计算力学前沿丛书

近场动力学
——理论、模型与应用

章青 顾鑫 著

科学出版社

北京

内 容 简 介

本书为"计算力学前沿丛书"之一。全书系统地论述了近场动力学的理论基础、建模方法、数值算法、软件技术和工程应用。

全书共 13 章,包括:绪论、近场动力学的基本理论、键型近场动力学模型及其改进、键型近场动力学在有限元中的实现、近场动力学的显式动力学解法、常规态型近场动力学模型、非常规态型近场动力学模型、非常规态型近场动力学模型的改进、近场动力学方法与有限单元法的混合模型、非均匀离散的近场动力学模型与自适应分析、冲击侵彻与爆炸问题的近场动力学模拟、热传导与热-力耦合问题的近场动力学模拟和混凝土材料与结构破坏的近场动力学建模分析。此外,本书还安排了两个附录,附录 A 介绍了近场动力学微分算子;附录 B 给出了近场动力学的显式动力学算法 FORTRAN 源程序。

本书可作为力学、土木、水利、交通、材料、航空航天、机械和能源等专业的研究生参考用书,亦可供从事损伤断裂力学、计算力学及工程模拟仿真的科技工作者参考和应用。

图书在版编目(CIP)数据

近场动力学:理论、模型与应用/章青,顾鑫著. —北京:科学出版社,2024.1
(计算力学前沿丛书)
ISBN 978-7-03-074291-9

Ⅰ.①近⋯　Ⅱ.①章⋯②顾⋯　Ⅲ.①动力学–研究　Ⅳ.①O313

中国版本图书馆 CIP 数据核字(2022)第 240361 号

责任编辑:赵敬伟　赵 颖／责任校对:彭珍珍

责任印制:张 伟　／封面设计:无极书装

科学出版社 出版
北京东黄城根北街 16 号
邮政编码:100717
http://www.sciencep.com
北京中科印刷有限公司 印刷
科学出版社发行　各地新华书店经销

*

2024 年 1 月第 一 版　开本:720×1000　1/16
2024 年 1 月第一次印刷　印张:34 3/4
字数:700 000
定价:298.00 元
(如有印装质量问题,我社负责调换)

编委会

主　　编：钟万勰　程耿东
副主编：李　刚　郭　旭　亢　战
编　　委（按姓氏音序排序）

　　　　　陈　震　崔俊芝　段宝岩　高效伟
　　　　　韩　旭　何颖波　李锡夔　刘书田
　　　　　欧阳华江　齐朝晖　申长雨　唐　山
　　　　　仝立勇　杨海天　袁明武　章　青
　　　　　郑　耀　庄　茁
秘书组：杨迪雄　阎　军　郑勇刚　高　强

丛 书 序

力学是工程科学的基础，是连接基础科学与工程技术的桥梁。钱学森先生曾指出，"今日的力学要充分利用计算机和现代计算技术去回答一切宏观的实际科学技术问题，计算方法非常重要"。计算力学正是根据力学基本理论，研究工程结构与产品及其制造过程分析、模拟、评价、优化和智能化的数值模型与算法，并利用计算机数值模拟技术和软件解决实际工程中力学问题的一门学科。它横贯力学的各个分支，不断扩大各个领域中力学的研究和应用范围，在解决新的前沿科学与技术问题以及与其他学科交叉渗透中不断完善和拓展其理论和方法体系，成为力学学科最具活力的一个分支。当前，计算力学已成为现代科学研究的重要手段之一，在计算机辅助工程（CAE）中占据核心地位，也是航空、航天、船舶、汽车、高铁、机械、土木、化工、能源、生物医学等工程领域不可或缺的重要工具，在科学技术和国民经济发展中发挥了日益重要的作用。

计算力学是在力学基本理论和重大工程需求的驱动下发展起来的。20 世纪 60 年代，计算机的出现促使力学工作者开始重视和发展数值计算这一与理论分析和实验并列的科学研究手段。在航空航天结构分析需求的强劲推动下，一批学者提出了有限元法的基本思想和方法。此后，有限元法短期内迅速得到了发展，模拟对象从最初的线性静力学分析拓展到非线性分析、动力学分析、流体力学分析等，也涌现了一批通用的有限元分析大型程序系统和可不断扩展的集成分析平台，在工业领域得到了广泛应用。时至今日，计算力学理论和方法仍在持续发展和完善中，研究对象已从结构系统拓展到多相介质和多物理场耦合系统，从连续介质力学行为拓展到损伤、破坏、颗粒流动等宏微观非连续行为，从确定性系统拓展到不确定性系统，从单一尺度分析拓展到时空多尺度分析。计算力学还出现了进一步与信息技术、计算数学、计算物理等学科交叉和融合的趋势。例如，数据驱动、数字孪生、人工智能等新兴技术为计算力学研究提供了新的机遇。

中国一直是计算力学研究最为活跃的国家之一。我国计算力学的发展可以追溯到近 60 年前。冯康先生 20 世纪 60 年代就提出"基于变分原理的差分格式"，被国际学术界公认为中国独立发展有限元法的标志。冯康先生还在国际上第一个给出了有限元法收敛性的严格的数学证明。早在 20 世纪 70 年代，我国计算力学的奠基人钱令希院士就致力于创建计算力学学科，倡导研究优化设计理论与方法，引领了中国计算力学走向国际舞台。我国学者在计算力学理论、方法和工程应用研究中都做出了贡献，其中包括有限元构造及其数学基础、结构力学与最优控制的相互模拟理论、结构拓扑优化基本理论等方向的先驱性工作。进入 21 世纪以来，我国计算力学研究队伍不断扩大，取得了一批有重要学术影响的研究成果，也为解决我国载人航天、高速列车、深海开发、核电装备等一批重大工程中的力学问题做出了突出贡献。

"计算力学前沿丛书"集中展现了我国计算力学领域若干重要方向的研究成果，广泛涉及计算力学研究热点和前瞻性方向。系列专著所涉及的研究领域，既包括计算力学基本理论体系和基础性数值方法，也包括面向力学与相关领域新的问题所发展的数学模型、高性能算法及其应用。例如，丛书纳入了我国计算力学学者关于 Hamilton 系统辛数学理论和保辛算法、周期材料和周期结构等效性能的高效数值预测、力学分析中对称性和守恒律、工程结构可靠性分析与风险优化设计、不确定性结构鲁棒性与非概率可靠性优化、结构随机振动与可靠度分析、动力学常微分方程高精度高效率时间积分、多尺度分析与优化设计等基本理论和方法的创新性成果，以及声学和声振问题的边界元法、计算颗粒材料力学、近场动力学方法、全速域计算空气动力学方法等面向特色研究对象的计算方法研究成果。丛书作者结合严谨的理论推导、新颖的算法构造和翔实的应用案例对各自专题进行了深入阐述。

本套丛书的出版，将为传播我国计算力学学者的学术思想、推广创新性的研究成果起到积极作用，也有助于加强计算力学向其他基础科学与工程技术前沿研究方向的交叉和渗透。丛书可为我国力学、计算数学、计算物理等相关领域的教学、科研提供参考，对于航空、航天、船舶、汽车、机械、土木、能源、化工等工程技术研究与开发的人员也将具有很好的借鉴价值。

"计算力学前沿丛书"从发起、策划到编著，是在一批计算力学同行的响应和支持下进行的。没有他们的大力支持，丛书面世是不可能的。同时，丛书的出版承蒙科学出版社全力支持。在此，对支持丛书编著和出版的全体同仁及编审人员表示深切谢意。

感谢大连理工大学工业装备结构分析优化与 CAE 软件全国重点实验室对"计算力学前沿丛书"出版的资助。

钟万勰　程耿东

2022 年 6 月

序

科学和工程中的众多问题都可归结为在给定初始和边界条件下的常微分或偏微分方程问题。基于经典连续介质力学和物理学的科学及工程计算方法经历多年发展，取得了极大的成功。但是，在分析固体材料及其结构的疲劳损伤、断裂破坏时，却存在诸多瓶颈性难题。究其原因：一是经典连续介质力学是基于连续性和局部接触假设建立微分型方程的，由于基本变量的空间导数在裂纹尖端不存在，或在材料界面处存在但不唯一，故不能精确描述固体材料及其结构的渐进破坏行为；二是经典连续介质力学缺少描述材料内部物质点之间含尺度参数的长程相互作用力，难以捕捉裂尖扩展过程的非局部效应，故不能分析长程相互作用力为主导的结构力学问题。

实际材料及其结构的断裂破坏过程具有明显的微-细-宏观特征，材料内部物质点之间存在非零距离的短程或长程作用力。对于材料断裂过程的微-细观结构变异特征及非局部效应，前人已经提出了多种非局部理论——统称为广义连续介质力学，包括强非局部理论和弱非局部理论。前者以 Eringen 等建立的积分型非局部理论为代表，摒弃了经典连续介质力学假设，在本构方程中引入了非局部的核函数，以考虑材料内部的长程相互作用。后者以各种应变梯度理论为代表，在经典连续介质模型中引入了场变量的梯度项，将尺度参数隐含于导数中，以描述物质点之间的非局部相互作用。上述非局部理论已在颗粒增强金属基复合材料、石墨烯和碳纳米管材料的性能分析中发挥了作用。但是，由于引入的一些参数难以通过实验测定，又没有行之有效的数值计算方法，以实现数值结果与工程实测的对比验证，故制约了它们应用于实际工程的计算分析。

值得指出的是，上述非局部理论在分析结构损伤、断裂、破坏时遇到困难，究其原因在于，当结构在极限载荷作用下发生损伤、断裂等不连续变形时，在物质点之间必然存在非局部的、与其距离相关的相互作用力。因此，许多学者一直在努力构建统一的、包含尺度参数相互作用力的非局部理论模型和高性能的计算方法。1998 年，美国圣地亚国家实验室 Silling 博士，基于 Eringen 和 Edelen 等提出的连续介质非局部积分理论，重构了积分型非局部连续介质力学的基本方程，提出了近场动力学（peridynamics，PD）的概念。近场动力学认为：材料由大量含有物性信息的物质点构成，每个物质点与其周围一定范围内的物质点之间存在相互作用力。基于非局部理论建立的积分型方程，回避了裂纹或界面处空间导数不存在而导致的奇性，能够自然地描述裂纹的萌生和扩展，可用于分析固体结构在极限载荷或长时间周期载荷作用下的损伤–断裂–失效–破坏的力学行为。

近场动力学问世二十多年来，吸引了世界各国的应用数学、力学、物理和工程科学家的广泛关注，积极投身于非局部理论、积分模型与高性能算法的研究。发展了多种形态的键型非局部积分模型，常规态、非常规态和改进的非常规态近场动力学模型，以及多种形式的近场动力学微分算子；将近场动力学扩展应用于复合材料结构、渗流、热扩散以及热–力耦合多物理场问题。同时，还发展了多种近场动力学的高性能算法，例如近场动力学自适应算法，与有限元法耦合算法等；研制了专门的近场动力学软件，与有限元法和计算流体力学软件相结合，开展了规模化的工程应用研究，例如水利水电工程、长大隧道和地下空间开发等。总之，过去二十年，近场动力学的理论、模型、算法、软件和工程应用均得到了迅速发展和完善。

该书作者章青教授长期从事水利水电工程和重大基础设施的工程力学研究，在服务于国家重大工程的同时，十分关注计算力学和近场动力学。他在非线性问题有限元法、加权残值法、界面元法、扩展有限元法和自然单元法等方面做出了出色成果，并将上述方法应用于工程实践。他参与解决了三峡、南水北调、向家坝、白鹤滩和江阴长江大桥、润扬长江大桥等多个重大工程的计算难题。2008年，章青教授与合作者将近场动力学引入国内，围绕着理论模型、建模技术、计算体系、软件实现和工程应用，持续开展了系统深入的研究；同时组织了多次关于近场动力学的研讨会，吸引了国内外众多学者投身于近场动力学的研究。

我与章青教授相识多年，受邀在他发起成立的"近场动力学与非局部理论国际研究中心"担任科学委员会主席。得知他与顾鑫博士的新著《近场动力学——理论、模型与应用》即将付梓，很是欣喜，这是中国学者的第一本全面系统反映近场动力学研究成果的专著。

该书系统地论述了近场动力学的理论基础、建模方法、数值算法、软件技术和工程应用，汇集了作者有关近场动力学的重要研究和应用成果。该书内容丰富，其特色是理论和应用相结合，书中通过算例详细阐述了近场动力学的实现方法、现有的开源代码，以及如何基于商用软件进行二次开发；还提供了FORTRAN源程序及其注释；这些都为读者深入理解近场动力学理论、模型、算法和应用提供了借鉴，是从事计算数学、计算力学及工程模拟仿真的研究人员、教师、研究生值得认真一读的专著。

我相信，该书的出版将进一步推进近场动力学的理论、建模、算法、软件及其工程应用在中国的发展。

是为序。

崔俊芝

中国工程院　院士

中国科学院数学与系统科学院　研究员

前　言

固体材料和结构的破坏分析是力学研究的经典难题，也是装备制造、航空航天、土木水利等领域的共性问题。由于固体材料和结构在破坏时必然会产生各种复杂的变形不连续特征，而传统连续介质力学理论及其相应的数值方法又要始终满足连续性条件以求解空间微分方程，这种根本性矛盾导致传统数值方法并不能很好地描述固体材料和结构复杂的损伤累积、宏观裂纹萌生与扩展、局部断裂乃至整体失稳的渐进破坏过程。此外，传统连续介质力学理论采用的局部接触作用假定，无法反映诸如裂尖断裂过程区的非局部效应。

近场动力学 (peridynamics, PD) 是一种积分型非局部连续介质力学理论，其将研究对象离散为大量包含所有物性信息的物质点，考虑近场范围内物质点间的非局部效应和长程相互作用进行建模。近场动力学积分型控制方程取代了传统连续介质力学的偏微分型控制方程，避免了裂纹等不连续处空间导数不存在而导致的奇异性，并通过"键"的断开和累积描述损伤与开裂，不需要预设裂纹路径，裂纹不受连续性和网格约束，可以自然萌生和扩展，适用于求解材料与结构非连续和非局部变形破坏问题。

近场动力学自 2000 年问世以来，备受数学、力学、物理和工程界的关注，已形成了较为完善的数学力学理论体系和数值计算方法，在宏、细、微观各个尺度的材料和结构的静动力变形和非连续力学问题中得到了广泛应用，并逐渐拓展至扩散问题、多物理场耦合问题、流体力学问题等研究领域，目前已成为国际计算力学界的研究热点和前沿课题之一。

本书作者课题组在国内最早开展近场动力学的研究，得到了多个国家自然科学基金项目和国家重点研发计划课题的资助，在近场动力学理论模型、计算体系、程序实现和工程应用等方面取得了较为丰硕的成果，协同成立了近场动力学与非局部理论国际研究中心，在国内外学术会议上组织了多次专题研讨会。本书以作者课题组十余年在近场动力学领域的主要成果为基础，系统地论述了近场动力学的理论模型、数值算法和工程应用；重点梳理了近场动力学的数值算法与程序实现方法、现有开源代码的应用与基于商用软件的二次开发，给读者提供详尽的算法流程图、源程序代码和一系列应用算例。本书内容论述完整，可作为近场动力学入门和进阶的教材和参考书。

本书共 13 章，包括：绪论、近场动力学的基本理论、键型近场动力学模型及其改进、键型近场动力学在有限元中的实现、近场动力学的显式动力学解法、常规态型近场动力学模型、非常规态型近场动力学模型、非常规态型近场动力学模型的改进、近场动力学方法与有限单元法的混合模型、非均匀离散的近场动力学模型与自适应分析、冲击侵彻与爆炸问题的近场动力学模拟、热传导与热–力耦合问题的近场动力学模拟和混凝土材料与结构破坏的近场动力学建模分析。此外，本书还安排了两个附录，附录 A 介绍了与本书有关的近场动力学微分算子；附录 B 给出了近场动力学的显式动力学算法 FORTRAN 源程序，并包含必要的注释和输入数据文件的说明，以便于读者更好地理解近场动力学理论和方法以及开展近场动力学相关问题的研究。

  本书是作者所在课题组多年来在近场动力学方面集体研究的成果，由章青教授和顾鑫副研究员共同执笔完成。

  作者感谢 Stewart A Silling 教授和 Erdogan Madenci 教授的关心和指教，感谢中国科学院系统与数学科学研究院研究员、近场动力学与非局部理论国际研究中心科学委员会主席崔俊芝院士的指导和为本书作序。

  本书的研究工作得到了国家重点基础研究发展计划课题 (2018YFC0406703、2017YFC1502603)、国家自然科学基金重点项目 (11932006、U1934206)、国家自然科学基金面上项目 (11672101、11372099、51179064、12172121)、国家自然科学基金青年科学基金项目 (12002118)、江苏省自然科学基金面上项目 (BK20151493)、河海大学中央高校基本科研业务费项目 (B210201031、B200202231) 的支持，大连理工大学工业装备结构分析优化与 CAE 软件全国重点实验室对本书的出版给予了经费资助，研究团队的夏晓舟副教授、第一著者指导的研究生 (沈峰博士、赵晶晶博士、赵世军博士、李天一博士、郁杨天硕士、郭士强硕士、杨思阳硕士和陈皖旱硕士等) 参加了相关研究工作，在此表示衷心的感谢。同时，对在写作过程中参考的国内外文献作者们一并表示感谢。

  本书的框架形成于 2018 年 12 月，2020 年 8 月完成初稿，后历经多次修改，于 2022 年 10 月定稿，并提交给出版社。由于近场动力学发展极为迅速，本书未能及时反映该领域国内外 (包括作者所在课题组) 众多最新研究成果，加之作者的学识和水平有限，书中难免存在不足之处，恳请读者不吝指出。

章 青 顾 鑫

2022 年 10 月

目 录

丛书序
序
前言

第1章 绪论 ·· 1
  1.1 引言 ·· 1
  1.2 连续介质力学的局部理论与非局部理论 ·· 3
  1.3 近场动力学的产生与发展 ·· 6
  1.4 近场动力学的理论特点 ··· 8
  1.5 近场动力学的研究现状 ··· 9
    1.5.1 固体力学问题 ·· 9
    1.5.2 流体力学问题 ·· 18
    1.5.3 输运扩散问题 ·· 19
    1.5.4 多物理场耦合问题 ·· 20
    1.5.5 近场动力学与其他数值方法的混合建模与多尺度分析 ····················· 25
    1.5.6 近场动力学相关应用 ·· 26
    1.5.7 近场动力学数理基础 ·· 30
    1.5.8 近场动力学程序和软件研发 ·· 30
  1.6 本书的主要内容 ··· 31
  参考文献 ·· 35

第2章 近场动力学的基本理论 ·· 58
  2.1 变形描述 ·· 58
    2.1.1 经典连续介质力学的变形描述 ··· 58
    2.1.2 近场动力学的变形描述 ··· 61
  2.2 受力描述与运动方程 ··· 64
    2.2.1 经典连续介质力学的受力描述与运动方程 ····································· 64
    2.2.2 近场动力学的受力描述与运动方程 ··· 66
  2.3 本构模型 ·· 70
    2.3.1 键型近场动力学的力密度矢量 ··· 71
    2.3.2 常规态型近场动力学的力密度矢量 ··· 71

2.3.3　非常规态型近场动力学的力密度矢量 ··· 72
　2.4　非局部边界条件 ··· 73
　2.5　守恒律的相容性条件 ·· 75
　　　2.5.1　线动量守恒的相容性条件 ··· 75
　　　2.5.2　角动量守恒的相容性条件 ··· 76
　　　2.5.3　能量守恒的相容性条件 ·· 77
　2.6　损伤与破坏：断键准则 ··· 78
　参考文献 ·· 81

第 3 章　键型近场动力学模型及其改进 ··· 83
　3.1　本构模型及其线性化 ·· 83
　　　3.1.1　键力密度函数的构建 ··· 83
　　　3.1.2　键力密度函数的线性化 ·· 86
　3.2　经典微弹脆性模型 ··· 88
　　　3.2.1　键力密度函数的具体形式 ··· 88
　　　3.2.2　参数率定 ·· 90
　3.3　经典微弹脆性模型的改进 ·· 95
　　　3.3.1　考虑长程力空间递减规律的修正 ·· 96
　　　3.3.2　考虑边界效应的修正 ··· 98
　3.4　其他键型近场动力学模型 ·· 100
　　　3.4.1　微梁模型 ·· 100
　　　3.4.2　共轭键模型 ··· 104
　　　3.4.3　另两种键型近场动力学模型简介 ·· 111
　参考文献 ·· 112

第 4 章　键型近场动力学在有限元中的实现 ··· 114
　4.1　近场动力学强形式方程的杆单元法 ·· 114
　　　4.1.1　基于杆单元离散的计算方法 ·· 114
　　　4.1.2　基于 ABAQUS 软件的二次开发 ··· 118
　4.2　近场动力学弱形式方程的非连续伽辽金有限元法 ··························· 121
　　　4.2.1　经典连续介质力学微分方程的弱形式 ···································· 121
　　　4.2.2　近场动力学积分方程的弱形式 ··· 123
　　　4.2.3　空间离散与矩阵装配 ··· 124
　4.3　键型近场动力学非连续伽辽金有限元法在 LS-DYNA 软件中的
　　　实现 ··· 128
　4.4　数值算例分析 ·· 131
　　　4.4.1　基于自编程序的近场动力学杆单元法算例 ······························ 131

4.4.2 基于 ABAQUS 二次开发的近场动力学杆单元法算例 ········ 132
4.4.3 基于近场动力学非连续伽辽金有限元法的算例 ············ 134
4.4.4 基于 LS-DYNA 的混凝土板爆炸冲击毁伤模拟 ············ 135
参考文献 ·· 137

第 5 章 近场动力学的显式动力学解法 ·· 140
5.1 控制方程的空间离散 ·· 140
5.2 时间离散与逐步积分法 ··· 142
    5.2.1 时间差分格式 ·· 142
    5.2.2 显式逐步积分法的稳定性条件 ····························· 143
5.3 显式拟静力方法 ·· 145
5.4 程序设计框架与计算流程 ·· 146
5.5 数值算例 ··· 147
    5.5.1 一维杆的弹性波传播 ·· 148
    5.5.2 二维简支梁的弹性变形 ····································· 149
    5.5.3 三维马氏体时效钢冲击动力裂纹扩展 ···················· 150
    5.5.4 混凝土板的冲击侵彻破坏 ·································· 152
5.6 近场动力学开源软件 PDLAMMPS ······························ 155
    5.6.1 PDLAMMPS 软件概述 ····································· 155
    5.6.2 安装与配置 ·· 156
    5.6.3 脚本文件示例与应用 ·· 156
参考文献 ·· 160

第 6 章 常规态型近场动力学模型 ·· 162
6.1 常规态型近场动力学弹性模型 ···································· 162
    6.1.1 基本变量说明 ·· 162
    6.1.2 弹性本构模型的建立过程 ·································· 165
6.2 常规态型近场动力学弹塑性模型 ································· 170
    6.2.1 弹塑性本构模型的一般形式 ······························· 170
    6.2.2 屈服函数与塑性流动法则 ·································· 171
    6.2.3 一致性条件与塑性乘子 ····································· 173
    6.2.4 材料强度参数的确定 ·· 174
    6.2.5 回映算法流程 ·· 176
6.3 数值算例 ··· 177
    6.3.1 含预制中心圆孔板的弹性变形 ····························· 177
    6.3.2 随机多孔脆性环氧板的拉伸开裂 ·························· 179
    6.3.3 方板的加卸载弹塑性响应 ·································· 180

参考文献 ··· 182
第 7 章 非常规态型近场动力学模型 ·· 184
7.1 非常规态型近场动力学建模方法 ··· 184
7.1.1 建模流程与基本方程 ··· 184
7.1.2 基本方程的离散 ·· 187
7.2 线性问题的隐式求解方法 ·· 188
7.2.1 线弹性小变形问题的求解方程 ··· 188
7.2.2 物质点劲度系数矩阵的构造 ·· 190
7.3 非线性问题的隐式求解方法 ··· 194
7.4 数值不稳定性分析与稳定控制方法 ·· 197
7.4.1 数值不稳定性的影响因素分析 ··· 198
7.4.2 稳定控制的计算策略 ··· 201
7.4.3 稳定控制计算策略的数值验证 ··· 202
7.5 晶体弹塑性变形与动态断裂的非常规态型近场动力学方法 ················· 209
7.5.1 单晶体弹塑性本构模型 ·· 209
7.5.2 塑性剪切变形增量与变形梯度张量更新 ····································· 211
7.5.3 晶体塑性本构的切线模量 ··· 214
7.5.4 平面多晶体弹塑性静力变形的算例分析 ····································· 216
7.5.5 平面多晶体弹塑性裂纹扩展的算例分析 ····································· 220
参考文献 ··· 225
第 8 章 非常规态型近场动力学模型的改进 ··· 228
8.1 几种无网格法与非常规态型近场动力学方法的对比 ·························· 228
8.1.1 光滑粒子流体动力学方法及其修正 ··· 228
8.1.2 再生核质点法和梯度再生核质点法 ··· 233
8.1.3 非常规态型近场动力学模型 ·· 237
8.1.4 几种无网格法与非常规态型近场动力学方法的对比 ······················· 238
8.2 高阶非常规态型近场动力学模型 ·· 240
8.2.1 高阶非局部变形梯度和力密度矢量状态 ····································· 240
8.2.2 线性隐式解法 ·· 242
8.2.3 非线性隐式解法 ··· 247
8.2.4 高阶非局部变形梯度的精度验证 ·· 248
8.2.5 弹性杆拉伸变形的算例分析 ·· 250
8.2.6 矩形板拉伸变形的算例分析 ·· 254
8.3 键关联高阶非常规态型近场动力学模型 ··· 257
8.3.1 键关联的非局部变形梯度 ··· 257

    8.3.2 键关联的力密度矢量状态和运动方程 · 259
    8.3.3 线性隐式静力解法 · 260
    8.3.4 矩形板拉伸变形的算例分析 · 262
    8.3.5 三维杆轴压变形的算例分析 · 265
    8.3.6 三维杆中准一维弹性波传播的算例分析 · 270
  8.4 一个新的非常规态型近场动力学模型 · 272
    8.4.1 基于近场动力学微分算子重构原非常规态型近场动力学模型 · · · · · · 272
    8.4.2 基于近场动力学微分算子构建新的非常规态型近场动力学模型 · · · · · · 276
    8.4.3 新的非常规态型近场动力学模型的隐式–显式混合解法 · · · · · · · · · · · · · 279
    8.4.4 矩形板拉伸变形的算例分析 · 283
    8.4.5 含线裂纹矩形板拉伸变形的算例分析 · 285
    8.4.6 含孔板拉伸变形的算例分析 · 286
    8.4.7 含孔板裂纹扩展的算例分析 · 290
  参考文献 · 292
第 9 章 近场动力学方法与有限单元法的混合模型 · 295
  9.1 近场动力学与有限元混合建模的几种方法 · 295
    9.1.1 力耦合方法与镶嵌单元技术 · 296
    9.1.2 位移协调约束与位移结合法 · 296
    9.1.3 力分解法 · 297
    9.1.4 混合函数方法 · 298
    9.1.5 子模型方法 · 298
  9.2 新的近场动力学与有限元混合模型 · 299
    9.2.1 重叠模型与接触模型 · 299
    9.2.2 重叠模型的定量分析 · 302
    9.2.3 接触模型的定量分析 · 304
  9.3 数值算例 · 305
    9.3.1 悬臂梁在端部受集中力作用 · 305
    9.3.2 简支梁在跨中受集中力作用 · 306
    9.3.3 含 I 型裂纹板的裂纹扩展分析 · 307
    9.3.4 含 I-II 复合型裂纹板的裂纹扩展分析 · 309
    9.3.5 多裂纹扩展分析 · 311
    9.3.6 含切口三点弯曲梁的裂纹扩展分析 · 313
  9.4 近场动力学有限元混合模型在重力坝稳定性分析中的应用 · · · · · · · · · · · · · 315
    9.4.1 典型重力坝的变形计算和分析 · 315
    9.4.2 典型重力坝的承载力评价 · 317

参考文献 ··· 319
第 10 章 非均匀离散的近场动力学模型与自适应分析 ······················· 322
10.1 近场动力学的空间离散方式与自适应分析 ····························· 322
10.2 基于 Voronoi 结构图离散的近场动力学自适应方法 ···················· 324
10.2.1 均匀/非均匀 Voronoi 胞元离散 ································ 324
10.2.2 键型近场动力学模型中参数的比例关系 ·························· 326
10.2.3 对偶双影响域近场动力学模型 ································· 327
10.2.4 基于 Voronoi 结构图的自适应方案 ····························· 329
10.3 数值算例 ··· 331
10.3.1 矩形板的拟静力弹性变形 ····································· 331
10.3.2 二维弹性波的传播 ··· 333
10.3.3 含预制裂纹板的动态裂纹扩展 ································· 339
10.3.4 三维马氏体时效钢冲击动力裂纹扩展 ··························· 345
10.4 主要结论 ··· 347
参考文献 ··· 348
第 11 章 冲击侵彻与爆炸问题的近场动力学模拟 ·························· 351
11.1 冲击接触算法与爆炸载荷的计算 ······································· 351
11.1.1 冲击接触算法 ··· 351
11.1.2 爆炸载荷施加方法 ··· 352
11.2 准脆性材料的 JH-2 本构模型 ··· 356
11.2.1 JH-2 本构关系 ·· 356
11.2.2 JH-2 本构的更新算法 ·· 359
11.2.3 JH-2 本构的基准验证 ·· 363
11.3 近场动力学的非局部色散特性与霍普金森压杆冲击试验的模拟 ········ 367
11.3.1 键型近场动力学的非局部色散特性 ····························· 367
11.3.2 分离式霍普金森压杆冲击巴西圆盘的近场动力学模拟 ············· 370
11.4 混凝土层裂与多重层裂的近场动力学模拟 ···························· 373
11.4.1 矩形冲击波作用下混凝土杆的单层层裂模拟 ····················· 373
11.4.2 三角冲击波作用下混凝土杆的多重层裂模拟 ····················· 376
11.5 钢筋混凝土板空中爆炸毁伤的近场动力学模拟 ························ 382
11.5.1 问题描述 ··· 382
11.5.2 计算结果与分析 ··· 384
参考文献 ··· 386
第 12 章 热传导与热-力耦合问题的近场动力学模拟 ······················· 388
12.1 热传导问题的近场动力学模型 ··· 388

12.1.1　键型近场动力学热传导模型 389
　　12.1.2　态型近场动力学热传导模型 390
　12.2　基于近场动力学微分算子的热传导模型 391
　12.3　基于近场动力学微分算子的热-力耦合模型 395
　12.4　热传导和热-力耦合问题近场动力学模型的数值计算 398
　　12.4.1　初始条件与边界条件 398
　　12.4.2　方程离散与求解 399
　　12.4.3　损伤区域导热系数的修正 403
　12.5　热传导问题近场动力学模型的算例分析 403
　　12.5.1　一维杆件热传导问题的算例分析 403
　　12.5.2　二维方板热传导问题的算例分析 405
　　12.5.3　三维厚板热传导问题的算例分析 408
　　12.5.4　混凝土试件热传导问题的细观分析 410
　12.6　热-力耦合问题近场动力学模型的算例分析 415
　　12.6.1　二维方板四边受给定温度荷载作用 415
　　12.6.2　二维方板三边绝热和一边受给定温度荷载作用 417
　　12.6.3　混凝土厚壁圆筒内壁温升后裂纹扩展的二维模拟 419
　　12.6.4　混凝土厚壁圆筒内壁温升后裂纹扩展的三维模拟 422
　参考文献 424

第13章　混凝土材料与结构破坏的近场动力学建模分析 427
　13.1　水泥水化产物的近场动力学建模与拉伸破坏分析 427
　　13.1.1　水泥水化产物微结构的生成 428
　　13.1.2　水泥水化微结构的近场动力学计算模型 434
　　13.1.3　水泥水化微结构单向拉伸的计算结果与分析 436
　13.2　混凝土细观破坏的近场动力学建模分析 441
　　13.2.1　混凝土骨料的富勒级配理论 442
　　13.2.2　基于蒙特卡罗法和随机游走法的骨料生成与投放 443
　　13.2.3　数值算例与分析 449
　13.3　混凝土冻融循环热-水-力耦合问题的近场动力学模型 452
　　13.3.1　多孔介质中孔隙水冻结规律与孔隙压力 453
　　13.3.2　多孔介质热-水-力耦合问题的近场动力学模型 456
　　13.3.3　水泥砂浆试件的冻融循环分析 458
　13.4　混凝土重力坝的侵彻与爆炸毁伤分析 461
　　13.4.1　混凝土重力坝的侵彻毁伤分析 461
　　13.4.2　空中爆炸致使混凝土重力坝的毁伤分析 465

13.5 向家坝水电站泄洪坝段的抗滑稳定分析 ··· 469
 13.5.1 工程概况 ·· 469
 13.5.2 泄④坝段近场动力学–有限元混合模型 ·· 469
 13.5.3 计算结果与分析 ··· 470
参考文献 ·· 472

附录 A 近场动力学微分算子 ··· 475
A.1 多维空间任意形状近场范围的任意阶近场动力学微分算子 ··················· 475
 A.1.1 近场动力学微分算子的绝对表达式 ·· 476
 A.1.2 近场动力学微分算子的相对表达式 ·· 477
A.2 三维球形对称近场范围的二阶近场动力学微分算子 ····························· 478
 A.2.1 近场动力学微分算子的绝对表达式 ·· 478
 A.2.2 近场动力学微分算子的相对表达式 ·· 483
A.3 二维圆形对称近场范围的二阶近场动力学微分算子 ····························· 488
 A.3.1 近场动力学微分算子的绝对表达式 ·· 488
 A.3.2 近场动力学微分算子的相对表达式 ·· 491
A.4 一维对称近场范围的二阶近场动力学微分算子 ···································· 493
参考文献 ·· 495

附录 B 近场动力学的显式动力学算法程序 ·· 496
B.1 键型近场动力学的显式动力学算法 FORTRAN 源程序 ························ 496
 B.1.1 公用模块：Module ·· 496
 B.1.2 主程序：Master ··· 500
 B.1.3 子程序 1：data_input ·· 502
 B.1.4 子程序 2：element_form ·· 506
 B.1.5 子程序 3：add_load_displacement ·· 508
 B.1.6 子程序 4：add_load_bc_force ·· 508
 B.1.7 子程序 5：add_load_impact_rigid ··· 510
 B.1.8 子程序 6：equation_solve ··· 513
 B.1.9 子程序 7：model_PMB ·· 515
 B.1.10 子程序 8：model_short_range_force ··· 517
 B.1.11 子程序 9：data_output ·· 518
B.2 输入数据文件 ·· 527
 B.2.1 输入数据文件的一般格式 ·· 527
 B.2.2 马氏体时效钢冲击破坏试验模拟的输入数据文件 ································· 528

近场动力学主要术语的中英文对照表 ··· 531
索引 ·· 533

第 1 章 绪 论

1.1 引 言

固体材料和结构的破坏分析是力学研究的经典难题,也是装备制造、航空航天、交通运输、能源矿业、土木水利等工程领域的共性问题。百余年来,研究者们开展了大量的理论和试验研究,发展了众多数值模型和计算方法,旨在揭示固体材料和结构的破坏机理,模拟得到材料和结构从损伤累积、裂纹萌生与扩展、局部断裂直至整体失效破坏的全过程,但由于问题的复杂性,该问题尚未得到很好的解决。

根据材料的组成方式和物质间的作用形式,在满足能量、质量、动量守恒定律和某些决定性原理的框架下,现有的固体力学理论可视为建立在两类基本假定的基础上,一是"连续性假定或非连续性假定",二是"局部接触作用假定或非局部长程作用假定",并由两类基本假定的不同组合,形成了如下理论和模型:

(1) 基于连续性假定与局部接触作用假定所建立的经典连续介质力学理论[1];

(2) 基于连续性假定和非局部长程作用假定所建立的弱非局部理论[2],如高阶梯度模型[3]等;

(3) 基于非连续性假定和局部接触作用假定所建立的接触作用类粒子动力学理论,如离散元模型等[4];

(4) 基于非连续性假定和非局部长程作用假定所建立的强非局部理论,如分子动力学[5]和本书涉及的近场动力学[6,7](peridynamics,PD)。

经典连续介质力学 (classical continuum mechanics, CCM) 理论和相应的数值方法历经多年的发展,已经在众多科学问题以及工程应用领域取得了极大成功,但在处理固体材料和结构破坏引起的各种复杂不连续变形时遇到瓶颈。其原因在于经典连续介质力学描述问题的偏微分方程的连续性要求与问题的不连续性之间存在根本矛盾,导致变量空间导数在裂纹等强不连续处不存在、在材料界面等弱不连续处不唯一[8-11],不能很好地描述固体材料和结构的渐进破坏过程。为此,往往需将控制方程和不连续条件分开处理,造成求解困难。此外,经典连续介质力学理论采用局部接触作用假定,缺乏描述材料长程相互作用或非局部效应的长度尺度参数,在分析长程力主导、非局部效应显著的力学问题时显现不足,如无法反映裂尖断裂过程区的非局部效应。

实际材料都是非均质的，具有特征细观结构，材料内部存在非零距离的相互作用力(短程力、长程力)，材料和结构在微细宏观尺度都表现出一定的非局部效应[12-16]。分子动力学(molecular dynamics, MD)理论和方法适用于涉及非局部长程相互作用的力学问题，可以预测材料的不连续演化过程，但会遭遇计算规模瓶颈和巨大计算量问题。考虑材料的细观结构特征和非局部效应等提出的各种改进理论模型[2]可归结为广义连续介质力学(generalized continuum mechanics, GCM)的范畴。积分型的强非局部模型显含相隔有限距离的物质点之间的相互作用力，与其相比，梯度型的弱非局部连续介质力学模型将结点有限邻域内的应变平均，位移导数项隐含着长度尺度，但不明确包含有限距离点间的相互作用，且在处理不连续变形破坏问题时仍存在不足[6]。

需要指出的是，对于具有潜在非局部效应的结构非连续变形问题，将局部或非局部连续介质力学方法与分子动力学方法进行耦合，进而提出的并发多尺度方法[17]是行之有效的分析手段，但该方法需要预判和设定潜在不连续区域和连续区域，且连续介质力学方法与分子动力学方法的数学框架并不一致，不同尺度间需要合理的界面过渡和参数传递，力学建模和求解过程较为复杂。

面对上述研究背景，将不连续变形特征和非局部长程相互作用包含在统一的数学力学模型框架内，发展新的物理概念清晰的力学理论和方法，以便捷地求解结构非连续和非局部变形问题，便成为学术界关注的重点。2000年，美国圣地亚国家实验室(Sandia National Laboratory)计算力学与物理学部的Stewart Silling博士重新构建了积分型非局部连续介质力学基本方程，提出了peridynamics(PD)理论[6]。PD将研究对象离散为大量包含所有物性信息的物质点，考虑近场范围内物质点间的非局部效应和长程相互作用进行建模，采用的积分型控制方程取代了传统连续介质力学的偏微分型控制方程，避免了裂纹等不连续处的空间导数不存在而导致的奇异性，可以自然地描述裂纹的萌生和扩展，适用于求解材料与结构非连续和非局部变形破坏问题。2010年，黄丹等[10]在《力学进展》发表了综述文章，系统地介绍了近场动力学理论和方法，并将peridynamics引入国内，译为"近场动力学"，2015年，杜强[18]从数学角度介绍了peridynamics，并称之为"毗域动力学"。

近场动力学自问世以来，备受数学、力学、物理学和工程界的关注，已形成了较为完善的数学力学理论框架和数值计算方法体系，在宏、细、微观各个尺度的材料和结构的静动力变形和非连续力学问题中得到广泛应用，并逐渐拓展至流体力学问题、扩散问题、多物理场耦合问题等研究领域，目前已成为国际计算力学界的研究热点和前沿课题之一。

本书作者所在的课题组在中国最早开展近场动力学的研究，在多个国家自然科学基金项目和国家重点研发计划项目的资助下，课题组围绕近场动力学理论模型、建模技术、计算体系、程序实现和工程应用等方面进行了较为系统和深入的研究,并在国内外学术会议上组织了多次专题研讨会。本书是作者十余年来在近场动力学领域的主要研究成果的汇集。

1.2 连续介质力学的局部理论与非局部理论

连续介质力学用统一的观点研究固体和流体的力学行为。一般认为,Euler 在 18 世纪中叶建立的理想流体的力学方程可看作连续介质力学的肇端,此后,Navier、Cauchy、Poisson、Stokes 等科学家在运动定律和物性定律方面做出了杰出贡献,促使弹性固体力学理论和黏性流体力学基本理论的问世。20 世纪 50~70 年代,Oldroyd[19]、Noll[20]、Truesdell 和 Noll[21] 和 Eringen[22] 等力学家提出了构造简单物质本构的基础公理,建立了力学公理化体系,奠定了现代连续介质力学体系的基础 [23]。

经典连续介质力学基于局部接触作用 (Cauchy 应力原理) 假定建立偏微分型控制方程,基于 Noll 原则建立简单材料的局部本构关系,以描述材料和结构的力学行为,其运动方程和本构方程分别为

$$\rho \ddot{\boldsymbol{u}}(\boldsymbol{x},t) = \nabla \cdot \boldsymbol{\sigma}(\boldsymbol{x}) + \boldsymbol{b}(\boldsymbol{x}), \quad t \geqslant 0 \tag{1.1}$$

$$\boldsymbol{\sigma}(\boldsymbol{x}) = \boldsymbol{C}(\boldsymbol{x}) : \boldsymbol{\varepsilon}(\boldsymbol{x}) \tag{1.2}$$

其中,ρ 为物质密度,$\ddot{\boldsymbol{u}}$ 为物质点加速度矢量,∇ 为散度算子,$\boldsymbol{b}$ 是单位体积物质所受的体力矢量,$\boldsymbol{\sigma}(\boldsymbol{x})$ 和 $\boldsymbol{\varepsilon}(\boldsymbol{x})$ 分别为 $\boldsymbol{x}$ 点处的 Cauchy 应力张量和应变张量,$\boldsymbol{C}$ 为材料的弹性张量。

经典连续介质力学所研究的均匀各向同性介质没有内部结构,也没有附加内部自由度和内部特征尺度,当考察问题的特征尺度始终大于连续介质力学模型的分辨率水平 (如代表性体积单元 RVE 尺度、单元尺度) 时,能够较准确地预测结构的宏观力学行为,在工程结构的力学分析中取得了极大成功。然而,当外部特征尺寸 (如裂纹宽度、波长等) 和内部特征尺寸 (如颗粒间距、晶格参数等) 相当时,材料内部的长程相互作用不可忽略。由于经典连续介质力学假设某一点的力学行为只和该点的状态相关,属于局部理论,无法反映材料微结构之间的长程作用效应。在这种背景下,广义连续介质力学的非局部理论应运而生,非局部理论假设材料某一点的物理力学行为与整个物体内部诸点的状态相关,能较好地描述尺寸效应或非局部效应显著的物理力学问题,已在颗粒增强金属基复合材料、石墨烯和碳纳米管材料的力学性能分析中发挥重要作用。

非局部作用思想在力学领域内已经有很长的历史,最早考虑连续介质力学的非局部性可以追溯到 19 世纪末,1893 年,Duhem 指出一点处的应力与整个物

体的应变状态有关,1918 年,Rayleigh 在讨论无限长滑动轴承的问题时提出了该问题的非局部性质,但非局部场论的真正形成是源于 20 世纪六七十年代的工作,自 20 世纪 60 年代末期以来,研究人员针对不同问题发展了多种非局部理论,Rogula[24] 以及 Bazant 和 Jirásek[2] 将它们归类为"强非局部理论"和"弱非局部理论"。前者主要是以 Eringen 建立的非局部理论为代表的积分型强非局部理论,后者主要是以应变梯度理论为代表的梯度型弱非局部理论。

积分型强非局部理论摒弃了经典连续介质力学中的"局部化假设",在本构方程中通过非局部核函数引入物体内部长程相互作用,由整体描述替代了局域描述。Kröner[25] 和 Di Paola 等[26] 考虑局部短程作用与非局部长程作用的共同作用,建立混合形式的控制方程为

$$\rho \ddot{\boldsymbol{u}}(\boldsymbol{x},t) = \nabla \cdot \boldsymbol{\sigma}(\boldsymbol{x}) + \int_{H_\infty} \boldsymbol{\varPhi}(\boldsymbol{x}'-\boldsymbol{x}) \cdot \boldsymbol{u}(\boldsymbol{x}') \mathrm{d}V_{\boldsymbol{x}'} + \boldsymbol{b} \tag{1.3}$$

其中,$\boldsymbol{\varPhi}$ 为微模量函数。Rogula[24] 和 Kunin[27,28] 所建立的积分型非局部模型,与近场动力学的线性化模型一致,但仅针对晶体微结构材料

$$\rho \ddot{\boldsymbol{u}}(\boldsymbol{x},t) = \int_{H_\infty} \boldsymbol{\varPhi}(\boldsymbol{x},\boldsymbol{x}') \left(\boldsymbol{u}(\boldsymbol{x}') - \boldsymbol{u}(\boldsymbol{x}) \right) \mathrm{d}V_{\boldsymbol{x}'} \tag{1.4}$$

Eringen 和 Edelen 仍采用经典连续介质力学控制方程,于 1972 年通过守恒定律、热力学定律和非局部变分原理系统地建立了非局部弹性理论[29],他们认为某一点的非局部应力张量 $\boldsymbol{t}$ 是全局经典应力张量 $\boldsymbol{\sigma}$ 的加权平均值,则某一点 $\boldsymbol{x}$ 的非局部应力张量可以表示为[29,30]

$$\boldsymbol{t} = \int_{H_\infty} K(|\boldsymbol{x}'-\boldsymbol{x}|,\tau) \boldsymbol{C}(\boldsymbol{x}') : \boldsymbol{\varepsilon}(\boldsymbol{x}') \mathrm{d}V_{\boldsymbol{x}'} \tag{1.5}$$

其中,$\boldsymbol{\varepsilon}(\boldsymbol{x}')$ 是 $\boldsymbol{x}'$ 点处的应变张量,$\boldsymbol{C}(\boldsymbol{x}') : \boldsymbol{\varepsilon}(\boldsymbol{x}')$ 为 $\boldsymbol{x}'$ 点处的应力张量,核函数 $K(|\boldsymbol{x}'-\boldsymbol{x}|,\tau)$ 为非局部系数,$|\boldsymbol{x}'-\boldsymbol{x}|$ 为 $\boldsymbol{x}$ 点和 $\boldsymbol{x}'$ 点间的距离,τ 是依赖于内部特征尺寸和外部特征尺寸的材料参数。积分形式的本构关系求解较为困难,通常等价为如下的微分方程,即

$$\left[1 - (e_0 a)^2 \nabla^2 \right] \boldsymbol{t} = \boldsymbol{C} : \boldsymbol{\varepsilon} \tag{1.6}$$

其中,e_0 为材料常数,a 为内禀特征尺寸。随后 Eringen 还建立了非局部热弹性理论[31]、非局部记忆弹性体理论[32] 和非局部位错连续理论[33] 等理论分支。在此基础上,许多研究者相继进行了大量深入细致的工作[34-38]。2011 年,Aifantis[39]

建立梯度型模型与 Eringen 非局部模型的关系,给出非局部–梯度线弹性理论的统一形式本构方程

$$[1 - c_1^2 \nabla^2] \, t = [1 - c_2^2 \nabla^2] \, C : \varepsilon \tag{1.7}$$

其中,$c_1 = e_0 a$ 为反映应力局部效应的材料内禀尺度,c_2 为反映应变梯度效应的材料内禀尺度,式 (1.7) 可以退化为式 (1.6)。

梯度型弱非局部理论在经典连续介质力学模型中引入场变量的梯度项来描述物质内部微观结构特性[40-43],非局部相互作用或长度尺度没有被明确地包含在本构方程中,而是隐含在导数项中。20 世纪 60 年代,Toupin[44]、Koiter[45] 和 Mindlin[46,47] 基于偶应力理论,在推导应变能密度时考虑了应变梯度的影响,发展了高阶弹性理论。1997 年,Fleck 和 Hutchinson[48] 简化了 Mindlin 的理论,并命名为"应变梯度理论"。2003 年,Lam 等[49] 消除了 Mindlin 理论中应变梯度张量的不对称部分,使得材料长度参数从 5 个减少为 3 个,并称其为"修正的应变梯度理论"。当只考虑一个长度参数时,即为 Aifantis 给出的最简单的应变梯度弹性理论[50,51],即

$$\boldsymbol{\sigma} = [1 - l^2 \nabla^2] \, \boldsymbol{C} : \boldsymbol{\varepsilon} \tag{1.8}$$

其中,l 是应变梯度的长度尺度参数,其不同于 Eringen 非局部理论中的参数 $e_0 a$。此外,应变梯度塑性理论也随应变梯度弹性理论不断发展。例如,Fleck 和 Hutchinson[52] 从几何必需位错及统计储存位错角度出发,发展了一种能够考虑旋转梯度影响的 CS(couple stress) 应变梯度塑性理论;在分析裂纹尖端场或微米压痕时,他们又同时考虑了拉伸梯度和旋转梯度 (stretch and rotation gradients, SG),提出了 SG 应变梯度塑性理论[48]。Gao 和 Huang 等[53,54] 于 1999 年发展了一种基于位错机制的应变梯度塑性理论 (MSG 理论)。最近,Lim 等[55] 将 Eringen 的非局部理论和应变梯度理论结合起来,发展了非局部应变梯度理论,将总应力张量 t 定义为

$$\boldsymbol{t} = \boldsymbol{\sigma} - \nabla \cdot \boldsymbol{\sigma}^{(1)} \tag{1.9}$$

其中,$\boldsymbol{\sigma}$ 为零阶非局部应力张量 (应变 ε 的功共轭量),$\boldsymbol{\sigma}^{(1)}$ 为高阶应力张量 (应变梯度 $\nabla \varepsilon$ 的功共轭量),具体表达式为

$$\boldsymbol{\sigma} = \int_{H_x} \alpha_0 \left(\boldsymbol{x}, \boldsymbol{x}', e_0 a \right) \boldsymbol{C}(\boldsymbol{x}') : \boldsymbol{\varepsilon}(\boldsymbol{x}') \mathrm{d} V_{\boldsymbol{x}'} \tag{1.10}$$

$$\boldsymbol{\sigma}^{(1)} = l^2 \int_{H_x} \alpha_1 \left(\boldsymbol{x}, \boldsymbol{x}', e_1 a \right) \boldsymbol{C}(\boldsymbol{x}') : \nabla \boldsymbol{\varepsilon}(\boldsymbol{x}') \mathrm{d} V_{\boldsymbol{x}'} \tag{1.11}$$

其中,$\alpha_0 \left(\boldsymbol{x}, \boldsymbol{x}', e_0 a \right)$ 和 $\alpha_1 \left(\boldsymbol{x}, \boldsymbol{x}', e_1 a \right)$ 分别是与应变张量 ε 以及应变梯度张量 $\nabla \varepsilon$ 相关的非局部核函数,l 为应变梯度的长度尺度参数。

1.3 近场动力学的产生与发展

长期以来,学术界一直致力于寻求和发展便于描述材料和结构非局部效应和非连续演化过程的力学理论和高效数值方法,以便捷地求解结构非连续和非局部变形问题。1998年,Stewart Silling在圣地亚国家实验室的一份研究报告中,首次提出近场动力学(peridynamics, PD)的概念,2000年,正式在 *Journal of the Mechanics and Physics of Solids* 上长文发表[6]。20多年来,近场动力学吸引了大量学者再次聚焦非局部理论和积分方程建模方法,推动了计算力学及相关领域的飞速发展,其发展历程大致可分为三个阶段:

2008年之前,主要是圣地亚国家实验室研究人员以及美国的波音公司、内布拉斯加大学林肯分校、新墨西哥大学、亚利桑那大学、麻省理工学院、加州理工学院、加州大学伯克利分校、洛斯阿拉莫斯国家实验室和德国的柏林工业大学等学者开展近场动力学理论和相关应用研究,另有匈牙利和印度等国家的少数研究者开始关注近场动力学。

2008～2012年,由于近场动力学已在固体破坏问题分析中显现出巨大潜力,并且非局部的积分建模思想为多尺度和多物理场问题研究提供了新思路,近场动力学的研究稳步增长,并向全球范围扩散,据不完全统计,这期间美国三十多家大学和德国、英国、法国、中国、意大利、日本、韩国、墨西哥、土耳其、印度、沙特阿拉伯等国的学者纷纷投入到近场动力学理论和方法的研究中。

2012年后,全球范围内众多领域的研究者开始关注近场动力学方法,掀起了近场动力学的研究热潮。美国国防部(DoD)将"Multiphysics and multiscale failure prediction through peridynamics theory"列为MURI(跨学科合作研究项目)23个主题之一,持续5年资助750万美元,进行近场动力学理论的多物理场与多尺度破坏分析研究;美国能源部(DOE)、自然科学基金会(NSF)、陆军研究局(ARO)、空军科学研究局(AFOSR)和海军研究总署(ONR)等都对近场动力学相关的研究项目予以了资助;英国的工程和自然科学研究委员会(EPSRC)、英国国防科技实验室(DSTL)、欧盟(EU)、中国国家自然科学基金委员会(NSFC)以及波音、三星、康宁等大型跨国公司也持续资助了多项近场动力学研究项目。从近场动力学研究团体的全球分布、资助项目的层次和数量、高水平国际学术期刊发表的研究论文数量等多方面观察,近年来,近场动力学研究已进入高速发展时期(见图1-1、图1-2)。

国内近场动力学的研究起步于2009年,河海大学、上海交通大学、南京航空航天大学、武汉理工大学、北京大学、重庆大学、哈尔滨工业大学、山东大学、郑州大学、大连理工大学、中国工程物理研究院、武汉大学、华中科技大学等多家高校和研究机构的学者先后开展了近场动力学理论模型、数值计算和应用研究。2015 年后,近场动力学逐渐成为国际计算力学的研究热点,国内也召开了多个以近场动力学为主题的研讨会和专题会议,促进了更多的高校和科研院所的研究人员从事近场动力学研究,发表的期刊论文、学位论文、专利和软件著作权数量增长很快,取得了一系列优秀成果。

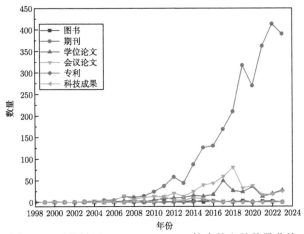

图 1-1　以关键词 "peridynamics" 检索的文献数量曲线

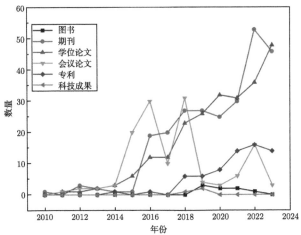

图 1-2　以关键词 "近场动力学" 检索的文献数量曲线

另外需要指出的是,美国亚利桑那大学 Madenci 教授于 2014 年出版了近场动力学首部专著 *Peridynamic Theory and Its Applications*[8],其中译本《近场动力学理论及其应用》于 2019 年由余音、胡祎乐翻译出版[56],内布拉斯加大学林肯分校 Bobaru 教授等于 2016 年编著出版了 PD 建模手册 *Handbook of Peridynamic*

Modeling[9]，新墨西哥大学 Gerstle 教授于 2015 年出版了 *Introduction to Practical Peridynamics: Computational Solid Mechanics Without Stress and Strain*[57] 一书，Madenci 教授又于 2019 年出版了专著 *Peridynamic Differential Operator for Numerical Analysis*[58]，其中译本《近场动力学微分算子——在数值分析中的应用》于 2021 年由韩非、张玲翻译出版[59]，由 Oterkus、Oterkus 和 Madenci 合著的 *Peridynamic Modeling, Numerical Techniques, and Applications*[60] 也于 2021 年出版，上述著作和手册极大地促进了近场动力学的国际传播与发展。

1.4 近场动力学的理论特点

近场动力学为固体材料和结构的破坏分析开辟了新的途径，在涉及非局部长程效应、裂纹扩展和多点损伤破碎的非连续破坏问题分析时具有独特优势。相较于以偏微分方程为基础的基于局部接触作用思想的经典连续介质力学，近场动力学基于非局部相互作用思想和积分方程建模，可归属于广义连续介质力学范畴，是经典连续介质力学和分子动力学的继承、发展与重构。

广义来说，近场动力学是假设每个物质点在承受一定范围内的非局部非接触相互作用下，研究整个物理系统变化过程的理论[61]。如图 1-3 所示，近场动力学借鉴并拓展了经典连续介质力学和分子动力学的建模思想，将研究对象离散为大量包含所有物性信息的物质点，考虑近场范围内物质点间的非局部效应和长程相互作用采用积分型控制方程进行问题的数学处理，避免了经典连续介质力学偏微分型控制方程在裂纹等不连续处空间导数不存在而导致的奇异性问题，适用于物体连续或不连续的任何区域。近场动力学采用"键的临界伸长率"等单一判据，通过断键的累积描述材料的损伤和开裂，不需要预设裂纹路径，裂纹不受连续性和网格约束，可以自然萌生和扩展，方便求解固体材料和结构损伤破坏等不连续力

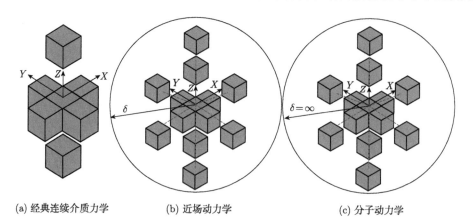

(a) 经典连续介质力学　　(b) 近场动力学　　(c) 分子动力学

图 1-3　经典连续介质力学、近场动力学和分子动力学的粒子间相互作用示意图

学问题。此外，非局部模型是对真实世界更合适和更准确的数学表达，其可被视为局部模型的推广而非简单近似[18]，当近场范围尺寸趋于零时，其理论解答收敛于经典局部理论解答[62-65]，故它是对传统连续介质力学理论的一种有益补充。

近场动力学又可视为一种宏观尺度的分子动力学方法[66]，其采用统一的模型和求解体系描述从原子尺度到宏观尺度的力学行为，突破了经典分子动力学模型在求解尺度上的局限，也避免了传统的多尺度模型在不同尺度力学量传递等方面的复杂性，方便进行跨尺度计算。此外，近场动力学固体力学模型的非局部长程相互作用的建模思想以及积分方程便于处理不连续问题的优势，已被拓展至流体力学问题、扩散问题 (包括热传导、水分传输、离子运移、渗流、电传导等) 和多物理场耦合问题的研究，为相关问题研究提供了新思路[67]。

1.5 近场动力学的研究现状

1.5.1 固体力学问题

近场动力学可分为键型近场动力学 (bond-based PD, BB PD)[6] 和态型近场动力学 (state-based PD, SB PD)[7,68-75]。键型近场动力学描述物质点之间相互作用的本构力函数 (响应函数) 只与近场范围内单一物质点对相关，而态型近场动力学的本构函数与两个物质点各自近场范围内的所有物质点相关 (即两物质点各自受其近场范围内的所有其他物质点作用)。进一步，态型近场动力学又分为常规态型近场动力学 (ordinary state-based PD, OSB PD)[70-73] 和非常规态型近场动力学 (non-ordinary state-based PD, NOSB PD)[74,75] 两类，其区别在于物质点对的力密度矢量状态与变形矢量状态是否共线。

OSB PD 的建模方法是：基于选定的建模变量，构造近场动力学中的弹性变形能密度函数，通过能量密度的弗雷歇 (Frechet) 导数等数学运算推导力状态与变形状态的本构关系式，最后将近场动力学中的弹性变形能密度函数与传统连续介质力学中的弹性应变能密度函数等效，确定 OSB PD 模型参数。NOSB PD 有两类建模方法，一类与 OSB PD 建模方法相似，另一类是与传统连续介质力学本构模型对应的非常规态型近场动力学模型 (non-ordinary state-based peridynamic model corresponding to the classical continuum model)，也称为近场动力学对应材料模型 (peridynamic correspondence material model, PD-CMM)。其中，SB PD 模型是 BB PD 模型的推广，BB PD 模型可视为 OSB PD 模型的一个特例。SB PD 模型突破了 BB PD 模型对材料泊松比的限制，加强了与传统连续介质力学理论模型的联系，拓宽了 PD 的材料和结构适用范围。本节将综述三类近场动力学模型的研究现状。

1.5.1.1 键型近场动力学

键型近场动力学模型具有概念清晰、易于理解和方便实施的特点，也是研究态型近场动力学模型的基础，得到了广泛研究和应用。

2000 年，Silling[6] 首先阐述了近场动力学的基本思想，就近场动力学中材料的各向同性、弹性、调和性、线性化、PD 方程理论解的稳定性条件以及近场动力学与传统固体力学理论之间的联系进行了详细的讨论和严格的数学证明，首次给出以"相对键长"表示的近场动力学微弹性模型，模拟了反平面剪切的裂纹扩展问题，并将分析结果与有限元方法结果比较，证明了近场动力学方法在裂纹尖端不存在应力场的奇异性问题。2004 年，Silling 和 Kahan[76] 在美国高速计算的学术会议上报告了在键层次上考虑微弹性、微弹脆性、微黏塑性和微弹塑性建模的可能性，指出了近场动力学在动力裂纹扩展、冲击侵彻、复合材料断裂失效、超弹性材料破碎、薄膜撕裂和纤维变形等领域的应用前景。随后，Silling 和 Askari[77] 在文献 [6] 定义的面力密度的基础上，给出了近场动力学应力张量的定义，并首次提出 PMB 模型。Gerstle 等 [78] 指出该 PMB 模型微观模量存在错误，并详细推导了三维、二维平面应变和平面应力问题的微观模量和临界伸长率表达，同时指出键型近场动力学存在泊松比限制，即三维和二维平面应变限制为 1/4，二维平面应力限制为 1/3。2005 年，Silling 和 Askari[79] 更正了 PMB 模型微观模量，正式提出 PMB 模型，给出物质点离散、近场范围尺寸和时间步长等参数的确定条件与方法，提出近场动力学显式动力学计算格式，并模拟了带预制中心裂缝的厚方板的裂纹扩展问题、刚性球体冲击下脆性固体生成 Hertzian 裂纹和脆性板的冲击破碎问题。在此之前，Silling 和 Askari[79,80] 还在键层次上考虑塑性效应，提出了微观塑性模型 (microplastic model, MP)，并模拟了 Kalthoff-Winkler 单裂纹动态扩展试验。2008 年，Ladányi 和 Jenei[81] 引入径向基函数 (RBF) 无网格方法和回映算法研究并发展了微观弹塑性模型。为反映工程材料的阻尼特性和黏性效应，Colavito 等在 PMB 模型中增加拉伸速率项和黏性阻尼系数 [82-84]。2013 年，Weckner 和 Mohamed[85] 研究了 PD 黏弹性材料模型，Azizi 等 [86] 将宏观 Burgers 弹簧–阻尼模型引入到 PD 框架中描述材料黏弹性行为，并模拟了蠕变、松弛等固体黏弹性现象。为描述材料弹性渐进损伤行为，Zaccariotto 等 [87] 于 2015 年提出键软化的双线性本构力函数。

Silling 和 Bobaru[88] 还于 2005 年提出适用于二维薄膜的多项式本构力函数，并引入 Lennard-Jones 势函数表示的范德瓦耳斯力长程作用，以描述一维纳米纤维材料质点间的相互作用，两者近场范围不必相同，并模拟了含初始缺陷薄膜的裂纹动力扩展、受内压橡胶气球在刚性碎片打击下的爆炸破碎、薄膜不对称撕裂、薄膜中的摆动裂纹路径、单根纳米纤维和纳米纤维网格的大变形问题，显示了应用 PD 方法模拟强非线性问题的可行性。2008 年，Kilic[84] 提出了考虑热力荷载的典型微观热弹性材料模型，为近场动力学热力耦合研究开辟了途径。针对 PMB 常微观模量不能反映长程力的空间递减特性，2010 年后，Kilic[84]、Ha 和 Bobaru[89] 以及 Huang 等 [90,91] 分别引入了高斯形式、线性递减、高次多项式非线性递减核函数修正 PMB 模型。2012~2014 年，Tao[92] 提出适合纤维增强复合材料的各向异性材料本构力函数，Ghajari 等 [93] 系统发展了适合正交各向异性材料动力断裂分析的键型近场动力学模型，Alali 和 Lipton[94] 发展了具有周期性微结构的非均质材料的近场动力学计算模型。2015 年，Taylor 和 Steigmann[95] 从三维键型近场动力学理论出发，在薄板参考平面的厚度方向上对控制方程进行泰勒展开和渐进分析，简化得到了参考平面内物质点的控制方程，并分析了带预制裂纹板的开裂破坏问题，还有学者提出了基于近场动力学的杆、梁、板理想模型。

2007 年，Demmie 等 [96,97] 为运用 PD 方法模拟爆炸现象，依据爆炸产物的气体内能随体积膨胀的变化规律建立了气体对势本构力函数，研究了弹头爆炸破碎问题，弹头典型时刻的破碎结果与 X 射线图像吻合良好，并研究了混凝土球壳与储水混凝土结构在炸药爆炸作用下的破坏过程，为爆炸问题的研究开辟了新渠道。2012~2014 年，Mikata[98] 和 Buryachenko[99] 提出近场静力学 (peristatics) 概念，专指采用近场动力学方法分析静力问题。2016 年，Ren 等 [100] 发展了对偶双影响域近场动力学理论模型，相互作用的两个物质点可以具有不同尺度的近场范围，有效解决虚假力和虚假应力波问题，适合自适应与多尺度分析；同年，Li 等 [101] 尝试将近场动力学建模思想推广到桁架和张拉整体结构非线性大变形分析中。

键型近场动力学的本构力函数类似于分子动力学方法中的中心对势函数，故由能量等效方法推导得到的单参数模型 (如 PMB 模型) 存在泊松比限制问题，为此，Gerstle 和 Sau 等 [102,103] 于 2007 年发展了微极近场动力学模型 (micropolar peridynamics)，将"微观杆"拓展到"微观梁"，微极模型将三维问题的泊松比扩展至 0~1/4，将平面应力问题的泊松比扩展至 0~1/3。2012 年，Rahman[104] 将微极近场动力学模型与格子模型思想相结合，计算中只考虑物质点最邻近的一圈物质点，在保证精度的同时提高了计算效率。由于原微极模型只考虑物质点的拉伸和物质点对间的相对旋转，2015 年，卢广达 [105] 进一步考虑物质点对的相对扭转，发展了 c-d 和 c-e 模型；同年，Prakash 和 Seidel[106] 考虑"径向键"与"切向键"共同作用，建立了二维双参数的线弹性本构模型，将平面问题的泊松比扩展为 0~1/3。2016~2018 年，Zhou 和 Shou[107] 进一步研究了径向键和切向键模型，Zhu 和 Ni[108] 借鉴虚内键方法，考虑单个键的转动克服泊松比限制，Wang 和 Zhou 等 [109,110] 同时考虑键的拉伸和键之间的转动效应建模，称为共轭键 PD 模型，本书作者随后改进了该共轭键模型，消除了原模型随离散构型变化的不足 [111]。需要指出的是，隐含在上述模型中的"拉伸弹簧和转动弹簧"的建模概念,最早由 Silling 提出[6],并被一些研究者应用建立了 PD 欧拉-伯努利梁模型[112,113]和铁摩辛柯梁模型[114]。Hu 和 Madenci[115]在均匀离散情况下将近场动力学键分为"正交键"和"斜交键",类比传统连续介质力学的拉伸应变和剪切应变的定义,重新定义了两种键的强度计算方式,并模拟了任意泊松比情况下弹性板的变形问题。此外,Liu[116]提出一种对势补偿格式,在微观模量中包含了泊松比,较为有效地解决了泊松比限制问题。

为了确定键型近场动力学模型中的参数,可以分别求得基于近场动力学和经典连续介质力学对应的应变能,再根据能量等效的原则得到相关参数。在这个过程中,物质点应具有完整对称的近场范围,但在边界区域,物质点的近场范围是不完整的,导致边界处的 PD 模型参数计算出现偏离,即存在"边界效应"或"表面效应"问题[8]。Ganzenmüller 等[117]和本书作者等[118]采用实际离散积分区域计算微观模量,提高了边界处参数的计算精度,有效降低了边界效应影响。Le 和 Bobaru 比较了几种边界修正方法对 PD 模型计算结果的影响[119]。

键型近场动力学问世最早,也得到了长足发展,但与经典连续介质力学的联系不够紧密。经典连续介质力学理论中的应力、应变和物性参数等在键型 PD 模型中缺乏描述和使用,虽然定义了 PD 应力和面力密度概念[6,77],但与传统应力概念存在较大差异。2008 年,Lehoucq 和 Silling 定义了积分型应力表达关系,可收敛于传统 PK 应力[120]。随后,研究者发展了非局部微积分理论[121],定义了非局部变形梯度和应变张量等[122,123],定义了应变不变量的 PD 积分表达[124],发展了 PD 微分算子[125],采用 PD 微分算子可以得到传统应力和应变的积分表达。这些研究工作大大增强了近场动力学与经典连续介质力学理论的联系。

1.5.1.2 常规态型近场动力学

态型近场动力学的核心概念是状态(state)算子,状态算子可以是非线性不连续的,张量算子可视为状态算子的一个特例。描述物质点间相互作用的力密度矢量状态(force density vector state)$\boldsymbol{T}$ 和描述物体变形的变形矢量状态(deformation vector state)$\boldsymbol{Y}$ 是一对功共轭量,$\boldsymbol{T}$ 和 $\boldsymbol{Y}$ 以及其他场变量的方程用来刻画材料的本构特征。常规态型近场动力学采用经典连续介质力学张量分解的思想进行建模,Silling 等[7]和 Madenci 等[8]分别构建了常规态型近场动力学模型的公式体系,虽然形式不同,但两套体系在本质上是一致的。

Silling 等[7]首次给出了常规态型近场动力学三维线弹性固体(LPS)模型和流体模型的建模方法,其主要步骤是:基于选定的建模变量,构造近场动力学弹性变形能密度函数,通过能量密度的弗雷歇导数等数学运算,推导力状态与变形状态的本构关系式,随后又建立了相应的线性化模型[68,69]。为减小计算量和提高计算效率,Le 等[70,126]推导了平面应力和平面应变条件下的二维线弹性固体本构模型,但模型不是最简形式。Sarego 等在三维状态线性化模型和二维模型的基础上,推导了二维线性化常规态型模型[71]。Merwe[127] 和 Ouchi 等[128] 利用常规态型平面应变模型研究了水力劈裂问题,其中 Merwe 的模型参数有待商榷。上述三个二维模型存在参数不一致、形式不完全统一的问题。为克服边界效应问题,Mitchell 等[129] 发展了位置依赖的线弹性固体模型 (PALS)。Le[126] 采用所建立的常规态型近场动力学二维线弹性模型,研究了带预制中心圆孔板的单轴拉伸问题,结果均与有限单元法 (finite element method, FEM) 模拟结果高度一致,并进而研究了缺陷对高温超导材料 $Bi_2Sr_2CaCu_2O_x$ 圆丝的局部应力集中等力学行为的影响。本书作者等[130] 采用平面应力模型研究了带中心预制孔板的单轴拉伸弹性变形与多孔环氧板的脆性开裂破坏问题。Sarego 等[131] 采用平面应力模型研究了半圆弯曲试验的不同角度初始裂纹的混合型扩展问题,并开展了 δ 和 m 收敛性分析。基于线弹性常规态型 PD 模型,Dipasquale 等[132] 研究了断键准则对混合型裂纹扩展路径的影响。Imachi 等[133] 计算了混合型动应力强度因子。Butt 等[134] 研究了常规态型 PD 的弹性波传播行为与色散特性。此外,Silling 和 Askari[135] 建立了常规态型近场动力学疲劳损伤模型,在该模型中,键损伤由键的残余寿命决定,残余寿命依循环应变随时间变化,很好地再现了典型材料的 S-N 曲线和帕里斯疲劳裂纹增长规律,模拟了二维紧凑拉伸试件和三维带螺旋裂纹铝杆的疲劳损伤与裂纹扩展问题,验证了该模型的适用性。Lejeune 和 Linder[136] 拓展了常规态型 PD 模型,用于分析细胞分裂导致的肿瘤生长问题。Zhang 和 Qiao[137] 推导了二维和三维常规态型 PD 热弹性模型。D'Antuono 和 Morandini[138] 在常规态型模型的力状态中加入热膨胀项,建立了弱耦合热力模型,分析了陶瓷结构的热冲击问题。Han 等[139] 发展了耦合线性化的常规态型近场动力学与经典连续介质力学模型的 Morphing 方法。Bie 等[140] 提出了基于常规态型 PD 线弹性模型的无网格粒子解法与光滑节点有限元法的耦合建模方法。Ren 等[141] 从总变形中减去刚体转动部分,发展了适用于小转动问题的修正的常规态型 PD 模型。

在弹塑性问题的常规态型 PD 建模研究方面,Silling 等[7] 提出了不可压缩固体的常规态型 PD 弹塑性模型,适用于金属和非多孔介质等材料的弹塑性分析。此后,于继东等[142] 建立了常规态型近场动力学弹塑性模型,并研究了冲击应力波传播问题。Mitchell[143] 类比传统 J_2 塑性理论建模方法,给出了包含矢量状态的弹塑性分解、本构关系、屈服准则、流动法则和加卸载一致性条件等的弹塑性不可压缩固体的近场动力学建模方法,并给出模型实现的回映算法与隐式实现方法。随后,Mitchell[144] 建立了常规态型 PD 黏弹性模型,将黏性效应包含于变形标量状态的偏量部分。Vogler 和 Lammi[145] 发展了适用于混凝土等塑性可压缩固体材料的三维常规态型近场动力学弹塑性模型。Zhou 等[146] 提出了塑性不可压缩材料的二维弹塑性模型,用于分析岩石材料的裂尖塑性区特征。Delorme

等[147]进一步发展了常规态黏弹性模型。张洪武和李辉等[148,149]建立水土混合物的平衡方程,构建了饱和土常规态型近场动力学隐式非线性整体解法的 u-p 格式,模拟了饱和土的固结变形问题。

Madenci 和 Oterkus[8]构造了以体应变、剪切变形量和温度表示的微势能函数,采用虚功原理建立了近场动力学的控制方程,其中的力密度矢量状态由物质点间的微势能函数确定,再对能量密度进行微分运算,得到线弹性固体材料的常规态型近场动力学本构关系式,并根据应变能等效原则推导了该模型的材料参数,此后又拓展至复合材料分析[150]。Dorduncu 等[151,152]采用杆单元,在 ABAQUS 软件中实现了弹性和黏弹性问题的常规态型近场动力学模拟,分析了方板的变形开裂问题。Madenci 等[153]和 Zhu 等[154]分别模拟了混合型裂纹扩展问题。Madenci 和 Oterkus[155]以及 Hu 等[156]分别基于常规态型近场动力学,发展了热黏弹性模型和热–力耦合模型。Madenci 和 Oterkus[157]建立了包含等向强化 Mises 屈服准则的常规态型 PD 弹塑性模型。Chen 等[158]也研究了 PD 弹塑性模型,模拟了材料变形和断裂过程。此外,Tuniki[159]、Richardson[160]和 McVey[161]将态型 PD 建模思想与点阵模型结合,建立了常规态型 PD 点阵模型 (state-based peridynamic lattice model, SPLM),该方法将构型离散为密排六方点阵,物质点与其相邻的所有物质点发生作用,以 PD 力密度矢量状态和变形矢量状态计算物质点间的作用力。

需要指出的是,与键型近场动力学的研究状况相比,常规态型近场动力学的研究成果相对较少,特别是在弹塑性模型的构建、扩散输运、多物理场耦合和应用等方面。

1.5.1.3 非常规态型近场动力学

2007 年,Silling 等[7]提出了基于转动弹簧概念建立非常规态型近场动力学模型的思路,但未给出具体模型及参数的确定方法。此后,Aguiar 和 Fosdick[162]借鉴该思路,考虑键的拉伸和相对转动,提出一个四参数的非常规态型 PD 三维线弹性模型,该模型中两个参数可通过与传统理论的等价关系求得,第三个参数为任意常数,第四个为键的拉伸与转动的耦合参数。Aguiar[163]给出了第三个任意参数的确定方法。O'Grady 和 Foster[112,164]基于转动弹簧概念,建立了非常规态型 PD 一维梁模型,并研究了欧拉–伯努利梁的弯曲问题,研究表明:采用较粗糙的物质点离散构型即可获得高精度弹性变形解答,采用较精细的物质点离散构型和相应较小的近场范围,可得到准确的塑性变形与残余变形解答;当材料满足位移可微条件时,该模型等价于 Eringen 非局部弹性理论,当近场范围趋于零时,得到传统欧拉–伯努利梁方程。O'Grady 和 Foster[165,166]在上述梁模型的基础上建立了板的 PD 模型,研究了基尔霍夫–勒夫板 (Kirchhoff-Love plate) 的弯曲问题, 结果表明非常规态型 PD 梁模型经过简单扩展, 可以与 PD 二维线弹性固体模型结合, 分析板的弯曲变形与开裂问题。

Silling 等[7] 在 2007 年同一篇文章中, 还提出与经典连续介质力学本构关系对应的非常规态型近场动力学模型 (NOSB PD), 也称为近场动力学对应材料模型 (peridynamic correspondence material model, PD-CMM)[9,167]。该模型可在近场动力学框架内使用传统材料的应力–应变 (率) 本构关系, 并保留近场动力学考虑长程力作用效应和自然描述断裂问题的优势, 因而 NOSB PD 模型被广泛应用于多种材料模型, 如线弹性、弹塑性、黏塑性、晶体弹黏塑性、各向同性损伤模型、修正的 Johnson-Cook 模型以及 Drucker-Prager 塑性模型等。2009 年, Warren 和 Silling 等[74] 具体给出了基于经典连续介质力学理论建立非常规态型 PD 模型的完整方法, 实现了弹性–理想塑性本构模型的 PD 建模, 并研究了含缺陷一维杆件静/动力拉伸问题, PD 数值解与解析解吻合良好。此后, Foster 等[75,168,169] 在 EMU 软件中实现了与传统黏塑性本构对应的 NOSB PD 模型, 采用基于能量密度的断键准则, 模拟验证了 Taylor 杆冲击变形与单边裂缝板的开裂问题。Khan[170] 在开源软件 Peridigm 中实现了等向强化弹塑性的 NOSB PD 模型, 分析了剪切带形成的应变局部化问题。Littlewood[171,172] 建立了与弹塑性硬化本构和晶体弹黏塑性本构对应的 NOSB PD 模型, 研究了受内部撞击钢管的动力断裂与单晶体的疲劳裂纹扩展问题。

Littlewood 在研究中首先注意到 NOSB PD 模型会产生由非局部变形梯度计算方法导致的零能模式与数值不稳定问题 (通常以物质点的快速振荡显现), 通过增加罚函数的方法, 引入修正键力密度可以抑制该类零能模式的发生。Tupek 等[173,174] 实现了修正的黏塑性 Johnson-Cook 连续损伤本构模型, 重点研究了 PD 模拟中的数值不稳定现象, 指出过大或过小的近场范围会导致非局部变形梯度张量的计算不精确, 允许物质渗透和物质坍塌等非物理可能变形的发生, 从而造成零能模式和能量消失模式等虚假模态, 指出可通过罚函数方法或监测键的变形, 以抑制数值不稳定现象, 基于常规态的 PD 模型同样不能确保物质渗透等现象不发生, 但基本不产生数值不稳定现象; 采用单高斯积分点的无网格粒子离散与多高斯点积分的有限元离散方法分别实现 NOSB PD 计算, 证明离散化和积分方法对 PD 数值稳定性影响较小, 增加积分点的增强积分能略微改进数值不稳定性。Breitenfeld 等[175,176] 给出了三维线弹性隐式静力分析方法, 并注意到该数值不稳定现象在裂尖等高应变区尤为显著, 进而提出在原力密度矢量状态中引入三种不同控制零能模式的力密度矢量状态函数, 以提高计算稳定性与计算精度, 具体为: ①弹簧键力; ②平均近场范围内位移场; ③沙漏力, 但其中的参数不易确定, 参数过小无法有效地抑制零能模式, 过大则易导致结果偏差较大, 需要通过多组数值试验确定合适的参数。Sun 和 Sundararaghavan[177] 进行了晶体弹塑性静力变形分析，他们的研究指出：对所有边界施加位移约束条件能有效抑制中等变形问题的零能模态产生，然而该方法可能仅适用于渐变位移边界条件问题，并仍然残留较为明显的数值振荡。Amani 等[178]采用修正的 JC 模型，以分析热塑性断裂问题，并通过 Taylor 冲击和 Kalthoff-Winkler 试验的模拟，对所建立的模型进行了验证，但他们未提及稳定控制的措施，数值结果也显现出轻微的数值不稳定现象。Wu 和 Ren[179,180]通过多个数值算例证实：非常规态型 PD 模型存在数值不稳定问题，且易造成问题的求解失效，并指出边界条件的不准确施加会加剧边界处数值计算的不稳定。他们利用凸核近似改进了影响函数，通过局部与非局部耦合方法实现了本质边界条件的准确施加，采用位移场加权平均技术改进了 PD 力密度矢量状态，较为成功地抑制了 PD 数值不稳定问题，并基于上述技术与塑性损伤本构模型，模拟了金属拉伸韧性断裂与金属切削问题。Yaghoobi 等[181,182]在 NOSB PD 框架内分别建立了混凝土基体的各向同性损伤模型和纤维的微机械模型，分析了纤维增强混凝土结构的力学响应与破坏特征。Yaghoobi 等[183]采用 Taylor 高阶多项式描述变形梯度，并通过给定不同位置物质点不同的影响函数值，以降低变形梯度的计算误差，较好地控制数值不稳定问题。Yaghoobi 和 Chorzepa[184]还提出了 NOSB PD 模型的对称边界条件施加方法，并采用 NOSB PD 与 FEM 耦合方法分析了结构线性静力变形问题。谷新保等[185,186]采用无稳定控制的线弹性模型，分析了含预制裂纹的三点弯曲梁和含中心预制裂纹的巴西圆盘压缩破坏问题。Wang 和 Zhou 等[187-190]发展了适用于混凝土和岩石类准脆性材料断裂分析的 NOSB PD 方法，建立了与线弹性本构模型对应的 NOSB PD 模型，采用基于应力的断键准则与线性弹簧类修正力状态抑制零能模态，获得了裂纹起裂和扩展路径的主要特征，但模拟得到的损伤区域较宽，采用更高效的稳定控制措施可以提高裂纹扩展路径的预测精度。Lai 等[191,192]建立了适用于饱和岩土体的 Drucker-Prager 塑性 PD 本构等模型，进行了裂纹扩展、侵彻冲击、剪切带演化、岩土破碎和边坡稳定等问题的数值模拟，所得结果与试验和有限元结果均吻合良好。Ren、Fan、Bergel 和 Li 等[193-195]发展了 NOSB PD 与光滑粒子流体动力学 (SPH) 的混合建模方法，模拟了内爆炸载荷作用下的土体破碎问题，并采用施加 Monaghan 人工阻尼和虚粒子方法，消除了冲击波传播的数值不稳定性。Bergel 和 Li[196]发展了物质坐标与空间坐标中的非局部微分算子，给出了 NOSB PD 模型的整体拉格朗日与更新式拉格朗日格式。Yu 等[197]建立了单向层压板的非线性弹塑性本构模型，模拟了单轴拉伸载荷作用下层压板的面内变形与渐进损伤过程，类似于在边界区域外设置虚拟区域，他们提出了一种补偿修正方法，以降低非完整近场范围导致的边界计算误差，但数值结果仍可见位移振荡。本书作者等[198]通过弹性静力变形与弹性波问题研究了罚函数方法的稳定控制效果，指出带有递减影响函数的罚函数方法能有效抑制数值不稳定问题，

1.5 近场动力学的研究现状

并指出在瞬时动力问题分析中,阻尼力与人工黏性项有利于降低零能模态导致的数值不稳定问题。Du 和 Tian 等 [199] 也指出影响函数对 NOSB PD 建模具有重要影响。Liu 等 [200] 采用各向同性弹性模型与临界等效应变断键准则,模拟了弹脆性冰的碰撞破碎问题,但其结果显现出数值不稳定,并且载荷和几何构型对称问题的计算结果不具对称性,这可能是缺少稳定控制的缘故。Sami [201] 建立了弹塑性线性硬化模型,预测了冷挤压孔的残余应力分布。Ganzenmüller 等 [202] 证明了 NOSB PD 与修正 SPH 的等价关系,认为 NOSB PD 中的零能模态源于粒子离散和单点积分,建议采用更精确的积分格式以提高数值稳定性。Bessa 等 [203] 证明了 NOSB PD 与再生核粒子法 (RKPM) 在均匀离散下的等价性和区别,指出非常规态型 PD 模型存在准确施加边界条件的困难和数值稳定性问题,但并未给出解决措施。Silling [167] 认为 NOSB PD 模型的数值不稳定根源在于材料不稳定,少部分源自无网格粒子离散,他认为近场范围内物质点的零能模态变形导致 PD-CMM 不满足材料稳定性条件,并通过最小能量原理导出关联材料模型,在应变能密度中引入额外项抑制均匀变形的位移偏离,该方法能够显著消除数值不稳定现象,但仍有轻微残余。李潘等 [204] 基于线性化 PD 键理论,推导非均匀变形状态对应弹性系数张量的具体形式,建立基于线性键理论的稳定关联材料模型,有效抑制了零能模式引起的数值不稳定现象,避免了参数调节过程。本书作者等 [205] 推导得到了 SPH、梯度修正的 SPH(CSPH)、再生核质点法 (RKPM)、梯度再生核质点法 (G-RKPM) 和 NOSB PD 的非局部核函数积分形式的统一表达,比较了各方法的完备性、计算复杂性和异同点,指出带有特定影响函数的 NOSB PD 与 SPH、CSPH、RKPM、G-RKPM 分别具有等价性,但可有效避免影响函数导数的复杂计算,并进一步提出了键关联高阶稳定的非常规态型近场动力学模型,极大地提高了原模型的计算精度与稳定性,但新模型所需内存与计算量增加较多。

此外,Chowdhury 等 [206,207] 建立了微极材料的 NOSB PD 三维模型,并推广为一维梁和二维壳模型,模拟了梁板壳的弹性变形与破坏问题。Song 等 [208,209] 采用非饱和土体弹塑性本构关系,提出了考虑力–化–水耦合效应的 NOSB PD 方法,分析了非饱和土的应变局部化变形。

综上所述,以非局部变形梯度和应力张量为媒介,NOSB PD 模型可以将经典连续介质力学发展的诸多本构关系转化到 PD 框架内,增强了近场动力学与经典连续介质力学的联系,传统的各种宏观力学参数可以直接使用,以分析各种材料复杂的力学响应。但 NOSB PD 模型存在的两个问题不容忽视,一是 NOSB PD 模型存在零能模式变形与数值不稳定性问题,往往表现为位移场的锯齿形振荡,在裂尖或加载区域等变形梯度较大区域尤为显著,可能导致计算结果偏离准确解,严重的甚至使计算结果发散。NOSB PD 的数值不稳定现象与变形梯度和力密度矢量状态的计算方法相关,少部分源自无网格粒子离散和单点积分方法,故需要深入分析 NOSB PD 数值不稳定的原因，提出行之有效的解决方案。二是 NOSB PD 模型中涉及形状张量和变形梯度张量的矩阵计算，需要基于近场范围内物质点的积分求和，然而裂纹扩展或局部严重损伤等大变形会造成拉格朗日型的近场范围内物质点数目不足，使得 NOSB PD 模型的矩阵条件数变差甚至奇异，导致变形梯度的计算误差较大，并且，由于近场动力学物质点受到的作用力是非局部的，近场范围内物质点不足意味着受力描述的不准确，可能导致数值计算不稳定，计算结果失真或计算终止。除此之外，NOSB PD 模型的显式动力学解法发展较为成熟，而隐式静力解法研究较少，缺乏相关的数值实施细节，需要系统研究 NOSB PD 的隐式静力解法。

1.5.2 流体力学问题

在流体运动模拟方面，Berge 和 Li[196] 首先研究了 NOSB PD 模型的完全拉格朗日格式与更新拉格朗日格式描述，Tu 和 Li[210] 发展了牛顿可压缩与不可压缩流体的 NOSB PD 的更新拉格朗日格式，模拟了溃坝和气泡运动问题。Wang 和 Zhang[211,212] 采用 NOSB PD 的非局部梯度张量和积分方程建模，数值模拟结果的计算精度和稳定性都较移动粒子半隐式方法 (MPS) 有所提高。Ren 和 Zhuang[213] 提出了弱可压缩流体的非局部积分公式，采用非局部散度和有限差分近似梯度推导得出内力矢量。Bazazzadeh 等[214] 基于不可压缩黏性流体的势流理论，采用近场动力学微分算子进行求解，模拟了二维、三维水池中液体晃荡问题。Gao 和 Oterkus[215,216] 采用近场动力学微分算子将 N-S 方程积分化求解，基于该非局部模型的整体与更新拉格朗日数值方法模拟了小雷诺数下多个层流运动实例，如：库埃特流、泊肃叶流、Taylor-Green 涡旋、剪切驱动空腔问题和溃坝问题。Katiyar 等[217] 基于质量守恒的变分原理，考虑多孔介质的物质点间存在的流动通道，提出了多孔介质渗流的常规态型近场动力学模型，建立了单相牛顿流体 (小常压缩系数液体) 在二维非均质各向异性多孔介质中渗流的 PD 模型，模拟了均质和非均质方形区域的五点井汇流问题，该模型后来被拓展应用于多相多组分的非牛顿可压缩流体[218]。

在流固耦合分析方面，Dalla 等[219] 提出了浸入边界方法 (IBM)–近场动力学混合方法，以求解流体–结构耦合问题，模拟了浸没二维圆柱体的绕流问题与槽流中弹性板弯曲问题。Liu 等[220] 发展了水–冰耦合作用的键型近场动力学与更新拉格朗日粒子水动力学方法，模拟了刚性柱体入水和冲击水面浮冰的过程。Zhang 等[221] 发展了耦合近场动力学–离散单元法–浸入边界法–级联格子玻尔兹曼方法 (PD-DEM-IB-CLBM)，采用 DEM-IB-CLBM 模拟黏性流体中的固体粒子运动，采用 PD 模拟固体表面损伤，通过黏性流体中单颗粒碰撞墙壁问题的模拟，验证了所提出方法的有效性，并进一步模拟计算了单/多个粒子从不同角度撞击固体

1.5 近场动力学的研究现状

表面的损伤过程，分析了撞击角度与流体性质对于固体损伤以及固体表面弹坑对于流体运动的影响。

1.5.3 输运扩散问题

热传导、水分传输、离子传输、渗流、电传导等输运扩散问题往往具有非局部效应，也是多物理场耦合问题的研究基础。研究者开展了基于近场动力学方法的扩散模型或传热传质研究，Gerstle 和 Read 等[222,223]首次给出了热传导、电传导，以及原子和空穴扩散的键型近场动力学非局部扩散模型，提出了温度场、电场、原子空穴浓度场变化的 PD 表述和数值求解方法，研究了一维电子迁移的多物理耦合过程，得到了机械变形、热传导、电势分布 (电荷流动) 和原子空穴扩散的结果，为扩散问题和多物理场耦合问题的 PD 建模研究奠定了基础。Oterkus、Fox 和 Madenci[224]将电迁移问题的近场动力学分析方法拓展至高维，研究了由电迁移、热迁移和应力迁移诱发原子空穴扩散而导致微结构材料性质劣化的问题。Bobaru 和 Duangpanya[225,226]从热传导 Fourier 定律和能量守恒法则出发，构建了键型 PD 一维、二维非稳态热传导方程，基于线性变化势场分布和局部与非局部等效方法，详细给出了两种微观热传导函数的推导方法，通过典型一维、二维热传导问题验证了 PD 解答，并研究了考虑含裂纹方板的热传导问题。Agwai、Guven 和 Madenci[227,228]通过典型一维热传导问题验证了该键型 PD 热传导模型的合理性，Agwai 和 Madenci、Oterkus[229]还采用 Euler-Lagrange 方程和热能守恒方法推导了态型近场动力学热传导方程，初步给出了非常规态型热传导模型，简化后得到键型近场动力学热传导方程，提出了 3 种热响应 (内核) 函数形式以及对应的微观热传导系数的计算方法，给出详细的空间离散积分、时间差分、初始条件和边界条件施加的数值实现方法，验证了键型 PD 模型在分析一维杆、二维板和三维块体在多种边界条件下热传导问题的有效性。Chen、Bobaru[230] 和 Jafari 等[231]分别研究了带有不同热响应 (内核) 函数键型 PD 热传导模型解答的 δ 收敛性和 m 收敛性问题。刘硕等[232]采用键型近场动力学无网格粒子法与有限元耦合方法研究了二维热传导问题。王飞和马玉娥等[233]给出了非均匀材料的键型 PD 热传导方程，通过键两端物质点热传导参数的加权平均描述非均匀特征，并分析 PD 各个计算参数对热传导结果的影响。刘英凯和程站起[234]研究了功能梯度材料的热传导问题。Liao 等[235]采用非常规态型近场动力学的缩减算子定义了非局部积分型温度梯度，构建了非常规态型近场动力学热传导模型。Wang 等[236,237]研究了上述 Fourier 型热传导的键型 PD 模型的 Green 函数解答，并进一步提出了非 Fourier 型热传导的 PD 模型，给出了广义态型热传导方程和键型热传导方程，并对比传统非 Fourier 结果和试验结果，验证了该键型 PD 模型的有效性。王超聪等[238]建立基于接触近邻的近场动力学热传导模型，引入热烧蚀断键准则, 模拟了热防护材料的热烧蚀温度场演化过程。Jabakhanji 和 Oterkus[239,240] 在他们的博士学位论文中系统建立了热传导、水分浓度扩散、渗流水分含量扩散和电传导问题的键型近场动力学模型, 并给出了当近场范围为完整对称线段、圆形或球形时, 不同物理场扩散问题的响应函数及其对应的一维、二维和三维 PD 微观传导/扩散系数。需要指出的是, 由于采用相同近场范围内的积分表达, 近场动力学运动方程和扩散方程可以统一在一个非局部微分算子框架内[121,241,242]。

Han 等[243] 与 Diyaroglu 等[244,245] 给出多种物理场的键型近场动力学扩散方程形式, 在大型商用软件 ANSYS 中, 通过 LINK33 三维热传导杆单元和 MASS71 热质量单元, 分别实现了 PD 热扩散、水分浓度扩散、电传导、归一化湿度场扩散问题的数值模拟, 充分利用了 ANSYS 软件的高效隐式求解器, 得到的一维杆和三维交叉堆叠封装结构浓度扩散问题的 PD 预测结果具有很高的精度。

Jabakhanji 和 Mohtar[246] 考虑 "均质单管" 水分运输通道, 建立了适用于饱和、均质、各向同性(或各向异性)土壤介质(多孔介质)中水分流动的键型近场动力学扩散模型, 包括水分流动的扩散方程和湿度通量表达, 给出基于传统水分流动模型确定一维、二维 PD 微观水力传导系数的方法。进而针对非均匀或非饱和土体不具有常量水力传导系数的特点, 考虑 "并行双管" 水分运输通道, 将该模型推广到非饱和、非均质土壤介质, 对比验证了一维均匀饱和土柱的排水问题、二维水平土层中的水分扩散问题以及初始饱和的二维垂直非均质土柱排水问题。此外, 同一阶段, Delgoshaie 等[247] 与 Jenny、Meyer[248] 从离散孔隙网络结点的流量守恒出发, 将多孔介质均匀化为连续介质, 建立了积分型非局部达西定律, 归一化后, 在空间均匀对称下, 能复现非局部扩散模型方程[241], 与局部连续达西定律相比, 积分型非局部达西定律得到的结果更接近孔隙网络流动模型的计算结果。

1.5.4 多物理场耦合问题

多物理耦合问题具有重要的科学价值和工程意义, 实际工程结构都处在力、水(水分、蒸汽、渗流)、热、电和磁等多物理场环境中, 涉及不同的多场耦合建模, 如: 微机电系统涉及力–热、力–电、电–热、力–电–热、力–热–电–磁、湿–热–力等多场耦合, 环境地质与岩石力学问题涉及热–水–力耦合、热–水–力–化耦合, 高寒地区冻土和混凝土结构、高温干旱和潮湿地区的混凝土结构开裂破坏均涉及水(湿)–热–力耦合, 页岩气等油藏开采与大坝的高水压水力劈裂问题与多孔弹性介质的水–力耦合问题密切相关。随着研究的不断深入, 各工程领域对多物理场耦合模型和仿真软件的需求大、精度和稳健性要求高[249,250], 发展基于近场动力学的多物理场耦合模型、数值计算方法和计算软件将为多场耦合问题的研究开辟新的途径。

1.5.4.1 考虑多物理场单向作用效应的非耦合 PD 模型

在近场动力学中，本构力函数或力密度矢量状态的构造是建模的关键，在单纯力场中，通常依赖于近场范围内物质点间初始和现时相对位置。分布规律已知的温度场、水分或离子浓度场、多孔介质渗流场等对固体结构变形场的影响可以反映在本构力函数或力密度矢量状态中，类似于传统的热弹性力学问题，是一种单向非耦合的多物理场作用下的近场动力学模型。

为研究热力载荷作用下的电子封装结构的力学响应，Kilic 和 Madenci[251,252] 首次提出了含温度项的键型 PD 热力本构函数，采用该非耦合 PD 热力模型分析了给定变温下物体的变形规律，研究了含初始裂纹的淬火玻璃在不同初始缺陷和温度条件下的裂纹扩展问题，捕捉到了裂纹偏转、分叉及连接现象。此后，Jeon 等[253]、Mella 和 Wenman[254]、Xu 等[255] 以及 Giannakeas 等[256] 分别采用非耦合键型 PD 热力模型研究了钢化玻璃、核燃料芯块、淬火玻璃板、陶瓷板块在常规热力载荷、冷水浴或热冲击载荷作用下的动态裂纹扩展。基于同样的建模思路，Madenci、Oterkus[155] 与 Zhang、Qiao[137] 在力密度矢量状态中加入物质点变温项，分别建立了非耦合的常规态型近场动力学热弹性模型和热黏弹性模型，分析了热力载荷下矩形板变形与双材料梁的变形开裂问题。Oterkus 等[257] 在键型 PD 本构力函数中考虑湿度增量、蒸气压或水压的影响，建立了非耦合的湿–热–力 PD 模型。苏伯阳[258] 等将该湿–热–力模型应用到非均匀复合材料中，模拟了不同湿热环境下复合材料的冲击损伤问题。

1.5.4.2 热–力耦合问题的 PD 模型

在热–力完全耦合的 PD 模型中，温度场对变形场的影响可通过热力本构模型反映在固体力学方程中，而结构变形会导致热传导方程中的热源项或边界条件有所变化。Agwai[227] 和 Oterkus 等[259,260] 基于热能和机械能量守恒方程以及热力学自由能函数[69]，建立了热–力完全耦合的态型近场动力学模型，并可简化得到热–力耦合键型 PD 模型与无量纲化的热–力耦合键型 PD 模型。他们采用交错差分格式求解耦合方程，分析验证了一维、二维和三维热传导问题，研究了一维杆件、二维均质板和单向纤维增强复合薄板的热力耦合变形问题。基于上述热–力完全耦合的键型 PD 模型，Chen 等[261,262] 忽略热传导方程中的变形影响项，采用 Newton-Raphson 方法通过 MOOSE 软件实现其隐式计算，分析了二维和三维核燃料芯块的断裂问题。Hu 等[156] 考虑空间离散为非规则、非均匀情形，建立了固体力学和热传导问题的键型及常规态型近场动力学模型，验证了非均匀离散的 PD 模型在分析结构变形和热传导问题中的有效性，并采用隐显式方法分析了三维核燃料芯块的断裂问题。D'Antuono 和 Morandini[138] 采用常规态型 PD 热–力本构方程描述结构变形，以键型 PD 热传导方程描述温度场变化，发展了热-力弱耦合的 PD 模型，采用多速率显式积分技术求解两类控制方程，在 Peridigm 中模拟了弹脆性陶瓷薄板与厚板的热冲击致裂行为，观察到了二维有序平行裂纹集和三维柱状节理蜂窝裂纹模式。Wang 等[263]采用热-力耦合键型 PD 模型分析了岩石的裂纹扩展问题。Oterkus[240]系统研究了多物理场耦合问题的近场动力学模型，除上述热扩散和完全耦合的热-力模型外，还涉及电子封装领域湿热蒸汽变形的湿-热-力耦合问题[257]、电子迁移致损问题[224]、高温环境下聚合物基复合材料表面氧化或老化的热-氧耦合问题[264]以及核燃料芯块高温开裂的热-力氧耦合问题[265]等。

1.5.4.3 力-电、热-电和热-力-电耦合问题的 PD 模型

针对柔性电子器件领域的多物理作用，Roy 等[266]提出了力-电（挠曲电）耦合的近场动力学模型，该模型能够描述纳米尺度挠曲电耦合导致的对称电介质非均匀变形问题。基于纳米尺度的电子跃迁将会引起材料的压电电阻效应，且电流具有非局部性。Prakash 和 Seidel[267,268]考虑变形对电子跃迁和电势分布影响的单向作用效应，建立了力-电耦合键型近场动力学模型，通过改变导电系数和压阻系数，分析了碳纳米管 (CNT) 增强聚合物纳米复合材料的压电电阻响应问题。

张振宇[269]基于 Seebeck 效应、Peltier 效应、Thomson 效应和高斯定理，建立了含缺陷热电材料热-电耦合的近场动力学模型，分析了三维热电器件的热电转换效率问题。Assefa 等[270,271]采用广义 Fourier 热传导定律和广义欧姆定律，考虑热-电耦合效应，基于热能守恒和电荷守恒，建立了态型近场动力学热传导和电传导方程，还建立了热-电耦合的键型 PD 方程，得到了一维和二维热-电耦合问题的键型 PD 模拟结果。

Wildman 和 Gazonas[272,273]采用近场动力学方法研究了固体电介质在热-力-电耦合作用下的脆性失效破坏问题，在他们的研究中，忽略热扩散效应，仅在 PD 本构模型中考虑由电流焦耳热效应导致的温度改变，将洛伦兹力和开尔文极化力等静电力加到 PD 本构力中，以反映静电场对机械变形场的影响，电传导系数是温度和电场的非线性函数，结构变形损伤将改变介电常数，并采用有限差分法或有限元法求解静电场方程，采用无网格粒子法求解近场动力学固体力学方程，介电材料方板在高电压下的裂纹扩展模式与试验观测到的脆性固体通道状和树状的介电击穿失效现象吻合良好。

1.5.4.4 多孔弹性介质流-固耦合问题的 PD 模型

鉴于传统多孔介质弹性理论不能很好地模拟土壤干缩致裂、油藏开采中水力劈裂等不连续变形问题，研究者发展了近场动力学框架内的多孔介质弹性理论，以描述多孔介质中液体流动与多孔介质固体变形的相互作用。

Turner[274]在力密度矢量状态中考虑流体孔隙压力效应,以反映流体压力对多孔介质变形的影响,建立了PD多孔介质弹性分析的常规态型近场动力学模型,并分析了地层中液体采掘导致的表面沉降和饱和土体固结问题,但该模型通过解析解或者其他数值方法预先给定孔隙压力,不考虑固体变形对液体流动和孔隙压力变化的影响,因而属于非耦合的流固相互作用分析模型。Jabakhanji[239]基于PD方法分析土壤干缩致裂问题,采用土壤收缩特征曲线(含水量减少导致土壤体积改变)以考虑水分流动对固体变形和开裂的影响,并与土约束试验的干缩开裂结果进行比较,但该模型只能分析含水率减少导致土壤体积改变引起的土壤收缩致裂问题,且也仅是包含含水量对固体变形作用效应的单向耦合模型。基于Katiyar等[217]提出的多孔介质渗流模型和常规态型近场动力学模型,Ouchi等[128,275−278]在常规态型PD弹性模型中进行孔压修正,以反映流体运动对固体变形的作用,将多孔介质孔隙率设为介质应变、孔压和总平均应力的函数,以反映固体变形对流体运动的影响,发展了适用于流体驱动断裂问题的非均质多孔弹性介质的流-固耦合模型,通过一维饱和岩体的固结理论结果对该流-固耦合模型进行了验证。进而采用含裂隙渗透系数和孔隙度的孔隙流动方程描述裂隙区域流体运动(裂隙渗透系数和孔隙度分别依赖于裂缝宽度和损伤情况),同时考虑裂隙流体和孔隙流体的对流输运,通过非常规态型PD的力密度矢量状态施加初始地应力场,分析了平面应变条件下单相流驱动单裂纹扩展问题,研究了水力裂缝与单/多自然裂缝的相互作用,模拟了天然裂隙储层中多水力裂纹扩展问题,探讨了微观小尺度非均匀性对岩体水力裂纹扩展模式的影响,以及储层非均质特性对垂直水力裂纹偏转、弯曲和分叉的影响。

Nadimi等[279]采用PDLAMMPS软件带有的常规态型近场动力学黏弹性模型,分析了不同注水速率下三维非均质地层材料的水力劈裂问题。Edmiston[280]联合运用损伤依赖的局部渗流方程和键型近场动力学方程,提出流体流动与固体变形的双向耦合分析方法,并用于求解水力劈裂问题,该模型通过形状函数插值类数值方法计算裂隙压力,并通过体力密度项施加到近场动力学运动方程中,通过变渗透系数反映变形和损伤对局部渗流方程的影响,但未考虑结构变形对渗流方程孔隙率的影响,也没有考虑孔压变化对结构变形的影响。Oterkus等[281]建立了双向流-固耦合问题的近场动力学多孔介质弹性模型,该模型在键型PD运动方程中引入流体孔压项,通过建立非局部形式的达西渗流方程,以反映固体变形对孔压变化的影响,分析验证了经典的五点井汇流问题、一维和二维固结问题和水力压裂引发的裂纹扩展问题,但同样没有考虑结构变形对孔隙率变化的影响。吴凡等[282]采用键型近场动力学模型,通过追踪裂纹面施加水压力载荷,初步模拟了射孔水平井剖面的水力压裂问题。此外,Ishimoto等[283]建立了无网格粒子类近场动力学方法与欧拉有限元法的混合模型,模拟了压力容器壁裂纹扩展时的氢气泄漏问题。

1.5.4.5 腐蚀损伤的力–化耦合以及湿–热–力耦合问题的 PD 模型

腐蚀环境中金属表面易发生点蚀现象，致使材料和结构发生腐蚀损伤开裂，点蚀过程可视为金属离子在固/液双相材料中的溶解扩散问题。研究者基于近场动力学非局部扩散方程，进行了金属材料的电化学腐蚀破坏分析。Chen 和 Bobaru[284]基于 PD 扩散方程和运动方程开展点蚀损伤研究，采用近场动力学方法描述腐蚀过程中阳极反应的浓度变化过程和相边界移动，建立了基于浓度与近场动力学"力键"损伤关系的腐蚀损伤模型，分别进行了一维、二维和三维问题的金属材料点蚀过程分析，结果与试验观察到的表面损伤现象具有一致性，证明该 PD 腐蚀模型能够模拟一定厚度范围内金属腐蚀对结构损伤的影响。Chen 等[285]进一步考虑点蚀过程中活化和扩散的影响，研究了表面钝化膜对结构点蚀损伤过程的影响。Jafarzadeh 等[286]在前述近场动力学腐蚀模型的基础上，计及再钝化膜和盐膜的形成过程，对不锈钢点蚀的再钝化处理进行了模拟分析。白小敏等[287]实现了不同过电位下一维点蚀的近场动力学数值模拟。De Meo 和 Oterkus[288]在商业有限元软件 ANSYS 中建立了近场动力学点蚀损伤模型，利用商业软件中的隐式求解算法模拟了二维点蚀的损伤开裂问题。De Meo 等[289]采用近场动力学方法，计算模拟了水溶液中含预制裂纹薄铁板的吸附氢应力腐蚀开裂问题，得到的裂纹扩展速度和分叉行为与试验结果吻合良好。

电子封装结构在制造过程中，由于湿热应力和蒸气压的存在，吸收的水分易使电子封装结构产生变形与损伤。Oterkus 等[257]在已知热量温度场和水分湿度场的条件下，构造包含变形、温度、湿度和蒸气压的键型 PD 本构力函数，建立了键型 PD 热传导模型、水分浓度扩散模型和湿–热–力耦合模型，给出了蒸气压的计算方法，分别验证了热–力模型和水 (湿)–力模型的预测结果，以及电子封装结构在湿–热–力共同作用下的结果。Wang 等[290]采用近场动力学研究了锂电池的断裂问题，采用近场动力学微分算子理论将裂尖局部形式的锂离子浓度方程转化为对应的非局部积分形式，同时在键型 PD 本构力函数中加入锂离子浓度的影响，初步模拟了含多裂纹电极板的开裂问题。

1.5.4.6 多物理场耦合问题 PD 模型的数值解法

采用显式动力学方法求解 PD 积分方程时，空间域常采用均匀无网格粒子离散，且每一物质点所属变量为常数，空间积分采用中心单高斯点积分法，时间域采用向前差分、中心差分或四阶龙格库塔差分等格式，以使得计算结果的稳定性、收敛性和精度得到保证[79]。通常来说，多物理场模型的控制方程是相互耦联的，常用的解法包括：

(1) 分离交错式解法：将所有控制方程分开考虑，按时间序列交错迭代求解，直到计算结果满足收敛条件[260,291]，发展合适的无条件稳定交错算法能较为高效稳定地求解耦合方程[292]；

(2) 整体解法：将所有控制方程作为一个系统，采用隐式方法整体求解方程变量[260]；

(3) 通过提供方程变量间的函数关系，将不同场方程解耦成非耦联的，分别求解各自场的独立积分方程。

目前，PD多物理场研究大多采用分离交错式解法，但不同控制方程显式时间积分的稳定时间步长可能存在较大的量级差距，交错计算方式极大地增加了计算时间，无条件稳定的隐式时间积分方法将具有优势。此外，常采用的基于空间均匀离散的PD计算格式，欲获得高精度结果通常需要匹配较精细的离散网格，从而导致计算量巨大，因此有必要发展非均匀离散的PD计算格式以及基于GPU等的并行计算方法。

1.5.5 近场动力学与其他数值方法的混合建模与多尺度分析

近场动力学兼具连续介质力学和分子动力学的特点，其数值实施方法可归结为无网格粒子类方法。为充分发挥近场动力学以及其他数值方法的各自优势，研究者发展了近场动力学与其他方法的混合建模方法，包括：PD-有限单元法[11]、PD-有限差分法[293]、PD-分子动力学方法[294,295]以及PD-SPH等其他无网格方法[193-195]。

非局部/局部模型的耦合理论以及相应的PD-FEM混合建模方法依然是目前计算力学界的研究热点。2016年，本书作者[11]综述了近场动力学的有限元实现方法，总结了近场动力学–有限元混合建模方法的研究进展，阐明了各种混合建模方法的基本原理和特点，并提出了一种基于力耦合的近场动力学–有限元混合建模的隐式分析方法。Wildman和Gazonas[293]发现PD方法的频散程度较为严重，提出了近场动力学–有限差分法(FDM)混合方法，可以减弱近场动力学计算结果的波频散程度，有效地模拟了波传播问题和固体动态裂纹问题。卢志堂等[296]研究了一维弹性波PD模拟的数值频散特性，并联合应用FDM-PD方法分析了岩石杆件的层裂问题。此外，研究者多采用拉格朗日型无网格粒子方法实现近场动力学计算，以避免各种常规无网格方法中不连续界面跨越判断和拉伸不稳定等问题，并提出了NOSB PD与光滑粒子流体动力学(SPH)的混合建模方法[193-195]。

在近场动力学多尺度分析方面，Askari等[297]综述了近场动力学从微观尺度到宏观尺度的应用情况，指出近场动力学属于原子尺度到宏观尺度的跨尺度模型，可以自然实现多尺度问题的统一建模和跨尺度数值模拟。Badia等[298]指出近场动力学有望成为原子尺度与连续介质尺度模型自然过渡的良好手段。Sears

和 Lehoucq[299,300] 从统计力学原理出发,推导出近场动力学的动量和能量守恒方程,指出近场动力学方法处于经典统计力学和传统连续介质力学的中间状态,有望采用近场动力学–连续介质力学的耦合方法代替原子理论–连续介质力学的耦合方法。Rahman 和 Foster[301] 引入经典统计力学,发展了 PD 方法的统计力学框架,并指出近场动力学的均匀化模型近似于传统连续介质力学模型。Lehoucq 和 Silling[302] 从微观分子动力学有限温度下的多体势函数出发,通过统计粗粒化过程,给出了 PD 宏观本构力函数的构建方法,该方法将原子集合视为 PD 物质点,通过统计力学方法均匀化物质点,进而通过比例缩放,由小尺度模型建立更大尺度的模型。Silling[15] 给出了近场动力学多尺度分析的多步多级粗化建模方法,可由小尺度模型经过不断粗化建立大尺度模型,一维算例的结果证明了所提出方法的有效性。Emmrich 和 Weckner[303] 以及 Seleson[304] 证明了近场动力学和分子动力学可以通过高阶梯度理论建立联系,数值论证了近场动力学方法可得到分子动力学的解答精度。Rahman、Haque 和 Foster[294,295] 建立了"自上而下"和"自下而上"两种 PD 分层多尺度模型以及 PD 和 MD 的分层多尺度模型,PD 和 MD 模型可以通过细化的小尺度 PD 模型自然连接,并分别采用 MD-FEM 耦合模型、PD-MD 分层多尺度耦合模型,分析了单边裂缝板拉伸的裂尖位移。Rahman 和 Foster[305,306] 修正了所提出的"自下而上"的近场动力学多尺度模型,以能够对不同尺度非局部作用进行比例缩放,并联合使用幂次定律和握手区,可以无缝桥接局部与非局部模型。针对非局部模型不同尺度的过渡区以及局部–非局部模型的过渡区易发生虚假力和虚假波反射问题,提出采用分数阶幂次核函数的方法,以有效抑制波反射问题。Silling[307] 发展了近场动力学分层多尺度方法,从关注的结构重点区域出发,最内层采用紧密粒子离散,依次对外围区域采用较稀疏粒子离散,并与 XFEM 和黏聚单元方法计算结果比较,模拟效果良好。Duzzi 等 [308] 基于四叉树的非均匀离散与自适应方法,建立了键型 PD 并发式多尺度模型,将重点关注区域离散至纳米尺度,并赋以纳米纤维的材料参数,求解分析了纳米复合材料的破坏问题。

1.5.6 近场动力学相关应用

近场动力学 (PD) 具有两个显著的特点:一是模型中显含非局部长程作用,另一是积分方程的建模思想 (避免了空间导数)。近场动力学在与上述特点有关的问题中自然能够发挥其作用,但不同的力学问题对上述两个方面的需求侧重也不同。如材料和结构的宏观失效破坏分析更侧重 PD 不连续分析的优势,在进行碳纳米管等微结构的弹性波传播、变形和振动特性分析时,更关注其非局部特点,而在复合材料的失效破坏问题中,非局部性和破坏引起的不连续特征均是重要的影响因素。迄今,近场动力学已在不同尺度诸多材料的静动力连续/不连续问题中得到

1.5 近场动力学的研究现状

广泛应用。涉及的主要材料包括：理想均质材料、纳米纤维、橡胶薄膜、多晶体微结构、金属、玻璃、陶瓷、木材、混凝土、岩石、天然油气储层、多种纤维或颗粒增强和层合复合材料、电极材料，以及颗粒材料等。应用的主要问题包括：结构静力弹塑性变形、弹性波色散与传播、静动力裂纹扩展、复合材料失效破坏、微结构非局部力学特性、多物理场耦合问题、结构局部化变形、高速碰撞、冲击侵彻、爆炸载荷作用下结构的剧烈破碎与动力响应以及数据和图像处理识别的非局部技术等。

1.5.6.1 弹性波传播与色散特性

近场动力学非局部特性决定了其具有波色散行为，准确描述弹性波的传播过程对于结构动力变形和裂纹扩展问题至关重要。Silling[6] 首次分析了平面弹性波问题，指出近场动力学的非局部长程力特性导致其具有弹性波频散/色散特性，计算的波速与波数相关。Silling 和 Zimmermann 等[309,310] 研究了自平衡荷载(双点集中荷载)作用下无限长一维杆的变形问题，采用 Fourier 变换和卷积定理方法，求得了近场动力学平衡方程(第一类 Fredholm 积分方程)的位移解析解，展示出传统局部解答未能获得的独特现象：荷载施加点区域以外出现位移场衰减振荡，同时变形不连续特征以渐变弱化方式从加载区域向远场传播，远离加载区位移逐渐趋于连续变化，其原因是 PD 存在由非局部特性决定的弹性波频散现象；并证明了当近场范围趋近于零时，PD 解答收敛于传统局部理论解析解。基于上述研究，Weckner 和 Abeyaratne[311]、Emmrich 和 Weckner[312,313] 提出了几种不同微弹性模型的本构力函数，研究了 PD 非局部长程力特性对无限长一维杆件动力学特性的影响，采用 Fourier 解析方法和数值积分方法分析了几种不同本构力函数对 PD 数值频散特性的影响。Weckner 等[314] 的研究表明，近场动力学的近场范围和物质点尺寸控制弹性波的数值频散程度。Silling[6] 和 Weckner[314] 将试验测得的实际材料非线性频散关系与近场动力学模型的频散曲线相匹配，确定了 PD 的本构模型，指出适用于材料的 PD 模型能够模拟出该材料的实际频散关系。Al-Barmany[315] 通过泰勒级数展开，获得了弹性小变形时近场动力学运动方程(波动方程)的级数解，证明了当近场范围趋于零时，近场动力学运动方程是收敛的，进而简略分析了弹性波频散和群速度特征。本书作者等[316] 研究了考虑长程力空间递减特性的改进 PMB 模型和原 PMB 模型对近场动力学频散特性的影响，分析了物质点尺寸和近场范围大小对数值频散程度的影响，并将一维弹性波传播问题的 PD 数值解与传统弹性波理论解进行了对比分析。Tao 和 Liu[317] 首次运用近场动力学方法研究了一维弹性波在 SHPB 杆件中的传播问题，为 PD 方法模拟 SHPB 冲击试验开辟了渠道，但未能关注近场动力学波频散的特征。Martowicz 等[318,319] 运用 PD 方法模拟了带预制裂纹板中振动与声波的相互作用规律，以及考虑用近场动力学非局部频散特性研究弹性波在石墨烯纳米带中的传播问题。Guan 和 Gunzburger[320] 给出带有分数阶拉普拉斯算子和分数阶导数算子的一维、二维非局部波动方程的一般形式,由波动方程一般解答推导得到两个非局部波动方程的频散关系,给出其泰勒级数表达,指出非局部波动方程在不同波数下发生不同程度的频散,且频散程度受近场范围尺寸影响。

1.5.6.2 裂纹扩展分析

如前所述,近场动力学便于分析复杂的裂纹扩展问题,吸引了大批研究者开展这方面的研究。Silling[77,80] 采用 EMU 程序模拟了 Kalthoff-Winkler 试验,即刚性柱体冲击马氏体时效钢导致的单裂纹动态扩展问题,模拟结果与试验结果吻合良好,验证了近场动力学方法和开发的 EMU 程序在动力裂纹扩展分析中的正确性与适用性。Aidun 和 Silling[321] 指出了动力断裂和破碎问题的典型特征,并给出近场动力学方法在动力破坏分析中的多个典型应用。Ha 和 Bobaru[89,322] 进行了均匀离散下的 m-收敛和 δ-收敛分析,研究了含单边裂纹脆性板的动态裂纹扩展问题,模拟得到了裂纹扩展、分叉、路径不稳定、裂纹不对称、连续分叉和次级分叉等试验中观测到的现象,且计算得到的裂纹传播速度与试验测试的结果相符,并指出边界反射应力波对动态裂纹的扩展路径具有强烈影响。鉴于动力问题的高应变率将会导致裂纹尖端出现软化现象,Mehrmashhadi 等[323] 在近场动力学模型中反映该裂尖软化机制,很好地模拟了 PMMA 玻璃板的动态断裂过程。Agwai 等[324] 基于 PD 方法模拟计算了两种不同的单边裂纹板的动态裂纹扩展路径和裂纹传播速度,并与试验、扩展有限单元元 (XFEM) 和黏聚力模型 (CZM) 的计算结果进行了对比分析,表明 PD 能自然模拟裂纹的扩展和分叉现象,且计算得到的裂纹扩展速度与其他数值方法相近;进而模拟了压头冲击含单个刚性夹杂板的动力裂纹扩展问题,不同预制裂纹位置得到的预测结果与试验结果吻合良好。Lipton[325,326] 基于数学推导和证明,系统研究了黏性动力响应和脆性断裂问题的近场动力学方法。Bobaru 和 Zhang[327] 综述了动态裂纹分叉的研究现状,结合含单边裂缝板的动态脆断问题,采用近场动力学脆性断裂分析方法进行了数值计算,分析了裂纹扩展和分叉的动力不稳定来源,研究了应力波对裂纹扩展的作用以及试件几何形状对裂纹分叉的影响;他们在研究中,还关注了裂纹扩展模式、裂纹分叉时的传播速度和裂纹分叉角的特征,研究表明,随着裂尖应力的增大和应力波堆积,导致最大应变能偏离裂纹对称线,产生偏离对称线的损伤,该损伤又反过来影响应变能分布,使得裂纹产生分叉和偏移。Dias 等[328] 综述了裂纹扩展问题的近场动力学研究现状。

值得说明的是,2012 年,美国圣地亚国家实验室联合美国国家科学基金会等单位,组织了圣地亚断裂竞赛 (Sandia Fracture Challenge),向全世界各研究团队发出参赛邀请。该竞赛目前已开展了三届,各参赛团队围绕主办方指定的断裂问题进行数值模拟计算,并提供研究报告。计算结果将与试验结果对比验证,以对相关模型和方法进行基准评估,每次竞赛的主题稍有不同。在第一次断裂竞赛中,PD 方法取得了良好效果[329],但没有 PD 课题组参加第二次断裂竞赛[330],PD 方法在第三次断裂竞赛中表现优异[331]。

1.5.6.3 高速碰撞与冲击侵彻问题

近场动力学,尤其是键型近场动力学方法非常适用于分析高速碰撞、冲击侵彻和爆炸载荷作用下,结构的剧烈损伤破碎和动力响应问题。Silling 和 Askari[79] 采用 PMB 模型,模拟了刚性球体冲击脆性固体并生成 Hertzian 裂纹的过程以及脆性板的冲击破碎问题。Macek 和 Silling[332] 提出了近场动力学杆单元与有限元耦合方法,并采用该方法研究了刚性球形弹丸撞击、贯穿韧性铝板的问题。Demmie 和 Silling[96,97] 研究了飞机撞击钢筋混凝土结构的问题,并发展了近场动力学气体模型,进而研究了炸药爆炸、弹头破碎等问题。Levine 等[333] 将 PD 的物质点及其相互作用视为弹簧质量系统,模拟了子弹冲击玻璃板和脆性固体跌落破碎的问题。Oterkus 等[334] 模拟了钢筋混凝土板在受到刚性弹丸冲击后再受压的问题,研究了损伤后建筑构件的残余强度。Bobaru 等[335] 模拟了弹丸高速冲击下多层玻璃的损伤演化问题,得到了许多试验观测到的损伤形态,但玻璃层间的接触算法较为简单,又受限于计算资源,尚有部分试验现象未能准确预测到。Hu 等[336] 采用试验和 PD 方法研究了球形弹丸冲击薄碳酸酯板支撑的玻璃板,分析了不同冲击速度对损伤模式的影响,模拟得到了试验观测到的大部分现象。Tupek[174] 采用 PD 方法对多个冲击破坏试验进行了数值模拟研究,包括:泰勒冲击试验、刚性球撞击夹芯板、含单边裂纹三维有机玻璃板的冲击破坏、Kalthoff 马氏体时效钢的冲击试验,计算得到了各种问题的裂纹扩展路径和最终破坏模式,在刚性球撞击夹芯板的模拟中,夹芯板破坏程度、弹丸与夹芯板的最终位置关系(贯穿、嵌入、反弹)等均与试验结果一致,为弹丸冲击侵彻破坏问题的研究提供了新的有效方法。本书作者等[337] 也采用键型近场动力学弹脆性模型,研究了刚性小球撞击混凝土板的问题。Diyaroglu 等[338] 基于 PD 方法研究了爆炸荷载作用下复合材料层压板的损伤破坏问题,模拟结果与试验现象吻合良好,显示了 PD 方法分析爆炸荷载作用下复合材料非线性大变形和损伤行为的有效性。

1.5.6.4 复合材料和结构的失效破坏

复合材料和结构的失效破坏分析是近场动力学的重点应用领域。研究者发展了多种复合材料层合板和颗粒增强复合材料的分析方法,取得了丰硕的成果,读者可参阅有关学位论文和研究综述[339−348]。

美国新墨西哥大学 Gerstle 课题组在混凝土 PD 建模和应用领域内开展了系统研究,并出版了专著 [57],他们采用 PD 方法对普通混凝土与钢筋混凝土结构进行了不同工况下的拉伸、压缩和剪切破坏模拟,分析了钢筋混凝土搭接接头的拉伸滑移破坏问题;还提出了微极 PD 模型和点阵 PD 模型,用于模拟混凝土和钢筋混凝土试块的拉压破坏等问题,计算结果可清晰地显示混凝土构件破坏过程中的裂纹扩展以及混凝土基体与钢筋间的相对滑移。但他们的研究没有考虑混凝土的细微观结构特征。此外,国内学者 Huang 等 [349]、Li 等 [350] 和 Lu 等 [351] 以及本书作者的课题组 [352,353] 也开展了近场动力学在混凝土材料和结构分析领域的研究工作,基于 PD 方法模拟分析了各类混凝土试件在多种荷载作用下的破坏问题,并应用于水利水电工程和交通工程。

1.5.7 近场动力学数理基础

数值方法的生命力有赖于其数理基础。近场动力学自问世以来,不仅吸引了力学工作者和相关技术领域专家的关注,也引起了数学家的兴趣。研究者们证明了键型和态型近场动力学模型边值、初值、初边值问题的适定性 [62,312,354−360],为近场动力学本构建模和数值求解提供了坚实的数学支撑。

近场动力学固体力学模型和扩散模型的研究也推动了非局部微积分理论的发展,美国宾夕法尼亚州立大学 Qiang Du(杜强) 课题组及其合作者 Lehoucq(圣地亚实验室) 和 Gunzburger(佛罗里达州立大学) 围绕多尺度建模的近场动力学数学基础,开展了大量的卓有成效的研究 [121,241,356−373]。他们指出了非局部模型及其数学理论与传统局部模型的相似之处,分析了非局部问题的特色和面临的挑战,参照传统矢量微积分的高斯定理和格林恒等式,发展了适合 PD 的非局部矢量微积分和非局部变分原理,定义了非局部梯度、散度、旋度算子及其共轭算子,给出变分弱形式的基本方程、方程基本解及其格林函数表达,研究了非局部模型渐进兼容的数值离散格式。

1.5.8 近场动力学程序和软件研发

目前,近场动力学程序和软件主要有:EMU、PDLAMMPS、PERIDIGM、Sierra/Solid Mechanics (Sierra/SM) 的 PD 模块、LS-DYNA 的内嵌求解模块、基于 ABAQUS 和 ANSYS 平台的二次开发模块,此外,还有各研究组针对具体问题的自编程序。

美国圣地亚国家实验室 (Sandia National Laboratory, SNL) 主持开发了近场动力学的三个主要计算程序,即 EMU、PDLAMMPS 和 PERIDIGM。其中,EMU[374] 是最早由 Silling 开发的 PD 计算程序,基于 FORTRAN90 编写,但 EMU 程序是 SNL 的自用程序,存在获取和使用限制,关于 EMU 的介绍资料较少。PDLAMMPS 是基于 LAMMPS 开发的 PD 计算程序,由 Parks 等利用

C++ 编写,平台扩展性良好,但是目前实现的近场动力学模型较为有限[375,376]。Parks 等开发了 PERIDIGM 开源程序[377,378],PERIDIGM 的稳定性好、适用性广,还在处于不断发展和完善的过程中。此外,供美国海军使用的固体结构分析软件 Sierra/Solid Mechanics (Sierra/SM)[379] 实现了常规态型和非常规态型近场动力学对应的材料模型。专著 [8,58] 及其中译本提供了各基准算例的 FORTRAN 源程序。本书的各章节也将给出本团队自编的 FORTRAN 源程序,并具体介绍开源软件 PDLAMMPS、商业软件 LS-DYNA 和 ABAQUS 平台二次开发的实现方法与数值应用,供不同需求的读者参考。

近场动力学由于其非局部性,具有计算量大的特点,需要大力发展相应的高性能计算技术。现有近场动力学程序软件 EMU、PDLAMMPS、PERIDIGM 等先后实现了基于 MPI 的并行计算功能,显著提高了计算规模和计算效率。新墨西哥大学 Gerstle 课题组开发了混凝土破坏分析的基于 MPI 并行的近场动力学程序。基于 OpenMP 的细粒度并行技术能将各物质点的计算映射到单个线程,实现多线程高效并行,并且易于实现[380,381]。Diehl 首先发表了近场动力学的 GPU 并行计算[382],上海交通大学的刘肃肃等[383] 开发了基于 CUDA 平台的近场动力学 GPU 并行化方法,模拟了纤维增强复合材料的拉伸破坏。Mossaiby 等[384] 实现了键型近场动力学的 OpenMP 并行与 OpenCL 平台的 GPU 并行计算。

需要说明,近场动力学的发展方兴未艾,成果众多,囿于作者能力和本书篇幅,本节综述的 PD 研究现状还有许多遗漏之处,一些最新的研究成果也未及进行评述,读者可关注国内外相关期刊发表的论文。

1.6 本书的主要内容

本书作者的课题组在中国国内最早开展的近场动力学的研究得到了多个国家自然科学基金项目和国家重点研发计划项目课题的资助,在近场动力学理论模型、计算体系、程序实现和工程应用等方面开展了较为系统的研究,在国内外学术会议上组织了多次专题研讨会,并于 2019 年 11 月协同成立了近场动力学与非局部理论国际研究中心。本书是以作者课题组十余年在近场动力学领域的主要成果为基础,并吸收了近场动力学最新的研究成果,系统地论述了近场动力学的建模方法、数值算法和应用技术。本书内容论述完整,可作为近场动力学入门和进阶的教材和参考书。本书共十三章,包括:绪论、近场动力学基本理论、键型近场动力学模型及其改进、键型近场动力学在有限元中的实现、近场动力学的显式动力学解法、常规态型近场动力学模型、非常规态型近场动力学模型、非常规态型近场动力学模型的改进、近场动力学方法与有限单元法的混合模型、非均匀离散的近场动力学模型与自适应分析、冲击侵彻与爆炸问题的近场动力学模拟、热传导与热力耦合问题的近场动力学建模分析、混凝土材料与结构破坏分析的近场动力学方法。此外，还安排两个附录，分别介绍与本书有关的近场动力学微分算子以及近场动力学的显式动力学算法的 FORTRAN 源程序和数据格式。

本书的主要内容如下：

(1) 近场动力学是一种以积分方程为基础的基于非局部相互作用思想的广义连续介质力学理论——积分型非局部理论，是进行固体材料和结构损伤累积、裂纹萌生、扩展、分叉和汇合等演化分析的一种有效方法。第 1 章作为绪论，简要介绍近场动力学问世的背景，概述了连续介质力学的局部理论与非局部理论，阐述了近场动力学的发展历程，分析了近场动力学的理论特点，较为详细地综述了近场动力学的研究现状。

(2) 近场动力学相较于以偏微分方程为基础的基于局部接触作用思想的经典连续介质力学，适用于求解连续体、离散体和含裂纹等不连续特征结构的力学问题。近场动力学与经典连续介质力学和分子动力学联系紧密，是对它们特点的继承、综合、发展与重构。第 2 章系统介绍近场动力学的基本理论，包括基于非局部思想的变形和受力描述、运动方程、本构力模型、守恒律、非局部边界条件与损伤断裂模型。

(3) 键型近场动力学模型是提出最早、研究最多、发展最快和应用最广的近场动力学模型，其建模方法概念清晰、简单易行，且方便拓展应用，并能作为态型近场动力学模型的基础，广受学术界和工程界的关注。第 3 章系统介绍键型近场动力学一般形式的键力本构模型及其线性化方法、经典微观弹脆性 PMB 模型、考虑长程力空间递减规律和边界效应的修正 PMB 模型，还介绍了突破固定泊松比限制的几类改进键型近场动力学模型。

(4) 近场动力学强形式的控制方程可以直接采用配点法离散求解，键型近场动力学微弹性线性模型的键力密度可以表示成有限元法中杆单元劲度矩阵模式，键型近场动力学弱形式方程的非连续 Galerkin 有限元方法能在有限元计算框架内充分发挥近场动力学的非连续分析能力。第 4 章将介绍近场动力学强形式方程的杆单元解法及其在 ABAQUS 软件中的二次开发技术，详细阐述键型近场动力学非连续 Galerkin 有限元法，给出其数值实施过程，并介绍 LS-DYNA 软件对于该方法的实现方法。

(5) 近场动力学运动方程通常无法得到解析解，需要通过空间和时间的数值积分技术获得相应的数值解，近场动力学数值算法具有分子动力学方法、无网格法与有限元法的特点。第 5 章主要介绍近场动力学的显式动力学无网格粒子法的算法和程序实现细节，包括空间离散、时间离散、体积修正、数值稳定性和算法流程图，给出若干算例以及相应的 FORTRAN 源程序，最后介绍近场动力学开源软件 PDLAMMPS 的安装使用。

1.6 本书的主要内容

(6) 经典键型近场动力学存在泊松比限制、不能区分几何形状改变和体积变形、不支持不可压缩材料本构建模以及与经典连续介质力学联系不够紧密等弱点，态型近场动力学可以有效克服上述经典键型近场动力学的局限性。在第 2 章介绍的近场动力学状态算子的基础上，第 6 章将系统介绍常规态型近场动力学线弹性模型和弹塑性模型的建模方法，详细给出线弹性固体本构建模过程，推导与传统 J_2 弹塑性及 Drucker-Prager 弹塑性模型关联的适用于体积变形不可压缩材料与可压缩材料的近场动力学弹塑性本构模型，并通过几个数值实例验证模型的有效性。

(7) 非常规态型近场动力学模型 (或近场动力学对应材料模型) 可以将经典连续介质力学已发展的各种应力–应变 (率) 本构模型转化为近场动力学的力密度矢量状态与变形矢量状态的本构关系，便于在近场动力学框架内分析各类材料和结构的复杂力学响应和破坏行为，但非常规态型近场动力学的数值求解体系尚未完善，并存在数值稳定性问题，影响计算精度甚至导致计算失效。第 7 章将给出非常规态型近场动力学的建模方法，介绍线弹性本构模型的线性隐式静力解法和非线性本构模型的非线性隐式静力解法，讨论数值不稳定性的原因，给出含稳定控制的数值计算策略，基于带预制裂纹矩形板的静力弹性变形分析，验证多种因素对数值计算稳定性的影响，最后发展了多晶体弹塑性静动力破坏分析的近场动力学方法。

(8) 非常规态型近场动力学具有描述材料与结构复杂的非线性变形和断裂行为的能力，其数值计算方法与基于经典连续介质力学的无网格方法具有一定的相似性。第 8 章将非常规态型近场动力学的数值计算方法与几种典型无网格方法进行对比分析，并在第 7 章数值不稳定性原因分析的基础上，为有效控制非常规态型近场动力学数值不稳定问题，重点阐述三类改进的高性能非常规态型近场动力学模型，包括高阶非常规态型近场动力学模型、键关联 (高阶) 非常规态型近场动力学模型和一种新的非常规态型近场动力学模型。

(9) 对以近场动力学方法与有限元方法为代表的传统数值方法进行混合建模，可以有效提高近场动力学的计算效率。第 9 章评述了近场动力学方法与传统有限单元法混合建模的研究进展，阐明了各种混合建模方法的基本原理与特点，重点介绍了作者课题组在近场动力学方法与有限元方法混合建模方面的研究工作，阐述了所提出的两种混合模型的建模方法以及算例验证情况，并应用于典型重力坝的抗滑稳定分析。

(10) 在近场动力学数值实施过程中，均匀离散要求整个结构采用高密度均匀网格或点阵，这将导致计算量急剧增大，计算效率低下。近场动力学数值实施并不限于均匀离散，况且非均匀离散也是自适应分析和多尺度建模的自然组成部分。第 10 章系统介绍基于 Voronoi 结构图的均匀或非均匀离散的近场动力学分析方法以及近场动力学自适应分析方法，包括：近场动力学数值计算的离散技术和自适应研究现状、基于非均匀离散的近场动力学与自适应分析的数值技术、非均匀离散近场动力学方法的若干计算细节、典型算例分析与验证等。

(11) 冲击侵彻与爆炸载荷作用下结构的动力响应与毁伤破坏问题涉及多学科交叉,近场动力学对强动载作用下结构的动态裂纹扩展和剧烈破碎毁伤模拟具有独特的优势。第 11 章首先给出适用于刚性体冲击侵彻可变形体的冲击接触算法以及一种经典的自由空中爆炸载荷施加方法,随后详细阐释了准脆性材料的 JH-2 动态损伤本构模型,分析了键型近场动力学方法的色散特性,建立了分离式霍普金森压杆 (SHPB) 冲击试验的近场动力学模拟方法,模拟了典型混凝土巴西盘的 SHPB 冲击破坏过程,最后采用非常规态型近场动力学方法模拟了混凝土的冲击层裂与爆炸毁伤破坏问题。

(12) 以粒子为载体的能量传输本质上是非局部过程,若粒子的平均自由程大于驱动力变化的距离尺度,则需要采用非局部传输理论进行研究。近场动力学框架下的积分非局部性为传输传质问题的深入研究创造了条件,扩散问题与扩散-变形耦合的近场动力学方法已引起研究者的日趋关注。第 12 章将简要概述热传导问题的近场动力学模型,发展基于近场动力学微分算子的近场动力学热传导模型和热-力耦合模型,构建相应的数值计算体系,开展混合热边界条件下完好或含裂纹结构的热传导分析,建立混凝土导热分析与热-力耦合分析的近场动力学方法,研究裂纹对于混凝土导热性能的影响,开展混凝土厚壁圆筒热冲击破裂模拟分析。

(13) 混凝土材料和结构广泛应用于各类工程,其损伤破坏过程和破坏机制是学术界和工程界长期关注的问题。近场动力学理论和方法能够模拟混凝土材料和结构中微缺陷的发展、损伤累积、宏观裂纹萌生、扩展到局部断裂直至整体失稳的渐进破坏全过程,揭示混凝土材料和结构的破坏机制。第 13 章分别从微观、细观和宏观尺度,进行水泥水化产物、混凝土材料和混凝土结构的近场动力学建模方法研究,建立水泥基材料热-力耦合和热-水-力耦合分析的近场动力学模型,模拟分析水泥基材料的冻融循环过程,采用近场动力学方法开展混凝土试件在各种荷载作用下的细观破坏分析,进行典型混凝土重力坝的冲击侵彻和爆炸毁伤模拟,并基于近场动力学-有限元混合方法研究向家坝水电站泄洪坝段的深层抗滑稳定问题。

(14) 在一点的局部空间域或时间域上,近场动力学微分算子可以定义该点任意阶局部导数的积分表达式,能将任意常 (偏) 微分方程转化为积分方程,避免导数奇异性问题。并且,近场动力学微分算子能够考虑时间进程的非对称性和空间离散的非均匀性,自然描述其对应的非局部特性。本书附录 A 具体介绍任意形状近场范围和完整对称近场范围下近场动力学微分算子的绝对和相对形式表达式的构建过程,有助于将任意阶局部微分转化为非局部积分。

(15) 近场动力学在某种意义上就是一种数值计算方法,其数值实现对于近场动力学的推广应用具有重要的作用。为了使读者更好地理解近场动力学理论和方法,也便于读者开展近场动力学研究,附录 B 给出与第 5 章算法配套的 FOR-

TRAN 源代码，并包含必要的注释；此外，还结合具体算例，给出输入数据文件的格式说明，供读者参考使用。

参 考 文 献

[1] 黄筑平. 连续介质力学基础 [M]. 北京: 高等教育出版社, 2003.

[2] Bazant Z P, Jirásek M. Nonlocal integral formulations of plasticity and damage: survey of progress [J]. Journal of Engineering Mechanics, 2002, 128(11): 1119-1149.

[3] Arndt M, Griebel M. Derivation of higher order gradient continuum models from atomistic models for crystalline solids [J]. Multiscale Modeling & Simulation, 2005, 4(2): 531-562.

[4] 刘凯欣, 高凌天. 离散元法研究的评述 [J]. 力学进展, 2003, 33(4): 483-490.

[5] Haile J M. Molecular Dynamics Simulation [M]. New York: Wiley, 1992.

[6] Silling S A. Reformulation of elasticity theory for discontinuities and long-range forces [J]. Journal of the Mechanics and Physics of Solids, 2000, 48(1): 175-209.

[7] Silling S A, Epton M, Weckner O, et al. Peridynamic states and constitutive modeling [J]. Journal of Elasticity, 2007, 88(2): 151-184.

[8] Madenci E, Oterkus E. Peridynamic Theory and Its Applications [M]. New York: Springer, 2014.

[9] Bobaru F, Foster J T, Geubelle P H, et al. Handbook of Peridynamic Modeling [M]. New York: Chapman and Hall/CRC Press, 2016.

[10] 黄丹, 章青, 乔丕忠, 等. 近场动力学方法及其应用 [J]. 力学进展, 2010, 40(4): 448-459.

[11] 章青, 郁杨天, 顾鑫. 近场动力学与有限元的混合建模方法 [J]. 计算力学学报, 2016, 33(4):441-448.

[12] 李杰, 吴建营, 陈建兵. 混凝土随机损伤力学 [M]. 北京：科学出版社, 2014.

[13] 杨桂通. 弹塑性动力学基础 [M]. 北京：科学出版社, 2008.

[14] 孙其诚, 刘晓星, 张国华, 等. 密集颗粒物质的介观结构 [J]. 力学进展, 2017, 47(1):263-308.

[15] Silling S A. Coarsening method for linear peridynamics [J]. International Journal of Multiscale Computational Engineering, 2011, 9(6):609-621.

[16] Silling S A. Origin and effect of nonlocality in a composite [J]. Journal of Mechanics of Materials and Structures, 2014, 9(2):245-258.

[17] 陈玉丽, 马勇, 潘飞, 等. 多尺度复合材料力学研究进展 [J]. 固体力学学报, 2018, 39(1):1-68.

[18] 杜强. 从毗域动力学到随机跳跃过程: 非局部平衡范例及非局部微积分框架 [J]. 中国科学: 数学, 2015, 45(7):939-952.

[19] Oldroyd J G. On the formulation of rheological equations of state [C]. Proceedings of the Royal Society A: Mathematical, Physical and Engineering Sciences, 1950, 200(1063):523-541.

[20] Noll W. A mathematical theory of the mechanical behavior of continuous media [J]. Archive for Rational Mechanics and Analysis, 1958, 2(1):197-226.

[21] Truesdell C, Noll W. The Non-linear Field Theories of Mechanics [M]. Berlin, Heidelberg: Springer, 2004.

[22] Eringen A C. Continuum Physica, 4 Volumes [M]. New York: Academic Press, 1974-1976.

[23] 赵亚溥. 近代连续介质力学 [M]. 北京：科学出版社，2016.

[24] Rogula D. Nonlocal Theory of Material Media [M]. Berlin: Springer-Verlag, 1982.

[25] Kröner E. Elasticity theory of materials with long range cohesive forces [J]. International Journal of Solids and Structures, 1967, 3(5):731-742.

[26] Di Paola M, Pirrotta A, Zingales M. Mechanically-based approach to non-local elasticity: variational principles [J]. International Journal of Solids and Structures, 2010, 47(5):539-548.

[27] Kunin I A. Elastic Media with Microstructure I: One-dimensional Models [M]. Berlin: Springer-Verlag, 1982.

[28] Kunin I A. Elastic Media with Microstructure II: Three-dimensional Models [M]. Berlin: Springer-Verlag, 1983.

[29] Eringen A C. Linear theory of nonlocal elasticity and dispersion of plane waves [J]. International Journal of Engineering Science, 1972, 10(5):425-435.

[30] Eringen A C. Nonlocal polar elastic continua [J]. International Journal of Engineering Science, 1972, 10(1):1-16.

[31] Eringen A C. Theory of nonlocal thermoelasticity [J]. International Journal of Engineering Science, 1974, 12(12):1063-1077.

[32] Eringen A C. Memory dependent nonlocal elastic solids [J]. Letters in Applied and Engineering Sciences, 1974, 2:145-159.

[33] Eringen A C. Screw dislocation in non-local elasticity [J]. Journal of Physics D: Applied Physics, 1977, 10(5):671-678.

[34] Atkinson C. On some recent crack tip stress calculations in nonlocal elasticity [J]. Archiwum Mechaniki Stosowanej, 1980, 32(2):317-328.

[35] Altan S B. Uniqueness of initial-boundary value problems in nonlocal elasticity [J]. International Journal of Solids and Structures, 1989, 25(11):1271-1278.

[36] Picu R C. On the functional form of non-local elasticity kernels [J]. Journal of the Mechanics and Physics of Solids, 2002, 50(9):1923-1939.

[37] 黄再兴. 关于非局部场论的两点注记 [J]. 固体力学学报, 1997, 18(2):158-162.

[38] Reddy J N. Nonlocal theories for bending, buckling and vibration of beams [J]. International Journal of Engineering Science, 2007, 45(2-8):288-307.

[39] Aifantis E C. On the gradient approach—relation to Eringen's nonlocal theory [J]. International Journal of Engineering Science, 2011, 49(12):1367-1377.

[40] 陈少华, 王自强. 应变梯度理论进展 [J]. 力学进展, 1900, 33(2):207-216.

[41] 黄克智, 邱信明. 应变梯度理论的新进展（一）：偶应力理论和 SG 理论 [J]. 机械强度, 1999, 21(2):81-87.

[42] 黄克智, 邱信明, 姜汉卿. 应变梯度理论的新进展 (二) ——基于细观机制的 MSG 应变梯度塑性理论 [J]. 机械强度, 1999, (3):161-165.

[43] 赵杰, 陈万吉, 冀宾. 关于两种二阶应变梯度理论 [J]. 力学学报, 2010(1):138-145.

[44] Toupin R A. Elastic materials with couple-stresses [J]. Archive for Rational Mechanics and Analysis, 1962, 11(1):385-414.

[45] Koiter W. Couple-stresses in the theory of elasticity [J]. Dictionary Geotechnical Engineering/Wörterbuch Geotechnik, 1969, 67:17-44.

[46] Mindlin R D, Tiersten H F. Effects of couple-stresses in linear elasticity [J]. Archive for Rational Mechanics and analysis, 1962, 11(1):415-448.

[47] Mindlin R D. Second gradient of strain and surface-tension in linear elasticity [J]. International Journal of Solids and Structures, 1965, 1(4):417-438.

[48] Fleck N A, Hutchinson J W. Strain gradient plasticity [J]. Advances in Applied Mechanics, 1997, 33:296-361.

[49] Lam D C, Yang F, Chong A C, et al. Experiments and theory in strain gradient elasticity [J]. Journal of the Mechanics and Physics of Solids, 2003, 51(8):1477-1508.

[50] Aifantis E C. On the role of gradients in the localization of deformation and fracture [J]. International Journal of Engineering Science, 1992, 30(10):1279-1299.

[51] Aifantis E C. Strain gradient interpretation of size effects [C]//Fracture Scaling. Springer, Dordrecht, 1999:299-314.

[52] Fleck N A, Hutchinson J W. A phenomenological theory for strain gradient effects in plasticity [J]. Journal of the Mechanics and Physics of Solids, 1993, 41(12):1825-1857.

[53] Gao H, Huang Y, Nix W D, et al. Mechanism-based strain gradient plasticity—I. Theory [J]. Journal of the Mechanics and Physics of Solids, 1999, 47(6):1239-1263.

[54] Huang Y, Gao H, Nix W D, et al. Mechanism-based strain gradient plasticity—II. Analysis [J]. Journal of the Mechanics and Physics of Solids, 2000, 48(1):99-128.

[55] Lim C W, Zhang G, Reddy J N. A higher-order nonlocal elasticity and strain gradient theory and its applications in wave propagation [J]. Journal of the Mechanics and Physics of Solids, 2015, 78:298-313.

[56] 埃尔多安·马德西, 额尔坎·奥特库斯. 近场动力学理论及其应用 [M]. 余音, 胡祎乐, 译. 上海: 上海交通大学出版社, 2019.

[57] Gerstle W H. Introduction to Practical Peridynamics: Computational Solid Mechanics Without Stress and Strain [M]. Singapore: World Scientific Publishing Co Pte Ltd, 2015.

[58] Madenci E, Barut A, Dorduncu M. Peridynamic Differential Operator for Numerical Analysis [M]. Switzerland: Springer International Publishing, 2019.

[59] Madenci E, Barut A, Dorduncu M. 近场动力学微分算子——在数值分析中的应用 [M]. 韩非, 张玲, 译. 北京: 科学出版社, 2021.

[60] Oterkus E, Oterkus S, Madenci E. Peridynamic Modeling, Numerical Techniques, and Applications [M]. 1st ed. Kidlington: Elsevier, 2021.

[61] 韩非. 体会近场动力学之动 [EB/OL]. 2017-01-21. http://blog.sciencenet.cn/blog-232936-1029029.html.

[62] Emmrich E, Weckner O. On the well-posedness of the linear peridynamic model and its convergence towards the Navier equation of linear elasticity [J]. Communications in Mathematical Sciences, 2007, 5(4):851-864.

[63] Silling S A, Lehoucq R B. Convergence of peridynamics to classical elasticity theory [J]. Journal of Elasticity, 2008, 93(1):13-37.

[64] Bobaru F, Yang M J, Alves L F, et al. Convergence, adaptive refinement, and scaling in 1D peridynamics [J]. International Journal for Numerical Methods in Engineering, 2009, 77(6):852-877.

[65] Tian X C, Du Q. Asymptotically compatible schemes and applications to robust discretization of nonlocal models [J]. SIAM Journal on Numerical Analysis, 2014, 52(4): 1641-1665.

[66] Seleson P, Parks M L, Gunzburger M, et al. Peridynamics as an upscaling of molecular dynamics [J]. Multiscale Modeling & Simulation, 2009, 8(1):204-227.

[67] 顾鑫, 章青, Madenci E. 多物理场耦合作用分析的近场动力学理论与方法 [J]. 力学进展, 2019, (201901):576-598.

[68] Silling S A. Linearized theory of peridynamic states [J]. Journal of Elasticity, 2010, 99(1):85-111.

[69] Silling S A, Lehoucq R B. Peridynamic theory of solid mechanics [J]. Advances in Applied Mechanics, 2010, 44(10):73-168.

[70] Le Q V, Chan W K, Schwartz J. A two-dimensional ordinary, state-based peridynamic model for linearly elastic solids [J]. International Journal for Numerical Methods in Engineering, 2014, 98(8):547-561.

[71] Sarego G, Le Q V, Bobaru F, et al. Linearized state-based peridynamics for 2-D problems [J]. International Journal for Numerical Methods in Engineering, 2016, 108(10): 1174-1197.

[72] Mousavi F, Jafarzadeh S, Bobaru F. An ordinary state-based peridynamic elastoplastic 2D model consistent with J2 plasticity [J]. International Journal of Solids and Structures, 2021, 229:111146.

[73] Liu Z, Bie Y, Cui Z, et al. Ordinary state-based peridynamics for nonlinear hardening plastic materials' deformation and its fracture process [J]. Engineering Fracture Mechanics, 2020, 223:106782.

[74] Warren T L, Silling S A, Askari A, et al. A non-ordinary state-based peridynamic method to model solid material deformation and fracture [J]. International Journal of Solids and Structures, 2009, 46(5):1186-1195.

[75] Foster J T, Silling S A, Chen W W. Viscoplasticity using peridynamics [J]. International Journal for Numerical Methods in Engineering, 2009, 81(10):1242-1258.

[76] Silling S A, Kahan S. Peridynamic modeling of structural damage and failure [C]. Conference on High Speed Computing, Gleneden Beach, Oregon, USA, 2004.

[77] Silling S A, Askari E. Peridynamic modeling of impact damage [C]. ASME/JSME 2004 Pressure Vessels and Piping Conference. American Society of Mechanical Engineers, 2004:197-205.

[78] Gerstle W H, Sau N, Silling S A. Peridynamic modeling of plain and reinforced concrete structures [C]. Proceedings of 18th International Conference on Structural Mechanics in Reactor Technology, 2005.

[79] Silling S A, Askari E. A meshfree method based on the peridynamic model of solid mechanics [J]. Computers & Structures, 2005, 83(17-18):1526-1535.

[80] Silling S A. Peridynamic modeling of the kalthoff-winkler experiment [R]. Submission for the 2001 Sandia Prize in Computational Science, 2002.

[81] Ladányi G, Jenei I. Analysis of plastic peridynamic material with RBF meshless method [J]. Pollack Periodica, 2008, 3(3):65-77.

[82] Colavito K, Kilic B, Celik E, et al. Effect of void content on stiffness and strength of composites by peridynamic analysis and static indentation test [C]. 48th AIAA/ASME/ASCE/AHS/ASC Structures, Structural Dynamics, and Materials Conference, 2007: 2251.

[83] Colavito K, Kilic B, Celik E, et al. Effect of nanoparticles on stiffness and impact strength of composites [C]. 48th AIAA/ASME/ASCE/AHS/ASC Structures, Structural Dynamics, and Materials Conference, 2007:2021.

[84] Kilic B. Peridynamic theory for progressive failure prediction in homogeneous and heterogeneous materials [D]. Tuscon: The University of Arizona, 2008.

[85] Weckner O, Mohamed N A. Viscoelastic material models in peridynamics [J]. Applied Mathematics and Computation, 2013, 219(11):6039-6043.

[86] Azizi M A, Mohd Ihsan A K, Nik Mohamed N A. The peridynamic model of viscoelastic creep and recovery [J]. Multidiscipline Modeling in Materials and Structures, 2015, 11(4):579-597.

[87] Zaccariotto M, Luongo F, Sarego G, et al. Examples of applications of the peridynamic theory to the solution of static equilibrium problems [J]. The Aeronautical Journal, 2015, 119(1216):677-700.

[88] Silling S A, Bobaru F. Peridynamic modeling of membranes and fibers [J]. Proposed for Publication in Peridynamic Modeling of Membranes & Fibers, 2005, 40(2):395-409.

[89] Ha Y D, Bobaru F. Studies of dynamic crack propagation and crack branching with peridynamics [J]. International Journal of Fracture, 2010, 162(1-2):229-244.

[90] Huang D, Lu G D, Wang C W, et al. An extended peridynamic approach for deformation and fracture analysis [J]. Engineering Fracture Mechanics, 2015, 141:196-211.

[91] Huang D, Lu G D, Qiao P Z. An improved peridynamic approach for quasi-static elastic deformation and brittle fracture analysis [J]. International Journal of Mechanical Sciences, 2015, 94:111-122.

[92] Tao J. Development and applications of new peridynamic models [D]. East Lansing: Michigan State University, 2012.

[93] Ghajari M, Iannucci L, Curtis P. A peridynamic material model for the analysis of dynamic crack propagation in orthotropic media [J]. Computer Methods in Applied Mechanics and Engineering, 2014, 276(7):431-452.

[94] Alali B, Lipton R. Multiscale dynamics of heterogeneous media in the peridynamic formulation [J]. Journal of Elasticity, 2012, 106(1):71-103.

[95] Taylor M, Steigmann D J. A two-dimensional peridynamic model for thin plates [J]. Mathematics and Mechanics of Solids, 2015, 20(8):998-1010.

[96] Demmie P N, Silling S A. An approach to modeling extreme loading of structures using peridynamics [J]. Journal of Mechanics of Materials and Structures, 2007, 2(10):1921-1945.

[97] Demmie P N, Preece D S, Silling S A. Warhead fragmentation modeling with peridynamics [R]. No.SAND2007-2184C. Sandia National Laboratories (SNL-NM), Albuquerque, NM (United States), 2007.

[98] Mikata Y. Analytical solutions of peristatic and peridynamic problems for a 1D infinite rod [J]. International Journal of Solids and Structures, 2012, 49(21):2887-2897.

[99] Buryachenko V A. Effective elastic modulus of heterogeneous peristatic bar of random structure [J]. International Journal of Solids and Structures, 2014, 51(17):2940-2948.

[100] Ren H L, Zhuang X Y, Cai Y C, et al. Dual-horizon peridynamics [J]. International Journal for Numerical Methods in Engineering, 2016, 108(12):1451-1476.

[101] Li H, Zhang H W, Zheng Y G, et al. A peridynamic model for the nonlinear static analysis of truss and tensegrity structures [J]. Computational Mechanics, 2016, 57(5):843-858.

[102] Gerstle W H, Sau N, Silling S A. Peridynamic modeling of concrete structures [J]. Nuclear Engineering and Design, 2007, 237(12-13):1250-1258.

[103] Sau N. Peridynamic modeling of quasibrittle structures [D]. Albuquerque: The University of New Mexico, 2008.

[104] Rahman A S M. Lattice-based peridynamic modeling of linear elastic solids [D]. Albuquerque: The University of New Mexico, 2012.

[105] 卢广达. 脆性材料和结构变形破坏的近场动力学模拟 [D]. 南京: 河海大学, 2015.

[106] Prakash N, Seidel G D. A novel two-parameter linear elastic constitutive model for bond based peridynamics [C]. 56th AIAA/ASCE/AHS/ASC Structures, Structural Dynamics, and Materials Conference, 2015.

[107] Zhou X P, Shou Y D. Numerical simulation of failure of rock-like material subjected to compressive loads using improved peridynamic method [J]. International Journal of Geomechanics, 2016, 17(3):04016086:1-12.

[108] Zhu Q Z, Ni T. Peridynamic formulations enriched with bond rotation effects [J]. International Journal of Engineering Science, 2017, 121:118-129.

[109] Wang Y T, Zhou X P, Wang Y, et al. A 3-D conjugated bond-pair-based peridynamic formulation for initiation and propagation of cracks in brittle solids [J]. International Journal of Solids and Structures, 2018, 134:89-115.

[110] Zhou X P, Wang Y T, Shou Y D, et al. A novel conjugated bond linear elastic model in

[110] (continued) bond-based peridynamics for fracture problems under dynamic loads [J]. Engineering Fracture Mechanics, 2018, 188:151-183.

[111] Gu X, Zhang Q. A modified conjugated bond-based peridynamic analysis for impact failure of concrete gravity dam [J]. Meccanica, 2020, 55(3):547-566.

[112] O'Grady J, Foster J. Peridynamic beams: a non-ordinary, state-based model [J]. International Journal of Solids and Structures, 2014, 51(18):3177-3183.

[113] Sarkar S, Nowruzpour M, Reddy J N, et al. A discrete Lagrangian based direct approach to macroscopic modelling [J]. Journal of the Mechanics and Physics of Solids, 2017, 98:172-180.

[114] Diyaroglu C, Oterkus E, Oterkus S, et al. Peridynamics for bending of beams and plates with transverse shear deformation [J]. International Journal of Solids and Structures, 2015, 69:152-168.

[115] Hu Y L, Madenci E. Bond-based peridynamic modeling of composite laminates with arbitrary fiber orientation and stacking sequence [J]. Composite Structures, 2016, 153:139-175.

[116] Liu W Y. Discretized bond-based peridynamics for solid mechanics [D]. East Lansing: Michigan State University, 2012.

[117] Ganzenmüller G C, Hiermaier S, May M. Improvements to the prototype micro-brittle model of peridynamics [C]. Meshfree Methods for Partial Differential Equations VII. Springer International Publishing, 2015: 163-183.

[118] 章青, 顾鑫, 郁杨天. 冲击载荷作用下颗粒材料动态力学响应的近场动力学模拟 [J]. 力学学报, 2016, 48(1):56-63.

[119] Le Q V, Bobaru F. Surface corrections for peridynamic models in elasticity and fracture [J]. Computational Mechanics, 2018, 61:499-518.

[120] Lehoucq R B, Silling S A. Force flux and the peridynamic stress tensor [J]. Journal of the Mechanics and Physics of Solids, 2008, 56(4):1566-1577.

[121] Du Q, Gunzburger M, Lehoucq R B, et al. A nonlocal vector calculus, nonlocal volume-constrained problems, and nonlocal balance laws [J]. Mathematical Models and Methods in Applied Sciences, 2012, 23(3):493-540.

[122] Turner D Z, Lehoucq R B, Reu P L. A nonlocal strain measure for DIC [C]//Advancement of Optical Methods in Experimental Mechanics, Springer, Cham, 2016, 3:79-83.

[123] Lehoucq R B, Reu P L, Turner D Z. A novel class of strain measures for digital image correlation [J]. Strain, 2015, 51(4):265-275.

[124] Madenci E. Peridynamic integrals for strain invariants of homogeneous deformation [J]. ZAMM-Journal of Applied Mathematics and Mechanics/Zeitschrift für Angewandte Mathematik and Mechanik, 2017, 97(10):1236-1251.

[125] Madenci E, Barut A, Futch M. Peridynamic differential operator and its applications [J]. Computer Methods in Applied Mechanics and Engineering, 2016, 304:408-451.

[126] Le Q V. Relationship between microstructure and mechanical properties in $Bi_2Sr_2CaCu_2O_x$ round wires using peridynamic simulation [D]. Raleigh: North Carolina State

University, 2014.
[127] Merwe C W V D. A peridynamic model for sleeved hydraulic fracture [D]. Stellenbosch: Stellenbosch University, 2014.
[128] Ouchi H, Katiyar A, York J, et al. A fully coupled porous flow and geomechanics model for fluid driven cracks: a peridynamics approach [J]. Computational Mechanics, 2015, 55(3): 561-576.
[129] Mitchell J A, Silling S A, Littlewood D J. A position-aware linear solid constitutive model for peridynamics [J]. Journal of Mechanics of Materials and Structures, 2015, 10(5):539-557.
[130] Zhang Q, Gu X, Huang D. Failure analysis of plate with non-uniform arrangement holes by ordinary state-based peridynamics [C]. ICCM2015, Auckland, New Zealand, 2015.
[131] Sarego G, Zaccariotto M, Galvanetto U. Mixed-mode crack patterns in ordinary state-based Peridynamics [J]. Key Engineering Materials, 2015, 665:53-56.
[132] Dipasquale D, Sarego G, Zaccariotto M, et al. A discussion on failure criteria for ordinary state-based peridynamics [J]. Engineering Fracture Mechanics, 2017, 186:378-398.
[133] Imachi M, Tanaka S, Bui T Q. Mixed-mode dynamic stress intensity factors evaluation using ordinary state-based peridynamics [J]. Theoretical and Applied Fracture Mechanics, 2018, 93:97-104.
[134] Butt S N, Timothy J J, Meschke G. Wave dispersion and propagation in state-based peridynamics [J]. Computational Mechanics, 2017, 60(5):725-738.
[135] Silling S A, Askari A. Peridynamic model for fatigue cracking [R]. SAND2014-18590. Albuquerque: Sandia National Laboratories, 2014.
[136] Lejeune E, Linder C. Modeling tumor growth with peridynamics [J]. Biomechanics and modeling in mechanobiology, 2017, 16(4):1141-1157.
[137] Zhang H, Qiao P Z. An extended state-based peridynamic model for damage growth prediction of bimaterial structures under thermomechanical loading [J]. Engineering Fracture Mechanics, 2018, 189:81-97.
[138] D′Antuono P, Morandini M. Thermal shock response via weakly coupled peridynamic thermo-mechanics [J]. International Journal of Solids and Structures, 2017, 129:74-89.
[139] Han F, Lubineau G, Azdoud Y, et al. A morphing approach to couple state-based peridynamics with classical continuum mechanics [J]. Computer Methods in Applied Mechanics and Engineering, 2016, 301:336-358.
[140] Bie Y H, Cui X Y, Li Z C. A coupling approach of state-based peridynamics with node-based smoothed finite element method [J]. Computer Methods in Applied Mechanics and Engineering, 2018, 331:675-700.
[141] Ren H L, Zhuang X Y, Rabczuk T. A new peridynamic formulation with shear deformation for elastic solid [J]. Journal of Micromechanics and Molecular Physics, 2016, 1(2):1650009.
[142] 于继东, 王文强, 李平. 弹塑性本构在 PeriDynamic 模型中的实现 [C]. 中国力学学会 2009

学术大会, 2009.

[143] Mitchell J A. A nonlocal, ordinary, state-based plasticity model for peridynamics [R]. Sandia Report SAND2011-3166, Albuquerque: Sandia National Laboratories, 2011.

[144] Mitchell J A. A non-local, ordinary-state-based viscoelasticity model for peridynamics [R]. Sandia National Lab Report, 2011, 8064:1-28.

[145] Vogler T, Lammi C J. A nonlocal peridynamic plasticity model for the dynamic flow and fracture of concrete [R]. Sandia National Laboratories (SNL-CA), Livermore, CA (United States), 2014.

[146] Zhou X P, Shou Y D, Berto F. Analysis of the plastic zone near the crack tips under the uniaxial tension using ordinary state-based peridynamics [J]. Fatigue & Fracture of Engineering Materials & Structures, 2018, 41(5):1159-1170.

[147] Delorme R, Tabiai I, Lebel L L, et al. Generalization of the ordinary state-based peridynamic model for isotropic linear viscoelasticity [J]. Mechanics of Time-Dependent Materials, 2017, 21(4):549-575.

[148] Zhang H W, Li H, Ye H F, et al. A coupling peridynamic approach for the consolidation and dynamic analysis of saturated porous media [J]. Computational Mechanics, 2019, 64(4):1097-1113.

[149] 李辉. 饱和多孔介质动力及断裂分析的多尺度有限元和近场动力学方法 [D]. 大连：大连理工大学, 2019.

[150] Oterkus E, Madenci E. Peridynamics for failure prediction in composites [C]. 53rd AIAA/ASME/ASCE/AHS/ASC Structures, Structural Dynamics and Materials Conference 20th AIAA/ASME/AHS Adaptive Structures Conference 14th AIAA 2012:1692.

[151] Dorduncu M, Barut A, Madenci E. Ordinary-state based peridynamic truss element [C]. 56th AIAA/ASCE/AHS/ASC Structures, Structural Dynamics, and Materials Conference, 2015.

[152] Dorduncu M, Barut A, Madenci E. Peridynamic truss element for viscoelastic deformation [C]. 57th AIAA/ASCE/AHS/ASC Structures, Structural Dynamics, and Materials Conference, 2016.

[153] Madenci E, Colavito K, Phan N. Peridynamics for unguided crack growth prediction under mixed-mode loading [J]. Engineering Fracture Mechanics, 2016, 167:34-44.

[154] Zhu N, De Meo D, Oterkus E. Modelling of granular fracture in polycrystalline materials using ordinary state-based peridynamics [J]. Materials, 2016, 9(12):977.

[155] Madenci E, Oterkus S. Ordinary state-based peridynamics for thermoviscoelastic deformation [J]. Engineering Fracture Mechanics, 2017, 175:31-45.

[156] Hu Y L, Chen H L, Spencer B W, et al. Thermomechanical peridynamic analysis with irregular non-uniform domain discretization [J]. Engineering Fracture Mechanics, 2018, 197:92-113.

[157] Madenci E, Oterkus S. Ordinary state-based peridynamics for plastic deformation according to von Mises yield criteria with isotropic hardening [J]. Journal of the Mechanics and Physics of Solids, 2016, 86:192-219.

[158] Chen W, Zhu F, Zhao J, et al. Peridynamics-based fracture animation for elastoplastic solids [C].Computer Graphics Forum, 2018, 37(1):112-124.

[159] Tuniki B K. Peridynamic constitutive model for concrete [D]. Albuquerque: The University of New Mexico, 2012.

[160] Richardson R. The state-based peridynamic lattice model [D]. Albuquerque: The University of New Mexico, 2014.

[161] McVey S. The state-based peridynamic lattice model and reinforced concrete structures [D]. Albuquerque: The University of New Mexico, 2016.

[162] Aguiar A R, Fosdick R. A constitutive model for a linearly elastic peridynamic body [J]. Mathematics and Mechanics of Solids, 2014, 19(5):502-523.

[163] Aguiar A R. On the determination of a peridynamic constant in a linear constitutive model [J]. Journal of Elasticity, 2016, 122(1):27-39.

[164] O'Grady J. Peridynamic beams, plates, and shells: a nonordinary, state-based model [D]. San Antonio: The University of Texas At San Antonio, 2014.

[165] O'Grady J, Foster J T. Peridynamic plates and flat shells: a non-ordinary, state-based model [J]. International Journal of Solids and Structures, 2014, 51(25-26):4572-4579.

[166] O'Grady J, Foster J T. A meshfree method for bending and failure in non-ordinary peridynamic shells [J]. Computational Mechanics, 2016, 57(6):921-929.

[167] Silling S A. Stability of peridynamic correspondence material models and their particle discretizations [J]. Computer Methods in Applied Mechanics and Engineering, 2017, 322:42-57.

[168] Foster J T. Dynamic crack initiation toughness: Experiments and peridynamic modeling [D]. West Lafayette: Purdue University, 2009.

[169] Foster J T, Silling S A, Chen W W. An energy based failure criterion for use with peridynamic states [J]. International Journal for Multiscale Computational Engineering, 2011, 9(6):675-688.

[170] Khan M I H. Shear band regularization with peridynamics [D]. San Antonio: The University of Texas at San Antonio, 2014.

[171] Littlewood D J. Simulation of dynamic fracture using peridynamics, finite element modeling, and contact [C]. ASME 2010 International Mechanical Engineering Congress and Exposition. American Society of Mechanical Engineers, 2010.

[172] Littlewood D J. A nonlocal approach to modeling crack nucleation in AA 7075-T651 [C]. ASME 2011 International Mechanical Engineering Congress and Exposition. American Society of Mechanical Engineers, 2011: 567-576.

[173] Tupek M R, Rimoli J J, Radovitzky R. An approach for incorporating classical continuum damage models in state-based peridynamics [J]. Computer Methods in Applied Mechanics and Engineering, 2013, 263:20-26.

[174] Tupek M R. Extension of the peridynamic theory of solids for the simulation of materials under extreme loadings [D]. Cambridge: Massachusetts Institute of Technology, 2014.

[175] Breitenfeld M S. Quasi-static non-ordinary state-based peridynamics for the modeling

of 3D fracture [D]. Urbana: University of Illinois at Urbana-Champaign, 2014.
[176] Breitenfeld M S, Geubelle P H, Weckner O, et al. Non-ordinary state-based peridynamic analysis of stationary crack problems [J]. Computer Methods in Applied Mechanics and Engineering, 2014, 272: 233-250.
[177] Sun S, Sundararaghavan V. A peridynamic implementation of crystal plasticity [J]. International Journal of Solids and Structures, 2014, 51: 3350-3360.
[178] Amani J, Oterkus E, Areias P, et al. A non-ordinary state-based peridynamics formulation for thermoplastic fracture [J]. International Journal of Impact Engineering, 2016, 87: 83-94.
[179] Wu C T. Kinematic constraints in the state-based peridynamics with mixed local/nonlocal gradient approximations [J]. Computational Mechanics, 2014, 54:1255-1267.
[180] Wu C T, Ren B. A stabilized non-ordinary state-based peridynamics for the nonlocal ductile material failure analysis in metal machining process [J]. Computer Methods in Applied Mechanics and Engineering, 2015, 291:197-215.
[181] Yaghoobi A, Mi G C. Meshless modeling framework for fiber reinforced concrete structures [J]. Computers & Structures, 2015, 161:43-54.
[182] Yaghoobi A, Chorzepa M G, Kim S S. Mesoscale fracture analysis of multiphase cementitious composites using peridynamics [J]. Materials, 2017, 10:162-1-21.
[183] Yaghoobi A, Chorzepa M G. Higher-order approximation to suppress the zero-energy mode in non-ordinary state-based peridynamics [J]. Computers & Structures, 2017, 188: 63-79.
[184] Yaghoobi A, Chorzepa M G. Formulation of symmetry boundary modeling in non-ordinary state-based peridynamics and coupling with finite element analysis [J]. Mathematics and Mechanics of Solids, 2018, 23(8):1156-1176.
[185] 谷新保, 周小平. 含圆孔拉伸板的近场动力学数值模拟 [J]. 固体力学学报, 2015, (5):376-383.
[186] Gu X B, Wu Q H. The application of nonordinary, state-based peridynamic theory on the damage process of the rock-like materials [J]. Mathematical Problems in Engineering, 2016, 2016:1-9.
[187] Wang Y T, Zhou X P, Xu X M. Numerical simulation of propagation and coalescence of flaws in rock materials under compressive loads using the extended non-ordinary state-based peridynamics [J]. Engineering Fracture Mechanics, 2016, 163: 248-273.
[188] Zhou X P, Wang Y T, Xu X M. Numerical simulation of initiation, propagation and coalescence of cracks using the non-ordinary state-based peridynamics [J]. International Journal of Fracture, 2016, 201:213-234.
[189] Zhou X P, Wang Y T, Qian Q H. Numerical simulation of crack curving and branching in brittle materials under dynamic loads using the extended non-ordinary state-based peridynamics [J]. European Journal of Mechanics - A/Solids, 2016, 60:277-299.
[190] Zhou X P, Wang Y T. Numerical simulation of crack propagation and coalescence in pre-cracked rock-like Brazilian disks using the non-ordinary state-based peridynamics

[J]. International Journal of Rock Mechanics and Mining Sciences, 2016, 89:235-249.
[191] Lai X, Ren B, Fan H F, et al. Peridynamics simulations of geomaterial fragmentation by impulse loads [J]. International Journal for Numerical and Analytical Methods in Geomechanics, 2015, 39:1304-1330.
[192] Lai X, Liu L S, Liu Q W, et al. Slope stability analysis by peridynamic theory [C]. Applied Mechanics and Materials, Trans. Tech. Publications, 2015,744:584-588.
[193] Ren B, Fan H F, Bergel G L, et al. A peridynamics-SPH coupling approach to simulate soil fragmentation induced by shock waves [J]. Computational Mechanics, 2015, 55:287-302.
[194] Fan H F, Bergel G L, Li S F. A hybrid peridynamics-SPH simulation of soil fragmentation by blast loads of buried explosive [J]. International Journal of Impact Engineering, 2016, 87:14-27.
[195] Fan H F, Li S F. A Peridynamics-SPH modeling and simulation of blast fragmentation of soil under buried explosive loads [J]. Computer Methods in Applied Mechanics and Engineering, 2017, 318: 349-381.
[196] Bergel G L, Li S F. The total and updated Lagrangian formulations of state-based peridynamics [J]. Computational Mechanics, 2016, 58(2):351-370.
[197] Yu Y, Liu S S, Zhao S L, et al. The nonlinear inplane behavior and progressive damage modeling for laminate by peridynamics [C]. ASME 2016 International Mechanical Engineering Congress and Exposition. American Society of Mechanical Engineers, 2016: V001T03A054-V001T03A054.
[198] Gu X, Zhang Q, Yu Y T. An effective way to control numerical instability of a nonordinary state-based peridynamic elastic model [J]. Mathematical Problems in Engineering, 2017, 2017:1-7.
[199] Du Q, Tian X C. Stability of nonlocal Dirichlet integrals and implications for peridynamic correspondence material modeling [J]. Journal on Applied Mathematics, 2018, 78(3):1536-1552.
[200] Liu M H, Wang Q, Lu W. Peridynamic simulation of brittle-ice crushed by a vertical structure [J]. International Journal of Naval Architecture and Ocean Engineering, 2017, 9: 209-218.
[201] Sami A. Peridynamic analysis of residual stress around a cold expanded fastener hole [J]. International Journal of Application or Innovation in Engineering and Management, 2017, 6:1-7
[202] Ganzenmüller G C, Hiermaier S, May M. On the similarity of meshless discretizations of peridynamics and smooth-particle hydrodynamics [J]. Computers & Structures, 2014, 150:71-78.
[203] Bessa M A, Foster J T, Belytschko T, et al. A meshfree unification: reproducing kernel peridynamics [J]. Computational Mechanics, 2014, 53:1251-1264.
[204] 李潘, 郝志明, 甄文强. 一种近场动力学非普通状态理论零能模式控制方法 [J]. 力学学报, 2018, 50(2): 329-338.

[205] Gu X, Zhang Q, Madenci E, et al. Possible causes of numerical oscillation in non-ordinary state-based peridynamics and a bond-associated higher-order stabilized model [J]. Computer Methods in Applied Mechanics and Engineering, 2019, 357:112592.

[206] Chowdhury S R, Rahaman M M, Roy D, et al. A micropolar peridynamic theory in linear elasticity [J]. International Journal of Solids and Structures, 2015, 59:171-182.

[207] Chowdhury S R, Roy P, Roy D, et al. A peridynamic theory for linear elastic shells [J]. International Journal of Solids and Structures, 2016, 84:110-132.

[208] Song X Y, Khalili N. A peridynamics model for strain localization analysis of geomaterials[J]. International Journal for Numerical and Analytical Methods in Geomechanics, 2019, 43(1):77-96.

[209] Song X Y, Menon S. Modeling of chemo-hydromechanical behavior of unsaturated porous media: a nonlocal approach based on integral equations [J]. Acta Geotechnica, 2019, 14(3):727-747.

[210] Tu Q S, Li S F. An updated Lagrangian particle hydrodynamics (ULPH) for Newtonian fluids [J]. Journal of Computational Physics, 2017, 348:493-513.

[211] Wang J Q, Zhang X B. Modified particle method with integral Navier-Stokes formulation for incompressible flows [J]. Journal of Computational Physics, 2018, 366:1-13.

[212] Wang J Q, Zhang X B. Improved moving particle semi-implicit method for multiphase flow with discontinuity [J]. Computer Methods in Applied Mechanics and Engineering, 2019, 346:312-331.

[213] Ren H L, Zhuang X Y. A nonlocal formulation for weakly compressible fluid [C]. International Conference on Advances in Computational Mechanics, Springer, Singapore, 2017:835-850.

[214] Bazazzadeh S, Shojaei A, Zaccariotto M, et al. Application of the peridynamic differential operator to the solution of sloshing problems in tanks [J]. Engineering Computations, 2018, 36(1):45-83.

[215] Gao Y, Oterkus S. Nonlocal numerical simulation of low Reynolds number laminar fluid motion by using peridynamic differential operator [J]. Ocean Engineering, 2019, 179:135-158.

[216] Gao Y, Oterkus S. Non-local modeling for fluid flow coupled with heat transfer by using peridynamic differential operator [J]. Engineering Analysis with Boundary Elements, 2019, 105:104-121.

[217] Katiyar A, Foster J T, Ouchi H, et al. A peridynamic formulation of pressure driven convective fluid transport in porous media [J].Journal of Computational Physics, 2014, 261:209-229.

[218] Katiyar A, Agrawal S, Ouchi H, et al. A general peridynamics model for multiphase transport of non-newtonian compressible fluids in porous media [J]. Journal of Computational Physics, 2019, 402:109075.

[219] Dalla Barba F, Campagnari P, Zaccariotto M, et al. A fluid-structure interaction model based on peridynamics and Navier-Stokes equations for hydraulic fracture problems

[C]. 6th European Conference on Computational Mechanics (ECCM 6), 7th European Conference on Computational Fluid Dynamics (ECFD 7), 2018.

[220] Liu R W, Yan J L, Li S F. Modeling and simulation of ice-water interactions by coupling peridynamics with updated Lagrangian particle hydrodynamics [J]. Computational Particle Mechanics, 2020, 7(2):241-255.

[221] Zhang Y, Pan G, Zhang Y, et al. A multi-physics peridynamics-DEM-IB-CLBM framework for the prediction of erosive impact of solid particles in viscous fluids [J]. Computer Methods in Applied Mechanics and Engineering, 2019, 352:675-690.

[222] Gerstle W H, Silling S A, Read D, et al. Peridynamic simulation of electromigration [J]. CMC-Computers Materials & Continua, 2008, 8(2):75-92.

[223] Read D T, Tewary V K, Gerstle W H. Modeling electromigration using the peridynamics approach [C]//Electromigration in Thin Films and Electronic Devices, 2011: 45-69.

[224] Oterkus S, Fox J, Madenci E. Simulation of electro-migration through peridynamics [C]. IEEE 63rd Electronic Components and Technology Conference (ECTC), 2013:1488-1493.

[225] Bobaru F, Duangpanya M. The peridynamic formulation for transient heat conduction [J]. International Journal of Heat and Mass Transfer, 2010, 53(19-20):4047-4059.

[226] Bobaru F, Duangpanya M. A peridynamic formulation for transient heat conduction in bodies with evolving discontinuities [J]. Journal of Computational Physics, 2012, 231(7):2764-2785.

[227] Agwai A. A peridynamic approach for coupled fields [D]. Tucson: The University of Arizona, 2011.

[228] Agwai A, Guven I, Madenci E. A new thermomechanical fracture analysis approach for 3D integration technology [C]. IEEE 61st Electronic Components and Technology Conference (ECTC), 2011:740-745.

[229] Oterkus S, Madenci E, Agwai A. Peridynamic thermal diffusion [J]. Journal of Computational Physics, 2014, 265(10):71-96.

[230] Chen Z G, Bobaru F. Selecting the kernel in a peridynamic formulation: A study for transient heat diffusion [J]. Computer Physics Communications, 2015,197:51-60.

[231] Jafari A, Bahaaddini R, Jahanbakhsh H. Numerical analysis of peridynamic and classical models in transient heat transfer, employing Galerkin approach [J]. Heat Transfer—Asian Research, 2018,47(3):531-555.

[232] 刘硕, 方国东, 王兵, 等. 近场动力学与有限元方法耦合求解热传导问题 [J]. 力学学报, 2018, 50(2):339-348.

[233] 王飞, 马玉娥, 郭妍宁. 近场动力学中内核参数对非均匀材料热传导数值解的影响研究 [J]. 西北工业大学学报, 2017, 35(2):203-207.

[234] 刘英凯, 程站起. 功能梯度材料热传导问题的近场动力学模型 [J]. 力学季刊, 2018, 39(1):82-89.

[235] Liao Y, Liu L S, Liu Q W, et al. Peridynamic simulation of transient heat conduction problems in functionally gradient materials with cracks [J]. Journal of Thermal Stresses,

2017,40(12):1484-1501.

[236] Wang L J, Xu J F, Wang J X. The Green's functions for peridynamic non-local diffusion [J]. Proceedings of the Royal Society A: Mathematical Physical and Engineering Sciences, 2016, 472(2193):20160185.

[237] Wang L J, Xu J F, Wang J X. A peridynamic framework and simulation of non-Fourier and nonlocal heat conduction [J]. International Journal of Heat and Mass Transfer, 2018, 118:1284-1292.

[238] 王超聪, 刘齐文, 刘立胜, 等. 热防护材料烧蚀温度场的近场动力学模拟 [J]. 科学技术与工程, 2017, 17(26):172-176.

[239] Jabakhanji R. Peridynamic modeling of coupled mechanical deformations and transient flow in unsaturated soils [D]. West Lafayette: Purdue University, 2013.

[240] Oterkus S. Peridynamics for the solution of multiphysics problems [D]. Tucson: The University of Arizona, 2015.

[241] Du Q, Gunzburger M, Lehoucq R B, et al. Analysis and approximation of nonlocal diffusion problems with volume constraints [J]. SIAM Review, 2012, 54(4): 667-696.

[242] Xu F F, Gunzburger M, Burkardt J. A multiscale method for nonlocal mechanics and diffusion and for the approximation of discontinuous functions [J]. Computer Methods in Applied Mechanics and Engineering, 2016, 307:117-143.

[243] Han S W, Diyaroglu C, Oterkus S, et al. Peridynamic direct concentration approach by using ANSYS [C]. IEEE 66th Electronic Components and Technology Conference (ECTC), 2016:544-549.

[244] Diyaroglu C, Oterkus S, Oterkus E, et al. Peridynamic modeling of diffusion by using finite-element analysis [J]. IEEE Transactions on Components, Packaging and Manufacturing Technology, 2017, 7(11):1823-1831.

[245] Diyaroglu C, Oterkus S, Oterkus E, et al. Peridynamic wetness approach for moisture concentration analysis in electronic packages [J]. Microelectronics Reliability, 2017, 70:103-111.

[246] Jabakhanji R, Mohtar R H. A peridynamic model of flow in porous media [J]. Advances in Water Resources, 2015, 78: 22-35.

[247] Delgoshaie A H, Meyer D W, Jenny P, et al. Non-local formulation for multiscale flow in porous media [J]. Journal of Hydrology, 2015, 531:649-654.

[248] Jenny P, Meyer D W. Non-local generalization of Darcy's law based on empirically extracted conductivity kernels [J]. Computational Geosciences, 2017, 21(5-6):1281-1288.

[249] Multiphysics C O. Comsol Multiphysics User's Guide, Version 4.3a [R]. COMSOL AB, 2012: 39.

[250] 孙培德, 杨东全, 陈奕柏. 多物理场耦合模型及数值模拟导论 [M]. 北京: 中国科学技术出版社, 2007.

[251] Kilic B, Madenci E. Prediction of crack paths in a quenched glass plate by using peridynamic theory [J]. International Journal of Fracture, 2009, 156(2):165-177.

[252] Kilic B, Madenci E. Peridynamic theory for thermomechanical analysis [J]. IEEE Transactions on Advanced Packaging, 2010, 33(1):97-105.

[253] Jeon B S, Stewart R J, Ahmed I Z. Peridynamic simulations of brittle structures with thermal residual deformation: Strengthening and structural reactivity of glasses under impacts [J]. Proceedings of the Royal Society A: Mathematical Physical and Engineering Sciences, 2015, 471(2183):20150231.

[254] Mella R, Wenman M R. Modelling explicit fracture of nuclear fuel pellets using peridynamics [J]. Journal of Nuclear Materials, 2015, 467:58-67.

[255] Xu Z P, Zhang G F, Chen Z G, et al. Elastic vortices and thermally-driven cracks in brittle materials with peridynamics [J]. International Journal of Fracture, 2018, 209:203-222.

[256] Giannakeas I N, Papathanasiou T K, Bahai H. Simulation of thermal shock cracking in ceramics using bond-based peridynamics and FEM [J]. Journal of the European Ceramic Society, 2018,38(8): 3037-3048.

[257] Oterkus S, Madenci E, Oterkus E, et al. Hygro-thermo-mechanical analysis and failure prediction in electronic packages by using peridynamics [C]. IEEE 64th Electronic Components and Technology Conference (ECTC), 2014:973-982.

[258] 苏伯阳, 李书欣, 刘立胜, 等. 湿热环境下复合材料冲击损伤的近场动力学模拟 [J]. 科学技术与工程, 2018,18(1):201-206.

[259] Oterkus S, Madenci E. Crack growth prediction in fully-coupled thermal and deformation fields using peridynamic theory [C]. 54th AIAA/ASME/ASCE/AHS/ASC Structures, Structural Dynamics, and Materials Conference, 2013:1477.

[260] Oterkus S, Madenci E, Agwai A. Fully coupled peridynamic thermomechanics [J]. Journal of the Mechanics and Physics of Solids, 2014, 64(1):1-23.

[261] Chen H L, Hu Y L, Spencer B W. A MOOSE-based implicit peridynamic thermomechanical model [C]. ASME 2016 International Mechanical Engineering Congress and Exposition, 2016:V009T12A072.

[262] Chen H L, Hu Y L, Spencer B W. Peridynamics using irregular domain discretization with moose-based implementation [C]. ASME 2017 International Mechanical Engineering Congress and Exposition, 2017: V009T12A067-V009T12A067.

[263] Wang Y T, Zhou X P, Kou M M. A coupled thermo-mechanical bond-based peridynamics for simulating thermal cracking in rocks [J]. International Journal of Fracture, 2018, 211(1-2):13-42.

[264] Madenci E, Oterkus S. Peridynamic modeling of thermo-oxidative damage evolution in a composite lamina [C]//58th AIAA/ASCE/AHS/ASC Structures, Structural Dynamics, and Materials Conference, 2017:0197.

[265] Oterkus S, Madenci E. Peridynamic modeling of fuel pellet cracking [J]. Engineering Fracture Mechanics, 2017, 176:23-37.

[266] Roy P, Roy D. A peridynamic approach to flexoelectricity [J/OL]. arXiv preprint arXiv:1603.03894,2016.

[267] Prakash N, Seidel G D. Electromechanical peridynamics modeling of piezoresistive response of carbon nanotube nanocomposites [J]. Computational Materials Science, 2016, 113:154-170.

[268] Prakash N, Seidel G D. Computational electromechanical peridynamics modeling of strain and damage sensing in nanocomposite bonded explosive materials (ncbx) [J]. Engineering Fracture Mechanics, 2017, 177:180-202.

[269] 张振宇. 基于 Voronoi 图方法的近场动力学键理论及热电耦合理论研究 [D]. 武汉：武汉理工大学, 2015.

[270] Assefa M, Lai X, Liu L S. Bond based peridynamic formulation for thermoelectric materials [J]. Materials Science Forum, 2017, 883: 51-59.

[271] Assefa M, Lai X, Liu L S, et al. Peridynamic formulation for coupled thermoelectric phenomena [J]. Advances in Materials Science and Engineering, 2017, 9836741:1-10.

[272] Wildman R A, Gazonas G A. A dynamic electro-thermo-mechanical model of dielectric breakdown in solids using peridynamics [J]. Journal of Mechanics of Materials and Structures, 2015, 10:613-630.

[273] Wildman R A, Gazonas G A. A multiphysics finite element and peridynamics model of dielectric breakdown [R]. US Army Research Laboratory Aberdeen Proving Ground United States, 2017.

[274] Turner D Z. A non-local model for fluid-structure interaction with applications in hydraulic fracturing [J]. International Journal for Computational Methods in Engineering Science and Mechanics, 2013, 14(5):391-400.

[275] Ouchi H, Katiyar A, Foster J T, et al. A peridynamics model for the propagation of hydraulic fractures in heterogeneous, naturally fractured reservoirs [C]//SPE Hydraulic Fracturing Technology Conference 2015. Society of Petroleum Engineers.

[276] Ouchi H. Development of peridynamics-based hydraulic fracturing model for fracture growth in heterogeneous reservoirs [D]. Austin: The University of Texas at Austin, 2016.

[277] Ouchi H, Agrawal S, Foster J T, et al. Effect of small scale heterogeneity on the growth of hydraulic fractures [C]//SPE Hydraulic Fracturing Technology Conference and Exhibition 2017. Society of Petroleum Engineers.

[278] Ouchi H, Foster J T, Sharma M M. Effect of reservoir heterogeneity on the vertical migration of hydraulic fractures [J]. Journal of Petroleum Science and Engineering, 2017, 151:384-408.

[279] Nadimi S, Miscovic I, McLennan J. A 3D peridynamic simulation of hydraulic fracture process in a heterogeneous medium [J]. Journal of Petroleum Science and Engineering, 2016, 145:444-452.

[280] Edmiston J K. Development of a geoperidynamic model for hydraulic fracture [C]. 49th US Rock Mechanics/Geomechanics Symposium, American Rock Mechanics Association, 2015.

[281] Oterkus S, Madenci E, Oterkus E. Fully coupled poroelastic peridynamic formulation for fluid-filled fractures [J]. Engineering Geology, 2017, 225:19-28.

[282] 吴凡, 李书卉, 段庆林, 等. 基于近场动力学方法的水力压裂过程数值模拟 [J]. 计算机辅助工程, 2017, 26(1):1-6.

[283] Ishimoto J, Sato T, Combescure A. Computational approach for hydrogen leakage with crack propagation of pressure vessel wall using coupled particle and Euler method [J]. International Journal of Hydrogen Energy, 2017, 42(15):10656-10682.

[284] Chen Z G, Bobaru F. Peridynamic modeling of pitting corrosion damage [J]. Journal of the Mechanics and Physics of Solids, 2015, 78:352-381.

[285] Chen Z G, Zhang G F, Bobaru F. The influence of passive film damage on pitting corrosion [J]. Journal of The Electrochemical Society, 2016, 163(2):C19-C24.

[286] Jafarzadeh S, Chen Z G, Bobaru F. Peridynamic modeling of repassivation in pitting corrosion of stainless steel [J]. Corrosion, 2017, 74(4):393-414.

[287] 白小敏, 唐建群, 巩建鸣. 不同过电位下一维点蚀的近场动力学数值模拟 [J]. 南京工业大学学报 (自科版), 2017, 6:91-98.

[288] De Meo D, Oterkus E. Finite element implementation of a peridynamic pitting corrosion damage model [J]. Ocean Engineering, 2017, 135:76-83.

[289] De Meo D, Diyaroglu C, Zhu N, et al. Modelling of stress-corrosion cracking by using peridynamics [J]. International Journal of Hydrogen Energy, 2016, 41(15):6593-6609.

[290] Wang H L, Oterkus E, Oterkus S. Predicting fracture evolution during lithiation process using peridynamics [J]. Engineering Fracture Mechanics, 2018,192:176-191.

[291] 戴旭东, 王义亮, 谢友柏. 以润滑油膜为动力耦合件的内燃机缸套–活塞系统中动力耦合方程的建立及求解方法 [J]. 润滑与密封, 2001, 5:5-8.

[292] Farhat C, Park K C, Dubois-Pelerin Y. An unconditionally stable staggered algorithm for transient finite element analysis of coupled thermoelastic problems [J]. Computer Methods in Applied Mechanics and Engineering, 1991, 85(3): 349-365.

[293] Wildman R A, Gazonas G A. A finite difference-augmented peridynamics method for reducing wave dispersion [J]. International Journal of Fracture, 2014, 190(1-2):39-52.

[294] Rahman R, Haque A. A peridynamics formulation based hierarchical multiscale modeling approach between continuum scale and atomistic scale [J]. International Journal of Computational Materials Science and Surface Engineering, 2013, 1(3):1-19.

[295] Rahman R, Foster J T, Haque A. A multiscale modeling scheme based on peridynamic theory [J]. International Journal for Multiscale Computational Engineering, 2014, 12(3):223-248.

[296] 卢志堂, 王志亮. 近场动力学法频散特性及其在岩石层裂分析中应用 [J]. 哈尔滨工业大学学报, 2016(2): 131-137, 151.

[297] Askari E, Bobaru F, Lehoucq R B, et al. Peridynamics for multiscale materials modeling [C]. Journal of Physics: Conference Series, IOP Publishing, 2008, 125(1):012078.

[298] Badia S, Bochev P, Gunzburger M, et al. Blending methods for coupling atomistic and continuum models [C]. 6th International Conference on Large-Scale Scientific Computing, 2007,4818:16-27.

[299] Sears M P, Lehoucq R B. The statistical mechanical foundations of peridynamics: I.

mass and momentum conservation laws [R].Technical Report SAND2009-0791J, Sandia National Laboratories, 2009.

[300] Lehoucq R B, Sears M P. Statistical mechanical foundation of the peridynamic nonlocal continuum theory: Energy and momentum conservation laws[J]. Physical Review E, 2011, 84(3):247-268.

[301] Rahman R, Foster J T. Peridynamic theory of solids from the perspective of classical statistical mechanics [J]. Physica A Statistical Mechanics & Its Applications, 2015, 437:162-183.

[302] Lehoucq R B, Silling S A. Statistical coarse-graining of molecular dynamics into peridynamics [R]. Report SAND2007-6410, Albuquerque: Sandia National Laboratories, 2007.

[303] Emmrich E, Weckner O. The peridynamic equation and its spatial discretisation [J]. Mathematical Modelling and Analysis, 2007, 12(1):17-27.

[304] Seleson P. Peridynamic multiscale models for the mechanics of materials: Constitutive relations, upscaling from atomistic systems, and interface problems [D]. Tallahassee: Florida State University, 2010.

[305] Rahman R, Foster J T. Bridging the length scales through nonlocal hierarchical multiscale modeling scheme [J]. Computational Materials Science, 2014, 92(5):401-415.

[306] Rahman R, Foster J T. Onto resolving spurious wave reflection problem with changing nonlocality among various length scales [J]. Communications in Nonlinear Science and Numerical Simulation, 2016, 34:86-122.

[307] Silling S A, James V C. Hierarchical multiscale method development for peridynamics [R]. SAND Report, 18565, 2014.

[308] Duzzi M, Zaccariotto M, Galvanetto U. Application of peridynamic theory to nanocomposite materials [J]. Advanced Materials Research, 2014, 1016:44-48.

[309] Silling S A, Zimmermann M, Abeyaratne R. Deformation of a peridynamic bar [J]. Journal of Elasticity, 2003, 73(1-3): 173-190.

[310] Zimmermann M. A continuum theory with long-range forces for solids [D]. Cambridge: Massachusetts Institute of Technology, 2005.

[311] Weckner O, Abeyaratne R. The effect of long-range forces on the dynamics of a bar [J]. Journal of the Mechanics and Physics of Solids, 2005, 53(3):705-728.

[312] Emmrich E, Weckner O. The peridynamic equation of motion in non-local elasticity theory [C]. III European Conference on Computational Mechanics-Solids, Structures and Coupled Problems in Engineering: Springer Netherlands, 2006:5-8.

[313] Emmrich E, Weckner O. Analysis and numerical approximation of an integro-differential equation modeling non-local effects in linear elasticity [J]. Mathematics and Mechanics of Solids, 2007, 4(4):363-384.

[314] Weckner O, Silling S A. Determination of nonlocal constitutive equations from phonon dispersion relations [J]. International Journal for Multiscale Computational Engineering, 2011, 9(6):623-634.

[315] Al-Barmany H M A. Dispersive standing waves in peridynamic model [J]. International Journal of Mathematics and Mathematical Sciences, 2013, 3(4):66-72.

[316] Gu X, Zhang Q, Huang D, et al. Wave dispersion analysis and simulation method for concrete SHPB test in peridynamics [J]. Engineering Fracture Mechanics, 2016, 160:124-137.

[317] Tao J, Liu D. Simulating wave propagation in SHPB with peridynamics [C]. Dynamic Behavior of Materials, Volume 1. Springer International Publishing, 2014: 195-200.

[318] Martowicz A, Staszewski W J, Ruzzene M, et al. Vibro-acoustic wave interaction in cracked plate modeled with peridynamics [C]. Proceedings of the WCCM XI, ECCM V, ECFD VI,(Barcelona, Spain), (2014): 20-25.

[319] Martowicz A, Staszewski W J, Ruzzene M, et al. Peridynamics as an analysis tool for wave propagation in graphene nanoribbons [C]. SPIE Smart Structures and Materials+ Nondestructive Evaluation and Health Monitoring. International Society for Optics and Photonics, 2015.

[320] Guan Q G, Gunzburger M. Stability and accuracy of time-stepping schemes and dispersion relations for a nonlocal wave equation [J]. Numerical Methods for Partial Differential Equations, 2014, 31(2):778-791.

[321] Aidun J B, Silling S A. Accurate prediction of dynamic fracture with peridynamics [C]. Joint US-Russian Conference on Advances in Materials Science, 2009.

[322] Ha Y D, Bobaru F. Characteristics of dynamic brittle fracture captured with peridynamics [J]. Engineering Fracture Mechanics, 2011, 78(6):1156-1168.

[323] Mehrmashhadi J, Wang L, Bobaru F. Uncovering the dynamic fracture behavior of PMMA with peridynamics: The importance of softening at the crack tip [J]. Engineering Fracture Mechanics, 2019, 219:106617.

[324] Agwai A, Guven I, Madenci E. Predicting crack propagation with peridynamics: A comparative study [J]. International Journal of Fracture, 2011, 171(1):65-78.

[325] Lipton R. Cohesive dynamics and brittle fracture [J]. Journal of Elasticity, 2016, 124(2): 143-191.

[326] Lipton R. Dynamic brittle fracture as a small horizon limit of peridynamics [J]. Journal of Elasticity, 2013, 117(1):21-50.

[327] Bobaru F, Zhang G F. Why do cracks branch? A peridynamic investigation of dynamic brittle fracture [J]. International Journal of Fracture, 2016, 196(1-2):59-98.

[328] Dias J P, Bazani M A, Paschoalini A T, et al. A review of crack propagation modeling using peridynamics [C]//Probabilistic Prognostics and Health Management of Energy Systems, Springer, Cham, 2017:111-126.

[329] Boyce B L, Kramer S L, Fang H E, et al. The Sandia fracture challenge: blind round robin predictions of ductile tearing [J]. International Journal of Fracture, 2014, 186(1-2):5-68.

[330] Boyce B L, Kramer S L, Bosiljevac T R, et al. The second Sandia fracture challenge: predictions of ductile failure under quasistatic and moderate-rate dynamic loading [J].

International Journal of Fracture, 2016, 198(1):5-100.

[331] Kramer S L, Jones A, Mostafa A, et al. The third Sandia fracture challenge: predictions of ductile fracture in additively manufactured metal [J]. International Journal of Fracture, 2019, 218(1):5-61.

[332] Macek R W, Silling S A. Peridynamics *via* finite element analysis [J]. Finite Elements in Analysis and Design, 2007, 43(15):1169-1178.

[333] Levine J A, Bargteil A W, Corsi C, et al. A peridynamic perspective on spring-mass fracture [C]. ACM Siggraph/eurographics Symposium on Computer Animation. Eurographics Association, 2014.

[334] Oterkus E, Guven I, Madenci E. Impact damage assessment by using peridynamic theory [J]. Open Engineering, 2012, 2(4):523-531.

[335] Bobaru F, Ha Y D, Hu W K. Damage progression from impact in layered glass modeled with peridynamics [J]. Central European Journal of Engineering, 2012, 2(4):551-561.

[336] Hu W K, Wang Y N, Yu J, et al. Impact damage on a thin glass plate with a thin polycarbonate backing [J]. International Journal of Impact Engineering, 2013, 62(24):152-165.

[337] 顾鑫, 章青, 黄丹. 基于近场动力学方法的混凝土板侵彻问题研究 [J]. 振动与冲击, 2016, 35(6): 52-58.

[338] Diyaroglu C, Oterkus E, Madenci E, et al. Peridynamic modeling of composite laminates under explosive loading [J]. Composite Structures, 2016, 144:14-23.

[339] 胡祎乐. 基于近场动力学的 FRP 层压板建模与分析 [D]. 上海: 上海交通大学, 2013.

[340] 王富伟. 近场动力学模拟复合材料层合板的冲击损伤 [D]. 南京: 南京航空航天大学, 2014.

[341] 刘贞谷. 基于近场动力学的复合材料单向板渐进损伤模拟与实验研究 [D]. 哈尔滨: 哈尔滨工业大学, 2017.

[342] Hu W K. Peridynamic models for dynamic brittle fracture [D]. Lincoln: University of Nebraska - Lincoln, 2012.

[343] Colavito K. Peridynamics for failure and residual strength prediction of fiber-reinforced composites [D]. Tucson: The University of Arizona, 2013.

[344] Hu Y L. Peridynamic modeling of fiber-reinforced composites with polymer and ceramic matrix [D]. Tucson: The University of Arizona, 2017.

[345] 顾继光. 基于近场动力学的复合材料结构损伤分析 [D]. 哈尔滨：哈尔滨工程大学, 2018.

[346] 姜晓伟. 常规态近场动力学复合材料球型域模型建立及应用研究 [D]. 上海：上海交通大学, 2019.

[347] 陈瑞. 改进的复材单向板键基近场动力学模型 [D]. 大连：大连理工大学, 2020.

[348] 郭帅, 焦学健, 李丽君, 等. 近场动力学方法研究复合材料失效的进展 [J]. 材料导报, 2019, 33(5):826-33.

[349] Huang D, Zhang Q, Qiao P Z. Damage and progressive failure of concrete structures using non-local peridynamic modeling [J]. Science China Technological Sciences, 2011, 54(3):591-596.

[350] Li W J, Guo L. Meso-fracture simulation of cracking process in concrete incorporating three-phase characteristics by peridynamic method [J]. Construction and Building Materials, 2018, 161:665-675.

[351] Lu J Z, Zhang Y T, Muhammad H, et al. Peridynamic model for the numerical simulation of anchor bolt pullout in concrete [J]. Mathematical Problems in Engineering, 2018, 2018:1-10.

[352] 沈峰. 混凝土材料和结构损伤破坏的近场动力学模拟 [D]. 南京: 河海大学, 2014.

[353] 李天一, 章青, 夏晓舟, 等. 考虑混凝土材料非均质特性的近场动力学模型 [J]. 应用数学和力学, 2018, 39(8):913-924.

[354] Emmrich E, Puhst D. Well-posedness of the peridynamic model with Lipschitz continuous pairwise force function [J]. Communications in Mathematical Sciences, 2013, 11(4):1039-1049.

[355] Hinds B, Radu P. Dirichlet's principle and wellposedness of solutions for a nonlocal p-Laplacian system [J]. Applied Mathematics and Computation, 2012, 219(4):1411-1419.

[356] Mengesha T, Du Q. Analysis of a scalar nonlocal peridynamic model with a sign changing kernel [J]. Discrete and Continuous Dynamical Systems-Series B (DCDS-B), 2017, 18(5):1415-1437.

[357] Mengesha T, Du Q. The bond-based peridynamic system with Dirichlet-type volume constraint [J]. Proceedings of the Royal Society of Edinburgh: Section A Mathematics, 2014, 144(1):161-186.

[358] Mengesha T, Du Q. Nonlocal constrained value problems for a linear peridynamic Navier equation [J]. Journal of Elasticity, 2014, 116(1):27-51.

[359] Aksoylu B, Mengesha T. Results on nonlocal boundary value problems [J]. Numerical Functional Analysis and Optimization, 2010, 31(12): 1301-1317.

[360] Gunzburger M, Lehoucq R B. A nonlocal vector calculus with application to nonlocal boundary value problems [J]. Multiscale Modeling & Simulation, 2010, 8(5):1581-1598.

[361] Du Q, Gunzburger M, Lehoucq R B, et al. Analysis of the volume-constrained peridynamic navier equation of linear elasticity [J]. Journal of Elasticity, 2013, 113(2):193-217.

[362] Zhou K, Du Q. Mathematical and numerical analysis of linear peridynamic models with nonlocal boundary conditions [J]. SIAM Journal on Numerical Analysis, 2010, 48(5):1759-1780.

[363] Du Q, Kamm J R. A new approach for a nonlocal, nonlinear conservation law [J]. Siam Journal on Applied Mathematics, 2012, 72(1):464-487.

[364] Tian X C, Du Q. Nonconforming discontinuous galerkin methods for nonlocal variational problems [J]. SIAM Journal on Numerical Analysis, 2015, 53(2):762-781.

[365] Du Q, Zhou K. Mathematical analysis for the peridynamic nonlocal continuum theory [J]. Esaim Mathematical Modelling and Numerical Analysis, 2011, 45(2):217-234.

[366] Mengesha T, Du Q. On the variational limit of a class of nonlocal functionals related to peridynamics [J]. Nonlinearity, 2015, 28(11):3999-4035.

[367] Mengesha T, Du Q. Characterization of function spaces of vector fields and an applica-

tion in nonlinear peridynamics [J]. Nonlinear Analysis: Theory, Methods and Applications, 2016, 140: 82-111.

[368] Du Q, Tian X C. Robust discretization of nonlocal models related to peridynamics [J]. Lecture Notes in Computational Science and Engineering, 2015, 100(5):97-113.

[369] Du Q, Tian L, Zhao X Y. A convergent adaptive finite element algorithm for nonlocal diffusion and peridynamic models [J]. SIAM Journal on Numerical Analysis, 2013, 51(2):1211-1234.

[370] Tian X C, Du Q. Analysis and comparison of different approximations to nonlocal diffusion and linear peridynamic equations [J]. SIAM Journal on Numerical Analysis, 2013, 51(6):3458-3482.

[371] D'Elia M, Gunzburger M. The fractional Laplacian operator on bounded domains as a special case of the nonlocal diffusion operator [J]. Computers & Mathematics with Applications, 2013, 66(7):1245-1260.

[372] D'Elia M, Gunzburger M. Identification of the diffusion parameter in nonlocal steady diffusion problems [J]. Applied Mathematics & Optimization, 2016,73(2): 227-249.

[373] Alali B, Liu K, Gunzburger M. A generalized nonlocal vector calculus [J]. Zeitschrift Für Angewandte Mathematik Und Physik Zamp, 2015, 66(5):2807-2828.

[374] Birkey J. Development of Visual EMU, a graphical user interface for the peridynamic EMU code [D]. Manhattan: Kansas State University, 2007.

[375] Parks M L, Rahman R, Foster J T. pair_style peri/pmb command [EB/OL]. https://docs.lammps.org/pair_peri.html.

[376] Parks M L, Plimpton S J, Lehoucq R B, et al. Peridynamics with LAMMPS: a user guide [R]. Sandia National Laboratory Report, SAND2008-0135.

[377] Parks M L, Littlewood D J, Mitchell J A, et al. Peridigm users' guide v1. 0.0. [R]. SAND Report, 2012: 7800.

[378] Rädel M, Willberg C. Peridigm installation guide[R]. DLR-Interner Bericht. DLR-IB-FA-BS-2019-33, 79S. 2019.

[379] Team S S. Sierra/solid mechanics 4.22 user's guide [R]. SAND2011-7597, Sandia National Laboratories, 2011.

[380] Fan H F, Li S F. Parallel peridynamics—SPH simulation of explosion induced soil fragmentation by using OpenMP [J]. Computational Particle Mechanics, 2017, 4(2):199-211.

[381] Lee J, Oh S E, Hong J W. Parallel programming of a peridynamics code coupled with finite element method [J]. International Journal of Fracture, 2017. 203(1-2):99-114.

[382] Diehl P. Implementierung eines peridynamik-verfahrens auf gpu [D]. Stuttgart: University of Stuttgart, 2012.

[383] 刘肃肃, 胡祎乐, 余音. 基于 GPU 的近场动力学模拟的并行化方法 [J]. 上海交通大学学报, 2016, 50(09):32-37, 45.

[384] Mossaiby F, Shojaei A, Zaccariotto M, et al. OpenCL implementation of a high performance 3D peridynamic model on graphics accelerators [J]. Computers & mathematics with applications, 2017, 74(8):1856-1870.

第 2 章 近场动力学的基本理论

近场动力学 (peridynamics, PD) 由 Silling[1] 在 1998 年首次提出，2000 年正式在 *Journal of the Mechanics and Physics of Solids* 上长文发表。二十多年来，近场动力学引领了大量学者再次聚焦非局部理论和积分方程建模方法 [2-6]，推动了计算力学及其相关领域的飞速发展。近场动力学是一种以积分方程为基础的基于非局部相互作用思想的广义连续介质力学理论——积分型非局部理论 [7]，相较于以偏微分方程为基础的基于局部接触作用思想的经典连续介质力学理论 [8-10]，近场动力学适用于求解连续体、离散体和含裂纹等不连续特征结构的力学问题，是进行固体材料和结构的损伤累积、裂纹萌生、扩展、分叉和汇合等演化分析的一种有效方法。近场动力学与经典连续介质力学和分子动力学联系紧密，是对它们特点的继承、发展与重构。本章系统介绍近场动力学的基本理论，包括基于非局部思想的变形和受力描述、运动方程、本构力模型、守恒律、非局部边界条件与损伤断裂模型。

2.1 变形描述

2.1.1 经典连续介质力学的变形描述

运动学用来描述物体或物体组成部分的空间位置随时间演进的改变，连续介质中的运动学，一般也被称为变形几何学，研究变形所引起物体各部分空间位置和方向的变化以及各邻近点相互距离的变化，位置、位移、速度、加速度、变形梯度、应变和应变率是经典连续介质力学的基本概念 [8-10]。

如图 2-1 所示，引入欧几里得空间中的笛卡儿直角坐标系，并采用 Lagrange 方法描述。参考构型中点 $\boldsymbol{x}$ 的位置与当前构型中点 $\boldsymbol{y}$ 的位置通过变形状函数 $\boldsymbol{\varphi}(\boldsymbol{x},t)$ 相联系，其矢量和分量表达式分别为

$$\boldsymbol{y} = \boldsymbol{\varphi}(\boldsymbol{x},t), \quad y_i = \varphi_i(x_K,t), \quad i=1,2,3; K=1,2,3 \tag{2.1}$$

其中，下标 K、i 分别指代参考构型和当前构型，变形状函数 $\boldsymbol{\varphi}$ 为连续或不连续的矢量函数，未必可逆。

2.1 变形描述

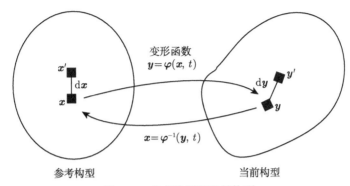

图 2-1 参考构型和当前构型

在参考构型中,邻近两点位置矢量可取为 $\boldsymbol{x}$ 和 $\boldsymbol{x}' = \boldsymbol{x}+\mathrm{d}\boldsymbol{x}$,其中两点间线微元矢量 $\mathrm{d}\boldsymbol{x}$ 为小量,当前构型中相应两点的位置矢量为 $\boldsymbol{y} = \boldsymbol{\varphi}(\boldsymbol{x},t)$ 和 $\boldsymbol{y}' = \boldsymbol{\varphi}(\boldsymbol{x}',t)$。略去泰勒级数展开式中 $\mathrm{d}\boldsymbol{x}$ 的二阶及以上高阶项,只保留一阶项,有

$$\mathrm{d}\boldsymbol{y} = \boldsymbol{y}' - \boldsymbol{y} \approx \boldsymbol{F} \cdot \mathrm{d}\boldsymbol{x} = \mathrm{d}\boldsymbol{x} \cdot \boldsymbol{F}^{\mathrm{T}} \tag{2.2}$$

其中,$\boldsymbol{F}$ 为变形梯度张量,是一个两点二阶张量,定义为

$$\boldsymbol{F} = F_{iK}\boldsymbol{e}_i \otimes \boldsymbol{e}_K = \frac{\partial y_i}{\partial x_K}\boldsymbol{e}_i \otimes \boldsymbol{e}_K \tag{2.3}$$

式中,$\otimes$ 为张量积或并矢积符号,$\boldsymbol{e}_i$ 为当前构型的基矢量,$\boldsymbol{e}_K$ 为参考构型的基矢量。变形梯度张量 $\boldsymbol{F}$ 的分量表达式为

$$\boldsymbol{F} = \begin{bmatrix} \dfrac{\partial y_1}{\partial x_1} & \dfrac{\partial y_1}{\partial x_2} & \dfrac{\partial y_1}{\partial x_3} \\ \dfrac{\partial y_2}{\partial x_1} & \dfrac{\partial y_2}{\partial x_2} & \dfrac{\partial y_2}{\partial x_3} \\ \dfrac{\partial y_3}{\partial x_1} & \dfrac{\partial y_3}{\partial x_2} & \dfrac{\partial y_3}{\partial x_3} \end{bmatrix} \tag{2.4}$$

变形梯度张量 $\boldsymbol{F}$ 的行列式为 $J = \det(\boldsymbol{F})$ $(0 < J < \infty)$,它给出了当前构型与参考构型中体积微元的转换关系或体积比,即 $\mathrm{d}v = J\mathrm{d}V$。式 (2.2) 给出了两个构型之间线元的转换关系,进一步可以得到两个构型之间面元的转换关系为 $\mathrm{d}\boldsymbol{a} = J\boldsymbol{F}^{-\mathrm{T}} \cdot \mathrm{d}\boldsymbol{A}$。

在当前构型中,点 $\boldsymbol{x}$ 和点 $\boldsymbol{x}'$ 的位移矢量分别为 $\boldsymbol{u} = \boldsymbol{y} - \boldsymbol{x}$ 和 $\boldsymbol{u}' = \boldsymbol{y}' - \boldsymbol{x}'$,因邻近两点的距离为小量 $\mathrm{d}\boldsymbol{x}$,略去泰勒级数展开式中的二阶及二阶以上高阶项,可以得到位移矢量的增量

$$\mathrm{d}\boldsymbol{u} = \boldsymbol{u}' - \boldsymbol{u} = (\boldsymbol{y}' - \boldsymbol{y}) - (\boldsymbol{x}' - \boldsymbol{x}) \approx (\boldsymbol{F} - \boldsymbol{I}) \cdot \mathrm{d}\boldsymbol{x} = \boldsymbol{H} \cdot \mathrm{d}\boldsymbol{x} \tag{2.5}$$

其中，H 为位移梯度张量。

对于有限变形，Green 应变张量定义为

$$E = \frac{1}{2}(F^{\mathrm{T}} \cdot F - I) \tag{2.6}$$

当变形微小时，Green 应变张量退化为 Cauchy 应变张量

$$\varepsilon = \frac{1}{2}(F + F^{\mathrm{T}}) - I \tag{2.7}$$

上述定义的两种应变张量都是对称的二阶张量。以平面问题的构型为例，采用正交的线微元的长度变化和夹角变化表征一点处的变形状态，如图 2-2 所示，则直角坐标系中 Cauchy 应变张量的分量为 $\varepsilon_{11} = \dfrac{\partial u_1}{\partial x_1}$、$\varepsilon_{22} = \dfrac{\partial u_2}{\partial x_2}$ 和 $\varepsilon_{12} = \dfrac{1}{2}\left(\dfrac{\partial u_2}{\partial x_1} + \dfrac{\partial u_1}{\partial x_2}\right)$，由此可计算获得任意方向的伸长和任意两方向的夹角改变。

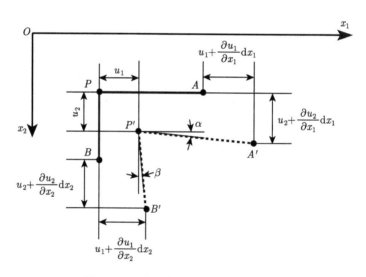

图 2-2　一点处线微元变形几何示意图

在当前构型中，设 ξ 表示点 x 处的一个线微元，则点 x 沿 ξ 的单位方向矢量为 $n = \xi/|\xi|$，该点沿 n 方向的单位伸长度和伸长量分别为

$$\varepsilon_n = n \cdot E \cdot n = n_i \varepsilon_{ij} n_j \tag{2.8}$$

$$\Delta l = \varepsilon_n |\xi| = \frac{\xi_i}{|\xi|} \varepsilon_{ij} \frac{\xi_j}{|\xi|} |\xi| = \frac{1}{|\xi|} \varepsilon_{ij} \xi_i \xi_j \tag{2.9}$$

2.1 变形描述

此外，对于动载荷作用下的应变率相关问题，需要进一步描述构型的应变率。定义点 $\boldsymbol{x}$ 的速度梯度张量 $\boldsymbol{L}$ 为当前构型中速度矢量 $\boldsymbol{v}$ 的右梯度，即

$$\boldsymbol{L} = \boldsymbol{v} \otimes \nabla = \frac{\partial v_i}{\partial x_j} \boldsymbol{e}_i \otimes \boldsymbol{e}_j \tag{2.10}$$

利用物质导数的定义、微分的连续性以及微分的链式法则，可以建立变形梯度张量 $\boldsymbol{F}$ 与速度梯度张量 $\boldsymbol{L}$ 的联系，即

$$\dot{\boldsymbol{F}} = \boldsymbol{L} \cdot \boldsymbol{F} \tag{2.11}$$

或

$$\boldsymbol{L} = \dot{\boldsymbol{F}} \cdot \boldsymbol{F}^{-1} \tag{2.12}$$

速度梯度张量 $\boldsymbol{L}$ 可分解为对称的变形率张量 $\boldsymbol{D}$ 和反对称的旋率张量 $\boldsymbol{W}$，即

$$\boldsymbol{L} = \boldsymbol{D} + \boldsymbol{W} \tag{2.13}$$

且有 $\boldsymbol{D} = \frac{1}{2}(\boldsymbol{L} + \boldsymbol{L}^{\mathrm{T}})$ 和 $\boldsymbol{W} = \frac{1}{2}(\boldsymbol{L} - \boldsymbol{L}^{\mathrm{T}})$。

2.1.2 近场动力学的变形描述

近场动力学的变形描述也是基于位置、位移、速度、加速度等基本概念，但不同于经典连续介质力学采用位移的导数描述构型的变形状态，近场动力学没有使用变量的空间微分，而是直接采用变量差值或变量差值的局部加权积分，避免了传统局部微分在空间不连续时的奇异性问题。

如图 2-3 所示，空间离散体或连续体 Ω 均可视为大量物质点 (material point) 的集合，物质点携带材料性质、体积、位移、速度、加速度、变形和受力等信息，近场动力学认为有限距离内 ($|\boldsymbol{x}' - \boldsymbol{x}| \leqslant \delta$) 的物质点之间存在非局部长程相互作用，这个有限距离内的区域称为近场范围 (peridynamic horizon)，δ 为近场范围半径，并记物质点 $\boldsymbol{x}$ 的近场范围内物质点 $\boldsymbol{x}'$ 的集合为 $H_{\boldsymbol{x}} = H(\boldsymbol{x}, \delta) := \{\boldsymbol{x}' \in R_0 : \{|\boldsymbol{x}' - \boldsymbol{x}| \leqslant \delta\}\}$。

为了考虑两个物质点之间在有限距离内的相互作用，近场动力学需要表征构型变形前后物质点之间的几何关系，进而描述构型的相对变形状态。近场动力学直接采用位置差和位移差表征构型的几何变形情况。如图 2-4 所示，在参考构型中，两物质点的位置矢量为 $\boldsymbol{x}$ 和 $\boldsymbol{x}'$，则物质点间初始相对位置矢量为

$$\boldsymbol{\xi} = \boldsymbol{x}' - \boldsymbol{x} \tag{2.14}$$

记构型变形引起的位移矢量分别为 $\boldsymbol{u}$ 和 $\boldsymbol{u}'$，则当前构型中物质点的位置矢量分别为 $\boldsymbol{y} = \boldsymbol{x} + \boldsymbol{u}$ 和 $\boldsymbol{y}' = \boldsymbol{x}' + \boldsymbol{u}'$，于是，相对位移矢量和当前相对位置矢量分别为

$$\boldsymbol{\eta} = \boldsymbol{u}' - \boldsymbol{u} \tag{2.15}$$

$$\boldsymbol{\xi} + \boldsymbol{\eta} = \boldsymbol{y}' - \boldsymbol{y} \tag{2.16}$$

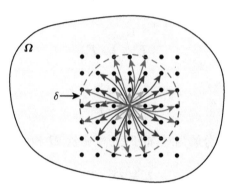

图 2-3 近场动力学中有限距离内的物质点间存在非局部长程作用

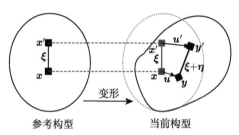

图 2-4 近场动力学中构型变形前后物质点间的几何特征

近场动力学的物质点之间通过非局部相互作用进行联系，在一定意义上，可以认为两点间存在一根虚拟的 "键"(bond)，这根 "键" 可以类比于 "弹簧、杆、梁、柱、板、壳等结构件" 发挥的承载功能。物质点 $\boldsymbol{x}$ 和 $\boldsymbol{x}'$ 间 "键" 的伸长率定义为

$$s = \frac{|\boldsymbol{y}' - \boldsymbol{y}| - |\boldsymbol{x}' - \boldsymbol{x}|}{|\boldsymbol{x}' - \boldsymbol{x}|} = \frac{|\boldsymbol{\xi} + \boldsymbol{\eta}| - |\boldsymbol{\xi}|}{|\boldsymbol{\xi}|} \tag{2.17}$$

对于小变形问题，有

$$s \approx \frac{\left(|\boldsymbol{\xi}| + \frac{\partial |\boldsymbol{\xi}|}{\partial \boldsymbol{\xi}} \boldsymbol{\eta}\right) - |\boldsymbol{\xi}|}{|\boldsymbol{\xi}|} = \frac{\partial |\boldsymbol{\xi}|}{\partial \boldsymbol{\xi}} \cdot \frac{\boldsymbol{\eta}}{|\boldsymbol{\xi}|} = \frac{\boldsymbol{\xi}}{|\boldsymbol{\xi}|} \cdot \frac{\boldsymbol{\eta}}{|\boldsymbol{\xi}|} \tag{2.18}$$

2.1 变形描述

$$s \approx \frac{\left(|\boldsymbol{\xi}| + \frac{\partial |\boldsymbol{\xi}+\boldsymbol{\eta}|}{\partial (\boldsymbol{\xi}+\boldsymbol{\eta})}\boldsymbol{\eta}\right) - |\boldsymbol{\xi}|}{|\boldsymbol{\xi}|} = \frac{\partial |\boldsymbol{\xi}+\boldsymbol{\eta}|}{\partial (\boldsymbol{\xi}+\boldsymbol{\eta})} \cdot \frac{\boldsymbol{\eta}}{|\boldsymbol{\xi}|} = \frac{\boldsymbol{\xi}+\boldsymbol{\eta}}{|\boldsymbol{\xi}+\boldsymbol{\eta}|} \cdot \frac{\boldsymbol{\eta}}{|\boldsymbol{\xi}|} \tag{2.19}$$

近场动力学关注近场范围 H_x 内所有点的变形情况，并定义当前 t 时刻物质点 $\boldsymbol{x}$ 的变形矢量状态 $\underline{\boldsymbol{Y}}$ 为

$$\underline{\boldsymbol{Y}} = \underline{\boldsymbol{Y}}[\boldsymbol{x},t] = \left\{ \begin{array}{c} (\boldsymbol{y}'_{(1)} - \boldsymbol{y}) \\ \vdots \\ (\boldsymbol{y}'_{(\infty)} - \boldsymbol{y}) \end{array} \right\} \tag{2.20}$$

上述定义表示，$\underline{\boldsymbol{Y}}[\boldsymbol{x},t]$ 是物质点 $\boldsymbol{x}$ 变形后，与其关联的各点相对位置矢量的集合，以描述物质点 $\boldsymbol{x}$ 的变形状态。同样地，定义初始时刻物质点 $\boldsymbol{x}$ 的变形矢量状态 $\underline{\boldsymbol{X}}$ 为

$$\underline{\boldsymbol{X}} = \underline{\boldsymbol{X}}[\boldsymbol{x},t] = \left\{ \begin{array}{c} (\boldsymbol{x}'_{(1)} - \boldsymbol{x}) \\ \vdots \\ (\boldsymbol{x}'_{(\infty)} - \boldsymbol{x}) \end{array} \right\} \tag{2.21}$$

进一步，定义初始变形矢量 $\underline{\boldsymbol{X}}\langle\boldsymbol{\xi}\rangle$ 和当前变形矢量 $\underline{\boldsymbol{Y}}\langle\boldsymbol{\xi}\rangle$ 为

$$\underline{\boldsymbol{X}}\langle\boldsymbol{\xi}\rangle = \underline{\boldsymbol{X}}[\boldsymbol{x},t]\langle\boldsymbol{\xi}\rangle = \boldsymbol{x}' - \boldsymbol{x} = \boldsymbol{\xi}$$

$$\underline{\boldsymbol{Y}}\langle\boldsymbol{\xi}\rangle = \underline{\boldsymbol{Y}}[\boldsymbol{x},t]\langle\boldsymbol{\xi}\rangle = \boldsymbol{y}' - \boldsymbol{y} = \boldsymbol{\xi}+\boldsymbol{\eta} \tag{2.22}$$

上式表示，变形矢量状态对初始相对位置矢量 $\boldsymbol{\xi}$ 进行映射，得到初始和当前的变形矢量 (实际上也是各个键的位置矢量)。

类似地，定义初始变形标量 $\underline{x}\langle\boldsymbol{\xi}\rangle$ 和当前变形标量 $\underline{y}\langle\boldsymbol{\xi}\rangle$ 为

$$\underline{x}\langle\boldsymbol{\xi}\rangle = \underline{x}[\boldsymbol{x},t]\langle\boldsymbol{\xi}\rangle = |\underline{\boldsymbol{X}}\langle\boldsymbol{\xi}\rangle|$$

$$\underline{y}\langle\boldsymbol{\xi}\rangle = \underline{y}[\boldsymbol{x},t]\langle\boldsymbol{\xi}\rangle = |\underline{\boldsymbol{Y}}\langle\boldsymbol{\xi}\rangle| \tag{2.23}$$

近场动力学要求变形协调性，即材料不发生相互渗透、坍缩等非物理变形，要求满足运动约束条件

$$\underline{\boldsymbol{Y}}\langle\boldsymbol{\xi}\rangle = \boldsymbol{0}, \quad \text{当且仅当} \quad \boldsymbol{\xi} = \boldsymbol{0} \tag{2.24}$$

需要说明的是，在上述变形描述和后续的受力描述中，近场动力学都定义了 "状态"(state)[2,3] 的概念，以表征 t 时刻某一物质点 $\boldsymbol{x}$ 的近场范围 H_x 内所有物质点 $\boldsymbol{x}'$ 对 $\boldsymbol{x}$ 的影响，通常是一个集合。具体地，定义 $\underline{\boldsymbol{A}} = \underline{\boldsymbol{A}}[\boldsymbol{x},t]$ 和 $\underline{\boldsymbol{B}} = \underline{\boldsymbol{B}}[\boldsymbol{x},t]$ 为 $m \geqslant 1$ 阶状态，一般将方括号 $[\boldsymbol{x},t]$ 省略，以简化公式表示，当 $m=1$ 时，$\underline{\boldsymbol{A}}$

和 $\underline{\boldsymbol{B}}$ 为矢量状态 (vector state); 还可类似定义 $\underline{a}$ 和 $\underline{b}$ 为 0 阶状态, 即标量状态 (scalar state)。

式 (2.22) 和式 (2.23) 的定义涉及近场动力学状态算子 (state operator) 的概念, 近场动力学状态算子是 $H_{\boldsymbol{x}}$ 上的函数或映射, 其完整记法为 $\underline{\boldsymbol{A}}\,[\boldsymbol{x},t]\,\langle\boldsymbol{\xi}\rangle$、$\underline{\boldsymbol{B}}\,[\boldsymbol{x},t]\,\langle\boldsymbol{\xi}\rangle$、$\underline{a}\,[\boldsymbol{x},t]\,\langle\boldsymbol{\xi}\rangle$ 和 $\underline{b}\,[\boldsymbol{x},t]\,\langle\boldsymbol{\xi}\rangle$, 可简记为 $\underline{\boldsymbol{A}}\,\langle\boldsymbol{\xi}\rangle$、$\underline{\boldsymbol{B}}\,\langle\boldsymbol{\xi}\rangle$、$\underline{a}\,\langle\boldsymbol{\xi}\rangle$ 和 $\underline{b}\,\langle\boldsymbol{\xi}\rangle$, 表示 t 时刻物质点 $\boldsymbol{x}$ 的某状态量对矢量 $\boldsymbol{\xi}$(键) 进行映射, 映射的结果可以是矢量或标量。读者需留意这里的特殊记号 $[\boldsymbol{x},t]$ 和 $\langle\boldsymbol{\xi}\rangle$ 的含义。

状态算子是态型近场动力学的核心概念, 它可以是非线性不连续的, 经典连续介质力学的张量算子可视为状态算子的一个特例 [2-4], 且标量状态对应零阶张量, 矢量状态对应二阶张量, 双状态 (double state) 对应四阶张量。

在上述状态算子的基础上, 近场动力学定义矢量状态间的点积运算为

$$\underline{\boldsymbol{A}}\cdot\underline{\boldsymbol{B}} = \int_{H_{\boldsymbol{x}}} \underline{\boldsymbol{A}}\,\langle\boldsymbol{\xi}\rangle \cdot \underline{\boldsymbol{B}}\,\langle\boldsymbol{\xi}\rangle \, \mathrm{d}V_{\boldsymbol{x}'} \tag{2.25}$$

定义标量状态间的点积运算为

$$\underline{a}\cdot\underline{b} = \int_{H_{\boldsymbol{x}}} \underline{a}\,\langle\boldsymbol{\xi}\rangle \, \underline{b}\,\langle\boldsymbol{\xi}\rangle \, \mathrm{d}V_{\boldsymbol{x}'} \tag{2.26}$$

定义矢量状态和标量状态的范数为

$$|\underline{\boldsymbol{A}}| = \sqrt{\underline{\boldsymbol{A}}\cdot\underline{\boldsymbol{A}}}, \quad |\underline{a}| = \sqrt{\underline{a}\cdot\underline{a}} \tag{2.27}$$

2.2 受力描述与运动方程

2.2.1 经典连续介质力学的受力描述与运动方程

连续介质力学假定构型介质在全体积内是连续分布的, 通常选取代表体积单元 (RVE), 要求物体中割取的最微小单元具有该物体相同的物理特性, 也即连续介质力学的唯象模型要求: ①空间尺度上的 "宏观充分小, 微观充分大"; ②时间尺度上的 "宏观充分短, 微观充分长"。初始无应力的考察体在外力作用下, 各部分之间将产生内力, 经典连续介质力学采用应力以表征相邻 RVE 之间单位接触面积上的内力。应力描述基于局部作用原理建模, 即物体中考察点的应力状态完全由充分邻近该点的微元的运动历史所决定, 而与离开该物质点有限距离的其他考察点的运动无关。局部作用原理概念直观、基础坚实①, 受到广泛认同。

---

① 微观原子、分子之间的相互作用并不限于第一近邻, 第二、第三近邻都会产生作用, 但原子、分子之间的短程相互作用随距离衰减迅速, 一般小于 3nm(约 10 个原子间距)[10], 宏观足够小。

2.2 受力描述与运动方程

如图 2-5 所示，真实的或有旋的 Cauchy 应力张量 $\boldsymbol{\sigma}$ 为定义在当前构型上的二阶对称应力张量，满足 $\boldsymbol{\sigma} = \boldsymbol{\sigma}^{\mathrm{T}}$。在单位外法线矢量为 $\boldsymbol{n}$ 的截面上，面力矢量为应力张量与单位外法线矢量的点积，即

$$\boldsymbol{t_n} = \boldsymbol{\sigma} \cdot \boldsymbol{n} = \sigma_{ij} n_j \boldsymbol{e}_i \tag{2.28}$$

该面力矢量在外法线方向的垂直分量 σ_n 和作用在该截面上的剪切矢量 $\boldsymbol{\sigma}_\tau$ 分别为

$$\sigma_n = \boldsymbol{t_n} \cdot \boldsymbol{n} = \boldsymbol{n} \cdot \boldsymbol{\sigma} \cdot \boldsymbol{n} = \sigma_{ij} n_i n_j \tag{2.29}$$

$$\boldsymbol{\sigma}_\tau = \boldsymbol{t_n} - \sigma_n \boldsymbol{n} = \boldsymbol{t_n} - \boldsymbol{t_n} \cdot (\boldsymbol{n} \otimes \boldsymbol{n}) = \boldsymbol{t_n} \cdot (\boldsymbol{I} - \boldsymbol{n} \otimes \boldsymbol{n}) \tag{2.30}$$

经典连续介质力学定义了多种应力度量方式，比较常用的还有无旋 Cauchy 应力、Kirchhoff 应力、第一类和第二类 Piola-Kirchhoff 应力等[4,5]。

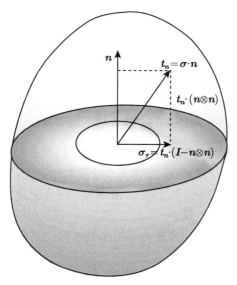

图 2-5 Cauchy 面力示意图

无旋 Cauchy 应力张量定义为

$$\boldsymbol{\sigma}_{\mathrm{unroated}} = \boldsymbol{R}^{\mathrm{T}} \cdot \boldsymbol{\sigma} \cdot \boldsymbol{R} \tag{2.31}$$

式中，$\boldsymbol{R}$ 为正交转动张量，描述构型的刚体旋转，可根据极分解定理计算获得。通过上式 (坐标变换)，可以将当前构型中的 Cauchy 应力变换到参考构型中的无旋 Cauchy 应力，便于引入标架不变的本构模型。

Kirchhoff 应力张量是作用在当前构型上的对称张量，定义为

$$\boldsymbol{\tau} = J\boldsymbol{\sigma} \tag{2.32}$$

其中，雅可比行列式 $J = \det(\boldsymbol{F})$，Kirchhoff 应力张量与变形率张量 $\boldsymbol{d}$ 是一对功共轭量。

第一类 Piola-Kirchhoff 应力张量用来表征作用在参考构型单位面积上的力，即"名义应力张量"，有

$$\boldsymbol{P} = J\boldsymbol{\sigma} \cdot \boldsymbol{F}^{-\mathrm{T}} \tag{2.33}$$

它是一个非对称的两点张量，与变形梯度张量 $\boldsymbol{F}$ 是一对功共轭量。

第二类 Piola-Kirchhoff 应力张量是作用在参考构型上的对称张量，定义为

$$\boldsymbol{T} = \boldsymbol{F}^{-1} \cdot \boldsymbol{P} = J\boldsymbol{F}^{-1} \cdot \boldsymbol{\sigma} \cdot \boldsymbol{F}^{-\mathrm{T}} = \boldsymbol{F}^{-1} \cdot \boldsymbol{\tau} \cdot \boldsymbol{F}^{-\mathrm{T}} \tag{2.34}$$

它与 Green 应变张量 $\boldsymbol{E}$ 是一对功共轭量。

采用虚功原理或动量守恒定律可建立偏微分型运动方程。在当前构型中，采用 Cauchy 应力张量表示的运动方程为

$$\rho \ddot{\boldsymbol{u}}(\boldsymbol{x}, t) = \nabla \cdot \boldsymbol{\sigma} + \boldsymbol{b}(\boldsymbol{x}, t) \tag{2.35}$$

在参考构型中，采用第一类 Piola-Kirchhoff 应力张量表示的运动方程为

$$\rho_0 \ddot{\boldsymbol{u}}(\boldsymbol{x}, t) = \nabla \cdot \boldsymbol{P} + \boldsymbol{b}_0(\boldsymbol{x}, t) \tag{2.36}$$

式中，ρ_0 和 ρ 分别为参考构型与当前构型中物质的质量密度，满足质量守恒定律或连续性方程 $\rho_0 = \rho J$；$\boldsymbol{u}$ 和 $\ddot{\boldsymbol{u}}$ 为物质点位移和加速度，∇ 为散度算子，$\boldsymbol{\sigma}$ 为 Cauchy 应力张量，$\boldsymbol{P}$ 为第一类 Piola-Kirchhoff 应力张量，$\boldsymbol{b}$ 表示单位体积物质所受的体力，即体力密度矢量。

对于弹性材料，经典连续介质力学给出其本构关系和应变能密度，分别为

$$\boldsymbol{T} = \mathbb{D} : \boldsymbol{E} \tag{2.37}$$

$$W_{\mathrm{CCM}} = \frac{1}{2} \boldsymbol{E} : \mathbb{D} : \boldsymbol{E} \tag{2.38}$$

式中，W_{CCM} 为应变能密度，下标 "CCM" 特指经典连续介质力学 (classical continuum mechanics)，$\mathbb{D}$ 为描述材料弹性性质的四阶张量，$\boldsymbol{E}$ 为 Green 应变张量。应力–应变本构模型、应变能密度和控制方程都是位移梯度的函数，涉及位移的空间导数，导致经典连续介质力学在处理断裂破坏等非连续问题时存在困难。

2.2.2 近场动力学的受力描述与运动方程

相较于经典连续介质力学，非局部连续介质力学具有特征空间长度和特征时间等尺度参数，能够较好地反映出材料的尺度效应和微结构效应，尤其适用于非

2.2 受力描述与运动方程

均质材料和结构材料。近场动力学就是这样一种 (位移) 积分型非局部理论,它通过近场范围内长程相互作用的消失与转移 (渐进断键) 描述非局部效应与非连续损伤断裂过程。

以平面问题为例,如图 2-6(a) 所示,经典连续介质力学采用一点处矩形微元的各表面力集度表征该点的受力状态,中心微元体只与最邻近的四个微元之间存在接触作用。在近场动力学中,中心物质点与近场范围内的其他物质点之间存在非局部长程相互作用,该相互作用可以跨越有限距离,达到第一近邻、第二近邻、第三近邻、第四近邻……,如图 2-6(b) 和 (c) 所示。

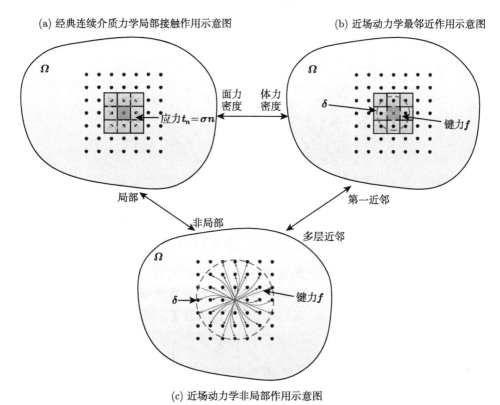

图 2-6 经典连续介质力学局部接触作用、近场动力学最邻近作用与非局部长程作用示意图

需要指出,从局部接触的应力 (面力密度) 到非局部非接触的键力 (体力密度) 的改变是连续介质力学建模范式的一个重要发展。此外,近场范围尺寸 δ 是一个重要的材料参数和计算参数,它控制非局部效应的影响程度。从图 2-6(a) 和 (b) 的直观显示可以看出,在足够光滑的条件下,当近场范围尺寸趋于零时,近场动力学可以收敛于经典连续介质力学,近场动力学数值方法 (第一近邻) 等价于局部数值方法,近场动力学解答收敛于传统局部解答[11-14],故 PD 是对经典连续介

质力学的一种有益补充。

考虑一连续体，初始时刻占据空间 R_0，t 时刻占据空间 R_t，该连续体初始时刻处于无应力状态，后受载变形，如图 2-7 所示。连续体内物质点 $\boldsymbol{x}$ 与其近场范围 $H_{\boldsymbol{x}} = H(\boldsymbol{x}, \delta) := \{\boldsymbol{x}' \in R_0 : \{|\boldsymbol{x}' - \boldsymbol{x}| \leqslant \delta\}\}$ 内其他物质点发生相互作用，处于某一受力状态。将物质点 $\boldsymbol{x}'$ 对物质点 $\boldsymbol{x}$ 的作用力表示为 $\boldsymbol{t}_{\boldsymbol{x}'\boldsymbol{x}}$，这些力密度矢量共同构成了点 $\boldsymbol{x}$ 的力密度矢量状态 $\underline{\boldsymbol{T}}$，即

$$\underline{\boldsymbol{T}} = \underline{\boldsymbol{T}}[\boldsymbol{x}, t] = \left\{ \begin{array}{c} \boldsymbol{t}_{\boldsymbol{x}'_{(1)}\boldsymbol{x}} \\ \vdots \\ \boldsymbol{t}_{\boldsymbol{x}'_{(\infty)}\boldsymbol{x}} \end{array} \right\} \tag{2.39}$$

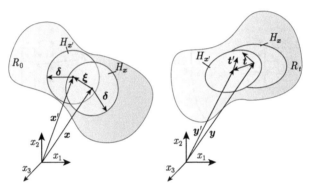

图 2-7 近场动力学物质点间的相互作用和变形前后构型信息

同样地，物质点 $\boldsymbol{x}'$ 与其近场范围 $H_{\boldsymbol{x}'} = H(\boldsymbol{x}', \delta) := \{\boldsymbol{x}'' \in R_0 : \{|\boldsymbol{x}'' - \boldsymbol{x}'| \leqslant \delta\}\}$ 内的物质点也有相互作用，将物质点 $\boldsymbol{x}''$ 对物质点 $\boldsymbol{x}'$ 的作用力记为 $\boldsymbol{t}_{\boldsymbol{x}''\boldsymbol{x}'}$，这些力密度矢量共同构成了 $\boldsymbol{x}'$ 点的力密度矢量状态 $\underline{\boldsymbol{T}}'$，即

$$\underline{\boldsymbol{T}}' = \underline{\boldsymbol{T}}[\boldsymbol{x}', t] = \left\{ \begin{array}{c} \boldsymbol{t}_{\boldsymbol{x}''_{(1)}\boldsymbol{x}'} \\ \vdots \\ \boldsymbol{t}_{\boldsymbol{x}''_{(\infty)}\boldsymbol{x}'} \end{array} \right\} \tag{2.40}$$

下面关注物质点 $\boldsymbol{x}$ 在键 $\boldsymbol{\xi} = \boldsymbol{x}' - \boldsymbol{x}$ 方向上的受力情况。为此，引入 $\boldsymbol{f}_{\boldsymbol{x}'\boldsymbol{x}}$ 表示单位体积的物质点 $\boldsymbol{x}'$ 施加给单位体积的物质点 $\boldsymbol{x}$ 的作用力，参考图 2-7，不难得到

$$\boldsymbol{f}_{\boldsymbol{x}'\boldsymbol{x}} = \boldsymbol{t}_{\boldsymbol{x}'\boldsymbol{x}} - \boldsymbol{t}_{\boldsymbol{x}\boldsymbol{x}'} = \boldsymbol{t} - \boldsymbol{t}' = \underline{\boldsymbol{T}}[\boldsymbol{x}, t]\langle \boldsymbol{x}' - \boldsymbol{x} \rangle - \underline{\boldsymbol{T}}[\boldsymbol{x}', t]\langle \boldsymbol{x} - \boldsymbol{x}' \rangle \tag{2.41}$$

显然，$\boldsymbol{f}_{\boldsymbol{x}'\boldsymbol{x}}$ 与 $H_{\boldsymbol{x}}$ 和 $H_{\boldsymbol{x}'}$ 两个近场范围内所有物质点的变形有关，称为键力密度 (bond force density) 矢量或本构力密度 (constitutive force density) 矢量，

单位为 N /(m³·m³)，它包含了材料的所有物性信息，且近场范围外的点与物质点 x 之间的相互作用力为零。此外，$t_{x'x}$ 和 $t_{xx'}$ 不必反向相等，由此引出了近场动力学的三类模型，将在 2.3 节中予以介绍。

将单位体积的物质点 x 受到的近场范围内所有物质点 x' 的作用力进行集成，并记集成后得到的合内力密度矢量为 L。对于物质点 x，在时刻 t 的内力密度矢量为

$$L(x,t) = \int_{H_\infty} f_{x'x} \mathrm{d}V_{x'} = \int_{H_\infty} \{\underline{T}[x,t]\langle x'-x\rangle - \underline{T}[x',t]\langle x-x'\rangle\} \mathrm{d}V_{x'} \quad (2.42)$$

于是，根据牛顿第二定律，获得参考构型中的近场动力学运动方程

$$\rho_0(x)\ddot{u}(x,t) = \int_{H_\infty} (\underline{T}[x,t]\langle x'-x\rangle - \underline{T}[x',t]\langle x-x'\rangle) \mathrm{d}V_{x'} + b_0(x,t) \quad (2.43)$$

在当前构型中的近场动力学运动方程为

$$\rho(x,t)\ddot{u}(x,t) = \int_{H_\infty} (\underline{T}[x,t]\langle x'-x\rangle - \underline{T}[x',t]\langle x-x'\rangle) \mathrm{d}V_{x'} + b(x,t) \quad (2.44)$$

其中，ρ_0 和 ρ 分别为参考构型与当前构型中的物质质量密度，b_0 和 b 分别为施加在参考构型与当前构型上的体力密度矢量，u 和 $\ddot{u}$ 为物质点位移矢量和加速度矢量，$\mathrm{d}V_{x'}$ 为物质点 x' 的体积微元。需要指出，当前构型的近场范围与参考构型的近场范围可以不一致。

对于静力问题，近场动力学的平衡方程为

$$\int_{H_\infty} (\underline{T}[x]\langle x'-x\rangle - \underline{T}[x']\langle x-x'\rangle) \mathrm{d}V_{x'} + b_0(x) = 0 \quad (2.45)$$

上述近场动力学积分型控制方程还可以通过虚功原理[5]推导建立。

由此可见，近场动力学采用非局部作用积分项，取代经典连续介质力学控制方程中的应力散度项，即得到了时间微分-空间积分型的 PD 控制方程。该积分方程不包含位移的局部导数而直接采用位移差建模，对位移场的连续性没有限制，避免了经典连续介质力学在裂纹等不连续处空间导数不存在而导致的奇异性问题，适用于变形连续或不连续问题的分析。此外，近场动力学通过非局部相互作用的消失与转移 (即渐进断键) 描述损伤和断裂问题，可以自然地模拟材料和结构的损伤累积、裂纹萌生、扩展、分叉和汇合的演化过程。

2.3 本构模型

本构关系 (constitutive relationship) 描述了材料的受力状态对变形状态的依赖关系，建立近场动力学本构模型的关键是确定力密度矢量或键力密度矢量的表达式。根据键力密度矢量的特点，近场动力学模型可分为键型近场动力学 (bond-based peridynamics, BB PD)、常规态型近场动力学 (ordinary state-based peridynamics, OSB PD) 与非常规态型近场动力学 (non-ordinary state-based peridynamics, NOSB PD) 三类，如图 2-8 所示。

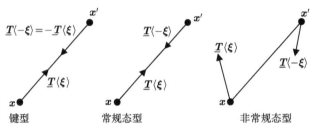

图 2-8　三类近场动力学模型的物质点间"键力"示意图

各种功共轭的应力张量和应变张量提供了经典连续介质力学的本构描述方式，在近场动力学中，力密度矢量状态 $\underline{T}$ 和变形矢量状态 $\underline{Y}$ 也是一对功共轭量，提供了材料本构关系的一般描述形式

$$\underline{T} = \underline{T}(\underline{Y}, \underline{A}, \underline{a}) \tag{2.46}$$

式中，$\underline{A}$ 和 $\underline{a}$ 为其他场变量的矢量状态或标量状态表达，如温度等。

若材料具有率相关性，则上述本构关系可以描述为

$$\underline{T} = \underline{T}(\underline{Y}, \underline{\dot{Y}}, \boldsymbol{x}, \underline{A}, \underline{a}) \tag{2.47}$$

其中，$\underline{\dot{Y}}$ 为变形矢量状态的时间导数，$\boldsymbol{x}$ 表示这种本构关系是位置依赖的。

若力密度矢量状态 $\underline{T}$ 仅依赖于变形矢量状态 $\underline{Y}$，则为通常的简单材料，在这种情况下，仅考虑材料弹性时，则存在可微标量函数 $W_{\mathrm{PD}}(\cdot)$，满足

$$\underline{T} = \underline{T}(\underline{Y}) = \nabla W_{\mathrm{PD}}(\underline{Y}) \tag{2.48}$$

式中，$W_{\mathrm{PD}}(\underline{Y})$ 为变形能密度函数，下标 "PD" 特指近场动力学 (peridynamics)，∇ 为弗雷歇导数 (Frechet derivative)[2-6]。

下面简要阐述键型近场动力学、常规态型近场动力学与非常规态型近场动力学相应的本构模型构建过程，不加推导地给出各自模型的力密度矢量表达式，具体细节可见本书后续相关章节。

2.3 本构模型

2.3.1 键型近场动力学的力密度矢量

在某一键方向上，当两物质点的力密度矢量大小相等、方向相反，且均与变形矢量 $\underline{Y}\langle\boldsymbol{\xi}\rangle$ 平行共线时，有 $\underline{T}\langle\boldsymbol{\xi}\rangle = -\underline{T}'\langle-\boldsymbol{\xi}\rangle$，并且力密度矢量只依赖于该物质点对的相对变形，即 $\underline{T}\langle\boldsymbol{\xi}\rangle = \underline{T}(\boldsymbol{\xi}+\boldsymbol{\eta})$，则称该模型为键型近场动力学模型[1,3,15]。

键型近场动力学的力密度矢量为

$$\underline{T}[\boldsymbol{x},t]\langle\boldsymbol{\xi}\rangle = -\underline{T}[\boldsymbol{x}',t]\langle-\boldsymbol{\xi}\rangle = \frac{1}{2}\boldsymbol{f}(\boldsymbol{\xi},\boldsymbol{\eta}) \tag{2.49}$$

经典微弹脆性 (prototype microelastic brittle, PMB) 模型是最常用的键型近场动力学模型，适用于均匀各向同性材料，其键力密度矢量为[15]

$$\boldsymbol{f}(\boldsymbol{\xi},\boldsymbol{\eta}) = cs\frac{\boldsymbol{\xi}+\boldsymbol{\eta}}{|\boldsymbol{\xi}+\boldsymbol{\eta}|}, \quad |\boldsymbol{\xi}| \leqslant \delta \tag{2.50}$$

式中，s 为物质点对伸长率，c 为微观模量，且有

$$c = \begin{cases} \dfrac{18K}{\pi\delta^4} = \dfrac{6E}{\pi\delta^4(1-2\nu)}, & \text{三维} \\[2mm] \dfrac{6E}{\pi\delta^3 h(1+\nu)(1-2\nu)}, & \text{平面应变} \\[2mm] \dfrac{6E}{\pi\delta^3 h(1-\nu)}, & \text{平面应力} \\[2mm] \dfrac{2E}{A\delta^2}, & \text{一维} \end{cases} \tag{2.51}$$

其中，E 为杨氏弹性模量，K 为体积模量，ν 为材料的泊松比，h 为平面问题考察体的厚度，A 为一维杆件的横截面面积。

2.3.2 常规态型近场动力学的力密度矢量

在某一键方向上，当物质点对的力密度矢量 $\underline{T}\langle\boldsymbol{\xi}\rangle$ 与变形矢量 $\underline{Y}\langle\boldsymbol{\xi}\rangle$ 共线，即满足 $\underline{T}\langle\boldsymbol{\xi}\rangle \times \underline{Y}\langle\boldsymbol{\xi}\rangle = 0$，并且 $\underline{T}\langle\boldsymbol{\xi}\rangle$ 还与两个物质点 $\boldsymbol{x}$、$\boldsymbol{x}'$ 的各自近场范围内所有物质点的整体变形状态相关，即 $\underline{T} = \underline{T}(\underline{Y},\underline{A},\underline{a})$，则称该模型为常规态型近场动力学模型。

常规态型近场动力学的建模方法是：确定近场动力学本构建模的变量，构造近场动力学的弹性变形能密度函数，通过能量密度的弗雷歇导数等运算，推导得出受力状态与变形状态的本构关系式，最后将近场动力学中的弹性变形能密度函数与经典连续介质力学中的弹性应变能密度函数等效，确定模型参数。

常规态型近场动力学弹性模型的力密度矢量 $\underline{\boldsymbol{T}}\langle\boldsymbol{\xi}\rangle$ 为 [2]

$$\underline{\boldsymbol{T}}[\boldsymbol{x},t]\langle\boldsymbol{\xi}\rangle = \underline{t}(\underline{e})\frac{\boldsymbol{\xi}+\boldsymbol{\eta}}{|\boldsymbol{\xi}+\boldsymbol{\eta}|} = \underline{t}(\theta,\underline{e}^d)\frac{\boldsymbol{\xi}+\boldsymbol{\eta}}{|\boldsymbol{\xi}+\boldsymbol{\eta}|} \tag{2.52}$$

其中，$\underline{t}$ 为力密度标量状态 [16]

$$\underline{t}(\theta,\underline{e}^d) = \left[\frac{\gamma k}{m} - \frac{(3-\gamma)\alpha}{9}\right]w\theta\underline{x} + \alpha w\underline{e}^d \tag{2.53}$$

且有

$$\begin{cases} k = K,\ \alpha = \dfrac{15G}{m},\ \gamma = 3, & \text{三维} \\[2mm] k = K + \dfrac{G}{9},\ \alpha = \dfrac{8G}{m},\ \gamma = 2, & \text{平面应变} \\[2mm] k = K + \dfrac{G(\nu+1)^2}{9(2\nu-1)^2},\ \alpha = \dfrac{8G}{m},\ \gamma = \dfrac{2(2\nu-1)}{\nu-1}, & \text{平面应力} \\[2mm] k = K,\ \alpha = \dfrac{3G}{m},\ \gamma = 1, & \text{一维} \end{cases} \tag{2.54}$$

上述三式中，$\underline{x}\langle\boldsymbol{\xi}\rangle = |\boldsymbol{\xi}|$ 为初始相对位置矢量的模；$\underline{e}$ 为拉伸标量状态，且 $\underline{e}\langle\boldsymbol{\xi}\rangle = |\boldsymbol{\xi}+\boldsymbol{\eta}| - |\boldsymbol{\xi}|$，表示变形前后键长的改变量；$\theta = \dfrac{3}{m}[w|\boldsymbol{\xi}|\cdot(|\boldsymbol{\xi}+\boldsymbol{\eta}|-|\boldsymbol{\xi}|)]$ 为膨胀量或体积应变，w 为影响函数，加权量 $m = w|\boldsymbol{\xi}|\cdot|\boldsymbol{\xi}| = \displaystyle\int_{H_x} w|\boldsymbol{\xi}|^2 \mathrm{d}V_{x'}$；$\underline{e}^d$ 为拉伸标量状态的偏量，其映射定义为 $\underline{e}^d\langle\boldsymbol{\xi}\rangle = \underline{e}\langle\boldsymbol{\xi}\rangle - \underline{e}^i\langle\boldsymbol{\xi}\rangle = (|\boldsymbol{\xi}+\boldsymbol{\eta}|-|\boldsymbol{\xi}|) - \dfrac{\theta|\boldsymbol{\xi}|}{3}$；$K$ 和 G 分别为材料的体积模量和剪切模量；ν 为材料的泊松比；γ 为问题的维度参数。

2.3.3 非常规态型近场动力学的力密度矢量

在某一键方向上，若物质点对的力密度矢量 $\underline{\boldsymbol{T}}\langle\boldsymbol{\xi}\rangle$ 与变形矢量 $\underline{\boldsymbol{Y}}\langle\boldsymbol{\xi}\rangle$ 不平行，并且 $\underline{\boldsymbol{T}}\langle\boldsymbol{\xi}\rangle$ 还与两个物质点 $\boldsymbol{x}$、$\boldsymbol{x}'$ 各自近场范围内所有物质点的整体变形状态相关，即 $\underline{\boldsymbol{T}} = \underline{\boldsymbol{T}}(\underline{\boldsymbol{Y}},\underline{\boldsymbol{A}},\underline{a})$，则称该模型为非常规态型近场动力学模型。

非常规态型近场动力学有两类建模方法：一类与上述常规态型近场动力学建模方法相似；另一类将经典连续介质力学的应力–应变 (率) 关系作为中间媒介，计算得到近场动力学的力密度矢量表达式，也称为近场动力学对应材料模型 (peridynamic correspondence material model, PD-CMM)。

非常规态型近场动力学的力密度矢量 $\boldsymbol{T}\langle\boldsymbol{\xi}\rangle$ 为[2]

$$\underline{\boldsymbol{T}}[\boldsymbol{x},t]\langle\boldsymbol{\xi}\rangle = w\boldsymbol{P}\boldsymbol{K}^{-1}\boldsymbol{\xi} \tag{2.55}$$

式中，w 为影响函数，$\boldsymbol{P}$ 为第一类 Piola-Kirchhoff 应力张量，物质点间初始相对位置矢量为 $\boldsymbol{\xi} = \boldsymbol{x}' - \boldsymbol{x}$，形状张量定义如下

$$\boldsymbol{K} = \int_{H_x} w(|\boldsymbol{\xi}|)(\boldsymbol{\xi}\otimes\boldsymbol{\xi})\mathrm{d}V_{\boldsymbol{x}'} \tag{2.56}$$

分析上述三种模型的特点，不难看出：键型近场动力学的"键力"沿键变形方向，键的径向拉压能反映物体的整体变形，但不能区分体积变形和剪切变形，且由于对键之间相对转动缺乏刻画机制，导致对剪切变形描述存在误差，只在特定泊松比时准确；常规态型近场动力学的"键力"也是沿键变形方向，但考虑了体积变形与剪切变形的贡献，对材料的泊松比没有限制；非常规态型近场动力学的"键力"可以沿任意方向，同样能考虑体积变形和剪切变形的影响，且借助了经典连续介质力学的本构模型，能更好地描述材料的复杂力学行为。值得指出的是，存在合适的力密度矢量，使得 BB PD 模型是 OSB PD 模型的一个特例，OSB PD 模型也是 NOSB PD 模型的一个特例。

近场动力学的三种模型各具特色，键型近场动力学概念清晰、简便易行，并能作为态型近场动力学研究的基础，但传统的键型近场动力学模型存在泊松比限制和塑性变形描述不足等问题。态型近场动力学是键型近场动力学的推广，对材料的泊松比没有限制，并加强了与经典连续介质力学的联系，拓展了近场动力学的应用范围。但常规态型近场动力学模型相关定义的物理意义不够明晰，力–位移本构描述与经典连续介质力学的应力–应变本构关系联系不够紧密，非常规态型近场动力学也存在数值稳定性问题。

2.4 非局部边界条件

近场动力学的非局部特性决定了其边界条件的施加方法与经典连续介质力学不同，称为非局部边界条件或体积约束条件[17]，一般施加在距边界 1~2 倍近场范围内的物质点上。对于自然边界条件或应力边界条件，面力、集中力需要转化为体力密度，施加在边界区域一定体积的物质点上，体力直接施加在所有物质点上。对于本质边界条件或位移边界条件，速度或位移施加在边界区域的物质点上。需要注意，施加边界条件的区域大小和物质点数目对计算结果的精确性存在一定影响。

如图 2-9 所示，某杆件左端承受均布面力 $\boldsymbol{t_n} = \boldsymbol{\sigma}\cdot\boldsymbol{n}$ 作用，右端受已知位移 $\boldsymbol{U}_0$ 或已知速度 $\boldsymbol{V}_0$ 约束，杆件横截面积为 A。通过体力密度施加应力边界条件，

如施加的体积为 V,则与局部应力边界条件等价的体力密度矢量为

$$b = \frac{\int_{\varGamma_f} t_n \mathrm{d}s}{\int_{R_f} \mathrm{d}V} = \frac{t_n A}{V} \tag{2.57}$$

如果右端受已知位移 U_0 约束,在 R_c 区域内的物质点上施加的位移矢量为

$$u(x,t) = U_0, \quad x \in R_c \tag{2.58}$$

如果右端受已知速度 V_0 约束,在 R_c 区域内的物质点上施加的速度矢量为

$$\dot{u}(x,t) = V_0, \quad x \in R_c \tag{2.59}$$

为了避免引入边界条件时物质点位置和速度状态的突变,可以按以下分步加载方式逐渐施加

$$u(x,t) = \begin{cases} U_0 \dfrac{t}{t_0}, & 0 \leqslant t \leqslant t_0 \\ U_0, & t > t_0 \end{cases} \quad \text{或} \quad \dot{u}(x,t) = \begin{cases} V_0 \dfrac{t}{t_0}, & 0 \leqslant t \leqslant t_0 \\ V_0, & t > t_0 \end{cases} \tag{2.60}$$

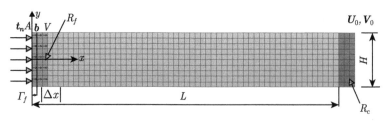

图 2-9 近场动力学中非局部边界条件的施加示意图

需要说明的是,为防止数值模拟中加载区域与自由区域间出现虚假断裂,除采用分级加载外,可在加载初始一定时间内,设定边界为无损区域[18]。近场动力学中不含局部接触力,可通过穿越界面的"混合键"来表征材料界面的力学行为,并可通过短程排斥体力或罚函数法等模拟无黏结接触问题。在冲击爆炸导致材料和结构剧烈破碎的问题中,断裂的物质点间没有"键"的相互作用,需要考虑短程排斥力,以防止出现物质点相互重叠等非物理变形现象。

2.5 守恒律的相容性条件

近场动力学方程需要满足线动量守恒、角动量守恒和能量守恒以及热力学定律，本节论述近场动力学方程与上述守恒律的相容性条件。

2.5.1 线动量守恒的相容性条件

首先，讨论近场动力学运动方程与线动量守恒的相容性条件。设 Ω 表示物体构型，H_x 为近场范围，在 t 时刻，物体构型的总线动量 L 为

$$L = \int_\Omega \rho(\boldsymbol{x})\dot{\boldsymbol{u}}(\boldsymbol{x},t)\mathrm{d}V_{\boldsymbol{x}} \tag{2.61}$$

根据近场动力学运动方程，物体系统所受合力 F 为

$$F = \int_\Omega \boldsymbol{b}(\boldsymbol{x},t)\mathrm{d}V_{\boldsymbol{x}} + \int_\Omega \int_{H_{\boldsymbol{x}}} \{\underline{\boldsymbol{T}}[\boldsymbol{x},t]\langle \boldsymbol{x}'-\boldsymbol{x}\rangle - \underline{\boldsymbol{T}}[\boldsymbol{x}',t]\langle \boldsymbol{x}-\boldsymbol{x}'\rangle\} \mathrm{d}V_{\boldsymbol{x}'}\mathrm{d}V_{\boldsymbol{x}} \tag{2.62}$$

由于 $H_x \in \Omega$，且力密度矢量状态 $\underline{\boldsymbol{T}}$ 具有紧支撑性，故上式的积分域可用 Ω 代替，则有

$$F = \int_\Omega \boldsymbol{b}(\boldsymbol{x},t)\mathrm{d}V_{\boldsymbol{x}} + \int_\Omega \int_\Omega \underline{\boldsymbol{T}}[\boldsymbol{x},t]\langle \boldsymbol{x}'-\boldsymbol{x}\rangle \mathrm{d}V_{\boldsymbol{x}'}\mathrm{d}V_{\boldsymbol{x}} - \int_\Omega \int_\Omega \underline{\boldsymbol{T}}[\boldsymbol{x}',t]\langle \boldsymbol{x}-\boldsymbol{x}'\rangle \mathrm{d}V_{\boldsymbol{x}'}\mathrm{d}V_{\boldsymbol{x}}$$

$$= \int_\Omega \boldsymbol{b}(\boldsymbol{x},t)\mathrm{d}V_{\boldsymbol{x}} + \int_\Omega \int_\Omega \underline{\boldsymbol{T}}[\boldsymbol{x},t]\langle \boldsymbol{x}'-\boldsymbol{x}\rangle \mathrm{d}V_{\boldsymbol{x}'}\mathrm{d}V_{\boldsymbol{x}} - \int_\Omega \int_\Omega \underline{\boldsymbol{T}}[\boldsymbol{x},t]\langle \boldsymbol{x}'-\boldsymbol{x}\rangle \mathrm{d}V_{\boldsymbol{x}}\mathrm{d}V_{\boldsymbol{x}'}$$

$$= \int_\Omega \boldsymbol{b}(\boldsymbol{x},t)\mathrm{d}V_{\boldsymbol{x}} \tag{2.63}$$

上式第二个等号后的第三项是通过交换变量 $\boldsymbol{x}$ 与 $\boldsymbol{x}'$ 得到的。同时需要指出，上式也隐含了键力具有反对称性的基本假定，即 $\boldsymbol{f}_{\boldsymbol{x}'\boldsymbol{x}} = -\boldsymbol{f}_{\boldsymbol{x}\boldsymbol{x}'}$，该式隐含了

$$\int_\Omega \int_\Omega \boldsymbol{f}_{\boldsymbol{x}'\boldsymbol{x}} \mathrm{d}V_{\boldsymbol{x}'}\mathrm{d}V_{\boldsymbol{x}} = \boldsymbol{0}$$

依据物体构型的总线动量守恒要求，应有 $\dot{\boldsymbol{L}} = \boldsymbol{F}$，即

$$\int_\Omega \rho(\boldsymbol{x})\ddot{\boldsymbol{u}}(\boldsymbol{x},t)\mathrm{d}V_{\boldsymbol{x}} = \int_\Omega \boldsymbol{b}(\boldsymbol{x},t)\mathrm{d}V_{\boldsymbol{x}}$$

$$\Rightarrow \int_\Omega [\rho(\boldsymbol{x})\ddot{\boldsymbol{u}}(\boldsymbol{x},t) - \boldsymbol{b}(\boldsymbol{x},t)]\mathrm{d}V_{\boldsymbol{x}} = \boldsymbol{0} \tag{2.64}$$

上式与近场动力学运动方程尤其是力密度矢量状态 $\underline{\boldsymbol{T}}$ 的形式无关，故对于任意的力密度矢量，近场动力学运动方程均自动满足线动量守恒定律。

2.5.2 角动量守恒的相容性条件

下面讨论近场动力学运动方程与角动量守恒的相容性条件。设 Ω 表示物体构型，$H_{\boldsymbol{x}}$ 为近场范围，在 t 时刻，物体构型的总角动量 $\boldsymbol{H}$ (关于坐标原点) 为

$$\boldsymbol{H} = \int_{\Omega} \boldsymbol{y}(\boldsymbol{x},t) \times \rho(\boldsymbol{x}) \dot{\boldsymbol{u}}(\boldsymbol{x},t) \mathrm{d}V_{\boldsymbol{x}} \tag{2.65}$$

根据近场动力学运动方程，物体系统所受合力矩 $\boldsymbol{M}$ 为

$$\boldsymbol{M} = \int_{\Omega} \boldsymbol{y}(\boldsymbol{x},t) \times \int_{H_{\boldsymbol{x}}} \underline{\boldsymbol{T}}[\boldsymbol{x},t]\langle \boldsymbol{x}'-\boldsymbol{x}\rangle \mathrm{d}V_{\boldsymbol{x}'}\mathrm{d}V_{\boldsymbol{x}}$$
$$- \int_{\Omega} \boldsymbol{y}(\boldsymbol{x},t) \times \int_{H_{\boldsymbol{x}}} \underline{\boldsymbol{T}}[\boldsymbol{x}',t]\langle \boldsymbol{x}-\boldsymbol{x}'\rangle \mathrm{d}V_{\boldsymbol{x}'}\mathrm{d}V_{\boldsymbol{x}} + \int_{\Omega} \boldsymbol{y}(\boldsymbol{x},t) \times \boldsymbol{b}(\boldsymbol{x},t)\mathrm{d}V_{\boldsymbol{x}} \tag{2.66}$$

由于 $H_{\boldsymbol{x}} \in \Omega$，且力密度矢量状态 $\underline{\boldsymbol{T}}$ 具有紧支撑性，故上式的积分域可用 Ω 代替，则有

$$\boldsymbol{M} = \int_{\Omega}\int_{\Omega} \boldsymbol{y}(\boldsymbol{x},t) \times \underline{\boldsymbol{T}}[\boldsymbol{x},t]\langle \boldsymbol{x}'-\boldsymbol{x}\rangle \mathrm{d}V_{\boldsymbol{x}'}\mathrm{d}V_{\boldsymbol{x}}$$
$$- \int_{\Omega}\int_{\Omega} \boldsymbol{y}(\boldsymbol{x},t) \times \underline{\boldsymbol{T}}[\boldsymbol{x}',t]\langle \boldsymbol{x}-\boldsymbol{x}'\rangle \mathrm{d}V_{\boldsymbol{x}'}\mathrm{d}V_{\boldsymbol{x}} + \int_{\Omega} \boldsymbol{y}(\boldsymbol{x},t) \times \boldsymbol{b}(\boldsymbol{x},t)\mathrm{d}V_{\boldsymbol{x}}$$
$$= \int_{\Omega}\int_{\Omega} \boldsymbol{y}(\boldsymbol{x},t) \times \underline{\boldsymbol{T}}[\boldsymbol{x},t]\langle \boldsymbol{x}'-\boldsymbol{x}\rangle \mathrm{d}V_{\boldsymbol{x}'}\mathrm{d}V_{\boldsymbol{x}}$$
$$- \int_{\Omega}\int_{\Omega} \boldsymbol{y}(\boldsymbol{x}',t) \times \underline{\boldsymbol{T}}[\boldsymbol{x},t]\langle \boldsymbol{x}'-\boldsymbol{x}\rangle \mathrm{d}V_{\boldsymbol{x}}\mathrm{d}V_{\boldsymbol{x}'} + \int_{\Omega} \boldsymbol{y}(\boldsymbol{x},t) \times \boldsymbol{b}(\boldsymbol{x},t)\mathrm{d}V_{\boldsymbol{x}}$$
$$= -\int_{\Omega}\int_{\Omega} [\boldsymbol{y}(\boldsymbol{x}',t) - \boldsymbol{y}(\boldsymbol{x},t)] \times \underline{\boldsymbol{T}}[\boldsymbol{x},t]\langle \boldsymbol{x}'-\boldsymbol{x}\rangle \mathrm{d}V_{\boldsymbol{x}'}\mathrm{d}V_{\boldsymbol{x}} + \int_{\Omega} \boldsymbol{y}(\boldsymbol{x},t) \times \boldsymbol{b}(\boldsymbol{x},t)\mathrm{d}V_{\boldsymbol{x}}$$
$$= -\int_{\Omega}\int_{\Omega} \underline{\boldsymbol{Y}}[\boldsymbol{x},t]\langle \boldsymbol{x}'-\boldsymbol{x}\rangle \times \underline{\boldsymbol{T}}[\boldsymbol{x},t]\langle \boldsymbol{x}'-\boldsymbol{x}\rangle \mathrm{d}V_{\boldsymbol{x}'}\mathrm{d}V_{\boldsymbol{x}} + \int_{\Omega} \boldsymbol{y}(\boldsymbol{x},t) \times \boldsymbol{b}(\boldsymbol{x},t)\mathrm{d}V_{\boldsymbol{x}}$$
$$\tag{2.67}$$

上式第二个等号后的第二项是通过交换变量 $\boldsymbol{x}$ 与 $\boldsymbol{x}'$ 得到的，第四个等号后的第一项采用式 (2.21) 替换，即 $\underline{\boldsymbol{Y}}\langle \boldsymbol{\xi}\rangle = \boldsymbol{y}' - \boldsymbol{y}$。

依据物体构型的总角动量守恒要求，总角动量的变化率等于外力力矩，即 $\dot{\boldsymbol{H}} = \boldsymbol{M}$，则有

$$\int_{\Omega} \boldsymbol{y}(\boldsymbol{x},t) \times \rho(\boldsymbol{x})\ddot{\boldsymbol{u}}(\boldsymbol{x},t) \mathrm{d}V_{\boldsymbol{x}}$$
$$= \int_{\Omega} \boldsymbol{y}(\boldsymbol{x},t) \times \boldsymbol{b}(\boldsymbol{x},t) \mathrm{d}V_{\boldsymbol{x}} - \int_{\Omega}\int_{\Omega} \underline{\boldsymbol{Y}}[\boldsymbol{x},t] \langle \boldsymbol{x}'-\boldsymbol{x} \rangle \times \underline{\boldsymbol{T}}[\boldsymbol{x},t] \langle \boldsymbol{x}'-\boldsymbol{x} \rangle \mathrm{d}V_{\boldsymbol{x}'} \mathrm{d}V_{\boldsymbol{x}} \tag{2.68}$$

上式左端项的计算利用了关系式 $\dot{\boldsymbol{y}}(\boldsymbol{x},t) \times \rho(\boldsymbol{x})\dot{\boldsymbol{y}}(\boldsymbol{x},t) = 0$；同时，该式右端第二项必须为零，则有

$$\int_{\Omega}\int_{\Omega} \underline{\boldsymbol{Y}}[\boldsymbol{x},t] \langle \boldsymbol{x}'-\boldsymbol{x} \rangle \times \underline{\boldsymbol{T}}[\boldsymbol{x},t] \langle \boldsymbol{x}'-\boldsymbol{x} \rangle \mathrm{d}V_{\boldsymbol{x}'} \mathrm{d}V_{\boldsymbol{x}} = 0 \tag{2.69}$$

如果有下式成立

$$\int_{H_{\boldsymbol{x}}} \underline{\boldsymbol{Y}}[\boldsymbol{x},t] \langle \boldsymbol{x}'-\boldsymbol{x} \rangle \times \underline{\boldsymbol{T}}[\boldsymbol{x},t] \langle \boldsymbol{x}'-\boldsymbol{x} \rangle \mathrm{d}V_{\boldsymbol{x}'} = 0 \tag{2.70}$$

则式 (2.69) 自动满足，近场动力学方程满足总角动量守恒定律，也即式 (2.70) 是近场动力学总角动量守恒的相容性条件。此时，近场动力学的键力密度场被称为是非极性的 (non-polar)，前述三类近场动力学模型均满足该条件。

式 (2.71) 是使近场动力学满足角动量守恒的更强条件，此时键型和常规态型近场动力学模型满足角动量守恒定律，但非常规态型近场动力学模型不满足角动量守恒定律。

$$\underline{\boldsymbol{Y}}[\boldsymbol{x},t] \langle \boldsymbol{x}'-\boldsymbol{x} \rangle \times \underline{\boldsymbol{T}}[\boldsymbol{x},t] \langle \boldsymbol{x}'-\boldsymbol{x} \rangle = 0 \tag{2.71}$$

2.5.3 能量守恒的相容性条件

忽略热效应和源汇项等影响，仅考虑机械能守恒，基于近场动力学参量表示的总能量守恒方程为

$$\int_{\Omega} (\dot{\kappa} + \dot{\varepsilon}) \mathrm{d}V_{\boldsymbol{x}} = \int_{\Omega} \dot{\boldsymbol{u}} \cdot \int_{H_{\boldsymbol{x}}} [\underline{\boldsymbol{T}}[\boldsymbol{x},t] \langle \boldsymbol{x}'-\boldsymbol{x} \rangle - \underline{\boldsymbol{T}}[\boldsymbol{x}',t] \langle \boldsymbol{x}-\boldsymbol{x}' \rangle] \mathrm{d}V_{\boldsymbol{x}'} \mathrm{d}V_{\boldsymbol{x}} + \int_{\Omega} \boldsymbol{b} \cdot \dot{\boldsymbol{u}} \mathrm{d}V_{\boldsymbol{x}} \tag{2.72}$$

式中，Ω 表示物体构型，$H_{\boldsymbol{x}}$ 为近场范围，$\kappa = \dfrac{\rho \dot{\boldsymbol{u}} \cdot \dot{\boldsymbol{u}}}{2}$ 为物质点动能密度，$\dot{\kappa} = \rho \dot{\boldsymbol{u}} \cdot \ddot{\boldsymbol{u}}$，$\varepsilon$ 为物质点的内能密度。由于 $H_{\boldsymbol{x}} \in \Omega$，且力密度矢量状态 $\underline{\boldsymbol{T}}$ 具有紧支撑性，故上式的积分域可用 Ω 代替，则有

$$\int_\Omega (\dot{\kappa}+\dot{\varepsilon})\mathrm{d}V_{\boldsymbol{x}}$$

$$=\int_\Omega \dot{\boldsymbol{u}}\cdot\int_\Omega [\underline{\boldsymbol{T}}[\boldsymbol{x},t]\langle \boldsymbol{x}'-\boldsymbol{x}\rangle-\underline{\boldsymbol{T}}[\boldsymbol{x}',t]\langle \boldsymbol{x}-\boldsymbol{x}'\rangle]\mathrm{d}V_{\boldsymbol{x}'}\mathrm{d}V_{\boldsymbol{x}} + \int_\Omega \boldsymbol{b}\cdot\dot{\boldsymbol{u}}\mathrm{d}V_{\boldsymbol{x}}$$

$$=\int_\Omega\int_\Omega \dot{\boldsymbol{u}}\cdot\underline{\boldsymbol{T}}[\boldsymbol{x},t]\langle \boldsymbol{x}'-\boldsymbol{x}\rangle\mathrm{d}V_{\boldsymbol{x}'}\mathrm{d}V_{\boldsymbol{x}}$$

$$-\int_\Omega\int_\Omega \dot{\boldsymbol{u}}\cdot\underline{\boldsymbol{T}}[\boldsymbol{x}',t]\langle \boldsymbol{x}-\boldsymbol{x}'\rangle\mathrm{d}V_{\boldsymbol{x}'}\mathrm{d}V_{\boldsymbol{x}} + \int_\Omega \boldsymbol{b}\cdot\dot{\boldsymbol{u}}\mathrm{d}V_{\boldsymbol{x}}$$

$$=\int_\Omega\int_\Omega \dot{\boldsymbol{u}}'\cdot\underline{\boldsymbol{T}}[\boldsymbol{x}',t]\langle \boldsymbol{x}-\boldsymbol{x}'\rangle\mathrm{d}V_{\boldsymbol{x}}\mathrm{d}V_{\boldsymbol{x}'} - \int_\Omega\int_\Omega \dot{\boldsymbol{u}}\cdot\underline{\boldsymbol{T}}[\boldsymbol{x}',t]\langle \boldsymbol{x}-\boldsymbol{x}'\rangle\mathrm{d}V_{\boldsymbol{x}'}\mathrm{d}V_{\boldsymbol{x}}$$

$$+\int_\Omega \boldsymbol{b}\cdot\dot{\boldsymbol{u}}\mathrm{d}V_{\boldsymbol{x}}$$

$$=\int_\Omega\int_\Omega [\dot{\boldsymbol{u}}'-\dot{\boldsymbol{u}}]\cdot\underline{\boldsymbol{T}}[\boldsymbol{x}',t]\langle \boldsymbol{x}-\boldsymbol{x}'\rangle\mathrm{d}V_{\boldsymbol{x}'}\mathrm{d}V_{\boldsymbol{x}} + \int_\Omega \boldsymbol{b}\cdot\dot{\boldsymbol{u}}\mathrm{d}V_{\boldsymbol{x}}$$

$$=\int_\Omega\int_\Omega \underline{\dot{\boldsymbol{Y}}}[\boldsymbol{x},t]\langle \boldsymbol{x}'-\boldsymbol{x}\rangle\cdot\underline{\boldsymbol{T}}[\boldsymbol{x}',t]\langle \boldsymbol{x}-\boldsymbol{x}'\rangle\mathrm{d}V_{\boldsymbol{x}'}\mathrm{d}V_{\boldsymbol{x}} + \int_\Omega \boldsymbol{b}\cdot\dot{\boldsymbol{u}}\mathrm{d}V_{\boldsymbol{x}} \tag{2.73}$$

上式第三个等号后的第二项是通过交换变量 $\boldsymbol{x}$ 与 $\boldsymbol{x}'$ 得到的,第五个等号后的第一项采用变形矢量的时间导数替换,即 $\underline{\dot{\boldsymbol{Y}}}\langle \boldsymbol{\xi}\rangle = \dot{\boldsymbol{y}}' - \dot{\boldsymbol{y}}$。

进而,局部化的能量守恒定律为

$$\dot{\kappa}+\dot{\varepsilon}=\underline{\boldsymbol{T}}\cdot\underline{\dot{\boldsymbol{Y}}}+\boldsymbol{b}\cdot\dot{\boldsymbol{u}} \tag{2.74}$$

2.6 损伤与破坏:断键准则

如图 2-10 所示,近场动力学将研究对象离散为大量粒子组成的系统,通过粒子间的非局部相互作用描述系统的运动和变形,采用粒子间 "键" 的断开表征损伤累积、裂纹萌生和扩展过程。损伤和不连续裂纹面通过物质点间的相互作用反映在积分方程中,判断 "键" 断开或相互作用丧失需要依据断键准则。合适的断键准则是近场动力学的重要组成部分,需要重点关注[19],常用的断键准则或参量有:键的临界伸长率、键的临界能量密度、非局部 J 积分、等效应变或等效体积应变、传统应力或能量失效准则、传统损伤本构模型的损伤内变量等。在模拟无损问题时,可将断键准则的阈值置为无穷大。

构型发生相对位移 $\boldsymbol{\eta}$ 时,近场动力学单个 "键" 储存的变形能量密度 (单位体积) 为 [20]

2.6 损伤与破坏：断键准则

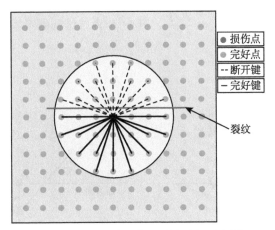

图 2-10　物质点损伤与裂纹路径示意图

$$\widehat{w}(\boldsymbol{\xi}, \boldsymbol{\eta}) = \int_0^{\boldsymbol{\eta}} \boldsymbol{f} \cdot \mathrm{d}\boldsymbol{\eta} = \int_0^{\boldsymbol{\eta}} \{\underline{\boldsymbol{T}}[\boldsymbol{x}, t]\langle \boldsymbol{x}' - \boldsymbol{x}\rangle - \underline{\boldsymbol{T}}[\boldsymbol{x}', t]\langle \boldsymbol{x} - \boldsymbol{x}'\rangle\} \cdot \mathrm{d}\boldsymbol{\eta} \quad (2.75)$$

于是，物质点 $\boldsymbol{x}$ 的近场动力学变形能密度为 [20]

$$W_{\mathrm{PD}} = \frac{1}{2}\int_{H_{\infty}} \widehat{w}(\boldsymbol{\xi}, \boldsymbol{\eta}) \mathrm{d}V_{\boldsymbol{x}'} = \frac{1}{2}\int_{H_{\infty}}\int_0^{\boldsymbol{\eta}} \{\underline{\boldsymbol{T}}[\boldsymbol{x}, t]\langle \boldsymbol{x}' - \boldsymbol{x}\rangle - \underline{\boldsymbol{T}}[\boldsymbol{x}', t]\langle \boldsymbol{x} - \boldsymbol{x}'\rangle\} \cdot \mathrm{d}\boldsymbol{\eta}\mathrm{d}V_{\boldsymbol{x}'} \quad (2.76)$$

其中，系数 $1/2$ 表示成键的两个物质点平均分配该键所具有的能量，W_{PD} 为近场动力学变形能密度。

依据能量守恒律，裂纹扩展单位面积所释放的能量与断开裂纹表面两边物质点间所有键所需的能量应保持相等。如图 2-11 所示，可建立键的临界变形能量密度与传统断裂力学能量释放率 G_F 的关系 [15,20]。

对于三维与二维问题，具体如下

$$G_F = \begin{cases} \int_0^{\delta}\int_0^{2\pi}\int_z^{\delta}\int_0^{\arccos(z/|\boldsymbol{\xi}|)} \widehat{w}_0 |\boldsymbol{\xi}|^2 \sin\phi \mathrm{d}\phi \mathrm{d}|\boldsymbol{\xi}|\mathrm{d}\varphi\mathrm{d}z, & \text{三维} \\ \int_0^{\delta}\int_z^{\delta}\int_{-\arccos(z/|\boldsymbol{\xi}|)}^{\arccos(z/|\boldsymbol{\xi}|)} \widehat{w}_0 |\boldsymbol{\xi}|\mathrm{d}\varphi\mathrm{d}|\boldsymbol{\xi}|\mathrm{d}z, & \text{二维} \end{cases} \quad (2.77)$$

式中，φ 和 ϕ 分别为极坐标系的方位角和极角。据此求得三维和二维临界键能密

度分别为 $\widehat{w}_0 = \dfrac{4G_F}{\pi\delta^4}$ 和 $\widehat{w}_0 = \dfrac{9G_F}{4\delta^3 h}$，其中 h 为二维问题考察体的厚度，可取单位 1。

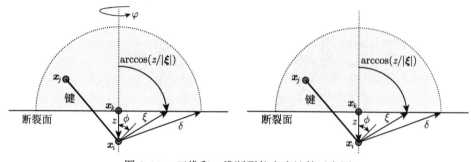

图 2-11　三维和二维断裂能密度计算示意图

特别地，在键型近场动力学中，以键的伸长率与能量密度的关系 $\widehat{w} = \dfrac{c|\boldsymbol{\xi}|}{2}s^{2[18]}$ 为媒介，可建立键的临界伸长率与能量释放率的关系

$$s_0 = \begin{cases} \sqrt{\dfrac{10G_F}{\pi c \delta^5}} = \sqrt{\dfrac{5G_F}{9K\delta}}, & \text{三维} \\ \sqrt{\dfrac{4G_F}{hc\delta^4}}, & \text{二维} \end{cases} \quad (2.78)$$

其中，c 为 PMB 模型的微模量常数。

当满足断键准则时，物质点间"键"断开，且断开后不再发生相互作用，采用历史依赖标量函数 χ 表征断键与否，其值为 0 或 1，即

$$\chi(\boldsymbol{\xi}, t) = \begin{cases} 1, & s < s_0 \text{ 或 } w < w_0 \\ 0, & \text{其他} \end{cases} \quad (2.79)$$

进而，定义近场动力学的损伤为物质点近场范围内断键数目与键总数目之比（见图 2-10）

$$D(\boldsymbol{x}, t) = 1 - \dfrac{\displaystyle\int_{H_\infty} \chi(\boldsymbol{x}, t, \boldsymbol{\xi}) \mathrm{d}V_{\boldsymbol{x}'}}{\displaystyle\int_{H_\infty} \mathrm{d}V_{\boldsymbol{x}'}} \quad (2.80)$$

该标量函数表征了一点 $\boldsymbol{x}$ 处的损伤度，取值范围为 0~1，0 表示该点未发生断键损伤，而 1 表示该点与其近场范围内的所有其他物质点之间的键全部断开。

注意到断键行为是成对 (pair-wise) 发生的，键的临界伸长率和临界变形能密度也是成对进行定义的，但临界 J 积分、应力和应变等物理量是逐点定义的 (point-wise)。材料的非均匀性要求临界伸长率和临界变形能密度也要逐点定义，但逐点定义的物理量与成对断键行为不匹配，故需要进行对称化处理，以选取唯一的断键阈值，如

$$s_{0ij} = \min(s_{0i}, s_{0j}) \quad 或 \quad s_{0ij} = \frac{1}{2}(s_{0i} + s_{0j}) \tag{2.81}$$

其中，s_{0i}、s_{0j} 分别为物质点 i、j 的临界伸长率，s_{0ij} 为最终临界伸长率。

另外指出，近场动力学也可以基于临界非局部 J 积分判断考察体是否起裂，态型 PD 和键型 PD 非局部 J 积分的计算式可参见文献 [3] 和 [21]。

近场动力学模型的键力密度函数中包含了损伤和断裂的描述，通过"键的临界伸长率"等单一判据判断损伤与开裂，不需要预设裂纹路径，裂纹可以自然萌生和扩展，材料的损伤和不连续性作为积分方程解的一部分，突破了经典连续介质力学理论和数值方法在求解变形不连续力学问题时的瓶颈。

参 考 文 献

[1] Silling S A. Reformulation of elasticity theory for discontinuities and long-range forces [J]. Journal of the Mechanics and Physics of Solids, 2000, 48(1):175-209.

[2] Silling S A, Epton M, Weckner O, et al. Peridynamic states and constitutive modeling [J]. Journal of Elasticity, 2007, 88(2):151-184.

[3] Silling S A, Lehoucq R B. Peridynamic theory of solid mechanics [J]. Advances in Applied Mechanics, 2010, 44(10):73-168.

[4] Silling S A. Linearized theory of peridynamic states [J]. Journal of Elasticity, 2010, 99(1):85-111.

[5] Madenci E, Oterkus E. Peridynamic Theory and Its Applications [M]. New York: Springer, 2014.

[6] Bobaru F, Foster J T, Geubelle P H, et al. Handbook of Peridynamic Modeling [M]. New York: Chapman and Hall CRC Press, 2016.

[7] Bazant Z P, Jirasek M. Nonlocal integral formulations of plasticity and damage: survey of progress [J]. Journal of Engineering Mechanics ASCE, 2002, 128(11):1119-1149.

[8] 徐芝纶. 弹性力学简明教程 [M]. 5 版. 北京：高等教育出版社, 2018.

[9] 王自强. 理性力学基础 [M]. 北京：科学出版社, 2000.

[10] 赵亚溥. 近代连续介质力学 [M]. 北京：科学出版社, 2016.

[11] Emmrich E, Weckner O. On the well-posedness of the linear peridynamic model and its convergence towards the Navier equation of linear elasticity [J]. Communications in Mathematical Sciences, 2007, 5(4):851-864.

[12] Silling S A, Lehoucq R B. Convergence of peridynamics to classical elasticity theory [J]. Journal of Elasticity, 2008, 93(1):13-37.

[13] Bobaru F, Yang M J, Alves L F, et al. Convergence, adaptive refinement, and scaling in 1D peridynamics [J]. International Journal for Numerical Methods in Engineering, 2009, 77(6):852-877.

[14] Tian X C, Du Q. Asymptotically compatible schemes and applications to robust discretization of nonlocal models [J]. SIAM Journal on Numerical Analysis, 2014, 52(4):1641-1665.

[15] Silling S A, Askari E. A meshfree method based on the peridynamic model of solid mechanics [J]. Computers & Structures, 2005, 83(17-18):1526-1535.

[16] 顾鑫. 非常规态型近场动力学建模及其微分算子重构 [D]. 南京：河海大学，2018.

[17] Du Q, Gunzburger M, Lehoucq R B, et al. Analysis of the volume-constrained peridynamic Navier equation of linear elasticity [J]. Journal of Elasticity, 2013, 113(2):193-217.

[18] Ha Y D, Bobaru F. Studies of dynamic crack propagation and crack branching with peridynamics [J]. International Journal of Fracture, 2010, 162(1-2):229-244.

[19] Yu H C, Li S F. On energy release rates in peridynamics [J]. Journal of the Mechanics and Physics of Solids, 2020,142:104024.

[20] Foster J T, Silling S A, Chen W. An energy based failure criterion for use with peridynamic states [J]. International Journal for Multiscale Computational Engineering, 2011, 9(6):675-688.

[21] Hu W K, Ha Y D, Bobaru F, et al. The formulation and computation of the nonlocal J-integral in bond-based peridynamics [J]. International Journal of Fracture, 2012, 176(2):195-206.

第 3 章 键型近场动力学模型及其改进

键型近场动力学 (bond-based peridynamics, BB PD) 模型是提出最早、研究最多、发展最快和应用最广的近场动力学模型, 其建模思想直观清晰、易于实施, 也是态型近场动力学建模研究的基础。本章系统介绍键型近场动力学模型的构建过程, 包括一般形式的键力密度本构模型及其线性化方法、经典微观弹脆性 PMB 模型、考虑长程力空间递减规律和边界效应修正的 PMB 模型、突破固定泊松比限制的几类改进模型等。

3.1 本构模型及其线性化

3.1.1 键力密度函数的构建

构造表征键力密度的函数 $f(x, x', u, u', t)$ 是近场动力学建模的关键。通常包含三个步骤: ①在纯机械变形场中, 如果不考虑温度等其他变量的影响, 键力密度函数 f 通常依赖于物质点间的初始相对位置与当前相对位置, 常以"键"的伸长量 $(|\xi+\eta|-|\xi|)$ 或伸长率 $s=(|\xi+\eta|-|\xi|)/|\xi|$ 为基本建模变量, 也可纳入描述"键"转动效应的建模变量, 并考虑弹性、塑性、黏性、各向同性、各向异性等材料特征与长程力空间变化特征, 构建键力密度函数的形式; ②根据所构建的键力密度函数, 求得近场动力学"键"中的微观弹性变形能密度, 进而积分求得物质点的弹性变形能密度的近场动力学表达式; ③基于近场动力学变形能密度和经典连续介质力学应变能密度等效的方法, 求得近场动力学键力密度函数的有关参数。

需要指出的是, 近场动力学本构建模过程对于三维问题、二维问题、一维问题和轴对称问题都是相似的, 只是对不同类型的问题, 近场范围形状和模型参数存在区别。此外, 基于统计力学与粗粒化方法, 将分子动力学的势函数唯象化, 以得到更高尺度的近场动力学键力密度函数[3-6], 也是一种有效的本构建模方法, 尤其适用于级联跨尺度建模, 但本章不具体叙述。

下面, 以均质材料的键力密度函数 $f(x, x', u, u', t)$(也称为本构力函数) 的构建为例, 阐述键型近场动力学建模的过程。

3.1.1.1 均质材料键力密度函数的一般形式

设考察体当前构型发生相对参考构型的某一位移场 u，近场范围内物质点之间的键力密度函数只与两点的相对位置 ξ 和相对位移 η 有关，即

$$f(x, x', u, u', t) = f(x' - x, u' - u) = f(\xi, \eta) \tag{3.1}$$

该函数表示了"键"的相对变形与"键力密度"之间的映射。

由牛顿第三定律，物质点 x' 对物质点 x 施加的作用力与物质点 x 对物质点 x' 施加的作用力，大小相等且方向相反，得到

$$f(-\xi, -\eta) = -f(\xi, \eta) \tag{3.2}$$

这一限制条件也被称为线容许条件 (linear admissibility condition) 或线动量守恒条件。

另外，角动量守恒定律要求键力密度函数 f 满足如下条件

$$(\xi + \eta) \times f(\xi, \eta) = 0 \tag{3.3}$$

上式表明，在物体构型变形过程中，两物质点之间的键力密度矢量 f 与当前构型中的相对位置矢量 $\eta + \xi$ 的方向始终保持一致，该限制条件也被称为角容许条件 (angular admissibility condition) 或角动量守恒条件。

满足上述两类容许条件的最一般化的键力密度函数 f 可以记为

$$f(\xi, \eta) = F(\xi, \eta)(\xi + \eta), \quad \forall \xi, \eta \tag{3.4}$$

其中，$F(\xi, \eta)$ 是一个标量函数，且满足

$$F(\xi, \eta) = F(-\xi, -\eta) \tag{3.5}$$

3.1.1.2 均质材料的微弹性势能密度

众所周知，质点在势力场中运动时，有势力所做的功与质点运动的路径无关，而只取决于质点的始末位置，而弹性力场是一类典型的有势力场。由此可知，在弹性力场中，物质点 x 近场范围内的任一物质点 x' 沿任意一个闭合回路 Γ 运动一周后，对物质点 x 所做的功为零，即存在如下关系

$$\oint_{\Gamma} f(\xi, \eta) \cdot \mathrm{d}\eta = 0, \quad \forall \Gamma \tag{3.6}$$

其中，Γ 为空间域 R 中任意闭合曲线，$\mathrm{d}\eta$ 是沿封闭路径 Γ 的积分路径长度的微分矢量。

3.1 本构模型及其线性化

满足上述关系的材料称为微观弹性 (micro-elastic) 材料。所谓微观弹性是指物质点间的相互作用是弹性的，而宏观弹性 (macro-elastic) 是指考察体整体的弹性行为。

若 $\boldsymbol{f}$ 关于 $\boldsymbol{\eta}$ 是连续可微的，根据斯托克斯定理 (Stokes' Theorem)，存在关系式 $\oint \boldsymbol{f} \cdot \mathrm{d}\boldsymbol{\eta} = \iint_S \nabla \times \boldsymbol{f} \cdot \mathrm{d}\boldsymbol{s}$，于是，式 (3.6) 成立的充分必要条件为

$$\nabla_{\boldsymbol{\eta}} \times \boldsymbol{f}(\boldsymbol{\xi}, \boldsymbol{\eta}) = 0, \quad \forall \boldsymbol{\xi} \neq \boldsymbol{0} \tag{3.7}$$

或

$$\nabla_{\boldsymbol{\eta}} \times \boldsymbol{f}(\boldsymbol{\xi}, \boldsymbol{\eta}) = \begin{vmatrix} \boldsymbol{e}_1 & \boldsymbol{e}_2 & \boldsymbol{e}_3 \\ \dfrac{\partial}{\partial \eta_1} & \dfrac{\partial}{\partial \eta_2} & \dfrac{\partial}{\partial \eta_3} \\ f_1 & f_2 & f_3 \end{vmatrix}$$

$$= \left(\frac{\partial f_3}{\partial \eta_2} - \frac{\partial f_2}{\partial \eta_3}\right) \boldsymbol{e}_1 + \left(\frac{\partial f_1}{\partial \eta_3} - \frac{\partial f_3}{\partial \eta_1}\right) \boldsymbol{e}_2 + \left(\frac{\partial f_2}{\partial \eta_1} - \frac{\partial f_1}{\partial \eta_2}\right) \boldsymbol{e}_3$$

$$= \boldsymbol{0} \tag{3.8}$$

其中，$\nabla_{\boldsymbol{\eta}} \times$ 是关于 $\boldsymbol{\eta}$ 的旋度算子，$\boldsymbol{e}_i$ 为当前构型的基矢量。此时 $\boldsymbol{f}$ 是保守力，保守力的旋度为零。

若微弹性材料是均质各向同性的，由式 (3.4)，一种可能的键力密度函数形式为

$$\boldsymbol{f}(\boldsymbol{\xi}, \boldsymbol{\eta}) = F(|\boldsymbol{\xi}|, |\boldsymbol{\xi} + \boldsymbol{\eta}|)(\boldsymbol{\xi} + \boldsymbol{\eta}), \quad \forall \boldsymbol{\xi}, \boldsymbol{\eta} \tag{3.9}$$

根据式 (3.6) 成立的充分必要条件，必然存在一个可微的标量函数 $\widehat{w}$，满足

$$\boldsymbol{f}(\boldsymbol{\xi}, \boldsymbol{\eta}) = \frac{\partial \widehat{w}(\boldsymbol{\xi}, \boldsymbol{\eta})}{\partial \boldsymbol{\eta}} \tag{3.10}$$

其中，标量函数 $\widehat{w}(\boldsymbol{\xi}, \boldsymbol{\eta})$ 为点对势能函数 (pairwise potential function)，也称微弹性势能密度，表示参考构型中物质点对 (物质点对的相对位置矢量为 $\boldsymbol{\xi}$) 发生相对位移 $\boldsymbol{\eta}$ 时，存储在该键中的能量密度。它类似于分子动力学的中心对势函数，描述近场范围内两物质点间相互作用的强弱，可根据 (半) 经验势函数或物理规律确定。对物质点近场范围内所有键的能量密度积分，可以求得该物质点的宏观能量密度 W_{PD}，再求得相同位移场下基于经典连续介质力学的能量密度 W_{CCM}，令两者等效，即可确定键力密度函数 $\boldsymbol{f}$。

3.1.2 键力密度函数的线性化

3.1.2.1 均质材料的线性键力密度函数

通过泰勒级数展开，可将 3.1.1 节给出的一般形式的非线性键力密度函数线性化，得到线性的键力密度函数。

假定 $|\boldsymbol{\eta}| \ll 1$，令 $\boldsymbol{\xi}$ 固定不变，将 $\boldsymbol{f}(\boldsymbol{\xi}, \boldsymbol{\eta})$ 在 $(\boldsymbol{\xi}, \boldsymbol{0})$ 点进行泰勒展开，可得

$$\boldsymbol{f}(\boldsymbol{\xi}, \boldsymbol{\eta}) = \boldsymbol{f}(\boldsymbol{\xi}, \boldsymbol{0}) + \boldsymbol{C}(\boldsymbol{\xi}) \cdot \boldsymbol{\eta} + o(|\boldsymbol{\eta}|^2), \quad \forall \boldsymbol{\xi}, \boldsymbol{\eta} \tag{3.11}$$

式中，$\boldsymbol{f}(\boldsymbol{\xi}, \boldsymbol{0})$ 表示未变形时的键力密度矢量，$\boldsymbol{C}(\boldsymbol{\xi})$ 为二阶微模量 (micro-modulus) 张量，且有

$$\boldsymbol{C}(\boldsymbol{\xi}) = \frac{\partial \boldsymbol{f}(\boldsymbol{\xi}, \boldsymbol{0})}{\partial \boldsymbol{\eta}} \tag{3.12}$$

略去式 (3.11) 中的高阶小量，并代入近场动力学运动方程中，则有

$$\rho \ddot{\boldsymbol{u}} = \int_{H_{\boldsymbol{x}}} [\boldsymbol{f}(\boldsymbol{\xi}, \boldsymbol{0}) + \boldsymbol{C}(\boldsymbol{\xi}) \cdot \boldsymbol{\eta}] \mathrm{d}V_{\boldsymbol{x}'} + \boldsymbol{b} = \int_{H_{\boldsymbol{x}}} \boldsymbol{C}(\boldsymbol{\xi}) \cdot \boldsymbol{\eta} \mathrm{d}V_{\boldsymbol{x}'} + \boldsymbol{b} \tag{3.13}$$

式中，$\int_{H_{\boldsymbol{x}}} \boldsymbol{f}(\boldsymbol{\xi}, \boldsymbol{0}) \mathrm{d}V_{\boldsymbol{x}'} = \boldsymbol{0}$ 是由于参考构型未变形时，物质点处于自平衡状态。

进而，根据式 (3.2) 给出的键力密度函数的特征，可知 $\boldsymbol{f}(\boldsymbol{\xi}, \boldsymbol{0}) = -\boldsymbol{f}(-\boldsymbol{\xi}, \boldsymbol{0})$，并注意到式 (3.12) 中的 $\boldsymbol{C}(\boldsymbol{\xi})$ 是关于物质点 $\boldsymbol{x}$ 定义的，对于物质点 $\boldsymbol{x}'$，则有下式成立

$$\boldsymbol{C}(-\boldsymbol{\xi}) = \frac{\partial \boldsymbol{f}(-\boldsymbol{\xi}, \boldsymbol{0})}{\partial (-\boldsymbol{\eta})} = \frac{\partial (-\boldsymbol{f}(\boldsymbol{\xi}, \boldsymbol{0}))}{\partial (-\boldsymbol{\eta})} = \frac{\partial \boldsymbol{f}(\boldsymbol{\xi}, \boldsymbol{0})}{\partial \boldsymbol{\eta}} = \boldsymbol{C}(\boldsymbol{\xi}) \tag{3.14}$$

表明这个微模量张量是关于 $\boldsymbol{\xi}$ 的偶函数。

将式 (3.4) 对 $\boldsymbol{\eta}$ 求微分，并令 $\boldsymbol{\eta} = \boldsymbol{0}$，则有

$$\boldsymbol{C}(\boldsymbol{\xi}) = \frac{\partial \boldsymbol{f}(\boldsymbol{\xi}, \boldsymbol{0})}{\partial \boldsymbol{\eta}} = \frac{\partial [F(\boldsymbol{\xi}, \boldsymbol{\eta})(\boldsymbol{\xi} + \boldsymbol{\eta})]}{\partial \boldsymbol{\eta}}\bigg|_{\boldsymbol{\eta}=\boldsymbol{0}} = \boldsymbol{\xi} \otimes \frac{\partial F(\boldsymbol{\xi}, \boldsymbol{0})}{\partial \boldsymbol{\eta}} + F(\boldsymbol{0}, \boldsymbol{\xi}) \boldsymbol{I}, \quad \forall \boldsymbol{\xi} \tag{3.15}$$

其中，$\otimes$ 表示张量积，$\boldsymbol{I}$ 表示单位张量。

将式 (3.15) 代入式 (3.11) 中，得到均质材料的线性键力密度函数

$$\begin{aligned} \boldsymbol{f}(\boldsymbol{\xi}, \boldsymbol{\eta}) &= \boldsymbol{f}(\boldsymbol{\xi}, \boldsymbol{0}) + \boldsymbol{C}(\boldsymbol{\xi}) \cdot \boldsymbol{\eta} \\ &= \boldsymbol{f}(\boldsymbol{\xi}, \boldsymbol{0}) + \left[\boldsymbol{\xi} \otimes \frac{\partial F(\boldsymbol{\xi}, \boldsymbol{0})}{\partial \boldsymbol{\eta}} + F(\boldsymbol{0}, \boldsymbol{\xi}) \boldsymbol{I} \right] \cdot \boldsymbol{\eta}, \quad \forall \boldsymbol{\xi}, \boldsymbol{\eta} \end{aligned} \tag{3.16}$$

3.1.2.2 均质微弹性材料的线性键力密度函数

近场动力学微弹性材料的充分必要条件要求键力密度函数满足式 (3.8),此时有

$$\frac{\partial f_i}{\partial \eta_j} = \frac{\partial f_j}{\partial \eta_i}, \quad i,j=1,2,3, \ i \neq j \tag{3.17}$$

根据微模量张量的定义 $C(\boldsymbol{\xi}) = \dfrac{\partial \boldsymbol{f}(\boldsymbol{\xi},0)}{\partial \boldsymbol{\eta}}$,可得到其分量形式,即

$$C(\boldsymbol{\xi}) = \frac{\partial \boldsymbol{f}(\boldsymbol{\xi},0)}{\partial \boldsymbol{\eta}} = \begin{bmatrix} \dfrac{\partial f_1}{\partial \eta_1} & \dfrac{\partial f_1}{\partial \eta_2} & \dfrac{\partial f_1}{\partial \eta_3} \\ \dfrac{\partial f_2}{\partial \eta_1} & \dfrac{\partial f_2}{\partial \eta_2} & \dfrac{\partial f_2}{\partial \eta_3} \\ \dfrac{\partial f_3}{\partial \eta_1} & \dfrac{\partial f_3}{\partial \eta_2} & \dfrac{\partial f_3}{\partial \eta_3} \end{bmatrix} \tag{3.18}$$

从式 (3.17) 和式 (3.18) 可知,微弹性材料的微模量张量为二阶对称张量,即有

$$C(\boldsymbol{\xi}) = C^{\mathrm{T}}(\boldsymbol{\xi}), \ \forall \boldsymbol{\xi} \tag{3.19}$$

需要说明的是,键力密度函数的线性化并未规定微模量张量的对称性,线性键力密度函数表征的材料不一定是微弹性的。材料的微弹性性质要求微模量张量具有对称性,式 (3.16) 已将微模量张量线性化,即 $C(\boldsymbol{\xi}) = \boldsymbol{\xi} \otimes \dfrac{\partial F(\boldsymbol{\xi},0)}{\partial \boldsymbol{\eta}} + F(\boldsymbol{\xi},0)\boldsymbol{I}$,微弹性材料进一步要求微模量张量满足对称性 $C(\boldsymbol{\xi}) = C^{\mathrm{T}}(\boldsymbol{\xi})$,则只需要 $\boldsymbol{\xi} \otimes \dfrac{\partial F(\boldsymbol{\xi},0)}{\partial \boldsymbol{\eta}}$ 满足对称性即可。根据该对称性的要求和张量性质可知,$\dfrac{\partial F(\boldsymbol{\xi},0)}{\partial \boldsymbol{\eta}}$ 必须和向量 $\boldsymbol{\xi}$ 平行,亦即必然存在某标量函数 $\lambda(\boldsymbol{\xi})$,使得 $\dfrac{\partial F(\boldsymbol{\xi},0)}{\partial \boldsymbol{\eta}} = \lambda(\boldsymbol{\xi})\boldsymbol{\xi}$。于是微模量张量可进一步表示为

$$C(\boldsymbol{\xi}) = \boldsymbol{\xi} \otimes \frac{\partial F(\boldsymbol{\xi},0)}{\partial \boldsymbol{\eta}} + F(\boldsymbol{\xi},0)\boldsymbol{I} = \lambda(\boldsymbol{\xi})\boldsymbol{\xi} \otimes \boldsymbol{\xi} + F_0(\boldsymbol{\xi})\boldsymbol{I} \tag{3.20}$$

其中

$$\lambda(\boldsymbol{\xi}) = \frac{1}{|\boldsymbol{\xi}|^2}\boldsymbol{\xi} \cdot \frac{\partial F(\boldsymbol{\xi},0)}{\partial \boldsymbol{\eta}}, \ F_0(\boldsymbol{\xi}) = F(\boldsymbol{\xi},0) \tag{3.21}$$

又由于线性材料的微模量张量满足 $C(-\boldsymbol{\xi}) = C(\boldsymbol{\xi})$,则 λ 和 F_0 还需要满足:

$$\lambda(-\boldsymbol{\xi}) = \lambda(\boldsymbol{\xi}), \ F_0(-\boldsymbol{\xi}) = F_0(\boldsymbol{\xi}) \tag{3.22}$$

最终，均质微弹性材料的线性键力密度函数为

$$f(\boldsymbol{\xi},\boldsymbol{\eta}) = f(\boldsymbol{\xi},\mathbf{0}) + \boldsymbol{C}(\boldsymbol{\xi}) \cdot \boldsymbol{\eta}$$
$$= f(\boldsymbol{\xi},\mathbf{0}) + [\lambda(\boldsymbol{\xi})\boldsymbol{\xi} \otimes \boldsymbol{\xi} + F_0(\boldsymbol{\xi})\boldsymbol{I}] \cdot \boldsymbol{\eta}, \quad \forall \boldsymbol{\xi},\boldsymbol{\eta} \quad (3.23)$$

3.1.2.3 均质各向同性微弹性材料的线性键力密度函数

对于各向同性材料，键力密度矢量仅是 $|\boldsymbol{\xi}+\boldsymbol{\eta}|$ 和 $|\boldsymbol{\xi}|$ 的函数，所以键力密度函数可写为

$$f(\boldsymbol{\xi},\boldsymbol{\eta}) = f(|\boldsymbol{\xi}|,\mathbf{0}) + \boldsymbol{C}(|\boldsymbol{\xi}|) \cdot \boldsymbol{\eta}$$
$$= f(|\boldsymbol{\xi}|,\mathbf{0}) + [\lambda(|\boldsymbol{\xi}|)\boldsymbol{\xi} \otimes \boldsymbol{\xi} + F_0(|\boldsymbol{\xi}|)\boldsymbol{I}] \cdot \boldsymbol{\eta}, \quad \forall \boldsymbol{\xi},\boldsymbol{\eta} \quad (3.24)$$

对于初始无应力构型，即 $f(|\boldsymbol{\xi}|,\mathbf{0}) = \mathbf{0}$，则有

$$f(\boldsymbol{\xi},\boldsymbol{\eta}) = \boldsymbol{C}(|\boldsymbol{\xi}|) \cdot \boldsymbol{\eta} = [\lambda(|\boldsymbol{\xi}|)\boldsymbol{\xi} \otimes \boldsymbol{\xi} + F_0(|\boldsymbol{\xi}|)\boldsymbol{I}] \cdot \boldsymbol{\eta}, \quad \forall \boldsymbol{\xi},\boldsymbol{\eta} \quad (3.25)$$

由式 (3.16)、式 (3.23)、式 (3.24) 和式 (3.25) 出发，考虑实际材料的各向同性、线弹性等特征和数学模型形式，结合材料频散行为的试验结果或经典连续介质力学模型，可构建不同材料的键力密度函数 $f(\boldsymbol{\xi},\boldsymbol{\eta})$。基于上述思路，提出了多种与经典弹性模型对应且满足守恒定律的线弹性键力密度函数，并在诸多力学问题分析中取得了理想的结果。此外，Silling 曾经指出 "一种宏观弹性本构模型对应多种近场动力学键力密度函数"，认为传统理论中给定的一种宏观材料模型可能有多个近场动力学材料模型 (当近场范围趋于零时) 与之对应，但这些近场动力学模型的小尺度力学行为却可能不同 [1]。

3.2 经典微弹脆性模型

3.2.1 键力密度函数的具体形式

3.1 节介绍了非线性和线性问题对应的键力密度函数的一般形式，但并未说明如何确定其中的材料参数。本节以 Silling 等提出的适用于均质各向同性弹脆性材料的经典微弹脆性 (prototype microelastic brittle, PMB) 模型 [2] 为例，详细阐释相关的建模细节。

对于均质各向同性微弹性材料，式 (3.9) 给出的非线性中心对势键力密度函数可改写为

$$f(\boldsymbol{\xi},\boldsymbol{\eta}) = f(|\boldsymbol{\xi}|,|\boldsymbol{\xi}+\boldsymbol{\eta}|)\frac{\boldsymbol{\xi}+\boldsymbol{\eta}}{|\boldsymbol{\xi}+\boldsymbol{\eta}|}, \quad \forall \boldsymbol{\xi},\boldsymbol{\eta} \quad (3.26)$$

3.2 经典微弹脆性模型

其中，f 为只与初始和当前相对距离相关的标量函数。

假定初始构型无应力，也即初始键力密度为零，键力密度满足反对称性要求，并考虑 PMB 模型中的键与拉压弹簧性质相似，则在构建 PMB 模型时假定：①键力密度函数与键伸长率 s 呈线性关系；②键受拉伸长率达到临界伸长率 s_0 时，该键断开且不可恢复，受压时键不会断裂，如图 3-1 所示。

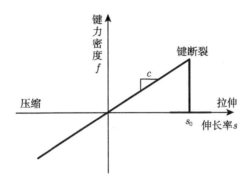

图 3-1 PMB 模型中键力与伸长率的线性关系

PMB 模型的键力密度函数可定义为

$$f(\boldsymbol{\xi},\boldsymbol{\eta}) = \chi cs\frac{\boldsymbol{\xi}+\boldsymbol{\eta}}{|\boldsymbol{\xi}+\boldsymbol{\eta}|}, \quad \forall \boldsymbol{\xi},\boldsymbol{\eta} \tag{3.27}$$

相当于式 (3.26) 中的标量函数 $f(|\boldsymbol{\xi}|,|\boldsymbol{\xi}+\boldsymbol{\eta}|) = \chi cs$；$c$ 为线性比例系数，也即微模量常数，其值和具体推导过程将在 3.2.2 节给出；χ 为表征键损伤与否的历史依赖标量函数，即

$$\chi(\boldsymbol{\xi},t) = \begin{cases} 1, & s < s_0 \\ 0, & \text{其他} \end{cases} \tag{3.28}$$

式中，s_0 为键的临界伸长率，键的伸长率定义为

$$s = \frac{|\boldsymbol{\xi}+\boldsymbol{\eta}|-|\boldsymbol{\xi}|}{|\boldsymbol{\xi}|} \tag{3.29}$$

对于小变形问题，有式 (2.19) 成立，即 $s \approx \dfrac{\boldsymbol{\xi}+\boldsymbol{\eta}}{|\boldsymbol{\xi}+\boldsymbol{\eta}|} \cdot \dfrac{\boldsymbol{\eta}}{|\boldsymbol{\xi}|}$，代入式 (3.27) 得到

$$f(\boldsymbol{\xi},\boldsymbol{\eta}) = \chi c\frac{\boldsymbol{\xi}+\boldsymbol{\eta}}{|\boldsymbol{\xi}+\boldsymbol{\eta}|} \cdot \frac{\boldsymbol{\eta}}{|\boldsymbol{\xi}|} \otimes \frac{\boldsymbol{\xi}+\boldsymbol{\eta}}{|\boldsymbol{\xi}+\boldsymbol{\eta}|} = \frac{\chi c}{|\boldsymbol{\xi}|}\frac{\boldsymbol{\xi}+\boldsymbol{\eta}}{|\boldsymbol{\xi}+\boldsymbol{\eta}|} \otimes \frac{\boldsymbol{\xi}+\boldsymbol{\eta}}{|\boldsymbol{\xi}+\boldsymbol{\eta}|} \cdot \boldsymbol{\eta} = C(|\boldsymbol{\xi}|) \cdot \boldsymbol{\eta} \tag{3.30}$$

其中，微模量张量为

$$C(|\bm{\xi}|) = \frac{\chi c}{|\bm{\xi}|} \frac{\bm{\xi}+\bm{\eta}}{|\bm{\xi}+\bm{\eta}|} \otimes \frac{\bm{\xi}+\bm{\eta}}{|\bm{\xi}+\bm{\eta}|} \tag{3.31}$$

此外，Macek 和 Silling[7] 还给出微模量张量的另一表达式为

$$C(|\bm{\xi}|) = \frac{\chi c}{|\bm{\xi}|^3} \bm{\xi} \otimes \bm{\xi} \tag{3.32}$$

亦即式 (3.25) 表示的线性键力本构函数 $\bm{f}(\bm{\xi},\bm{\eta}) = [\lambda(|\bm{\xi}|)\bm{\xi} \otimes \bm{\xi} + F_0(|\bm{\xi}|)\bm{I}] \cdot \bm{\eta}$ 中，有

$$\lambda(|\bm{\xi}|) = \frac{\chi c}{|\bm{\xi}|^3}, \quad F_0(|\bm{\xi}|) = 0 \tag{3.33}$$

需要指出的是，对于近场范围完整对称的内部物质点，在小变形假定下，式 (3.31) 与式 (3.32) 中的微模量张量具有等价性。

3.2.2 参数率定

由上一节可知，键型近场动力学 PMB 模型有两个重要的材料参数，即微模量常数 c 和临界伸长率 s_0，以下详述这两个参数的确定方法。

在键力密度函数形式确定后，对物质点近场范围内所有"键"能量密度积分，可求得该物质点处的宏观能量密度 W_{PD}，再求出相同均匀变形①下经典连续介质力学的能量密度 W_{CCM}，令两者等效，即近场动力学非局部意义下的应变能和断裂能与经典连续介质力学局部意义下的应变能和断裂能相等，即可确定 PMB 模型参数与杨氏模量、泊松比和能量释放率等的关系。

3.2.2.1 微模量常数

三维情况下，以三个主应变分量表示的弹性应变能密度为

$$W_{\text{CCM}} = \frac{1}{2}[\lambda(\varepsilon_1+\varepsilon_2+\varepsilon_3)^2 + 2\mu(\varepsilon_1^2+\varepsilon_2^2+\varepsilon_3^2)] \tag{3.34}$$

式中，$\lambda = \dfrac{E\nu}{(1+\nu)(1-2\nu)}$ 和 $\mu = \dfrac{E}{2(1+\nu)}$ 为拉梅常数，E 为杨氏模量，ν 为泊松比；ε_1、ε_2 和 ε_3 为三个主应变分量。

1) 均匀拉伸变形状态

小变形假设下，考虑一无限大体，在各向均匀的拉伸荷载作用下产生各向同性均匀膨胀变形 ε_0，此时单元体仅发生体积变形而无形状改变，则不同维度问题

---

① 均匀变形是指具有常仿射边界条件的变形，也即应变张量在整个物体上为常量的情形。

3.2 经典微弹脆性模型

的主应变分别为

$$\begin{cases} \varepsilon_1 = \varepsilon_2 = \varepsilon_3 = \varepsilon_0, & \text{三维} \\ \varepsilon_1 = \varepsilon_2 = \varepsilon_0, \ \varepsilon_3 = 0, & \text{平面应变} \\ \varepsilon_1 = \varepsilon_2 = \varepsilon_0, \ \varepsilon_3 = \dfrac{\nu(\varepsilon_1 + \varepsilon_2)}{\nu - 1}, & \text{平面应力} \\ \varepsilon_1 = \varepsilon_0, \ \varepsilon_2 = \varepsilon_3 = 0, & \text{一维} \end{cases} \quad (3.35)$$

对于该应变状态下的不同维度问题,经典连续介质力学的应变能密度为

$$W_{\text{CCM}} = \begin{cases} \dfrac{3E}{2(1-2\nu)}\varepsilon_0^2, & \text{三维} \\ \dfrac{E}{(1+\nu)(1-2\nu)}\varepsilon_0^2, & \text{平面应变} \\ \dfrac{E}{1-\nu}\varepsilon_0^2, & \text{平面应力} \\ \dfrac{1}{2}E\varepsilon_0^2, & \text{一维} \end{cases} \quad (3.36)$$

在该变形状态下,近场动力学中物质点之间键的伸长率为 $s = \varepsilon_0$,此时 PMB 模型的点对势能函数 $\widehat{w}$ 为

$$\widehat{w} = \int_0^{\boldsymbol{\eta}} cs\dfrac{\boldsymbol{\xi}+\boldsymbol{\eta}}{|\boldsymbol{\xi}+\boldsymbol{\eta}|} \cdot \mathrm{d}\boldsymbol{\eta} = \int_0^{\varepsilon_0} cs|\boldsymbol{\xi}|\, \mathrm{d}\left(\dfrac{\boldsymbol{\xi}+\boldsymbol{\eta}}{|\boldsymbol{\xi}+\boldsymbol{\eta}|} \cdot \dfrac{\boldsymbol{\eta}}{|\boldsymbol{\xi}|}\right) = \int_0^{\varepsilon_0} cs|\boldsymbol{\xi}|\, \mathrm{d}s$$

$$= \dfrac{cs^2|\boldsymbol{\xi}|}{2}\bigg|_{\varepsilon_0} = \dfrac{c\varepsilon_0^2|\boldsymbol{\xi}|}{2} \quad (3.37)$$

则不同维度问题的近场动力学弹性变形能密度为

$$W_{\text{PD}} = \dfrac{1}{2}\int_{H_{\boldsymbol{x}}} \widehat{w}\, \mathrm{d}V_{\boldsymbol{x}'} = \begin{cases} \dfrac{1}{2}\int_0^\delta \int_0^{2\pi} \int_0^\pi \dfrac{c\varepsilon_0^2|\boldsymbol{\xi}|}{2}|\boldsymbol{\xi}|^2 \sin\varphi\, \mathrm{d}\varphi \mathrm{d}\phi \mathrm{d}|\boldsymbol{\xi}| = \dfrac{\pi\delta^4\varepsilon_0^2}{4}c, & \text{三维} \\ \dfrac{1}{2}\int_0^\delta \int_0^{2\pi} \dfrac{c\varepsilon_0^2|\boldsymbol{\xi}|}{2}(|\boldsymbol{\xi}|\,\mathrm{d}\phi \mathrm{d}|\boldsymbol{\xi}|h) = \dfrac{\pi\delta^3\varepsilon_0^2 h}{6}c, & \text{二维} \\ \dfrac{A}{2}\int_{-\delta}^\delta \dfrac{c\varepsilon_0^2|\boldsymbol{\xi}|}{2}\,\mathrm{d}|\boldsymbol{\xi}| = \dfrac{A\delta^2\varepsilon_0^2}{4}c, & \text{一维} \end{cases}$$

$$(3.38)$$

其中，h 为二维问题考察体的厚度，A 为一维杆件的横截面面积，ϕ 和 φ 分别表示方位角和极角，如图 3-2 所示。

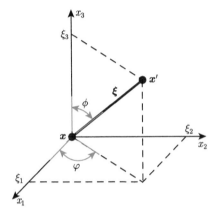

图 3-2　球坐标系中物质点的相对位置关系

对不同维度的问题，令 W_{PD} 与 W_{CCM} 相等，则可以得到 PMB 模型的微模量常数为

$$c = \begin{cases} \dfrac{18K}{\pi\delta^4} = \dfrac{6E}{\pi\delta^4(1-2\nu)}, & \text{三维} \\[2mm] \dfrac{6E}{\pi\delta^3 h(1+\nu)(1-2\nu)}, & \text{平面应变} \\[2mm] \dfrac{6E}{\pi\delta^3 h(1-\nu)}, & \text{平面应力} \\[2mm] \dfrac{2E}{A\delta^2}, & \text{一维} \end{cases} \quad (3.39)$$

式中，$K = E/(3(1-2\nu))$ 为体积模量。

2) 纯剪切变形状态

纯剪切变形可以认为是物体的三维均匀展平变形，变形过程无刚体转动，一种典型情况为：变形体沿一个方向拉伸，而在垂直面内缩短。如图 3-3 所示，设构型发生一微小纯剪切变形，剪应变值为 γ，则相应的变形梯度和 Cauchy 应变分别为

$$\boldsymbol{F} = \begin{bmatrix} 1 & \gamma & 0 \\ \gamma & 1 & 0 \\ 0 & 0 & 1 \end{bmatrix}, \quad \boldsymbol{E} = \begin{bmatrix} 0 & \gamma & 0 \\ \gamma & 0 & 0 \\ 0 & 0 & 0 \end{bmatrix} \quad (3.40)$$

3.2 经典微弹脆性模型

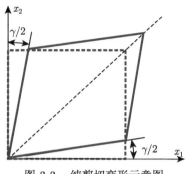

图 3-3 纯剪切变形示意图

经典连续介质力学的弹性应变能密度可分解为体应变能密度和畸变能密度，对不同维度的问题，分别为

$$W_{\text{CCM}} = \begin{cases} \dfrac{K}{2}\varepsilon_{ii}^2 + \mu \sum\limits_{i,j=1,2,3} \varepsilon_{ij}^d \varepsilon_{ij}^d, & \text{三维} \\[2mm] \left(\dfrac{K}{2} + \dfrac{\mu}{9}\right)\varepsilon_{ii}^2 + \mu \sum\limits_{i,j=1,2} \varepsilon_{ij}^d \varepsilon_{ij}^d, & \text{平面应变} \\[2mm] \left[\dfrac{K}{2} + \dfrac{\mu}{9}\left(\dfrac{\nu+1}{2\nu-1}\right)^2\right]\varepsilon_{ii}^2 + \mu \sum\limits_{i,j=1,2} \varepsilon_{ij}^d \varepsilon_{ij}^d, & \text{平面应力} \end{cases} \quad (3.41)$$

式中，K 为体积模量，$\mu = E/(2(1+\nu))$ 为剪切模量，E 为杨氏模量，ν 为泊松比，$\varepsilon_{ii} = \varepsilon_{11} + \varepsilon_{22} + \varepsilon_{33}$ 为体应变，ε_{ij}^d 为应变偏张量分量，且有 $\varepsilon_{ij}^d = \varepsilon_{ij} - \dfrac{1}{3}\varepsilon_{mm}\delta_{ij}$，其中 δ_{ij} 为克罗内克符号，则该纯剪切变形下的应变能密度为

$$W_{\text{CCM}} = 2\mu\gamma^2 \quad (3.42)$$

关注物质点 $\boldsymbol{x}$ 及其近场范围内相互作用物质点 $\boldsymbol{x}'$，设物质点 $\boldsymbol{x}$ 为坐标原点，在球坐标系中，参考构型中 $\boldsymbol{x}'$ 相对于 $\boldsymbol{x}$ 点的坐标为 (ξ_1, ξ_2, ξ_3)，且 $\xi_1 = |\boldsymbol{\xi}|\sin\varphi\cos\phi$，$\xi_2 = |\boldsymbol{\xi}|\sin\varphi\sin\phi$ 和 $\xi_3 = |\boldsymbol{\xi}|\cos\varphi$，则键的伸长量为

$$\underline{e} = |\underline{\boldsymbol{Y}}| - |\underline{\boldsymbol{X}}| = \sqrt{(\xi_1 + \gamma\xi_2)^2 + (\xi_2 + \gamma\xi_1)^2 + \xi_3^2} - |\boldsymbol{\xi}|$$
$$\approx 2\cos\phi\sin\phi\sin^2\varphi|\boldsymbol{\xi}|\gamma \quad (3.43)$$

同理，在二维极坐标系中，参考构型中 $\boldsymbol{x}'$ 相对于 $\boldsymbol{x}$ 点的坐标为 (ξ_1, ξ_2)，且 $\xi_1 = \cos\phi|\boldsymbol{\xi}|$，$\xi_2 = \sin\phi|\boldsymbol{\xi}|$，则键的伸长量为

$$\underline{e} = |\underline{\boldsymbol{Y}}| - |\underline{\boldsymbol{X}}| = \sqrt{(\xi_1 + \gamma\xi_2)^2 + (\xi_2 + \gamma\xi_1)^2} - |\boldsymbol{\xi}|$$

$$\approx 2\cos\phi\sin\phi|\boldsymbol{\xi}|\gamma \tag{3.44}$$

于是，在该变形状态下，近场动力学中物质点之间键的伸长率为

$$s_\gamma = \frac{e}{|\boldsymbol{\xi}|} = \begin{cases} 2\cos\phi\sin\phi\sin^2\varphi\gamma, & 三维 \\ 2\cos\phi\sin\phi\gamma, & 二维 \end{cases} \tag{3.45}$$

此时，PMB 模型的点对势函数 $\widehat{w}$ 为

$$\widehat{w} = \int_0^{\boldsymbol{\eta}} cs\frac{\boldsymbol{\xi}+\boldsymbol{\eta}}{|\boldsymbol{\xi}+\boldsymbol{\eta}|}\cdot\mathrm{d}\boldsymbol{\eta} = \int_0^{s_\gamma} cs|\boldsymbol{\xi}|\mathrm{d}s = \left.\frac{cs^2|\boldsymbol{\xi}|}{2}\right|_{s_\gamma}$$

$$= \begin{cases} \dfrac{c|\boldsymbol{\xi}|}{2}(2\cos\phi\sin\phi\sin^2\varphi\gamma)^2, & 三维 \\[2mm] \dfrac{c|\boldsymbol{\xi}|}{2}(2\cos\phi\sin\phi\gamma)^2, & 二维 \end{cases} \tag{3.46}$$

进而，不同维度问题的近场动力学弹性变形能密度为

$$W_{\mathrm{PD}} = \frac{1}{2}\int_{H_x}\widehat{w}\mathrm{d}V_{x'}$$

$$= \begin{cases} \dfrac{1}{2}\int_0^\delta\int_0^{2\pi}\int_0^\pi \dfrac{c|\boldsymbol{\xi}|}{2}(2\cos\phi\sin\phi\sin^2\varphi\gamma)^2|\boldsymbol{\xi}|^2\sin\varphi\mathrm{d}\varphi\mathrm{d}\phi\mathrm{d}|\boldsymbol{\xi}| = \dfrac{\pi\delta^4}{15}c\gamma^2, & 三维 \\[2mm] \dfrac{1}{2}\int_0^\delta\int_0^{2\pi}\dfrac{c|\boldsymbol{\xi}|}{2}(2\cos\phi\sin\phi\gamma)^2(|\boldsymbol{\xi}|\mathrm{d}\phi\mathrm{d}|\boldsymbol{\xi}|h) = \dfrac{\pi\delta^3 h}{12}c\gamma^2, & 二维 \end{cases}$$

式中，h 为二维问题考察体的厚度，ϕ 和 φ 分别表示方位角和极角。

对不同维度问题，令 W_{PD} 与 W_{CCM} 相等，则可以得到微模量常数为

$$c = \begin{cases} \dfrac{30\mu}{\pi\delta^4} = \dfrac{15E}{\pi\delta^4(1+\nu)}, & 三维 \\[2mm] \dfrac{24\mu}{\pi\delta^3 h} = \dfrac{12E}{\pi\delta^3 h(1+\nu)}, & 二维 \end{cases} \tag{3.47}$$

考虑到两种不同的变形场假设确定的 PMB 模型微模量参数应当一致，故令式 (3.39) 与式 (3.47) 相等，则有

$$\nu = \begin{cases} 1/4, & 三维 \\ 1/4, & 平面应变 \\ 1/3, & 平面应力 \end{cases} \tag{3.48}$$

从上述推导过程可见，键型近场动力学模型存在泊松比限制问题，即只能用来模拟材料泊松比为 1/4 的三维问题和平面应变问题以及泊松比为 1/3 的平面应力问题，这也是键型近场动力学的一个不足。

3.2.2.2 临界伸长率

如 2.6 节所述，在断裂面完全分离和忽略裂纹尖端其他损耗的假设下，基于裂纹扩展单位面积所释放的能量与断开裂纹表面两边物质点间所有键所需的能量相等的原则，可建立键的临界变形能量密度与传统断裂力学能量释放率 G_F 的关系。对于三维与二维问题，分别有下式成立

$$G_F = \begin{cases} \int_0^\delta \int_0^{2\pi} \int_z^\delta \int_0^{\arccos(z/|\boldsymbol{\xi}|)} \widehat{w}_0 |\boldsymbol{\xi}|^2 \sin\phi \mathrm{d}\phi \mathrm{d}|\boldsymbol{\xi}| \mathrm{d}\varphi \mathrm{d}z, & \text{三维} \\ \int_0^\delta \int_z^\delta \int_{-\arccos(z/|\boldsymbol{\xi}|)}^{\arccos(z/|\boldsymbol{\xi}|)} \widehat{w}_0 |\boldsymbol{\xi}| \mathrm{d}\varphi \mathrm{d}|\boldsymbol{\xi}| \mathrm{d}z, & \text{二维} \end{cases} \quad (3.49)$$

其中，φ 和 ϕ 分别为极坐标系的方位角和极角。据此，可求得三维和二维临界键能密度，分别为 $\widehat{w}_0 = \dfrac{4G_F}{\pi\delta^4}$ 和 $\widehat{w}_0 = \dfrac{9G_F}{4\delta^3 h}$，其中的二维考察体厚度 h 可取单位 1。

需要指出，在进行积分计算时，考虑到 $|z/|\boldsymbol{\xi}|| < 1$，在对 $\arccos(z/|\boldsymbol{\xi}|)$ 进行泰勒级数展开时，可取其前三项，即

$$\arccos(z/|\boldsymbol{\xi}|) = \pi/2 - z/|\boldsymbol{\xi}| - (z/|\boldsymbol{\xi}|)^3/6 + O((z/|\boldsymbol{\xi}|)^5) \quad (3.50)$$

特别地，在键型近场动力学 PMB 模型中，以键的伸长率与能量密度的关系 $\widehat{w} = \dfrac{c|\boldsymbol{\xi}|}{2}s^2$ 为媒介，可建立键的临界伸长率与能量释放率的关系

$$s_0 = \begin{cases} \sqrt{\dfrac{10G_F}{\pi c \delta^5}} = \sqrt{\dfrac{5G_F}{9K\delta}}, & \text{三维} \\ \sqrt{\dfrac{4G_F}{hc\delta^4}}, & \text{二维} \end{cases} \quad (3.51)$$

其中，c 为 PMB 模型的微模量常数。

3.3 经典微弹脆性模型的改进

经典 PMB 模型中微模量 c 为常数，无法反映长程力的空间分布规律，影响 PD 的计算精度，需要提出能反映物质点间长程力基本特性的键力密度函数。此

外，在材料边界或不同材料的交界面区域，由于积分区域不完整或不对称，近场动力学会产生"边界效应"或"表面效应"问题，使得 PD 在边界区域的计算结果不准确，亦需要发展边界效应的修正方法。

3.3.1 考虑长程力空间递减规律的修正

近场动力学的键力密度函数需要满足以下三个基本性质：①递减变化规律，即键力密度值随物质点间相对距离的增大而减小；②有限值，即当点对极限距离趋于零时，键力密度到达有限最大值，点对距离趋于近场范围半径时，键力密度到达零值，点对距离超过近场范围半径时，键力密度值为零；③对称性，即遵从作用力与反作用力规律。

不再假设近场动力学键的微模量为常数，引入反映物质点间长程力强度随点对距离变化的核函数 $g(\boldsymbol{\xi},\delta)$，则均匀各向同性微弹性材料的键力密度函数可进一步构造为

$$f(\boldsymbol{\xi},\boldsymbol{\eta}) = \chi c(\boldsymbol{\xi},\delta) s \frac{\boldsymbol{\xi}+\boldsymbol{\eta}}{|\boldsymbol{\xi}+\boldsymbol{\eta}|} = \chi s \frac{\boldsymbol{\xi}+\boldsymbol{\eta}}{|\boldsymbol{\xi}+\boldsymbol{\eta}|} c(0,\delta) g(\boldsymbol{\xi},\delta) \tag{3.52}$$

式中，$c(\boldsymbol{\xi},\delta) = c(0,\delta) g(\boldsymbol{\xi},\delta)$ 称为微模量函数，表征物质点对的刚度，为相对位置与近场范围尺寸的函数，它由集中函数 $c(0,\delta)$ 和核函数 $g(\boldsymbol{\xi},\delta)$ 组成，且集中函数 $c(0,\delta)$ 用来描述两物质点在无限靠近时的点对刚度，其他各参数含义同前文。根据长程作用力随距离增加而递减的特性，核函数 $g(\boldsymbol{\xi},\delta)$ 需要满足以下基本规律

$$\begin{cases} \text{(a):} \ g(\boldsymbol{\xi},\delta) = g(-\boldsymbol{\xi},\delta) \\ \text{(b):} \ \lim_{\boldsymbol{\xi}\to 0} g(\boldsymbol{\xi},\delta) = \max g \\ \text{(c):} \ \lim_{\boldsymbol{\xi}\to \delta} g(\boldsymbol{\xi},\delta) = 0 \\ \text{(d):} \ \int_{-\infty}^{\infty} \lim_{\delta\to 0} g(\boldsymbol{\xi},\delta) = \int_{-\infty}^{\infty} \Delta(\boldsymbol{\xi}) \mathrm{d}x = 1 \end{cases} \tag{3.53}$$

式中，$\Delta(\boldsymbol{\xi})$ 为狄拉克函数 (此处为了区分近场范围尺寸 δ，采用大写 $\Delta(\boldsymbol{\xi})$ 表示)。

对于经典的 PMB 模型，键力密度函数 $f(\boldsymbol{\xi},\boldsymbol{\eta}) = \chi cs \dfrac{\boldsymbol{\xi}+\boldsymbol{\eta}}{|\boldsymbol{\xi}+\boldsymbol{\eta}|}$，相当于微模量取常数，即式 (3.52) 中的核函数 $g(\boldsymbol{\xi},\delta) = \begin{cases} 1, & |\boldsymbol{\xi}| \leqslant \delta \\ 0, & |\boldsymbol{\xi}| > \delta \end{cases}$，但常数微模量忽略了长程力的空间递减变化特性，影响计算结果的精度。

可采用四次多项式型的空间变化函数构建核函数，即

3.3 经典微弹脆性模型的改进

$$g(\boldsymbol{\xi},\delta) = \begin{cases} \left(1-\left(\dfrac{|\boldsymbol{\xi}|}{\delta}\right)^2\right)^2, & |\boldsymbol{\xi}| \leqslant \delta \\ 0, & |\boldsymbol{\xi}| > \delta \end{cases} \quad (3.54)$$

该核函数充分考虑了近场范围尺寸 δ 对键力密度函数的影响,计算精度较原 PMB 模型有所提高。

当构型在各向均匀的拉伸荷载作用下产生各向同性均匀膨胀变形 ε_0 时,而同样变形状态对应于近场动力学中,有 $s = \varepsilon_0$,则不同维度问题的近场动力学弹性变形能密度为

$$W_{\text{PD}} = \begin{cases} \dfrac{1}{2}\displaystyle\int_0^\delta\int_0^{2\pi}\int_0^{\pi}\left(\dfrac{c(\boldsymbol{\xi},\delta)\varepsilon_0^2|\boldsymbol{\xi}|}{2}\right)|\boldsymbol{\xi}|^2\sin\varphi\mathrm{d}\varphi\mathrm{d}\phi\mathrm{d}|\boldsymbol{\xi}| = \dfrac{\pi\delta^4\varepsilon_0^2}{24}c(0,\delta) \\[2mm] \dfrac{1}{2}\displaystyle\int_0^\delta\int_0^{2\pi}\left(\dfrac{c(\boldsymbol{\xi},\delta)\varepsilon_0^2|\boldsymbol{\xi}|}{2}\right)(|\boldsymbol{\xi}|\mathrm{d}\phi\mathrm{d}|\boldsymbol{\xi}|h) = \dfrac{4\pi\delta^3\varepsilon_0^2 h}{105}c(0,\delta) \\[2mm] \dfrac{A}{2}\displaystyle\int_0^\delta\left(\dfrac{c(\boldsymbol{\xi},\delta)\varepsilon_0^2|\boldsymbol{\xi}|}{2}\right)\mathrm{d}|\boldsymbol{\xi}| = \dfrac{A\delta^2\varepsilon_0^2}{12}c(0,\delta) \end{cases} \quad (3.55)$$

根据近场动力学变形能量密度与经典连续介质力学应变能密度相等的原则,令 W_{PD} 与 W_{CCM} 相等,则可以得到微模量常数 $c(0,\delta)$,为

$$c(0,\delta) = \begin{cases} \dfrac{36E}{\pi\delta^4(1-2\nu)}, & \text{三维} \\[2mm] \dfrac{105E}{4\pi h\delta^3(1+\nu)(1-2\nu)}, & \text{平面应变} \\[2mm] \dfrac{105E}{4\pi h\delta^3(1-\nu)}, & \text{平面应力} \\[2mm] \dfrac{6E}{A\delta^2}, & \text{一维} \end{cases} \quad (3.56)$$

以平面应力问题为例,图 3-4 给出了改进的 PMB 模型和经典 PMB 模型的微模量函数与两个物质点对相对位置 $\boldsymbol{\xi}$ 的分布图。从图中曲线可以看出,经典模型将近场范围内物质点点对的刚度设为常数,忽略了物质点对间的距离对相互作用力的影响,并在物质点间距趋于近场范围时突然截断;而改进的 PMB 模型以连续函数的形式反映长程力强度随物质点间的距离增大而减弱的物理特性,更为合理。

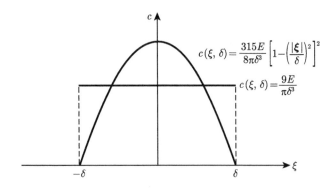

图 3-4　经典 PMB 模型与改进 PMB 模型的微模量函数比较

此外，当选取不同的核函数 $g(\pmb{\xi},\delta)$ 时，可以获得不同的微模量 $c(0,\delta)$，表 3-1 给出了平面应力情况下的模型参数。

表 3-1　不同核函数 $g(\pmb{\xi},\delta)$ 对应的平面应力微模量常数 $c(0,\delta)$

序号	1	2	3	4
核函数$g(\pmb{\xi},\delta)$	$1-\dfrac{\|\pmb{\xi}\|}{\delta}$	$1-\left(\dfrac{\|\pmb{\xi}\|}{\delta}\right)^2$	$1-3\left(\dfrac{\|\pmb{\xi}\|}{\delta}\right)^2 + 2\left(\dfrac{\|\pmb{\xi}\|}{\delta}\right)^3$	$\left[1-\left(\dfrac{\|\pmb{\xi}\|}{\delta}\right)^2\right]^2$
微模量$c(0,\delta)$	$\dfrac{24E}{\pi\delta^3(1-\nu)}$	$\dfrac{15E}{\pi\delta^3(1-\nu)}$	$\dfrac{30E}{\pi\delta^3(1-\nu)}$	$\dfrac{105E}{4\pi\delta^3(1-\nu)}$
序号	5	6	7	
核函数$g(\pmb{\xi},\delta)$	$1-10\left(\dfrac{\|\pmb{\xi}\|}{\delta}\right)^3 + 15\left(\dfrac{\|\pmb{\xi}\|}{\delta}\right)^4 - 6\left(\dfrac{\|\pmb{\xi}\|}{\delta}\right)^5$	$\dfrac{\delta^2\left[1-\left(\dfrac{\|\pmb{\xi}\|}{\delta}\right)^2\right]}{\|\pmb{\xi}\|^2+\delta^2}$	$\cos\left(\dfrac{\pi\|\pmb{\xi}\|}{2\delta}\right)$	
微模量$c(0,\delta)$	$\dfrac{168E}{5\pi\delta^3(1-\nu)}$	$\dfrac{12E}{(10-3\pi)\pi\delta^3(1-\nu)}$	$\dfrac{\pi^2 E}{(\pi^2-8)\delta^3(1-\nu)}$	

3.3.2　考虑边界效应的修正

在上述考虑长程力特性的改进 PMB 模型中，近场范围是假设完整对称的，在物体内部和边界处均取相同的常数 $c(0,\delta)$，弱化了边界处的 $c(0,\delta)$ 值，导致计算结果存在误差，即存在"边界效应"问题。需要对边界处 $c(0,\delta)$ 的计算方法进行改进。

改进的基本原则是：在计算边界区域的应变能密度时，仅考虑边界区域近场范

围内物质点的贡献。当发生空间或平面内各向均匀变形 ε_0 时，PD 应变能密度为

$$W_{\mathrm{PD}}(\boldsymbol{x}) = \frac{1}{2} \int_{H_{\boldsymbol{x}}} \widehat{w}(\boldsymbol{\xi}, \boldsymbol{\eta}) \mathrm{d}V_{\boldsymbol{x}'} = \frac{1}{2} \sum_{j \in H_i} \widehat{w}(\boldsymbol{\xi}, \boldsymbol{\eta}) V_j$$

$$= \frac{1}{4} \sum_{j \in H_i} c(0, \delta) \left(1 - \left(\frac{|\boldsymbol{\xi}_{ij}|}{\delta}\right)^2\right)^2 \varepsilon_0^2 |\boldsymbol{\xi}_{ij}| V_j \tag{3.57}$$

式中，H_i 为 $H_{\boldsymbol{x}}$ 的离散形式，为边界处物质点 $\boldsymbol{x}_i$ 近场范围内的物质点 $\boldsymbol{x}_j$ 的集合。令 W_{PD} 与 W_{CCM} 相等，即令式 (3.57) 与式 (3.36) 相等，可推得

$$c_i = \begin{cases} \dfrac{6E}{(1-2\nu) \sum\limits_{j \in H_i} \left[1 - \left(\dfrac{|\boldsymbol{\xi}_{ij}|}{\delta}\right)^2\right]^2 |\boldsymbol{\xi}_{ij}| V_j}, & \text{三维} \\[2ex] \dfrac{4E}{(1+\nu)(1-2\nu) \sum\limits_{j \in H_i} \left[1 - \left(\dfrac{|\boldsymbol{\xi}_{ij}|}{\delta}\right)^2\right]^2 |\boldsymbol{\xi}_{ij}| V_j}, & \text{平面应变} \\[2ex] \dfrac{4E}{(1-\nu) \sum\limits_{j \in H_i} \left[1 - \left(\dfrac{|\boldsymbol{\xi}_{ij}|}{\delta}\right)^2\right]^2 |\boldsymbol{\xi}_{ij}| V_j}, & \text{平面应力} \\[2ex] \dfrac{2E}{\sum\limits_{j \in H_i} \left[1 - \left(\dfrac{|\boldsymbol{\xi}_{ij}|}{\delta}\right)^2\right]^2 |\boldsymbol{\xi}_{ij}| V_j}, & \text{一维} \end{cases} \tag{3.58}$$

在改进 (或原)PMB 模型中，$\boldsymbol{f}_{ij} = \boldsymbol{f}_{ji}$ 必然成立，但在采用离散求和方式计算时，H_i 与 H_j 不一定相同，为保证动量守恒，取

$$c_{ij} = \frac{c_i + c_j}{2} \tag{3.59}$$

图 3-5 给出了弹性模量 $E = 1\mathrm{GPa}$，泊松比 $\nu = 1/3$，尺寸 $50\mathrm{mm} \times 50\mathrm{mm}$ 的方形薄板在均匀离散情况下，不同 PMB 模型计算得到的 $c(0, \delta)$ 值分布情况，结果表明：改进模型中的边界处 $c(0, \delta)$ 值不再与内部 $c(0, \delta)$ 值相同，有效改善了表面效应问题。

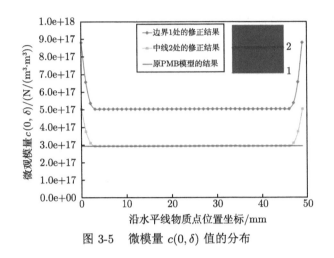

图 3-5 微模量 $c(0,\delta)$ 值的分布

3.4 其他键型近场动力学模型

经典键型近场动力学模型的"键"可以视为一种"微弹簧"或"微杆"模型[1]，即以径向键力密度函数，传递沿键方向的拉压载荷，这也是可以通过 ABAQUS 或 ANSYS 等商业有限元软件的杆单元实现近场动力学计算的根本原因，但仅考虑径向键力作用的键型近场动力学模型，不仅存在泊松比取值的限制，在描述材料的力学行为方面也有所欠缺。于是，研究者在考虑径向键力密度的基础上，自然想到要引入切向键力密度，以克服泊松比取值的限制，并能更好地刻画材料的力学行为。本节主要介绍两种突破固定泊松比限制的近场动力学模型，即微梁模型[8,9]和共轭键模型[10,11]，并简要评述考虑单键转动的增强模型[12]和考虑正交键和斜交键的键型 PD 模型[13,14]，上述模型都含有"切向键力密度"项，仅存在引入思路和计算模式的差异。

3.4.1 微梁模型

通过引入转角自由度，可以将微杆模型拓展为微梁模型，典型的如"微欧拉梁"和"微铁摩辛柯梁"模型，从而引入垂直于"键"的切向键力密度。将两物质点间的"键"视为一个微梁，且两物质点位于梁端点位置。对于三维问题，物质点对具有 3 个平动自由度和 3 个转动自由度，即 $\{u\ v\ w\ \theta_x\ \theta_y\ \theta_z\}$，在局部坐标系下，物质点对的变形为轴向拉压、两个主平面内的弯曲和一个扭转变形的组合。对于二维问题，物质点对具有 2 个平动自由度和 1 个转动自由度，即 $\{u\ v\ \theta\}$，在局部坐标系下，物质点对的变形为轴向拉压和平面内的弯曲组合。

以二维问题为例，根据式 (3.30)，局部坐标系中的广义力 $\underset{\sim}{f'}$、广义位移 $\underset{\sim}{u'}$ 和

3.4 其他键型近场动力学模型

键劲度系数矩阵 $\boldsymbol{K}'$ 的关系为

$$\underset{\sim}{\boldsymbol{f}'} = \boldsymbol{K}' \underset{\sim}{\boldsymbol{u}'} \tag{3.60}$$

且有

$$\begin{cases} \underset{\sim}{\boldsymbol{f}'} = \begin{bmatrix} \boldsymbol{f}'_{ji} & \boldsymbol{f}'_{ij} \end{bmatrix}^{\mathrm{T}} = \begin{bmatrix} f'_{ji(1)} & f'_{ji(2)} & m'_{ji} & f'_{ij(1)} & f'_{ij(2)} & m'_{ij} \end{bmatrix}^{\mathrm{T}} \\ \underset{\sim}{\boldsymbol{u}'} = \begin{bmatrix} \boldsymbol{u}'_i & \boldsymbol{u}'_j \end{bmatrix}^{\mathrm{T}} = \begin{bmatrix} u'_{i(1)} & u'_{i(2)} & \theta'_i & u'_{j(1)} & u'_{j(2)} & \theta'_j \end{bmatrix}^{\mathrm{T}} \end{cases} \tag{3.61}$$

其中，下标 i, j 表示键两端的物质点，$f'_{ji(1)}$ 和 $f'_{ij(1)}$ 为径向键力密度，$f'_{ji(2)}$ 和 $f'_{ij(2)}$ 为切向键力密度，$f'_{ji(12)}$ 和 $f'_{ij(12)}$ 为微梁力矩密度。因此，除了描述物质点平动的运动方程外，尚需建立描述物质点转动的运动方程，为

$$\boldsymbol{I}\frac{\partial^2 \boldsymbol{\theta}(\boldsymbol{x}_i, t)}{\partial t^2} = \int_{H_{xi}} \boldsymbol{m}'_{ji}(\boldsymbol{x}_i, \boldsymbol{x}_j, \boldsymbol{u}_i, \boldsymbol{u}_j, \boldsymbol{\theta}_i, \boldsymbol{\theta}_j, t) \mathrm{d}V_{x_j} + \boldsymbol{m}_{\mathrm{ext}}(\boldsymbol{x}_i, t) \tag{3.62}$$

式中，$\boldsymbol{I}$ 为微梁横截面转动惯量，$\boldsymbol{m}_{\mathrm{ext}}$ 为物质点所受外力矩密度矢量，$\boldsymbol{m}_{ji}$ 为微梁横向变形时物质点 i 所受的力矩密度。

2007 年，Gerstle 等发展的微极近场动力学 (micro-polar peridynamic) 模型[8] 是一类"微欧拉梁"模型。欧拉梁弯曲理论考虑 Kirchhoff 平截面假定，梁截面转角等于挠度曲线切线的斜率，借鉴有限元法的欧拉梁单元，可以构造近场动力学中的"微欧拉梁"模型，即假设物质点间的键具有径向微模量 c 和切向微模量 d，则物质点对在局部坐标系下的劲度矩阵为

$$\boldsymbol{K}'_E = \begin{bmatrix} c/|\boldsymbol{\xi}_{ij}| & & & & & \\ 0 & 12d/|\boldsymbol{\xi}_{ij}|^3 & & \text{对称} & & \\ 0 & 6d/|\boldsymbol{\xi}_{ij}|^2 & 4d/|\boldsymbol{\xi}_{ij}| & & & \\ -c/|\boldsymbol{\xi}_{ij}| & 0 & 0 & c/|\boldsymbol{\xi}_{ij}| & & \\ 0 & -12d/|\boldsymbol{\xi}_{ij}|^3 & -6d/|\boldsymbol{\xi}_{ij}|^2 & 0 & 12d/|\boldsymbol{\xi}_{ij}|^3 & \\ 0 & 6d/|\boldsymbol{\xi}_{ij}|^2 & 2d/|\boldsymbol{\xi}_{ij}| & 0 & -6d/|\boldsymbol{\xi}_{ij}|^2 & 4d/|\boldsymbol{\xi}_{ij}| \end{bmatrix} \tag{3.63}$$

由于物质点对所构成的微梁单元的最大"跨高比"$m = \delta/\Delta x$ (一般 $m \leqslant 5$)，在很多情况下属于深梁范围，因此梁的横向剪切变形不应该被忽略。因此，可以借鉴有限元法的铁摩辛柯梁单元，构建近场动力学中的"微铁摩辛柯梁"模型[9]。在铁摩辛柯梁理论中，梁的横截面仍满足平截面假定，但不再与梁中面垂直，挠度与截面转角各自独立。假设物质点间的键具有径向微模量 c 和切向微模量 e (以

区分微欧拉梁模型中的切向微模量 d), 则可得到物质点对在局部坐标系下的劲度矩阵, 为

$$\boldsymbol{K}_T' = \begin{bmatrix} c/|\boldsymbol{\xi}_{ij}| & & & & & \\ 0 & e/|\boldsymbol{\xi}_{ij}| & & & 对称 & \\ 0 & e/2 & e|\boldsymbol{\xi}_{ij}|/3 & & & \\ -c/|\boldsymbol{\xi}_{ij}| & 0 & 0 & c/|\boldsymbol{\xi}_{ij}| & & \\ 0 & -e/|\boldsymbol{\xi}_{ij}| & -e/2 & 0 & e/|\boldsymbol{\xi}_{ij}| & \\ 0 & e/2 & e|\boldsymbol{\xi}_{ij}|/6 & 0 & -e/2 & e|\boldsymbol{\xi}_{ij}|/3 \end{bmatrix} \tag{3.64}$$

下面, 以平面应力问题为例给出"微梁模型"的微模量系数的确定方法。式 (3.34) 已经给出了以主应变表示的应变能密度表达式, 将应变关系 $\varepsilon_3 = -\dfrac{\nu}{1-\nu} \times (\varepsilon_1 + \varepsilon_2)$ 代入式 (3.34) 中, 并写为矩阵形式

$$W_{\text{CCM}} = (\varepsilon_1 \ \varepsilon_2) \begin{bmatrix} \dfrac{E}{2(1-\nu^2)} & \dfrac{E\nu}{2(1-\nu^2)} \\ \dfrac{E\nu}{2(1-\nu^2)} & \dfrac{E}{2(1-\nu^2)} \end{bmatrix} \begin{pmatrix} \varepsilon_1 \\ \varepsilon_2 \end{pmatrix} \tag{3.65}$$

沿应变主轴方向选取整体坐标系, 以物质点 $\boldsymbol{x}_i$ 为坐标原点, 则两物质点的相对位移矢量与主应变的关系为

$$\boldsymbol{u} = \begin{pmatrix} u_1 \\ u_2 \end{pmatrix} = \begin{pmatrix} u_{j(1)} - u_{i(1)} \\ u_{j(2)} - u_{i(2)} \end{pmatrix} = \boldsymbol{\xi} \begin{pmatrix} \varepsilon_1 l \\ \varepsilon_2 m \end{pmatrix} \tag{3.66}$$

则局部坐标系下两物质点的相对位移矢量为

$$\boldsymbol{u}' = \boldsymbol{T}\boldsymbol{u} = \begin{bmatrix} l & m \\ -m & l \end{bmatrix} \begin{pmatrix} u_1 \\ u_2 \end{pmatrix} \tag{3.67}$$

其中, $l = \cos\theta, m = \sin\theta$。

进而, 近场动力学中单一键存储的能量密度和物质点的变形能密度分别为

$$\widehat{w} = \frac{1}{2} \boldsymbol{u}'^{\text{T}} \boldsymbol{K}' \boldsymbol{u}' \tag{3.68}$$

$$W_{\text{PD}} = \frac{1}{2} \int_{H_{\boldsymbol{x}}} \frac{1}{2} \boldsymbol{u}'^{\text{T}} \boldsymbol{K}' \boldsymbol{u}' \text{d}V_{\boldsymbol{x}'} \tag{3.69}$$

对于微欧拉梁模型,取 $\boldsymbol{K}' = \boldsymbol{K}'_E$,则有

$$\begin{aligned}
W_{\mathrm{PD}} &= \frac{1}{2}\int_0^\delta \int_0^{2\pi} \frac{1}{2}|\boldsymbol{\xi}|^3 \begin{pmatrix} \varepsilon_1 & \varepsilon_2 \end{pmatrix} \begin{bmatrix} l^2 & -lm \\ m^2 & lm \end{bmatrix} \begin{bmatrix} c/|\boldsymbol{\xi}| & 0 \\ 0 & 12d/|\boldsymbol{\xi}|^3 \end{bmatrix} \\
&\quad \times \begin{bmatrix} l^2 & m^2 \\ -lm & lm \end{bmatrix} \begin{pmatrix} \varepsilon_1 \\ \varepsilon_2 \end{pmatrix} \mathrm{d}\phi \mathrm{d}|\boldsymbol{\xi}| \\
&= \begin{pmatrix} \varepsilon_1 & \varepsilon_2 \end{pmatrix} \begin{bmatrix} \dfrac{\pi}{4}\left(\dfrac{c\delta^3}{4}+3d\delta\right) & \dfrac{\pi}{4}\left(\dfrac{c\delta^3}{12}-3d\delta\right) \\ \dfrac{\pi}{4}\left(\dfrac{c\delta^3}{12}-3d\delta\right) & \dfrac{\pi}{4}\left(\dfrac{c\delta^3}{4}+3d\delta\right) \end{bmatrix} \begin{pmatrix} \varepsilon_1 \\ \varepsilon_2 \end{pmatrix}
\end{aligned} \quad (3.70)$$

令式 (3.70) 与式 (3.65) 相等,即 $W_{\mathrm{PD}} = W_{\mathrm{CCM}}$,则有

$$c = \frac{6E}{\pi\delta^3(1-\nu)}, \quad d = \frac{E(1-3\nu)}{6\pi\delta(1-\nu^2)} \quad (3.71)$$

对于铁摩辛柯梁模型,取 $\boldsymbol{K}' = \boldsymbol{K}'_T$,则有

$$W_{\mathrm{PD}} = \begin{pmatrix} \varepsilon_1 & \varepsilon_2 \end{pmatrix} \begin{bmatrix} \dfrac{\pi\delta^3(3c+e)}{48} & \dfrac{\pi\delta^3(c-e)}{48} \\ \dfrac{\pi\delta^3(c-e)}{48} & \dfrac{\pi\delta^3(3c+e)}{48} \end{bmatrix} \begin{pmatrix} \varepsilon_1 \\ \varepsilon_2 \end{pmatrix} \quad (3.72)$$

令式 (3.72) 与式 (3.65) 相等,即 $W_{\mathrm{PD}} = W_{\mathrm{CCM}}$,则有

$$c = \frac{6E}{\pi\delta^3(1-\nu)}, \quad e = \frac{6E(1-3\nu)}{\pi\delta^3(1-\nu^2)} \quad (3.73)$$

对于平面应变问题和三维问题,同样可以导出相应的结果。经比较后发现,只需将平面应力问题的径向微模量 c 中对应的 $(1-\nu)$ 替换为 $(1-2\nu)(1+\nu)$,d 和 e 对应的 $\dfrac{1-3\nu}{1-\nu^2}$ 替换为 $\dfrac{1-4\nu}{(1-2\nu)(1+\nu)}$,就可以得到平面应变问题的微模量参数。

由式 (3.71) 和式 (3.73) 可见,要保证微模量系数为正,对于平面应力问题,这两种模型的泊松比不得大于 1/3。同样,对于平面应变问题和三维问题,不难知道微梁模型的泊松比不得大于 1/4。因此,微梁模型虽然在一定程度上拓宽了泊松比的取值范围,但并未突破泊松比上限限制。

此外，为提高计算精度，还可引入近场域内可积的核函数 $g(\boldsymbol{\xi},\delta)$ 来反映近场范围内物质点的微模量系数的强度分布特征。具体可令 $\begin{cases} c(\boldsymbol{\xi},\delta) = c(0,\delta)g(\boldsymbol{\xi},\delta) \\ d(\boldsymbol{\xi},\delta) = d(0,\delta)g(\boldsymbol{\xi},\delta) \end{cases}$ 或 $\begin{cases} c(\boldsymbol{\xi},\delta) = c(0,\delta)g(\boldsymbol{\xi},\delta) \\ e(\boldsymbol{\xi},\delta) = e(0,\delta)g(\boldsymbol{\xi},\delta) \end{cases}$，当给出核函数的具体形式后，可类似 3.3.1 节的处理方法求得相关参数。表 3-2 给出了平面应力问题下几个核函数对应的微梁模型的参数。

表 3-2 不同核函数 $g(\boldsymbol{\xi},\delta)$ 对应的等效刚度系数 $c(0,\delta)$、$d(0,\delta)$ 和 $e(0,\delta)$

序号	1	2	3	4
核函数 $g(\boldsymbol{\xi},\delta)$	$1 - \dfrac{|\boldsymbol{\xi}|}{\delta}$	$1 - \left(\dfrac{|\boldsymbol{\xi}|}{\delta}\right)^2$	$1 - 3\left(\dfrac{|\boldsymbol{\xi}|}{\delta}\right)^2 + 2\left(\dfrac{|\boldsymbol{\xi}|}{\delta}\right)^3$	$\left[1 - \left(\dfrac{|\boldsymbol{\xi}|}{\delta}\right)^2\right]^2$
拉压微模量 $c(0,\delta)$	$\dfrac{24E}{\pi\delta^3(1-\nu)}$	$\dfrac{15E}{\pi\delta^3(1-\nu)}$	$\dfrac{30E}{\pi\delta^3(1-\nu)}$	$\dfrac{105E}{4\pi\delta^3(1-\nu)}$
弯曲微模量 $d(0,\delta)$	$\dfrac{E(1-3\nu)}{3\pi\delta(1-\nu^2)}$	$\dfrac{E(1-3\nu)}{4\pi\delta(1-\nu^2)}$	$\dfrac{E(1-3\nu)}{3\pi\delta(1-\nu^2)}$	$\dfrac{15E(1-3\nu)}{48\pi\delta(1-\nu^2)}$
扭转微模量 $e(0,\delta)$	$\dfrac{24E(1-3\nu)}{\pi\delta^3(1-\nu^2)}$	$\dfrac{15E(1-3\nu)}{\pi\delta^3(1-\nu^2)}$	$\dfrac{30E(1-3\nu)}{\pi\delta^3(1-\nu^2)}$	$\dfrac{105E(1-3\nu)}{4\pi\delta^3(1-\nu^2)}$
序号	5	6	7	
核函数 $g(\boldsymbol{\xi},\delta)$	$1 - 10\left(\dfrac{|\boldsymbol{\xi}|}{\delta}\right)^3 + 15\left(\dfrac{|\boldsymbol{\xi}|}{\delta}\right)^4 - 6\left(\dfrac{|\boldsymbol{\xi}|}{\delta}\right)^5$	$\dfrac{\delta^2\left[1-\left(\dfrac{|\boldsymbol{\xi}|}{\delta}\right)^2\right]}{|\boldsymbol{\xi}|^2+\delta^2}$	$\cos\left(\dfrac{\pi|\boldsymbol{\xi}|}{2\delta}\right)$	
拉压微模量 $c(0,\delta)$	$\dfrac{168E}{5\pi\delta^3(1-\nu)}$	$\dfrac{12E}{(10-3\pi)\pi\delta^3(1-\nu)}$	$\dfrac{\pi^2 E}{(\pi^2-8)\delta^3(1-\nu)}$	
弯曲微模量 $d(0,\delta)$	$\dfrac{E(1-3\nu)}{3\pi\delta(1-\nu^2)}$	$\dfrac{E(1-3\nu)}{3(\pi-2)\pi\delta(1-\nu^2)}$	$\dfrac{E(1-3\nu)}{24\delta(1-\nu^2)}$	
扭转微模量 $e(0,\delta)$	$\dfrac{168E(1-3\nu)}{5\pi\delta^3(1-\nu^2)}$	$\dfrac{12E(1-3\nu)}{(10-3\pi)\pi\delta^3(1-\nu^2)}$	$\dfrac{\pi^2 E(1-3\nu)}{(\pi^2-8)\delta^3(1-\nu^2)}$	

3.4.2 共轭键模型

本节介绍共轭键型近场动力学模型 (conjugated BB PD)，该模型基于"拉压弹簧"和"旋转弹簧"或"两体对势"和"三体势"共同建模，即考虑单一键的径向拉伸或压缩变形对径向键力密度的贡献时，同时考虑近场范围内键与键之间的夹角变化对垂直于当前键方向的切向键力密度的贡献，总的键力密度不再沿着当前键的方向，获得的双参数模型突破了原单参数模型的泊松比限制问题。同样地，

3.4 其他键型近场动力学模型

还可采用表 3-2 给出的核函数反映长程作用力的空间递减特性，以提高计算精度。

3.4.2.1 键力密度函数

在共轭键型近场动力学模型中，物质点 x_i 与 x_j 之间的键力密度 f_{ji} 可分解为径向键力密度 f_n 和垂直于当前键方向的切向键力密度 f_t，如图 3-6 所示，即

$$f_{ji} = f_n + f_t \tag{3.74}$$

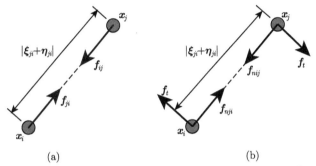

图 3-6 经典键型近场动力学与共轭键型近场动力学的键力密度示意图

径向键力密度可与经典键型近场动力学的键力密度一致，则径向键力密度与点对势能密度分别为

$$f_n = c_n(\boldsymbol{\xi}_{ji},\delta)s(\boldsymbol{\xi}_{ji},\boldsymbol{\eta}_{ji})\frac{\boldsymbol{\xi}_{ji}+\boldsymbol{\eta}_{ji}}{|\boldsymbol{\xi}_{ji}+\boldsymbol{\eta}_{ji}|} = c_n s \boldsymbol{e}_n \tag{3.75}$$

$$\widehat{w}_n(\boldsymbol{\xi}_{ji},\boldsymbol{\eta}_{ji}) = \frac{c_n(\boldsymbol{\xi}_{ji},\delta)s^2(\boldsymbol{\xi}_{ji},\boldsymbol{\eta}_{ji})|\boldsymbol{\xi}_{ji}|}{2} \tag{3.76}$$

式中，$c_n(\boldsymbol{\xi}_{ji},\delta)$ 为径向微模量系数，$\boldsymbol{e}_n$ 是径向键力密度的单位方向矢量。

以下将介绍切向键力密度 f_t 的计算方法与物理解释。

如图 3-7(a) 所示，物质点 x_i 近场范围内共有 N 个物质点 (含 x_i 自身)，共有 $N-1$ 个键 $\boldsymbol{\xi}_{ji}$，定义与键 $\boldsymbol{\xi}_{ji}$ 共同形成键角 θ_{jik_m} 的键 $\boldsymbol{\xi}_{k_mi}$ 为 $\boldsymbol{\xi}_{ji}$ 的共轭键，因而每个键共有 $N-2$ 个共轭键，即 $\boldsymbol{\xi}_{k_mi}(m=1,2,\cdots,N-2)$，共有 $N-2$ 个键角，即 $\theta_{jik_m}(m=1,2,\cdots,N-2)\in[0,\pi]$。也可以认为存在 $N-2$ 个旋转弹簧 (rotation spring) 阻抗某一键与其共轭键之间的相对转动。于是，可定义微转动势函数或微三体势函数为

$$w_t^{k_m}(\boldsymbol{\xi}_{ji},\boldsymbol{\xi}_{k_mi},\theta_{jik_m}) = \frac{c_t(\theta_{jik_0},\delta)\left[\Delta\theta_{jik_m}(\boldsymbol{\xi}_{ji},\boldsymbol{\xi}_{k_mi},\theta_{jik_m})\right]^2}{2} \tag{3.77}$$

其中，$c_t(\theta_{jik_m0},\delta)$ 为键之间的抗旋转微模量 (rotation resistance micro-modulus)。

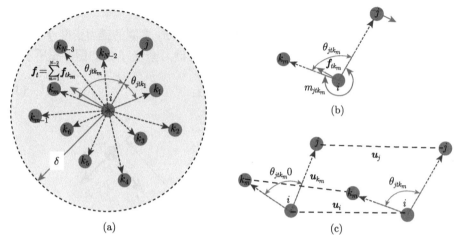

图 3-7　键 $\boldsymbol{\xi}_{ji}$ 与其共轭键 $\boldsymbol{\xi}_{k_mi}$ 形成相对键角 $\theta_{jik_m}(m=1,2,\cdots,N-2)$(a)、转动弹簧产生的弯矩和切向键力密度的示意图 (b) 和变形时键角 θ_{jik_m} 发生相对转动示意图 (c)

如图 3-7(c) 所示，键 $\boldsymbol{\xi}_{ji}$ 与其共轭键 $\boldsymbol{\xi}_{k_mi}$ 的初始夹角、当前夹角和相对转动之间的关系如下：

$$\Delta\theta_{jik_m}(\boldsymbol{\xi}_{ji},\boldsymbol{\xi}_{k_mi},\theta_{jik_m}) = \theta_{jik_m} - \theta_{jik_m0} \tag{3.78}$$

类似地，考虑图 3-7(b)，旋转弹簧的微转动势函数相对物质点 $\boldsymbol{x}_i$ 产生的力矩和相应的切应力分别为

$$m_{jik_m} = \frac{\partial w_t^{k_m}(\boldsymbol{\xi}_{ji},\boldsymbol{\xi}_{k_mi},\theta_{jik_m})}{\partial \Delta\theta_{jik_m}}V_{k_m} = c_t(\theta_{jik_m0},\delta)\Delta\theta_{jik_m}(\boldsymbol{\xi}_{ji},\boldsymbol{\xi}_{k_mi},\theta_{jik_m})V_{k_m} \tag{3.79}$$

$$\begin{aligned}\boldsymbol{f}_t^{k_m}(\boldsymbol{\xi}_{ji},\boldsymbol{\xi}_{k_mi},\theta_{jik_m}) &= \frac{m_{jik_m}}{|\boldsymbol{\xi}_{ji}+\boldsymbol{\eta}_{ji}|}\cdot \boldsymbol{e}_t \\ &= \frac{m_{jik_m}}{|\boldsymbol{\xi}_{ji}+\boldsymbol{\eta}_{ji}|}\cdot\frac{\boldsymbol{\xi}_{ji}+\boldsymbol{\eta}_{ji}}{|\boldsymbol{\xi}_{ji}+\boldsymbol{\eta}_{ji}|}\times\left[\frac{\boldsymbol{\xi}_{ji}+\boldsymbol{\eta}_{ji}}{|\boldsymbol{\xi}_{ji}+\boldsymbol{\eta}_{ji}|}\times\frac{\boldsymbol{\xi}_{k_mi}+\boldsymbol{\eta}_{k_mi}}{|\boldsymbol{\xi}_{k_mi}+\boldsymbol{\eta}_{k_mi}|}\right]\end{aligned} \tag{3.80}$$

式中，V_{k_m} 为物质点 $\boldsymbol{x}_{k_m}$ 的体积，$\boldsymbol{e}_t$ 是切向键力密度的单位方向矢量。对于所有共轭键对产生的切向键力密度累加求和，则切向键力密度为

$$\boldsymbol{f}_t = \sum_{m=1}^{N-2}\boldsymbol{f}_t^{k_m}(\boldsymbol{\xi}_{ji},\boldsymbol{\xi}_{k_mi},\theta_{jik_m}) = \sum_{m=1}^{N-2}\frac{c_t(\theta_{jik_m0},\delta)\Delta\theta_{jik_m}(\boldsymbol{\xi}_{ji},\boldsymbol{\xi}_{k_mi},\theta_{jik_m})V_{k_m}}{|\boldsymbol{\xi}_{ji}+\boldsymbol{\eta}_{ji}|}\cdot\boldsymbol{e}_t \tag{3.81}$$

3.4 其他键型近场动力学模型

至此，建立了共轭键型近场动力学弹性模型的键力密度表达式，尚需确定径向和切向微模量参数。

3.4.2.2 径向和切向微模量系数

假定抗旋转微模量 $c_t(\theta_{jik},\delta)$ 为常数，而在径向微模量 $c_n(\boldsymbol{\xi}_{ji},\delta)$ 中引入递减核函数，并以表 3-2 中第一种核函数为例，即有

$$c_n(\boldsymbol{\xi}_{ji},\delta) = c_{n0}(0,\delta)g_n(\boldsymbol{\xi}_{ji},\delta) = c_{n0}\left(1 - \frac{|\boldsymbol{\xi}_{ji}|}{\delta}\right) \tag{3.82}$$

其中，c_t 和 c_{n0} 需要通过应变能等效方法确定。

近场动力学弹性变形能密度可分解为拉压应变能密度和转动应变能密度

$$W^{\mathrm{PD}} = W_n^{\mathrm{PD}} + W_t^{\mathrm{PD}} \tag{3.83}$$

且有

$$W_n^{\mathrm{PD}} = \frac{1}{2}\int_{H_{\boldsymbol{x}_i}} w_n(\boldsymbol{\xi}_{ji},\boldsymbol{\eta}_{ji})\mathrm{d}V_{\boldsymbol{x}_j} = \frac{1}{4}\int_{H_{\boldsymbol{x}_i}} c_{n0}g_n(\boldsymbol{\xi}_{ji},\delta)s^2(\boldsymbol{\xi}_{ji},\boldsymbol{\eta}_{ji})|\boldsymbol{\xi}_{ji}|\mathrm{d}V_{\boldsymbol{x}_j} \tag{3.84}$$

$$\begin{aligned}W_t^{\mathrm{PD}} &= \frac{1}{2}\int_{H_{\boldsymbol{x}_i}}\frac{1}{2}\int_{H_{\boldsymbol{x}_i}} w_t^{k_m}(\boldsymbol{\xi}_{ji},\boldsymbol{\xi}_{k_mi},\theta_{jik_m})\mathrm{d}V_{\boldsymbol{x}_{k_m}}\mathrm{d}V_{\boldsymbol{x}_j}\\ &= \frac{1}{8}\int_{H_{\boldsymbol{x}_i}}\int_{H_{\boldsymbol{x}_i}} c_t(\theta_{jik0},\delta)\left[\Delta\theta_{jik_m}(\boldsymbol{\xi}_{ji},\boldsymbol{\xi}_{k_mi},\theta_{jik_m})\right]^2\mathrm{d}V_{\boldsymbol{x}_{k_m}}\mathrm{d}V_{\boldsymbol{x}_j}\end{aligned} \tag{3.85}$$

式中，第一个 1/2 表示键两端物质点各占有一半总键能，第二个 1/2 表示共轭键对中两个键各占一半转动键能。

1) 拉压应变能密度

在小变形假设下，由任意变形场的变形梯度张量 $\boldsymbol{F}$ 和 Green 应变张量 $\boldsymbol{E}$，通过 Cauchy-Born 准则可以确定变形后键 $\boldsymbol{\xi}_{ji}$ 和键 $\boldsymbol{\xi}_{k_mi}$ 的键长，为

$$|\boldsymbol{\xi}_{ji}+\boldsymbol{\eta}_{ji}| = |\boldsymbol{\xi}_{ji}|\sqrt{\boldsymbol{\chi}^{\mathrm{T}}\boldsymbol{F}^{\mathrm{T}}\boldsymbol{F}\boldsymbol{\chi}} = |\boldsymbol{\xi}_{ji}|\sqrt{\boldsymbol{\chi}^{\mathrm{T}}(2\boldsymbol{E}+\boldsymbol{I})\boldsymbol{\chi}} = (\varepsilon_{mn}\chi_m\chi_n+1)|\boldsymbol{\xi}_{ji}| \tag{3.86}$$

$$\begin{aligned}|\boldsymbol{\xi}_{k_mi}+\boldsymbol{\eta}_{k_mi}| &= |\boldsymbol{\xi}_{k_mi}|\sqrt{\boldsymbol{\psi}^{\mathrm{T}}\boldsymbol{F}^{\mathrm{T}}\boldsymbol{F}\boldsymbol{\psi}}\\ &= |\boldsymbol{\xi}_{k_mi}|\sqrt{\boldsymbol{\psi}^{\mathrm{T}}(2\boldsymbol{E}+\boldsymbol{I})\boldsymbol{\psi}} = (\varepsilon_{mn}\psi_m\psi_n+1)|\boldsymbol{\xi}_{k_mi}|\end{aligned} \tag{3.87}$$

其中，$\boldsymbol{I}$ 为二阶单位张量，ε_{mn} 为 Green 应变张量分量，χ_m 和 χ_n 分别为参考构型中键 $\boldsymbol{\xi}_{ji}$ 的方向矢量 $\boldsymbol{\chi}$ 的分量，ψ_m 和 ψ_n 分别为参考构型中键 $\boldsymbol{\xi}_{k_mi}$ 的方向矢

量 ψ 的分量,m 和 n 为哑标,参考构型中的方向矢量分别为

$$\chi = \begin{cases} \begin{bmatrix} \sin\varphi\cos\phi & \sin\varphi\sin\phi & \cos\varphi \end{bmatrix}^{\mathrm{T}}, & \text{三维} \\ \begin{bmatrix} \cos\phi & \sin\phi \end{bmatrix}^{\mathrm{T}}, & \text{二维} \end{cases} \quad (3.88)$$

$$\psi = \begin{cases} \begin{bmatrix} \sin\tilde{\varphi}\cos\tilde{\phi} & \sin\tilde{\varphi}\sin\tilde{\phi} & \cos\tilde{\varphi} \end{bmatrix}^{\mathrm{T}}, & \text{三维} \\ \begin{bmatrix} \cos\tilde{\phi} & \sin\tilde{\phi} \end{bmatrix}^{\mathrm{T}}, & \text{二维} \end{cases} \quad (3.89)$$

式中,ϕ 和 φ 分别为方位角和极角。

将式 (3.86) 代入伸长率公式,或根据式 (2.9),有下式成立

$$s(\boldsymbol{\xi}_{ji}, \boldsymbol{\eta}_{ji}) = (|\boldsymbol{\xi}_{ji} + \boldsymbol{\eta}_{ji}| - |\boldsymbol{\xi}_{ji}|)/|\boldsymbol{\xi}_{ji}| = \varepsilon_{mn}\chi_m\chi_n \quad (3.90)$$

再将核函数 $g_n(\boldsymbol{\xi}_{ji}, \delta) = 1 - |\boldsymbol{\xi}_{ji}|/\delta$ 和伸长率 $s(\boldsymbol{\xi}_{ji}, \boldsymbol{\eta}_{ji}) = \varepsilon_{mn}\chi_m\chi_n$ 代入式 (3.83) 的拉压应变能密度项中,则可得到三维情况下的拉压应变能密度

$$\begin{aligned} W_n^{\mathrm{PD}} &= \frac{1}{4}\int_0^\delta \int_0^{2\pi} \int_0^\pi c_{n0}\left(1 - \frac{|\boldsymbol{\xi}_{ji}|}{\delta}\right)(\varepsilon_{mn}\chi_m\chi_n)^2 |\boldsymbol{\xi}_{ji}|^3 \sin\varphi \mathrm{d}\varphi \mathrm{d}\phi \mathrm{d}|\boldsymbol{\xi}_{ji}| \\ &= c_{n0}\frac{\pi\delta^4}{300}\begin{pmatrix} 3\varepsilon_{11}^2 + 3\varepsilon_{22}^2 + 3\varepsilon_{33}^2 + 4\varepsilon_{12}^2 + 4\varepsilon_{13}^2 \\ +4\varepsilon_{23}^2 + 2\varepsilon_{11}\varepsilon_{22} + 2\varepsilon_{11}\varepsilon_{33} + 2\varepsilon_{22}\varepsilon_{33} \end{pmatrix} \\ &= c_{n0}\frac{\pi\delta^4}{300}(3I_1^2 - 4I_2) \end{aligned} \quad (3.91)$$

其中,$I_1 = \varepsilon_{11} + \varepsilon_{22} + \varepsilon_{33}$ 和 $I_2 = \varepsilon_{11}\varepsilon_{22} + \varepsilon_{11}\varepsilon_{33} + \varepsilon_{22}\varepsilon_{33} - \varepsilon_{12}^2 - \varepsilon_{13}^2 - \varepsilon_{23}^2$ 为主应变不变量。

二维情况下的拉压应变能密度为

$$\begin{aligned} W_n^{\mathrm{PD}} &= \frac{1}{4}\int_0^\delta \int_0^{2\pi} c_{n0}\left(1 - \frac{|\boldsymbol{\xi}_{ji}|}{\delta}\right)(\varepsilon_{mn}\chi_m\chi_n)^2 |\boldsymbol{\xi}_{ji}|^2 \mathrm{d}\phi \mathrm{d}|\boldsymbol{\xi}_{ji}| h \\ &= c_{n0}\frac{\pi h\delta^3}{192}(3\varepsilon_{11}^2 + 3\varepsilon_{22}^2 + 4\varepsilon_{12}^2 + 2\varepsilon_{11}\varepsilon_{22}) \\ &= c_{n0}\frac{\pi h\delta^3}{192}(3I_1^2 - 4I_2) \end{aligned} \quad (3.92)$$

式中,$I_1 = \varepsilon_{11} + \varepsilon_{22}$ 和 $I_2 = \varepsilon_{11}\varepsilon_{22} - \varepsilon_{12}^2$ 为二维情况下的主应变不变量。

3.4 其他键型近场动力学模型

2) 转动应变能密度

如图 3-7(c) 所示，参考构型和当前构型中共轭键对 $\boldsymbol{\xi}_{ji}$、$\boldsymbol{\xi}_{k_m i}$ 的键角分别为 $\theta_{jik_m 0}$、θ_{jik_m}，并有

$$\theta_{jik_m 0} = \arccos\left(\frac{\boldsymbol{\xi}_{ji} \cdot \boldsymbol{\xi}_{k_m i}}{|\boldsymbol{\xi}_{ji}| \cdot |\boldsymbol{\xi}_{k_m i}|}\right), \quad \theta_{jik_m} = \arccos\left(\frac{(\boldsymbol{\xi}_{ji} + \boldsymbol{\eta}_{ji}) \cdot (\boldsymbol{\xi}_{k_m i} + \boldsymbol{\eta}_{k_m i})}{|\boldsymbol{\xi}_{ji} + \boldsymbol{\eta}_{ji}| \cdot |\boldsymbol{\xi}_{k_m i} + \boldsymbol{\eta}_{k_m i}|}\right) \tag{3.93}$$

进一步，采用单位方向矢量和应变描述，上述键角可重写为

$$\theta_{jik_m 0} = \arccos\left(\frac{\boldsymbol{\chi}^{\mathrm{T}} \boldsymbol{\psi}}{\sqrt{\boldsymbol{\chi}^{\mathrm{T}} \boldsymbol{\chi}} \cdot \sqrt{\boldsymbol{\psi}^{\mathrm{T}} \boldsymbol{\psi}}}\right) \tag{3.94}$$

$$\begin{aligned}
\theta_{jik_m} &= \arccos\left(\frac{\boldsymbol{\chi}^{\mathrm{T}} \boldsymbol{F}^{\mathrm{T}} \boldsymbol{F} \boldsymbol{\psi}}{\sqrt{\boldsymbol{\chi}^{\mathrm{T}} \boldsymbol{F}^{\mathrm{T}} \boldsymbol{F} \boldsymbol{\chi}} \cdot \sqrt{\boldsymbol{\psi}^{\mathrm{T}} \boldsymbol{F}^{\mathrm{T}} \boldsymbol{F} \boldsymbol{\psi}}}\right) \\
&= \arccos\left(\frac{\boldsymbol{\chi}^{\mathrm{T}}(2\boldsymbol{E}+\boldsymbol{I})\boldsymbol{\psi}}{\sqrt{\boldsymbol{\chi}^{\mathrm{T}}(2\boldsymbol{E}+\boldsymbol{I})\boldsymbol{\chi}} \cdot \sqrt{\boldsymbol{\psi}^{\mathrm{T}}(2\boldsymbol{E}+\boldsymbol{I})\boldsymbol{\psi}}}\right) \\
&= \arccos\left(\frac{2\boldsymbol{\chi}^{\mathrm{T}}\boldsymbol{E}\boldsymbol{\psi} + \boldsymbol{\chi}^{\mathrm{T}}\boldsymbol{\psi}}{\sqrt{2\boldsymbol{\chi}^{\mathrm{T}}\boldsymbol{E}\boldsymbol{\chi}+1} \cdot \sqrt{2\boldsymbol{\psi}^{\mathrm{T}}\boldsymbol{E}\boldsymbol{\psi}+1}}\right) = \arccos(Q)
\end{aligned} \tag{3.95}$$

由上式可知，键角 θ_{jik_m} 是关于应变张量 $\boldsymbol{E}$ 的函数，因此，θ_{jik_m} 的一阶泰勒级数展开式为

$$\theta_{jik_m}(\boldsymbol{E}) \approx \theta_{jik_m 0}(\boldsymbol{0}) + \left.\frac{\partial \theta_{jik_m}(\boldsymbol{E})}{\partial \boldsymbol{E}}\right|_{\boldsymbol{E}=0} \boldsymbol{E} \tag{3.96}$$

则当前构型中相对键角为

$$\begin{aligned}
\Delta\theta_{jik_m} &= \theta_{jik_m}(\boldsymbol{E}) - \theta_{jik_m 0}(\boldsymbol{0}) \approx \left.\frac{\partial \theta_{jik_m}(\boldsymbol{E})}{\partial \boldsymbol{E}}\right|_{\boldsymbol{E}=0} \boldsymbol{E} \\
&= \left.\left(\frac{\partial \arccos(Q)}{\partial Q}\frac{\partial Q}{\partial \boldsymbol{E}}\right)\right|_{\boldsymbol{E}=0} \boldsymbol{E} = \left.\left(-\frac{1}{\sqrt{1-Q^2}}\frac{\partial Q}{\partial \boldsymbol{E}}\right)\right|_{\boldsymbol{E}=0} \boldsymbol{E} \\
&= -\frac{1}{\sqrt{1-(\boldsymbol{\chi}^{\mathrm{T}}\boldsymbol{\psi})^2}}(2\chi_m\psi_n - \boldsymbol{\chi}^{\mathrm{T}}\boldsymbol{\psi}(\chi_m\chi_n + \psi_m\psi_n))\varepsilon_{mn} \\
&= -\frac{1}{\sqrt{1-(\chi_i\psi_i)^2}}(2\chi_m\psi_n - \chi_i\psi_i(\chi_m\chi_n + \psi_m\psi_n))\varepsilon_{mn}
\end{aligned} \tag{3.97}$$

再将上式代入式 (3.85) 中，并利用 Wolfram Mathematica 软件，可计算得到二维问题的转动应变能密度为

$$W_t^{\text{PD}} = \frac{1}{8}\int\limits_{H_{x_i}}\int\limits_{H_{x_i}} c_t(\theta_{jik0},\delta)\left[\Delta\theta_{jik_m}(\boldsymbol{\xi}_{ji},\boldsymbol{\xi}_{k_mi},\theta_{jik_m})\right]^2 \mathrm{d}V_{\boldsymbol{x}_{k_m}}\mathrm{d}V_{\boldsymbol{x}_j}$$

$$= \frac{1}{8}\int_0^\delta\int_0^{2\pi}\left\{\int_0^\delta\int_0^{2\pi} c_t\frac{1}{1-(\chi_i\psi_i)^2}\left[\begin{pmatrix} 2\chi_m\psi_n - \chi_i\psi_i \\ (\chi_m\chi_n + \psi_m\psi_n) \end{pmatrix}\varepsilon_{mn}\right]^2 |\boldsymbol{\xi}_{k_mi}|\mathrm{d}\tilde{\phi}\mathrm{d}|\boldsymbol{\xi}_{k_mi}|\right\}$$

$$\times |\boldsymbol{\xi}_{ji}|\mathrm{d}\phi\mathrm{d}|\boldsymbol{\xi}_{ji}|$$

$$= \frac{c_t\pi^2\delta^4 h}{32}(\varepsilon_{11}^2 + \varepsilon_{22}^2 + 4\varepsilon_{12}^2 - 2\varepsilon_{11}\varepsilon_{22})$$

$$= \frac{c_t\pi^2\delta^4 h}{32}(I_1^2 - 4I_2) \tag{3.98}$$

需要指出的是，在三维情况下，采用式 (3.77) 给出的微转动势函数不能求得转动应变能密度的解析表达式，此处仅给出三维问题的转动应变能密度的计算式

$$W_t^{\text{PD}} = \frac{1}{8}\int\limits_{H_{x_i}}\int\limits_{H_{x_i}} c_t(\theta_{jik0},\delta)\left[\Delta\theta_{jik_m}(\boldsymbol{\xi}_{ji},\boldsymbol{\xi}_{k_mi},\theta_{jik_m})\right]^2 \mathrm{d}V_{\boldsymbol{x}_{k_m}}\mathrm{d}V_{\boldsymbol{x}_j}$$

$$= \frac{1}{8}\int_0^\delta\int_0^{2\pi}\int_0^\pi\left\{\int_0^\delta\int_0^{2\pi}\int_0^\pi c_t\frac{1}{1-(\chi_i\psi_i)^2}\left[\begin{pmatrix} 2\chi_m\psi_n - \chi_i\psi_i \\ (\chi_m\chi_n + \psi_m\psi_n) \end{pmatrix}\varepsilon_{mn}\right]^2\right.$$

$$\left.\times |\boldsymbol{\xi}_{k_mi}|^2\sin\tilde{\varphi}\mathrm{d}\tilde{\varphi}\mathrm{d}\tilde{\phi}\mathrm{d}|\boldsymbol{\xi}_{k_mi}|\right\}$$

$$\times |\boldsymbol{\xi}_{ji}|^2\sin\varphi\mathrm{d}\varphi\mathrm{d}\phi\mathrm{d}|\boldsymbol{\xi}_{ji}| \tag{3.99}$$

并可推测其最终表达式为 $A(I_1^2 - 4I_2)$，其中 A 与 c_t, π, δ 相关，且 $I_1 = \varepsilon_{11} + \varepsilon_{22} + \varepsilon_{33}, I_2 = \varepsilon_{11}\varepsilon_{22} + \varepsilon_{11}\varepsilon_{33} + \varepsilon_{22}\varepsilon_{33} - \varepsilon_{12}^2 - \varepsilon_{13}^2 - \varepsilon_{23}^2$。

3) 应变能等效方法确定微模量系数

在二维情况下，近场动力学应变能密度为拉伸应变能密度与转动应变能密度之和，即为

$$W^{\text{PD}} = W_n^{\text{PD}} + W_t^{\text{PD}} = c_{n0}\frac{\pi h\delta^3}{192}(3\varepsilon_{11}^2 + 3\varepsilon_{22}^2 + 4\varepsilon_{12}^2 + 2\varepsilon_{11}\varepsilon_{22})$$

$$+ \frac{c_t\pi^2\delta^4 h}{16}(\varepsilon_{11}^2 + \varepsilon_{22}^2 + 4\varepsilon_{12}^2 - 2\varepsilon_{11}\varepsilon_{22}) \tag{3.100}$$

根据式 (3.41)，对于平面应变和平面应力问题，经典连续介质力学对应的弹性应变能密度分别为

$$W_{\text{CCM}} = \frac{E(1-\nu)}{2(1+\nu)(1-2\nu)}(\varepsilon_{11}^2 + \varepsilon_{22}^2) + 2\mu\varepsilon_{12}^2 + \frac{E\nu}{(1+\nu)(1-2\nu)}\varepsilon_{11}\varepsilon_{22} \tag{3.101}$$

$$W_{\mathrm{CCM}} = \frac{E}{2(1+\nu)(1-\nu)}(\varepsilon_{11}^2 + \varepsilon_{22}^2) + 2\mu\varepsilon_{12}^2 + \frac{E\nu}{(1+\nu)(1-\nu)}\varepsilon_{11}\varepsilon_{22} \quad (3.102)$$

令近场动力学应变能密度与经典连续介质力学应变能密度相等，即 $W_{\mathrm{PD}} = W_{\mathrm{CCM}}$，则对于平面应变问题，有

$$\begin{cases} c_{n0}\dfrac{\pi h \delta^3}{64} + \dfrac{c_t \pi^2 \delta^4 h}{32} = \dfrac{E(1-\nu)}{2(1+\nu)(1-2\nu)} \\ c_{n0}\dfrac{\pi h \delta^3}{48} + \dfrac{c_t \pi^2 \delta^4 h}{8} = \dfrac{E}{(1+\nu)} \\ c_{n0}\dfrac{\pi h \delta^3}{96} - \dfrac{c_t \pi^2 \delta^4 h}{16} = \dfrac{E\nu}{(1+\nu)(1-2\nu)} \end{cases} \Rightarrow \begin{cases} c_{n0} = \dfrac{24E}{\pi h \delta^3 (1+\nu)(1-2\nu)} \\ c_t = \dfrac{4E(1-4\nu)}{\pi^2 \delta^4 h (1+\nu)(1-2\nu)} \end{cases}$$
(3.103)

对于平面应力问题，有

$$\begin{cases} c_{n0}\dfrac{\pi h \delta^3}{64} + \dfrac{c_t \pi^2 \delta^4 h}{32} = \dfrac{E}{2(1+\nu)(1-\nu)} \\ c_{n0}\dfrac{\pi h \delta^3}{48} + \dfrac{c_t \pi^2 \delta^4 h}{8} = \dfrac{E}{(1+\nu)} \\ c_{n0}\dfrac{\pi h \delta^3}{96} - \dfrac{c_t \pi^2 \delta^4 h}{16} = \dfrac{E\nu}{(1+\nu)(1-\nu)} \end{cases} \Rightarrow \begin{cases} c_{n0} = \dfrac{24E}{\pi h \delta^3 (1-\nu)} \\ c_t = \dfrac{4E(1-3\nu)}{\pi^2 \delta^4 h (1-\nu^2)} \end{cases}$$
(3.104)

3.4.3 另两种键型近场动力学模型简介

不同于考虑"共轭键对之间夹角变化"建立的共轭键型近场动力学模型，朱其志等[12]考虑当前构型中键相对于参考构型中"键自身的转角变化"，建立了转角增强的键型近场动力学模型，如图 3-8 所示，具体建模细节不再详述。需要指出的是，该模型仍未突破三维问题和平面应变问题泊松比为 1/4 以及平面应力问题泊松比为 1/3 的上限。

此外，胡祎乐和 Madenci[13,14] 将键分为"正交键"(normal bond) 和"斜交键"(shear bond) 两类，提出了无泊松比限制的键型近场动力学模型，即 $\nu = [-0.5, 0.5]$，如图 3-9 所示。该模型能直接使用经典连续介质力学的弹性本构和相关的工程参数建模，并应用于任意铺层角和铺层顺序的复合材料层合板的力学分析，但该模型的数值实施限于构型的均匀正交离散。

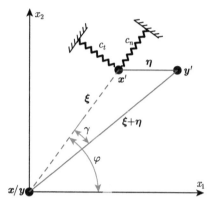

图 3-8　键拉伸与键转动产生径向和切向键力的概念图

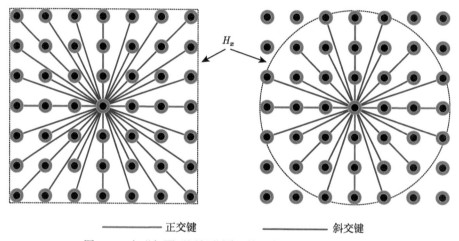

图 3-9　方形与圆形近场范围下的正交键和斜交键示意图

参 考 文 献

[1] Silling S A. Reformulation of elasticity theory for discontinuities and long-range forces [J]. Journal of the Mechanics and Physics of Solids, 2000, 48(1):175-209.

[2] Silling S A, Askari E. A meshfree method based on the peridynamic model of solid mechanics [J]. Computers & Structures, 2005, 83(17-18):1526-1535.

[3] Rahman R, Foster J T. Peridynamic theory of solids from the perspective of classical statistical mechanics [J]. Physica A: Statistical Mechanics and its Applications, 2015, 437:162-183.

[4] Gur S, Sadat M R, Frantziskonis G N, et al. The effect of grain-size on fracture of polycrystalline silicon carbide: a multiscale analysis using a molecular dynamics-peridynamics framework [J]. Computational Materials Science, 2019, 159:341-348.

[5] Xu W, Jiao Y, Fish J. An atomistically-informed multiplicative hyper-elasto-plasticity-damage model for high-pressure induced densification of silica glass [J]. Computational Mechanics, 2020, 66(1):155-187.

[6] Zhan J M, Yao X H, Han F. An approach of peridynamic modeling associated with molecular dynamics for fracture simulation of particle reinforced metal matrix composites [J]. Composite Structures, 2020:112613.

[7] Macek R W, Silling S A. Peridynamics via finite element analysis [J]. Finite Elements in Analysis and Design, 2007, 43(15):1169-1178.

[8] Gerstle W, Sau N, Silling S A. Peridynamic modeling of concrete structures [J]. Nuclear Engineering and Design, 2007, 237(12-13):1250-1258.

[9] 秦洪远, 黄丹, 刘一鸣, 等. 基于改进型近场动力学方法的多裂纹扩展分析 [J]. 工程力学, 2017, 34(12):31-38.

[10] Zhou X P, Wang Y T, Shou Y D, et al. A novel conjugated bond linear elastic model in bond-based peridynamics for fracture problems under dynamic loads [J]. Engineering Fracture Mechanics, 2018, 188:151-183.

[11] Gu X, Zhang Q. A modified conjugated bond-based peridynamic analysis for impact failure of concrete gravity dam [J]. Meccanica, 2020, 55(3):547-566.

[12] Zhu Q Z, Ni T. Peridynamic formulations enriched with bond rotation effects [J]. International Journal of Engineering Science, 2017, 121:118-129.

[13] Hu Y L, Madenci E. Bond-based peridynamics with an arbitrary Poisson's ratio [C]. 57th AIAA/ASCE/AHS/ASC Structures, Structural Dynamics, and Materials Conference 2016:1722.

[14] Hu Y L, Madenci E. Bond-based peridynamic modeling of composite laminates with arbitrary fiber orientation and stacking sequence [J]. Composite Structures, 2016, 153:139-175.

第 4 章 键型近场动力学在有限元中的实现

近场动力学强形式的控制方程可以直接采用配点法离散求解,在一定程度上,键型近场动力学微弹性线性模型中物质点间的"键"可以视为不同尺度的"杆",其中的键力密度可以采用有限元法中杆单元劲度矩阵进行表示。于是,对有限元软件的杆单元功能稍作修改,即可基于近场动力学方法进行计算分析。除强形式方程之外,键型近场动力学弱形式方程对应的非连续 Galerkin 有限元方法,能在有限元计算框架内充分发挥近场动力学特有的非连续变形分析能力,且便于施加传统边界条件和减轻表面效应问题,近年来广受关注。本章将介绍近场动力学强形式方程的杆单元解法及其在 ABAQUS 软件中的二次开发,详细阐述键型近场动力学非连续 Galerkin 有限元法,给出其数值实施过程,并介绍 LS-DYNA 软件对于该方法的实现。

4.1 近场动力学强形式方程的杆单元法

4.1.1 基于杆单元离散的计算方法

近场动力学强形式运动方程可以直接采用无网格配点法求解,物质点 x_i 的离散形式运动方程为

$$V_i \rho_i \ddot{u}_i = \sum_{j=1}^{N_j} f_{ji}(\xi_{ji}, \eta_{ji}) V_j V_i + b_i V_i \tag{4.1}$$

其中,ρ_i 为物质点 x_i 的质量密度,$\ddot{u}_i$ 为物质点 x_i 的加速度矢量,f_{ji} 为物质点 x_i 受物质点 x_j 的键力密度 (本构力函数) 矢量,b_i 为物质点 x_i 所受的外体力密度矢量,N_j 为物质点 x_i 近场范围内物质点 x_j 的总数,$\xi_{ji} = x_j - x_i$,$\eta_{ji} = u_j - u_i$,V_i 和 V_j 分别为物质点 x_i 和 x_j 的体积。需要注意的是,为使近场动力学方程与有限元方程形式一致,上式两端同乘以体积 V_i,也即 PD 与传统 FEM 一样,都采用集中质量矩阵表示惯性项。

在整体坐标系中,两个物质点 x_i 和 x_j 的键力矢量 $f_{ji}V_jV_i$ 和 $f_{ij}V_iV_j$ 可以重写成矩阵形式

$$\underline{f} = \chi \underline{k}\,\underline{u} \tag{4.2}$$

4.1 近场动力学强形式方程的杆单元法

式中，χ 为表征断键与否的历史依赖标量，$\underset{\sim}{\boldsymbol{k}}$ 为键的劲度系数矩阵，整体坐标系下物质点对 i,j 的作用力列阵 $\underset{\sim}{\boldsymbol{f}}$ 和位移列阵 $\underset{\sim}{\boldsymbol{u}}$ 分别为

$$\begin{cases} \underset{\sim}{\boldsymbol{f}} = V_i V_j \begin{bmatrix} \boldsymbol{f}_{ji} & \boldsymbol{f}_{ij} \end{bmatrix}^{\mathrm{T}} = V_i V_j \begin{bmatrix} f_{ji(1)} & f_{ji(2)} & f_{ji(3)} & f_{ij(1)} & f_{ij(2)} & f_{ij(3)} \end{bmatrix}^{\mathrm{T}} \\ \underset{\sim}{\boldsymbol{u}} = \begin{bmatrix} \boldsymbol{u}_i & \boldsymbol{u}_j \end{bmatrix}^{\mathrm{T}} = \begin{bmatrix} u_{i(1)} & u_{i(2)} & u_{i(3)} & u_{j(1)} & u_{j(2)} & u_{j(3)} \end{bmatrix}^{\mathrm{T}} \end{cases} \tag{4.3}$$

其中，$u_{i(1)}$ 表示物质点 i 的第一个自由度位移分量，其他以此类推。

进一步，建立局部直角坐标系 $x'y'z'$，如图 4-1 所示。令 x' 轴与两物质点的相对位置矢量 $\boldsymbol{\xi}_{ji}$ 或 $\boldsymbol{\xi}_{ji} + \boldsymbol{\eta}_{ji}$ 的方向平行，考虑微弹脆性 PMB 模型的键力密度矢量 $\boldsymbol{f}(\boldsymbol{\xi},\boldsymbol{\eta}) = \boldsymbol{C}(|\boldsymbol{\xi}|) \cdot \boldsymbol{\eta} = \dfrac{\chi c}{|\boldsymbol{\xi}|} \boldsymbol{I} \cdot (\boldsymbol{u}' - \boldsymbol{u})$，则在局部坐标系下，物质点对 i,j 的键力列阵 $\underset{\sim}{\boldsymbol{f}}'$、位移列阵 $\underset{\sim}{\boldsymbol{u}}'$ 和键的劲度系数矩阵 $\underset{\sim}{\boldsymbol{k}}'$ 满足下列关系式

$$\underset{\sim}{\boldsymbol{f}}' = \chi \underset{\sim}{\boldsymbol{k}}' \underset{\sim}{\boldsymbol{u}}' \tag{4.4}$$

且有

$$\begin{cases} \underset{\sim}{\boldsymbol{f}}' = V_i V_j \begin{bmatrix} \boldsymbol{f}'_{ji} & \boldsymbol{f}'_{ij} \end{bmatrix}^{\mathrm{T}} = V_i V_j \begin{bmatrix} f'_{ji(1)} & f'_{ji(2)} & f'_{ji(3)} & f'_{ij(1)} & f'_{ij(2)} & f'_{ij(3)} \end{bmatrix}^{\mathrm{T}} \\ \underset{\sim}{\boldsymbol{u}}' = \begin{bmatrix} \boldsymbol{u}'_i & \boldsymbol{u}'_j \end{bmatrix}^{\mathrm{T}} = \begin{bmatrix} u'_{i(1)} & u'_{i(2)} & u'_{i(3)} & u'_{j(1)} & u'_{j(2)} & u'_{j(3)} \end{bmatrix}^{\mathrm{T}} \end{cases} \tag{4.5}$$

$$\underset{\sim}{\boldsymbol{k}}' = \dfrac{c V_i V_j}{|\boldsymbol{\xi}_{ji}|} \begin{bmatrix} 1 & 0 & 0 & -1 & 0 & 0 \\ 0 & 0 & 0 & 0 & 0 & 0 \\ 0 & 0 & 0 & 0 & 0 & 0 \\ -1 & 0 & 0 & 1 & 0 & 0 \\ 0 & 0 & 0 & 0 & 0 & 0 \\ 0 & 0 & 0 & 0 & 0 & 0 \end{bmatrix} \tag{4.6}$$

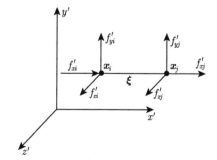

图 4-1　局部坐标系下物质点间的相互作用示意图

注意到整体坐标系和局部坐标系的变量之间存在相应的转换关系，即

$$\begin{cases} \underset{\sim}{\boldsymbol{f}} = \boldsymbol{T}\underset{\sim}{\boldsymbol{f}}' \\ \underset{\sim}{\boldsymbol{u}} = \boldsymbol{T}\underset{\sim}{\boldsymbol{u}}' \\ \underset{\sim}{\boldsymbol{k}} = \boldsymbol{T}\underset{\sim}{\boldsymbol{k}}'\boldsymbol{T}^{\mathrm{T}} \end{cases} \quad (4.7)$$

式中，$\boldsymbol{T}$ 为转换矩阵，并具有正交性 $\boldsymbol{T}^{-1} = \boldsymbol{T}^{\mathrm{T}}$，具体为

$$\boldsymbol{T} = \begin{bmatrix} l & \dfrac{-lm}{\sqrt{l^2+n^2}} & \dfrac{-n}{\sqrt{l^2+n^2}} & 0 & 0 & 0 \\ m & \sqrt{l^2+n^2} & 0 & 0 & 0 & 0 \\ n & \dfrac{-mn}{\sqrt{l^2+n^2}} & \dfrac{l}{\sqrt{l^2+n^2}} & 0 & 0 & 0 \\ 0 & 0 & 0 & l & \dfrac{-lm}{\sqrt{l^2+n^2}} & \dfrac{-n}{\sqrt{l^2+n^2}} \\ 0 & 0 & 0 & m & \sqrt{l^2+n^2} & 0 \\ 0 & 0 & 0 & n & \dfrac{-mn}{\sqrt{l^2+n^2}} & \dfrac{l}{\sqrt{l^2+n^2}} \end{bmatrix} \quad (4.8)$$

其中，l、m 和 n 分别为局部坐标系 x' 轴 (即键的相对位置矢量) 与整体坐标系 x 轴、y 轴和 z 轴的方向余弦。如图 4-2 所示，不难求得：$l = \cos\alpha = \dfrac{x_j - x_i}{|\boldsymbol{\xi}|}$、$m = \cos\beta = \dfrac{y_j - y_i}{|\boldsymbol{\xi}|}$ 和 $n = \cos\gamma = \dfrac{z_j - z_i}{|\boldsymbol{\xi}|}$。

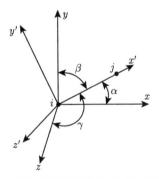

图 4-2 整体坐标系下的物质点对空间位置

对局部坐标系下物质点对的劲度系数矩阵 $\underset{\sim}{\boldsymbol{k}}'$ 进行坐标变换，则整体坐标系

下物质点对的劲度系数矩阵 $\underset{\sim}{k}$ 为

$$\underset{\sim}{k} = \frac{cV_iV_j}{|\boldsymbol{\xi}_{ji}|} \begin{bmatrix} l^2 & & & & & \\ lm & m^2 & & & \text{对称} & \\ ln & mn & n^2 & & & \\ -l^2 & -lm & -ln & l^2 & & \\ -lm & -m^2 & -mn & lm & m^2 & \\ -ln & -mn & -n^2 & ln & mn & n^2 \end{bmatrix} \quad (4.9)$$

另一方面，在有限单元法的杆单元分析中，考虑横截面为 A、长度为 $|\boldsymbol{\xi}_{ji}|$ 的拉压杆，在局部坐标系下，该杆的单元劲度矩阵为

$$\underset{\sim}{k}'_{\text{FEM}} = \frac{EA}{|\boldsymbol{\xi}_{ji}|} \begin{bmatrix} 1 & 0 & 0 & -1 & 0 & 0 \\ 0 & 0 & 0 & 0 & 0 & 0 \\ 0 & 0 & 0 & 0 & 0 & 0 \\ -1 & 0 & 0 & 1 & 0 & 0 \\ 0 & 0 & 0 & 0 & 0 & 0 \\ 0 & 0 & 0 & 0 & 0 & 0 \end{bmatrix} \quad (4.10)$$

其中，E 为弹性模量。进一步，可求得整体坐标系下该杆的单元劲度系数矩阵，即

$$\underset{\sim}{k}_{\text{FEM}} = \frac{EA}{|\boldsymbol{\xi}_{ji}|} \begin{bmatrix} l^2 & & & & & \\ lm & m^2 & & & \text{对称} & \\ ln & mn & n^2 & & & \\ -l^2 & -lm & -ln & l^2 & & \\ -lm & -m^2 & -mn & lm & m^2 & \\ -ln & -mn & -n^2 & ln & mn & n^2 \end{bmatrix} \quad (4.11)$$

令键的劲度系数矩阵与杆的单元劲度系数矩阵相等，即令式 (4.9) 和式 (4.11) 相等，则有

$$cV_iV_j = EA \quad (4.12)$$

若令 $A = (V_iV_j)^{1/3}$，则键型近场动力学 PMB 模型对应的杆单元的弹性模量参数为[1]

$$E = c(V_iV_j)^{2/3} \quad (4.13)$$

采用均匀正交离散时，物质点间距为 Δx，物质点体积为 $V_i = V_j = \Delta x^3$，则有[1]

$$A = \Delta x^2, \ E = c\Delta x^4 \quad (4.14)$$

其中，A 和 E 即为一般杆件有限元法中的几何参数和材料参数。需要注意，式 (4.13) 和式 (4.14) 的定义并不唯一。

此外，由键型近场动力学的相关定义可知，键的伸长率就是杆的工程应变，于是可以通过求得的工程应变值，并与材料的临界伸长率相比较，进行断键判别。在计算中，断键的具体做法可通过有限元软件的生死单元技术或单元删除技术实现。

进而，对物质点 x_i 的离散运动方程 (4.1) 进行组集，即可形成体系整体的运动方程

$$M\ddot{U} = KU + F_{\text{ext}} \tag{4.15}$$

其中，M 为集中质量矩阵，主对角元素为 $V_i\rho_i$；K 为由各键的劲度矩阵式 (4.9) 组集后形成的整体劲度矩阵；U 为整体位移列阵；F_{ext} 为外荷载列阵，各分量为 b_iV_i。

令运动方程中的惯性力项为零，则得到近场动力学静力平衡方程

$$KU + F_{\text{ext}} = 0 \tag{4.16}$$

施加边界条件后，即可求解代数方程组获得静力问题的解答。

上述求解方法可归属为隐式非局部有限元方法，由于该方法考虑了非局部效应，对应的劲度矩阵带宽大于传统有限法，存在计算机内存需求大和组装整体劲度矩阵耗时多等弱点，但随着高性能代数方程组求解器的普遍使用，且通过研究高效的矩阵封装算法，可提高计算效率。相较于拟静力问题的显式动力学求解方法 (施加人工阻尼)[2] 和动态松弛方法 [3]，这种方法是基于 PD 求解拟静力问题的一种合适方法。

4.1.2 基于 ABAQUS 软件的二次开发

Macek 和 Silling[1] 首次在 ABAQUS/Explicit 软件中采用杆单元，进行了键型近场动力学 PMB 模型的计算，同时采用杆单元与实体单元镶嵌方法，实现了近场动力学与经典连续介质力学的耦合建模。随后，有一些学者基于 ABAQUS 平台，开展了近场动力学建模以及近场动力学与其他数值方法耦合建模等研究[4-13]，用户单元子程序 UEL、VUEL 和用户材料子程序 UMAT、VUMAT 都可以实现近场动力学计算。为结合 ABAQUS 的高效求解器和近场动力学方法的优势，本书作者团队[7,8] 将上述键型近场动力学模型在 ABAQUS 软件中进行 VUMAT 二次开发，并利用 Python 开发了便捷的应用插件，实现了基于近场动力学方法的结构开裂破坏计算模拟。

4.1.2.1 ABAQUS 二次开发的环境配置

ABAQUS 提供了一系列诸如建模、载荷、材料、单元等用户自定义子程序 (user subroutine)，以满足用户解决特定问题的需求。用户子程序支持 C++ 和

FORTRAN 语言，将 ABAQUS 同 VISUAL STUDIO 和 VISUAL FORTRAN 进行关联后，方可以运行子程序。关联方法为：在计算机中查找 vcvarsall.bat 和 ifortvars.bat 文件，得到其路径后在 ABAQUS 安装目录 Commands 文件夹中编辑 abq—.bat 文档 ("—"为 ABAQUS 软件版本号)，具体添加 bat 文件路径、VS 和 IVF 的版本号并保存。

以 ABAQUS2020、VS2013 和 IVF2013 为测试版本，则添加示例如下：

```
@call "C:\Program Files (x86)\Microsoft Visual Studio 12.0\VC
    \vcvarsall.bat" x86_amd64
@call "C:\Program Files (x86)\Intel\Composer XE 2013 SP1\bin
    \ifortvars.bat" intel64 vs2013
@echo off"C:\SIMULIA\EstProducts\2020\win_b64\code\bin
    \ABQLauncher.exe" %*
```

随后，打开 ABAQUS Verification 进行验证，待所有验证通过后启动 ABAQUS CAE，弹出以下界面内容 (图 4-3) 即表示关联成功，则可使用户子程序。

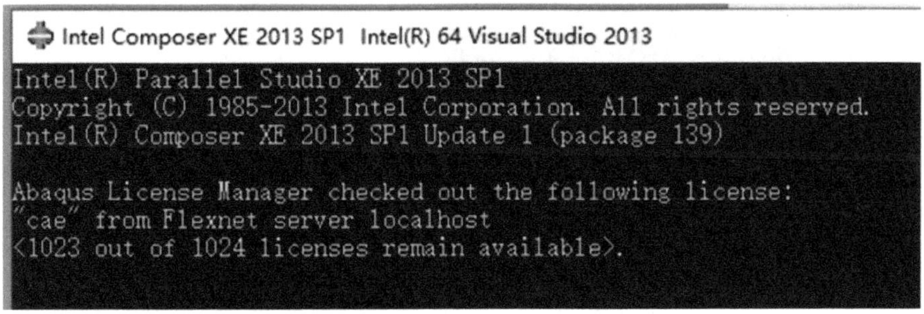

图 4-3 ABAQUS 与 VS 和 IVF 关联成功后的 ABAQUS 启动界面

4.1.2.2　ABAQUS 中用户材料子程序的简要说明

ABAQUS/Standard 通用分析模块 (准静态问题) 和 ABAQUS/Explicit 显式求解模块 (动力问题) 分别提供了两类用户材料子程序 UMAT 和 VUMAT，二者功能类似，UMAT 需要用户提供材料本构的雅可比矩阵 (DDSDDE，即弹性系数矩阵)，而 VUMAT 则不需要。此外，VUMAT 可以定义变量 STATENEW(K, I)(其中行 K 表示材料点，列 I 为用户指定的变量数)，譬如用户要指定 STATENEW 中第 2 个变量为删除单元的状态量，则在 ABAQUS 的 inp 文件的材料定义中，添加 delete=2 即可。需要说明的是，ABAQUS 默认该删除单元的状态量初始值为 1，要删除某单元，则指定其值为 0，一旦该单元被删除就不可被恢复。

4.1.2.3 键型近场动力学的 ABAQUS 插件简介

本书作者团队编写了键型近场动力学的用户材料子程序"PDBondVumat.for",并利用 Python 语言开发了应用插件"Peri_Dynamics",包含的程序有:"PDModule.py""peri_DynamicsDB.py""peri_Dynamics_plugin.py"和"PDBondVumat.for",利用该插件可进行固体材料和结构的破坏问题计算。

插件程序需要添加至 plugins 文件夹(位于安装目录 C:\SIMULIA\CAE\plugins\2020)中,并将用户子程序"PDBondVumat.for"拷贝至 ABAQUS 工作目录下。启动 ABAQUS CAE 软件,在菜单栏 Plug-ins 下拉选项里能够看到创建的插件 Peri_Dynamics,如图 4-4(a) 所示,即表示插件嵌入成功。

插件的运行界面如图 4-4(b)(c)(d) 所示,包括 About 启动页面、PD_Module 输入页面和 Help 帮助页面。开始 PD 计算前,用户需准备构型的物质点坐标、边界约束和外力加载等文件,并输入几何、材料和求解参数,操作无误后点击按钮"OK"即可运行。

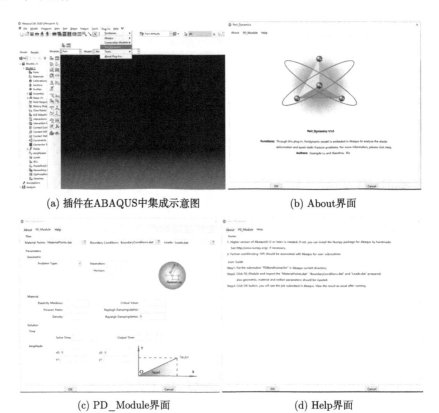

(a) 插件在ABAQUS中集成示意图　　(b) About界面

(c) PD_Module界面　　(d) Help界面

图 4-4　Peri_Dyanmics V1.0 运行界面

4.2 近场动力学弱形式方程的非连续伽辽金有限元法

如前所述,近场动力学在涉及断裂破坏等非连续变形力学问题的分析中显现出独特优势,但近场动力学方法也存在计算量大、表面效应和自然边界条件施加困难等问题,制约其应用于复杂的工程实际问题。继上述键型近场动力学的杆单元方法之后,研究者又采用 Galerkin 有限元方法,实现了近场动力学的计算求解[14-21],以充分发挥近场动力学和有限元方法的特点。在此基础上发展的近场动力学与传统有限元的耦合建模方法,不仅显著提高了近场动力学的计算效率,又有效弥补了经典连续介质力学在分析非局部和非连续变形问题的不足。

基于经典连续介质力学的传统 Galerkin 有限单元法要求位移场是一阶连续可微的,近似解属于 H^1 函数空间。近场动力学不要求位移场一阶连续可微,仅要求位移场是有限间断、平方可积的,近似解属于 L^2 函数空间[18-21]。因此,传统 Galerkin 有限元法在模拟结构裂纹扩展、剧烈破碎等非连续变形问题中遇到瓶颈和挑战,如:不连续处导数的奇异性、需预设裂纹路径、需引入裂纹扩展分叉准则、网格重构工作量大、网格依赖性显著等。非连续 Galerkin 有限元法是弱形式变分方程的一种离散解法,采用分段连续场代替连续近似场,允许单元交界面之间存在不连续[19],但需要在单元边界上增加数值通量的限制条件[17]。将近场动力学与非连续 Galerkin 有限元法相结合,通过各积分点之间的键,建立单元之间的连通性,则无需增加额外的限制条件,既放松了有限元法对位移解连续性的要求,又能保持有限元格式,继承了有限元法能直接施加局部边界条件和接触条件以及适应非均匀网格等优点。

4.2.1 经典连续介质力学微分方程的弱形式

经典连续介质力学边值问题的强形式通常采用微分方程描述,可表示为

$$\boldsymbol{A}(\boldsymbol{x},\boldsymbol{u}) = \boldsymbol{0}, \quad \boldsymbol{x} \in \Omega$$
$$\boldsymbol{B}(\boldsymbol{x},\boldsymbol{u}) = \boldsymbol{0}, \quad \boldsymbol{x} \in \Gamma \tag{4.17}$$

其中,$\boldsymbol{A}$、$\boldsymbol{B}$ 为微分算子,Ω 为计算域,Γ 为区域边界,位移函数 $\boldsymbol{u}$ 需满足控制方程和边界条件。

若对于任意的函数矩阵 $\boldsymbol{v}$ 和 $\bar{\boldsymbol{v}}$,有下式成立,即

$$\int_\Omega \boldsymbol{v}^{\mathrm{T}} \boldsymbol{A}(\boldsymbol{u}) \mathrm{d}\Omega + \int_\Gamma \bar{\boldsymbol{v}}^{\mathrm{T}} \boldsymbol{B}(\boldsymbol{u}) \mathrm{d}\Gamma = \boldsymbol{0}, \quad \forall \boldsymbol{v}, \bar{\boldsymbol{v}} \tag{4.18}$$

则称该式为原强形式方程的等效积分方程。在上述等效积分方程中,需要保证函

数矩阵 $\boldsymbol{v}$ 和 $\bar{\boldsymbol{v}}$ 在对应积分域上可积，位移函数 $\boldsymbol{u}$ 以偏导数形式出现，要求位移场是一阶连续可微的。

对上式进行分部积分，可以得到

$$\int_\Omega \boldsymbol{C}^\mathrm{T}(\boldsymbol{v})\boldsymbol{D}(\boldsymbol{u})\mathrm{d}\Omega + \int_\Gamma \boldsymbol{E}^\mathrm{T}(\bar{\boldsymbol{v}})\boldsymbol{F}(\boldsymbol{u})\mathrm{d}\Gamma = \boldsymbol{0} \tag{4.19}$$

式中，$\boldsymbol{C}$、$\boldsymbol{D}$、$\boldsymbol{E}$、$\boldsymbol{F}$ 为微分算子，它们包含的导数阶数低于 $\boldsymbol{A}$ 和 $\boldsymbol{B}$，故上式降低了对位移函数 $\boldsymbol{u}$ 的连续性要求，但提高了对检验函数 (权函数) $\boldsymbol{v}$ 和 $\bar{\boldsymbol{v}}$ 的连续性要求，由于 $\boldsymbol{v}$ 和 $\bar{\boldsymbol{v}}$ 为任意函数矩阵，容易满足低阶连续性要求。上式放松了对原控制方程中位移的连续性要求，可称之为强形式微分方程的等效弱形式积分方程。

若位移函数 $\boldsymbol{u}$ 为精确解，则强形式方程、等效积分方程和等效弱形式积分方程都严格满足且具有等价性。由于难以获得严格的精确解，在实际求解时，转而寻求满足精度要求的近似解。近似解 $\boldsymbol{u}^h$ 要满足本质边界条件和连续性要求，对于有限元方法，节点位移要满足等效弱形式的积分方程，并通过形状函数插值将单元位移场表示为单元节点位移的组合 (如图 4-5 所示)，具体为

$$\boldsymbol{u}(\boldsymbol{x}) \approx \boldsymbol{u}^h(\boldsymbol{x}) = \boldsymbol{N}(\boldsymbol{x})\boldsymbol{d} = \sum_{i=1}^n N_i(\boldsymbol{x})\boldsymbol{d}_i \tag{4.20}$$

且有

$$\boldsymbol{u} = \begin{bmatrix} u_{(1)} & u_{(2)} & u_{(3)} \end{bmatrix}_{3\times 1}^\mathrm{T}$$

$$\boldsymbol{N} = \begin{bmatrix} N_1 & 0 & 0 & \cdots & N_n & 0 & 0 \\ 0 & N_1 & 0 & \cdots & 0 & N_n & 0 \\ 0 & 0 & N_1 & \cdots & 0 & 0 & N_n \end{bmatrix}_{3\times 3n}$$

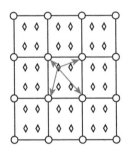

图 4-5　有限元形状函数插值示意图 (以四节点单元为例)

$$\boldsymbol{d} = \begin{bmatrix} u_{1(1)} & u_{1(2)} & u_{1(3)} & \cdots & u_{n(1)} & u_{n(2)} & u_{n(3)} \end{bmatrix}^{\mathrm{T}}_{3n \times 1} \quad (4.21)$$

其中，$\boldsymbol{u}$ 为单元内任意点 $\boldsymbol{x}$ 的位移精确解，$\boldsymbol{u}^h$ 为单元内任意点的位移近似解，$\boldsymbol{d}$ 为单元节点位移列阵，$u_{1(1)}$ 为第一个节点的第一个自由度位移分量，其他类推，n 为单元节点总数，$\boldsymbol{N}$ 为形状函数矩阵，N_i 为第 i 个节点的形状函数。对于平面问题常用的四节点等参单元，形状函数为 $N_i = \dfrac{1}{4}(1+\zeta_i\zeta)(1+\psi_i\psi)$，其中 ζ,ψ 为定义在标准单元上的局部坐标，$\zeta_i,\psi_i(i=1,2,3,4)$ 分别代表标准单元 4 个节点处的局部坐标值。

由此，等效弱形式的积分方程可进一步表述为

$$\int_{\Omega} \boldsymbol{C}^{\mathrm{T}}(\boldsymbol{v})\boldsymbol{D}(\boldsymbol{N}\boldsymbol{u}^e)\mathrm{d}\Omega + \int_{\Gamma} \boldsymbol{E}^{\mathrm{T}}(\bar{\boldsymbol{v}})\boldsymbol{F}(\boldsymbol{N}\boldsymbol{u}^e)\mathrm{d}\Gamma = \boldsymbol{0} \quad (4.22)$$

如果选取近似解 $\boldsymbol{u}^h$ 的形状函数作为权函数，不难推导得出有限元法求解的支配方程，这种方法也称为 Galerkin 有限元法。

4.2.2 近场动力学积分方程的弱形式

考虑三维空间中的计算域 $\Omega \in R^3$，物质点 $\boldsymbol{x} \in \Omega$ 的近场动力学运动方程为

$$\rho(\boldsymbol{x})\ddot{\boldsymbol{u}}(\boldsymbol{x},t) = \int_{H_{\boldsymbol{x}}} \boldsymbol{f}(\boldsymbol{\eta},\boldsymbol{\xi})\mathrm{d}V_{\boldsymbol{x}'} + \boldsymbol{b}(\boldsymbol{x},t) \quad (4.23)$$

边界区域记为 S_u，定义在 L^2 空间内的位移解 $\boldsymbol{u}$ 和检验函数 $\boldsymbol{v}$ 均需满足边界条件，具体有

$$S(\Omega) = \left\{\boldsymbol{u}(\boldsymbol{x}) \in L^2(\Omega) | \boldsymbol{u}(\boldsymbol{x}^g) = \boldsymbol{g}(\boldsymbol{x}^g), \forall \boldsymbol{x}^g \in S_u\right\}$$

$$S'(\Omega) = \left\{v(\boldsymbol{x}) \in L^2(\Omega) | v(\boldsymbol{x}^g) = 0, \forall \boldsymbol{x}^g \in S_u\right\} \quad (4.24)$$

式中，$\boldsymbol{x}^g$ 为高斯点坐标，$\boldsymbol{g}$ 为位移边界的已知函数。当选定检验函数 $\boldsymbol{v}$ 时，本质边界条件的加权余量恒为零，因而在构建弱形式积分方程时可以忽略。

在方程 (4.23) 两端同乘以检验函数矩阵 $\boldsymbol{v}(\boldsymbol{x})$ 后，在求解域内积分，得到相应的弱形式积分控制方程

$$\int_{\Omega} \boldsymbol{v}(\boldsymbol{x}) \cdot \rho\ddot{\boldsymbol{u}}(\boldsymbol{x})\mathrm{d}V_{\boldsymbol{x}} = \int_{\Omega} \boldsymbol{v}(\boldsymbol{x}) \cdot \int_{H_{\boldsymbol{x}}} \boldsymbol{f}(\boldsymbol{\eta},\boldsymbol{\xi})\mathrm{d}V_{\boldsymbol{x}'}\mathrm{d}V_{\boldsymbol{x}} + \int_{\Omega} \boldsymbol{v}(\boldsymbol{x}) \cdot \boldsymbol{b}(\boldsymbol{x})\mathrm{d}V_{\boldsymbol{x}},$$

$$\forall \boldsymbol{u}(\boldsymbol{x}) \in S(\Omega), v(\boldsymbol{x}) \in S'(\Omega) \quad (4.25)$$

在 L^2 空间内求解上述弱形式控制方程,则可得到相应解答。与常规 Galerkin 有限元方法一样,选取近似解 $\boldsymbol{u}^h$ 的形状函数作为检验函数,则待求位移近似解和检验函数矩阵分别为

$$\boldsymbol{u}^h(\boldsymbol{x}) = \boldsymbol{N}(\boldsymbol{x})\boldsymbol{d}, \ \boldsymbol{v}(\boldsymbol{x}) = \boldsymbol{N}(\boldsymbol{x}) \tag{4.26}$$

将求解域 Ω 划分为若干个单元,并选择单元体积 Ω_e 作为积分域,将式 (4.26) 的近似解和检验函数代入弱形式方程 (4.25) 中,则单元弱形式积分方程为

$$\int_{\Omega_e} \boldsymbol{N}^{\mathrm{T}}(\boldsymbol{x})\rho\boldsymbol{N}(\boldsymbol{x})\ddot{\boldsymbol{d}}\mathrm{d}V_{\boldsymbol{x}} = \int_{\Omega_e} \boldsymbol{N}^{\mathrm{T}}(\boldsymbol{x}) \left(\int_{H_{\boldsymbol{x}}} \boldsymbol{f}(\boldsymbol{\eta},\boldsymbol{\xi})\mathrm{d}V_{\boldsymbol{x}'}\right) \mathrm{d}V_{\boldsymbol{x}} + \int_{\Omega_e} \boldsymbol{N}^{\mathrm{T}}(\boldsymbol{x})\boldsymbol{b}(\boldsymbol{x})\mathrm{d}V_{\boldsymbol{x}} \tag{4.27}$$

4.2.3 空间离散与矩阵装配

如图 4-6 所示,不同于传统有限元采用连续网格离散构型 (相邻单元间共享节点),构造非连续 Galerkin 有限元的解空间需采用非连续网格离散构型,相邻单元间不共用节点,则节点总数等于所有单元的节点数之和。建模时可先生成连续网格,再通过复制共享节点的方法生成非连续单元网格。

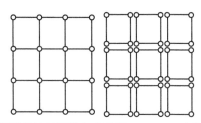

图 4-6 传统有限元的连续网格与非连续 Galerkin 有限元的非连续网格

将式 (4.27) 离散化并进行高斯积分,即可得到离散形式的近场动力学非连续伽辽金有限元法的弱形式方程

$$\sum_{g=1}^{ng} \boldsymbol{N}^{\mathrm{T}}(\boldsymbol{x}^g)\rho\boldsymbol{N}(\boldsymbol{x}^g)\ddot{\boldsymbol{d}}V^g = \sum_{g=1}^{ng} \boldsymbol{N}^{\mathrm{T}}(\boldsymbol{x}^g) \left(\sum_{g'=1}^{ng'} \boldsymbol{f}(\boldsymbol{\eta}(\boldsymbol{x}^g,\boldsymbol{x}^{g'}),\boldsymbol{\xi}(\boldsymbol{x}^g,\boldsymbol{x}^{g'}))V^{g'}\right) V^g$$

$$+ \sum_{g=1}^{ng} \boldsymbol{N}^{\mathrm{T}}(\boldsymbol{x}^g)\boldsymbol{b}(\boldsymbol{x}^g)V^g \tag{4.28}$$

式中,$\sum_{g'=1}^{ng'} \boldsymbol{f}(\boldsymbol{\eta}(\boldsymbol{x}^g,\boldsymbol{x}^{g'}),\boldsymbol{\xi}(\boldsymbol{x}^g,\boldsymbol{x}^{g'}))V^{g'}$ 为式 (4.27) 右端第一项的第一重积分求和,反映了近场域内所有高斯点的相互作用,也体现高斯点间存在跨越单元的非局部

作用效应，并可通过高斯点间键的断开实现损伤开裂模拟，如图 4-7 所示。外层积分的积分域为单元自身，表征单元行为，进而可扩展到在整个构型积分，以表征系统的力学行为。形状函数矩阵 $\boldsymbol{N}$ 和单元节点位移列阵 $\boldsymbol{d}$ 如公式 (4.21) 所示，ng 为单元 Ω_e 的高斯点总数，ng' 是高斯点 $\boldsymbol{x}^g$ 的近场域内的高斯点数，V^g 是高斯点 $\boldsymbol{x}^g$ 对应的体积，$\boldsymbol{\xi}(\boldsymbol{x}^g,\boldsymbol{x}^{g'})$ 是高斯点 $\boldsymbol{x}^g$ 与其近场域内的高斯点 $\boldsymbol{x}^{g'}$ 之间键的相对位置，$\boldsymbol{\eta}(x^g,x^{g'})$ 为两高斯点间的相对位移。

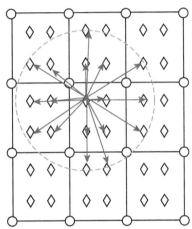

图 4-7　近场范围内的高斯点间存在非局部相互作用

类似于有限元法，将式 (4.28) 中的相关项记为单元质量矩阵 $\boldsymbol{M}^e$ 和单元荷载列阵 $\boldsymbol{F}^e$，具体为

$$\begin{cases} \boldsymbol{M}^e = \sum_{i=1}^{ng} \rho \boldsymbol{N}^{\mathrm{T}}(\boldsymbol{x}^g)\boldsymbol{N}(\boldsymbol{x}^g)V^g \\ \boldsymbol{F}^e = \sum_{i=1}^{ng} \boldsymbol{N}^{\mathrm{T}}(\boldsymbol{x}^g)bV^g \end{cases} \quad (4.29)$$

对于边界单元，单元荷载列阵采用如下形式

$$\boldsymbol{F}^e = \sum_{i=1}^{ng} \boldsymbol{N}^{\mathrm{T}}(\boldsymbol{x}^g)bV^g + \sum_{i=1}^{ng} \boldsymbol{N}^{\mathrm{T}}(\boldsymbol{x}^g)\bar{\boldsymbol{T}}S_{\boldsymbol{x}} \quad (4.30)$$

其中，$\bar{\boldsymbol{T}}$ 为边界上受到的面力矢量，$S_{\boldsymbol{x}}$ 为单元表面积。需要指出的是，近场动力学的自然边界 (应力边界) 条件的施加不同于传统有限元方法，需要转化为体力密度施加于边界物质点上，此处通过形状函数插值实现单元表面面力密度到高斯点的体力密度的转换，并保持有限元格式，继承了有限元法能直接施加局部边界条件的特点。

同样地，引入与式 (4.28) 中右端第二项有关的单元劲度矩阵 $\boldsymbol{K}^e$，可根据高斯点受其近场域内其他物质点的键力进行计算，并组装而成，具体说明如下。

为简便计，高斯点 $\boldsymbol{x}^g$ 的初始位置矢量也记为 $\boldsymbol{x}^g$，这个位置矢量和位移矢量 $\boldsymbol{u}^g$ 可由所属单元的形状函数插值得到，分别为

$$\boldsymbol{x}^g = \boldsymbol{N}(\boldsymbol{x}^g)\boldsymbol{X}, \ \boldsymbol{u}^g = \boldsymbol{N}(\boldsymbol{x}^g)\boldsymbol{d} \tag{4.31}$$

其中，$\boldsymbol{X}$ 为该高斯点所在单元的节点坐标列阵，即 $\boldsymbol{X} = [X_{1(1)} \quad X_{1(2)} \quad X_{1(3)} \quad \cdots \quad X_{n(1)} \quad X_{n(2)} \quad X_{n(3)}]_{3n \times 1}^{\mathrm{T}}$。高斯点的相对位置和相对位移分别为

$$\begin{cases} \boldsymbol{\xi}(\boldsymbol{x}^g, \boldsymbol{x}^{g'}) = \boldsymbol{x}^{g'} - \boldsymbol{x}^g = \boldsymbol{N}'(\boldsymbol{x}^{g'})\boldsymbol{X}' - \boldsymbol{N}(\boldsymbol{x}^g)\boldsymbol{X} \\ \boldsymbol{\eta}(\boldsymbol{x}^g, \boldsymbol{x}^{g'}) = \boldsymbol{u}^{g'} - \boldsymbol{u}^g = \boldsymbol{N}'(\boldsymbol{x}^{g'})\boldsymbol{d}' - \boldsymbol{N}(\boldsymbol{x}^g)\boldsymbol{d} \end{cases} \tag{4.32}$$

进而，以 i $(i = 1, 2, \cdots, n)$ 指示高斯点 $\boldsymbol{x}^g$、j $(j = 1, 2, \cdots, ng')$ 指示高斯点 $\boldsymbol{x}^{g'}$，采用式 (4.21) 的形状函数矩阵，将相对位移改写为

$$\boldsymbol{\eta}_{ji}(\boldsymbol{x}^g, \boldsymbol{x}^{g'}) = \begin{bmatrix} -N_1^i & 0 & 0 & \cdots & N_1^j & 0 & 0 & \cdots \\ 0 & -N_1^i & 0 & \cdots & 0 & N_1^j & 0 & \cdots \\ 0 & 0 & -N_1^i & \cdots & 0 & 0 & N_1^j & \cdots \end{bmatrix}_{3 \times 6n} \begin{bmatrix} u_{1(1)}^i \\ u_{1(2)}^i \\ u_{1(3)}^i \\ \vdots \\ u_{1(1)}^j \\ u_{1(2)}^j \\ u_{1(3)}^j \\ \vdots \end{bmatrix}_{6n \times 1}$$

$$= \begin{bmatrix} -\boldsymbol{N}^i & \boldsymbol{N}^j \end{bmatrix} \begin{bmatrix} \boldsymbol{d}^i & \boldsymbol{d}^j \end{bmatrix}^{\mathrm{T}} \tag{4.33}$$

其中，n 为单元节点数，N_1^i 表示第 i 个高斯点所属单元第一个节点的形状函数，$u_{1(1)}^i$ 表示第 i 个高斯点所属单元第一个节点的第一个自由度位移，其他物理量的含义以此类推；$\boldsymbol{N}^i$ 和 $\boldsymbol{N}^j$ 分别为高斯点 i 和 j 所属单元的形状函数矩阵，$\boldsymbol{d}^i$ 和 $\boldsymbol{d}^j$ 分别为高斯点 i 和 j 所属单元的节点位移列阵。

第 3 章给出了 PMB 模型线性化的键力密度函数 $\boldsymbol{f}_{ji} = \dfrac{\chi c}{|\boldsymbol{\xi}_{ji}|}\boldsymbol{I} \cdot (\boldsymbol{u}_j^g - \boldsymbol{u}_i^g)$，本章 4.1 节给出了整体坐标下两个物质点之间键的劲度矩阵，根据式 (4.2) 和式 (4.9)，可获得高斯点 $\boldsymbol{x}_i^g$ 受高斯点 $\boldsymbol{x}_j^g$ 作用的键力矩阵形式表达式

$$\boldsymbol{f}_{ji} V_j^g = [\boldsymbol{T}_{ji}] \left[\dfrac{\chi c}{|\boldsymbol{\xi}_{ji}|}\boldsymbol{\eta}_{ji}\right] V_j^g = \dfrac{\chi c V_j^g}{|\boldsymbol{\xi}_{ji}|} [\boldsymbol{T}_{ji}] [\boldsymbol{\eta}_{ji}]$$

4.2 近场动力学弱形式方程的非连续伽辽金有限元法

$$= \frac{\chi c V_j^g}{|\boldsymbol{\xi}_{ji}|} [\boldsymbol{T}_{ji}] \begin{bmatrix} -\boldsymbol{N}^i & \boldsymbol{N}^j \end{bmatrix} \begin{bmatrix} \boldsymbol{d}^i & \boldsymbol{d}^j \end{bmatrix}$$

$$= \left[\boldsymbol{K}_{ji}^{\boldsymbol{\xi}}\right]_{3\times 6n} \begin{bmatrix} \boldsymbol{d}^i & \boldsymbol{d}^j \end{bmatrix}_{6n\times 1}, \ (i=1,2,\cdots,n, j=1,2,\cdots,ng') \quad (4.34)$$

式中,坐标变换矩阵为 $[\boldsymbol{T}_{ji}] = \begin{bmatrix} l^2 & lm & ln \\ lm & m^2 & mn \\ ln & mn & n^2 \end{bmatrix}$,键的劲度矩阵为 $[\boldsymbol{K}_{ji}^{\boldsymbol{\xi}}] = \frac{\chi c V_j^g}{|\boldsymbol{\xi}_{ji}|} [\boldsymbol{T}_{ji}] \begin{bmatrix} -\boldsymbol{N}^i & \boldsymbol{N}^j \end{bmatrix}$。于是,式 (4.28) 的右端第一项可重写为

$$\sum_{g=1}^{ng} \boldsymbol{N}^{\mathrm{T}}(\boldsymbol{x}^g) \left(\sum_{g'=1}^{ng'} \boldsymbol{f}(\boldsymbol{\eta}(\boldsymbol{x}^g,\boldsymbol{x}^{g'}), \boldsymbol{\xi}(\boldsymbol{x}^g,\boldsymbol{x}^{g'})) V^{g'} \right) V^g$$

$$= \sum_{i=1}^{ng} \left(\sum_{j=1}^{ng'} \left[\boldsymbol{N}^i\right]^{\mathrm{T}} \left[\boldsymbol{K}_{ji}^{\boldsymbol{\xi}}\right] \begin{bmatrix} \boldsymbol{d}^i & \boldsymbol{d}^j \end{bmatrix} \right) V^g \quad (4.35)$$

至此,根据式 (4.31)~ 式 (4.35) 的推导,对高斯点 $i=1,2,\cdots,n$ 的近场范围内的所有高斯点 $j=1,2,\cdots,ng'$ 进行两重循环求和,可以获得单元劲度矩阵,再按总体节点编号进行组装,即可形成整体劲度矩阵 $[\boldsymbol{K}]_{3N\times 3N}$,$N$ 为系统节点总数。同时,对单元质量矩阵与单元载荷列阵进行整体组装,得到整个系统的运动方程为

$$[\boldsymbol{M}]_{3N\times 3N} \left[\ddot{\boldsymbol{U}}(t)\right]_{3N\times 1} = [\boldsymbol{K}]_{3N\times 3N} [\boldsymbol{U}(t)]_{3N\times 1} + [\boldsymbol{F}(t)]_{3N\times 1} \quad (4.36)$$

相应的静力平衡方程为

$$[\boldsymbol{K}]_{3N\times 3N} [\boldsymbol{U}]_{3N\times 1} + [\boldsymbol{F}]_{3N\times 1} = \boldsymbol{0} \quad (4.37)$$

施加边界条件后即可求解系统方程。边界条件可按有限元法的标准形式施加,对于应力边界条件,面力通过形状函数直接转化为等效节点载荷。

对于位移边界条件,可将其表示为矩阵形式 $\boldsymbol{GU} + \boldsymbol{U}^* = \boldsymbol{0}$,引入拉格朗日乘子 $\boldsymbol{\lambda}$,对系统平衡方程和约束方程的混合变分形式进行重构,得到求解未知量 $\boldsymbol{U}$ 和 $\boldsymbol{\lambda}$ 的代数方程组为

$$\begin{bmatrix} \boldsymbol{K} & \boldsymbol{G}^T \\ \boldsymbol{G} & \boldsymbol{0} \end{bmatrix} \begin{Bmatrix} \boldsymbol{U} \\ \boldsymbol{\lambda} \end{Bmatrix} = -\begin{Bmatrix} \boldsymbol{F} \\ \boldsymbol{U}^* \end{Bmatrix} \quad (4.38)$$

式中，G 为约束方程系数矩阵，与未知位移向量 U 相关，U^* 为已知位移约束值构成的列阵。上式可采用 MKL 库提供的大型稀疏方程组求解器 PARDISO 或 GMRES 等进行求解。

4.3 键型近场动力学非连续伽辽金有限元法在 LS-DYNA 软件中的实现

DYNA 程序最初由美国 Lawrence Livermore National Lab 的 John O. Hallquist 博士于 1976 年开发完成，主要用于求解高速碰撞和爆炸冲击作用下三维非弹性结构的大变形动力响应。后经不断的功能扩充和改进，形成 LS-DYNA 软件，现已成为国际著名的非线性动力分析软件，并于 2019 年被 ANSYS 软件公司收购。LS-DYNA 以 Lagrange 算法为主，兼有 ALE 和 Euler 算法；以显式求解为主，兼有隐式求解功能；以非线性动力分析为主，兼有静力分析功能；以结构分析为主，兼有热分析、流体–结构耦合分析功能。LS-DYNA 作为显式瞬态动力分析软件的权威，加上其开放的结构体系，很多公司为其开发了通用的前后处理器，如：LSTC 公司开发的预处理器 LS-PrePost、ANSYS 公司针对 LS-DYNA 开发的前后处理器、ALTAIR 公司开发的 Hypermesh 以及 XYZ Scientific 公司开发的 TrueGrid 等。

R10 以上版本的 LS-DYNA 软件中包含了多种固体结构分析的高性能有限元和无网格方法，如：Smoothed Particle Galerkin Method (SPGM) 和键型近场动力学的非连续 Galerkin 有限元方法等。键型近场动力学的非连续 Galerkin 方法根据临界能量释放率判定失效，裂纹扩展模拟过程中无需删除单元，且能够适应非均匀网格，并可以直接施加边界条件和约束。2015 年 6 月，该方法被植入 LS-DYNA 中，研究员们相继开发了适用于脆性材料破坏[21]和纤维增强复合材料破坏[22]的计算模拟模块，并用于挡风玻璃和塑料嵌板等的冲击损伤分析，用户可以在计算模型参数主控文件 (K 文件) 中定义 PD 相关关键字，开展计算分析。

下面具体介绍相关关键字的含义和相应参数卡的填写。

1) *SECTION_SOLID_PERI

该关键字包含两个参数卡，见表 4-1 和表 4-2。

表 4-1 参数卡 1

变量	SECID	ELEFORM
类型	整型 I	整型 I
缺省值		

表 4-2　参数卡 2

变量	DR	PTYPE
类型	浮点型 F	整型 I
缺省值	1.01	1

表 4-1 参数卡用于定义计算模型采用的单元算法。其中，SECID 表示部件号，ELEFORM 表示单元公式号，取值为 48，48 号算法在 LS-DYNA 关键字手册中表示近场动力学模型算法，算法支持四节点、六节点和八节点实体单元建模。标有 Type 的行给出了变量类型，F 表示浮点数，I 表示整数，如果指定了 0 或者该卡片留空，则表示变量将采用 Default 指定的默认值。

表 4-2 参数卡用于定义使用近场动力学模型的类型和近场范围尺寸。其中，PTYPE 表示近场动力学模型类型：1 表示键型近场动力学模型 (目前已导入软件)，2 表示态型近场动力学 (已通过测试)。DR 为归一化的邻域大小，表示近场域大小与单元最长对角线长度的比值，如图 4-8 所示，一般推荐使用 1.0。实际上，DR 为用户自定义的物质点近场域大小 (推荐值为 $0.6 \leqslant DR \leqslant 1.2$)，如果用户定义的网格极端不规则，LS-DYNA 将自动调整 DR 值的大小，使每个材料点的相邻材料点数为 $10 \leqslant ng' \leqslant 136$。

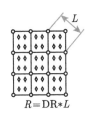

图 4-8　参数 DR 的定义

2) *MAT_ELASTIC_PERI

该关键字用于定义材料模型和参数，目前，LS-DYNA 中只引入了键型 PD 的微弹性脆性材料模型，此关键字包含一个参数卡，见表 4-3。

表 4-3　参数卡 1

变量	MID	RO	E	G_T	G_S
类型	整型 I	浮点型 F	浮点型 F	浮点型 F	浮点型 F
缺省值				1.0E20	1.0E20

表 4-3 中，MID 为用户定义的材料号，RO 表示材料密度，E 表示弹性模量，G_T 表示拉伸状态下的断裂能量释放率，主要针对脆性材料 (玻璃、水泥和硬塑料等)，G_S 表示压缩状态下的断裂能量释放率，该参数值由人为设定，对大多数以

压缩为主的问题，$G_S = 2.0G_T$，对于其他问题，直接采用缺省值即可。泊松比为 0.25，故予以缺省。

3)*MAT_ELASTIC_PERI_LAMINATE

该关键字用于定义层合复合材料模型（一个铺层的横观各向同性弹性材料），仅适用于近场动力学层压板模型（peridynamic laminate model），是一个二维材料模型，此关键字包含两个参数卡，见表 4-4 和表 4-5。

表 4-4 参数卡 1

变量	MID	RO	Ef	Em	vfm	FOPT	FCf	FCm
类型	整型 I	浮点型 F	浮点型 F	浮点型 F	浮点型 F	整型 I	浮点型 F	浮点型 F
缺省值								

表 4-5 参数卡 2

变量	V1	V2	V3
类型	浮点型 F	浮点型 F	浮点型 F
缺省值			

表 4-4 中的 MID 和 RO 含义同表 4-3，Ef 表示薄层纤维轴向的弹性模量 E_1，Em 表示横向弹性模量 E_2，vfm 表示薄层面内泊松比 ν_{12}，FOPT 表示失效准则类型，取值 1 为能量释放率，取值 2 为拉伸破坏伸长比，FCf 表示轴向失效准则，FCm 表示横向失效准则。通常推荐采用能量释放率破坏准则。表 4-5 中的 V1、V2、V3 表示局部正交的材料主轴，由单元面内矢量与单元法向量叉乘确定。

4)*SET_PERI_LAMINATE

该关键字将由 *MAT_ELASTIC_PERI_LAMINATE 定义的单层复合材料薄层/薄板组装为层合复合结构，包含 2 个以上参数卡，见表 4-6 和表 4-7。

表 4-6 参数卡 1

变量	SID
类型	整型 I
缺省值	

表 4-7 参数卡 2

变量	PID1	A1	T1	PID2	A2	T2
类型	整型 I	浮点型 F	浮点型 F	整型 I	浮点型 F	浮点型 F
缺省值						

表 4-6 中的 SID 表示 Set 部件号。表 4-7 中的 PID1 和 PID2 表示第一铺层和第二铺层的部件号；A1 和 A2 表示相应铺层中的纤维角度，0 度纤维角度由表

4-5 中关键字 (V1,V2,V3) 定义；T1 和 T2 表示单层铺层厚度。如果有 n 个铺层，则以此类推，在参数卡 2 中可以定义 n 组 PID*、A*、T*。

5)*ELEMENT_SOLID_PERI

不同于常规三维实体单元，近场动力学层合板模型中的单元是三节点或四节点表面单元 (surface element)，该关键字包含 1 个参数卡，见表 4-8。

表 4-8　参数卡 1

变量	EID	PID	n1	n2	n3	n4
类型	整型 I	整型 I	整型 I	整型 I	整型 I	整型 I
缺省值						

表 4-8 中，EID 表示单元编号，PID 表示部件编号，n1、n2、n3 和 n4 表示单元节点编号，近场动力学层合板模型采用四节点网格，几何形状为壳单元形状，但不采用壳单元的模型公式。

4.4　数值算例分析

4.4.1　基于自编程序的近场动力学杆单元法算例

如图 4-9 所示，考虑一受水平均匀单轴拉伸的各向同性矩形板，发生静力弹性变形，采用线弹性 BB PD 模型的隐式静力程序求解该问题。板的长度、宽度和厚度分别为 L=1m、W=0.5m 和 h=0.01m；材料的杨氏弹性模量为 E=200GPa，泊松比为 ν=1/3，质量密度为 ρ=7850kg/m^3；上下水平边界自由，水平方向拉伸最大位移值为 d=0.0005m，线性变化施加在左右竖直边界上，边界层数与近场范围半径一致。x_1、x_2 和 x_3 方向分别离散为 100、50 和 1 个物质点，共 5000 个物质点，物质点间距 $|\Delta x|=0.01$m，圆形近场范围半径固定为 $\delta=m|\Delta x|=0.03$m，影响函数选取 $w(|\boldsymbol{\xi}|)=1$。

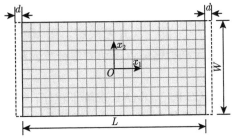

图 4-9　矩形板单轴拉伸变形的计算模型

图 4-10 给出了 BB PD 模型计算得到的水平位移和竖向位移云图，图中上层和下层分别给出未做表面修正和进行表面修正的计算结果。以竖向最大位移的

解析解 $d_y^{\max} = \dfrac{W}{L} v d_x^{\max} = 8.3333 \times 10^{-5}\text{m}$ 为基准比较计算误差, 未作表面修正的 BB PD 计算最大值为 $8.5403 \times 10^{-5}\text{m}$, 相对误差为 2.484%, 进行表面修正的 BB PD 计算最大值为 $8.3701 \times 10^{-5}\text{m}$, 相对误差为 0.4416%。计算结果表明, 键型近场动力学弹性模型的强形式求解器可以准确分析板结构的弹性变形问题, 且表面修正技术能明显降低计算误差。

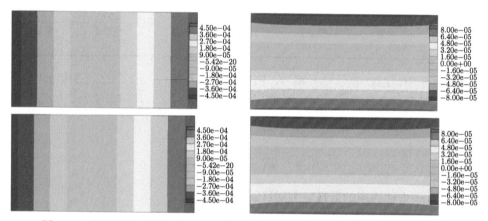

图 4-10 矩形板的水平位移和竖向位移云图: (上) 无表面修正; (下) 表面修正

4.4.2 基于 ABAQUS 二次开发的近场动力学杆单元法算例

仍以上述矩形板为研究对象, 几何尺寸与 4.4.1 节中相同, 材料的杨氏弹性模量为 $E=20\text{GPa}$, 泊松比为 $\nu=1/3$; 边界条件有所区别, 板左端受固定约束, 右端施加均布拉力 $P = 1 \times 10^5 \text{N}/\text{m}$, 两类边界条件均施加在边界一层的物质点上, 如图 4-11 所示。x_1、x_2 和 x_3 方向分别离散为 100、50 和 1 个物质点, 共 5000 个物质点, 物质点间距 $|\Delta x| = 0.01\text{m}$, 圆形近场范围半径固定为 $\delta = m|\Delta x| = 0.03\text{m}$, 共有 69396 个杆单元, 影响函数选取 $w(|\boldsymbol{\xi}|) = 1$。通过 ABAQUS 软件中的 VUMAT 材料子程序, 实现了 BB PD 的二次开发求解。图 4-12 给出了基于 CPS4 平面四节点线性单元的有限元法和键型近场动力学杆单元法的计算结果, 二者吻合良好。

进一步, 采用 ABAQUS 的 VUMAT 求解子程序, 模拟了含裂纹板的拟静力开裂问题。板几何参数和边界条件同上, 在底部预置切口, 切口长 $a=20\text{mm}$, 如图 4-13 所示。其他材料参数同上例, 质量密度为 $\rho = 5 \times 10^9 \text{ kg/m}^3$, 以使用大时间步长, 临界伸长率为 $s_0 = 2 \times 10^{-5}$, 计算总时长为 4×10^{-2} s。图 4-14 和图 4-15 分别给出了裂纹扩展过程中的水平位移和裂纹路径, 较好地模拟了裂纹扩展问题。

4.4 数值算例分析

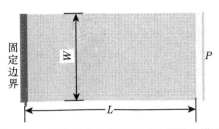

图 4-11　矩形板单轴拉伸变形的杆单元模型

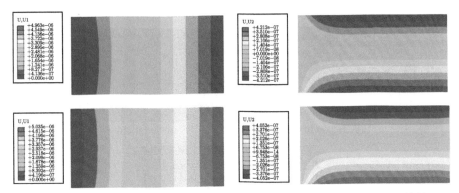

图 4-12　矩形板的水平和竖向位移云图：(上)FEM 结果；(下)BB PD 结果

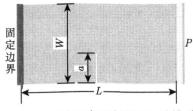

图 4-13　含切口矩形板单轴拉伸模型

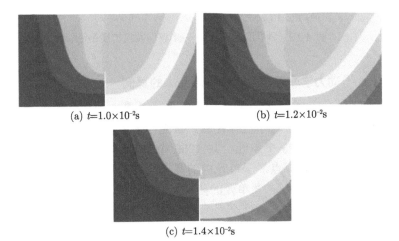

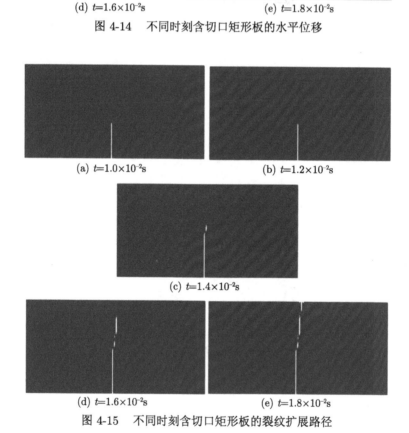

(d) $t=1.6\times10^{-2}$s   (e) $t=1.8\times10^{-2}$s

图 4-14   不同时刻含切口矩形板的水平位移

(a) $t=1.0\times10^{-2}$s   (b) $t=1.2\times10^{-2}$s

(c) $t=1.4\times10^{-2}$s

(d) $t=1.6\times10^{-2}$s   (e) $t=1.8\times10^{-2}$s

图 4-15   不同时刻含切口矩形板的裂纹扩展路径

4.4.3   基于近场动力学非连续伽辽金有限元法的算例

仍以 4.4.1 节的二维矩形板静力变形算例为基准，基于自编的键型 PD 非连续 Galerkin 有限元方法程序进行计算。首先，采用传统四边形单元划分网格，x、y 方向各划分 50 和 25 个单元，共 1250 个单元，单元边长为 0.02m。每个单元内部采用 2×2 个高斯点，子单元内的高斯点位置即作为 PD 方法中的物质点，根据有限元平面四节点等参单元的高斯点位置，物质点的间距约为单元尺寸的 1/2，即 $|\Delta x| \approx 0.01$m，材料参数等与 4.4.1 节算例相同，近场范围半径为 $\delta = m|\Delta x| \approx 0.03$m。进而，复制各单元的共享节点并重新编号，形成新的非连

续网格，提交计算。键型近场动力学非连续 Galerkin 有限元方法计算得到的水平位移和竖向位移如图 4-16 所示，变形特征与解析解一致，量值除边界处因表面效应误差较大外，其他处误差较小。

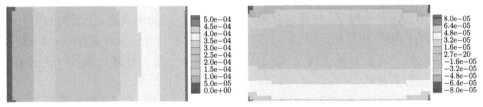

图 4-16 矩形板的水平位移和竖向位移云图

4.4.4 基于 LS-DYNA 的混凝土板爆炸冲击毁伤模拟

考虑一混凝土板受爆炸荷载作用，板的长度和宽度均为 1m，厚度为 40mm，爆炸中心距离混凝土板上表面中心 0.4m，TNT 炸药当量为 0.15kg。材料参数为：混凝土密度 $\rho = 2750\text{kg/m}^3$，弹性模量 $E = 38.2\text{GPa}$，临界断裂能释放率 $G_T = 120\text{J/m}^2$，泊松比为 0.25。图 4-17 为研究两对边夹持和四边夹持两种约束条件下混凝土板的毁伤破坏问题的几何模型。

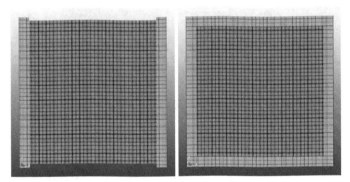

图 4-17 LS-ProPost 显示的两对边夹持和四边夹持约束的混凝土板几何模型

采用 TrueGrid 网格划分软件进行几何建模[23,24]，沿混凝土板长、宽、厚度方向网格尺寸分别为 10mm、10mm 和 8mm，基于 TrueGrid 软件逐个生成单元以形成非连续网格。采用 LS-DYNA 软件中集成的键型近场动力学非连续 Galerkin 有限元求解器进行混凝土板的近场空中爆炸毁伤模拟，通过 *MAT_ELASTIC_PERI 定义混凝土材料模型，并使用命令 lsdyna keyword 生成可被 LS-DYNA 软件识别的网格模型文件 trugrdo。

为实现夹持约束效果，采用固定边框进行边界约束建模，使用 *SECTION_SOLID 和 *MAT_RIGID 两个关键字进行定义。*SECTION_SOLID 关键字用于定义边框采用的单元算法，取值 elform=1(常应力体单元算法)；*MAT_RIGID 关键字用于定义边框为刚体，不发生破坏。*MAT_20 材料关键字用于定义边框的材料参数。此外，使用 *CONTACT_AUTOMATIC_SURFACE_TO_SURFACE 关键字，将混凝土板与边框定义为自动面面接触，无需做额外设定。随后，通过 *LOAD_BLAST 关键字，将爆炸载荷加载到混凝土板的上表面，上表面的所有单元表面构造为爆炸载荷加载面。最后，将生成的网格模型 trugrdo 文件和 main.k(K 文件) 放入同一工作目录，提交 LS-DYNA 求解器计算。计算完成后，使用 LS-ProPost 进行后处理。

以炸药起爆时刻为零点，图 4-18 和图 4-19 分别给出了不同约束条件下，混凝土板受爆炸冲击载荷作用后不同时刻的毁伤情况。

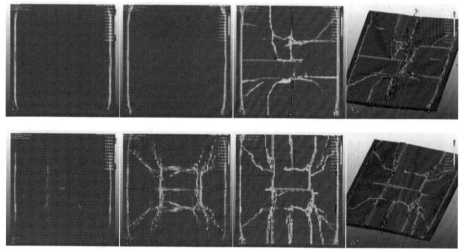

图 4-18　起爆后不同时刻两对边夹持混凝土板迎爆面 (上) 与背爆面 (下) 的毁伤情况 (自左到右依次为 0.9ms、1.5ms、5.0ms、11.5ms)

对于两对边夹持情况，在 0.9ms 时刻，混凝土板迎爆面和背爆面的左右两边固定边界内缘处均出现裂纹，迎爆面的裂纹更长，已逐渐延伸至混凝土板的上下边缘处，但背爆面中心处开始出现不规则裂纹。在 1.5ms 时刻，混凝土板背爆面左右两边裂纹扩展至板的上下边缘，中心出现较大范围的环向裂纹，并向边界方向扩展；相较于背爆面，迎爆面上的破损现象没有明显变化，仅固定边界内缘处的裂纹进一步扩展，并出现一定程度的偏折。在 5.0ms 时刻，混凝土板迎爆面出现较多的裂纹，垂直约束方向出现了横向裂纹，中部沿约束方向出现贯穿裂纹；背爆面裂纹继续向四周扩展至边界。在 11.5ms 时刻，混凝土板出现较多的贯穿裂纹，在中部和约束内缘处出现断裂现象，混凝土板脱离左右边框的固定约束，整体从中间发生断裂破坏。

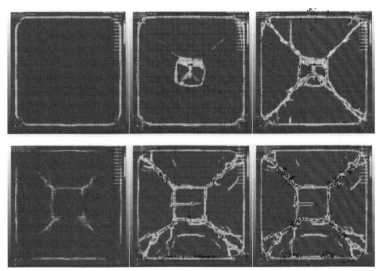

图 4-19　起爆后不同时刻四边夹持混凝土板迎爆面（上）与背爆面（下）的毁伤情况（自左到右依次为 0.9ms、5.0ms、11.5ms）

对于四边夹持情况，在 0.9ms 时刻，混凝土板迎爆面裂纹相较于两边固定板更为均匀，在边框内缘处形成了明显的环状裂纹角；背爆面边框内缘处出现环向裂纹，中心处出现环向裂纹并开始向四周扩展。在 5.0ms 时刻，混凝土板迎爆面中心出现环形裂纹，四边内缘处的裂纹也有微小扩展；背爆面开裂显著，中心处的环形裂纹沿对角线扩展，并与边缘裂纹贯通，出现多条放射状裂纹，背爆面的毁伤情况要比迎爆面更为严重，反映了反射拉伸波的作用效果。在 11.5ms 时刻，混凝土板的毁伤情况进一步恶化，迎爆面出现较大范围破碎现象，且有向外飞溅的趋势，中间环形裂纹沿对角线扩展贯通至边缘裂纹处；背爆面形成了对角贯穿裂纹，且有碎片向外飞溅。相比于两边固定混凝土板，四周固定的混凝土板受力更为均匀，虽然板的破损程度更为严重，且有碎片向外飞溅现象，但四周固定混凝土板并未从中间断裂，能保持更好的稳定性。

参 考 文 献

[1] Macek R W, Silling S A. Peridynamics via finite element analysis [J]. Finite Elements in Analysis and Design, 2007, 43(15):1169-1178.

[2] Huang D, Lu G D, Qiao P Z. An improved peridynamic approach for quasi-static elastic deformation and brittle fracture analysis [J]. International Journal of Mechanical Sciences, 2015, 94:111-122.

[3] Kilic B, Madenci E. An adaptive dynamic relaxation method for quasi-static simulations using the peridynamic theory [J]. Theoretical and Applied Fracture Mechanics, 2010, 53(3):194-204.

[4] Lall P, Shantaram S, Locker D. Reliability modeling of electronic systems subjected to high strain rates [C]. 13th International Thermal, Mechanical and Multi-Physics Simulation and Experiments in Microelectronics and Microsystems, IEEE, 2012:1-17.

[5] Beckmann R, Mella R, Wenman M R. Mesh and timestep sensitivity of fracture from thermal strains using peridynamics implemented in Abaqus [J]. Computer Methods in Applied Mechanics and Engineering, 2013, 263:71-80.

[6] Yolum U, Taştan A, Güler M A. A peridynamic model for ductile fracture of moderately thick plates [J]. Procedia Structural Integrity, 2016, 2:3713-3720.

[7] 卢广达, 夏晓舟. 近场动力学 Peri_Dynamics 软件 (简称：Peri_Dynamics V1.0). 软件著作权登记号 2016SR030649. 2016.

[8] 李天一, 章青, 夏晓舟, 等. 考虑混凝土材料非均质特性的近场动力学模型 [J]. 应用数学和力学, 2018, 39(8):913-924.

[9] Li W J, Guo L. Meso-fracture simulation of cracking process in concrete incorporating three-phase characteristics by peridynamic method [J]. Construction and Building Materials, 2018, 161:665-675.

[10] Madenci E, Dorduncu M, Barut A, et al. A state-based peridynamic analysis in a finite element framework [J]. Engineering Fracture Mechanics, 2018, 195:104-128.

[11] Huang X H, Bie Z W, Wang L, et al. Finite element method of bond-based peridynamics and its ABAQUS implementation [J]. Engineering Fracture Mechanics, 2019, 206:408-426.

[12] 别业辉. 近场动力学与连续介质力学耦合算法的研究 [D]. 长沙：湖南大学，2019.

[13] Bie Y H, Li S, Hu X, et al. An implicit dual-based approach to couple peridynamics with classical continuum mechanics [J]. International Journal for Numerical Methods in Engineering, 2019, 120(12):1349-1379.

[14] Chen X, Gunzburger M. Continuous and discontinuous finite element methods for a peridynamics model of mechanics [J]. Computer Methods in Applied Mechanics and Engineering, 2011, 200(9-12):1237-1250.

[15] Aksoy H G, Şenocak E. Discontinuous Galerkin method based on peridynamic theory for linear elasticity [J]. International Journal for Numerical Methods in Engineering, 2011, 88(7):673-692.

[16] Lubineau G, Azdoud Y, Han F, et al. A morphing strategy to couple non-local to local continuum mechanics [J]. Journal of the Mechanics and Physics of Solids, 2012, 60(6):1088-1102.

[17] Azdoud Y. A hybrid local/non-local framework for the simulation of damage and fracture [D]. Thuwal: King Abdullah University of Science and Technology, 2014.

[18] Han F, Lubineau G, Azdoud Y. Adaptive coupling between damage mechanics and peridynamics: a route for objective simulation of material degradation up to complete failure [J]. Journal of the Mechanics and Physics of Solids, 2016, 94: 453-472.

[19] Tian X C, Du Q. Nonconforming discontinuous Galerkin methods for nonlocal variational problems [J]. SIAM Journal on Numerical Analysis, 2015, 53(2):762-781.

[20] Xu F F, Gunzburger M, Burkardt J. A multiscale method for nonlocal mechanics and diffusion and for the approximation of discontinuous functions [J]. Computer Methods in Applied Mechanics and Engineering, 2016, 307:117-143.

[21] Ren B, Wu C T, Askari E. A 3D discontinuous Galerkin finite element method with the bond-based peridynamics model for dynamic brittle failure analysis [J]. International Journal of Impact Engineering, 2017, 99:14-25.

[22] Ren B, Wu C T, Seleson P, et al. A peridynamic failure analysis of fiber-reinforced composite laminates using finite element discontinuous Galerkin approximations [J]. International Journal of Fracture, 2018, 214(1):49-68.

[23] 辛春亮, 薛再清, 涂建, 等. TrueGrid 和 LS-DYNA 动力学数值计算详解 [M]. 北京: 机械工业出版社, 2019.

[24] 辛春亮, 涂建, 王俊林, 等. 由浅入深精通 LS-DYNA [M]. 北京: 中国水利水电出版社, 2019.

第 5 章　近场动力学的显式动力学解法

近场动力学运动方程通常无法得到解析解，需要进行空间域和时间域的离散，并采用数值积分技术获得相应数值解，近场动力学数值算法具有分子动力学方法、无网格粒子类方法和有限单元法的特点。根据空间域的离散方式，数值积分可基于粒子类无网格的单点积分实现[1,2]或基于网格的多高斯点积分实现[3-5]，按照显式/隐式求解格式，有显式直接积分法[1,2]或隐式非局部有限元求解方法[6-8]。其中，显式动力无网格粒子方法便于描述材料和结构的动力破坏问题，且引入阻尼项的拟静力法或动态松弛法可以分析拟静力问题[3,4,9]。本章主要介绍近场动力学的显式动力无网格粒子法的算法和程序实现细节，包括：空间离散、时间离散、体积修正、数值稳定性和算法流程图，给出几个算例和相应的 FORTRAN 源程序，最后介绍近场动力学开源软件 PDLAMMPS 的安装使用。

5.1　控制方程的空间离散

如图 5-1 所示，采用无网格粒子法求解近场动力学运动方程时，常将结构离散为中心正交的均匀粒子点阵，且近场范围尺寸取为常数，空间积分采用单高斯点积分方法。值得注意的是，近场动力学的空间离散不限于均匀离散，近场范围尺寸可以是非定常的，并可以采用多高斯点积分，但本章并不涉及。

采用无网格配点法求解近场动力学强形式运动方程。物质点 $\boldsymbol{x}_i$ 受其近场范围内 ($|\boldsymbol{x}_j - \boldsymbol{x}_i| \leqslant \delta$) 其他物质点 $\boldsymbol{x}_j$ 作用，对近场动力学的运动方程 (2.44) 在键型 PD 和态型 PD 框架下进行空间离散，得到

$$\rho_i \ddot{\boldsymbol{u}}_i^n = \sum_j^{N_i} \boldsymbol{f}(\boldsymbol{u}_j^n - \boldsymbol{u}_i^n, \boldsymbol{x}_j - \boldsymbol{x}_i) V_j + \boldsymbol{b}(\boldsymbol{x}_i^n) \tag{5.1}$$

$$\rho_i \ddot{\boldsymbol{u}}_i^n = \sum_j^{N_i} \left\{ \underline{\boldsymbol{T}}[\boldsymbol{x}_i^n, t] \langle \boldsymbol{x}_j^n - \boldsymbol{x}_i^n \rangle - \underline{\boldsymbol{T}}[\boldsymbol{x}_j^n, t] \langle \boldsymbol{x}_i^n - \boldsymbol{x}_j^n \rangle \right\} V_j + \boldsymbol{b}(\boldsymbol{x}_i^n) \tag{5.2}$$

式中，n 为时间步号，相应时刻为 t 时刻，ρ_i 为物质点 $\boldsymbol{x}_i$ 的质量密度，$\boldsymbol{b}$ 为体力密度矢量，$\ddot{\boldsymbol{u}}_i^n$ 为物质点 $\boldsymbol{x}_i$ 的加速度矢量，$\boldsymbol{f}$ 为物质点 $\boldsymbol{x}_i$ 受物质点 $\boldsymbol{x}_j$ 作用的键力密度矢量，$\underline{\boldsymbol{T}}$ 为力密度矢量状态，N_i 为物质点 $\boldsymbol{x}_i$ 近场范围内物质点 $\boldsymbol{x}_j$ 的总数，$\boldsymbol{L} = \sum_j^{N_i}$ 为物质点 $\boldsymbol{x}_i$ 受其近场范围内其他物质点作用的合力，V_j 为物

5.1 控制方程的空间离散

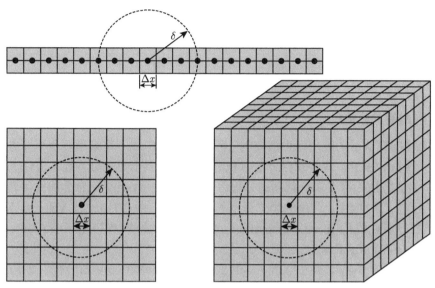

图 5-1　一维、二维与三维结构的中心正交均匀离散

质点 $\boldsymbol{x}_j$ 的积分体积，采用正交均匀离散时，三维物质点实际体积为 $\bar{V}_j = (\Delta x)^3$，二维物质点实际体积为 $\bar{V}_j = (\Delta x)^2$，Δx 为物质点 (或晶格) 长度。

注意到在边界附近，近场范围并不完整，物质点的积分体积不同于其实际体积，如图 5-2 所示。为保证积分精度，根据物质点坐标位置的不同，可采用 Parks 等 [2] 提出的一维插值方法修正物质点的积分体积 V_j，即

$$V_j = \begin{cases} \left(\dfrac{\delta - |\boldsymbol{\xi} + \boldsymbol{\eta}|}{2r_j} + \dfrac{1}{2}\right) \bar{V}_j, & (\delta - r_j) \leqslant |\boldsymbol{\xi} + \boldsymbol{\eta}| \leqslant \delta \\ \bar{V}_j, & |\boldsymbol{\xi} + \boldsymbol{\eta}| \leqslant \delta - r_j \\ 0, & \delta > |\boldsymbol{\xi} + \boldsymbol{\eta}| \end{cases} \quad (5.3)$$

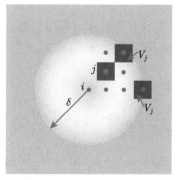

图 5-2　物质点的积分域与积分体积示意图

其中，$r_j = \Delta x/2$ 为物质点半边长。为进一步提高积分精度，还可采用 "以直代曲" 的方法 [10]，对近场范围边界上的物质点积分体积进行修正。

5.2 时间离散与逐步积分法

5.2.1 时间差分格式

对空间离散后的近场动力学控制方程 (5.1) 和 (5.2)，可采用逐步积分方法进行时间离散和求解。逐步积分方法从积分格式的形式上可划分为显式方法和隐式方法，显式方法不需要求解耦联的方程组，计算量较小，且更适合非线性问题的求解，得到了广泛应用。常用的显式逐步积分方法有：中心差分法、Velocity-Verlet 差分法、预测–校正 (predictor-corrector) 法、高阶龙格–库塔 (Runge-Kutta) 差分法、Lax-Wendroff 差分法等。

以上标 n、$n+1$、$n+1/2$ 和 $n-1/2$ 分别表示 t 时刻、$t+\Delta t$ 时刻、$t+0.5\Delta t$ 时刻和 $t-0.5\Delta t$ 时刻，Δt 为时间步长，下面给出几种典型算法的格式。

在中心差分法中，根据 $\dot{\boldsymbol{u}}^{n+1/2} = \dfrac{\boldsymbol{u}^{n+1} - \boldsymbol{u}^n}{\Delta t}$ 和 $\ddot{\boldsymbol{u}}^n = \dfrac{\dot{\boldsymbol{u}}^{n+1/2} - \dot{\boldsymbol{u}}^{n-1/2}}{\Delta t}$，并结合空间离散后的 PD 运动方程，可以得到 $n+1$ 时步的物质点速度和位移为

$$\begin{cases} \dot{\boldsymbol{u}}_i^{n+1/2} = \dot{\boldsymbol{u}}_i^{n-1/2} + \dfrac{\Delta t}{\rho}(\boldsymbol{L}+\boldsymbol{b})^n \\ \boldsymbol{u}_i^{n+1} = \boldsymbol{u}_i^n + \dot{\boldsymbol{u}}_i^{n+1/2}\Delta t \end{cases} \tag{5.4}$$

该格式的初始条件为：初始坐标 $\boldsymbol{u}(\boldsymbol{x},0)$ 及初始速度 $\dot{\boldsymbol{u}}(\boldsymbol{x},0)$，并根据初始构型计算初始加速度，其中起步条件 $\dot{\boldsymbol{u}}^{-1/2}$ 可根据向后差分 $\ddot{\boldsymbol{u}}^0 = 2\dfrac{\dot{\boldsymbol{u}}^0 - \dot{\boldsymbol{u}}^{-1/2}}{\Delta t}$ 求得

$$\dot{\boldsymbol{u}}_i^{-1/2} = \dot{\boldsymbol{u}}_i^0 - \dfrac{\Delta t}{2\rho}(\boldsymbol{L}+\boldsymbol{b})^0 \tag{5.5}$$

Velocity-Verlet 算法具有计算量适中且精度较高的特点，由 n 时步的速度以及 n 和 $n+1$ 时步的加速度，可以确定 $n+1$ 时步的物质点速度和位移

$$\begin{cases} \dot{\boldsymbol{u}}_i^{n+1/2} = \dot{\boldsymbol{u}}_i^n + \dfrac{\Delta t}{2\rho}(\boldsymbol{L}+\boldsymbol{b})^n \\ \dot{\boldsymbol{u}}_i^{n+1} = \dot{\boldsymbol{u}}_i^{n+1/2} + \dfrac{\Delta t}{2\rho}(\boldsymbol{L}+\boldsymbol{b})^{n+1} = \dot{\boldsymbol{u}}_i^n + \dfrac{\Delta t}{2\rho}(\boldsymbol{L}+\boldsymbol{b})^n + \dfrac{\Delta t}{2\rho}(\boldsymbol{L}+\boldsymbol{b})^{n+1} \\ \boldsymbol{u}_i^{n+1} = \boldsymbol{u}_i^n + \dot{\boldsymbol{u}}_i^{n+1/2}\Delta t = \boldsymbol{u}_i^n + \dot{\boldsymbol{u}}_i^n \Delta t + \dfrac{(\Delta t)^2}{2\rho}(\boldsymbol{L}+\boldsymbol{b})^n \end{cases} \tag{5.6}$$

该格式的初始条件为：初始坐标 $u(x,0)$ 及初始速度 $\dot{u}(x,0)$，并根据初始构型和第一次加载构型计算 0 和 1 时步的加速度，继而依时间序列开展逐步积分求解。

预测–校正法更新物质点速度和位置的算法如下

$$\underbrace{\begin{cases} \dot{u}^{n+1/2} = \dot{u}_i^n + \dfrac{\Delta t}{2\rho}(L+b)^n \\ u_i^{n+1/2} = u_i^n + \dot{u}_i^n \dfrac{\Delta t}{2} \end{cases}}_{(n+1/2) \text{ 时步的预测值}} \Rightarrow \underbrace{\begin{cases} \dot{u}^{n+1/2} = \dot{u}_i^n + \dfrac{\Delta t}{2\rho}(L+b)^{n+1/2} \\ u_i^{n+1/2} = u_i^n + \dot{u}_i^{n+1/2}\dfrac{\Delta t}{2} \end{cases}}_{(n+1/2) \text{ 时步的校正值}} \quad (5.7)$$

$$\Rightarrow \underbrace{\begin{cases} \dot{u}_i^{n+1} = 2\dot{u}^{n+1/2} - \dot{u}_i^n \\ u_i^{n+1} = 2u^{n+1/2} - u_i^n \end{cases}}_{\text{由 } n \text{ 和 } (n+1/2) \text{ 时步值获得 } (n+1) \text{ 时步值}}$$

该格式的初始条件为：初始坐标 $u(x,0)$ 及初始速度 $\dot{u}(x,0)$，并根据初始构型计算初始加速度。

5.2.2 显式逐步积分法的稳定性条件

显式逐步积分法是条件稳定的，时间步长必须小于控制方程所决定的某个临界值 Δt_{cr}。时间步长 Δt 选择过大，会导致粒子间相互作用过程描述不准确，使得位移、速度和加速度的累计误差不断扩大，并可能引起数值计算发散至"爆炸"失稳。时间步长选择过小，会导致计算量急剧增大。故需要进行稳定性分析，得到最佳的计算时间步长，以保证计算精度和稳定性，并节省计算资源和提高计算效率。可根据冯·诺依曼稳定性分析方法、传统动力学的简谐振动法和瑞利波法等，综合确定临界时间步长。

以单自由度体系的振动为例，其动力平衡方程可以表示为

$$\ddot{u} + C\dot{u} + \omega^2 u = r \quad (5.8)$$

其中，C 为阻尼系数，ω 为系统振动的固有频率，r 为动力荷载与体系质量的比值。讨论解的稳定性实质上是讨论误差引起的响应，为简便计，可令上式中的 $r=0$。由于阻尼对于解的稳定性是有利的，故在讨论解的稳定性问题时，一般不考虑阻尼项，即令 $C=0$。因此，上式简化为无阻尼单自由度体系的自由振动方程，即为

$$\ddot{u} + \omega^2 u = 0 \quad (5.9)$$

采用中心差分格式，上式化为

$$u^{n+1} = -(\Delta t^2 \omega^2 - 2)u^n - u^{n-1} \quad (5.10)$$

假设解的形式为 $u^{n+1} = \lambda u^n, u^n = \lambda u^{n-1}$，代入上式，可得特征方程为

$$\lambda^2 + (\Delta t^2 \omega^2 - 2)\lambda + 1 = 0 \tag{5.11}$$

该方程的根为

$$\lambda_{1,2} = \frac{2 - \Delta t^2 \omega^2 \pm \sqrt{(\Delta t^2 \omega^2 - 2)^2 - 4}}{2} \tag{5.12}$$

λ 的根与解的性质有关，为使其满足自由振动的特性，λ 须为复数，即要求 $(\Delta t^2 \omega^2 - 2)^2 < 4$，又因为 $\omega = 2\pi/T$，所以有

$$\Delta t < \frac{T}{\pi} = \Delta t_{cr} \tag{5.13}$$

其中，T 为材料的最小固有振动周期，为 $T = 2\pi\sqrt{\rho/k}$，k 为体系的劲度系数。为使解答不会无限增长，还应满足 $|\lambda| \leqslant 1$，而式 (5.12) 中的 $\lambda_{1,2}$ 的模 $|\lambda| = 1$，已自动满足该要求。

考虑式 (3.13) 给出的键型近场动力学运动方程，对于一维无体力问题，先进行空间离散，再采用中心差分格式对时间离散，得到

$$\rho_i \frac{u_i^{n+1} - 2u_i^n + u_i^{n-1}}{(\Delta t)^2} = \sum_{j=-\infty}^{\infty} C_{ji}(u_j - u_i)V_j \tag{5.14}$$

式中，j 表示 i 点两侧近场范围内物质点的编号，微模量 C_{ji} 与近场范围尺寸 δ 相关。

根据冯·诺依曼稳定性分析方法，假设位移场的形式为

$$u_i^n = \zeta^n e^{(\kappa i \sqrt{-1})} \tag{5.15}$$

式中，κ 为正实数，ζ 为复数。解的稳定性要求对于任意正实数 κ，确定稳定时间步长 Δt，满足 $|\zeta| \leqslant 1$，也即保证解答不会无限增长。将式 (5.15) 代入式 (5.14)，有

$$\frac{\rho_i}{(\Delta t)^2}(\zeta - 2 + \zeta^{-1}) = \sum_{j=-\infty}^{\infty} C_{ji}(e^{(\kappa(j-i)\sqrt{-1})} - 1)V_j$$

$$= \sum_{j=1}^{\infty} 2C_{ji}(\cos \kappa(j-i) - 1)V_j$$

$$= -2M_\kappa \tag{5.16}$$

在上述推导中，利用了欧拉公式 $e^{x\sqrt{-1}} = \cos x + \sqrt{-1}\sin x$，并令 $M_\kappa = \sum_{j=1}^{\infty} C_{ji}(1 - \cos \kappa(j-i))V_j$。

将上式改写为
$$\zeta^2 + \left(2\frac{M_\kappa(\Delta t)^2}{\rho_i} - 2\right)\zeta + 1 = 0 \tag{5.17}$$
得到该方程的根
$$\zeta = 1 - \frac{M_\kappa(\Delta t)^2}{\rho_i} \pm \sqrt{\left(\frac{M_\kappa(\Delta t)^2}{\rho_i} - 1\right)^2 - 1} \tag{5.18}$$
当 ζ 为复数时,条件 $|\zeta| \leqslant 1$ 恒成立,则有
$$\left(\frac{M_\kappa(\Delta t)^2}{\rho_i} - 1\right)^2 - 1 < 0 \Rightarrow \Delta t < \sqrt{\frac{2\rho_i}{M_\kappa}} \tag{5.19}$$
考虑到 $M_\kappa = \sum\limits_{j=1}^{\infty} C_{ji}(1 - \cos\kappa(j-i))V_j \leqslant \sum\limits_{j=1}^{\infty} 2C_{ji}V_j$,进而有
$$\Delta t < \sqrt{\frac{\rho_i}{\sum\limits_{j=1}^{\infty} C_{ji}V_j}} = \Delta t_{cr} \tag{5.20}$$

从上式可知,近场动力学的稳定时间步长除与材料性能有关以外,还取决于数值实施时人为选取的近场范围大小和物质点离散情况,为了提高计算效率,可采用自适应方法,在计算过程中自动调整时间步长。

5.3 显式拟静力方法

采用显式逐步积分法求解动力学方程时,可通过增加人工黏滞阻尼 (artificial damping) 的方法,获得系统的稳态解,也称为显式拟静力方法。此外,当通过低加载速率使动力学系统的惯性效应满足拟静力判据时,显式逐步积分法也可视为是显式拟静力方法。动态松弛法 (dynamic relaxation method, DRM) 是通过构造和求解包含阻尼的虚拟动力系统而获得静力解[4],采用虚时间步长在虚时间域更新变量值;与其不同的是,显式拟静力法则通过对考察体的真实动力系统施加阻尼,采用真实时间步长在真实时间域内更新变量值。

增加了人工阻尼项后,空间离散后的近场动力学控制方程 (5.1) 和 (5.2) 为
$$\rho_i \ddot{\boldsymbol{u}}_i^n + C\dot{\boldsymbol{u}}_i^n = \sum_j^{N_i} \boldsymbol{f}(\boldsymbol{u}_j^n - \boldsymbol{u}_i^n, \boldsymbol{x}_j - \boldsymbol{x}_i)V_j + \boldsymbol{b}(\boldsymbol{x}_i^n) \tag{5.21}$$
$$\rho_i \ddot{\boldsymbol{u}}_i^n + C\dot{\boldsymbol{u}}_i^n = \sum_j^{N_i} \left\{\underline{\boldsymbol{T}}[\boldsymbol{x}_i^n,t]\langle \boldsymbol{x}_j^n - \boldsymbol{x}_i^n\rangle - \underline{\boldsymbol{T}}[\boldsymbol{x}_j^n,t]\langle \boldsymbol{x}_i^n - \boldsymbol{x}_j^n\rangle\right\}V_j + \boldsymbol{b}(\boldsymbol{x}_i^n) \tag{5.22}$$

在 Velocity-Verlet 算法中，由 n 时步速度以及 n 和 $n+1$ 时步的加速度，可以确定 $n+1$ 时步的物质点速度和位移，具体为

$$\begin{cases} \dot{u}_i^{n+1/2} = \dfrac{\rho - 0.5C\Delta t}{\rho}\dot{u}_i^n + \dfrac{\Delta t}{2\rho}(L+b)^n \\ \dot{u}_i^{n+1} = \dfrac{\rho}{\rho + 0.5C\Delta t}\dot{u}_i^{n+1/2} + \dfrac{\Delta t}{2(\rho + 0.5C\Delta t)}(L+b)^{n+1} \\ u_i^{n+1} = u_i^n + \dot{u}_i^{n+1/2}\Delta t = u_i^n + \dfrac{\rho - 0.5C\Delta t}{\rho}\dot{u}_i^n\Delta t + \dfrac{(\Delta t)^2}{2\rho}(L+b)^n \end{cases} \quad (5.23)$$

需要指出的是，显式拟静力方法尚存在一些局限性，包括：计算成本随物质点尺寸和质量减小而急剧增大、人工阻尼系数确定依赖经验、时间差分格式会导致解的偏离等。

5.4 程序设计框架与计算流程

现有的近场动力学分析程序和软件包括 EMU、PDLAMMPS、PERIDIGM、Sierra/Solid Mechanics 的 PD 模块、LS-DYNA 以及各研究组针对具体问题的自编程序。通常采用均匀空间离散，并取定常近场范围，导致近场动力学计算量巨大。由于近场动力学非局部特性，决定其计算效率低于有限元等传统数值方法，因此，开发高效实用的近场动力学计算软件，必须着力提高近场动力学方法的计算效率，具体措施有：空间非均匀离散与自适应网格技术，多时间步长技术，基于 MPI、OpenMP 和 GPU-CPU 并行等高性能计算技术，近场动力学与有限元混合分析技术，隐式静力与显式动力联合分析方法等。

与其他方法类似，近场动力学的数值分析也包括前处理、PD 计算和后处理三个步骤，相应的程序软件需要包括三个模块：①几何建模和计算文件生成的前处理模块；②PD 计算模块；③数据图形显示的后处理模块。对于前处理，可采用 ANSYS 或 ABAQUS 等成熟商业软件，实现物质点的离散化并导出所需的物质点坐标，编写可供 PD 程序计算的数据文件。对于 PD 计算，可在统一框架下开发包含"键型"和"态型"多种模型的计算程序，实现一维、二维和三维静动力问题的计算，程序可施加位移和力边界条件，可实现位移加载、速度加载、力加载和冲击加载等。对于计算结果的后处理，可输出供 Ensight、Tecplot 和 Paraview 等成熟图形显示软件读取的结果数据，或基于图形库编写图像显示程序。

数据文件包含下列信息：计算总控信息，物体的坐标信息、材料信息、约束信息、加载信息，对于冲击问题，还包括冲击弹丸的坐标信息、材料信息、加载信息，冲击接触算法信息等。采用全局搜索方法形成某物质点的近场物质点集合信息，包括近场范围内其他物质点的编号、"键"的类型、个数和求解区域。施加

5.5 数值算例

初始条件、约束条件和加载条件后,通过显式或隐式方法求解近场动力学控制方程。计算结果包括物质点的位置、位移、速度、加速度和损伤等信息,进而通过运算获得应变、应力和能量密度等物理量。图 5-3 给出了近场动力学算法流程图。

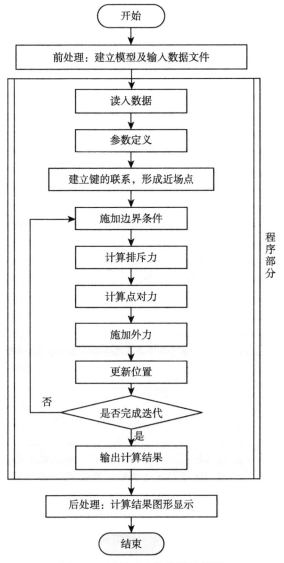

图 5-3 近场动力学算法流程图

5.5 数 值 算 例

本节采用键型近场动力学方法,给出一维杆的弹性波传播、二维简支梁的弹性变形、三维马氏体时效钢冲击动力裂纹扩展和混凝土板的冲击侵彻破坏四个算例,相应的 FORTRAN 源程序和数据文件可见本书附录 B。

5.5.1 一维杆的弹性波传播

如图 5-4 所示,长为 0.5m 两端自由的 347 不锈钢杆,受初始均匀应变 $\varepsilon_0 = 0.0001$,研究初始应变释放后,杆上 A 点的应变随时间变化特征。杆的弹性模量 $E=193$GPa,质量密度 $\rho = 8027$kg/m^3。为方便与经典连续介质力学的结果进行比较,定义如下工程应变

$$\varepsilon_{P2} = \frac{u_{P3} - u_{P1}}{|x_{P3} - x_{P1}|} = \frac{\eta_{13}}{|\xi_{13}|} \tag{5.24}$$

固定坐标系于杆上,初始条件为 $\begin{cases} u(x,0) = \varepsilon_0 x \\ \dot{u}(x,0) = 0 \end{cases}$,相应的经典连续介质力学的应变解析解为 [11,12]

$$\varepsilon = \frac{\partial u}{\partial x} = \frac{2}{L}\varepsilon_0 \sum_{i=0}^{\infty}\left[1-(-1)^i\right]\frac{1}{\lambda_i}\sin\lambda_i x \cdot \cos\omega_i t \tag{5.25}$$

式中,$\lambda_i = \frac{i\pi}{L}$,$\omega_i = \sqrt{\frac{E}{\rho}}\lambda_i$,$L$ 为杆长。

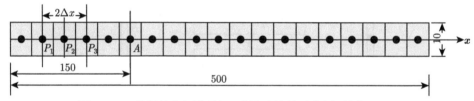

图 5-4　一维杆件几何模型与工程应变计算示意图 (单位:mm)

采用本书 3.3 节的键型近场动力学改进 PMB 模型进行数值计算,施加初始条件后杆件自由振动,近场范围取为 $\delta = 3\Delta x$,物质点间距考虑了 7 种情况,由大到小分别为 10^{-3}m、0.5×10^{-3}m、0.2×10^{-3}m、10^{-4}m、0.5×10^{-4}m、0.2×10^{-4}m 和 10^{-5}m,时间步长为 $\Delta t = 1\times 10^{-7}$s $\sim 1\times 10^{-9}$s。

计算得到了杆中 A 点的应变历程,并与解析解进行对比分析,如图 5-5 所示。此外,还选取 $\Delta x = 10^{-5}$m 和 $\Delta t = 1\times 10^{-9}$s 为计算条件,比较了原 PMB 模型与改进 PMB 模型的计算结果,如图 5-6 所示。图 5-5 显示:改进 PMB 模型数值解与经典理论解具有良好的一致性,随着物质点间距的减小,PD 解 (应变幅值、相位、周期) 逐渐收敛于经典解,如 PD 计算得到的波传播周期的数值解为 2.107×10^{-4}s,稍大于 2.038×10^{-4}s 的理论解,这是由于近场动力学的非局部

色散特性导致其计算波速低于理论波速，计算周期略大于理论周期。物质点间距较大时，方波转角处 PD 解存在过振幅现象，随着物质点间距逐渐减小，该现象迅速减少。图 5-6 显示：原 PMB 模型的计算结果偏差较大，改进 PMB 模型在求解波的传播问题时具有良好的精度，且当物质点间距满足频散关系后，减小其间距能增加应变的计算精度，但不能显著提高波速和周期的计算精度。

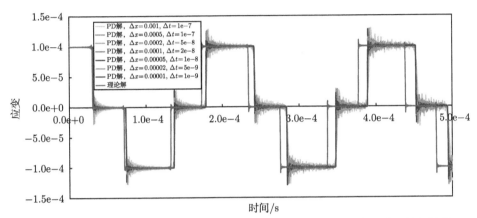

图 5-5　钢杆上 A 点的应变-时间曲线：改进 PMB 模型数值解与经典理论解比较

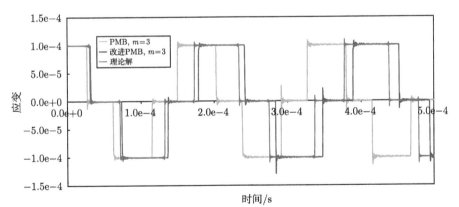

图 5-6　钢杆上 A 点的应变-时间曲线：原 PMB 模型与改进 PMB 模型的计算结果比较 ($\Delta x = 0.00001\text{m}$, $\Delta t = 1 \times 10^{-9}\text{s}$)

5.5.2　二维简支梁的弹性变形

如图 5-7 所示，矩形截面简支梁跨中受 1kN 集中力作用，梁长 1m、高 0.1m，按平面应力问题考虑，取单位厚度。材料的弹性模量 $E = 20\text{GPa}$，泊松比 $\nu = 1/3$。采用 $2\text{mm} \times 2\text{mm}$ 方形物质点均匀离散，共有 25551 个物质点；梁两端下部角隅处各约束 1 个物质点，左端在水平向和竖直向均为固定，右端只约束竖直向位移；荷载施加在梁的上表面中部 4 个物质点上，每个物质点施加集中力 0.25kN[13]。在拟静力分析中，人工阻尼系数 $C = 5\times 10^6$，时间步长 $\Delta t = 4\times 10^{-7}$s，共计算 4 万步。

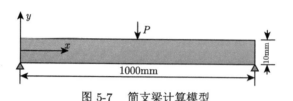

图 5-7　简支梁计算模型

分别采用改进的 PMB 模型、原 PMB 模型、FEM 方法对该问题进行了求解，图 5-8 给出了各种方法计算得到的梁中性轴挠度曲线，图中还附上理论解的结果。由图中结果可知，对于所考虑的算例，有限元解大于理论解，PD 解则小于理论解，改进的 PMB 模型计算结果更接近于理论解。

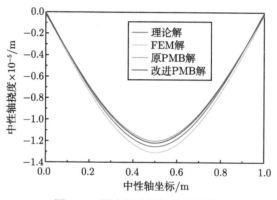

图 5-8　简支梁中性轴挠度曲线

5.5.3　三维马氏体时效钢冲击动力裂纹扩展

Kalthoff-Winkler 试验[14,15]是指以不同速度的刚性柱体，撞击马氏体时效钢试件致其发生动力断裂的冲击破坏试验。如图 5-9 所示，试验结果表明，刚性柱体撞击马氏体时效钢靶体时，产生压缩应力波，这个压缩波与预制的裂缝相互作用，靶体出现 II 型开裂现象。如果选取恰当的撞击速度，可从预制裂缝的裂尖起裂，发生剪切与拉伸混合型开裂，形成约 68° 的裂纹扩展角。能否成功模拟该试验已成为评价数值方法对脆性材料动力断裂问题适用性的重要手段。

基于键型近场动力学改进 PMB 模型，并采用显式动力学算法对该问题进行了计算模拟。材料的弹性模量 E=191GPa，泊松比 $\nu = 1/4$，质量密度 $\rho =$ 8000kg/m³，材料临界伸长率取为 s_0=0.015；刚性柱体弹丸质量 m=1.57kg，初速

度为 $v_0=-32\text{m/s}$。在 PD 模拟中，按三维问题考虑，马氏体时效钢靶体共均匀离散成 178200 个物质点，物质点各方向间距均为 1mm，物质点体积为 $1\times10^{-9}\text{m}^3$，显式动力计算的时间步长取 $\Delta t=8\times10^{-8}\text{s}$，采用了文献 [16] 中提出的冲击接触算法，共计算 4 千步。

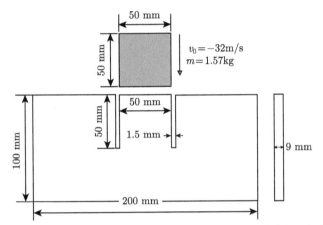

图 5-9　Kalthoff-Winkler 试验：刚性柱体冲击马氏体时效钢几何模型

PD 模拟结果见图 5-10 和图 5-11。计算结果表明，在冲击荷载作用下，预制裂缝的裂尖首先出现损伤，损伤不断累积致使裂尖起裂，后续在剪切与拉伸的共同作用下，新裂纹形成约 67° 的扩展角并不断演化发展。PD 计算中，以损伤值 $\varphi=0.35$ 作为开裂判据，得到的裂纹起裂时间为 28.56μs，扩展完成时间为 91.52μs，试验给出的起裂时间略低于 29μs，PD 模拟与试验所得的起裂时间和扩展角高度一致。此外，该问题的瑞利波波速 (Rayleigh wave speed) 为 2799.2m/s，由于计算中物质点间距设置较大，导致预测的波速波动较大，但 PD 模拟的裂纹传播速度与其他两种数值方法 [17,18] 总体趋势一致。该例较好地验证了近场动力学方法对动力断裂问题的适用性。

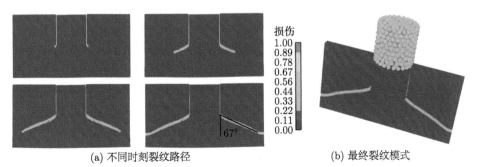

(a) 不同时刻裂纹路径　　　　　　　(b) 最终裂纹模式

图 5-10　Kalthoff-Winkler 试验的 PD 模拟结果

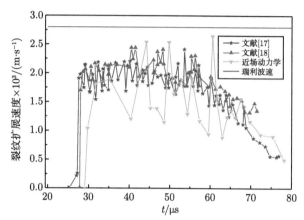

图 5-11　Kalthoff-Winkler 试验中马氏体时效钢裂尖传播速度

5.5.4　混凝土板的冲击侵彻破坏

考虑一刚性小球 (弹丸) 垂直撞击混凝土板表面中心的问题。混凝土板的长度 200mm、宽度 100mm、厚度 20mm，混凝土板沿宽度方向垂直放置于两辊轴支座上，同时上部也受两辊轴支座的抵持约束，抵持压力与横向摩擦力较小，可以忽略，混凝土板可以在垂直板面方向自由移动。刚性小球半径 10mm，质量密度 $\rho = 7800 \text{kg/m}^3$，小球初速度分别为 10m/s、20m/s、50m/s、100m/s，从一侧撞击混凝土板。混凝土为均匀各向同性的脆弹性材料，弹性模量 E=30GPa，泊松比 ν=1/4，能量释放率 $G_F = 175 \text{N/m}$，质量密度 $\rho = 2400 \text{kg/m}^3$。

基于键型近场动力学改进 PMB 模型进行计算模拟，采用文献 [16] 中的冲击接触算法描述刚性小球冲击侵彻混凝土板，物质点在三个方向上的间距均为 2mm，均匀离散混凝土板，共有 56661 个物质点，物质点体积 $8 \times 10^{-9} \text{m}^3$，时间步长 $\Delta t = 1 \times 10^{-7}$s，共计算 5 千步。

图 5-12～图 5-15 给出了四种不同冲击速度下，刚性小球冲击侵彻混凝土板的损伤累积和渐进破坏过程 (从侵彻初始至 2200 时间时步，间隔 200 时步)。图中红色表示损伤度为 1，蓝色表示损伤度为 0。由图中所示结果可以看出，弹丸撞击混凝土靶板后，产生向靶板背面传播的压缩应力波，弹着点附近发生混凝土压缩破碎现象，出现表面弹坑，并伴随碎片粒子反向飞溅；反射拉伸应力波与入射压缩波共同作用致物质点间作用弱化，产生锥形体 (圆台体) 损伤区域，并伴随大量微裂纹产生，随后裂纹扩展、贯通，形成锥形体冲切表面和冲切破坏区域，冲切区域内部由于拉压应力波共同作用，混凝土破碎，出现粒子飞溅脱离现象。此外，不同冲击速度对应的冲切区域大小无显著差别，但对弹坑的大小和深度、弹坑周围混凝土破碎程度、冲击接触区域碎片反向飞溅的数量和剧烈程度有很大影响，均随着弹丸的撞击速度增大而趋于增大或严重。

5.5 数 值 算 例

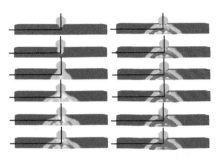

图 5-12　速度 10m/s 弹丸侵彻下混凝土板损伤破坏过程

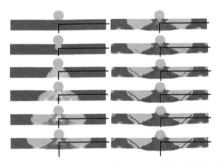

图 5-13　速度 20m/s 弹丸侵彻下混凝土板损伤破坏过程

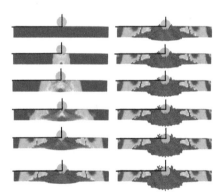

图 5-14　速度 50m/s 弹丸侵彻下混凝土板损伤破坏过程

定义结构的累积总损伤量为物质点损伤值总和与物质点数目的比值，图 5-16 给出了不同初始速度冲击下的混凝土板累积损伤变化过程，图 5-17 给出了混凝土板总损伤量和刚性小球残余速度与弹丸初速度的关系曲线。从图中结果可见，在弹丸侵彻混凝土板不久，混凝土板的总损伤量迅速增大，弹丸速度越大，总损伤量越大，损伤量的增幅亦越大；当混凝土板的总损伤量增加到一定程度后，增量趋于平缓。此外，弹丸的残余速度与弹丸初速度成正比，基本呈线性变化。

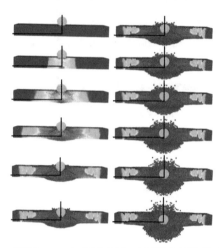

图 5-15　速度 100m/s 弹丸侵彻下混凝土板损伤破坏过程

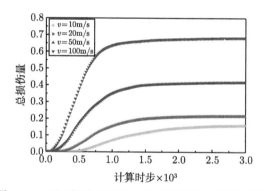

图 5-16　不同初始速度弹丸侵彻下混凝土板累积损伤

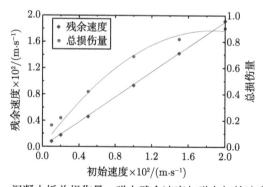

图 5-17　混凝土板总损伤量、弹丸残余速度与弹丸初始速度的关系

5.6 近场动力学开源软件 PDLAMMPS

5.6.1 PDLAMMPS 软件概述

美国圣地亚国家实验室 (Sandia National Laboratory, SNL) 主持开发了近场动力学的三个主要计算程序，即 EMU、PDLAMMPS 和 PERIDIGM。其中，EMU[19] 是最早由 Silling 开发的 PD 计算程序，基于 FORTRAN90 编写，但 EMU 是圣地亚国家实验室的自用程序，存在获取和使用限制，相关的介绍资料也较少。PDLAMMPS 是基于分子动力学通用软件 LAMMPS 开发的 PD 计算程序，由 Parks、Lehoucq、Plimpton、Silling 和 Seleson 等采用 C++ 编写，程序的扩展性良好，但可实现的近场动力学模型较为有限。Parks、Littlewood、Mitchell、Silling 和 Foster 等联合开发了 PERIDIGM 开源程序[20]，Rädel 和 Willberg 发布了软件安装教程[21]，该程序适用性广、稳定性好，目前还处于不断发展和完善的过程中。此外，供美国海军使用的固体结构分析软件 Sierra/Solid Mechanics (Sierra/SM)[22] 实现了常规态型和非常规态型近场动力学对应材料模型。本节着重介绍 PDLAMMPS 的安装和使用。

LAMMPS[23](Large-scale Atomic/Molecular Massively Parallel Simulator) 直译为大规模原子分子并行模拟器，它是由美国圣地亚国家实验室开发的经典开源的分子动力学计算软件，可以在任何一个安装了 C++ 编译器和 MPI 的平台上运算，包括 Linux、Mac 和 Windows 操作系统，具有良好的并行扩展性。LAMMPS 支持包括气态、液态和固态相形态下的各种系综，支持多种势函数以描述不同粒子间的短程力和长程力相互作用，可以采用不同的力场模型和边界条件对全原子、聚合物、生物、金属、粒状和粗料化体系进行计算模拟。

LAMMPS 是一个可以修改和扩展的开源计算程序，使用者可以自行加上一些新的力场、原子模型、边界条件和诊断功能等。LAMMPS 采用相邻链表跟踪粒子，这些链表是根据粒子间短程互斥力的大小进行优化的，以防止局部区域的粒子密度过高。LAMMPS 采用空间分解技术分配模拟区域，将整个模拟空间分成较小的三维小空间，每一个小空间分配在一个处理器上，各个处理器之间互相通信并且存储每一个小空间边界上的 "ghost" 原子的信息，进而实现并行计算。LAMMPS 不具有图形输出及可视化功能，需要借助其他软件来实现高质量的可视化，常用软件有 VMD、AtomEye、OVITO、ParaView、PyMol、Raster3d 和 RasMol 等。

PDLAMMPS 是基于 LAMMPS 平台开发的 PD 计算程序，也即在 LAMMPS 中加入了 Peri 模块，支持几乎所有边界条件。PDLAMMPS 通过势函数 peri/pmb、peri/pmb/omp、peri/lps、peri/lps/omp、peri/eps 和 peri/ves 等[23−25,28−30] 实

现了键型近场动力学和常规态型近场动力学的四个模型, 分别是

(1) 线性近场动力学固体模型 (LPS), 该常规态型近场动力学模型适用于线弹性各向同性材料;

(2) 经典微弹脆性模型 (PMB), 该模型是典型的键型近场动力学模型, 也是 LPS 模型的特例;

(3) 常规态型弹塑性模型 (EPS)[26];

(4) 常规态型黏弹塑性模型 (VES)[27]。

在 PDLAMMPS 中, 运行程序需要输入脚本文件 (input script), 其具体格式可见文献 [2, 24, 25], 其中各语句含义可见文献 [23], 本节也将结合具体算例给出脚本文件。

5.6.2 安装与配置

以 Windows10 64 位操作系统为例, 对 LAMMPS 的安装与配置做简要说明。首先, 从 http://packages.lammps.org/windows.html 下载适合本机的安装包, 如最新 64 位 MPI 的并行版本可执行文件 LAMMPS-64Bit-latest-MPI.exe, 或串行版本可执行文件 LAMMPS-64bit-latest, 并下载实现 MPI 并行功能的 mpich2-1.4.1p1-win-x86-64.msi 文件, 若电脑上尚无支持 mpich 的.Net Framework, 需要在微软开发者界面下载 NetFx64.exe 文件。随后直接点击相关安装文件, 可便捷地实现 MPICH、LAMMPS 串行或并行版本的安装。脚本运行时, 需要将脚本 in. 文件、lmp_serial.exe 或 lmp_mpi.exe 文件和 wmpiexec.exe 文件 (配合 lmp_mpi.exe) 放在同一路径下。

5.6.3 脚本文件示例与应用

如图 5-18 所示, 直径为 5mm 的刚性或非刚性小球以 100m/s 的速度撞击素

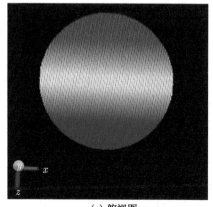
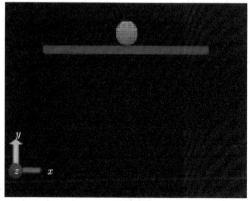

(a) 俯视图　　　　　　　　　　　(b) 正视图

图 5-18　刚性或非刚性小球撞击混凝土薄板的计算模型

5.6 近场动力学开源软件 PDLAMMPS

混凝土圆形薄板中心,薄板直径 74mm、厚度 2.5mm。假定混凝土是均匀、各向同性的弹脆性材料,密度 2400kg/m³、体积模量 K=40GPa、临界伸长 s_{00}=0.0005、计算参数 α=0.25[2]。物质点各方向的间距 Δx=0.0005m,近场范围 δ 取 3 倍物质点间距,为 0.0015m。小球为刚性时,离散后的物质点总数为 103110;小球非刚性时,离散后的物质点总数为 107274。

- **刚性小球撞击混凝土板 LAMMPS 模拟的脚本文件如下**

```
# small Peridynamic thin circular plate hit by a rigid projectile
units           si
boundary        s s s
atom_style      peri
atom_modify     map array
neighbor        0.0010 bin
lattice         sc 0.0005
region          target cylinder y 0.0 0.0 0.037 -0.0025 0.0 units box
create_box      1 target
create_atoms    1 region target
pair_style      peri/pmb
pair_coeff      * * 2.2635e22 0.0015001 0.0005 0.25
set             group all density 2400
set             group all volume 1.25e-10
velocity        all set 0.0 0.0 0.0 sum no units box
fix             1 all nve
fix             2 all indent 1e17 sphere 0 -100 0 0.005 units box
compute         1 all damage/atom
timestep        1.0e-7
thermo          300
dump            1 all custom 20 dump_rigid.lammpstrj id type x y z c_1
run             8000
```

- **非刚性小球撞击混凝土板 LAMMPS 模拟的脚本文件如下**

```
# small Peridynamic thin circular plate hit by a unrigid projectile
units           si
boundary        s s s
atom_style      peri
atom_modify     map array
neighbor        0.0010 bin
lattice         sc 0.0005
region          target1 cylinder y 0.0 0.0 0.037 -0.0025 0.0 units box
region          target2 sphere 0.0 0.005 0.0 0.005 side in units box
```

```
region          target union 2 target1 target2
create_box      1 target
create_atoms    1 region target
pair_style      peri/pmb
pair_coeff      * * 2.2635e22 0.0015001 0.0005 0.25
set             group all density 2400
set             group all volume 1.25e-10
velocity        all set 0.0 0.0 0.0 sum no units box
group           indenter region target2
velocity        indenter set 0.0 -100 0.0 sum no units box
fix             1 all nve
compute         1 all damage/atom
timestep        1.0e-7
thermo          300
dump            1 all custom 20 dump_unrigid.lammpstrj id type x y z c_1
run             8000
```

图 5-19 给出了非刚性小球撞击混凝土薄板过程中, 圆板和小球的破坏演变过程, 图 5-20 给出了混凝土板在刚性小球和非刚性小球撞击后的最终裂纹模式。刚性小球撞击后, 混凝土板径向裂纹较环向裂纹更为明显; 非刚性小球撞击后, 混凝土薄板的破坏面积更大, 碎块更多更细小, 其原因可能是非刚性小球撞击后自身破碎, 产生更大的撞击面积, 颗粒的动量交换也更多。

将所得到的 PD 模拟结果与试验结果[31]相比较, 破坏程度和裂纹图样符合程度较高, 也表明 PD 方法能较好地反映撞击过程和物理图景。

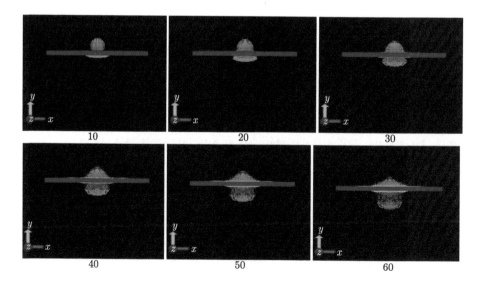

5.6 近场动力学开源软件 PDLAMMPS

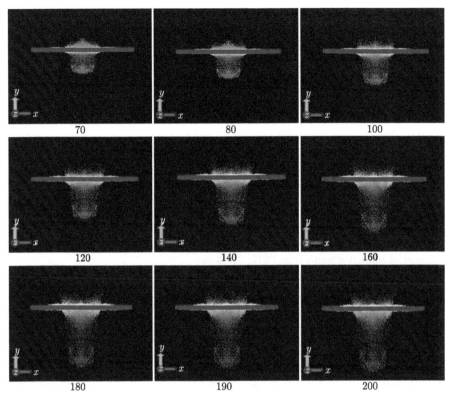

图 5-19 混凝土小球 (非刚性) 撞击混凝土薄板的冲击过程 (单位: 步, 每时间步长 10^{-7}s)

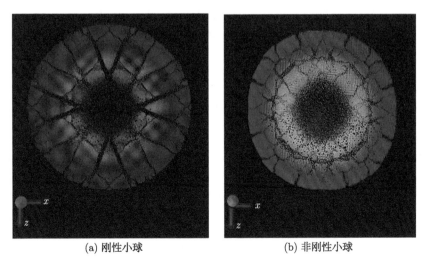

(a) 刚性小球　　　　　　　(b) 非刚性小球

图 5-20 小球撞击后薄板的破坏情况

参 考 文 献

[1] Silling S A, Askari E. A meshfree method based on the peridynamic model of solid mechanics [J]. Computers & Structures, 2005, 83(17-18): 1526-1535.

[2] Parks M L, Lehoucq R B, Plimpton S J, et al. Implementing peridynamics within a molecular dynamics code [J]. Computer Physics Communications, 2008, 179(11): 777-783.

[3] Kilic B. Peridynamic theory for progressive failure prediction in homogeneous and heterogeneous materials [D]. Tuscon: The University of Arizona, 2008.

[4] Kilic B, Madenci E. An adaptive dynamic relaxation method for quasi-static simulations using the peridynamic theory [J]. Theoretical and Applied Fracture Mechanics, 2010, 53(3):194-204.

[5] Ren B, Wu C T, Askari E. A 3D discontinuous Galerkin finite element method with the bond-based peridynamics model for dynamic brittle failure analysis [J]. International Journal of Impact Engineering, 2017, 99:14-25.

[6] Breitenfeld M S. Quasi-static non-ordinary state-based peridynamics for the modeling of 3D fracture [D]. Urbana-Champaign: University of Illinois at Urbana-Champaign, 2014.

[7] 郁杨天, 章青, 顾鑫. 近场动力学与有限单元法的混合模型与隐式求解格式 [J]. 浙江大学学报 (工学版), 2017, 51(7):1324-1330.

[8] Gu X, Zhang Q, Madenci E. Non-ordinary state-based peridynamic simulation of elasto-plastic deformation and dynamic cracking of polycrystal [J]. Engineering Fracture Mechanics, 2019, 218:106568.

[9] Huang D, Lu G D, Qiao P Z. An improved peridynamic approach for quasi-static elastic deformation and brittle fracture analysis [J]. International Journal of Mechanical Sciences, 2015, 94:111-122.

[10] Seleson P. Improved one-point quadrature algorithms for two-dimensional peridynamic models based on analytical calculations [J]. Computer Methods in Applied Mechanics and Engineering, 2014, 282:184-217.

[11] Tao J. Development and applications of new peridynamic models[D]. East Lansing: Michigan State University, 2012.

[12] Gu X, Zhang Q, Huang D, et al. Wave dispersion analysis and simulation method for concrete SHPB test in peridynamics [J]. Engineering Fracture Mechanics, 2016, 160:124-137.

[13] 顾鑫, 章青, 黄丹. 基于近场动力学方法的混凝土板侵彻问题研究 [J]. 振动与冲击, 2016, 35(6):52-58.

[14] Kalthoff J F, Winkler S. Failure mode transition at high rates of shear loading [C]. DGM Informationsgesellschaft mbH, Impact Loading and Dynamic Behavior of Materials, 1988, 1:185-195.

[15] Kalthoff J F, Bürgel A. Influence of loading rate on shear fracture toughness for failure mode transition [J]. International Journal of Impact Engineering, 2004, 30(8):957-971.

[16] Madenci E, Oterkus E. Peridynamic Theory and Its Applications [M]. Berlin: Springer, 2014.

[17] Belytschko T, Chen H, Xu J, et al. Dynamic crack propagation based on loss of hyperbolicity and a new discontinuous enrichment [J]. International Journal for Numerical Methods in Engineering, 2003, 58(12):1873-1905.

[18] Song J H, Areias P, Belytschko T. A method for dynamic crack and shear band propagation with phantom nodes [J]. International Journal for Numerical Methods in Engineering, 2006, 67(6): 868-893.

[19] Birkey J. Development of Visual EMU, a graphical user interface for the peridynamic EMU code [D]. Manhattan: Kansas State University, 2007.

[20] Parks M L, Littlewood D J, Mitchell J A, et al. Peridigm Users' Guide v1.0.0 [R]. Sandia National Lab Report, 2012, 7800.

[21] Rädel M, Willberg C. Peridigm Installation Guide [R]. DLR-Interner Bericht, DLR-IB-FA-BS-2019-33, 79S. 2019.

[22] Team S S. Sierra/Solid Mechanics 4.22 Users' Guide [R]. Sandia National Laboratory Report, SAND2011-7597.

[23] Parks M L, Rahman R, Foster J T. pair_style peri/pmb command [EB/OL]. https://docs.lammps.org/pair_peri.html.

[24] Parks M L, Plimpton S J, Lehoucq R B, et al. Peridynamics with LAMMPS: A user guide [R]. Sandia National Laboratory Report, SAND2008-0135.

[25] Parks M L, Seleson P, Plimpton S J, et al. Peridynamics with lammps: a user guide, V0.3 beta [R]. Sandia National Laboratory Report, SAND2011-8253.

[26] Mitchell J A. A nonlocal ordinary state-based plasticity model for peridynamics [R]. Sandia National Lab Report, 2011, 3166:1-34.

[27] Mitchell J A. A non-local, ordinary-state-based viscoelasticity model for peridynamics [R]. Sandia National Lab Report, 2011, 8064:1-28.

[28] Rahman R, Foster J T. Implementation of elastic-plastic model in PDLAMMPS [R]. University of Texas at San Antonio, Sandia National Laboratory, 2013.

[29] Rahman R, Foster J T. Implementation of linear viscoelasticity model in pdlammps [R]. University of Texas at San Antonio, Sandia National Laboratory, 2014.

[30] Rahman R, Foster J T, Plimpton S J. PDLAMMPS-made easy [R]. University of Texas at San Antonio, Sandia National Laboratory, 2014.

[31] Bouzid S, Nyoungue A, Azari Z, et al. Fracture criterion for glass under impact loading [J]. International Journal of Impact Engineering, 2001, 25(9):831-845.

第 6 章 常规态型近场动力学模型

针对键型近场动力学模型的局限性,如:固定泊松比限制、不能区分形状变形和体积变形、不支持不可压缩材料本构建模以及与经典连续介质力学联系不够紧密等,Silling 在 2000 年即提出考虑体积变形能密度的建模思路[1],并于 2007 年正式提出近场动力学状态算子的概念,构建了常规态型近场动力学和非常规态型近场动力学两类模型[2],拓宽了近场动力学的适用范围。本章在第 2 章介绍的近场动力学状态算子的基础上,系统介绍常规态型近场动力学弹性模型和弹塑性模型的建模方法,详细给出固体材料近场动力学弹性本构模型的构建过程,建立与传统 J_2 弹塑性模型和 Drucker-Prager 弹塑性模型关联的近场动力学弹塑性本构模型,并通过几个数值实例验证所建立的常规态型近场动力学模型的有效性。

6.1 常规态型近场动力学弹性模型

通过前文介绍可知,键型近场动力学模型的键力密度函数只与近场范围内单一物质点对相关,态型近场动力学模型的键力密度函数与两个物质点的各自近场范围内的所有物质点相关。本节将结合弹性问题,具体介绍常规态型近场动力学的力密度矢量状态与变形状态的关联方法。

6.1.1 基本变量说明

在第 2 章中,基于式 (2.22) 给出了初始构型和当前构型中变形矢量 $\underline{\boldsymbol{X}}\langle\boldsymbol{\xi}\rangle$ 与 $\underline{\boldsymbol{Y}}\langle\boldsymbol{\xi}\rangle$ 的定义,即

$$\underline{\boldsymbol{X}}\langle\boldsymbol{\xi}\rangle = \underline{\boldsymbol{X}}[\boldsymbol{x},t]\langle\boldsymbol{\xi}\rangle = \boldsymbol{x}' - \boldsymbol{x} = \boldsymbol{\xi}$$
$$\underline{\boldsymbol{Y}}\langle\boldsymbol{\xi}\rangle = \underline{\boldsymbol{Y}}[\boldsymbol{x},t]\langle\boldsymbol{\xi}\rangle = \boldsymbol{y}' - \boldsymbol{y} = \boldsymbol{\xi} + \boldsymbol{\eta}$$

由此可以决定初始构型和当前构型中的变形矢量状态。类似地,还通过式 (2.23),定义了初始变形标量 $\underline{x}\langle\boldsymbol{\xi}\rangle$ 和当前变形标量 $\underline{y}\langle\boldsymbol{\xi}\rangle$,具体为

$$\underline{x}\langle\boldsymbol{\xi}\rangle = \underline{x}[\boldsymbol{x},t]\langle\boldsymbol{\xi}\rangle = |\underline{\boldsymbol{X}}\langle\boldsymbol{\xi}\rangle| = |\boldsymbol{\xi}|$$
$$\underline{y}\langle\boldsymbol{\xi}\rangle = \underline{y}[\boldsymbol{x},t]\langle\boldsymbol{\xi}\rangle = |\underline{\boldsymbol{Y}}\langle\boldsymbol{\xi}\rangle| = |\boldsymbol{\xi} - \boldsymbol{\eta}|$$

顺便指出,近场动力学中定义的变形矢量和变形标量其实就是键矢量和键长度,通过受载后键矢量和键长度的变化,可以描述键的变形状态,这也是将其称

6.1 常规态型近场动力学弹性模型

为变形矢量 (变形标量) 的缘故。在本章中，根据上述初始和当前变形标量 (也是变形前后键的长度) 的含义，先定义拉伸标量 $\underline{e}\langle\boldsymbol{\xi}\rangle$ 为

$$\underline{e}\langle\boldsymbol{\xi}\rangle = \underline{y}\langle\boldsymbol{\xi}\rangle - \underline{x}\langle\boldsymbol{\xi}\rangle = |\boldsymbol{\xi}+\boldsymbol{\eta}| - |\boldsymbol{\xi}| \tag{6.1}$$

其含义是变形前后键长的改变量，也可视为拉伸标量状态 (extension state) $\underline{e}$ 的映射。

再根据第 2 章定义的点积运算，定义加权量 (weighted volume) 为

$$m = w\underline{x} \cdot \underline{x} = \int_{H_{\infty}} w(\boldsymbol{\xi})\underline{x}\langle\boldsymbol{\xi}\rangle \cdot \underline{x}\langle\boldsymbol{\xi}\rangle \, dV_{\boldsymbol{x}'} = \int_{H_{\infty}} w(\boldsymbol{\xi})|\boldsymbol{\xi}|^2 \, dV_{\boldsymbol{x}'} \tag{6.2}$$

式中，w 为影响函数，是关于 $\boldsymbol{\xi}$ 的函数，该函数形式的选取以能够反映长程相互作用的空间分布特征为佳。

式 (6.2) 定义的加权标量反映了近场域内的物质点集或键集的空间分布特征。类似地，式 (2.56) 定义的形状张量是反映近场域内物质点集或键集空间分布特征的二阶张量。

进一步，定义与传统体积应变等价的膨胀量 (dilatation) 为

$$\theta(\underline{e}) = \gamma\frac{w\underline{x} \cdot \underline{e}}{m} = \gamma\frac{\displaystyle\int_{H_{\infty}} w(|\boldsymbol{\xi}|)|\boldsymbol{\xi}|\underline{e}\langle\boldsymbol{\xi}\rangle \, dV_{\boldsymbol{x}'}}{\displaystyle\int_{H_{\infty}} w(|\boldsymbol{\xi}|)|\boldsymbol{\xi}|^2 \, dV_{\boldsymbol{x}'}} \tag{6.3}$$

式中，维度参数 γ 取为[3,4]

$$\gamma = \begin{cases} 3, & \text{三维} \\ 2, & \text{平面应变} \\ \dfrac{2(2\nu-1)}{\nu-1}, & \text{平面应力} \\ 1, & \text{一维} \end{cases} \tag{6.4}$$

其中，ν 为泊松比。当构型在各个方向发生均匀的小应变 $|\varepsilon_0| \ll 1$ 时，变形矢量为 $\underline{Y}\langle\boldsymbol{\xi}\rangle = (1+\varepsilon_0)\underline{X}\langle\boldsymbol{\xi}\rangle$，在这种情况下，影响函数形式不影响计算结果，根据式 (6.3) 和式 (6.4)，不难求出膨胀量分别为：三维 $\theta = 3\varepsilon_0$、平面应变 $\theta = 2\varepsilon_0$、平面应力 $\theta = \dfrac{2(2\nu-1)}{\nu-1}\varepsilon_0$ 和一维 $\theta = \varepsilon_0$，均与经典弹性理论中应变张量矩阵的迹一致，表明上述关于膨胀量的定义具有合理性。

借鉴连续介质力学张量分解思想，拉伸标量状态可分解为球量部分 $\underline{e}^{\text{i}}$ 和偏量部分 $\underline{e}^{\text{d}}$，分别表示体积变形状态和剪切变形状态，具体为

$$\underline{e}\langle\boldsymbol{\xi}\rangle = \underline{e}^{\text{i}}\langle\boldsymbol{\xi}\rangle + \underline{e}^{\text{d}}\langle\boldsymbol{\xi}\rangle \tag{6.5}$$

其中，拉伸标量的球量和偏量分别为

$$\underline{e}^{\mathrm{i}}\langle\boldsymbol{\xi}\rangle = \frac{\theta\underline{x}\langle\boldsymbol{\xi}\rangle}{3} \tag{6.6}$$

$$\underline{e}^{\mathrm{d}}\langle\boldsymbol{\xi}\rangle = |\boldsymbol{\xi}+\boldsymbol{\eta}| - |\boldsymbol{\xi}| - \frac{\theta\underline{x}\langle\boldsymbol{\xi}\rangle}{3} \tag{6.7}$$

需要说明的是，不同维度下，拉伸标量的球量与膨胀量的关系是固定的。

对于常规态型近场动力学模型，由于物质点对的力密度矢量 $\underline{\boldsymbol{T}}\langle\boldsymbol{\xi}\rangle$ 与变形矢量 $\underline{\boldsymbol{Y}}\langle\boldsymbol{\xi}\rangle$ 共线，于是，由力密度矢量状态 $\underline{\boldsymbol{T}}$ 和变形矢量状态 $\underline{\boldsymbol{Y}}$ 确定的本构方程 $\underline{\boldsymbol{T}} = \underline{\boldsymbol{T}}(\underline{\boldsymbol{Y}})$ 可简化为标量状态形式，即 $\underline{t} = \underline{t}(\underline{e})$，通过对拉伸标量状态 $\underline{e}$ 进行分解，可将上述本构方程进一步表示为 $\underline{t} = \underline{t}(\theta, \underline{e}^{\mathrm{d}})$ 或 $\underline{t} = \underline{t}(\underline{e}^{\mathrm{i}}, \underline{e}^{\mathrm{d}})$ 的形式。

考虑所研究的材料为简单材料，即其力学行为只依赖于物质点 $\boldsymbol{x}$ 邻域表征的变形梯度历史，亦即近场动力学中的力密度矢量状态 $\underline{\boldsymbol{T}}$ 仅依赖于变形矢量状态 $\underline{\boldsymbol{Y}}$，在这种情况下，当仅考虑材料弹性时，常规态型近场动力学中力密度矢量状态 $\underline{\boldsymbol{T}}$ 可表示为

$$\underline{\boldsymbol{T}} = \underline{\boldsymbol{T}}(\underline{\boldsymbol{Y}}) = \nabla W_{\mathrm{PD}}(\underline{\boldsymbol{Y}}) = \underline{t}\underline{\boldsymbol{M}} \tag{6.8}$$

式中，$W_{\mathrm{PD}}(\underline{\boldsymbol{Y}})$ 为近场动力学变形能密度函数，∇ 为弗雷歇导数，$\underline{\boldsymbol{M}}\langle\boldsymbol{\xi}\rangle = (\boldsymbol{\xi}+\boldsymbol{\eta})/|\boldsymbol{\xi}+\boldsymbol{\eta}|$ 为键的单位方向矢量，$\underline{t}$ 为力密度标量状态。

借鉴连续介质力学张量分解思想，力密度标量可分解为球量 $\underline{t}^{\mathrm{i}}$ 和偏量 $\underline{t}^{\mathrm{d}}$，即

$$\underline{t}\langle\boldsymbol{\xi}\rangle = \underline{t}^{\mathrm{i}}\langle\boldsymbol{\xi}\rangle + \underline{t}^{\mathrm{d}}\langle\boldsymbol{\xi}\rangle \tag{6.9}$$

类似连续介质力学，存在下述关系式

$$\underline{t} = \frac{\partial W_{\mathrm{PD}}}{\partial \underline{e}}, \quad \underline{t}^{\mathrm{i}} = \frac{\partial W_{\mathrm{PD}}}{\partial \underline{e}^{\mathrm{i}}}, \quad \underline{t}^{\mathrm{d}} = \frac{\partial W_{\mathrm{PD}}}{\partial \underline{e}^{\mathrm{d}}} \tag{6.10}$$

此外，定义压力为

$$p = -\frac{\underline{t} \cdot \underline{x}}{3} = -\frac{\partial W}{\partial \theta} \tag{6.11}$$

进而，定义力密度标量的球量为

$$\underline{t}^{\mathrm{i}}\langle\boldsymbol{\xi}\rangle = -3p\frac{w\underline{x}\langle\boldsymbol{\xi}\rangle}{m} \tag{6.12}$$

式中，w、m 为前文所述的影响函数和加权量。

6.1.2 弹性本构模型的建立过程

常规态型近场动力学模型的本构方程 $\underline{t} = \underline{t}(\theta, \underline{e}^{\mathrm{d}})$ 易于直接获得, 而本构方程 $\underline{t} = \underline{t}(\underline{e}^{\mathrm{i}}, \underline{e}^{\mathrm{d}})$ 是塑性建模的基础, 需要以 $\underline{t}(\theta, \underline{e}^{\mathrm{d}})$ 本构方程为媒介求得, 以下建立一维、二维和三维形式统一的近场动力学弹性固体本构模型 (linear peridynamic solid, LPS)。具体的本构建模步骤如下 [2-4]:

(1) 以体积应变和偏应变张量为参量, 确定连续介质力学中的弹性应变能密度;
(2) 选择近场动力学本构建模变量, 构造近场动力学中的弹性变形能密度函数;
(3) 利用近场动力学弹性变形能密度函数与经典连续介质力学弹性应变能密度函数等效, 确定近场动力学的模型参数;
(4) 通过能量密度的弗雷歇导数等运算, 导出力密度矢量状态与变形矢量状态的本构关系式。

6.1.2.1 连续介质力学的弹性应变能密度

经典连续介质力学的弹性应变能密度可分解为体积应变能密度和畸变能密度, 三维问题的弹性应变能密度为

$$W_{\mathrm{CCM}} = \frac{K}{2}\varepsilon_{ii}^2 + G\sum_{i,j=1,2,3}\varepsilon_{ij}^{\mathrm{d}}\varepsilon_{ij}^{\mathrm{d}} \tag{6.13}$$

式中, K 为体积模量, $G = E/(2(1+\nu))$ 为剪切模量, E 为杨氏模量, ν 为泊松比, ε_{ii} 为体积应变, $\varepsilon_{ij}^{\mathrm{d}}$ 为应变偏张量分量, 且有 $\varepsilon_{ij}^{\mathrm{d}} = \varepsilon_{ij} - \frac{1}{3}\varepsilon_{mm}\delta_{ij}$, 其中 δ_{ij} 为克罗内克符号。

表 6-1 给出了三维、平面应变、平面应力和一维问题对应的应力张量和应变张量等, 将这些物理量代入式 (6.13), 可以推得不同维度的弹性应变能密度, 为

$$W_{\mathrm{CCM}} = \begin{cases} \dfrac{K}{2}\varepsilon_{ii}^2 + G\displaystyle\sum_{i,j=1,2,3}\varepsilon_{ij}^{\mathrm{d}}\varepsilon_{ij}^{\mathrm{d}}, & \text{三维} \\[2mm] \left(\dfrac{K}{2} + \dfrac{G}{9}\right)\varepsilon_{ii}^2 + G\displaystyle\sum_{i,j=1,2}\varepsilon_{ij}^{\mathrm{d}}\varepsilon_{ij}^{\mathrm{d}}, & \text{平面应变} \\[2mm] \left[\dfrac{K}{2} + \dfrac{G}{9}\left(\dfrac{\nu+1}{2\nu-1}\right)^2\right]\varepsilon_{ii}^2 + G\displaystyle\sum_{i,j=1,2}\varepsilon_{ij}^{\mathrm{d}}\varepsilon_{ij}^{\mathrm{d}}, & \text{平面应力} \\[2mm] \left(\dfrac{K}{2} + \dfrac{2G}{3}\right)\varepsilon_{ii}^2, & \text{一维} \end{cases} \tag{6.14}$$

6.1.2.2 近场动力学的弹性变形能密度

考虑三维问题, 根据式 (6.13) 给出的连续介质力学应变能密度的表达式, 选取近场动力学中的膨胀量和拉伸标量状态的偏量为建模变量, 可以类似地构建如

表 6-1 三维、二维和一维应变、应力状态变量列表

	三维	平面应变	平面应力	一维
应变张量	$\boldsymbol{\varepsilon} = \begin{pmatrix} \varepsilon_{11} & \varepsilon_{12} & \varepsilon_{13} \\ \varepsilon_{21} & \varepsilon_{22} & \varepsilon_{23} \\ \varepsilon_{31} & \varepsilon_{32} & \varepsilon_{33} \end{pmatrix}$	$\boldsymbol{\varepsilon} = \begin{pmatrix} \varepsilon_{11} & \varepsilon_{12} & 0 \\ \varepsilon_{21} & \varepsilon_{22} & 0 \\ 0 & 0 & 0 \end{pmatrix}$	$\boldsymbol{\varepsilon} = \begin{pmatrix} \varepsilon_{11} & \varepsilon_{12} & 0 \\ \varepsilon_{21} & \varepsilon_{22} & 0 \\ 0 & 0 & \varepsilon_{33} \end{pmatrix}$	$\boldsymbol{\varepsilon} = \begin{pmatrix} \varepsilon_{11} & 0 & 0 \\ 0 & 0 & 0 \\ 0 & 0 & 0 \end{pmatrix}$
应力张量	$\boldsymbol{\sigma} = \begin{pmatrix} \sigma_{11} & \sigma_{12} & \sigma_{13} \\ \sigma_{21} & \sigma_{22} & \sigma_{23} \\ \sigma_{31} & \sigma_{32} & \sigma_{33} \end{pmatrix}$	$\boldsymbol{\sigma} = \begin{pmatrix} \sigma_{11} & \sigma_{12} & 0 \\ \sigma_{21} & \sigma_{22} & 0 \\ 0 & 0 & \sigma_{33} \end{pmatrix}$	$\boldsymbol{\sigma} = \begin{pmatrix} \sigma_{11} & \sigma_{12} & 0 \\ \sigma_{21} & \sigma_{22} & 0 \\ 0 & 0 & 0 \end{pmatrix}$	$\boldsymbol{\sigma} = \begin{pmatrix} \sigma_{11} & 0 & 0 \\ 0 & 0 & 0 \\ 0 & 0 & 0 \end{pmatrix}$
胡克定律	$\varepsilon_{11} = \frac{1}{E}(\sigma_{11} - \nu\sigma_{22} - \nu\sigma_{33})$ $\varepsilon_{22} = \frac{1}{E}(\sigma_{22} - \nu\sigma_{11} - \nu\sigma_{33})$ $\varepsilon_{33} = \frac{1}{E}(\sigma_{33} - \nu\sigma_{11} - \nu\sigma_{22})$	$\varepsilon_{11} = \frac{1}{E}(\sigma_{11} - \nu\sigma_{22} - \nu\sigma_{33})$ $\varepsilon_{22} = \frac{1}{E}(\sigma_{22} - \nu\sigma_{11} - \nu\sigma_{33})$ $\varepsilon_{33} = 0$	$\varepsilon_{11} = \frac{1}{E}(\sigma_{11} - \nu\sigma_{22})$ $\varepsilon_{22} = \frac{1}{E}(\sigma_{22} - \nu\sigma_{11})$ $\varepsilon_{33} = -\frac{\nu}{E}(\sigma_{11} + \sigma_{22})$	$\varepsilon_{11} = \frac{\sigma_{11}}{E}$ $\varepsilon_{22} = 0$ $\varepsilon_{33} = 0$
体积应变	$\varepsilon_{ii} = \varepsilon_{11} + \varepsilon_{22} + \varepsilon_{33}$	$\varepsilon_{ii} = \varepsilon_{11} + \varepsilon_{22}$	因 $\varepsilon_{33} = \frac{\nu}{\nu-1}(\varepsilon_{11}+\varepsilon_{22})$, 故 $\varepsilon_{ii} = \frac{2\nu-1}{\nu-1}(\varepsilon_{11}+\varepsilon_{22})$, 且 $\varepsilon_{33} = \frac{\nu}{2\nu-1}\varepsilon_{ii}$	$\varepsilon_{ii} = \varepsilon_{11} + \varepsilon_{22} + \varepsilon_{33}$ $= \varepsilon_{11}$
应变偏张量	$\varepsilon^{\mathrm{d}}_{ij} = \varepsilon_{ij} - \frac{1}{3}\varepsilon_{mm}\delta_{ij}$	$\boldsymbol{\varepsilon}^{\mathrm{d}} = \begin{pmatrix} \varepsilon^{\mathrm{d}}_{11} & \varepsilon^{\mathrm{d}}_{12} & 0 \\ \varepsilon^{\mathrm{d}}_{21} & \varepsilon^{\mathrm{d}}_{22} & 0 \\ 0 & 0 & \varepsilon^{\mathrm{d}}_{33} \end{pmatrix}$ 其中 $\varepsilon^{\mathrm{d}}_{33} = -\frac{1}{3}\varepsilon_{ii}$	$\boldsymbol{\varepsilon}^{\mathrm{d}} = \begin{pmatrix} \varepsilon^{\mathrm{d}}_{11} & \varepsilon^{\mathrm{d}}_{12} & 0 \\ \varepsilon^{\mathrm{d}}_{21} & \varepsilon^{\mathrm{d}}_{22} & 0 \\ 0 & 0 & \varepsilon^{\mathrm{d}}_{33} \end{pmatrix}$ 其中 $\varepsilon^{\mathrm{d}}_{33} = \frac{\nu+1}{3(2\nu-1)}\varepsilon_{ii}$	$\boldsymbol{\varepsilon}^{\mathrm{d}} = \begin{pmatrix} \varepsilon^{\mathrm{d}}_{11} & 0 & 0 \\ 0 & \varepsilon^{\mathrm{d}}_{22} & 0 \\ 0 & 0 & \varepsilon^{\mathrm{d}}_{33} \end{pmatrix}$ 其中 $\nu=0$, $\varepsilon^{\mathrm{d}}_{11} = \frac{2}{3}\varepsilon_{ii}$, $\varepsilon^{\mathrm{d}}_{22} = -\frac{1}{3}\varepsilon_{ii}$, $\varepsilon^{\mathrm{d}}_{33} = -\frac{1}{3}\varepsilon_{ii}$

下形式的近场动力学弹性变形能密度 $W_{\mathrm{PD}}(\theta, \underline{e}^{\mathrm{d}})$,为 [2]

$$W_{\mathrm{PD}}(\theta, \underline{e}^{\mathrm{d}}) = \frac{k\theta^2}{2} + \frac{\alpha}{2}(w\underline{e}^{\mathrm{d}}) \cdot \underline{e}^{\mathrm{d}} \tag{6.15}$$

式中,k 和 α 为与体积模量 K 和剪切模量 G 等宏观材料参数相关的待定参数。当材料在各方向发生均匀变形 (纯体积变形) 时,有 $W_{\mathrm{PD}}(\theta, \underline{e}^{\mathrm{d}}) = \frac{k\theta^2}{2}$;当材料发生纯剪切变形时,有 $W_{\mathrm{PD}}(\theta, \underline{e}^{\mathrm{d}}) = \frac{\alpha}{2}(w\underline{e}^{\mathrm{d}}) \cdot \underline{e}^{\mathrm{d}}$;对于一般的变形状态,弹性变形能密度为两者的叠加。

注意到式 (6.1) 实质上是两物质点构成的键在变形过程中伸长量的计算关系式,参考基于连续介质力学得到的任意方向微段伸长量的关系式 (2.9),式 (6.1) 可以进一步表示为

$$\underline{e}\langle\boldsymbol{\xi}\rangle = |\boldsymbol{\xi} + \boldsymbol{\eta}| - |\boldsymbol{\xi}| = |\boldsymbol{F}\boldsymbol{\xi}| - |\boldsymbol{\xi}| = \varepsilon_{ij}\xi_i\xi_j/|\boldsymbol{\xi}| \tag{6.16}$$

式中,$\boldsymbol{F}$ 为变形梯度张量,考虑到上式右端是应变张量 ε_{ij} 的线性关系,则拉伸

标量的偏量与应变偏量亦具有如下关系[2]

$$\underline{e}^{\mathrm{d}}\langle\boldsymbol{\xi}\rangle = \varepsilon_{ij}^{\mathrm{d}}\xi_i\xi_j/|\boldsymbol{\xi}| \tag{6.17}$$

下面以二维问题为例，给出弹性变形能密度的计算方法。对于一维问题，可类似处理，对于三维问题，可参考文献 [2] 推导得到。

当发生纯剪切变形时，将式 (6.17) 代入式 (6.15) 中，得到弹性变形能密度

$$W_{\mathrm{PD}}(\theta,\underline{e}^{\mathrm{d}}) = \frac{\alpha}{2}(w\underline{e}^{\mathrm{d}})\cdot\underline{e}^{\mathrm{d}} = \frac{\alpha}{2}\int_{H_\infty} w(\underline{e}^{\mathrm{d}}\langle\boldsymbol{\xi}\rangle)^2 \mathrm{d}V_{\boldsymbol{\xi}}$$

$$= \frac{\alpha}{2}\int_{H_\infty} w\left(\frac{\varepsilon_{ij}^{\mathrm{d}}\xi_i\xi_j}{|\boldsymbol{\xi}|}\right)\left(\frac{\varepsilon_{kl}^{\mathrm{d}}\xi_k\xi_l}{|\boldsymbol{\xi}|}\right) \mathrm{d}V_{\boldsymbol{\xi}}$$

$$= \frac{\alpha}{2}\int_{H_\infty} \frac{w}{|\boldsymbol{\xi}|^2}\left[\sum_{i,j=1,2}\varepsilon_{ij}^{\mathrm{d}}\xi_i\xi_j\right]\left[\sum_{k,l=1,2}\varepsilon_{kl}^{\mathrm{d}}\xi_k\xi_l\right] \mathrm{d}V_{\boldsymbol{\xi}}$$

$$= \frac{\alpha}{2}\int_{H_\infty} \frac{w}{|\boldsymbol{\xi}|^2}\left[\begin{array}{l}(\varepsilon_{11}^{\mathrm{d}})^2(\xi_1)^4+(\varepsilon_{22}^{\mathrm{d}})^2(\xi_2)^4\\ +4(\varepsilon_{12}^{\mathrm{d}})^2(\xi_1)^2(\xi_2)^2+2\varepsilon_{11}^{\mathrm{d}}\varepsilon_{22}^{\mathrm{d}}(\xi_1)^2(\xi_2)^2\\ +4\varepsilon_{11}^{\mathrm{d}}\varepsilon_{12}^{\mathrm{d}}(\xi_1)^3\xi_2+4\varepsilon_{22}^{\mathrm{d}}\varepsilon_{12}^{\mathrm{d}}\xi_1(\xi_2)^3\end{array}\right] \mathrm{d}V_{\boldsymbol{\xi}}$$

$$= \frac{\alpha}{2}\left[\frac{3m}{8}(\varepsilon_{11}^{\mathrm{d}})^2 + \frac{3m}{8}(\varepsilon_{22}^{\mathrm{d}})^2 + \frac{4m}{8}(\varepsilon_{12}^{\mathrm{d}})^2 + \frac{2m}{8}\varepsilon_{11}^{\mathrm{d}}\varepsilon_{22}^{\mathrm{d}}\right]$$

$$= \frac{\alpha}{2}\left[\frac{2m}{8}\sum_{i,j=1,2}\varepsilon_{ij}^{\mathrm{d}}\varepsilon_{ij}^{\mathrm{d}} + \frac{m}{8}\left(\sum_{i,j=1,2}\varepsilon_{ii}^{\mathrm{d}}\right)^2\right]$$

$$= \frac{\alpha m}{16}(\varepsilon_{33}^{\mathrm{d}})^2 + \frac{\alpha m}{8}\sum_{i,j=1,2}\varepsilon_{ij}^{\mathrm{d}}\varepsilon_{ij}^{\mathrm{d}}$$

$$= \begin{cases} \dfrac{\alpha m}{144}\theta^2 + \dfrac{\alpha m}{8}\sum_{i,j=1,2}\varepsilon_{ij}^{\mathrm{d}}\varepsilon_{ij}^{\mathrm{d}}, & \text{平面应变} \\ \dfrac{\alpha m}{144}\dfrac{(\nu+1)^2}{(2\nu-1)^2}\theta^2 + \dfrac{\alpha m}{8}\sum_{i,j=1,2}\varepsilon_{ij}^{\mathrm{d}}\varepsilon_{ij}^{\mathrm{d}}, & \text{平面应力} \end{cases} \tag{6.18}$$

式中，第四行到第五行的推导利用了如下关系：任意奇数的指标分量只出现一次时，其圆形区域积分为零，故只需计算含有 $(\xi_1)^4$、$(\xi_2)^4$ 和 $(\xi_1)^2(\xi_2)^2$ 项的积分，即 $\int_{H_\infty}\dfrac{w}{|\boldsymbol{\xi}|^2}(\xi_1)^4\mathrm{d}V_{\boldsymbol{\xi}} = \dfrac{3m}{8}$、$\int_{H_\infty}\dfrac{w}{|\boldsymbol{\xi}|^2}(\xi_2)^4\mathrm{d}V_{\boldsymbol{\xi}} = \dfrac{3m}{8}$、$\int_{H_\infty}\dfrac{w}{|\boldsymbol{\xi}|^2}(\xi_1)^2(\xi_2)^2\mathrm{d}V_{\boldsymbol{\xi}} = \dfrac{m}{8}$，其中 $m = 2\pi l_z\int_0^\delta w\langle r\rangle r^3\mathrm{d}r$。在平面极坐标系中，记 $r = |\boldsymbol{\xi}|$、$\xi_1 = r\cos\varphi$、$\xi_2 = r\sin\varphi$

和 $dV_\xi = l_z r dr d\varphi$，其中的 l_z 为考察体厚度，对于平面问题取为单位长度 1，此处显含 l_z 是为了保持量纲一致性。此外，上式中第六行到第七行的推导利用了纯剪切变形时的关系 $\varepsilon_{11}^d + \varepsilon_{22}^d + \varepsilon_{33}^d = 0$；第七行到八行的推导利用了表 6-1 中给出的应变偏量与体积应变的关系 $\varepsilon_{33}^d = -\frac{1}{3}\theta$ 和 $\varepsilon_{33}^d = \frac{\nu+1}{3(2\nu-1)}\theta$。

于是，考虑一般的变形状态，并将一维和三维问题对应的结果直接引入，汇总后得到近场动力学弹性变形能密度为

$$W_{\mathrm{PD}}(\theta, \underline{e}^d) = \begin{cases} \dfrac{k}{2}\theta^2 + \dfrac{\alpha m}{15}\sum_{i,j=1,2,3}\varepsilon_{ij}^d \varepsilon_{ij}^d, & \text{三维} \\[2mm] \left(\dfrac{k}{2} + \dfrac{\alpha m}{144}\right)\theta^2 + \dfrac{\alpha m}{8}\sum_{i,j=1,2}\varepsilon_{ij}^d \varepsilon_{ij}^d, & \text{平面应力} \\[2mm] \left[\dfrac{k}{2} + \dfrac{\alpha m}{144}\dfrac{(\nu+1)^2}{(2\nu-1)^2}\right]\theta^2 + \dfrac{\alpha m}{8}\sum_{i,j=1,2}\varepsilon_{ij}^d \varepsilon_{ij}^d, & \text{平面应力} \\[2mm] \left(\dfrac{k}{2} + \dfrac{2\alpha m}{9}\right)\theta^2, & \text{一维} \end{cases} \quad (6.19)$$

进一步，基于近场动力学弹性变形能密度与经典连续介质力学弹性应变能密度相等的方法，即令式 (6.19) 与式 (6.14) 相等，确定模型参数 k 和 α 为

$$\begin{cases} k = K, \ \alpha = \dfrac{15G}{m}, & \text{三维} \\[2mm] k = K + \dfrac{G}{9}, \ \alpha = \dfrac{8G}{m}, & \text{平面应力} \\[2mm] k = K + \dfrac{G(\nu+1)^2}{9(2\nu-1)^2}, \ \alpha = \dfrac{8G}{m}, & \text{平面应力} \\[2mm] k = K, \ \alpha = \dfrac{3G}{m}, & \text{一维} \end{cases} \quad (6.20)$$

6.1.2.3 本构方程的推导

在常规态型近场动力学模型中，可以通过能量密度的弗雷歇导数等运算，导出受力状态与变形状态的本构关系式，具体如下。

将能量密度函数 W_{PD} 对拉伸标量状态 $\underline{e}$ 进行弗雷歇导数计算，并参照式 (6.10)，得到

$$\Delta W_{\mathrm{PD}}(\underline{e}) = \frac{\partial W_{\mathrm{PD}}}{\partial \underline{e}} \cdot \Delta \underline{e} = \underline{t} \cdot \Delta \underline{e} = (\underline{t}^i + \underline{t}^d) \cdot \Delta \underline{e} \quad (6.21)$$

同样，将能量密度函数 W_{PD} 对膨胀量 θ 和拉伸标量状态的偏量 $\underline{e}^d$ 进行弗雷歇导数计算，得到

$$\Delta W_{\mathrm{PD}}(\theta, \underline{e}^{\mathrm{d}}) = \frac{\partial W_{\mathrm{PD}}}{\partial \theta} \frac{\partial \theta}{\partial \underline{e}} \cdot \Delta \underline{e} + \frac{\partial W_{\mathrm{PD}}}{\partial \underline{e}^{\mathrm{d}}} \cdot \Delta \underline{e}^{\mathrm{d}} \qquad (6.22)$$

结合式 (6.3) 和式 (6.6)，进行弗雷歇求导运算，有下式成立

$$\Delta \underline{e}^{\mathrm{d}} = \Delta \underline{e} - \Delta \underline{e}^{\mathrm{i}} = \Delta \underline{e} - \frac{\partial \underline{e}^{\mathrm{i}}}{\partial \theta} \frac{\partial \theta}{\partial \underline{e}} \Delta \underline{e} = \Delta \underline{e} - \frac{\gamma}{3} \underline{x} \frac{w\underline{x}}{m} \cdot \Delta \underline{e} \qquad (6.23)$$

下面，先计算 $(w\underline{e}^{\mathrm{d}}) \cdot \underline{x}$，以备后续推导使用，具体为

$$\begin{aligned}
(w\underline{e}^{\mathrm{d}}) \cdot \underline{x} &= w(\underline{e} - \underline{e}^{\mathrm{i}}) \cdot \underline{x} = w\left(\underline{e} - \frac{\theta\underline{x}}{3}\right) \cdot \underline{x} = (w\underline{e}) \cdot \underline{x} - \frac{\theta(w\underline{x}) \cdot \underline{x}}{3} \\
&= (w\underline{e}) \cdot \underline{x} - \frac{\theta m}{3} = \frac{\theta m}{\gamma} - \frac{\theta m}{3} = \left(\frac{1}{\gamma} - \frac{1}{3}\right)\theta m \qquad (6.24)
\end{aligned}$$

将式 (6.23) 和式 (6.24) 代入式 (6.22) 中，进行如下推导

$$\begin{aligned}
\Delta W_{\mathrm{PD}}(\theta, \underline{e}^{\mathrm{d}}) &= \frac{\partial W_{\mathrm{PD}}}{\partial \theta} \frac{\partial \theta}{\partial \underline{e}} \cdot \Delta \underline{e} + \frac{\partial W_{\mathrm{PD}}}{\partial \underline{e}^{\mathrm{d}}} \cdot \Delta \underline{e}^{\mathrm{d}} \\
&= \frac{\partial W_{\mathrm{PD}}}{\partial \theta} \frac{\partial \theta}{\partial \underline{e}} \cdot \Delta \underline{e} + \frac{\partial W_{\mathrm{PD}}}{\partial \underline{e}^{\mathrm{d}}} \cdot \left(\Delta \underline{e} - \frac{\gamma}{3}\underline{x}\frac{w\underline{x}}{m} \cdot \Delta \underline{e}\right) \\
&= \left\{\frac{\partial W_{\mathrm{PD}}}{\partial \theta} \frac{\partial \theta}{\partial \underline{e}} + \frac{\partial W_{\mathrm{PD}}}{\partial \underline{e}^{\mathrm{d}}} - \frac{\partial W_{\mathrm{PD}}}{\partial \underline{e}^{\mathrm{d}}} \cdot \frac{\gamma}{3}\underline{x}\frac{w\underline{x}}{m}\right\} \cdot \Delta \underline{e} \\
&= \left\{(k\theta)\left(\gamma\frac{w\underline{x}}{m}\right) + \alpha w\underline{e}^{\mathrm{d}} - (\alpha w\underline{e}^{\mathrm{d}}) \cdot \frac{\gamma}{3}\underline{x}\frac{w\underline{x}}{m}\right\} \cdot \Delta \underline{e} \\
&= \left\{(k\theta)\gamma\left(\frac{w\underline{x}}{m}\right) + \alpha w\underline{e}^{\mathrm{d}} - (\alpha m\theta)\frac{3-\gamma}{9}\left(\frac{w\underline{x}}{m}\right)\right\} \cdot \Delta \underline{e} \\
&= \left\{\left(\gamma k - \frac{(3-\gamma)\alpha m}{9}\right)\frac{w\theta\underline{x}}{m} + \alpha w\underline{e}^{\mathrm{d}}\right\} \cdot \Delta \underline{e} \qquad (6.25)
\end{aligned}$$

比较式 (6.25) 与式 (6.21)，可得 $\underline{t}(\theta, \underline{e}^{\mathrm{d}})$ 形式的力密度标量状态，即

$$\underline{t}(\underline{e}) = \underline{t}(\theta, \underline{e}^{\mathrm{d}}) = \left[\frac{\gamma k}{m} - \frac{(3-\gamma)\alpha}{9}\right]w\theta\underline{x} + \alpha w\underline{e}^{\mathrm{d}} \qquad (6.26)$$

依据力密度标量状态分解的思想，上式第二个等号右端第一项 $\left[\dfrac{\gamma k}{m} - \dfrac{(3-\gamma)\alpha}{9}\right] \times w\theta\underline{x} = \underline{t}^{\mathrm{i}}$，第二项 $\alpha w\underline{e}^{\mathrm{d}} = \underline{t}^{\mathrm{d}}$。进而，根据已知拉伸标量状态的球量与膨胀量的关系，将式 (6.6) 的关系代入上式，则可得到 $\underline{t}(\underline{e}^{\mathrm{i}}, \underline{e}^{\mathrm{d}})$ 形式的力密度标量状态，即

$$\underline{t}(\underline{e}) = \underline{t}(\underline{e}^{\mathrm{i}}, \underline{e}^{\mathrm{d}}) = \left[\frac{3\gamma k}{m} - \frac{(3-\gamma)\alpha}{3}\right]w\underline{e}^{\mathrm{i}} + \alpha w\underline{e}^{\mathrm{d}} \qquad (6.27)$$

将上式代入式 (6.8), 即可得到力密度矢量状态与变形矢量状态的本构关系式。

至此, 得到一维、二维、三维形式统一的常规态型近场动力学弹性本构模型的两种表达式。

需要说明的是, 此处基于弗雷歇导数运算得到的本构关系式 (6.26) 和式 (6.27) 只是微弹性的, 并不是线性的。考虑小变形假定, 常规态型近场动力学弹性模型的线性化近似可参见文献 [6,7], 此处不再赘述。

由于常规态型近场动力学模型采用与经典连续介质力学能量等效方法确定参数, 要求物质点具有完整对称的近场范围, 但边界区域物质点的近场范围并不完整, 导致边界处的参数计算存在误差, 同样也存在 "表面效应" 问题[5]。

6.2 常规态型近场动力学弹塑性模型

本节将介绍常规态型近场动力学的弹塑性模型的建模方法, 给出弹塑性本构模型、屈服准则和塑性流动法则、一致性条件和塑性乘子计算方法、屈服函数中的材料强度参数确定方法和弹塑性分析算法流程。综合而言, 常规态型近场动力学的弹塑性、黏弹性等非线性模型的研究尚不充分[4,8-14], 还在发展完善中。

6.2.1 弹塑性本构模型的一般形式

近场动力学假定固体的永久塑性变形由 "键" 层次变形决定, 但屈服准则和流动法则既可以在 "键" 上定义, 也可以在 "点" 上定义。键型近场动力学塑性模型在分析纯体积变形问题时, 体积不可压缩材料会出现永久塑性变形现象, 与经典塑性不可压缩理论相矛盾, 常规态型近场动力学的塑性模型可以较好地解决该问题。

在弹性阶段, 键力密度函数可由考察体的变形状态唯一确定, 与变形历史无关; 但在塑性阶段, 键力密度函数不仅与当前变形状态有关, 还依赖于整个键族的变形历史, 即键力密度是键族当前变形与变形历史的函数。在材料内一点的键力密度进入塑性状态后, 对式 (6.5) 中的拉伸标量状态的球量 $\underline{e}^i$ 和偏量 $\underline{e}^d$ 进行弹塑性加法分解, 即[8,9]

$$\underline{e}^i = \underline{e}^{ie} + \underline{e}^{ip}, \quad \underline{e}^d = \underline{e}^{de} + \underline{e}^{dp} \tag{6.28}$$

对于金属等塑性变形体积不可压缩材料, 球量部分不分解, 可只将偏量部分进行弹塑性分解。对于混凝土和岩土 (多孔介质) 等塑性变形体积可压缩材料, 需同时进行拉伸标量状态的球量和偏量的弹塑性分解。其中, 弹性部分服从常规态型近场动力学弹性固体本构模型, 即式 (6.26) 或式 (6.27), 同时考虑松弛构型 (即塑性变形时无应力贡献), 则可以假定仅键的弹性变形对力密度标量状态的球量与偏量部分有贡献,则 $\underline{t} = \underline{t}(\underline{e})$ 形式的本构模型可化为 $\underline{t} = \underline{t}(\underline{e}^{ie}, \underline{e}^{de})$ 或 $\underline{t} = \underline{t}(\underline{e}^{i}, \underline{e}^{ip}, \underline{e}^{d}, \underline{e}^{dp})$ 形式。

类似式 (6.15),近场动力学的弹性变形能密度可表示为 [9]

$$W_{\mathrm{PD}}(\underline{e}^{ie}, \underline{e}^{de}) = \frac{k'}{2}(w\underline{e}^{ie} \cdot \underline{e}^{ie}) + \frac{\alpha'}{2}(w\underline{e}^{de} \cdot \underline{e}^{de}) \quad (6.29a)$$

或

$$W_{\mathrm{PD}}(\underline{e}^{i}, \underline{e}^{ip}, \underline{e}^{d}, \underline{e}^{dp}) = \frac{k'}{2}\left[w(\underline{e}^{i} - \underline{e}^{ip}) \cdot (\underline{e}^{i} - \underline{e}^{ip})\right] + \frac{\alpha'}{2}\left[w(\underline{e}^{d} - \underline{e}^{dp}) \cdot (\underline{e}^{d} - \underline{e}^{dp})\right] \quad (6.29b)$$

其中,k' 和 α' 是与体积模量 K 和剪切模量 G 相关的待定参数。

对式 (6.29a) 给出的弹性变形能密度关于 $\underline{e}^{ie}$ 和 $\underline{e}^{de}$ 求弗雷歇导数,即采用与式 (6.21) ~ 式 (6.27) 类似的推导过程,便可得到适用于塑性变形体积可压缩材料的近场动力学弹塑性 (elastoplastic peridynamic solid, EPS) 本构方程

$$\underline{t} = k'w(\underline{e}^{i} - \underline{e}^{ip}) + \alpha'w(\underline{e}^{d} - \underline{e}^{dp}) \quad (6.30)$$

令拉伸标量状态的球量塑性部分 $\underline{e}^{ip} = 0$,则适用于塑性变形体积不可压缩材料的近场动力学弹塑性本构方程为

$$\underline{t} = k'w\underline{e}^{i} + \alpha'w(\underline{e}^{d} - \underline{e}^{dp}) \quad (6.31)$$

当仅发生弹性变形时,式 (6.29b) 和式 (6.30) 中 $\underline{e}^{ip} = 0$ 和 $\underline{e}^{dp} = 0$,两式即退化为常规态型近场动力学弹性固体模型的应变能密度和本构方程,因此弹塑性本构模型的参数与弹性模型参数相同。在上述情况下,令式 (6.30) 中的 $\underline{e}^{ip} = 0$,并与式 (6.27) 进行对比,得到 $k' = \dfrac{3\gamma k}{m} - \alpha\left(1 - \dfrac{\gamma}{3}\right)$,$\alpha' = \alpha$。于是,塑性变形体积可压缩材料和不可压缩材料的弹塑性本构方程的一般形式分别为

$$\underline{t}(\underline{e}^{i}, \underline{e}^{d}) = \left[\frac{3\gamma k}{m} - \frac{(3-\gamma)\alpha}{3}\right]w(\underline{e}^{i} - \underline{e}^{ip}) + \alpha w(\underline{e}^{d} - \underline{e}^{dp}) \quad (6.32)$$

$$\underline{t}(\underline{e}^{i}, \underline{e}^{d}) = \left[\frac{3\gamma k}{m} - \frac{(3-\gamma)\alpha}{3}\right]w\underline{e}^{i} + \alpha w(\underline{e}^{d} - \underline{e}^{dp}) \quad (6.33)$$

其中,参数 k, α, γ 如式 (6.4) 和式 (6.20) 所示。

6.2.2 屈服函数与塑性流动法则

受传统 J_2 塑性理论的启发,Mitchell[8] 首先建立常规态型近场动力学弹塑性模型。在力密度标量状态空间 S 中,考虑松弛构型 (塑性变形时无应力贡献),假定仅键的弹性变形对力密度标量状态有贡献,故需依据屈服准则来判定键所处的弹塑性状态。定义键处于弹性变形时的力密度标量状态集合为 N,即

$$N = \{\underline{t} \in S \,|\, F(\underline{t}, \kappa) = f(\underline{t}) - \kappa \leqslant 0\} \quad (6.34)$$

其中，$F(\underline{t},\kappa)$ 为屈服函数，表示力密度标量状态空间中的屈服面；$f(\underline{t})$ 为力密度标量状态 $\underline{t}$ 的函数，可包含偏量力状态 $\underline{t}^{\mathrm{d}}$ 和静水压力 p；κ 为材料强度参数，可反映加载塑性硬化和软化效应。弹性变形时 $F \leqslant 0$，塑性变形时 $F > 0$，此时需给定塑性流动势函数和流动法则，采用回映算法将超出屈服面的力状态重新拉回屈服面上，并计算塑性变形增量。

对于塑性变形体积不可压缩的金属材料，类比 Mises 屈服函数 $F = J_2 - \kappa$，定义常规态型近场动力学的屈服函数为 [8]

$$F(\underline{t}^{\mathrm{d}},\kappa) = \frac{\underline{t}^{\mathrm{d}} \cdot \underline{t}^{\mathrm{d}}}{2} - \kappa \tag{6.35}$$

对于塑性变形体积可压缩的混凝土和岩土类等多孔介质材料，类比 Drucker-Prager 屈服函数 $F = \sqrt{J_2} + \alpha I_1 - \kappa$，定义常规态型近场动力学的屈服函数 $F(\underline{t}^{\mathrm{d}}, \underline{t}^{\mathrm{i}}, \kappa)$ 或 $F(\underline{t}^{\mathrm{d}}, p, \kappa)$ 为 [9]

$$F = (\underline{t}^{\mathrm{d}} \cdot \underline{t}^{\mathrm{d}})^{1/2} - \beta p - \kappa \tag{6.36}$$

式中，β 为与内摩擦角相关的参数。

弹塑性材料的本构关系一般以增量形式给出，以反映变形的历史，称为塑性增量理论，也称塑性流动理论。为了描述材料发生塑性变形时的变化规律，引入塑性流动势函数 $\mathcal{G}$ 表征塑性应变增量与加载曲面 (后继屈服面) 的关系。对于塑性变形体积不可压缩材料，取 $\mathcal{G} = F$，即采用屈服函数 F 作为塑性流动势函数，称之为与屈服函数相关联的流动法则，在这种情况下，塑性应变增量的矢量与屈服面正交，故也称为正交法则。对于塑性变形体积可压缩材料，一般 $\mathcal{G} \neq F$，称之为非关联流动法则，即塑性应变增量的矢量与屈服面不正交，混凝土、岩土材料的塑性本构关系一般认为服从非关联流动法则。

在近场动力学中，非关联的塑性流动势函数 $\mathcal{G}$ 可以表示为

$$\mathcal{G} = (\underline{t}^{\mathrm{d}} \cdot \underline{t}^{\mathrm{d}})^{1/2} + \psi p \tag{6.37}$$

式中，ψ 为与膨胀角相关的参数。

拉伸标量状态的塑性增量可由关联或非关联塑性流动法则确定，具体为

$$\underline{\dot{e}}^{\mathrm{p}} = \lambda \nabla \mathcal{G} \tag{6.38}$$

式中，λ 为塑性乘子 (plasticity multiplier)；$\underline{\dot{e}}^{\mathrm{p}}$ 包含球量 $\underline{\dot{e}}^{\mathrm{ip}}$ 和偏量 $\underline{\dot{e}}^{\mathrm{dp}}$ 贡献，并有下式成立

$$\underline{\dot{e}}^{\mathrm{p}} = \underline{\dot{e}}^{\mathrm{ip}} + \underline{\dot{e}}^{\mathrm{dp}} = \lambda \nabla \mathcal{G} = \lambda \left(\frac{\partial \mathcal{G}}{\partial \underline{t}^{\mathrm{i}}} + \frac{\partial \mathcal{G}}{\partial \underline{t}^{\mathrm{d}}} \right) = \lambda (\nabla^{\mathrm{i}} \mathcal{G} + \nabla^{\mathrm{d}} \mathcal{G})$$

$$\Rightarrow \underline{\dot{e}}^{\mathrm{ip}} = \lambda \nabla^{\mathrm{i}} \mathcal{G}, \underline{\dot{e}}^{\mathrm{dp}} = \lambda \nabla^{\mathrm{d}} \mathcal{G} \tag{6.39}$$

其中，$\nabla^{\mathrm{i}} \mathcal{G}$ 和 $\nabla^{\mathrm{d}} \mathcal{G}$ 分别为塑性流动势函数对力密度标量状态的球量和偏量的弗雷歇导数；两类材料分别取相应屈服函数和塑性流动势函数时，$\nabla^{\mathrm{i}}(\cdot)$ 和 $\nabla^{\mathrm{d}}(\cdot)$ 的具体值如表 6-2 所示。

表 6-2 塑性变形体积不可压缩和可压缩材料的屈服函数与塑性流动法则

材料	屈服函数	塑性流动势函数	$\nabla^{\mathrm{i}} \mathcal{G}$ 与 $\nabla^{\mathrm{d}} \mathcal{G}$
塑性变形体积不可压缩（关联流动法则）	$F = \dfrac{\underline{t}^{\mathrm{d}} \cdot \underline{t}^{\mathrm{d}}}{2} - \kappa$	$\mathcal{G} = \dfrac{\underline{t}^{\mathrm{d}} \cdot \underline{t}^{\mathrm{d}}}{2} - \kappa$	$\nabla^{\mathrm{i}} \mathcal{G} = 0, \quad \nabla^{\mathrm{d}} \mathcal{G} = \underline{t}^{\mathrm{d}}$
塑性变形体积可压缩（关联流动法则）	$F = (\underline{t}^{\mathrm{d}} \cdot \underline{t}^{\mathrm{d}})^{1/2} - \beta p - \kappa$	$\mathcal{G} = (\underline{t}^{\mathrm{d}} \cdot \underline{t}^{\mathrm{d}})^{1/2} - \beta p - \kappa$	$\nabla^{\mathrm{i}} \mathcal{G} = \dfrac{\beta \underline{x}}{3},$ $\nabla^{\mathrm{d}} \mathcal{G} = \dfrac{\underline{t}^{\mathrm{d}}}{(\underline{t}^{\mathrm{d}} \cdot \underline{t}^{\mathrm{d}})^{1/2}}$
塑性变形体积可压缩（非关联流动法则）	$F = (\underline{t}^{\mathrm{d}} \cdot \underline{t}^{\mathrm{d}})^{1/2} - \beta p - \kappa$	$\mathcal{G} = (\underline{t}^{\mathrm{d}} \cdot \underline{t}^{\mathrm{d}})^{1/2} + \psi p$	$\nabla^{\mathrm{i}} \mathcal{G} = -\dfrac{\psi \underline{x}}{3},$ $\nabla^{\mathrm{d}} \mathcal{G} = \dfrac{\underline{t}^{\mathrm{d}}}{(\underline{t}^{\mathrm{d}} \cdot \underline{t}^{\mathrm{d}})^{1/2}}$

6.2.3 一致性条件与塑性乘子

类似于经典弹塑性理论，近场动力学也使用一致性条件 (consistency condition) $\lambda \dot{F} = 0$ 判断考察体加载和卸载过程的弹塑性状态，具体为

$$\begin{cases} \lambda = 0, & F(\underline{t}, \kappa) < 0, & \text{弹性加卸载} \\ \lambda > 0, & F(\underline{t}, \kappa) = 0, & \dot{F}(\underline{t}, \kappa) = 0, & \text{塑性加载} \\ \lambda = 0, & F(\underline{t}, \kappa) = 0, & \dot{F}(\underline{t}, \kappa) < 0, & \text{塑性卸载} \end{cases} \tag{6.40}$$

弹塑性变形分析的关键是确定不同受力状态下的塑性乘子。已知一致性条件、屈服函数、非关联塑性流动势函数和力密度矢量状态，对屈服函数进行弗雷歇导数运算，得到

$$\lambda \dot{F} = \lambda \left(\frac{\partial F}{\partial \underline{t}^{\mathrm{i}}} \frac{\partial \underline{t}^{\mathrm{i}}}{\partial \underline{e}^{\mathrm{ie}}} \cdot \frac{\partial (\underline{e}^{\mathrm{i}} - \underline{e}^{\mathrm{ip}})}{\partial t} + \frac{\partial F}{\partial \underline{t}^{\mathrm{d}}} \frac{\partial \underline{t}^{\mathrm{d}}}{\partial \underline{e}^{\mathrm{de}}} \cdot \frac{\partial (\underline{e}^{\mathrm{d}} - \underline{e}^{\mathrm{dp}})}{\partial t} \right) = 0$$

$$\Rightarrow \lambda \dot{F} = \lambda \left(\frac{\partial F}{\partial \underline{t}^{\mathrm{i}}} \frac{\partial \underline{t}^{\mathrm{i}}}{\partial \underline{e}^{\mathrm{ie}}} \cdot (\underline{\dot{e}}^{\mathrm{i}} - \underline{\dot{e}}^{\mathrm{ip}}) + \frac{\partial F}{\partial \underline{t}^{\mathrm{d}}} \frac{\partial \underline{t}^{\mathrm{d}}}{\partial \underline{e}^{\mathrm{de}}} \cdot (\underline{\dot{e}}^{\mathrm{d}} - \underline{\dot{e}}^{\mathrm{dp}}) \right) = 0$$

$$\Rightarrow \lambda \dot{F} = \lambda \left(\nabla^{\mathrm{i}} F(k'w) \cdot (\underline{\dot{e}}^{\mathrm{i}} - \underline{\dot{e}}^{\mathrm{ip}}) + \nabla^{\mathrm{d}} F(\alpha' w) \cdot (\underline{\dot{e}}^{\mathrm{d}} - \underline{\dot{e}}^{\mathrm{dp}}) \right) = 0$$

$$\Rightarrow k'w \nabla^{\mathrm{i}} F \cdot (\underline{\dot{e}}^{\mathrm{i}} - \underline{\dot{e}}^{\mathrm{ip}}) + \alpha' w \nabla^{\mathrm{d}} F \cdot (\underline{\dot{e}}^{\mathrm{d}} - \underline{\dot{e}}^{\mathrm{dp}}) = 0$$

$$\Rightarrow k'w \nabla^{\mathrm{i}} F \cdot (\underline{\dot{e}}^{\mathrm{i}} - \lambda \nabla^{\mathrm{i}} \mathcal{G}) + \alpha' w \nabla^{\mathrm{d}} F \cdot (\underline{\dot{e}}^{\mathrm{d}} - \lambda \nabla^{\mathrm{d}} \mathcal{G}) = 0$$

$$\Rightarrow \lambda = \frac{k'w\nabla^{\mathrm{i}}F \cdot \underline{\dot{e}}^{\mathrm{i}} + \alpha'w\nabla^{\mathrm{d}}F \cdot \underline{\dot{e}}^{\mathrm{d}}}{k'w\nabla^{\mathrm{i}}F \cdot \nabla^{\mathrm{i}}\mathcal{G} + \alpha'w\nabla^{\mathrm{d}}F \cdot \nabla^{\mathrm{d}}\mathcal{G}} \tag{6.41}$$

相似地，当使用关联塑性流动势函数 F 时，塑性乘子为

$$\lambda = \frac{k'w\nabla^{\mathrm{i}}F \cdot \underline{\dot{e}}^{\mathrm{i}} + \alpha'w\nabla^{\mathrm{d}}F \cdot \underline{\dot{e}}^{\mathrm{d}}}{k'w\nabla^{\mathrm{i}}F \cdot \nabla^{\mathrm{i}}F + \alpha'w\nabla^{\mathrm{d}}F \cdot \nabla^{\mathrm{d}}F} \tag{6.42}$$

其中，$k' = \dfrac{3\gamma k}{m} - \alpha\left(1 - \dfrac{\gamma}{3}\right)$ 和 $\alpha' = \alpha$，模型参数 k, α, γ 如式 (6.4) 和式 (6.20) 所示。

6.2.4 材料强度参数的确定

Mitchell 给出了三维情况下塑性变形体积不可压缩材料强度参数 κ 与 J_2 塑性屈服函数联系的计算方法 [8]，Lammi 和 Vogler 建立了塑性变形体积可压缩材料的屈服函数与传统 Drucker-Prager 塑性屈服函数的联系 [9]。本书作者团队推导得到了二维模型的强度参数 [4]，以下给出具体推导过程。

对于塑性变形体积不可压缩材料，塑性屈服函数和流动势函数一致。关注物质点 $\boldsymbol{x}$ 及其近场范围内相互作用物质点 $\boldsymbol{x}'$，设物质点 $\boldsymbol{x}$ 为坐标原点，参考构型中 $\boldsymbol{x}'$ 相对于 $\boldsymbol{x}$ 点的平面直角坐标为 (ξ_1, ξ_2)，且在平面极坐标系中，有 $\xi_1 = \cos\varphi|\boldsymbol{\xi}|$，$\xi_2 = \sin\varphi|\boldsymbol{\xi}|$。考虑构型发生纯剪切变形，且当切应变达到 γ 时发生屈服，忽略刚体位移后，可求得相对位移 $\boldsymbol{u}(\boldsymbol{\xi}) = \begin{bmatrix} \xi_2\gamma & \xi_1\gamma \end{bmatrix}$，则键的伸长量为

$$\begin{aligned} \underline{e} = |\underline{\boldsymbol{Y}}| - |\underline{\boldsymbol{X}}| &= \sqrt{(\xi_1 + \gamma\xi_2)^2 + (\xi_2 + \gamma\xi_1)^2} - |\boldsymbol{\xi}| \\ &= \left(\sqrt{1 + 4\cos\varphi\sin\varphi\gamma + \gamma^2} - 1\right)|\boldsymbol{\xi}| \\ &\approx 2\cos\varphi\sin\varphi\gamma|\boldsymbol{\xi}| = \underline{e}^{\mathrm{d}} \end{aligned} \tag{6.43}$$

为保持量纲的一致性，在下面的推导中引入厚度 l_z(对于平面问题可取为单位 1)，则有

$$\begin{aligned} \underline{e}^{\mathrm{d}} \cdot \underline{e}^{\mathrm{d}} = |\underline{e}^{\mathrm{d}}|^2 &= \int_{H_x} \underline{e}^{\mathrm{d}}\underline{e}^{\mathrm{d}}\mathrm{d}V_{\boldsymbol{\xi}} = \int_{H_x} (2\cos\varphi\sin\varphi\gamma|\boldsymbol{\xi}|)^2 \mathrm{d}V_{\boldsymbol{\xi}} \\ &= \int_0^{\delta}\int_0^{2\pi} (2\cos\varphi\sin\varphi\gamma|\boldsymbol{\xi}|)^2 l_z |\boldsymbol{\xi}|\,\mathrm{d}|\boldsymbol{\xi}|\,\mathrm{d}\varphi \\ &= 4l_z \int_0^{\delta} \gamma^2|\boldsymbol{\xi}|^3\,\mathrm{d}|\boldsymbol{\xi}| \int_0^{2\pi} (\cos\varphi\sin\varphi)^2\mathrm{d}\varphi = \frac{\pi}{4}\gamma^2\delta^4 l_z \end{aligned} \tag{6.44}$$

6.2 常规态型近场动力学弹塑性模型

根据式 (6.27) 可知 $\underline{t}^{\mathrm{d}} = \alpha w \underline{e}^{\mathrm{d}}$，并考虑屈服时 $F = \underline{t}^{\mathrm{d}} \cdot \underline{t}^{\mathrm{d}}/2 - \kappa = 0$，当取影响函数为 $w=1$ 时，材料的强度参数为

$$\begin{aligned}\kappa &= \left(\frac{\underline{t}^{\mathrm{d}} \cdot \underline{t}^{\mathrm{d}}}{2}\right)_0 = \frac{1}{2}\alpha^2 \left|w\underline{e}^{\mathrm{d}}\right|^2 = \frac{1}{2}\alpha^2 \left|\underline{e}^{\mathrm{d}}\right|^2 \\ &= \frac{1}{2}\left(\frac{8G}{m}\right)^2 \frac{\pi}{4}\gamma^2\delta^4 l_z = \frac{8\pi\delta^4}{m^2}(G\gamma)^2 l_z \\ &= \frac{8\pi\delta^4\tau_{\mathrm{s}}^2}{m^2}l_z \\ &= \frac{32\tau_{\mathrm{s}}^2}{\pi l_z \delta^4} = \frac{32\sigma_{\mathrm{s}}^2}{3\pi l_z \delta^4}\end{aligned} \qquad (6.45)$$

式中，$(\cdot)_0$ 表示括号内项为材料屈服时的对应值，加权量 $m = 2\pi l_z \int\limits_0^\delta w\langle r\rangle r^3 \mathrm{d}r = \frac{\pi l_z \delta^4}{2}$，$\delta$ 为近场范围尺寸，G 为剪切模量，τ_{s} 为纯剪切屈服应力，其与单轴屈服应力的关系为 $\sigma_{\mathrm{s}} = \sqrt{3}\tau_{\mathrm{s}}$。

这里顺便给出 Mitchell 针对三维问题，求得的塑性变形体积不可压缩材料的强度参数 [8]，具体为

$$\kappa = \frac{75\tau_{\mathrm{s}}^2}{8\pi\delta^5} = \frac{25\sigma_{\mathrm{s}}^2}{8\pi\delta^5} \qquad (6.46)$$

对于塑性变形体积可压缩材料，采用屈服函数 $F = (\underline{t}^{\mathrm{d}} \cdot \underline{t}^{\mathrm{d}})^{1/2} - \beta p - \kappa$ 和非关联塑性流动势函数 $\mathcal{G} = (\underline{t}^{\mathrm{d}} \cdot \underline{t}^{\mathrm{d}})^{1/2} + \psi p$，建立其与传统 Drucker-Prager 塑性模型的联系，以确定参数 β、κ 和 ψ。

当构型发生纯剪切变形并导致材料屈服时，有静水压力 $p=0$ 和屈服函数 $F=0$，取影响函数为 $w=1$，则屈服面方程为

$$\kappa = (\underline{t}^{\mathrm{d}} \cdot \underline{t}^{\mathrm{d}})_0^{\frac{1}{2}} = \sqrt{\alpha^2 \left|w\underline{e}^{\mathrm{d}}\right|^2} = \sqrt{\alpha^2 \left|\underline{e}^{\mathrm{d}}\right|^2} = \tau_{\mathrm{s}}\sqrt{\frac{16\pi\delta^4}{m^2}l_z} \qquad (6.47)$$

再考虑各向同性加载 (纯体积变形) 导致材料屈服的情形，此时，力密度矢量状态的偏量相关函数为零，根据屈服函数 $F=0$ 的条件，并参考式 (6.47)，则有

$$\kappa = -\beta p_0 \Rightarrow \beta = -\frac{\kappa}{p_0} = \frac{\tau_{\mathrm{s}}}{-p_0}\sqrt{\frac{16\pi\delta^4}{m^2}l_z} \qquad (6.48)$$

式中，p_0 为屈服失效压强 (含义与经典连续介质力学的静水压力一致，以拉为正)，τ_{s} 为最大容许剪应力，反映材料的抗剪黏聚强度。注意到 τ_{s} 与 $-p_0$(通常受压，

$-p_0$ 为正值) 的比值即为传统 Draucker-Prager 塑性模型内摩擦角 $\beta_{\rm DP}$ 的正切, 则有

$$\beta = -\frac{\kappa}{p_0} = \tan\beta_{\rm DP}\sqrt{\frac{16\pi\delta^4}{m^2}l_z} \tag{6.49}$$

进而, 类比近场动力学模型的内摩擦系数 β 与传统 Draucker-Prager 塑性模型中的内摩擦角 $\beta_{\rm DP}$ 的关系, 直接给出近场动力学模型的膨胀系数 ψ 与传统 Drucker-Prager 模型中的膨胀角 $\psi_{\rm DP}$ 的关系

$$\psi = \tan\psi_{\rm DP}\sqrt{\frac{16\pi\delta^4}{m^2}l_z} \tag{6.50}$$

Lammi 和 Vogler 针对三维问题, 求出了塑性变形体积可压缩材料的强度参数 [9], 为

$$\kappa = \tau_{\rm s}\sqrt{\frac{12\pi\delta^5}{m^2}},\ \beta = \tan(\beta_{\rm DP})\sqrt{\frac{12\pi\delta^5}{m^2}},\ \psi = \tan(\psi_{\rm DP})\sqrt{\frac{12\pi\delta^5}{m^2}} \tag{6.51}$$

6.2.5 回映算法流程

采用增量法求解弹塑性问题, 要求在每一增量步中, 迭代计算收敛。具体实施时, 采用向后欧拉数值积分法, 计算拉伸标量状态球量和偏量的塑性部分, 采用回映算法 (return mapping), 使得超出屈服面的力密度状态沿着垂直于流动势函数的路径, 映射返回屈服面。下面给出具体算法流程, 其中的脚标 n、$n+1$ 分别表示时刻 t 和 $t+\Delta t$。

步骤 1: 初始时刻 $t=0$, 设置一些变量初始值为 0, 包括: $\theta, \underline{e}^{\rm i}, \underline{e}^{\rm ip}, \underline{e}^{\rm d}, \underline{e}^{\rm dp}, \underline{t}, \underline{t}^{\rm i}$, $\underline{t}^{\rm d}, p, \Delta\lambda$。

步骤 2: t 时刻, 已完成 n 次计算, 进行第 $n+1$ 次分析。由塑性变形历史, 已知 $\underline{e}^{\rm i}_n, \underline{e}^{\rm ip}_n, \underline{e}^{\rm d}_n, \underline{e}^{\rm dp}_n$ 等变量值, 由第 $n+1$ 次计算之初的物质点坐标, 可求得

$$\underline{e}_{n+1} = |\boldsymbol{\xi}+\boldsymbol{\eta}|_{n+1} - |\boldsymbol{\xi}|_0,\ \theta_{n+1} = \frac{\gamma}{m}(w\boldsymbol{x})\cdot\underline{e}_{n+1},\ \underline{e}^{\rm i}_{n+1} = \frac{\theta_{n+1}\boldsymbol{x}}{3},\ \underline{e}^{\rm d}_{n+1} = \underline{e}_{n+1} - \underline{e}^{\rm i}_{n+1} \tag{6.52}$$

步骤 3: 进行第 $n+1$ 时步的弹性试算。假定除历史塑性变形外, 不产生新的塑性变形, 即 $\Delta\lambda = 0$、$\underline{e}^{\rm ie}_{n+1} = \underline{e}^{\rm i}_{n+1} - \underline{e}^{\rm ip}_n$ 和 $\underline{e}^{\rm de}_{n+1} = \underline{e}^{\rm d}_{n+1} - \underline{e}^{\rm dp}_n$。此时, 力密度标量状态的球量、偏量和静水压力的试算值分别为

$$\begin{cases} \underline{t}^{\rm i}_{\rm trial} = \left[\frac{3\gamma k}{m} - \alpha\left(1-\frac{\gamma}{3}\right)\right]w\underline{e}^{\rm ie}_{n+1} = \left[\frac{3\gamma k}{m} - \alpha\left(1-\frac{\gamma}{3}\right)\right]w(\underline{e}^{\rm i}_{n+1} - \underline{e}^{\rm ip}_n) \\ \underline{t}^{\rm d}_{\rm trial} = \alpha w\underline{e}^{\rm de}_{n+1} = \alpha w(\underline{e}^{\rm d}_{n+1} - \underline{e}^{\rm dp}_n) \\ p^{\rm trial}_{n+1} = -k\theta_{n+1} \end{cases} \tag{6.53}$$

步骤 4：计算屈服函数 $F(\underline{t}_{\text{trial}}^{\text{d}}, \kappa)$ 或 $F(\underline{t}_{\text{trial}}^{\text{d}}, p_{\text{trial}}, \kappa)$，用以判断 $n+1$ 时步物质点是否屈服。

步骤 5：判断物质点所处的弹塑性状态，并计算第 $n+1$ 次真实的变形状态和力状态。

(1) 若 $F(\underline{t}_{\text{trial}}, \kappa) \leqslant 0$ 或 $F(\underline{t}_{\text{trial}}^{\text{d}}, p_{\text{trial}}, \kappa) \leqslant 0$，则步骤 3 中弹性试算正确，即有 $\underline{e}_{n+1}^{\text{ip}} = \underline{e}_n^{\text{ip}}$、$\underline{e}_{n+1}^{\text{dp}} = \underline{e}_n^{\text{dp}}$ 和式 (6.53) 成立。

(2) 若 $F(\underline{t}_{\text{trial}}, \kappa) > 0$ 或 $F(\underline{t}_{\text{trial}}^{\text{d}}, p_{\text{trial}}, \kappa) > 0$，则物质点处于非弹性阶段，则塑性乘子和相应的力密度标量状态和变形标量状态分别为

$$\begin{cases} \Delta \lambda = \dfrac{1}{\alpha' w}\left[\dfrac{\|\underline{t}_{\text{trial}}^{\text{d}}\|}{\sqrt{2\kappa}} - 1\right], & \text{关联流动法则} \\ \Delta \lambda = \dfrac{\sqrt{\underline{t}_{\text{trial}}^{\text{d}} \cdot \underline{t}_{\text{trial}}^{\text{d}}} - \beta p_{\text{trial}} - \kappa}{\alpha' w + k'\beta\psi}, & \text{非关联流动法则} \end{cases} \quad (6.54)$$

$$\begin{cases} \underline{t}_{n+1}^{\text{i}} = \underline{t}_{\text{trial}}^{\text{i}} - k'w\Delta\lambda\nabla^{\text{i}}\mathcal{G} \\ \underline{t}_{n+1}^{\text{d}} = \underline{t}_{\text{trial}}^{\text{d}} - \alpha'w\Delta\lambda\nabla^{\text{d}}\mathcal{G} \\ p_{n+1} = p_{\text{trial}} + k'\Delta\lambda\psi \\ \underline{e}_{n+1}^{\text{ip}} = \underline{e}_n^{\text{ip}} + \Delta\lambda\nabla^{\text{i}}\mathcal{G} \\ \underline{e}_{n+1}^{\text{dp}} = \underline{e}_n^{\text{dp}} + \Delta\lambda\nabla^{\text{d}}\mathcal{G} \end{cases} \quad (6.55)$$

步骤 6：根据近场动力学运动方程，更新物质点位置坐标；若满足收敛性条件，则进入步骤 7，否则重复步骤 2 至步骤 6 的过程。

步骤 7：获得 $n+1$ 次加载后的物质点位置坐标，即 $n+2$ 计算之初的坐标情况；进入下一增量步计算，直至计算完成。

6.3 数值算例

本节分别采用常规态型近场动力学弹性模型、弹脆性模型和弹塑性模型，对三个典型算例进行计算分析，具体为：含预制中心圆孔板的拉伸弹性变形[15]、随机多孔环氧板的拉伸脆性开裂[15] 和二维方板的弹塑性响应[4]。

6.3.1 含预制中心圆孔板的弹性变形

如图 6-1 所示，几何尺寸 100mm×50mm 的二维矩形板中心含一直径为 20mm 的孔洞，板左端固定约束、右端受均布拉力 $\sigma = 1$MPa，板为均匀各向同性弹性材料，相关参数为：质量密度$\rho = 1100$kg/m³、弹性模量 $E = 3.26$GPa、泊松比 $\nu = 0.38$、抗拉强度 $f_t = 62.89$MPa。近场动力学中的计算参数为：物质点边长 $\Delta x = 0.5$mm、物质点总数 19501、近场范围半径 $\delta = 3\Delta x$、人工阻尼系数 $C = 4 \times 10^6$、时间步长 $\Delta t = 2 \times 10^{-7}$s、计算总时步 6×10^4。

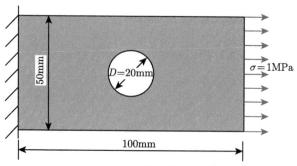

图 6-1　含中心圆孔板受拉伸荷载作用

图 6-2 给出了基于常规态型近场动力学弹性模型计算得到的板水平中线上的水平位移和垂直中线上的竖向位移的曲线,并同时绘出了采用有限单元法计算所得的结果。比较两种方法的计算结果可以看出,变形规律一致,数值差异很小,验证了常规态型近场动力学弹性模型的合理性和可靠性。

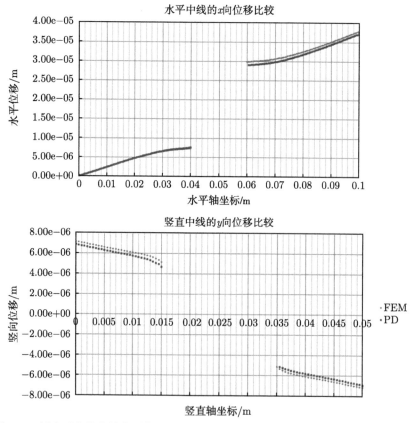

图 6-2　板水平中线上的水平位移和垂直中线上的竖向位移:PD 和 FEM 结果比较

6.3.2 随机多孔脆性环氧板的拉伸开裂

如图 6-3 所示，含多个随机分布圆形孔的二维方形环氧板受竖向单轴拉伸载荷作用，通过速度边界条件实现加载，板的几何参数为 82.5mm×82.5mm，共有 31 个直径 $D = 6.4$mm 的孔洞，孔洞率为 23%，采用二维泊松过程生成孔洞圆心位置，孔洞互不重合。假定环氧板为均匀各向同性弹脆性材料，材料参数为：质量密度 $\rho = 1100$kg/m^3、弹性模量 $E = 3.26$GPa、泊松比 $\nu = 0.38$、抗拉强度 $f_t = 62.89$MPa、临界伸长率 $s_0 = f_t/E = 0.01929$。

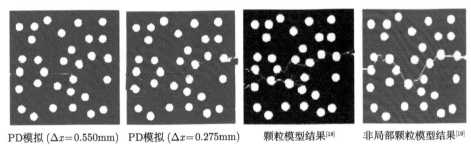

PD模拟 (Δx=0.550mm)　　PD模拟 (Δx=0.275mm)　　颗粒模型结果[18]　　非局部颗粒模型结果[19]

图 6-3　多孔环氧板的最终裂纹模式

对于该问题，Al-Ostaz 等 [17] 和 Ostoja-Starzewski 等 [18] 进行了试验研究，Ostoja-Starzewski 等 [18] 和 Lin 等 [19] 还分别采用颗粒模型和非局部颗粒模型进行了计算模拟。在应用常规态型近场动力学弹脆性模型进行该问题求解时，共建立了两个均匀离散的几何模型，物质点间距分别为 $\Delta x = 0.550$mm 和 $\Delta x = 0.275$mm，物质点总数分别为 19517 和 77432；采用显式算法进行计算，时间步长分别为 $\Delta t = 2 \times 10^{-7}$s 和 $\Delta t = 1 \times 10^{-7}$s，共迭代 1×10^5 步和 2×10^5 步。试验 [17] 中的加载速率约为 1.7×10^{-5}m/s，由于在数值模拟中，加载速率小意味着时间步长小，过小的时间步长将使得计算效率偏低，还会导致计算截断误差过大，经综合考虑，计算中采用 $v = 0.025$m/s 的速度，比拟准静态加载。

文献 [18] 和 [19] 的计算模型均存在泊松比为 1/3 的限制，为便于比较，在二维常规态型近场动力学计算中，也选取泊松比 $\nu = 1/3$，并将近场动力学与文献 [18] 和 [19] 的计算结果共同列入图 6-3 中。结果表明：采用近场动力学两种物质点间距得到的多孔板裂纹模式一致性良好，包括裂纹主路径和裂纹集中分布区域；近场动力学与颗粒模型和非局部颗粒模型得到的裂纹主路径基本相同，均能反映裂纹扩展过程中的偏转、分叉、局部破碎等现象，但不同方法所得的最终裂纹模式在部分区域还是有所差异的。

进一步，选取材料的真实泊松比 $\nu = 0.38$，采用常规态型近场动力学进行计算模拟，并将计算结果与试验结果 [17,18] 一并列入图 6-4 中。Ostoja-Starzewski

等[18]的试验结果显示，由于材料性质的空间变异性与随机性，试验给出的多孔环氧板的开裂模式并不完全相同，数值模拟结果与试验结果的严格匹配难度很大，但 PD 预测的裂纹模式符合 Ostoja-Starzewski 等[18]试验得到的 7 种主要裂纹分布特征，且预测的路径与 Al-Ostaz 等[17]的试验结果高度一致，显示近场动力学在预测结构复杂裂纹扩展模式方面具有独特的优势。

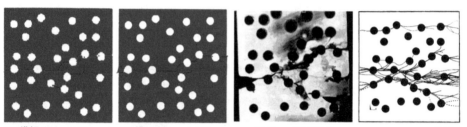

PD模拟 ($\Delta x = 0.550$mm)　　PD模拟 ($\Delta x = 0.275$mm)　　Al-Ostaz等试验[17]　　Ostoja-Starzewski 等[18]试验

图 6-4　多孔环氧板的最终裂纹模式

图 6-5 给出了基于近场动力学方法计算得到的多孔环氧板的损伤累积与裂纹扩展过程的若干细节，首先在孔洞边缘薄弱处产生损伤，继而损伤累积与裂纹扩展，损伤不断演化发展，先后在多处产生新的裂纹，最后形成贯穿性裂纹导致板的断裂破坏，并呈现出多裂纹同时扩展、裂纹偏转、裂纹汇聚和裂纹止裂等现象。

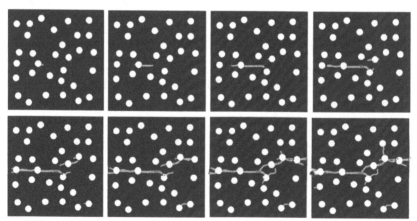

图 6-5　近场动力学预测的多孔环氧板的渐进破坏过程 ($\Delta x = 0.275$mm，$\nu = 0.38$，$v = 0.025$m/s)

6.3.3　方板的加卸载弹塑性响应

一均匀各向同性的二维方板，边长 $L = 1$m，质量密度 $\rho = 4428$kg/m^3，杨氏模量 $E = 113$GPa，泊松比 $\nu = 0.342$，方板左边界设置为固定约束，右边界施加水平位移。考虑如图 6-6 所示的两种位移梯度 (位移梯度 $\varepsilon_i = \partial u/\partial x$) 加载路径, 在线性加载路径 1 中, 位移持续线性增加, 并使得方板的应力超出材料的屈服强度, 在非线性加卸载路径 2 中, 历经加载、卸载、再加载的过程。

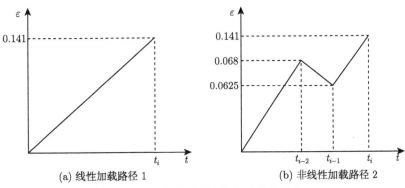

图 6-6 方板右边界上施加的位移梯度

采用常规态型近场动力学弹塑性模型模拟方板的力学行为, 均匀点阵离散方板, 物质点间距为 $\Delta x = \Delta y = 0.1 \text{m}$, 共离散 100×100 个物质点, 近场半径为 $\delta = 3\Delta x$, 时间步长取 $\Delta t = 1 \times 10^{-7}$ s。计算中, 选取式 (6.35) 的屈服准则, 屈服强度为 $\sigma_s = 1.017 \text{GPa}$, 不考虑塑性强化过程。

采用增量方式施加位移边界条件, 位移载荷等分为 304 步施加, 获得了每一加载步的稳态结果。常规态型近场动力学中没有应力概念, 为便于分析, 采用近场动力学中定义的等效应力 $\bar{\sigma} = \sqrt{6\alpha W^k}$[20], 其中, $W^k = \dfrac{k\theta^2}{2}$, α 和 k 是近场动力学的模型参数, 具体可见式 (6.20)。图 6-7 给出了两种路径下的支反力-水平

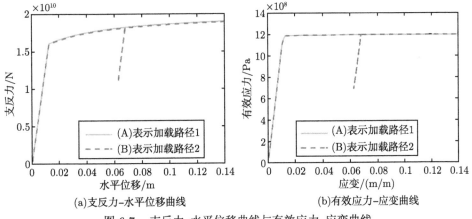

图 6-7 支反力-水平位移曲线与有效应力-应变曲线

位移曲线和有效应力-应变曲线。从图中所示结果可以看出，在加载的第一阶段，随着拉伸的增大，支反力与有效应力均线性增加；随着施加的位移不断增加，呈现了超出屈服应力后的强化现象；卸载与重新加载段保持线性变化规律，且与弹性段的直线平行。总体来看，采用常规态型近场动力学弹塑性模型计算得到的方板弹塑性力学行为符合客观规律。

参 考 文 献

[1] Silling S A. Reformulation of elasticity theory for discontinuities and long-range forces [J]. Journal of the Mechanics and Physics of Solids, 2000, 48(1):175-209.

[2] Silling S A, Epton M, Weckner O, et al. Peridynamic states and constitutive modeling [J]. Journal of Elasticity, 2007, 88(2):151-184.

[3] Le Q V, Chan W K, Schwartz J. A two-dimensional ordinary, state-based peridynamic model for linearly elastic solids [J]. International Journal for Numerical Methods in Engineering, 2014, 98(8):547-561.

[4] Li T Y, Gu X, Zhang Q, et al. Elastoplastic constitutive modeling for reinforced concrete in ordinary state-based peridynamics [J]. Journal of Mechanics, 2020, 36(6):799-811.

[5] Mitchell J, Silling S A, Littlewood D. A position-aware linear solid constitutive model for peridynamics [J]. Journal of Mechanics of Materials and Structures, 2015, 10(5):539-557.

[6] Silling S A. Linearized theory of peridynamic states [J]. Journal of Elasticity, 2010, 99(1):85-111.

[7] Sarego G, Le Q V, Bobaru F, et al. Linearized state-based peridynamics for 2-D problems [J]. International Journal for Numerical Methods in Engineering, 2016, 108(10): 1174-1197.

[8] Mitchell J A. A nonlocal ordinary state-based plasticity model for peridynamics [R]. Sandia National Lab Report, 2011.

[9] Lammi C J, Vogler T J. A nonlocal peridynamic plasticity model for the dynamic flow and fracture of concrete [R]. Sandia National Lab Report, 2014:18257.

[10] Asgari M, Kouchakzadeh M A. An equivalent von Mises stress and corresponding equivalent plastic strain for elastic–plastic ordinary peridynamics [J]. Meccanica, 2019, 54(7):1001-1014.

[11] Liu Z, Bie Y H, Cui Z, et al. Ordinary state-based peridynamics for nonlinear hardening plastic materials' deformation and its fracture process [J]. Engineering Fracture Mechanics, 2020, 223:106782.

[12] Mitchell J A. A non-local, ordinary-state-based viscoelasticity model for peridynamics [R]. Sandia National Lab Report, 2011:8064.

[13] Delorme R, Tabiai I, Lebel L L, et al. Generalization of the ordinary state-based peridynamic model for isotropic linear viscoelasticity [J]. Mechanics of Time-Dependent Materials, 2017, 21(4):549-575.

[14] Kulkarni S S, Tabarraei A. An ordinary state based peridynamic correspondence model for metal creep [J]. Engineering Fracture Mechanics, 2020, 233:107042.

[15] Zhang Q, Gu X, Huang D. Failure analysis of plate with non-uniform arrangement holes by ordinary state-based peridynamics [C].//Proceedings of the International Conference on Computational Methods, 2015, 2:1-10.

[16] Li T Y, Gu X, Zhang Q, et al. Coupled digital image correlation and peridynamics for full-field deformation measurement and local damage prediction [J]. CMES-Computer Modeling in Engineering & Sciences, 2019, 121(2):425-444.

[17] Al-Ostaz A, Jasiuk I. Crack initiation and propagation in materials with randomly distributed holes [J]. Engineering Fracture Mechanics, 1997, 58(5):395-420.

[18] Ostoja-Starzewski M, Wang G. Particle modeling of random crack patterns in epoxy plates [J]. Probabilistic Engineering Mechanics, 2006, 21(3):267-275.

[19] Lin E, Chen H L, Liu Y. Finite element implementation of a non-local particle method for elasticity and fracture analysis [J]. Finite Elements in Analysis and Design, 2015, 93:1-11.

[20] Madenci E, Oterkus S. Ordinary state-based peridynamics for plastic deformation according to Von Mises yield criteria with isotropic hardening [J]. Journal of the Mechanics and Physics of Solids, 2016, 86:192-219.

第 7 章 非常规态型近场动力学模型

如前所述，相比于键型近场动力学模型，常规态型近场动力学模型可以考虑体积变形和剪切变形，并突破了材料的泊松比限制，拓展了近场动力学的应用范围。但常规态型近场动力学模型相关定义的物理意义不够明晰，力–位移的本构描述与经典连续介质力学的本构关系缺乏联系，对于材料复杂力学行为的刻画尚有不足。于是，非常规态型近场动力学模型 (non-ordinary state-based peridynamics) 应运而生。作为一类典型的非常规态型近场动力学模型，近场动力学对应材料模型 (peridynamic correspondence material model)[1-3] 问世以后便广受关注。近场动力学对应材料模型以非局部变形梯度张量、应变张量、变形率张量、转动张量和应力张量等为媒介，可以将经典连续介质力学已发展的各种应力-应变 (率) 本构模型转化为近场动力学的力密度矢量状态与变形矢量状态之间的本构关系，便于在近场动力学框架内分析各类材料和结构的复杂力学响应和变形破坏行为，加强了近场动力学与经典连续介质力学的联系，大大拓宽了近场动力学的应用领域。本章给出非常规态型近场动力学的建模方法和模型的构建过程，系统介绍非常规态型近场动力学模型的隐式静力解法，以作为第 5 章显式动力学解法的补充。鉴于非常规态型近场动力学模型存在的数值不稳定性问题[4-6]，研究给出含稳定控制的数值计算策略，结合带预制裂纹矩形板的静力弹性变形分析，验证多种因素对数值计算稳定性的影响，并应用于多晶体弹塑性静动力变形破坏分析中。

7.1 非常规态型近场动力学建模方法

7.1.1 建模流程与基本方程

成对的向量并置形成并矢张量 (dyadic tensor)，局部理论采用六面体微元描述构型变化，如矢量微元 $\mathrm{d}\boldsymbol{\xi} = (\mathrm{d}\xi_1, \mathrm{d}\xi_2, \mathrm{d}\xi_3)$ 的并矢 $\mathrm{d}\boldsymbol{\xi} \otimes \mathrm{d}\boldsymbol{\xi}$ 能够描述微元体的几何形状改变。区别于局部理论，近场动力学采用近场域内点集的共同变形来描述构型的变化，显然，近场动力学的变形描述依赖于近场范围 H_x(积分体积) 的形状。考虑物质点 $\boldsymbol{x}$ 与物质点 $\boldsymbol{x}'$ 构成的相对位置矢量 $\boldsymbol{\xi}$，定义物质点 $\boldsymbol{x}$ 的非局部形状张量 (nonlocal shape tensor) $\boldsymbol{K}$ 为 [1]

$$\boldsymbol{K} = \underline{\boldsymbol{X}} * \underline{\boldsymbol{X}} = \int_{H_x} w(|\boldsymbol{\xi}|)(\boldsymbol{\xi} \otimes \boldsymbol{\xi}) \mathrm{d}V_{x'} \qquad (7.1)$$

其中，$\underline{\boldsymbol{X}}\langle\boldsymbol{\xi}\rangle=\boldsymbol{\xi}=\boldsymbol{x}'-\boldsymbol{x}$ 为点对初始相对位置矢量，w 为影响函数，$\otimes$ 为两矢量的张量积符号或并矢符号，$\mathrm{d}V_{\boldsymbol{x}'}$ 为物质点 $\boldsymbol{x}'$ 的积分体积。

定义状态缩减算子 (reduction operator) 为 [1]

$$\Re\{\underline{\boldsymbol{A}}\}=(\underline{\boldsymbol{A}}*\underline{\boldsymbol{X}})\boldsymbol{K}^{-1}=\int_{H_{\infty}}w(|\boldsymbol{\xi}|)(\underline{\boldsymbol{A}}\otimes\underline{\boldsymbol{X}})\mathrm{d}V_{\boldsymbol{x}'}\boldsymbol{K}^{-1} \tag{7.2}$$

其中，$\underline{\boldsymbol{A}}$ 为状态变量，该式表明，状态缩减算子作用在任意状态变量上，可获得对应的二阶梯度张量。

进而，定义物质点 $\boldsymbol{x}$ 的非局部变形梯度张量 $\boldsymbol{F}$ 和位移梯度张量 $\boldsymbol{H}$ [1]

$$\boldsymbol{F}=\nabla\underline{\boldsymbol{Y}}=\Re\{\underline{\boldsymbol{Y}}\}=(\underline{\boldsymbol{Y}}*\boldsymbol{X})\boldsymbol{K}^{-1}=\int_{H_{\infty}}w(|\boldsymbol{\xi}|)(\underline{\boldsymbol{Y}}\langle\boldsymbol{\xi}\rangle\otimes\boldsymbol{\xi})\mathrm{d}V_{\boldsymbol{x}'}\boldsymbol{K}^{-1}$$

$$=\int_{H_{\infty}}w(|\boldsymbol{\xi}|)((\boldsymbol{y}'-\boldsymbol{y})\otimes\boldsymbol{\xi})\mathrm{d}V_{\boldsymbol{x}'}\boldsymbol{K}^{-1} \tag{7.3}$$

$$\boldsymbol{H}=\nabla\underline{\boldsymbol{U}}=\Re\{\underline{\boldsymbol{U}}\}=(\underline{\boldsymbol{U}}*\boldsymbol{X})\boldsymbol{K}^{-1}=\int_{H_{\infty}}w(|\boldsymbol{\xi}|)(\underline{\boldsymbol{U}}\langle\boldsymbol{\xi}\rangle\otimes\boldsymbol{\xi})\mathrm{d}V_{\boldsymbol{x}'}\boldsymbol{K}^{-1}$$

$$=\int_{H_{\infty}}w(|\boldsymbol{\xi}|)((\boldsymbol{u}'-\boldsymbol{u})\otimes\boldsymbol{\xi})\mathrm{d}V_{\boldsymbol{x}'}\boldsymbol{K}^{-1} \tag{7.4}$$

式中，$\underline{\boldsymbol{Y}}$ 为当前相对位置状态，$\underline{\boldsymbol{U}}$ 为当前相对位移状态，且有 $\underline{\boldsymbol{U}}\langle\boldsymbol{\xi}\rangle=\boldsymbol{u}'-\boldsymbol{u}$。$\boldsymbol{F}$ 对近场范围内所有键的变形进行加权平均，以描述物质点的变形状态。

考虑到 $\boldsymbol{y}'-\boldsymbol{y}=\boldsymbol{\xi}+\boldsymbol{\eta}=(\boldsymbol{x}'-\boldsymbol{x})+(\boldsymbol{u}'-\boldsymbol{u})$，代入式 (7.3)，注意到式 (7.4) 中的 $(\boldsymbol{u}'-\boldsymbol{u})$ 替换为 $(\boldsymbol{x}'-\boldsymbol{x})$ 后，运算结果为单位张量 $\boldsymbol{I}$。进一步，对非局部变形梯度张量 $\boldsymbol{F}$ 进行变分运算，有

$$\delta\boldsymbol{F}=\delta\boldsymbol{H}+\delta\boldsymbol{I}=\int_{H_{\infty}}w(|\boldsymbol{\xi}|)((\delta\boldsymbol{u}'-\delta\boldsymbol{u})\otimes\boldsymbol{\xi})\mathrm{d}V_{\boldsymbol{x}'}\boldsymbol{K}^{-1} \tag{7.5}$$

将非局部变形梯度张量 $\boldsymbol{F}$ 近似作为局部变形梯度张量，根据经典连续介质力学的应变分析过程，计算变形率张量 $\boldsymbol{d}$、正交转动张量 $\boldsymbol{R}$ 和应变张量 $\boldsymbol{E}$ 等，根据某一客观①应力-应变本构模型，获得相应的应力张量。典型应力张量具有如下关系

$$\boldsymbol{P}=\det(\boldsymbol{F})\boldsymbol{\sigma}\boldsymbol{F}^{-\mathrm{T}}=\boldsymbol{F}\boldsymbol{T},\quad\boldsymbol{\sigma}=\boldsymbol{R}\boldsymbol{\tau}\boldsymbol{R}^{\mathrm{T}} \tag{7.6}$$

其中，$\det(\cdot)$ 为矩阵行列式值，$\boldsymbol{\sigma}$ 为现时构型的真实/有旋 Cauchy 应力张量，$\boldsymbol{\tau}$ 为参考构型的无旋 Cauchy 应力张量，$\boldsymbol{R}$ 为正交转动张量，可由极分解定理确定，

---

① 标架不变性原理要求建模变量及本构关系律均是客观的 [3,4]。

P 为第一类 Piola-Kirchhoff 应力张量，T 为第二类 Piola-Kirchhoff 应力张量。第一类 Piola-Kirchhoff 应力张量 P 用于传统线性本构或非线性本构的表征，是变形梯度张量 F 的函数，P 和 F 是一对功共轭量。

以下给出力密度矢量状态的建立过程，即由应力张量到力密度矢量状态的转换。Silling 在其文中通过式 (134)~式 (142)，给出了非常规态型近场动力学力密度矢量状态的推导方法[1]，再将经典连续介质力学应变能密度 W_CCM 和近场动力学变形能密度 W_PD 等效，并对变形能密度进行弗雷歇导数运算，获得近场动力学力密度矢量状态的表达式，即

$$\underline{T} = \nabla W_\text{PD}(\underline{Y}) = \nabla W_\text{CCM}(F(\underline{Y})) \tag{7.7}$$

变形梯度张量 F 的弗雷歇导数分量 ∇F_{ijk}，可通过变形梯度张量的分量 F_{ij} 关于变形矢量状态 $\underline{Y}$ 的增量计算获得[1]

$$\begin{aligned}
F_{ij}(\underline{Y}+\Delta\underline{Y}) &= \int_{H_x} w(\underline{y}_i\langle\boldsymbol{\xi}\rangle + \Delta\underline{y}_i\langle\boldsymbol{\xi}\rangle)\xi_p K_{pj}^{-1}\mathrm{d}V_{x'} \\
&= F_{ij}(\underline{Y}) + \int_{H_x} w\Delta\underline{y}_i\langle\boldsymbol{\xi}\rangle \xi_p K_{pj}^{-1}\mathrm{d}V_{x'} \\
&= F_{ij}(\underline{Y}) + \int_{H_x} w\Delta\underline{y}_k\langle\boldsymbol{\xi}\rangle \delta_{ik}\xi_p K_{pj}^{-1}\mathrm{d}V_{x'} \\
&= F_{ij}(\underline{Y}) + (\delta_{ik}wK_{pj}^{-1}\xi_p)\cdot\Delta\underline{y}_k \\
&= F_{ij}(\underline{Y}) + \nabla F_{ijk}\cdot\Delta\underline{y}_k
\end{aligned} \tag{7.8}$$

根据上述推导，并引入 ∇F_{ijk}，可见 $\nabla F_{ijk} = \delta_{ik}wK_{pj}^{-1}\xi_p$。

近场动力学变形能密度是位置矢量状态的函数，由变形矢量状态 Y 的增量 $\Delta\underline{y}_k$ 引起的变形能密度增量 ΔW_PD 表示为[1]

$$\Delta W_\text{PD} = \nabla_{\underline{y}_k}W_\text{PD}\cdot\Delta\underline{y}_k = \underline{T}_k\cdot\Delta\underline{y}_k \tag{7.9}$$

式中，最后一个等号利用了式 (7.7) 的分量形式。

另一方面，经典连续介质力学的应变能密度增量可以表示为

$$\begin{aligned}
\Delta W_\text{CCM} &= \frac{\partial W_\text{CCM}}{\partial F_{ij}}\Delta F_{ij} = P_{ij}\nabla F_{ijk}\cdot\Delta\underline{y}_k \\
&= (P_{ij}\delta_{ik}wK_{pj}^{-1}\xi_p)\cdot\Delta\underline{y}_k = (wP_{kj}K_{pj}^{-1}\xi_p)\cdot\Delta\underline{y}_k
\end{aligned} \tag{7.10}$$

在上式的推导过程中，第一个等号表示经典连续介质力学的应变能密度增量是关于变形梯度张量的全微分；第二个等号基于格林公式引入变形梯度张量的功

共轭量，即第一类 Piola-Kirchhoff 应力张量 P_{ij}，并利用式 (7.8) 给出的非局部变形梯度张量的增量表达式；第三、第四个等号是将式 (7.8) 得到的 ∇F_{ijk} 的具体形式代入，再进行哑标归并的自然结果。

根据近场动力学对应的变形能密度增量与经典连续介质力学相应的应变能密度增量相等的原则，比较式 (7.9) 和式 (7.10)，得到非常规态型近场动力学力密度矢量的分量 $\underline{T}_k = w P_{kj} K_{pj}^{-1} \xi_p$，则力密度矢量为 [1]

$$\underline{T}[\boldsymbol{x},t]\langle\boldsymbol{\xi}\rangle = w\boldsymbol{P}\boldsymbol{K}^{-1}\boldsymbol{\xi} \tag{7.11}$$

值得注意的是，$\boldsymbol{P}$ 和 $\boldsymbol{K}$ 的值与物质点的整体变形状态相关，且物质点对的力密度矢量 $\underline{\boldsymbol{T}}\langle\boldsymbol{\xi}\rangle$ 与变形矢量 $\underline{\boldsymbol{Y}}\langle\boldsymbol{\xi}\rangle$ 不平行共线。

将力密度矢量表达式代入到近场动力学运动方程中，有 [1]

$$\rho(\boldsymbol{x})\ddot{\boldsymbol{u}}(\boldsymbol{x},t) = \int_{H_x}\left(w(|\boldsymbol{\xi}|)\boldsymbol{P}\boldsymbol{K}^{-1}\boldsymbol{\xi} - w(|\boldsymbol{\xi}'|)\boldsymbol{P}'\boldsymbol{K}'^{-1}\boldsymbol{\xi}'\right)\mathrm{d}V_{\boldsymbol{x}'} + \boldsymbol{b}(\boldsymbol{x},t) \tag{7.12}$$

式中，物质点间初始相对位置矢量为 $\boldsymbol{\xi} = \boldsymbol{x}' - \boldsymbol{x}$ 和 $\boldsymbol{\xi}' = \boldsymbol{x} - \boldsymbol{x}'$，上标 "'" 表示物质点 $\boldsymbol{x}'$ 所属变量。

至此，完成了非常规态型近场动力学模型构建和求解过程，即现时构型的变形矢量状态 $\underline{\boldsymbol{Y}}$ → 求出非局部变形梯度张量 $\boldsymbol{F}$ → 将 $\boldsymbol{F}$ 近似作为局部变形梯度张量，进行连续介质力学的应变分析 → 由选定的本构关系求无旋 Cauchy 应力张量 $\boldsymbol{\tau}$ → 参照式 (7.6) 求有旋 Cauchy 应力张量 $\boldsymbol{\sigma}$ → 求第一类 Piola-Kirchhoff 应力张量 $\boldsymbol{P}$ → 力密度矢量状态 $\underline{\boldsymbol{T}}$ → 根据 NOSB PD 运动方程，进行数值求解，更新物质点位置，得到现时构型的变形矢量状态。

7.1.2 基本方程的离散

以下给出非常规态型近场动力学运动方程的空间离散形式。对于物质点 $\boldsymbol{x}_i$，离散后的静力平衡方程为

$$\sum_{j=1}^{N^i}\{\underline{\boldsymbol{T}}[\boldsymbol{x}_i]\langle\boldsymbol{x}_j - \boldsymbol{x}_i\rangle - \underline{\boldsymbol{T}}[\boldsymbol{x}_j]\langle\boldsymbol{x}_i - \boldsymbol{x}_j\rangle\}V_j + \boldsymbol{b}(\boldsymbol{x}_i) = \boldsymbol{0} \tag{7.13}$$

亦可表示为下面的形式

$$\boldsymbol{\psi}(\boldsymbol{u}_p) = \boldsymbol{L}(\boldsymbol{u}_p) + \boldsymbol{b} = \boldsymbol{0} \tag{7.14}$$

其中，N^i 为物质点 $\boldsymbol{x}_i$ 近场范围内其他物质点 $\boldsymbol{x}_j$ 的总数，V_j 为物质点 $\boldsymbol{x}_j$ 的体积，$\boldsymbol{b}$ 为外体力密度矢量，$\boldsymbol{L}$ 为内力密度矢量，是关于位移的函数，$\boldsymbol{u}_p$ 为与物质点 $\boldsymbol{x}_i$ 成键的物质点的位移列阵 (含 $\boldsymbol{x}_i$ 自身)。

将力密度矢量状态代入离散平衡方程中，有

$$\sum_{j=1}^{N^i} \left\{ w_{ji} \bm{P}_i \bm{K}_i^{-1} \bm{\xi}_{ji} - w_{ij} \bm{P}_j \bm{K}_j^{-1} \bm{\xi}_{ij} \right\} V_j + \bm{b}(\bm{x}_i) = \bm{0} \tag{7.15}$$

其中，下标 i、j 表示变量分属于物质点 $\bm{x}_i$ 和 $\bm{x}_j$，$\bm{\xi}_{ji} = \bm{x}_j - \bm{x}_i$、$\bm{\xi}_{ij} = \bm{x}_i - \bm{x}_j$，$w_{ji} = w(|\bm{\xi}_{ji}|)$ 和 $w_{ij} = w(|\bm{\xi}_{ij}|)$ 为影响函数。非局部形状张量 $\bm{K}_i$ 为

$$\bm{K}_i = \int_{H_{\bm{x}_i}} w_{ji}(\bm{\xi}_{ji} \otimes \bm{\xi}_{ji}) \mathrm{d}V_{\bm{x}_j} \approx \sum_{j=1}^{N^i} w_{ji} \begin{bmatrix} \xi_1^2 & \xi_1\xi_2 & \xi_1\xi_3 \\ \xi_1\xi_2 & \xi_2^2 & \xi_2\xi_3 \\ \xi_1\xi_3 & \xi_2\xi_3 & \xi_3^2 \end{bmatrix} V_j \tag{7.16}$$

式中，$\xi_1 = x_{j1} - x_{i1}$、$\xi_2 = x_{j2} - x_{i2}$ 和 $\xi_3 = x_{j3} - x_{i3}$。

非局部变形梯度张量 $\bm{F}_i$ 为

$$\bm{F}_i \approx \sum_{j=1}^{N^i} w_{ji} \left((\bm{u}_j - \bm{u}_i) \otimes (\bm{x}_j - \bm{x}_i) \right) \bm{K}_i^{-1} V_j + \bm{I} = \nabla \bm{u}_i + \bm{I} \tag{7.17}$$

进而，可求得变形梯度张量的变分

$$\delta \bm{F}_i \approx \sum_{j=1}^{N^i} w_{ji} \left((\delta\bm{u}_j - \delta\bm{u}_i) \otimes (\bm{x}_j - \bm{x}_i) \right) \bm{K}_i^{-1} V_j \tag{7.18}$$

据此，可计算拉格朗日-格林应变 $\bm{E} = \dfrac{1}{2}(\bm{F}^\mathrm{T}\bm{F} - \bm{I})$、无旋/有旋 Cauchy 应力 $\bm{\tau}$ 和 $\bm{\sigma}$，以及第一类/第二类 Piola-Kirchhoff 应力 $\bm{P}$ 和 $\bm{T}$ 等。

需要着重指出，$\bm{K}$ 与 $\bm{F}$ 必须正定可逆，以确保逆矩阵 $\bm{K}^{-1}$ 及 $\bm{F}^{-1}$ 的存在，三维问题的近场范围内至少要有 3 个其他物质点 (非共面或非共线)，二维问题的近场范围内至少有 2 个其他物质点。但局部损伤或裂纹扩展等易造成近场范围内物质点数目不足，过少的物质点会导致矩阵条件数变差甚至矩阵奇异，致使 $\bm{K}$ 和 $\bm{F}$ 的计算误差较大，甚至计算终止；另一方面，近场范围内物质点不足意味着受力描述的精度不够，也会影响数值计算的精度和稳定性。

7.2 线性问题的隐式求解方法

7.2.1 线弹性小变形问题的求解方程

在各向同性线弹性材料的本构模型中，第二类 Piola-Kirchhoff 应力张量 $\bm{T}$ 为

$$\bm{T} = \mathcal{L}(\bm{E}) = \lambda \mathrm{tr}(\bm{E})\bm{I} + 2\mu \bm{E} \tag{7.19}$$

其中，$\mathcal{L}$ 为四阶各向同性弹性张量，E 为拉格朗日-格林应变张量，$\lambda = \dfrac{E\nu}{(1+\nu)(1-2\nu)}$ 和 $\mu = G = \dfrac{E}{2(1+\nu)}$ 为拉梅常数，E 为杨氏模量，G 为剪切模量，ν 为泊松比，tr 为矩阵的迹，$\boldsymbol{I}$ 为二阶单位矩阵。

在小变形假定下，Cauchy 应力张量、第一类/第二类 Piola-Kirchhoff 应力张量相等，即 $\boldsymbol{\sigma} = \boldsymbol{T} = \boldsymbol{P}$，则非常规态型近场动力学的平衡方程变为

$$\sum_{j=1}^{N^i} \left\{ w_{ji}\boldsymbol{\sigma}_i \boldsymbol{K}_i^{-1} \boldsymbol{\xi}_{ji} - w_{ij}\boldsymbol{\sigma}_j \boldsymbol{K}_j^{-1} \boldsymbol{\xi}_{ij} \right\} V_j + \boldsymbol{b}(\boldsymbol{x}_i) = \boldsymbol{0} \tag{7.20}$$

此时，$\boldsymbol{L}(\boldsymbol{u}_p) + \boldsymbol{b} = \boldsymbol{0}$ 中内力密度矢量 $\boldsymbol{L}(\boldsymbol{u}_p)$ 可分解为系数矩阵 $\boldsymbol{H}_0$ 与未知位移列阵 $\boldsymbol{u}_p$，即

$$\boldsymbol{H}_0 \boldsymbol{u}_p + \boldsymbol{b} = \boldsymbol{0} \tag{7.21}$$

其中，$\boldsymbol{H}_0$ 为 $n_d \times n_d(N^i+1)$ 矩阵，表示单个物质点的劲度系数，$\boldsymbol{u}_p$ 为 $n_d(N^i+1) \times 1$ 列阵，$\boldsymbol{b}$ 为 $n_d \times 1$ 列阵，n_d 为问题维度。

进而，整个系统的 PD 平衡方程可组装为代数方程组

$$\boldsymbol{H}\boldsymbol{U} + \boldsymbol{b}^* = \boldsymbol{0} \tag{7.22}$$

式中，$\boldsymbol{U}$ 为 $n_d N_{\text{total}} \times 1$ 的列阵，由系统所有未知位移分量集合而成，$\boldsymbol{H}$ 为 $n_d N_{\text{total}} \times n_d N_{\text{total}}$ 的整体劲度矩阵，$\boldsymbol{b}^*$ 为 $n_d N_{\text{total}} \times 1$ 的列阵，由已知的外体力密度组成，其中的 N_{total} 为系统物质点总数。

应力边界条件中的面力需转化成外体力密度，施加在靠近边界一定区域的物质点上，其作用体现在 $\boldsymbol{b}^*$ 向量中；位移边界条件也施加在靠近边界一定区域的物质点上，其矩阵形式表示为

$$\boldsymbol{A}\boldsymbol{U} + \boldsymbol{U}^* = \boldsymbol{0} \tag{7.23}$$

式中，$\boldsymbol{A}$ 为约束方程已知的系数矩阵，$\boldsymbol{U}$ 为待求的未知位移向量，$\boldsymbol{U}^*$ 为给定位移约束值。

约束方程可通过拉格朗日乘子法 (Lagrange multipliers)[7]、乘大数法、罚函数法等引入到系统控制方程中，形成定解的代数方程组系统。引入拉格朗日乘子 $\boldsymbol{\lambda}$，系统平衡方程和约束方程可以统一在下述变分中

$$\delta \boldsymbol{U}^{\text{T}} (\boldsymbol{H}\boldsymbol{U} + \boldsymbol{b}^*) + \delta \left[\boldsymbol{\lambda}^{\text{T}} (\boldsymbol{A}\boldsymbol{U} + \boldsymbol{U}^*) \right] = 0 \tag{7.24}$$

式中，$\delta \boldsymbol{U}$ 为位移矢量的任意变分。对上式第二项进行一阶变分运算，有

$$\delta \boldsymbol{U}^{\text{T}} (\boldsymbol{H}\boldsymbol{U} + \boldsymbol{b}^*) + \delta \boldsymbol{\lambda}^{\text{T}} (\boldsymbol{A}\boldsymbol{U} + \boldsymbol{U}^*) + \delta \boldsymbol{U}^{\text{T}} \boldsymbol{A}^{\text{T}} \boldsymbol{\lambda} = 0 \tag{7.25}$$

上式可重构为

$$\left\{\begin{array}{c}\delta U\\ \delta\lambda\end{array}\right\}^{\mathrm{T}}\left\{\left[\begin{array}{cc}H & A^{\mathrm{T}}\\ A & 0\end{array}\right]\left(\begin{array}{c}U\\ \lambda\end{array}\right)+\left(\begin{array}{c}b^{*}\\ U^{*}\end{array}\right)\right\}=0 \quad (7.26)$$

因此，对任意变分 δU 和 $\delta\lambda$，供求解未知量 U 和 λ 的代数方程组为

$$\left[\begin{array}{cc}H & A^{\mathrm{T}}\\ A & 0\end{array}\right]\left(\begin{array}{c}U\\ \lambda\end{array}\right)=-\left(\begin{array}{c}b^{*}\\ U^{*}\end{array}\right) \quad (7.27)$$

该方程组具有稀疏系数矩阵，其带宽大于传统有限元方法，并依赖于近场范围尺寸选取和粒子编号方法，可采用 Intel 数学核心函数库 MKL 提供的大型稀疏方程组求解器 PARDISO 或 GMRES 等进行求解。

7.2.2 物质点劲度系数矩阵的构造

隐式静力求解方法的关键是构造物质点的劲度系数矩阵 H_0，进而组装成系统整体劲度矩阵。对于小变形弹性问题，Cauchy 应力张量与第一类/第二类 Piola-Kirchhoff 应力张量相等，式 (7.11) 给出的本构关系 $\boldsymbol{T}\langle\boldsymbol{\xi}\rangle = w\boldsymbol{PK}^{-1}\boldsymbol{\xi}$ 即为 $\boldsymbol{T}\langle\boldsymbol{\xi}\rangle = w\boldsymbol{\sigma K}^{-1}\boldsymbol{\xi}$。下面，将此式改写为矩阵形式，给出物质点劲度系数矩阵的计算方法[6,8-10]。

小变形问题的应变张量为 $\boldsymbol{E} = \frac{1}{2}(\boldsymbol{F}+\boldsymbol{F}^{\mathrm{T}})-\boldsymbol{I}$，将式 (7.17) 的非局部变形梯度张量 $\boldsymbol{F}_i$ 代入其中，可得

$$\begin{aligned}\boldsymbol{E}_i &\approx \frac{1}{2}\left(\begin{array}{c}\sum_{j=1}^{N^i}w_{ji}((\boldsymbol{u}_j-\boldsymbol{u}_i)\otimes(\boldsymbol{x}_j-\boldsymbol{x}_i))\boldsymbol{K}_i^{-1}V_j+\boldsymbol{I}\\ +\sum_{j=1}^{N^i}w_{ji}((\boldsymbol{x}_j-\boldsymbol{x}_i)\otimes(\boldsymbol{u}_j-\boldsymbol{u}_i))\boldsymbol{K}_i^{-1}V_j+\boldsymbol{I}\end{array}\right)-\boldsymbol{I}\\ &= \frac{1}{2}\sum_{j=1}^{N^i}w_{ji}\left[(\boldsymbol{u}_j-\boldsymbol{u}_i)\otimes(\boldsymbol{x}_j-\boldsymbol{x}_i)+(\boldsymbol{x}_j-\boldsymbol{x}_i)\otimes(\boldsymbol{u}_j-\boldsymbol{u}_i)\right]\boldsymbol{K}_i^{-1}V_j \quad (7.28)\end{aligned}$$

在下面的推导过程中，对物理含义相同但表现形式不同的张量及其重构的矩阵，采用帽子符号"^"加以区分。

7.2.2.1 三维情况

对于三维问题，将系数矩阵与位移向量解耦，式 (7.28) 的应变张量可重写为如下 Vogit 向量形式

$$\hat{\boldsymbol{E}} = \left[\begin{array}{cccccc}\varepsilon_{11} & \varepsilon_{22} & \varepsilon_{33} & \varepsilon_{23} & \varepsilon_{13} & \varepsilon_{12}\end{array}\right]^{\mathrm{T}} = \hat{\boldsymbol{K}}\boldsymbol{N}\boldsymbol{u}_p \quad (7.29)$$

式中，$\hat{E}$ 为 6×1 列向量，$\hat{K}$ 为 6×9 矩阵，N 为 $9\times(3N^i+3)$ 矩阵，u_p 为 $(3N^i+3)\times1$ 列向量，且有

$$[\hat{K}]_{6\times9} = \begin{bmatrix} K_{11}^{-1} & 0 & 0 & K_{12}^{-1} & 0 & 0 & K_{13}^{-1} & 0 & 0 \\ 0 & K_{12}^{-1} & 0 & 0 & K_{22}^{-1} & 0 & 0 & K_{23}^{-1} & 0 \\ 0 & 0 & K_{13}^{-1} & 0 & 0 & K_{23}^{-1} & 0 & 0 & K_{33}^{-1} \\ K_{12}^{-1} & K_{11}^{-1} & 0 & K_{22}^{-1} & K_{12}^{-1} & 0 & K_{23}^{-1} & K_{13}^{-1} & 0 \\ K_{13}^{-1} & 0 & K_{11}^{-1} & K_{23}^{-1} & 0 & K_{21}^{-1} & K_{33}^{-1} & 0 & K_{13}^{-1} \\ 0 & K_{13}^{-1} & K_{12}^{-1} & 0 & K_{23}^{-1} & K_{22}^{-1} & 0 & K_{33}^{-1} & K_{23}^{-1} \end{bmatrix} \tag{7.30}$$

$$[N]_{9\times(3N^i+3)} = \begin{bmatrix} N^1 & 0 & 0 & \cdots & w_{ji}V_j(x_{j1}-x_{i1}) & 0 & 0 & \cdots \\ 0 & N^1 & 0 & \cdots & 0 & w_{ji}V_j(x_{j1}-x_{i1}) & 0 & \cdots \\ 0 & 0 & N^1 & \cdots & 0 & 0 & w_{ji}V_j(x_{j1}-x_{i1}) & \cdots \\ N^2 & 0 & 0 & \cdots & w_{ji}V_j(x_{j2}-x_{i2}) & 0 & 0 & \cdots \\ 0 & N^2 & 0 & \cdots & 0 & w_{ji}V_j(x_{j2}-x_{i2}) & 0 & \cdots \\ 0 & 0 & N^2 & \cdots & 0 & 0 & w_{ji}V_j(x_{j2}-x_{i2}) & \cdots \\ N^3 & 0 & 0 & \cdots & w_{ji}V_j(x_{j3}-x_{i3}) & 0 & 0 & \cdots \\ 0 & N^3 & 0 & \cdots & 0 & w_{ji}V_j(x_{j3}-x_{i3}) & 0 & \cdots \\ 0 & 0 & N^3 & \cdots & 0 & 0 & w_{ji}V_j(x_{j3}-x_{i3}) & \cdots \end{bmatrix} \tag{7.31}$$

$$[u_p]_{(3N^i+3)\times1} = \begin{bmatrix} u_{11} & u_{12} & u_{13} & \cdots & u_{N^i1} & u_{N^i2} & u_{N^i3} & u_{(N^i+1)1} & u_{(N^i+1)2} & u_{(N^i+1)3} \end{bmatrix}^{\mathrm{T}} \tag{7.32}$$

其中，K_{ij}^{-1} 为形状张量逆矩阵对应 ij 位置的值；矩阵 N 的前三列值 $N^1 = -\sum_{j=1}^{N^i} w_{ji}V_j(x_{j1}-x_{i1})$、$N^2 = -\sum_{j=1}^{N^i} w_{ji}V_j(x_{j2}-x_{i2})$ 和 $N^3 = -\sum_{j=1}^{N^i} w_{ji}V_j(x_{j3}-x_{i3})$ 对应于源点 x_i，其余各列对应于场点 x_j；u_{11}、u_{12} 和 u_{13} 为源点 x_i 的位移值，其余各列为场点 x_j 的位移值。

于是，小变形线弹性问题本构方程的 Vogit 向量形式为

$$\hat{\boldsymbol{\sigma}} = \boldsymbol{C}\hat{\boldsymbol{E}} \Rightarrow \begin{bmatrix} \sigma_{11} \\ \sigma_{22} \\ \sigma_{33} \\ \sigma_{23} \\ \sigma_{31} \\ \sigma_{12} \end{bmatrix} = \begin{bmatrix} C_{11} & C_{12} & C_{12} & 0 & 0 & 0 \\ C_{12} & C_{11} & C_{12} & 0 & 0 & 0 \\ C_{12} & C_{12} & C_{11} & 0 & 0 & 0 \\ 0 & 0 & 0 & 2C_{44} & 0 & 0 \\ 0 & 0 & 0 & 0 & 2C_{44} & 0 \\ 0 & 0 & 0 & 0 & 0 & 2C_{44} \end{bmatrix} \begin{bmatrix} \varepsilon_{11} \\ \varepsilon_{22} \\ \varepsilon_{33} \\ \varepsilon_{23} \\ \varepsilon_{31} \\ \varepsilon_{12} \end{bmatrix}$$
(7.33)

其中，$\boldsymbol{C}$ 为材料的弹性矩阵，且 $C_{11} = E(1-\nu)/[(1+\nu)(1-2\nu)]$、$C_{12} = E\nu/[(1+\nu)(1-2\nu)]$ 和 $C_{44} = G = E/[2(1+\nu)]$。

Cauchy 应力张量和力密度矢量的矩阵表达式分别为

$$\hat{\boldsymbol{\sigma}} = \boldsymbol{C}\hat{\boldsymbol{E}} = \boldsymbol{C}\hat{\boldsymbol{K}}\boldsymbol{N}\boldsymbol{u}_p \tag{7.34}$$

$$\boldsymbol{T}\langle\boldsymbol{\xi}\rangle = \boldsymbol{G}^l \boldsymbol{C}\hat{\boldsymbol{K}}\boldsymbol{N}\boldsymbol{u}_p \tag{7.35}$$

式中，$\boldsymbol{G}^l$ 为 3×6 矩阵，可依据力密度矢量表达式 $\boldsymbol{T}\langle\boldsymbol{\xi}\rangle = w\boldsymbol{\sigma}\boldsymbol{K}^{-1}\boldsymbol{\xi}$ 求得，即

$$\left[\boldsymbol{G}^l\right]_{3\times6} = w_{ji} \begin{bmatrix} G_1 & 0 & 0 & G_2 & G_3 & 0 \\ 0 & G_2 & 0 & G_1 & 0 & G_3 \\ 0 & 0 & G_3 & 0 & G_1 & G_2 \end{bmatrix} \tag{7.36}$$

其中，$G_1 = K_{11}^{-1}\xi_1 + K_{12}^{-1}\xi_2 + K_{13}^{-1}\xi_3$、$G_2 = K_{21}^{-1}\xi_1 + K_{22}^{-1}\xi_2 + K_{23}^{-1}\xi_3$、$G_3 = K_{31}^{-1}\xi_1 + K_{32}^{-1}\xi_2 + K_{33}^{-1}\xi_3$，$\xi_1 = x_{j1} - x_{i1}$、$\xi_2 = x_{j2} - x_{i2}$ 和 $\xi_3 = x_{j3} - x_{i3}$。

则物质点的劲度系数矩阵为

$$\boldsymbol{H}_0 = \boldsymbol{G}^l \boldsymbol{C}\hat{\boldsymbol{K}}\boldsymbol{N} \tag{7.37}$$

最终，可组装获得式 (7.22) 的整体劲度系数矩阵 $\boldsymbol{H}$，求解线性代数方程组即可得到解答。

7.2.2.2 二维情况

对于二维问题，将系数矩阵与位移向量解耦，式 (7.28) 的应变张量也可类似地重写为如下 Vogit 向量形式

$$\hat{\boldsymbol{E}} = \begin{bmatrix} \varepsilon_{11} & \varepsilon_{22} & \varepsilon_{12} \end{bmatrix}^{\mathrm{T}} = \hat{\boldsymbol{K}}\boldsymbol{N}\boldsymbol{u}_p \tag{7.38}$$

式中，$\hat{\boldsymbol{E}}$ 为 3×1 列向量，$\hat{\boldsymbol{K}}$ 为 3×4 矩阵，$\boldsymbol{N}$ 为 $4\times(2N^i+2)$ 矩阵，$\boldsymbol{u}_p$ 为 $(2N^i+2)\times1$ 列向量，且有

$$\left[\hat{\boldsymbol{K}}\right]_{3\times4} = \begin{bmatrix} K_{11}^{-1} & 0 & K_{12}^{-1} & 0 \\ 0 & K_{12}^{-1} & 0 & K_{22}^{-1} \\ K_{12}^{-1} & K_{11}^{-1} & K_{22}^{-1} & K_{12}^{-1} \end{bmatrix} \tag{7.39}$$

7.2 线性问题的隐式求解方法

$$[\boldsymbol{N}]_{4\times(2N^i+2)} = \begin{bmatrix} N^1 & 0 & \cdots & w_{ji}V_j(x_{j1}-x_{i1}) & 0 & \cdots \\ 0 & N^1 & \cdots & 0 & w_{ji}V_j(x_{j1}-x_{i1}) & \cdots \\ N^2 & 0 & \cdots & w_{ji}V_j(x_{j2}-x_{i2}) & 0 & \cdots \\ 0 & N^2 & \cdots & 0 & w_{ji}V_j(x_{j2}-x_{i2}) & \cdots \end{bmatrix} \tag{7.40}$$

$$[\boldsymbol{u}_p]_{(2N^i+2)\times 1} = \begin{bmatrix} u_{11} & u_{12} & \cdots & u_{N^i 1} & u_{N^i 2} & u_{(N^i+1)1} & u_{(N^i+1)2} \end{bmatrix}^{\mathrm{T}} \tag{7.41}$$

其中，K_{ij}^{-1} 为形状张量逆矩阵对应 ij 位置的值；矩阵 $\boldsymbol{N}$ 的前两列值 $N^1 = -\sum_{j=1}^{N^i} w_{ji}V_j(x_{j1}-x_{i1})$ 和 $N^2 = -\sum_{j=1}^{N^i} w_{ji}V_j(x_{j2}-x_{i2})$ 对应于源点 $\boldsymbol{x}_i$，其余各列相应于场点 $\boldsymbol{x}_j$；u_{11} 和 u_{12} 为源点 $\boldsymbol{x}_i$ 的位移值，其余各列为场点 $\boldsymbol{x}_j$ 的位移值。

于是，小变形线弹性问题本构方程的 Vogit 向量形式为

$$\hat{\boldsymbol{\sigma}} = \boldsymbol{C}\hat{\boldsymbol{E}} \Rightarrow \begin{bmatrix} \sigma_{11} \\ \sigma_{22} \\ \sigma_{12} \end{bmatrix} = \begin{bmatrix} C_{11} & C_{12} & 0 \\ C_{12} & C_{11} & 0 \\ 0 & 0 & 2C_{44} \end{bmatrix} \begin{bmatrix} \varepsilon_{11} \\ \varepsilon_{22} \\ \varepsilon_{12} \end{bmatrix} \tag{7.42}$$

其中，$\boldsymbol{C}$ 为材料弹性矩阵。对于平面应力问题，可采用缩减弹性矩阵 $\boldsymbol{Q}$ 代替矩阵 $\boldsymbol{C}$，且有 $Q_{11}=C_{11}-C_{12}^2/C_{11}$、$Q_{12}=C_{12}-C_{12}^2/C_{11}$、$Q_{44}=C_{44}$、$Q_{11}-Q_{12}=2Q_{44}$ 或 $\boldsymbol{Q} = \dfrac{E}{1-\nu^2}\begin{bmatrix} 1 & \nu & 0 \\ \nu & 1 & 0 \\ 0 & 0 & 1-\nu \end{bmatrix}$。

Cauchy 应力和力密度矢量的矩阵表达分别为

$$\hat{\boldsymbol{\sigma}} = \boldsymbol{C}\hat{\boldsymbol{E}} = \boldsymbol{C}\hat{\boldsymbol{K}}\boldsymbol{N}\boldsymbol{u}_p \tag{7.43}$$

$$\underline{\boldsymbol{T}}\langle\boldsymbol{\xi}\rangle = \boldsymbol{G}^l\boldsymbol{C}\hat{\boldsymbol{K}}\boldsymbol{N}\boldsymbol{u}_p \tag{7.44}$$

式中，$\boldsymbol{G}^l$ 为 2×3 矩阵，即

$$[\boldsymbol{G}^l]_{2\times 3} = w_{ji}\begin{bmatrix} G_1 & 0 & G_2 \\ 0 & G_2 & G_1 \end{bmatrix} \tag{7.45}$$

其中，$G_1 = K_{11}^{-1}\xi_1 + K_{12}^{-1}\xi_2$，$G_2 = K_{21}^{-1}\xi_1 + K_{22}^{-1}\xi_2$，$\xi_1 = x_{j1}-x_{i1}$ 和 $\xi_2 = x_{j2}-x_{i2}$。

则物质点的劲度系数矩阵为

$$\boldsymbol{H}_0 = \boldsymbol{G}^l\boldsymbol{C}\hat{\boldsymbol{K}}\boldsymbol{N} \tag{7.46}$$

最终，可组装获得式 (7.22) 的整体劲度系数矩阵 $\boldsymbol{H}$，求解线性代数方程组即可得到解答。

7.2.2.3 一维情况

对于一维问题，可按前述类似方法构造物质点劲度系数矩阵和系统整体劲度系数矩阵，其中涉及的相关物理量表达式如下：

形状张量及其逆分别：$K_i = \sum_{j=1}^{N^i} w_{ji}\xi_{ji}\xi_{ji}V_j$ 和 $K_i^{-1} = 1/K_i$

变形梯度：$F_i = K_i^{-1} \sum_{j=1}^{N^i} w_{ji}(u_j - u_i)\xi_{ji}V_j + 1$

应变：$\varepsilon_i = K_i^{-1} \sum_{j=1}^{N^i} w_{ji}(u_j - u_i)\xi_{ji}V_j = K_i^{-1}\boldsymbol{N}\boldsymbol{u}_p$

应力：$\sigma_i = E\varepsilon_i = EK_i^{-1}\boldsymbol{N}\boldsymbol{u}_p$（$E$ 为杨氏弹性模量）

力密度标量：$\underline{T}(x_i)\langle\xi_{ji}\rangle = w_{ji}\sigma(x_i)K_i^{-1}\xi_{ji} = w_{ji}EK_i^{-1}\boldsymbol{N}\boldsymbol{u}_pK_i^{-1}\xi_{ji}$

其中，行向量 $[\boldsymbol{N}]_{1\times(N^i+1)} = \begin{bmatrix} N_i^1 & w_{j_1i}V_{j_1}\boldsymbol{\xi}_{j_1i} & \cdots & w_{j_{N^i}i}V_{j_{N^i}}\boldsymbol{\xi}_{j_{N^i}i} \end{bmatrix}$，$N_i^1 = -\sum_{j=1}^{N^i} w_{ji}V_j(x_j - x_i)$；位移列向量 $[\boldsymbol{u}_p]_{(N^i+1)\times 1} = \begin{bmatrix} u_i & u_{j_1} & \cdots & u_{j_{N^i}} \end{bmatrix}^\mathrm{T}$，$u_i$ 为源点 x_i 的位移值，其余为场点 x_j 的位移值。

7.3 非线性问题的隐式求解方法

对于线性与非线性问题，均可以按照非线性格式求解。此外，在非常规态型近场动力学中，采用非线性格式求解线性问题，还可以方便地施加罚力的稳定性条件。为求解非线性平衡方程 $\boldsymbol{L}(\boldsymbol{u}_p) + \boldsymbol{b} = \boldsymbol{0}$，采用 Newton Raphson 迭代法求解位移增量 $\delta\boldsymbol{u}_p$，当位移增量收敛于 $\boldsymbol{0}$ 时，便可获得问题解答。

Newton Raphson 迭代法求解步骤如下：

步骤 1：初始时刻 $t = 0$，加载步 $n = 0$，令初始位移为 $\boldsymbol{u}_p^0 = \boldsymbol{0}$，同时初始化各状态变量；

步骤 2：t 时刻，完成 n 加载步后，已知稳态构型各变量值；进入 $t + \Delta t$ 时刻计算，加载步 $n+1$，更新边界条件；

步骤 3：赋位移试探初值 $\boldsymbol{u}_p^{\mathrm{trial}} = \boldsymbol{u}_p^n$，令各变量当前值为迭代计算的试探初值；

步骤 4：采用 Newton Raphson 迭代法求解

(a) 在当前构型下，构造雅可比 (Jacobian) 矩阵或切线刚度矩阵 $\boldsymbol{J}$；

(b) 求解线性方程组 $\boldsymbol{J}\delta\boldsymbol{u}_p = -(\boldsymbol{L} + \boldsymbol{b})$，获得牛顿迭代步内的位移增量 $\delta\boldsymbol{u}_p$；

7.3 非线性问题的隐式求解方法

(c) 令 $\boldsymbol{u}_p^{\text{trial}} = \boldsymbol{u}_p^{\text{trial}} + \delta \boldsymbol{u}_p$；

(d) 更新 $\boldsymbol{u}_p^{\text{trial}}$ 后，基于收敛准则判断解答是否收敛 (如位移收敛准则 $\|\delta \boldsymbol{u}_p\| < \varepsilon_1 \|\boldsymbol{u}_p\|$，$\varepsilon_1$ 为相对容差)；若收敛，则执行步骤 5；

(e) 若不满足收敛条件，则重新执行步骤 4(a)；不断迭代求解，直至收敛；

步骤 5：令 $\boldsymbol{u}_p^{n+1} = \boldsymbol{u}_p^{\text{trial}}$，并令各变量值为下一步计算的试探值；

步骤 6：如果加载未完成，则进入步骤 2，继续计算；若加载完成，则结束计算。

记平衡方程 (7.14) 的残余矢量为 $\mathrm{d}\boldsymbol{L}$，可表示为

$$\mathrm{d}\boldsymbol{L} = \boldsymbol{J}\delta \boldsymbol{u}_p = -[\boldsymbol{L}(\boldsymbol{u}_p) + \boldsymbol{b}] \tag{7.47}$$

式中，$\boldsymbol{J} = \partial \boldsymbol{L}/\partial \boldsymbol{u}_p$ 为雅可比矩阵或切线刚度矩阵；$\boldsymbol{u}_p$ 为 $n_\mathrm{d}(N^i+1)\times 1$ 的位移列向量，n_d 为问题维度。

对照式 (7.13) 和式 (7.14)，可得到 [11,12]

$$\mathrm{d}\boldsymbol{L}(\boldsymbol{x}_i) = \boldsymbol{J}\delta \boldsymbol{u}_p = \frac{\partial \boldsymbol{L}}{\partial \boldsymbol{u}_p}\delta \boldsymbol{u}_p = \left[\sum_{j=1}^{N^i}\left(\frac{\partial \underline{\boldsymbol{T}}[\boldsymbol{x}_i]\langle \boldsymbol{\xi}_{ji}\rangle}{\partial \boldsymbol{u}_p} - \frac{\partial \underline{\boldsymbol{T}}[\boldsymbol{x}_j]\langle \boldsymbol{\xi}_{ij}\rangle}{\partial \boldsymbol{u}_p}\right)V_j\right]\delta \boldsymbol{u}_p \tag{7.48}$$

将力密度矢量状态 $\underline{\boldsymbol{T}}[\boldsymbol{x}_i]\langle \boldsymbol{\xi}_{ji}\rangle = w_{ji}\boldsymbol{P}_i \boldsymbol{K}_i^{-1}\boldsymbol{\xi}_{ji}$ 与 $\underline{\boldsymbol{T}}[\boldsymbol{x}_j]\langle \boldsymbol{\xi}_{ij}\rangle = w_{ij}\boldsymbol{P}_j \boldsymbol{K}_j^{-1}\boldsymbol{\xi}_{ij}$ 代入上式，并注意第一类 Piola-Kirchhoff 应力张量 $\boldsymbol{P}$ 是变形梯度张量 $\boldsymbol{F}$ 的函数，两者是一组功共轭量，且有 $\delta \boldsymbol{P} = \frac{\partial \boldsymbol{P}}{\partial \boldsymbol{F}}\delta \boldsymbol{F}$，则可得

$$\frac{\partial \underline{\boldsymbol{T}}[\boldsymbol{x}_i]\langle \boldsymbol{\xi}_{ji}\rangle}{\partial \boldsymbol{u}_p} = \frac{\partial [w_{ji}\boldsymbol{P}_i \boldsymbol{K}_i^{-1}\boldsymbol{\xi}_{ji}]}{\partial \boldsymbol{u}_p} = w_{ji}\frac{\partial \boldsymbol{P}_i}{\partial \boldsymbol{u}_p}\boldsymbol{K}_i^{-1}\boldsymbol{\xi}_{ji} = w_{ji}\frac{\partial \boldsymbol{P}_i}{\partial \boldsymbol{F}_i}\frac{\partial \boldsymbol{F}_i}{\partial \boldsymbol{u}_p}\boldsymbol{K}_i^{-1}\boldsymbol{\xi}_{ji} \tag{7.49}$$

$$\frac{\partial \underline{\boldsymbol{T}}[\boldsymbol{x}_j]\langle \boldsymbol{\xi}_{ij}\rangle}{\partial \boldsymbol{u}_p} = \frac{\partial [w_{ij}\boldsymbol{P}_j \boldsymbol{K}_j^{-1}\boldsymbol{\xi}_{ij}]}{\partial \boldsymbol{u}_p} = w_{ij}\frac{\partial \boldsymbol{P}_j}{\partial \boldsymbol{u}_p}\boldsymbol{K}_j^{-1}\boldsymbol{\xi}_{ij} = w_{ij}\frac{\partial \boldsymbol{P}_j}{\partial \boldsymbol{F}_j}\frac{\partial \boldsymbol{F}_j}{\partial \boldsymbol{u}_p}\boldsymbol{K}_j^{-1}\boldsymbol{\xi}_{ij} \tag{7.50}$$

将上述两式代入式 (7.48) 中，得到 [11,12]

$$\begin{aligned}&\mathrm{d}\boldsymbol{L}(\boldsymbol{x}_i)\\&= \left[\sum_{i=1}^{N^i}\left(w_{ji}\frac{\partial \boldsymbol{P}_i}{\partial \boldsymbol{F}_i}\frac{\partial \boldsymbol{F}_i}{\partial \boldsymbol{u}_p}\boldsymbol{K}_i^{-1}\boldsymbol{\xi}_{ji} - w_{ij}\frac{\partial \boldsymbol{P}_j}{\partial \boldsymbol{F}_j}\frac{\partial \boldsymbol{F}_j}{\partial \boldsymbol{u}_p}\boldsymbol{K}_j^{-1}\boldsymbol{\xi}_{ij}\right)V_j\right]\delta \boldsymbol{u}_p\\&= \sum_{i=1}^{N^i}\left(w_{ji}\frac{\partial \boldsymbol{P}_i}{\partial \boldsymbol{F}_i}\delta \boldsymbol{F}_i \boldsymbol{K}_i^{-1}\boldsymbol{\xi}_{ji}\right)V_j - \sum_{i=1}^{N^i}\left(w_{ij}\frac{\partial \boldsymbol{P}_j}{\partial \boldsymbol{F}_j}\delta \boldsymbol{F}_j \boldsymbol{K}_j^{-1}\boldsymbol{\xi}_{ij}\right)V_j\end{aligned}$$

$$= \sum_{i=1}^{N^i} \left(w_{ji} \frac{\partial \boldsymbol{P}_i}{\partial \boldsymbol{F}_i} \left(\left(\sum_{k=1}^{N^i} (w_{ki} V_k (\delta \boldsymbol{u}_k - \delta \boldsymbol{u}_i) \otimes (\boldsymbol{x}_k - \boldsymbol{x}_i)) \right) \boldsymbol{K}_i^{-1} \right) \boldsymbol{K}_i^{-1} \boldsymbol{\xi}_{ji} \right) V_j$$

$$- \sum_{i=1}^{N^i} \left(w_{ij} \frac{\partial \boldsymbol{P}_j}{\partial \boldsymbol{F}_j} \left(\left(\sum_{k=1}^{N^j} (w_{kj} V_k (\delta \boldsymbol{u}_k - \delta \boldsymbol{u}_j) \otimes (\boldsymbol{x}_k - \boldsymbol{x}_j)) \right) \boldsymbol{K}_j^{-1} \right) \boldsymbol{K}_j^{-1} \boldsymbol{\xi}_{ij} \right) V_j$$

(7.51)

其中，$w_{ji} = w(|\boldsymbol{x}_j - \boldsymbol{x}_i|)$、$w_{ij} = w(|\boldsymbol{x}_i - \boldsymbol{x}_j|)$、$w_{ki} = w(|\boldsymbol{x}_k - \boldsymbol{x}_i|)$ 和 $w_{kj} = w(|\boldsymbol{x}_k - \boldsymbol{x}_j|)$，$\boldsymbol{\xi}_{ji} = \boldsymbol{x}_j - \boldsymbol{x}_i$ 和 $\boldsymbol{\xi}_{ij} = \boldsymbol{x}_i - \boldsymbol{x}_j$。切线模量 $\partial \boldsymbol{P}/\partial \boldsymbol{F}$ 为物质点劲度系数矩阵的唯一未知量，一旦求得切线模量 $\partial \boldsymbol{P}/\partial \boldsymbol{F}$ 后，即可获得雅可比矩阵 $\boldsymbol{J}$ 表达式，进而可组装整体劲度矩阵 $\boldsymbol{H}$，最后采用广义最小残量法 (GMRES) 求解，得到系统方程组的解答。

有两种方法确定 $\partial \boldsymbol{P}/\partial \boldsymbol{F}$ 的值：① $\partial \boldsymbol{P}/\partial \boldsymbol{F}$ 本身与 $\delta \boldsymbol{P}$ 和 $\delta \boldsymbol{F}$ 的具体值无关，可以通过本构模型获得 $\delta \boldsymbol{P}$ 和 $\delta \boldsymbol{F}$ 的关系，进而对 $\delta \boldsymbol{P} = \dfrac{\partial \boldsymbol{P}}{\partial \boldsymbol{F}} \delta \boldsymbol{F}$ 进行矩阵重构，获得对应的 $\partial \boldsymbol{P}/\partial \boldsymbol{F}$ 表达式；② 在迭代过程中，根据上一步的数据信息 $\delta \boldsymbol{F}$ 和 $\delta \boldsymbol{P}$，计算 $\dfrac{\partial \boldsymbol{P}}{\partial \boldsymbol{F}} = \delta \boldsymbol{P} (\delta \boldsymbol{F})^{-1}$。两种处理方法的计算结果会有轻微差异，第二种方法简单易于实现，本书算例采用第二种方法。

对于一般的本构模型，考虑 $\boldsymbol{P} = \boldsymbol{F}\boldsymbol{T}$，则有如下变分关系

$$\delta \boldsymbol{P} = \delta(\boldsymbol{F}\boldsymbol{T}) = \delta \boldsymbol{F}\boldsymbol{T} + \boldsymbol{F}\delta \boldsymbol{T} \tag{7.52}$$

进而，对于经典的线弹性本构模型，第二类 Piola-Kirchhoff 应力的变分 $\delta \boldsymbol{T}$ 为

$$\delta \boldsymbol{T} = \delta \mathcal{L}(\boldsymbol{E}) = \delta \mathcal{L}\left(\frac{1}{2}(\boldsymbol{F}^\mathrm{T}\boldsymbol{F} - \boldsymbol{I})\right) = \frac{1}{2}\mathcal{L}\left(\delta(\boldsymbol{F}^\mathrm{T}\boldsymbol{F} - \boldsymbol{I})\right) = \frac{1}{2}\mathcal{L}(\delta \boldsymbol{F}^\mathrm{T}\boldsymbol{F} + \boldsymbol{F}^\mathrm{T}\delta \boldsymbol{F})$$

(7.53)

将式 (7.53) 给出的 $\delta \boldsymbol{T}$ 代入式 (7.52) 中，即可获得线弹性本构模型的变分 $\delta \boldsymbol{P}$ 和 $\delta \boldsymbol{F}$ 的关系。

以二维问题为例，为求出物质点劲度矩阵并组装系统整体劲度矩阵，需要将式 (7.51) 中的 $w_{ji} V_j \dfrac{\partial \boldsymbol{P}_i}{\partial \boldsymbol{F}_i} \delta \boldsymbol{F}_i \boldsymbol{K}_i^{-1} \boldsymbol{\xi}_{ji}$ 重写为如下矩阵形式，以分离雅可比矩阵 $\boldsymbol{J}$ 与位移增量列向量 $\delta \boldsymbol{u}_p$[12]

$$w_{ji} V_j \frac{\partial \boldsymbol{P}_i}{\partial \boldsymbol{F}_i} \delta \boldsymbol{F}_i \boldsymbol{K}_i^{-1} \boldsymbol{\xi}_{ji} = [\boldsymbol{G}^n]_{2\times 4} [\delta \boldsymbol{F}]_{4\times 1}$$

$$= [\boldsymbol{G}^n]_{2\times 4} \left[\hat{\boldsymbol{K}}\right]_{4\times 4} [\mathbf{N}]_{4\times 2(N^i+1)} [\delta \boldsymbol{u}_p]_{2(N^i+1)\times 1} \tag{7.54}$$

其中

$$[\delta \boldsymbol{F}]_{4\times 1} = \left[\hat{\boldsymbol{K}}\right]_{4\times 4} [\boldsymbol{N}]_{4\times 2(N^i+1)} [\delta \boldsymbol{u}_p]_{2(N^i+1)\times 1} \tag{7.55}$$

$$[\delta \boldsymbol{F}]_{4\times 1} = \left[\begin{array}{cccc} \delta F_{11} & \delta F_{22} & \delta F_{12} & \delta F_{21} \end{array}\right]^{\mathrm{T}} \tag{7.56}$$

$$\left[\hat{\boldsymbol{K}}\right]_{4\times 4} = \left[\begin{array}{cccc} K_{11}^{-1} & 0 & K_{12}^{-1} & 0 \\ 0 & K_{12}^{-1} & 0 & K_{22}^{-1} \\ K_{12}^{-1} & 0 & K_{22}^{-1} & 0 \\ 0 & K_{11}^{-1} & 0 & K_{12}^{-1} \end{array}\right] \tag{7.57}$$

$$[\boldsymbol{N}]_{4\times 2(N^i+1)} = \left[\begin{array}{cccccc} N^1 & 0 & \cdots & w_{ji}V_j(x_{j1}-x_{i1}) & 0 & \cdots \\ 0 & N^1 & \cdots & 0 & w_{ji}V_j(x_{j1}-x_{i1}) & \cdots \\ N^2 & 0 & \cdots & w_{ji}V_j(x_{j2}-x_{i2}) & 0 & \cdots \\ 0 & N^2 & \cdots & 0 & w_{ji}V_j(x_{j2}-x_{i2}) & \cdots \end{array}\right] \tag{7.58}$$

$$[\delta \boldsymbol{u}_p]_{2(N^i+1)\times 1} = \left[\begin{array}{cccccc} \delta u_{11} & \delta u_{12} & \cdots & \delta u_{N^i1} & \delta u_{N^i2} & \delta u_{(N^i+1)1} & \delta u_{(N^i+1)2} \end{array}\right]^{\mathrm{T}} \tag{7.59}$$

$$[\boldsymbol{G}^n]_{2\times 4} = w_{ji}V_j \left[\begin{array}{cccc} G_1\left(\dfrac{\partial P}{\partial F}\right)_{11} & G_2\left(\dfrac{\partial P}{\partial F}\right)_{12} & G_2\left(\dfrac{\partial P}{\partial F}\right)_{11} & G_1\left(\dfrac{\partial P}{\partial F}\right)_{12} \\ G_1\left(\dfrac{\partial P}{\partial F}\right)_{21} & G_2\left(\dfrac{\partial P}{\partial F}\right)_{22} & G_2\left(\dfrac{\partial P}{\partial F}\right)_{21} & G_1\left(\dfrac{\partial P}{\partial F}\right)_{22} \end{array}\right] \tag{7.60}$$

式中，K_{ij}^{-1} 为形状张量的逆矩阵对应 ij 位置的值；矩阵 $\boldsymbol{N}$ 的前两列值 $N^1 = -\sum_{j=1}^{N^i} w_{ji}V_j(x_{j1}-x_{i1})$ 和 $N^2 = -\sum_{j=1}^{N^i} w_{ji}V_j(x_{j2}-x_{i2})$ 对应于源点 $\boldsymbol{x}_i$，其余各列相应于场点 $\boldsymbol{x}_j$；δu_{11} 和 δu_{12} 为源点 $\boldsymbol{x}_i$ 的位移增量，其余各列为场点 $\boldsymbol{x}_j$ 的位移值增量；$G_1 = K_{11}^{-1}\xi_1 + K_{12}^{-1}\xi_2$，$G_2 = K_{21}^{-1}\xi_1 + K_{22}^{-1}\xi_2$，$\xi_1 = x_{j1}-x_{i1}$ 和 $\xi_2 = x_{j2}-x_{i2}$。至此，根据上述所获得的结果，便可求得物质点劲度矩阵并组集得到系统的整体劲度矩阵。

7.4 数值不稳定性分析与稳定控制方法

非常规态型近场动力学模型的计算范式导致数值计算中出现数值不稳定性问题，通常表现为粒子位移的锯齿形振荡，继而引起应力场和应变场的数值振荡，在加载区域或裂尖等变形梯度变化较为剧烈的区域尤为显著，影响计算精度甚至导

致计算失效 [5,6,12,13]。即使是对于分析弹性变形等变形梯度变化平缓的问题，仍有轻微锯齿形的位移振荡。本节将探讨数值不稳定性的来源，简述现有稳定控制算法和模型，给出考虑罚函数稳定控制的数值计算方案。

7.4.1 数值不稳定性的影响因素分析

7.4.1.1 非局部变形梯度的理论精度不高

考察由物质点位置函数求解非局部变形梯度张量的过程，归纳如下

$$\boldsymbol{y}(\boldsymbol{x}_j) = \boldsymbol{y}(\boldsymbol{x}_i + \boldsymbol{\xi}_{ji}) = \boldsymbol{y}(\boldsymbol{x}_i) + \nabla \boldsymbol{y}(\boldsymbol{x}_i) \cdot \boldsymbol{\xi}_{ji} + \frac{1}{2}(\nabla\nabla \boldsymbol{y}(\boldsymbol{x}_i) \cdot \boldsymbol{\xi}_{ji}) \cdot \boldsymbol{\xi}_{ji}$$

$$\Rightarrow \boldsymbol{y}(\boldsymbol{x}_j) - \boldsymbol{y}(\boldsymbol{x}_i) = \boldsymbol{F}(\boldsymbol{x}_i) \cdot \boldsymbol{\xi}_{ji} + \frac{1}{2}(\nabla \boldsymbol{F}(\boldsymbol{x}_i) \cdot \boldsymbol{\xi}_{ji}) \cdot \boldsymbol{\xi}_{ji}$$

$$\Rightarrow (\boldsymbol{y}(\boldsymbol{x}_j) - \boldsymbol{y}(\boldsymbol{x}_i)) \otimes \boldsymbol{\xi}_{ji} = \boldsymbol{F}(\boldsymbol{x}_i) \cdot \boldsymbol{\xi}_{ji} \otimes \boldsymbol{\xi}_{ji} + \frac{1}{2}(\nabla \boldsymbol{F}(\boldsymbol{x}_i) \cdot \boldsymbol{\xi}_{ji}) \cdot \boldsymbol{\xi}_{ji} \otimes \boldsymbol{\xi}_{ji}$$

$$\Rightarrow w_{ji}(\boldsymbol{y}(\boldsymbol{x}_j) - \boldsymbol{y}(\boldsymbol{x}_i)) \otimes \boldsymbol{\xi}_{ji}$$

$$= w_{ji}\boldsymbol{F}(\boldsymbol{x}_i) \cdot \boldsymbol{\xi}_{ji} \otimes \boldsymbol{\xi}_{ji} + w_{ji}\frac{1}{2}(\nabla \boldsymbol{F}(\boldsymbol{x}_i) \cdot \boldsymbol{\xi}_{ji}) \cdot \boldsymbol{\xi}_{ji} \otimes \boldsymbol{\xi}_{ji}$$

$$\Rightarrow \int_{H_{\boldsymbol{x}_i}} w_{ji}(\boldsymbol{y}(\boldsymbol{x}_j) - \boldsymbol{y}(\boldsymbol{x}_i)) \otimes \boldsymbol{\xi}_{ji} \mathrm{d}V_{\boldsymbol{x}_j}$$

$$= \boldsymbol{F}(\boldsymbol{x}_i) \cdot \int_{H_{\boldsymbol{x}_i}} w_{ji}\boldsymbol{\xi}_{ji} \otimes \boldsymbol{\xi}_{ji} \mathrm{d}V_{\boldsymbol{x}_j} + \nabla \boldsymbol{F}(\boldsymbol{x}_i) \cdot \int_{H_{\boldsymbol{x}_i}} \frac{1}{2}w_{ji}\boldsymbol{\xi}_{ji} \cdot \boldsymbol{\xi}_{ji} \otimes \boldsymbol{\xi}_{ji} \mathrm{d}V_{\boldsymbol{x}_j}$$

$$\Rightarrow \boldsymbol{F}(\boldsymbol{x}_i) = \int_{H_{\boldsymbol{x}_i}} w_{ji}(\boldsymbol{y}(\boldsymbol{x}_j) - \boldsymbol{y}(\boldsymbol{x}_i)) \otimes \boldsymbol{\xi}_{ji} \mathrm{d}V_{\boldsymbol{x}_j} \cdot \boldsymbol{K}_i^{-1}$$

$$- \nabla \boldsymbol{F}(\boldsymbol{x}_i) \cdot \int_{H_{\boldsymbol{x}_i}} \frac{1}{2}w_{ji}\boldsymbol{\xi}_{ji} \cdot \boldsymbol{\xi}_{ji} \otimes \boldsymbol{\xi}_{ji} \mathrm{d}V_{\boldsymbol{x}_j} \cdot \boldsymbol{K}_i^{-1} \tag{7.61}$$

从上述推导过程可见，将物质点位置函数进行泰勒级数展开，经过平移、乘以同一变量、在局部范围内加权积分、矩阵求逆等运算，可以得到非局部变形梯度张量 [1,6,7,14]。对于内部区域 (距离边界一个近场范围以上) 的物质点，由于完整圆形或球体近场范围或积分域具有对称性，则与位移二阶导数项关联的积分项 (即最后一式等号右边的第二项) 为零，因此变形梯度张量具有二阶精度。但对于边界区域的物质点，其近场范围非完整和非对称，与位移二阶导数项关联的积分项不为零，则变形梯度张量具有一阶精度。对于具有任意形状近场范围的物质点，其变形梯度的精度也是一阶的。

在具体计算过程中，如果物质点是均匀离散的，边界区域物质点变形梯度张量的低阶精度必然影响 NOSB PD 的整体精度，也在一定程度上影响数值计算的稳定性。如果是非均匀离散的，整个构型所有物质点的变形梯度张量精度都较低，进而造成全局计算误差与数值不稳定性。同样地，由应力变换得到的力密度矢量状态具有同阶精度，也会影响整体计算精度和稳定性。

7.4.1.2 点关联的非局部变形梯度导致"键变形协调性"无法严格满足

一般来说，近场范围是一点周围径向对称的球域或圆域，可称之为"点关联的近场范围 (point-associated horizon)"，对应的变形梯度可称之为"点关联的非局部变形梯度 (point-associated nonlocal deformation gradient)"。然而，由于非局部变形梯度的点关联性，对于任意非均匀 (连续) 变形状态，下式的"键变形协调性"条件不能严格满足 [6]

$$\boldsymbol{\xi} + \boldsymbol{\eta} = \boldsymbol{F}(\boldsymbol{x})\boldsymbol{\xi} = \boldsymbol{F}(\boldsymbol{x}')\boldsymbol{\xi} \tag{7.62}$$

该式只在均匀变形场时，即 $\boldsymbol{F}(\boldsymbol{x}) = \boldsymbol{F}(\boldsymbol{x}')$ 时严格成立。值得注意的是，相比于经典连续介质力学的局部变形梯度张量，点关联的非局部变形梯度通过加权平均能包含更多的局部变形信息、表征更为复杂的变形状态，但在一定程度上忽略了局部变形的协调性特征、取向性特征和局部峰值特征。

7.4.1.3 点关联的非局部变形梯度导致零能模式和伪能量模式问题

在 NOSB PD 方法中，基于非局部变形梯度张量 $\boldsymbol{F}$ 描述物体变形状态时存在误差，导致某些非物理变形不能避免，并存在某些物理可能变形状态下的零能模式 (zero-energy mode) 和伪能量模式 (pseudo-energy mode) 等缺陷 [6,15,16]。

首先，NOSB PD 方法放松了相容性要求，没有严格满足"键的变形协调性"，导致一些不该发生的非物理变形状态出现，其对应的非局部变形梯度值貌似合理，如近场域内当发生局部多个物质点凝聚为一点时，其变形梯度值并没有出现异常现象 [16]。

其次，计算物理可能变形状态时，一些变形状态得不到有效描述，变形梯度计算不准确，导致零能模式或伪能量模式的发生。下面以单点移动为例，说明点关联的非局部变形梯度张量导致零能模式和伪能量模式问题。

如图 7-1 所示，当前构型中，物质点 $\boldsymbol{x}$ 和 $\boldsymbol{x}'$ 的变形梯度张量分别为 $\boldsymbol{F}_{\text{old}}(\boldsymbol{x})$ 和 $\boldsymbol{F}_{\text{old}}(\boldsymbol{x}')$，此时给物质点 $\boldsymbol{x}$ 施加载荷使其发生单点移动 $\boldsymbol{u}_{\text{d}}(\boldsymbol{x})$，则物质点 $\boldsymbol{x}$ 和 $\boldsymbol{x}'$ 的变形梯度张量分别为

$$\boldsymbol{F}_{\text{new}}(\boldsymbol{x}) = \int_{H_{\boldsymbol{x}}} w[\boldsymbol{y}_{\text{new}}(\boldsymbol{x}') - \boldsymbol{y}_{\text{new}}(\boldsymbol{x})] \otimes (\boldsymbol{x}' - \boldsymbol{x}) \mathrm{d}V_{\boldsymbol{x}'} \boldsymbol{K}^{-1}(\boldsymbol{x})$$

$$= \int_{H_x} w[\boldsymbol{y}_{\text{old}}(\boldsymbol{x}') - \boldsymbol{y}_{\text{old}}(\boldsymbol{x}) - \boldsymbol{u}_{\text{d}}(\boldsymbol{x})] \otimes (\boldsymbol{x}' - \boldsymbol{x}) \mathrm{d}V_{\boldsymbol{x}'} \boldsymbol{K}^{-1}(\boldsymbol{x})$$

$$= \boldsymbol{F}_{\text{old}}(\boldsymbol{x}) - \boldsymbol{u}_{\text{d}}(\boldsymbol{x}) \otimes \int_{H_x} w(\boldsymbol{x}' - \boldsymbol{x}) \mathrm{d}V_{\boldsymbol{x}'} \boldsymbol{K}^{-1}(\boldsymbol{x}) \tag{7.63}$$

$$\boldsymbol{F}_{\text{new}}(\boldsymbol{x}') = \int_{H_{x'}} w[\boldsymbol{y}_{\text{new}}(\boldsymbol{x}'') - \boldsymbol{y}_{\text{new}}(\boldsymbol{x}')] \otimes (\boldsymbol{x}'' - \boldsymbol{x}') \mathrm{d}V_{\boldsymbol{x}''} \boldsymbol{K}^{-1}(\boldsymbol{x}')$$

$$= \int_{H_{x'}} w[\boldsymbol{y}_{\text{old}}(\boldsymbol{x}'') - \boldsymbol{y}_{\text{old}}(\boldsymbol{x}')] \otimes (\boldsymbol{x}'' - \boldsymbol{x}') \mathrm{d}V_{\boldsymbol{x}''} \boldsymbol{K}^{-1}(\boldsymbol{x}') +$$

$$w \boldsymbol{u}_{\text{d}}(\boldsymbol{x}) \otimes (\boldsymbol{x} - \boldsymbol{x}') V_{\boldsymbol{x}} \boldsymbol{K}^{-1}(\boldsymbol{x}')$$

$$= \boldsymbol{F}_{\text{old}}(\boldsymbol{x}') + w \boldsymbol{u}_{\text{d}}(\boldsymbol{x}) \otimes (\boldsymbol{x} - \boldsymbol{x}') V_{\boldsymbol{x}} \boldsymbol{K}^{-1}(\boldsymbol{x}') \tag{7.64}$$

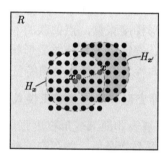

图 7-1 构型中单点移动时,点关联的非局部变形梯度张量导致零能模式和伪能量模式

当采用均匀正交点阵离散构型和球形对称影响函数 w 时,式 (7.63) 右端积分项为零,表明物质点 $\boldsymbol{x}$ 的变形梯度张量增量为零,对应的变形能增量的计算值也为零,不符合客观事实,即为零能模式问题。式 (7.64) 右端第二项不为零,因而物质点 $\boldsymbol{x}'$ 的变形能增量不为零,符合加载点一定范围内的物质点能量状态变化的物理实际,但从数值计算的角度分析,零能模式变形状态导致非零能的变形状态并不准确,称为伪能量模式。

7.4.1.4 构型变形信息丢失与键力密度矢量计算存在的一对多映射问题

如前所述,由变形矢量求键力密度矢量的计算流程为 $\underline{\boldsymbol{Y}}[x,t] \rightarrow \boldsymbol{F}(\boldsymbol{x}) \rightarrow \boldsymbol{P}(\boldsymbol{x}) \rightarrow \underline{\boldsymbol{T}}[x,t]\langle\boldsymbol{\xi}\rangle$。其中,由变形矢量计算点关联的非局部变形梯度张量时 ($\underline{\boldsymbol{Y}}[x,t] \rightarrow \boldsymbol{F}(\boldsymbol{x})$),由于采用了近场域内加权积分的模式,存在构型变形信息丢失问题。此外,根据式 (7.11),由点关联变形梯度张量或点关联应力张量计算键力

密度矢量时 $(P(x) \to \underline{T}[x,t]\langle\xi\rangle)$，存在一对多映射的问题，并且当选定不同的影响函数时，该映射也是不唯一的 [6,17,18]。

7.4.1.5 双重积分放大误差与单点数值积分误差

非常规态型近场动力学模型在近场范围开展双重积分建模，先后获得积分型应变、应力和积分型内力密度矢量，双重积分放大了计算误差 [5]。此外，基于粒子离散的单高斯点积分方法 (或不恰当的数值方法) 引入了额外的计算误差，也会加剧数值计算的不稳定性 [19,20]。

7.4.2 稳定控制的计算策略

当前，非常规态型近场动力学的稳定控制方法或修正模型主要包括：
(1) 使用增强积分方法 [19,20]；
(2) 增加额外力密度矢量状态或阻尼项 [4,15,13,21−23]；
(3) 选用合理影响函数 [5,6,24−26]；
(4) 位移场光滑化或平均化 [24,26]；
(5) 高阶修正模型 [6,14]；
(6) 应力点法 [27]；
(7) 子域平均方法 [28]；
(8) 考虑键层次或键关联近场范围的变量修正方法 [6,17,18,29,30]；
(9) 采用近场动力学微分算子重构 Navier 位移平衡方程建立的新 NOSB PD 模型 [5]。

这些方法在很大程度上提高了计算精度和稳定性。本书不再赘述各种方法的细节，读者可参阅相关文献资料。

本节提出含罚函数稳定控制的计算策略 [13]，具体如下：
(1) 将边界条件施加到一个近场范围的物质点上，实现非局部边界条件，确保加载区域不发生突变；
(2) 采用较精细的离散网格/点阵，有利于抑制数值不稳定性；
(3) 选用特定的空间递减形式的影响函数，如 $w(|\xi|) = \delta^3/|\xi|^3$；
(4) 对于动力或基于隐式非线性静力求解的问题，采用罚函数方法引入沙漏力，以抑制数值不稳定性。

下面具体介绍沙漏力的算法，其基本思想是通过引入与物质点变形梯度相关的沙漏力，使得近场动力学的计算能客观反映常变形梯度、构型线性变化，满足"键变形协调性"。

考虑当前构型中物质点 x 的非局部变形梯度张量 F，则 t 时刻物质点 x' 的理论预测位置 y'^* 为

$$y'^* = y + F(x' - x) \tag{7.65}$$

记沙漏矢量为

$$\boldsymbol{h} = \boldsymbol{y'}^* - \boldsymbol{y'} \tag{7.66}$$

表示根据"点关联变形梯度"预测的物质点位置与根据运动方程计算的物质点位置的差异。若 $\boldsymbol{h} = \boldsymbol{0}$,则近场动力学模拟结果正确反映常变形梯度,满足"键变形协调性";若 $\boldsymbol{h} \neq \boldsymbol{0}$,则近场动力学模拟结果未能满足"键变形协调性",此时 $\boldsymbol{y'}^*$、$\boldsymbol{y'}$ 与 $\boldsymbol{y}$ 构成了漏斗形 $\angle \boldsymbol{y'yy'}^*$。

进而,引入与沙漏矢量 $\boldsymbol{h}$ 在 $\boldsymbol{x'} - \boldsymbol{x}$ 上的投影 $h_{\text{proj}} = \boldsymbol{h} \cdot (\boldsymbol{x'} - \boldsymbol{x})$ 成比例的罚函数 $\boldsymbol{f}_{\text{hg}}$(也称为沙漏力密度矢量),该沙漏力密度矢量可定义为

$$\boldsymbol{f}_{\text{hg}} = f_{\text{hg}} \frac{\boldsymbol{y'} - \boldsymbol{y}}{|\boldsymbol{y'} - \boldsymbol{y}|} = -C_{\text{hg}} C_{\text{bmodulus}} \frac{h_{\text{proj}}}{|\boldsymbol{x'} - \boldsymbol{x}|} \frac{\boldsymbol{y'} - \boldsymbol{y}}{|\boldsymbol{y'} - \boldsymbol{y}|} \tag{7.67}$$

式中,C_{bmodulus} 为 PMB 模型中的微观模量;C_{hg} 为沙漏力常数,可取 $10^{-3} \sim 10^1$,该值需要通过多次数值试验确定,且在非线性问题中需要不断调整,具有一定人为性。在发生沙漏零能模态时,沙漏力沿键伸长方向做功,通过引入非零功改变了沙漏模式导致的零能量状态,从而可有效抑制非常规态型近场动力学运动方程的虚假解答。

7.4.3 稳定控制计算策略的数值验证

本节以带预制裂纹的矩形板的静力弹性变形分析为例,验证多种因素对数值计算稳定性的影响,以最大限度控制数值不稳定现象,并验证隐式线性与非线性算法的正确性。

如图 7-2 所示,含有中心预制线裂纹的各向同性矩形板受单轴拉伸载荷作用。矩形板的长度、宽度和厚度分别为 $L = 1\text{m}$、$W = 0.5\text{m}$ 和 $h = 0.01\text{m}$,中心竖直裂纹长度为 $2a = 0.1\text{m}$;材料质量密度为 $\rho = 7850\text{kg/m}^3$,杨氏模量 $E = 200\text{GPa}$,泊松比 $\nu = 1/3$,则与式 (7.42) 对应的 $Q_{11} = 225\text{GPa}$、$Q_{12} = 75\text{GPa}$ 和 $Q_{44} = 75\text{GPa}$。分别采用线弹性 BB PD 和线弹性 NOSB PD 的隐式方法求解该问题,在左右两端竖直边界的一定层数物质点上,施加均匀的水平方向拉伸位移 $d = 0.0005\text{m}$,位移场线性变化,并固定两竖直边界中点的竖向位移为 0,以消除刚体位移,上下水平边界自由。

首先,采用键型近场动力学线弹性隐式分析方法求解该问题,以获得该问题的参考解答。进而,采用非常规态型近场动力学方法求解该问题,以研究数值稳定控制方案中相关因素对计算结果稳定性和精度的影响。沿长度、宽度和厚度方向分别采用 100×50×1 点阵离散该矩形板,物质点间距、物质点体积和近场范围分别为 $\Delta x = 0.01\text{m}$、$\Delta V = (\Delta x)^3$ 和 $\delta = 3\Delta x$,影响函数 $w(|\xi|) = 1$,水平方向拉伸位移按线性变化施加在三层物质点上。BB PD 获得的水平位移与竖向位移云图 (三维与二维视图) 如图 7-3 所示。

7.4 数值不稳定性分析与稳定控制方法 · 203 ·

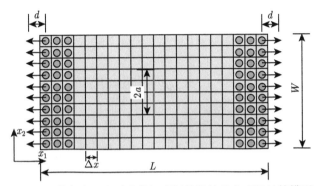

图 7-2 带有中心线裂纹的矩形板单轴拉伸变形的计算模型

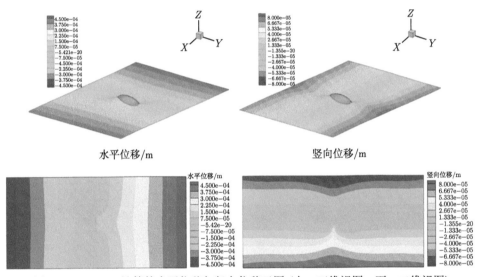

图 7-3 BB PD 计算的水平位移与竖向位移云图 (上: 三维视图, 下: 二维视图)

7.4.3.1 边界条件的影响

为研究边界条件的影响，采用线弹性 NOSB PD 的隐式方法求解该问题。离散格式与前面相同，不考虑沙漏力作用，水平方向拉伸位移按线性变化分别施加在一层、二层和三层物质点上。图 7-4 给出基于 NOSB PD 获得的水平位移与竖向位移的三维视图，从图中给出的计算结果可以看出，对于 NOSB PD 模型，将位移载荷仅施加在一层物质点上，将导致计算结果严重失真，这是由于变形梯度变化剧烈时，边界处近场范围不完整导致变形梯度张量和力状态计算不准确；当位移载荷施加在二层物质点上时，能够获得趋势正确的近场动力学解答，但变形的锯齿形振荡现象较为显著；当位移载荷施加在三层物质点上时，计算精度和稳定性有所提高，数值振荡程度减弱。该算例分析说明，NOSB PD 对于边界条件的施加方式较为敏感，要严格满足边界条件至少施加在一个近场范围的物质点上。

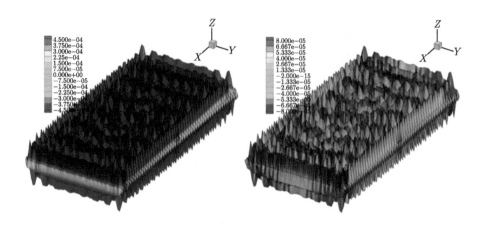

(a) 位移载荷施加在一层物质点上

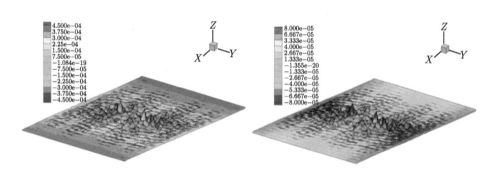

(b) 位移载荷施加在二层物质点上

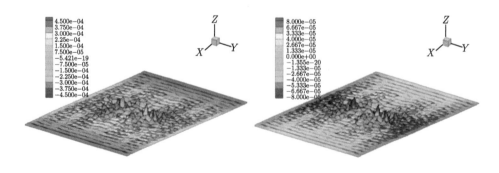

(c) 位移载荷施加在三层物质点上

图 7-4　位移载荷不同处理方式对应的水平位移 (左) 与竖向位移 (右) 云图 (单位：m)

7.4.3.2 网格/点阵密度的影响

为研究离散网格/点阵密度对 NOSB PD 计算结果的影响，依然基于线弹性 NOSB PD 隐式方法求解该问题，分别采用 100×50×1、160×80×1 和 200×100×1 三组点阵离散矩形板，物质点间距分别为 $\Delta x = 0.005\text{m}$、$\Delta x = 0.00625\text{m}$、$\Delta x = 0.01\text{m}$，影响函数 $w(|\boldsymbol{\xi}|) = 1$，不考虑沙漏力，水平方向拉伸位移按线性变化施加在三层物质点上。不同点阵密度下，水平位移与竖向位移的三维视图如图 7-5 所

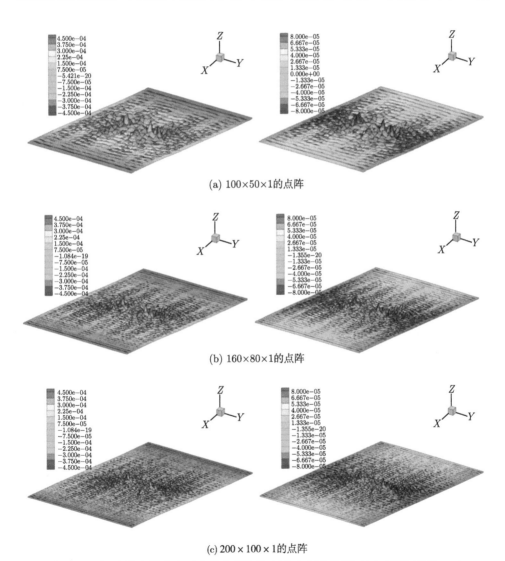

(a) 100×50×1 的点阵

(b) 160×80×1 的点阵

(c) 200×100×1 的点阵

图 7-5 不同离散点阵密度对应的水平位移 (左) 与竖向位移 (右) 云图 (单位：m)

示。由图可见，点阵密度越大，NOSB PD 的解答越精确，数值振荡程度越低，增加点阵密度有利于抑制数值不稳定现象。

7.4.3.3 影响函数的影响

分别考虑三种影响函数 $w(|\boldsymbol{\xi}|)=1$、$w(|\boldsymbol{\xi}|)=\mathrm{e}^{-|\boldsymbol{\xi}|^2/\delta^2}$ 与 $w(|\boldsymbol{\xi}|)=\delta^3/|\boldsymbol{\xi}|^3$，研究影响函数对 NOSB PD 计算结果的影响。还是基于线弹性 NOSB PD 隐式分析方法进行求解，采用 100×50×1 点阵离散矩形板，物质点间距、体积和近场范围分别为 $\Delta x=0.01\mathrm{m}$、$\Delta V=(\Delta x)^3$ 和 $\delta=3\Delta x$，不考虑沙漏力，水平方向拉伸位移按线性变化施加在三层物质点上。三种不同影响函数得到的水平位移与竖向位移的三维视图如图 7-6 所示。由图可见，相比于常影响函数，$w(|\boldsymbol{\xi}|)=\mathrm{e}^{-|\boldsymbol{\xi}|^2/\delta^2}$ 高斯递减形式的影响函数轻微加剧了数值振荡，而 $w(|\boldsymbol{\xi}|)=\delta^3/|\boldsymbol{\xi}|^3$ 三次样条递减形式的影响函数有效地降低了常影响函数的数值振荡程度。一般来说，递减形式的影响函数考虑了近场范围内不同点的影响权重，加大了非局部变形梯度张量计算中的局部影响，有利于降低数值不稳定性；但高斯形式的影响函数递减规律可能不匹配本例中变形和应力的变化规律，使得变形梯度张量的计算误差较大，也表明需要合理地确定递减函数。

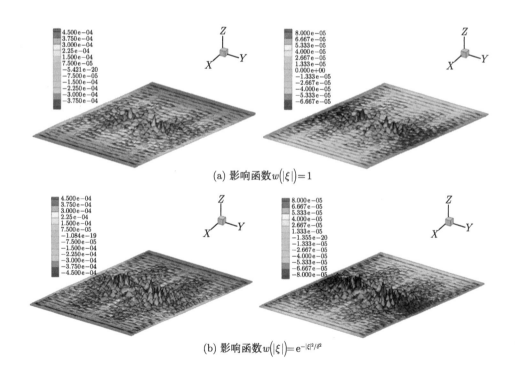

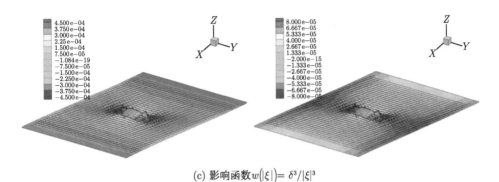

(c) 影响函数 $w(|\xi|)=\delta^3/|\xi|^3$

图 7-6 不同影响函数对应的水平位移 (左) 与竖向位移 (右) 云图 (单位：m)

7.4.3.4 沙漏力的影响

由于采用线性静力求解方式不便于施加沙漏力，本节采用非线性方法求解该问题，研究沙漏力对 NOSB PD 计算结果的影响。分别模拟沙漏力常数取 $C_{\rm hg}=0$、$C_{\rm hg}=0.1$、$C_{\rm hg}=1$、$C_{\rm hg}=10$ 和 $C_{\rm hg}=100$ 的五种情况。采用 $100\times50\times1$ 网格离散矩形板，物质点间距、物质点体积和近场范围分别为 $\Delta x=0.01{\rm m}$、$\Delta V=(\Delta x)^3$ 和 $\delta=3\Delta x$，影响函数取为 $w(|\boldsymbol{\xi}|)=1$，水平方向拉伸位移按线性变化施加在三层物质点上。图 7-7 分别给出基于非线性 NOSB PD 方法求得的不同沙漏力对应的水平位移与竖向位移三维云图。结果显示，未施加沙漏力时 ($C_{\rm hg}=0$)，NOSB PD 解答虽然趋势和分布规律正确，但存在明显的位移振荡现象；施加沙漏力后，数值不稳定现象得到很大程度的抑制，NOSB PD 解答的光滑性与连续性明显提高。当沙漏力常数较小时 ($C_{\rm hg}=0.1$)，虽然能一定程度上减小数值不稳定性，但对于高应变梯度附近 (如裂尖) 抑制效果不佳；当沙漏力常数大小合适时 ($C_{\rm hg}=1$ 与 $C_{\rm hg}=10$)，数值不稳定性得到极大削弱，解答光滑性与连续性显著提高，NOSB PD 解答与 BB-PD 解答吻合良好，但仔细观察裂尖仍然残留微小位移振荡与不连续现象，显示 NOSB PD 方法在变形梯度较大的区域，计算误差依然较大；当沙漏力常数过大时 ($C_{\rm hg}=100$)，结果偏差较大 (未给出具体视图)。综合来看，引入沙漏力的罚函数方法能够有效抑制 NOSB PD 的数值不稳定现象，但不能完全消除，合理选择沙漏力常数能够获得较高精度且稳定的解答；沙漏力常数取值范围较广，选择难度较大，需要多次数值试验确定；过大的沙漏常数使结果偏离精确解，过小的沙漏力对数值不稳定的抑制效果不足[13]。

通过本算例的研究，可以得到以下认识：非局部边界条件的准确施加是解答正确的基本保证；增加沙漏力可使结果光滑性显著提高；三次样条形式的影响函数在很大程度上降低了数值计算的不稳定性；增大网格密度可以降低数值计算的不稳定性；采用系统控制方法，能够最大程度地抑制数值不稳定现象，从而获得高精度与稳定的解答。

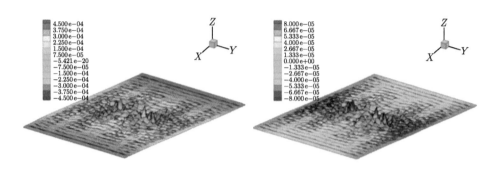

(a) 不考虑沙漏力，沙漏力常数为 $C_{hg}=0$

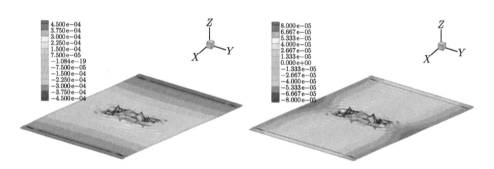

(b) 考虑沙漏力，沙漏力常数为 $C_{hg}=0.1$

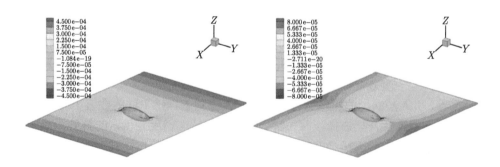

(c) 考虑沙漏力，沙漏力常数为 $C_{hg}=1$

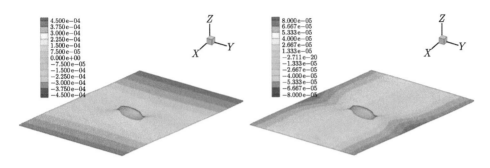

(d) 考虑沙漏力，沙漏力常数为 $C_{hg}=10$

图 7-7　不同沙漏力对应的水平位移 (左) 与竖向位移 (右) 云图 (单位：m)

7.5　晶体弹塑性变形与动态断裂的非常规态型近场动力学方法

本节在非常规态型近场动力学的框架下建立晶体的弹塑性模型，采用晶体滑移系滑动引起塑性变形的单晶体塑性本构模型[31,32]，基于罚函数方法控制零能模式等数值不稳定现象，采用隐式静力和显式动力方法求解晶体弹塑性变形与动态断裂问题。

7.5.1　单晶体弹塑性本构模型

多晶体 (polycrystal) 是由单晶体 (single-crystal) 组成的，因此，可基于单晶本构模型研究多晶体的力学响应。本节介绍 Anand 与 Kothari[31,32] 建立的单晶本构模型，并在 NOSB PD 框架内，求解多晶体弹塑性变形与动力裂纹扩展问题。

Anand 与 Kothari 所建立的单晶本构模型认为，晶体滑移系 (crystal slip systems) 的滑移导致结构的塑性变形，适用于单晶晶粒内的每一个物质点。不失一般性，假设晶体有 N 个滑移系，采用正交矢量 (m_0^α, n_0^α) 描述第 $\alpha(\alpha=1,\cdots,N)$ 个滑移系，m_0^α 和 n_0^α 表示 $t=0$ 时滑移面的滑移方向和滑移面法向。

如图 7-8 所示，xOy 为平面晶体的全局参考坐标系，每个晶粒中的正交箭头为晶体的局部坐标系 $x'O'y'$，参考坐标系逆时针旋转 θ 后与晶体的局部坐标系重合，夹角以逆时针为正，θ 为单晶体晶向 (orientation)。对于二维问题，晶体的局部坐标系与全局参考坐标系的坐标变换矩阵为

$$\boldsymbol{R} = \begin{bmatrix} \cos\theta & -\sin\theta \\ \sin\theta & \cos\theta \end{bmatrix} \qquad (7.68)$$

在进行本构建模时，应力、应变等物理量均在晶体的局部坐标系中表示。在

晶体的局部坐标系中，相对位置矢量、变形梯度张量、形状张量、Cauchy 应力张量、第一类 Piola-Kirchhoff 应力张量等与全局参考坐标系中对应的变量关系为

$$\boldsymbol{x}^{\text{ref}} = \boldsymbol{R}\boldsymbol{x}, \quad \boldsymbol{F}^{\text{ref}} = \boldsymbol{R}\boldsymbol{F}\boldsymbol{R}^{\text{T}}, \quad \boldsymbol{K}^{\text{ref}} = \boldsymbol{R}\boldsymbol{K}\boldsymbol{R}^{\text{T}}, \quad \boldsymbol{\sigma}^{\text{ref}} = \boldsymbol{R}\boldsymbol{\sigma}\boldsymbol{R}^{\text{T}}, \quad \boldsymbol{P}^{\text{ref}} = \boldsymbol{R}\boldsymbol{P}\boldsymbol{R}^{\text{T}} \tag{7.69}$$

其中，上标 ref 表示全局参考坐标系。

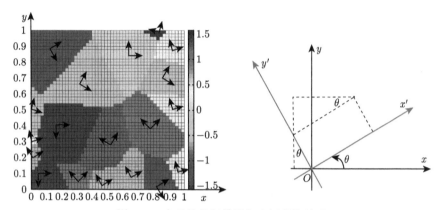

图 7-8　多晶体几何模型与坐标变换关系

当构型发生变形时，变形梯度张量 $\boldsymbol{F}$ 可通过对中间构型应用微分链式法则，进行弹塑性乘法分解[33]，化为弹性变形梯度张量 $\boldsymbol{F}^{\text{e}}$ 和塑性变形梯度张量 $\boldsymbol{F}^{\text{p}}$ 的乘积，即

$$\boldsymbol{F} = \boldsymbol{F}^{\text{e}}\boldsymbol{F}^{\text{p}} \tag{7.70}$$

其中，$\boldsymbol{F}^{\text{e}}$ 表示由拉伸和旋转的晶格畸变导致的弹性变形梯度张量，且 $\det(\boldsymbol{F}^{\text{e}}) > 0$；$\boldsymbol{F}^{\text{p}}$ 表示由滑移系滑移导致的塑性变形梯度张量，且 $\det(\boldsymbol{F}^{\text{p}}) = 1$ 始终成立，当 $\det(\boldsymbol{F}^{\text{p}}) = 1$ 不成立时，采用下式对 $\boldsymbol{F}^{\text{p}}$ 进行更新[32]

$$\boldsymbol{F}^{\text{p}} = \det(\boldsymbol{F}^{\text{p}})^{-1/3}\boldsymbol{F}^{\text{p}} \tag{7.71}$$

在松弛构型中 (塑性变形时无应力)，格林弹性应变张量为 $\boldsymbol{E}^{\text{e}} = \dfrac{1}{2}(\boldsymbol{F}^{\text{e T}}\boldsymbol{F}^{\text{e}} - \boldsymbol{I})$，与其共轭的第二类 Piola-Kirchhoff 应力张量为

$$\boldsymbol{T} = \mathcal{L}(\boldsymbol{E}^{\text{e}}) = \det(\boldsymbol{F}^{\text{e}})(\boldsymbol{F}^{\text{e}})^{-1}\boldsymbol{\sigma}(\boldsymbol{F}^{\text{e}})^{-\text{T}} \tag{7.72}$$

式中，$\boldsymbol{\sigma}$ 为 Cauchy 应力张量，$\mathcal{L}$ 为四阶弹性张量，所涉及的材料参数不受变形状态的影响。

假定晶体滑移或位错导致塑性变形，则塑性速度梯度张量定义为

$$\boldsymbol{L}^{\text{p}} = \dot{\boldsymbol{F}}^{\text{p}}(\boldsymbol{F}^{\text{p}})^{-1} = \sum_{\alpha}\dot{\gamma}^{\alpha}\boldsymbol{S}_{0}^{\alpha}\text{sign}(\tau^{\alpha}) \tag{7.73}$$

其中，$S_0^\alpha = m_0^\alpha \otimes n_0^\alpha$ 为施密特张量 (Schmidt tensor)，$\dot{\gamma}^\alpha$ 为第 α 个滑移系的塑性剪切变形率，sign($\cdot$) 是符号函数，则驱动滑移系滑动的分切应力 (resolved shear stress)τ^α 为

$$\tau^\alpha = T : S_0^\alpha = \text{tr}(T^T S_0^\alpha) \tag{7.74}$$

式中，tr 为矩阵的迹。当第 α 个滑移系滑移时，分切应力 τ^α 到达临界值，这个临界值即为滑移阻抗应力 s^α(slip resistances)，且 $s^\alpha > 0$。因而，塑性剪切变形率由下式确定

$$\begin{cases} \dot{\gamma}^\alpha = 0, & \tau^\alpha \leqslant s^\alpha, \quad \text{未滑动滑移系} \\ \dot{\gamma}^\alpha > 0, & \tau^\alpha > s^\alpha, \quad \text{已滑动滑移系} \end{cases} \tag{7.75}$$

采用塑性硬化模型[31]，即滑移系无滑移时阻抗应力不变，滑移系滑移时阻抗应力增加。于是，滑移阻抗应力采用下式进行更新

$$\begin{aligned} s^\alpha(t + \Delta t) &= s^\alpha(t) + \Delta t \cdot \dot{s}^\alpha(t) \\ \dot{s}^\alpha(t) &= \sum_\beta h^{\alpha\beta} \dot{\gamma}^\beta(t), \quad s^\alpha(0) = \tau_0^\alpha \end{aligned} \tag{7.76}$$

式中，上标 α 和 β 表示滑移系编号，$\dot{\gamma}^\beta$ 为第 β 个滑移系的塑性剪切变形率，τ_0^α 为第 α 个滑移系的初始滑移阻抗应力值，$h^{\alpha\beta}$ 为滑移系塑性硬化项，采用如下模型[31] 计算

$$h^{\alpha\beta} = h_0^\beta \left(q + (1-q)\delta^{\alpha\beta}\right) \left(1 - \frac{s^\beta(t)}{s_s^\beta}\right)^a \quad (\beta\text{不是哑标求和符号}) \tag{7.77}$$

式中，h_0 为硬化系数，q 为潜在硬化系数，$\delta^{\alpha\beta}$ 为克罗内克-δ 常数，s_s 为滑移阻抗的饱和值，a 为硬化指数。在后文数值模拟中，硬化模型各个参数取为

$$h_0 = 10\text{MPa}, \quad s_s = 200\text{MPa}, \quad a = 2, \quad \tau_0 = 10\text{MPa}, \quad q = 1.4 \tag{7.78}$$

7.5.2 塑性剪切变形增量与变形梯度张量更新

对于显式或隐式求解方法，需通过前一加载时步的变量值获得当前加载时步的变量值。上标 $n+1$ 和 n 分别表示当前加载时刻 $t + \Delta t$ 与上一加载时刻 t。当前时步加载完成后，得到 $n+1$ 时步的初始变形梯度张量 F_{n+1}，在非线性求解方法的 Newton Raphson 迭代法中，F_{n+1} 不断更新，直至位移场收敛。

对速度变形梯度张量表达式 (7.73)，右乘矩阵 F^p，再采用时间向后欧拉差分格式，则有

$$\dot{F}^\text{p}(F^\text{p})^{-1}F^\text{p} = \sum_\alpha \dot{\gamma}^\alpha S_0^\alpha \text{sign}(\tau^\alpha) F^\text{p}$$

$$\Rightarrow \frac{\bm{F}_{n+1}^{\mathrm{p}} - \bm{F}_n^{\mathrm{p}}}{\Delta t} = \sum_\alpha \dot{\gamma}^\alpha \bm{S}_0^\alpha \mathrm{sign}(\tau^\alpha) \bm{F}_n^{\mathrm{p}}$$

$$\Rightarrow \bm{F}_{n+1}^{\mathrm{p}} = \Delta t \sum_\alpha \dot{\gamma}^\alpha \bm{S}_0^\alpha \mathrm{sign}(\tau^\alpha) \bm{F}_n^{\mathrm{p}} + \bm{F}_n^{\mathrm{p}}$$

$$\Rightarrow \bm{F}_{n+1}^{\mathrm{p}} = \left(\bm{I} + \sum_\alpha \Delta \gamma^\alpha \bm{S}_0^\alpha \mathrm{sign}(\tau^\alpha) \right) \bm{F}_n^{\mathrm{p}} \tag{7.79}$$

将上式代入乘法分解 $\bm{F} = \bm{F}^{\mathrm{e}} \bm{F}^{\mathrm{p}}$ 中，有下式成立

$$\bm{F}_{n+1}^{\mathrm{e}} = \bm{F}_{n+1} (\bm{F}_{n+1}^{\mathrm{p}})^{-1}$$

$$\Rightarrow \bm{F}_{n+1}^{\mathrm{e}} = \bm{F}_{n+1} \left(\left(\bm{I} + \sum_\alpha \Delta \gamma^\alpha \bm{S}_0^\alpha \mathrm{sign}(\tau^\alpha) \right) \bm{F}_n^{\mathrm{p}} \right)^{-1}$$

$$\Rightarrow \bm{F}_{n+1}^{\mathrm{e}} = \bm{F}_{n+1} (\bm{F}_n^{\mathrm{p}})^{-1} \left(\bm{I} + \sum_\alpha \Delta \gamma^\alpha \bm{S}_0^\alpha \mathrm{sign}(\tau^\alpha) \right)^{-1}$$

$$\Rightarrow \bm{F}_{n+1}^{\mathrm{e}} \approx \bm{F}_{n+1} (\bm{F}_n^{\mathrm{p}})^{-1} \left(\bm{I} - \sum_\alpha \Delta \gamma^\alpha \bm{S}_0^\alpha \mathrm{sign}(\tau^\alpha) \right)$$

$$\Rightarrow \bm{F}_{n+1}^{\mathrm{e}} = \bm{F}_{\mathrm{trial}}^{\mathrm{e}} \left(\bm{I} - \sum_\alpha \Delta \gamma^\alpha \bm{S}_0^\alpha \mathrm{sign}(\tau^\alpha) \right) \tag{7.80}$$

其中，$\bm{I}$ 为单位矩阵，$\bm{F}_{\mathrm{trial}}^{\mathrm{e}} = \bm{F}_{n+1} (\bm{F}_n^{\mathrm{p}})^{-1}$ 为试探弹性变形梯度张量，$\bm{F}_n^{\mathrm{p}}$ 为上一加载步完成后的塑性变形梯度张量，当前加载步保持不变。

进而，格林弹性应变张量为

$$\bm{E}_{n+1}^{\mathrm{e}} = \frac{1}{2} \left((\bm{F}_{n+1}^{\mathrm{e}})^{\mathrm{T}} \bm{F}_{n+1}^{\mathrm{e}} - \bm{I} \right) = \bm{E}_{\mathrm{trial}}^{\mathrm{e}} - \frac{1}{2} \sum_\alpha \Delta \gamma^\alpha \mathrm{sign}(\tau^\alpha) \bm{B}^\alpha \tag{7.81}$$

上式忽略了与 $\Delta \gamma^\alpha$ 相关的高阶项，试探弹性格林应变张量 $\bm{E}_{\mathrm{trial}}^{\mathrm{e}} = ((\bm{F}_{\mathrm{trial}}^{\mathrm{e}})^{\mathrm{T}} \bm{F}_{\mathrm{trial}}^{\mathrm{e}} - \bm{I})/2$ 和 $\bm{B}^\alpha = (\bm{S}_0^\alpha)^{\mathrm{T}} (\bm{F}_{\mathrm{trial}}^{\mathrm{e}})^{\mathrm{T}} \bm{F}_{\mathrm{trial}}^{\mathrm{e}} + (\bm{F}_{\mathrm{trial}}^{\mathrm{e}})^{\mathrm{T}} \bm{F}_{\mathrm{trial}}^{\mathrm{e}} \bm{S}_0^\alpha$ 均为对称张量。利用本构关系 $\bm{T} = \mathcal{L}[\bm{E}^{\mathrm{e}}]$，则可获得第二类 Piola-Kirchhoff 应力张量

$$\bm{T} = \mathcal{L}\left[\bm{E}_{n+1}^{\mathrm{e}}\right] = \mathcal{L}\left[\bm{E}_{\mathrm{trial}}^{\mathrm{e}} - \frac{1}{2} \sum_\alpha \Delta \gamma^\alpha \mathrm{sign}(\tau^\alpha) \bm{B}^\alpha \right]$$

$$= \bm{T}_{\mathrm{trial}} - \frac{1}{2} \sum_\alpha \Delta \gamma^\alpha \mathrm{sign}(\tau^\alpha) \mathcal{L}[\bm{B}^\alpha] \tag{7.82}$$

对于塑性变形的迭代分析过程，采用试探值分析应力应变状态的过程是必须的，通过试探值计算获得塑性剪切变形增量 $\Delta\gamma^\alpha$，再代入上述各式，获得当前牛顿迭代步的各个变量值。

在当前加载时步的初始时刻，完成加载后，已知当前时刻第一次牛顿迭代步的 $\boldsymbol{F}_{n+1}$ 与上一时刻塑性变形 $\boldsymbol{F}_n^{\mathrm{p}}$，依次计算试探弹性变形梯度张量 $\boldsymbol{F}_{\mathrm{trial}}^{\mathrm{e}} = \boldsymbol{F}_{n+1}(\boldsymbol{F}_n^{\mathrm{p}})^{-1}$、试探格林弹性应变张量 $\boldsymbol{E}_{\mathrm{trial}}^{\mathrm{e}}$、试探第二类 Piola-Kirchhoff 应力张量 $\boldsymbol{T}_{\mathrm{trial}}$ 与试探分切应力 $\tau_{\mathrm{trial}}^\alpha$。进而，通过试探分切应力判断潜在滑动滑移系的集合，即当 $|\tau_{\mathrm{trial}}^\alpha| - s^\alpha > 0$ 时，该滑移系为潜在滑动滑移系。晶体塑性变形中，滑动滑移系遵循一致性条件 $|\tau^\alpha| = s^\alpha$，故有下式成立

$$|\tau^\alpha| = s_{n+1}^\alpha = |\tau_{\mathrm{trial}}^\alpha| - \frac{1}{2}\sum_{\beta}^M \mathrm{sign}(\tau_{\mathrm{trial}}^\alpha)\mathrm{sign}(\tau_{\mathrm{trial}}^\beta)\Delta\gamma^\beta \mathcal{L}\left[\boldsymbol{B}^\beta\right]:\boldsymbol{S}_0^\alpha \quad (7.83)$$

式中，M 表示某一物质点的潜在滑动滑移系总数。进而考虑到滑移阻抗具有关系 $s_{n+1}^\alpha = s_n^\alpha + \Delta t \cdot \dot{s}^\alpha(t) = s_n^\alpha + \sum_\beta h^{\alpha\beta}\Delta\gamma^\beta(t)$，则上式变换为

$$|\tau_{\mathrm{trial}}^\alpha| - s_n^\alpha = \frac{1}{2}\sum_{\beta}^M \mathrm{sign}(\tau_{\mathrm{trial}}^\alpha)\mathrm{sign}(\tau_{\mathrm{trial}}^\beta)\Delta\gamma^\beta \mathcal{L}\left[\boldsymbol{B}^\beta\right]:\boldsymbol{S}_0^\alpha + \sum_\beta^M h^{\alpha\beta}\Delta\gamma^\beta$$

$$= \sum_\beta^M \left\{\frac{1}{2}\mathrm{sign}(\tau_{\mathrm{trial}}^\alpha)\mathrm{sign}(\tau_{\mathrm{trial}}^\beta)\mathcal{L}\left[\boldsymbol{B}^\beta\right]:\boldsymbol{S}_0^\alpha + h^{\alpha\beta}\right\}\Delta\gamma^\beta \quad (7.84)$$

上式可以写为矩阵形式，即

$$\boldsymbol{A}\Delta\boldsymbol{\gamma} = \boldsymbol{b} \Rightarrow \begin{bmatrix} A^{11} & \cdots & A^{1\beta} & \cdots & A^{1M} \\ \vdots & \cdots & \vdots & \cdots & \vdots \\ A^{\alpha 1} & \cdots & A^{\alpha\beta} & \cdots & A^{\alpha M} \\ \vdots & \cdots & \vdots & \cdots & \vdots \\ A^{M1} & \cdots & A^{M\beta} & \cdots & A^{MM} \end{bmatrix} \begin{bmatrix} \Delta\gamma^1 \\ \vdots \\ \Delta\gamma^\alpha \\ \vdots \\ \Delta\gamma^M \end{bmatrix} = \begin{bmatrix} b^1 \\ \vdots \\ b^\alpha \\ \vdots \\ b^M \end{bmatrix} \quad (7.85)$$

其中，$A^{\alpha\beta} = \frac{1}{2}\mathrm{sign}(\tau_{\mathrm{trial}}^\alpha)\mathrm{sign}(\tau_{\mathrm{trial}}^\beta)\mathcal{L}^{\mathrm{e}}\left[\boldsymbol{B}^\beta\right]:\boldsymbol{S}_0^\alpha + h^{\alpha\beta}$，$b^\alpha = |\tau_{\mathrm{trial}}^\alpha| - s_n^\alpha$。求解该代数方程或方程组，可获得该加载步的物质点滑移系的塑性剪切变形增量 $\Delta\gamma^\beta$；且若 $\Delta\gamma^\beta \leqslant 0$，则该潜在滑动滑移系不发生滑移，从集合中移除，重新求解代数方程组，直至所有潜在滑动滑移系的 $\Delta\gamma^\beta > 0$ 都成立。

至此，已由试探值出发，获得相应的塑性剪切变形增量 $\Delta\gamma^\beta$，进而获得弹性变形梯度 $\boldsymbol{F}_{n+1}^{\mathrm{e}}$、格林弹性应变张量 $\boldsymbol{E}_{n+1}^{\mathrm{e}}$、第二类 Piola-Kirchhoff 应力张量 $\boldsymbol{T}_{n+1}$ 与试探分切应力 τ_{n+1}^α；计算得到 Cauchy 应力张量 $\boldsymbol{\sigma} = \dfrac{1}{\det(\boldsymbol{F}^{\mathrm{e}})}\boldsymbol{F}^{\mathrm{e}}\boldsymbol{T}(\boldsymbol{F}^{\mathrm{e}})^{\mathrm{T}}$，第

一类 Piola-Kirchhoff 应力 $\boldsymbol{P} = \det(\boldsymbol{F})\boldsymbol{\sigma}\boldsymbol{F}^{-\mathrm{T}} = \boldsymbol{FT}$ 和力密度矢量状态 $\boldsymbol{T}\langle\boldsymbol{\xi}\rangle = w\boldsymbol{PK}^{-1}\boldsymbol{\xi}$。将上述各变量代入增量形式的平衡方程中，获得位移增量 $\delta\boldsymbol{u}_p$；若满足收敛条件，则此时 $\Delta\gamma^\beta$ 即为当前加载步的最终塑性剪切变形增量，对应的各变量值为当前加载步的状态变量值。采用下式 [32] 计算该时步的等效塑性应变增量

$$\mathrm{d}\varepsilon^{\mathrm{P}} = \frac{\sum_\beta |\tau^\beta|\Delta\gamma^\beta}{\sigma_{\mathrm{eff}}} \tag{7.86}$$

其中，等效应力 $\sigma_{\mathrm{eff}} = \sqrt{\dfrac{3}{2}S_{ij}S_{ij}}$，$S_{ij}$ 为应力偏张量的分量。

7.5.3 晶体塑性本构的切线模量

当采用隐式非线性求解方法时，需要获得 $\dfrac{\partial \boldsymbol{P}}{\partial \boldsymbol{F}} = \delta\boldsymbol{P}(\delta\boldsymbol{F})^{-1}$。对 $\boldsymbol{P} = \det(\boldsymbol{F})\boldsymbol{\sigma}\boldsymbol{F}^{-\mathrm{T}}$ 进行变分运算，有 [11]

$$\begin{aligned}\delta\boldsymbol{P} &= \det(\boldsymbol{F})\mathrm{tr}(\delta\boldsymbol{F}\boldsymbol{F}^{-1})\boldsymbol{\sigma}\boldsymbol{F}^{-\mathrm{T}} + \det(\boldsymbol{F})\delta(\boldsymbol{\sigma})\boldsymbol{F}^{-\mathrm{T}} - \det(\boldsymbol{F})\boldsymbol{\sigma}(\delta\boldsymbol{F}\boldsymbol{F}^{-1})^{\mathrm{T}}\boldsymbol{F}^{-\mathrm{T}} \\ &= \det(\boldsymbol{F})\left[\mathrm{tr}(\delta\boldsymbol{F}\boldsymbol{F}^{-1})\boldsymbol{\sigma} + \delta\boldsymbol{\sigma} - \boldsymbol{\sigma}(\delta\boldsymbol{F}\boldsymbol{F}^{-1})^{\mathrm{T}}\right]\boldsymbol{F}^{-\mathrm{T}}\end{aligned} \tag{7.87}$$

式中，$\boldsymbol{\sigma} = \dfrac{1}{\det(\boldsymbol{F}^{\mathrm{e}})}\boldsymbol{F}^{\mathrm{e}}\boldsymbol{T}(\boldsymbol{F}^{\mathrm{e}})^{\mathrm{T}}$。

对于每一个牛顿迭代步，已知 $\boldsymbol{F}$，进而可知 $\Delta\gamma^\beta$、$\boldsymbol{F}^{\mathrm{e}}$、$\boldsymbol{T}$、$\boldsymbol{\sigma}$，还需要求出 $\delta\boldsymbol{\sigma}$。为此，对 Cauchy 应力进行变分运算，有

$$\begin{aligned}\boldsymbol{\sigma} &= \frac{1}{\det(\boldsymbol{F}^{\mathrm{e}})}\boldsymbol{F}^{\mathrm{e}}\boldsymbol{T}(\boldsymbol{F}^{\mathrm{e}})^{\mathrm{T}} \\ \Rightarrow \delta\boldsymbol{\sigma} &= \delta\left[\frac{1}{\det(\boldsymbol{F}^{\mathrm{e}})}\right]\boldsymbol{F}^{\mathrm{e}}\boldsymbol{T}(\boldsymbol{F}^{\mathrm{e}})^{\mathrm{T}} + \frac{1}{\det(\boldsymbol{F}^{\mathrm{e}})}\delta\boldsymbol{F}^{\mathrm{e}}\boldsymbol{T}(\boldsymbol{F}^{\mathrm{e}})^{\mathrm{T}} \\ &\quad + \frac{1}{\det(\boldsymbol{F}^{\mathrm{e}})}\boldsymbol{F}^{\mathrm{e}}\delta\boldsymbol{T}(\boldsymbol{F}^{\mathrm{e}})^{\mathrm{T}} + \frac{1}{\det(\boldsymbol{F}^{\mathrm{e}})}\boldsymbol{F}^{\mathrm{e}}\boldsymbol{T}(\delta\boldsymbol{F}^{\mathrm{e}})^{\mathrm{T}} \\ \Rightarrow \delta\boldsymbol{\sigma} &= -\frac{1}{\det(\boldsymbol{F}^{\mathrm{e}})}\mathrm{tr}(\delta\boldsymbol{F}^{\mathrm{e}}\boldsymbol{F}^{\mathrm{e}^{-1}})\boldsymbol{F}^{\mathrm{e}}\boldsymbol{T}(\boldsymbol{F}^{\mathrm{e}})^{\mathrm{T}} + \frac{1}{\det(\boldsymbol{F}^{\mathrm{e}})}\delta\boldsymbol{F}^{\mathrm{e}}\boldsymbol{T}(\boldsymbol{F}^{\mathrm{e}})^{\mathrm{T}} \\ &\quad + \frac{1}{\det(\boldsymbol{F}^{\mathrm{e}})}\boldsymbol{F}^{\mathrm{e}}\delta\boldsymbol{T}(\boldsymbol{F}^{\mathrm{e}})^{\mathrm{T}} + \frac{1}{\det(\boldsymbol{F}^{\mathrm{e}})}\boldsymbol{F}^{\mathrm{e}}\boldsymbol{T}(\delta\boldsymbol{F}^{\mathrm{e}})^{\mathrm{T}} \\ \Rightarrow \delta\boldsymbol{\sigma} &= \frac{1}{\det(\boldsymbol{F}^{\mathrm{e}})}[-\mathrm{tr}(\delta\boldsymbol{F}^{\mathrm{e}}\boldsymbol{F}^{\mathrm{e}^{-1}})\boldsymbol{F}^{\mathrm{e}}\boldsymbol{T}(\boldsymbol{F}^{\mathrm{e}})^{\mathrm{T}} + \delta\boldsymbol{F}^{\mathrm{e}}\boldsymbol{T}(\boldsymbol{F}^{\mathrm{e}})^{\mathrm{T}} \\ &\quad + \boldsymbol{F}^{\mathrm{e}}\delta\boldsymbol{T}(\boldsymbol{F}^{\mathrm{e}})^{\mathrm{T}} + \boldsymbol{F}^{\mathrm{e}}\boldsymbol{T}(\delta\boldsymbol{F}^{\mathrm{e}})^{\mathrm{T}}]\end{aligned} \tag{7.88}$$

其中，需要求出 $\delta \boldsymbol{F}^{\mathrm{e}}$ 和 $\delta \boldsymbol{T}$，分别表述如下

$$\begin{aligned}
\delta \boldsymbol{F}_{n+1}^{\mathrm{e}} &= \delta \left\{ \boldsymbol{F}_{\mathrm{trial}}^{\mathrm{e}} \left(\boldsymbol{I} - \sum_{\alpha} \Delta \gamma^{\alpha} \boldsymbol{S}_{0}^{\alpha} \mathrm{sign}(\tau^{\alpha}) \right) \right\} \\
&= \delta \boldsymbol{F}_{\mathrm{trial}}^{\mathrm{e}} \left(\boldsymbol{I} - \sum_{\alpha} \Delta \gamma^{\alpha} \boldsymbol{S}_{0}^{\alpha} \mathrm{sign}(\tau^{\alpha}) \right) + \boldsymbol{F}_{\mathrm{trial}}^{\mathrm{e}} \delta \left(\boldsymbol{I} - \sum_{\alpha} \Delta \gamma^{\alpha} \boldsymbol{S}_{0}^{\alpha} \mathrm{sign}(\tau^{\alpha}) \right) \\
&= \delta \boldsymbol{F}_{n+1} (\boldsymbol{F}_{n}^{\mathrm{p}})^{-1} \left(\boldsymbol{I} - \sum_{\alpha} \Delta \gamma^{\alpha} \boldsymbol{S}_{0}^{\alpha} \mathrm{sign}(\tau^{\alpha}) \right) \\
&\quad - \boldsymbol{F}_{n+1} (\boldsymbol{F}_{n}^{\mathrm{p}})^{-1} \sum_{\alpha} \delta \Delta \gamma^{\alpha} \boldsymbol{S}_{0}^{\alpha} \mathrm{sign}(\tau^{\alpha})
\end{aligned} \tag{7.89}$$

$$\begin{aligned}
\delta \boldsymbol{T} &= \delta \mathcal{L}^{\mathrm{e}} \left[\boldsymbol{E}_{\mathrm{trial}}^{\mathrm{e}} - \frac{1}{2} \sum_{\alpha} \Delta \gamma^{\alpha} \mathrm{sign}(\tau^{\alpha}) \boldsymbol{B}^{\alpha} \right] \\
&= \mathcal{L}^{\mathrm{e}} [\delta \boldsymbol{E}_{\mathrm{trial}}^{\mathrm{e}}] - \frac{1}{2} \sum_{\alpha} \mathrm{sign}(\tau^{\alpha}) \delta \Delta \gamma^{\alpha} \mathcal{L}^{\mathrm{e}} [\boldsymbol{B}^{\alpha}] - \frac{1}{2} \sum_{\alpha} \mathrm{sign}(\tau^{\alpha}) \Delta \gamma^{\alpha} \mathcal{L}^{\mathrm{e}} [\delta \boldsymbol{B}^{\alpha}]
\end{aligned} \tag{7.90}$$

式中，

$$\begin{aligned}
\delta \boldsymbol{E}_{\mathrm{trial}}^{\mathrm{e}} &= \delta \left(\frac{1}{2} \left((\boldsymbol{F}_{\mathrm{trial}}^{\mathrm{e}})^{\mathrm{T}} \boldsymbol{F}_{\mathrm{trial}}^{\mathrm{e}} - \boldsymbol{I} \right) \right) = \frac{1}{2} \delta (\boldsymbol{F}_{\mathrm{trial}}^{\mathrm{e}})^{\mathrm{T}} \boldsymbol{F}_{\mathrm{trial}}^{\mathrm{e}} + \frac{1}{2} (\boldsymbol{F}_{\mathrm{trial}}^{\mathrm{e}})^{\mathrm{T}} \delta \boldsymbol{F}_{\mathrm{trial}}^{\mathrm{e}} \\
&= \frac{1}{2} (\boldsymbol{F}_{n}^{\mathrm{p}})^{-\mathrm{T}} \delta (\boldsymbol{F}_{n+1}^{\mathrm{T}}) \boldsymbol{F}_{n+1} (\boldsymbol{F}_{n}^{\mathrm{p}})^{-1} + \frac{1}{2} \left(\boldsymbol{F}_{n+1} (\boldsymbol{F}_{n}^{\mathrm{p}})^{-1} \right)^{\mathrm{T}} \delta \boldsymbol{F}_{n+1} (\boldsymbol{F}_{n}^{\mathrm{p}})^{-1}
\end{aligned} \tag{7.91}$$

$$\begin{aligned}
\delta \boldsymbol{B}^{\alpha} &= \delta \left[(\boldsymbol{S}_{0}^{\alpha})^{\mathrm{T}} (\boldsymbol{F}_{\mathrm{trial}}^{\mathrm{e}})^{\mathrm{T}} \boldsymbol{F}_{\mathrm{trial}}^{\mathrm{e}} + (\boldsymbol{F}_{\mathrm{trial}}^{\mathrm{e}})^{\mathrm{T}} \boldsymbol{F}_{\mathrm{trial}}^{\mathrm{e}} \boldsymbol{S}_{0}^{\alpha} \right] \\
&= (\boldsymbol{S}_{0}^{\alpha})^{\mathrm{T}} \delta \left((\boldsymbol{F}_{\mathrm{trial}}^{\mathrm{e}})^{\mathrm{T}} \boldsymbol{F}_{\mathrm{trial}}^{\mathrm{e}} \right) + \delta \left((\boldsymbol{F}_{\mathrm{trial}}^{\mathrm{e}})^{\mathrm{T}} \boldsymbol{F}_{\mathrm{trial}}^{\mathrm{e}} \right) \boldsymbol{S}_{0}^{\alpha} \\
&= 2 (\boldsymbol{S}_{0}^{\alpha})^{\mathrm{T}} \delta \boldsymbol{E}_{\mathrm{trial}}^{\mathrm{e}} + 2 \delta \boldsymbol{E}_{\mathrm{trial}}^{\mathrm{e}} \boldsymbol{S}_{0}^{\alpha}
\end{aligned} \tag{7.92}$$

对于上述各式，还需要确定 $\delta \Delta \gamma^{\alpha}$，可采用如下方法进行计算。

注意到 $|\tau_{\mathrm{trial}}^{\alpha}| = \mathrm{sign}(\tau_{\mathrm{trial}}^{\alpha}) \mathcal{L} \left[\bar{\boldsymbol{E}}_{\mathrm{trial}}^{\mathrm{e}} \right] : \boldsymbol{S}_{0}^{\alpha}$，并对式 (7.84) 进行变分运算，则有下式成立

$$\delta(|\tau_{\mathrm{trial}}^{\alpha}| - s_{n}^{\alpha}) = \delta \left\{ \sum_{\beta}^{M} \left\{ \frac{1}{2} \mathrm{sign}(\tau_{\mathrm{trial}}^{\alpha}) \mathrm{sign}(\tau_{\mathrm{trial}}^{\beta}) \mathcal{L}^{\mathrm{e}} \left[\boldsymbol{B}^{\beta} \right] : \boldsymbol{S}_{0}^{\alpha} + h^{\alpha\beta} \right\} \Delta \gamma^{\beta} \right\}$$

$$\Rightarrow \delta b^{\alpha} = \delta (A^{\alpha\beta} \Delta \gamma^{\beta})$$

$$\Rightarrow \delta\Delta\gamma^\beta = A_{\alpha\beta}^{-1}(\delta b^\alpha - \delta A^{\alpha b}\Delta\gamma^b) \tag{7.93}$$

式中，重复指标表示 Einstein 求和约定，并且有

$$\delta b^\alpha = \text{sign}(\tau_{\text{trial}}^\alpha)\mathcal{L}^{\text{e}}\left[\delta \boldsymbol{E}_{\text{trial}}^{\text{e}}\right] : \boldsymbol{S}_0^\alpha \tag{7.94}$$

$$A^{\alpha\beta} = \frac{1}{2}\text{sign}(\tau_{\text{trial}}^\alpha)\text{sign}(\tau_{\text{trial}}^\beta)\mathcal{L}^{\text{e}}\left[\boldsymbol{B}^\beta\right] : \boldsymbol{S}_0^\alpha + h^{\alpha\beta} \tag{7.95}$$

$$\delta A^{\alpha\beta} = \delta\left\{\frac{1}{2}\text{sign}(\tau_{\text{trial}}^\alpha)\text{sign}(\tau_{\text{trial}}^\beta)\mathcal{L}^{\text{e}}\left[\boldsymbol{B}^\beta\right] : \boldsymbol{S}_0^\alpha + h^{\alpha\beta}\right\}$$

$$= \frac{1}{2}\text{sign}(\tau_{\text{trial}}^\alpha)\text{sign}(\tau_{\text{trial}}^\beta)\mathcal{L}^{\text{e}}\left[\delta\boldsymbol{B}^\beta\right] : \boldsymbol{S}_0^\alpha \tag{7.96}$$

则从初始的 $\boldsymbol{F}$ 和 $\delta\boldsymbol{F}$ 出发，可计算获得 $\delta\Delta\gamma^\alpha$。

上述从 $\boldsymbol{P} = \det(\boldsymbol{F})\boldsymbol{\sigma}\boldsymbol{F}^{-\text{T}}$ 出发的推导过程较为复杂，可考虑从另一种角度对 $\boldsymbol{P} = \boldsymbol{FT}$ 进行变分运算，得到

$$\delta\boldsymbol{P} = \delta(\boldsymbol{FT}) = \delta\boldsymbol{F}\boldsymbol{T} + \boldsymbol{F}\delta\boldsymbol{T} \tag{7.97}$$

如前所述，由 $\boldsymbol{F}$ 可获得 $\boldsymbol{T}$，参照式 (7.90)～式 (7.92)，可得到 $\delta\boldsymbol{T}$ 及其相关变量的表达式。

至此，采用假设初值的计算方法，通过 $\dfrac{\partial\boldsymbol{P}}{\partial\boldsymbol{F}} = \delta\boldsymbol{P}(\delta\boldsymbol{F})^{-1}$ 即可获得 $\partial\boldsymbol{P}/\partial\boldsymbol{F}$。

7.5.4 平面多晶体弹塑性静力变形的算例分析

采用 NOSB PD 模型的非线性隐式静力解法，在平面应变条件下，模拟平面多晶体的双轴拉压弹塑性变形问题[11]。如图 7-9 所示，考虑具有 π 角度对称轴的一般平面晶体，1mm×1mm 方形微观结构含 19 个该晶粒，共有 12 个不同晶向，晶体取向 θ 在 $[-\pi/2, \pi/2)$ 范围内，每隔 $\pi/6$ 变化。单晶具有两个滑移系，滑移方向为 $-\pi/6$ 和 $\pi/6$。各向同性线弹性材料的参数为：质量密度 2700kg/m^3，弹性模量 $E = 8/3$GPa，剪切模量 $G = 1$GPa，泊松比 $\nu = 1/3$，与式 (7.33) 对应的 $C_{11} = 2$GPa、$C_{12} = C_{44} = 1$GPa。位移边界条件通过速度梯度张量 $\boldsymbol{L} = \eta\begin{bmatrix}1 & 0\\ 0 & -1\end{bmatrix}$ 施加在所有边界的近场范围内的粒子上，相应的位移边界为

$$\begin{bmatrix}u_x\\ u_y\end{bmatrix} = \begin{bmatrix}\eta t & 0\\ 0 & -\eta t\end{bmatrix}\begin{bmatrix}x\\ y\end{bmatrix} = \begin{bmatrix}x\eta t\\ -y\eta t\end{bmatrix} \tag{7.98}$$

其中，$\eta = 0.002$ 为应变率常数，共施加 30 个加载时步 ($t = 1\sim 30$)。

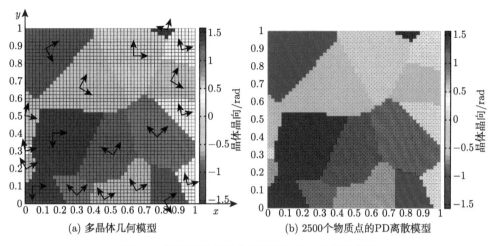

图 7-9 平面多晶体的几何模型与 PD 的离散模型

NOSB PD 模拟的计算条件均与文献 [11] 相同。采用 50×50 点阵离散，近场范围为 $\delta = \Delta x$，影响函数为 $w(|\boldsymbol{\xi}|) = \delta^3/|\boldsymbol{\xi}|^3 = 1$，物质点积分体积为常数。基于不含罚函数稳定控制的 NOSB PD 模型，计算得到应力分量值、晶体晶向变化、等效塑性应变增量、两个滑移系的塑性切应变增量、水平位移与竖向位移等，如图 7-10 所示。结果表明，本节计算得到的应力、应变以及晶体的织态结构的分布规律等均与文献给出的 NOSB PD 结果、晶体塑性有限元 (crystal plasticity finite element, CPFE) 结果高度相似，但应力分量 σ_{xx} 客观存在负值，各个变量的最大值和最小值数值与文献结果存在轻微差别。模拟结果验证了该晶体弹塑性 NOSB PD 模型和隐式求解程序的正确性，NOSB PD 模型能够很好地模拟晶体弹塑性非线性变形过程中精细剪切带的产生与演化过程，但也呈现出一定的数值振荡现象。

图 7-11 给出了含有罚函数稳定控制的 NOSB PD 计算结果，相比之下，含稳定控制的 NOSB PD 结果更为光滑连续，粒子的锯齿形位移得到有效避免。

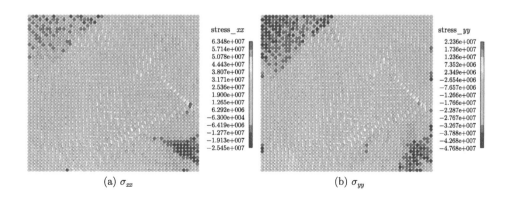

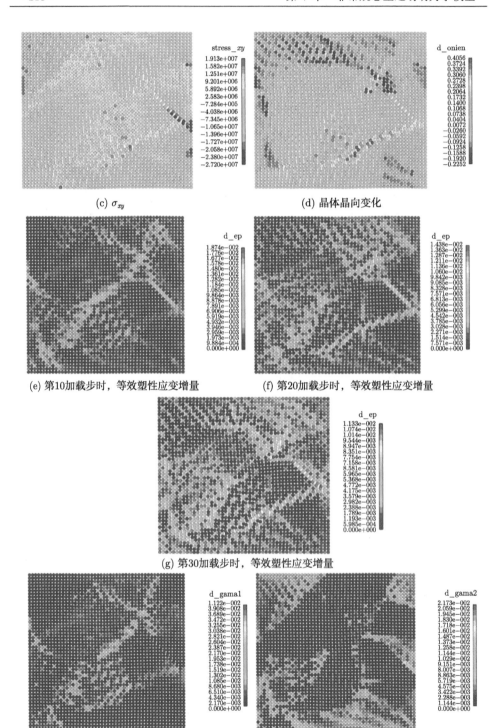

(c) σ_{xy}

(d) 晶体晶向变化

(e) 第10加载步时，等效塑性应变增量

(f) 第20加载步时，等效塑性应变增量

(g) 第30加载步时，等效塑性应变增量

(h) 第10加载步时，两个滑移系的塑性剪切变形增量

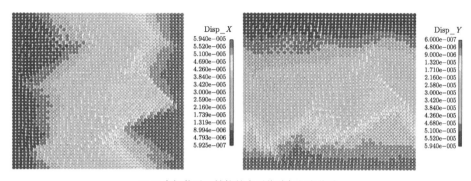

(i) 30步加载后，结构的水平位移与竖向位移

图 7-10 不含罚函数稳定控制的 NOSB PD 模型隐式非线性计算结果

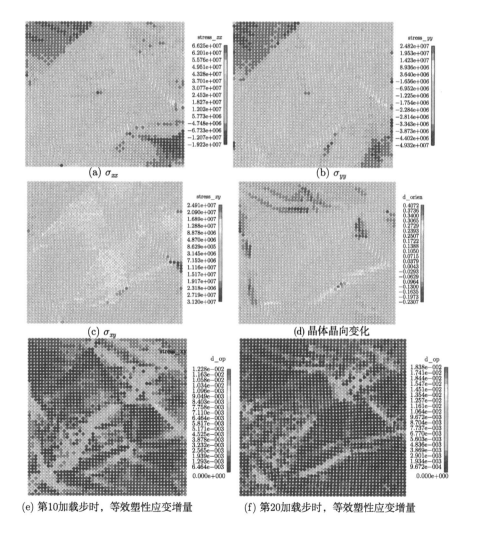

(a) σ_{xx} (b) σ_{yy}

(c) σ_{xy} (d) 晶体晶向变化

(e) 第10加载步时，等效塑性应变增量 (f) 第20加载步时，等效塑性应变增量

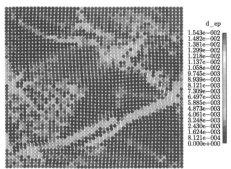

(g) 第30加载步时，等效塑性应变增量

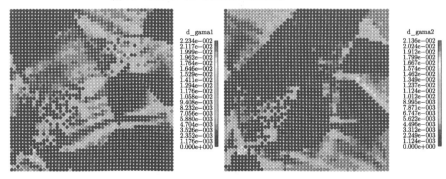

(h) 第10加载步时，两个滑移系的塑性剪切变形增量

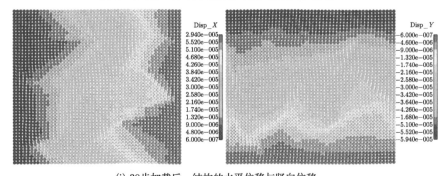

(i) 30步加载后，结构的水平位移与竖向位移

图 7-11　含罚函数稳定控制的 NOSB PD 模型隐式非线性计算结果

7.5.5　平面多晶体弹塑性裂纹扩展的算例分析

采用求解弹塑性问题的 NOSB PD 显式动力学方法，并考虑数值稳定性控制措施，模拟计算多晶体的裂纹扩展问题。

如图 7-12 所示，承受单轴拉伸载荷作用的 4mm×4mm 的多晶体方板含有 68 个单晶晶粒，采用沃罗诺伊图 (Voronoi tessellation) 方法生成该 68 个 Voronoi 胞元。粒子晶向随机分布，采用均匀分布随机数算法生成 [0,360] 的任意晶向角度，

7.5 晶体弹塑性变形与动态断裂的非常规态型近场动力学方法

假定单晶的两个滑移方向为 $-\pi/6$ 和 $\pi/6$。各向同性线弹性材料的参数与上例相同。在板的上下边缘中部，各预制一条长 $a = 0.4$mm 的线裂纹，裂纹生成方法为：(1) 删除裂纹线两侧共一个近场范围的粒子，生成的几何模型记为模型 I；(2) 删除线裂纹处的物质点，并断开所有穿越裂纹的键，生成的几何模型记为模型 II。在左右竖直边界上施加速度 $V = 5$m/s，边界设置无损区域，确保荷载能够传递到板的内部。影响函数取为 $w(|\pmb{\xi}|) = \delta^3/|\pmb{\xi}|^3$。计算中，施加沙漏力控制数值振荡，沙漏力常数取为 $C_{hg} = 100$。显式动力学算法的时间步长取为 $\Delta t = 0.02$μs，共计算 1×10^5 步。采用临界伸长率作为断键准则，即与抗拉强度对应的临界伸长率 $s_0 = \dfrac{\sigma_{\max}}{2E} = 0.1$。

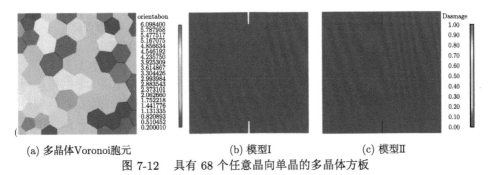

(a) 多晶体Voronoi胞元　　　(b) 模型I　　　(c) 模型II

图 7-12　具有 68 个任意晶向单晶的多晶体方板

为避免裂纹扩展过程中，近场范围内物质点减少可能导致的矩阵条件数增大甚至矩阵奇异，在计算中，当键两端任意物质点的损伤值大于 0.80 时，该键断开，不再对变形状态与力密度状态有贡献；当源点物质点损伤值大于 0.80 时，一次断开所有剩余键，此时物质点不再采用本构力描述；对于完全损伤粒子，采用短程排斥力避免非物理侵入。选取损伤值 0.80 的依据是：在近场动力学中，当物质点损伤到达 0.50 左右时，认为已形成宏观裂纹，损伤超过 0.50 的物质点集中于损伤带附近，当损伤达到 0.80 时，仍与该物质点存在相互作用的点已经很少，可以忽略它们对系统变形与受力的贡献。计算实践表明，这种处理方式牺牲的精度要低于矩阵奇异带来的计算误差。

首先开展 m 收敛和 δ 收敛分析。选取模型 II，固定物质点间距为 $\Delta x = 1.6 \times 10^{-2}$mm，选取不同的近场范围尺寸 $\delta = 2\Delta x$、$\delta = 3\Delta x$、$\delta = 4\Delta x$ 和 $\delta = 5\Delta x$，开展 m 收敛性分析。固定 $m = 3$，分别采用 $\Delta x = 0.04$mm、$\Delta x = 0.025$mm、$\Delta x = 0.02$mm 和 $\Delta x = 0.016$mm 离散构型，开展 δ 收敛性分析。图 7-13 给出了计算得到的多晶体板的载荷-位移曲线 (横坐标为最大水平位移值，纵坐标为 $x = 0.14$mm 处横截面合力)，从图中结果可以看出，除了 $m = 2$ 时的开裂下降段，图 7-13(a) 中各条载荷-位移曲线吻合良好，计算结果稳定可靠，也表明过小的近场范围 (非局部性或长度尺度) 不利于获得稳定的动态开裂结果；随着离散的精细化，图 7-13(b) 中载荷峰值和曲线下降段收敛，验证了 δ 收敛行为良好。

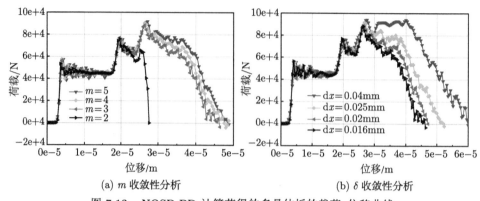

(a) m 收敛性分析　　　(b) δ 收敛性分析

图 7-13　NOSB PD 计算获得的多晶体板的载荷–位移曲线

从计算得到的载荷–位移曲线还可以看出，在曲线的第一个水平段和直线上升段，弹性波从边界开始向板内传播，并至 $x=0.14$mm 处的截面，该截面持续发生弹性变形；在曲线随后的第一个平台，发生了塑性滑移变形；在曲线随后的上升、下降与上升段，交替发生塑性滑移强化和局部损伤开裂；在最后的下降段，裂纹不断扩展直至完全断裂。基于 NOSB PD 的模拟很好地捕捉了受荷多晶体的弹性波传播、线弹性变形、塑性滑移变形、屈服强化和局部开裂以及最终失效的动态开裂过程。

进一步，基于模型 Ⅰ 和 Ⅱ 开展多晶体的动力弹塑性变形与开裂分析，不同时刻的裂纹扩展路径 (损伤分布)、水平位移与等效塑性应变分布云图分别见图 7-14 和图 7-15。比较两组计算结果可见，两种离散方式获得的裂纹路径相似度较高，

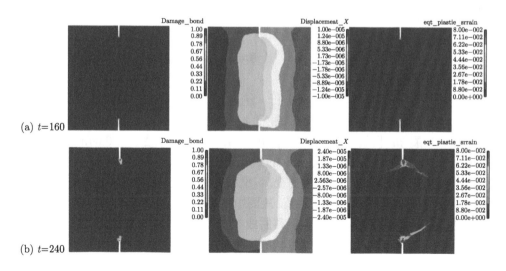

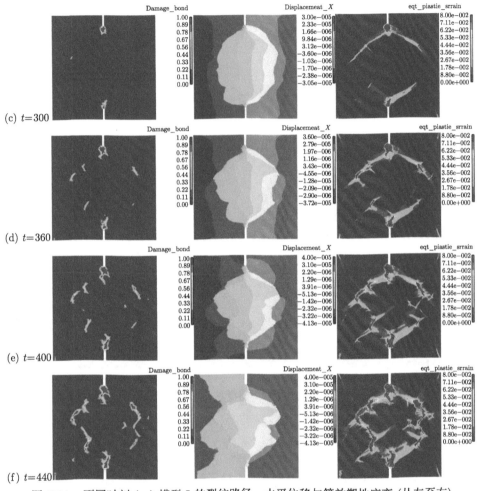

(c) $t=300$

(d) $t=360$

(e) $t=400$

(f) $t=440$

图 7-14　不同时刻 (μs) 模型 I 的裂纹路径、水平位移与等效塑性应变 (从左至右)

裂纹扩展速度基本一致，水平位移与等效塑性应变分布规律吻合良好，表明所提出的方法能够较好地模拟多晶体弹塑性变形和开裂问题。但也应看到，两种几何模型的计算结果存在轻微差别，表明几何模型对计算结果也有一定的影响。

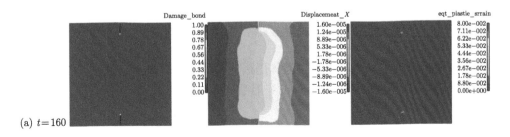

(a) $t=160$

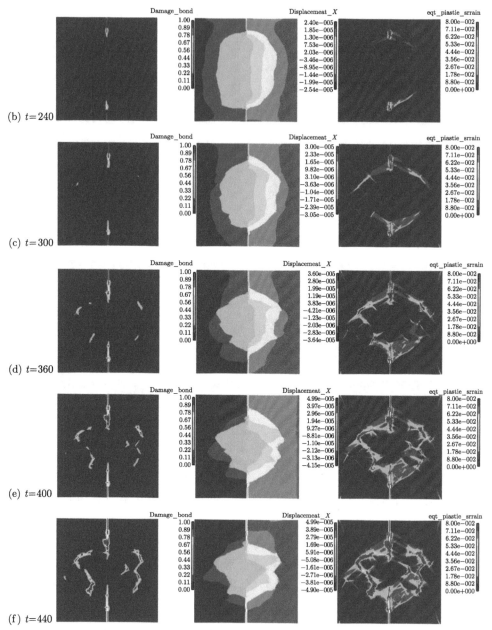

图 7-15　不同时刻 (μs) 模型 II 的裂纹路径、水平位移与等效塑性应变 (从左至右)

图 7-16 还给出了受荷平面多晶体的最终裂纹模式，显示 NOSB PD 模拟可以清晰地捕捉到晶间断裂为主、局部穿晶断裂的破坏特征。

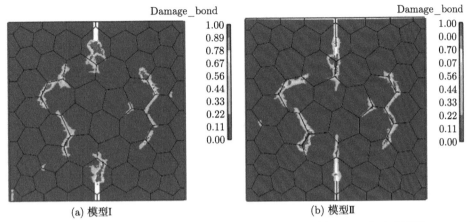

图 7-16　受荷平面多晶体的最终裂纹模式：晶间断裂为主、局部穿晶断裂

参 考 文 献

[1] Silling S A, Epton M, Weckner O, et al. Peridynamic states and constitutive modeling [J]. Journal of Elasticity, 2007, 88(2):151-184.

[2] Warren T L, Silling S A, Askari A, et al. A non-ordinary state-based peridynamic method to model solid material deformation and fracture [J]. International Journal of Solids and Structures, 2009, 46(5):1186-1195.

[3] Foster J T, Silling S A, Chen W W. Viscoplasticity using peridynamics [J]. International Journal for Numerical Methods in Engineering, 2010, 81(10):1242-1258.

[4] Silling S A. Stability of peridynamic correspondence material models and their particle discretizations [J]. Computer Methods in Applied Mechanics and Engineering, 2017, 322:42-57.

[5] Gu X, Madenci E, Zhang Q. Revisit of non-ordinary state-based peridynamics [J]. Engineering Fracture Mechanics, 2018, 190:31-52.

[6] Gu X, Zhang Q, Madenci E, et al. Possible causes of numerical oscillations in non-ordinary state-based peridynamics and a bond-associated higher-order stabilized model [J]. Computer Methods in Applied Mechanics and Engineering, 2019, 357:112592.

[7] Madenci E, Barut A, Futch M. Peridynamic differential operator and its applications [J]. Computer Methods in Applied Mechanics and Engineering, 2016, 304:408-451.

[8] Breitenfeld M S, Geubelle P H, Weckner O, et al. Non-ordinary state-based peridynamic analysis of stationary crack problems [J]. Computer Methods in Applied Mechanics and Engineering, 2014, 272:233-250.

[9] Breitenfeld M S. Quasi-static non-ordinary state-based peridynamics for the modeling of 3D fracture [D]. Urbana-Champaign: The University of Illinois at Urbana-Champaign, 2014.

[10] Yaghoobi A, Chorzepa M G. Meshless modeling framework for fiber reinforced concrete structures [J]. Computers & Structures, 2015, 161:43-54.

[11] Sun S, Sundararaghavan V. A peridynamic implementation of crystal plasticity [J]. International Journal of Solids and Structures, 2014, 51(19-20):3350-3360.

[12] Gu X, Zhang Q, Madenci E. Non-ordinary state-based peridynamic simulation of elasto-plastic deformation and dynamic cracking of polycrystal [J]. Engineering Fracture Mechanics, 2019, 218:106568.

[13] Gu X, Zhang Q, Yu Y T. An effective way to control numerical instability of a non-ordinary state-based peridynamic elastic model [J]. Mathematical Problems in Engineering, 2017, 2017:1-7.

[14] Yaghoobi A, Chorzepa M G. Higher-order approximation to suppress the zero-energy mode in non-ordinary state-based peridynamics [J]. Computers & Structures, 2017, 188:63-79.

[15] Littlewood D J. A nonlocal approach to modeling crack nucleation in AA 7075-T651 [C]. ASME International Mechanical Engineering Congress & Exposition, 2011:567-576.

[16] Tupek M R. Extension of the peridynamic theory of solids for the simulation of materials under extreme loadings [D]. Cambridge: Massachusetts Institute of Technology, 2014.

[17] Chen H L. Bond-associated deformation gradients for peridynamic correspondence model [J]. Mechanics Research Communications, 2018, 90:34-41.

[18] Chen H L, Spencer B W. Peridynamic bond-associated correspondence model: Stability and convergence properties [J]. International Journal for Numerical Methods in Engineering, 2019, 117(6):713-727.

[19] Ganzenmüller G C, Hiermaier S, May M. On the similarity of meshless discretizations of peridynamics and smooth-particle hydrodynamics [J]. Computers & Structures, 2014, 150:71-78.

[20] Bessa M A, Foster J T, Belytschko T, et al. A meshfree unification: reproducing kernel peridynamics [J]. Computational Mechanics, 2014, 53:1251-1264.

[21] Bergel G L, Li S F. The total and updated Lagrangian formulations of state-based peridynamics [J]. Computational Mechanics, 2016, 58(2):351-370.

[22] Ren B, Fan H F, Bergel G L, et al. A peridynamics–SPH coupling approach to simulate soil fragmentation induced by shock waves [J]. Computational Mechanics, 2015, 55(2):287-302.

[23] Li P, Hao Z M, Zhen W Q. A stabilized non-ordinary state-based peridynamic model [J]. Computer Methods in Applied Mechanics and Engineering, 2018, 339:262-280.

[24] Queiruga A F, Moridis G. Numerical experiments on the convergence properties of state-based peridynamic laws and influence functions in two-dimensional problems [J]. Computer Methods in Applied Mechanics and Engineering, 2017, 322:97-122.

[25] Du Q, Tian X C. Stability of nonlocal Dirichlet integrals and implications for peridynamic correspondence material modeling [J]. Journal on Applied Mathematics, 2018, 78(3):1536-1552.

[26] Wu C T, Ren B. A stabilized non-ordinary state-based peridynamics for the nonlocal ductile material failure analysis in metal machining process [J]. Computer Methods in

Applied Mechanics and Engineering, 2015, 291:197-215.

[27] Luo J Y, Sundararaghavan V. Stress-point method for stabilizing zero-energy modes in non-ordinary state-based peridynamics [J]. International Journal of Solids and Structures, 2018, 150:197-207.

[28] Chowdhury S R, Roy P, Roy D, et al. A modified peridynamics correspondence principle: Removal of zero-energy deformation and other implications [J]. Computer Methods in Applied Mechanics and Engineering, 2019, 346:530-549.

[29] Breitzman T, Dayal K. Bond-level deformation gradients and energy averaging in peridynamics [J]. Journal of the Mechanics and Physics of Solids, 2018, 110:192-204.

[30] Madenci E, Dorduncu M, Phan N, et al. Weak form of bond-associated non-ordinary state-based peridynamics free of zero energy modes with uniform or non-uniform discretization [J]. Engineering Fracture Mechanics, 2019, 218:106613.

[31] Anand L, Kothari M. A computational procedure for rate-independent crystal plasticity [J]. Journal of the Mechanics and Physics of Solids, 1996, 44(4):525-558.

[32] Anand L, Kalidindi S R. The process of shear band formation in plane strain compression of fcc metals: Effects of crystallographic texture [J]. Mechanics of Materials, 1994, 17(2):223-243.

[33] 赵亚溥. 近代连续介质力学 [M]. 北京：科学出版社, 2016.

第 8 章 非常规态型近场动力学模型的改进

非常规态型近场动力学具有描述材料与结构复杂非线性变形与断裂行为的能力，相应的数值计算方法可归属为无网格粒子类方法，与现有基于经典连续介质力学的无网格方法具有一定相似性。本章首先将非常规态型近场动力学方法与几种典型的无网格方法进行对比分析。考虑到数值不稳定性是影响非常规态型近场动力学模型广泛应用的重要原因，在第 7 章数值不稳定性原因分析的基础上，为有效控制数值不稳定问题，本章重点介绍三类改进的非常规态型近场动力学模型，包括高阶非常规态型近场动力学模型、键关联 (高阶) 非常规态型近场动力学模型和一种新的非常规态型近场动力学模型。

8.1 几种无网格法与非常规态型近场动力学方法的对比

本节以参考构型中的运动方程和变形梯度张量为重点关注对象，分析它们在不同方法中的连续和离散形式特点，讨论非常规态型近场动力学 (non-ordinary state-based peridynamics, NOSB PD) 与光滑粒子流体动力学 (smoothed particle hydrodynamics, SPH)、修正 SPH(corrected SPH, CSPH)、再生核质点法 (reproducing kernel particle method, RKPM) 和梯度再生核质点法 (gradient-RKPM) 的异同点。

8.1.1 光滑粒子流体动力学方法及其修正

M 维空间中，函数 $f(\boldsymbol{x}') = f(\boldsymbol{x}+\boldsymbol{\xi})$ 关于 $\boldsymbol{x}$ 点函数 $f(\boldsymbol{x})$ 的泰勒级数展开 (Taylor series expansion, TSE) 为

$$f(\boldsymbol{x}+\boldsymbol{\xi}) = \sum_{n_1=0}^{N}\sum_{n_2=0}^{N-n_1}\cdots\sum_{n_M=0}^{N-n_1-\cdots-n_{M-1}} \frac{1}{n_1!n_2!\cdots n_M!}\xi_1^{n_1}\xi_2^{n_2}\cdots\xi_M^{n_M}$$
$$\cdot\frac{\partial^{n_1+n_2+\cdots+n_M}f(\boldsymbol{x})}{\partial x_1^{n_1}\partial x_2^{n_2}\cdots\partial x_M^{n_M}} + R(N,\boldsymbol{x}) \tag{8.1}$$

式中，N 表示泰勒级数展开的阶数，R 为残余项，$n_i(n_i = 0, 1, \cdots, N)$ 表示关于变量 $x_i(i = 1, 2, \cdots, M)$ 的微分阶数，ξ_i 为两点相对位置矢量 $\boldsymbol{\xi} = \boldsymbol{x}' - \boldsymbol{x}$ 的第 i 维分量。

根据式 (8.1)，三维函数的 N 阶泰勒级数展开式为

$$f(\bm{x}+\bm{\xi})=\sum_{n_1=0}^{N}\sum_{n_2=0}^{N-n_1}\sum_{n_3=0}^{N-n_1-n_2}\frac{1}{n_1!n_2!n_3!}\xi_1^{n_1}\xi_2^{n_2}\xi_3^{n_3}\frac{\partial^{n_1+n_2+n_3}f(\bm{x})}{\partial x_1^{n_1}\partial x_2^{n_2}\partial x_3^{n_3}} \qquad (8.2)$$

光滑粒子流体动力学[1-3]是基于核函数近似 (kernel approximation) 的拉格朗日无网格粒子法，概念简单直观且适用于大变形问题。物质点 $\bm{x}$ 的任意标量函数 $f(\bm{x})$ 的核近似为

$$f(\bm{x})=\int_{H_{\bm{x}}} f(\bm{x}')w(\bm{x}'-\bm{x},\delta)\mathrm{d}V_{\bm{x}'} \qquad (8.3)$$

其中，$H_{\bm{x}} = H(\bm{x},\delta) = \{\bm{x}' \in R_0 : \{|\bm{x}'-\bm{x}| \leqslant \delta\}\}$ 为 SPH 和 RKPM 的紧支域 (compact support) 或 PD 的近场范围 (horizon)，δ 为积分域半径，积分域可以是不限于对称圆形或球形的任意形状。如图 8-1 所示，积分域 $H_{\bm{x}}$ 是欧拉型的 (Eulerian) 或拉格朗日型的 (Lagrangian)，相应的数值方法是更新拉格朗日格式 (updated Lagrangian, UL) 和完全拉格朗日格式 (total Lagrangian, TL)[4]。$w(\bm{x}'-\bm{x},\delta)$ 为核函数 (或光滑函数、窗函数、权函数和影响函数等)，通常是指数型、高斯型和样条型的径向对称函数。

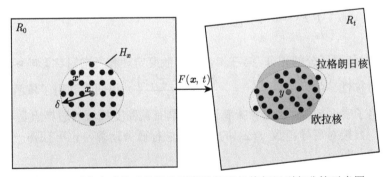

图 8-1 受载变形前后的欧拉型积分核和拉格朗日型积分核示意图

类似于文献 [5] 的工作，将 N 阶泰勒级数展开代入核函数近似中，即将式 (8.2) 代入式 (8.3) 中，得到

$$f(\bm{x})=\int_{H_{\bm{x}}}\sum_{n_1=0}^{N}\sum_{n_2=0}^{N-n_1}\sum_{n_3=0}^{N-n_1-n_2}\frac{1}{n_1!n_2!n_3!}\xi_1^{n_1}\xi_2^{n_2}\xi_3^{n_3}\frac{\partial^{n_1+n_2+n_3}f(\bm{x})}{\partial x_1^{n_1}\partial x_2^{n_2}\partial x_3^{n_3}}w(\bm{x}'-\bm{x},\delta)\mathrm{d}V_{\bm{x}'}$$

$$(8.4)$$

该核函数积分的完备性 (completeness) 阶数[6]可认为是该积分能够精确近似的多项式阶数。通常情况下，求解偏微分方程的数值方法至少需具备一阶完备性，这

样常变形梯度的线性变形场能够得到正确反映。因此，为实现 N 阶完备性，函数自身的一致性条件 (consistency conditions) 或再生条件 (reproducing conditions)[7,8] 要求下式成立

$$\begin{cases} \int_{H_{\boldsymbol{x}}} w \mathrm{d} V_{\boldsymbol{x}'} = 1 \\ \int_{H_{\boldsymbol{x}}} w \xi_1^{n_1} \xi_2^{n_2} \xi_3^{n_3} \mathrm{d} V_{\boldsymbol{x}'} = 0 \ (0 < n_1 + n_2 + n_3 \leqslant N) \end{cases} \quad (8.5)$$

进而，函数 $f(\boldsymbol{x})$ 的一阶导数的核近似可由如下推导过程获得

$$\begin{aligned}
\nabla_{\boldsymbol{x}} f(\boldsymbol{x}) &= \int_{H_{\boldsymbol{x}}} \nabla_{\boldsymbol{x}'} f(\boldsymbol{x}') w(\boldsymbol{x}' - \boldsymbol{x}, \delta) \mathrm{d} V_{\boldsymbol{x}'} \\
&= \int_{H_{\boldsymbol{x}}} \left[\frac{\partial (f(\boldsymbol{x}') w(\boldsymbol{x}' - \boldsymbol{x}, \delta))}{\partial \boldsymbol{x}'} - f(\boldsymbol{x}') \frac{\partial w(\boldsymbol{x}' - \boldsymbol{x}, \delta)}{\partial \boldsymbol{x}'} \right] \mathrm{d} V_{\boldsymbol{x}'} \\
&= \int_{\partial H_{\boldsymbol{x}}} f(\boldsymbol{x}') w(\boldsymbol{x}' - \boldsymbol{x}, \delta) \boldsymbol{n} \mathrm{d} S + \int_{H_{\boldsymbol{x}}} f(\boldsymbol{x}') \frac{\partial w(\boldsymbol{x}' - \boldsymbol{x}, \delta)}{\partial \boldsymbol{x}} \mathrm{d} V_{\boldsymbol{x}'} \\
&= \int_{H_{\boldsymbol{x}}} f(\boldsymbol{x}') \frac{\partial w(\boldsymbol{x}' - \boldsymbol{x}, \delta)}{\partial \boldsymbol{x}} \mathrm{d} V_{\boldsymbol{x}'}
\end{aligned} \quad (8.6)$$

其中，第二式利用了分部积分，第三步使用了散度定理和径向对称影响函数的空间导数的反对称性 $\dfrac{\partial w(\boldsymbol{x}' - \boldsymbol{x}, \delta)}{\partial \boldsymbol{x}} = -\dfrac{\partial w(\boldsymbol{x} - \boldsymbol{x}', \delta)}{\partial \boldsymbol{x}'}$，第四步利用了核函数的紧支撑性并忽略了表面效应，这也意味着该式对具有截断支撑域的边界点是不准确的。

此外，直接对可导函数 $f(\boldsymbol{x})$ 的核近似进行微分运算，也可获得一阶导数的核近似形式，即

$$\nabla f(\boldsymbol{x}) = \frac{\partial f(\boldsymbol{x})}{\partial \boldsymbol{x}} = \int_{H_{\boldsymbol{x}}} \left[\frac{\partial f(\boldsymbol{x}')}{\partial \boldsymbol{x}} w(\boldsymbol{x}' - \boldsymbol{x}, \delta) + f(\boldsymbol{x}') \frac{\partial w(\boldsymbol{x}' - \boldsymbol{x}, \delta)}{\partial \boldsymbol{x}} \right] \mathrm{d} V_{\boldsymbol{x}'} \quad (8.7)$$

将函数 $f(\boldsymbol{x})$ 的 N 阶泰勒级数展开代入上式，得到

$$\begin{aligned}
\nabla f(\boldsymbol{x}) &= \int_{H_{\boldsymbol{x}}} \left[\frac{\partial f(\boldsymbol{x}')}{\partial \boldsymbol{x}} w(\boldsymbol{x}' - \boldsymbol{x}, \delta) + f(\boldsymbol{x}') \frac{\partial w(\boldsymbol{x}' - \boldsymbol{x}, \delta)}{\partial \boldsymbol{x}} \right] \mathrm{d} V_{\boldsymbol{x}'} \\
&= \int_{H_{\boldsymbol{x}}} \left[\frac{\partial}{\partial \boldsymbol{x}} \left(\sum_{n_1=0}^{N} \sum_{n_2=0}^{N-n_1} \sum_{n_3=0}^{N-n_1-n_2} \frac{1}{n_1! n_2! n_3!} \xi_1^{n_1} \xi_2^{n_2} \xi_3^{n_3} \frac{\partial^{n_1+n_2+n_3} f(\boldsymbol{x})}{\partial x_1^{n_1} \partial x_2^{n_2} \partial x_3^{n_3}} \right) w \right.
\end{aligned}$$

$$+ f(\boldsymbol{x}')\frac{\partial w}{\partial \boldsymbol{x}}\bigg] \mathrm{d}V_{\boldsymbol{x}'}$$

$$= \int_{H_{\boldsymbol{x}}} \left[\left\{ \begin{array}{l} \dfrac{1}{n_1!n_2!n_3!}\xi_1^{n_1}\xi_2^{n_2}\xi_3^{n_3}\dfrac{\partial^{n_1+1+n_2+n_3}f(\boldsymbol{x})}{\partial x_1^{n_1+1}\partial x_2^{n_2}\partial x_3^{n_3}} \\ \dfrac{1}{n_1!n_2!n_3!}\xi_1^{n_1}\xi_2^{n_2}\xi_3^{n_3}\dfrac{\partial^{n_1+n_2+1+n_3}f(\boldsymbol{x})}{\partial x_1^{n_1}\partial x_2^{n_2+1}\partial x_3^{n_3}} \\ \dfrac{1}{n_1!n_2!n_3!}\xi_1^{n_1}\xi_2^{n_2}\xi_3^{n_3}\dfrac{\partial^{n_1+n_2+n_3+1}f(\boldsymbol{x})}{\partial x_1^{n_1}\partial x_2^{n_2}\partial x_3^{n_3+1}} \end{array} \right\} w + f(\boldsymbol{x}')\frac{\partial w}{\partial \boldsymbol{x}} \right] \mathrm{d}V_{\boldsymbol{x}'}$$

(8.8)

上式的最后一步中,有 $0 < n_1 + n_2 + n_3 = N$。考虑零阶导数的一致性条件,即式 (8.5) 中的 $\int_{H_{\boldsymbol{x}}} w\xi_1^{n_1}\xi_2^{n_2}\xi_3^{n_3}\mathrm{d}V_{\boldsymbol{x}}' = 0 \ (0 < n_1 + n_2 + n_3 \leqslant N)$,则一阶导数的核近似可以简化为

$$\nabla f(\boldsymbol{x}) = \int_{H_{\boldsymbol{x}}} f(\boldsymbol{x}')\frac{\partial w(\boldsymbol{x}' - \boldsymbol{x}, \delta)}{\partial \boldsymbol{x}}\mathrm{d}V_{\boldsymbol{x}'} \tag{8.9}$$

将函数 $f(\boldsymbol{x})$ 的 N 阶泰勒级数展开代入上式,有

$$\nabla f(\boldsymbol{x}) = \int_{H_{\boldsymbol{x}}} \left(\sum_{n_1=0}^{N} \sum_{n_2=0}^{N-n_1} \sum_{n_3=0}^{N-n_1-n_2} \frac{1}{n_1!n_2!n_3!} \xi_1^{n_1}\xi_2^{n_2}\xi_3^{n_3}\frac{\partial^{n_1+n_2+n_3}f(\boldsymbol{x})}{\partial x_1^{n_1}\partial x_2^{n_2}\partial x_3^{n_3}} \right)$$

$$\times \frac{\partial w(\boldsymbol{x}' - \boldsymbol{x}, \delta)}{\partial \boldsymbol{x}}\mathrm{d}V_{\boldsymbol{x}'} \tag{8.10}$$

因此,使一阶导数满足 N 阶完备性的一致性条件或再生性条件为

$$\begin{array}{l} \displaystyle\int_{H_{\boldsymbol{x}}} \frac{\partial w}{\partial \boldsymbol{x}}\mathrm{d}V_{\boldsymbol{x}'} = \boldsymbol{0} \\ \displaystyle\int_{H_{\boldsymbol{x}}} \boldsymbol{\xi} \otimes \frac{\partial w}{\partial \boldsymbol{x}}\mathrm{d}V_{\boldsymbol{x}'} = \boldsymbol{I} \\ \displaystyle\int_{H_{\boldsymbol{x}}} \xi_1^{n_1}\xi_2^{n_2}\xi_3^{n_3}\frac{\partial w}{\partial \boldsymbol{x}}\mathrm{d}V_{\boldsymbol{x}'} = \boldsymbol{0} \quad (1 < n_1 + n_2 + n_3 \leqslant N) \end{array} \tag{8.11}$$

具体地,函数的零阶导数和一阶导数满足一阶完备性的一致性条件为

$$\begin{array}{l} \displaystyle\int_{H_{\boldsymbol{x}}} w\mathrm{d}V_{\boldsymbol{x}'} = 1 \\ \displaystyle\int_{H_{\boldsymbol{x}}} w\xi_1^{n_1}\xi_2^{n_2}\xi_3^{n_3}\mathrm{d}V_{\boldsymbol{x}'} = 0 \quad (n_1 + n_2 + n_3 = 1) \end{array} \tag{8.12}$$

$$\int_{H_{\boldsymbol{x}}} \frac{\partial w}{\partial \boldsymbol{x}} \mathrm{d} V_{\boldsymbol{x}'} = \boldsymbol{0}$$
$$\int_{H_{\boldsymbol{x}}} \boldsymbol{\xi} \otimes \frac{\partial w}{\partial \boldsymbol{x}} \mathrm{d} V_{\boldsymbol{x}'} = \boldsymbol{I} \qquad (8.13)$$

对于典型径向对称核函数 w 和任意形状支撑域, 式 (8.5) 和式 (8.11) 的一致性条件过强, 难以严格满足。对于对称圆形或球形支撑域, 含有特定系数的径向对称核函数能使一致性条件成立, 因此 SPH 的连续核积分形式满足零阶和一阶完备性。但 SPH 的离散形式方程并不能满足零阶完备性的一致性条件[1,2], 尤其是对边界区域粒子或支撑域内具有非均匀粒子分布的情形, 将会显著影响 SPH 离散数值方法的精度和稳定性。

对式 (8.13) 的第一式乘以 $f(\boldsymbol{x})$, 并利用式 (8.9), 得到函数一阶导数的核近似为

$$\nabla f(\boldsymbol{x}) = \int_{H_{\boldsymbol{x}}} [f(\boldsymbol{x}') - f(\boldsymbol{x})] \frac{\partial w(\boldsymbol{x}' - \boldsymbol{x}, \delta)}{\partial \boldsymbol{x}} \mathrm{d} V_{\boldsymbol{x}'} \qquad (8.14)$$

该式能够保证 SPH 离散格式的零阶完备性[1]。

第 2 章式 (2.3) 给出了经典连续介质力学的变形梯度张量定义 $\boldsymbol{F} = \partial \boldsymbol{y}/\partial \boldsymbol{x}$, 式 (2.36) 给出了参考构型中的第一类 Piola-Kirchhoff 应力 $\boldsymbol{P}$ 表示的运动方程 $\rho_0 \ddot{\boldsymbol{u}}(\boldsymbol{x}, t) = \nabla \cdot \boldsymbol{P} + \boldsymbol{b}$。根据式 (8.14), 对于任意矢量函数 $\boldsymbol{y}(\boldsymbol{x})$, 其变形梯度张量可表示为

$$\boldsymbol{F}(\boldsymbol{x}) = \frac{\partial \boldsymbol{y}(\boldsymbol{x})}{\partial \boldsymbol{x}} = \int_{H_{\boldsymbol{x}}} [\boldsymbol{y}(\boldsymbol{x}') - \boldsymbol{y}(\boldsymbol{x})] \otimes \frac{\partial w(\boldsymbol{x}' - \boldsymbol{x}, \delta)}{\partial \boldsymbol{x}} \mathrm{d} V_{\boldsymbol{x}'} \qquad (8.15)$$

类似地, 局部运动方程可以转化为

$$\rho_0 \ddot{\boldsymbol{u}}(\boldsymbol{x}, t) = \int_{H_{\boldsymbol{x}}} [\boldsymbol{P}(\boldsymbol{x}') + \boldsymbol{P}(\boldsymbol{x})] \cdot \frac{\partial w(\boldsymbol{x}' - \boldsymbol{x}, \delta)}{\partial \boldsymbol{x}} \mathrm{d} V_{\boldsymbol{x}'} + \boldsymbol{b}(\boldsymbol{x}, t) \qquad (8.16)$$

为了保证 SPH 离散格式梯度插值的一阶完备性插值, 基于式 (8.13) 的第二式, 引入修正矩阵 $\boldsymbol{L}$[9-11], 具体为

$$\boldsymbol{L}(\boldsymbol{x}) = \int_{H_{\boldsymbol{x}}} (\boldsymbol{x}' - \boldsymbol{x}) \otimes \frac{\partial w(\boldsymbol{x}' - \boldsymbol{x}, \delta)}{\partial \boldsymbol{x}} \mathrm{d} V_{\boldsymbol{x}'} \qquad (8.17)$$

因此, 保证 SPH 离散格式一阶完备性的变形梯度张量和运动方程为

$$\boldsymbol{F}(\boldsymbol{x}) = \int_{H_{\boldsymbol{x}}} [\boldsymbol{y}(\boldsymbol{x}') - \boldsymbol{y}(\boldsymbol{x})] \otimes \left[\boldsymbol{L}^{-1}(\boldsymbol{x}) \cdot \frac{\partial w(\boldsymbol{x}' - \boldsymbol{x}, \delta)}{\partial \boldsymbol{x}} \right] \mathrm{d} V_{\boldsymbol{x}'} \qquad (8.18)$$

$$\rho_0\ddot{\boldsymbol{u}}(\boldsymbol{x},t) = \int_{H_{\boldsymbol{x}}} [\boldsymbol{P}(\boldsymbol{x}') + \boldsymbol{P}(\boldsymbol{x})] \cdot \left[\boldsymbol{L}^{-1}(\boldsymbol{x}) \cdot \frac{\partial w(\boldsymbol{x}'-\boldsymbol{x},\delta)}{\partial \boldsymbol{x}}\right] \mathrm{d}V_{\boldsymbol{x}'} + \boldsymbol{b}(\boldsymbol{x},t) \quad (8.19)$$

上述带有梯度修正的 SPH 格式被称为修正 SPH(CSPH)，它可以有效提高传统 SPH 方法的一致性、精度和稳定性。此外，考虑到影响函数导数的反对称性 $\dfrac{\partial w(\boldsymbol{x}'-\boldsymbol{x},\delta)}{\partial \boldsymbol{x}} = -\dfrac{\partial w(\boldsymbol{x}-\boldsymbol{x}',\delta)}{\partial \boldsymbol{x}'}$，方程 (8.19) 可以改写为

$$\rho_0\ddot{\boldsymbol{u}}(\boldsymbol{x},t) = \int_{H_{\boldsymbol{x}}} \left[\boldsymbol{P}(\boldsymbol{x})\boldsymbol{L}^{-1}(\boldsymbol{x})\frac{\partial w(\boldsymbol{x}'-\boldsymbol{x},\delta)}{\partial \boldsymbol{x}} \right.$$
$$\left. - \boldsymbol{P}(\boldsymbol{x}')\boldsymbol{L}^{-1}(\boldsymbol{x}')\frac{\partial w(\boldsymbol{x}'-\boldsymbol{x},\delta)}{\partial \boldsymbol{x}'}\right] \mathrm{d}V_{\boldsymbol{x}'} + \boldsymbol{b}(\boldsymbol{x},t) \quad (8.20)$$

8.1.2 再生核质点法和梯度再生核质点法

研究表明，采用结点积分的再生核粒子法 (RKPM) 与 CSPH 具有等价性[12]，RKPM 在经典 SPH 的核近似格式中，引入修正函数 (correction function)，以保证一致性条件成立。物质点 $\boldsymbol{x}$ 处任意标量函数 $f(\boldsymbol{x})$ 的再生核近似为[7,8]

$$f(\boldsymbol{x}) = \int_{H_{\boldsymbol{x}}} f(\boldsymbol{x}') C_N^0(\boldsymbol{x}',\boldsymbol{x}) w(\boldsymbol{x}'-\boldsymbol{x},\delta) \mathrm{d}V_{\boldsymbol{x}'} \quad (8.21)$$

若修正函数为单位函数，即 $C_N^0(\boldsymbol{x}',\boldsymbol{x}) = 1$，则再生核近似退化为经典核近似。修正函数通常可用 N 阶多项式定义，如以下的二阶多项式

$$C_{N=2}^0(\boldsymbol{x}',\boldsymbol{x}) = c_0 + c_1\xi_1 + c_2\xi_2 + c_3\xi_3 + c_4\xi_1^2 + c_5\xi_2^2 + c_6\xi_3^2 + c_7\xi_1\xi_2 + c_8\xi_1\xi_3 + c_9\xi_2\xi_3 \quad (8.22)$$

其中，ξ_1、ξ_2 和 ξ_3 为相对位置矢量 $\boldsymbol{\xi} = \boldsymbol{x}' - \boldsymbol{x}$ 的分量，$c_i(i=0,\cdots,9)$ 为待定修正系数，需根据一致性条件或再生性条件确定。

采用前述 SPH 一致性条件的推导方法，将函数的 N 阶泰勒级数展开代入式 (8.21) 中，有

$$f(\boldsymbol{x}) = \int_{H_{\boldsymbol{x}}} \sum_{n_1=0}^{N} \sum_{n_2=0}^{N-n_1} \sum_{n_3=0}^{N-n_1-n_2} \frac{1}{n_1!n_2!n_3!} \xi_1^{n_1}\xi_2^{n_2}\xi_3^{n_3}$$
$$\times \frac{\partial^{n_1+n_2+n_3} f(\boldsymbol{x})}{\partial x_1^{n_1} \partial x_2^{n_2} \partial x_3^{n_3}} C_N^0(\boldsymbol{x}',\boldsymbol{x}) w(\boldsymbol{x}'-\boldsymbol{x},\delta) \mathrm{d}V_{\boldsymbol{x}'} \quad (8.23)$$

为满足 N 阶完备性条件，函数自身 (零阶导数) 的再生条件要求下式成立

$$\int_{H_\infty} C_N^0 w \mathrm{d}V_{x'} = 1$$
$$\int_{H_\infty} C_N^0 w \xi_1^{n_1} \xi_2^{n_2} \xi_3^{n_3} \mathrm{d}V_{x'} = 0 \quad (0 < n_1 + n_2 + n_3 \leqslant N) \tag{8.24}$$

进而，直接对可导函数的再生核近似进行微分运算，则一阶导数的再生核近似为

$$\frac{\partial f(\boldsymbol{x})}{\partial \boldsymbol{x}} = \int_{H_\infty} \left[\frac{\partial f(\boldsymbol{x}')}{\partial \boldsymbol{x}} \left(C_N^0(\boldsymbol{x}', \boldsymbol{x}) w(\boldsymbol{x}' - \boldsymbol{x}, \delta) \right) + f(\boldsymbol{x}') \frac{\partial}{\partial \boldsymbol{x}} \left(C_N^0(\boldsymbol{x}', \boldsymbol{x}) w(\boldsymbol{x}' - \boldsymbol{x}, \delta) \right) \right] \mathrm{d}V_{x'} \tag{8.25}$$

代入函数 $f(\boldsymbol{x}')$ 的 N 阶泰勒级数展开，上式可重写为

$$\nabla f(\boldsymbol{x})$$
$$= \int_{H_\infty} \left[\frac{\partial f(\boldsymbol{x}')}{\partial \boldsymbol{x}} \left(C_N^0(\boldsymbol{x}', \boldsymbol{x}) w(\boldsymbol{x}' - \boldsymbol{x}, \delta) \right) + f(\boldsymbol{x}') \frac{\partial}{\partial \boldsymbol{x}} \left(C_N^0(\boldsymbol{x}', \boldsymbol{x}) w(\boldsymbol{x}' - \boldsymbol{x}, \delta) \right) \right] \mathrm{d}V_{x'}$$
$$= \int_{H_\infty} \left[\frac{\partial}{\partial \boldsymbol{x}} \left(\sum_{n_1=0}^{N} \sum_{n_2=0}^{N-n_1} \sum_{n_3=0}^{N-n_1-n_2} \frac{1}{n_1! n_2! n_3!} \xi_1^{n_1} \xi_2^{n_2} \xi_3^{n_3} \frac{\partial^{n_1+n_2+n_3} f(\boldsymbol{x})}{\partial x_1^{n_1} \partial x_2^{n_2} \partial x_3^{n_3}} \right) (C_N^0 w) \right.$$
$$\left. + f(\boldsymbol{x}') \frac{\partial (C_N^0 w)}{\partial \boldsymbol{x}} \right] \mathrm{d}V_{x'}$$
$$= \int_{H_\infty} \left[\left\{ \begin{array}{l} \dfrac{1}{n_1! n_2! n_3!} \xi_1^{n_1} \xi_2^{n_2} \xi_3^{n_3} \dfrac{\partial^{n_1+1+n_2+n_3} f(\boldsymbol{x})}{\partial x_1^{n_1+1} \partial x_2^{n_2} \partial x_3^{n_3}} \\ \dfrac{1}{n_1! n_2! n_3!} \xi_1^{n_1} \xi_2^{n_2} \xi_3^{n_3} \dfrac{\partial^{n_1+n_2+1+n_3} f(\boldsymbol{x})}{\partial x_1^{n_1} \partial x_2^{n_2+1} \partial x_3^{n_3}} \\ \dfrac{1}{n_1! n_2! n_3!} \xi_1^{n_1} \xi_2^{n_2} \xi_3^{n_3} \dfrac{\partial^{n_1+n_2+n_3+1} f(\boldsymbol{x})}{\partial x_1^{n_1} \partial x_2^{n_2} \partial x_3^{n_3+1}} \end{array} \right\} (C_N^0 w) + f(\boldsymbol{x}') \dfrac{\partial (C_N^0 w)}{\partial \boldsymbol{x}} \right] \mathrm{d}V_{x'} \tag{8.26}$$

上式的最后一步中，有 $0 < n_1 + n_2 + n_3 = N$。考虑零阶导数的再生条件，即式 (8.24) 中的 $\int_{H_\infty} C_N^0 w \xi_1^{n_1} \xi_2^{n_2} \xi_3^{n_3} \mathrm{d}V_{x'} = 0 \ (0 < n_1 + n_2 + n_3 \leqslant N)$，则一阶导数的再生核近似可以简化为

$$\nabla f(\boldsymbol{x}) = \int_{H_\infty} f(\boldsymbol{x}') \frac{\partial}{\partial \boldsymbol{x}} (C_N^0(\boldsymbol{x}', \boldsymbol{x}) w(\boldsymbol{x}' - \boldsymbol{x}, \delta)) \mathrm{d}V_{x'} \tag{8.27}$$

8.1 几种无网格法与非常规态型近场动力学方法的对比

再将函数 $f(x)$ 的 N 阶泰勒级数展开代入上式,有

$$\nabla f(x) = \int_{H_x} \left(\sum_{n_1=0}^{N} \sum_{n_2=0}^{N-n_1} \sum_{n_3=0}^{N-n_1-n_2} \frac{1}{n_1!n_2!n_3!} \xi_1^{n_1} \xi_2^{n_2} \xi_3^{n_3} \frac{\partial^{n_1+n_2+n_3} f(x)}{\partial x_1^{n_1} \partial x_2^{n_2} \partial x_3^{n_3}} \right)$$
$$\times \frac{\partial}{\partial x}(C_N^0(x', x)w(x'-x, \delta)) \mathrm{d}V_{x'} \tag{8.28}$$

因此,使一阶导数满足 N 阶完备性的再生性条件为

$$\begin{aligned}
&\int_{H_x} \frac{\partial (C_N^0 w)}{\partial x} \mathrm{d}V_{x'} = \mathbf{0} \\
&\int_{H_x} \xi \otimes \frac{\partial (C_N^0 w)}{\partial x} \mathrm{d}V_{x'} = \mathbf{I} \\
&\int_{H_x} \xi_1^{n_1} \xi_2^{n_2} \xi_3^{n_3} \frac{\partial (C_N^0 w)}{\partial x} \mathrm{d}V_{x'} = \mathbf{0} \quad (1 < n_1+n_2+n_3 \leqslant N)
\end{aligned} \tag{8.29}$$

具体地,函数的零阶导数和一阶导数满足一阶完备性的一致性条件为

$$\begin{aligned}
&\int_{H_x} C_N^0 w \mathrm{d}V_{x'} = 1 \\
&\int_{H_x} C_N^0 w \xi_1^{n_1} \xi_2^{n_2} \xi_3^{n_3} \mathrm{d}V_{x'} = 0 \quad (n_1+n_2+n_3=1)
\end{aligned} \tag{8.30}$$

$$\begin{aligned}
&\int_{H_x} \frac{\partial (C_N^0 w)}{\partial x} \mathrm{d}V_{x'} = \mathbf{0} \\
&\int_{H_x} \xi \otimes \frac{\partial (C_N^0 w)}{\partial x} \mathrm{d}V_{x'} = \mathbf{I}
\end{aligned} \tag{8.31}$$

与 SPH 相比,对于径向对称核函数 w 和任意形状支撑域,修正函数保证了一致性条件成立,可有效提高边界区域物质点或非均匀分布点数值计算结果的精度和稳定性。

对式 (8.31) 的第一式乘以 $f(x)$,并利用式 (8.27),得到函数一阶导数的核近似为

$$\nabla f(x) = \frac{\partial f(x)}{\partial x} = \int_{H_x} [f(x') - f(x)] \frac{\partial C_N^0 w(x'-x, \delta)}{\partial x} \mathrm{d}V_{x'} \tag{8.32}$$

对于任意矢量函数 $\boldsymbol{y}(\boldsymbol{x})$，其变形梯度张量可表示为

$$\boldsymbol{F}(\boldsymbol{x}) = \int_{H_\infty} [\boldsymbol{y}(\boldsymbol{x}') - \boldsymbol{y}(\boldsymbol{x})] \otimes \frac{\partial C_N^0 w(\boldsymbol{x}' - \boldsymbol{x}, \delta)}{\partial \boldsymbol{x}} \mathrm{d}V_{\boldsymbol{x}'} \tag{8.33}$$

局部运动方程可以转化为

$$\begin{aligned}\rho_0 \ddot{\boldsymbol{u}}(\boldsymbol{x}, t) &= \int_{H_\infty} [\boldsymbol{P}(\boldsymbol{x}') + \boldsymbol{P}(\boldsymbol{x})] \cdot \frac{\partial C_N^0 w(\boldsymbol{x}' - \boldsymbol{x}, \delta)}{\partial \boldsymbol{x}} \mathrm{d}V_{\boldsymbol{x}'} + \boldsymbol{b}(\boldsymbol{x}, t) \\ &= \int_{H_\infty} \left[\boldsymbol{P}(\boldsymbol{x}) \cdot \frac{\partial C_N^0 w(\boldsymbol{x}' - \boldsymbol{x}, \delta)}{\partial \boldsymbol{x}} - \boldsymbol{P}(\boldsymbol{x}') \frac{\partial C_N^0 w(\boldsymbol{x}' - \boldsymbol{x}, \delta)}{\partial \boldsymbol{x}'} \right] \mathrm{d}V_{\boldsymbol{x}'} + \boldsymbol{b}(\boldsymbol{x}, t) \end{aligned}$$
$$\tag{8.34}$$

由于对再生核近似直接求导非常复杂，文献 [13] 提出了梯度再生核近似 (G-RK) 的概念，以降低一阶微分的计算复杂性，一阶导数的梯度再生核近似为

$$\nabla f(\boldsymbol{x}) = \int_{H_\infty} f(\boldsymbol{x}') C_N^1(\boldsymbol{x}', \boldsymbol{x}) w(\boldsymbol{x}' - \boldsymbol{x}, \delta) \mathrm{d}V_{\boldsymbol{x}'} \tag{8.35}$$

式中，$\boldsymbol{C}_N^1 = [\ C_N^{100}\ \ C_N^{010}\ \ C_N^{001}\ ]^\mathrm{T}$ 为修正矢量，C_N^{100}、C_N^{010} 和 C_N^{001} 为坐标轴 x_1、x_2 和 x_3 方向的修正函数，构造方法类似于式 (8.22)，上标 "1" 表示一阶导数的修正函数。

将函数 $f(\boldsymbol{x}')$ 的 N 阶泰勒级数展开代入式 (8.35)，得到满足 N 阶完备性的再生条件

$$\begin{aligned} &\int_{H_\infty} \boldsymbol{C}_N^1 w \mathrm{d}V_{\boldsymbol{x}'} = \boldsymbol{0} \\ &\int_{H_\infty} \boldsymbol{\xi} \otimes (\boldsymbol{C}_N^1 w) \mathrm{d}V_{\boldsymbol{x}'} = \boldsymbol{I} \\ &\int_{H_\infty} \xi_1^{n_1} \xi_2^{n_2} \xi_3^{n_3} (\boldsymbol{C}_N^1 w) \mathrm{d}V_{\boldsymbol{x}'} = \boldsymbol{0} \quad (1 < n_1 + n_2 + n_3 \leqslant N) \end{aligned} \tag{8.36}$$

将式 (8.36) 的第一式乘以 $f(\boldsymbol{x})$，并利用式 (8.35)，得到函数一阶导数的积分近似为

$$\nabla f(\boldsymbol{x}) = \int_{H_\infty} [f(\boldsymbol{x}') - f(\boldsymbol{x})] C_N^1(\boldsymbol{x}', \boldsymbol{x}) w(\boldsymbol{x}' - \boldsymbol{x}, \delta) \mathrm{d}V_{\boldsymbol{x}'} \tag{8.37}$$

进而，G-RKPM 的对称近似格式的变形梯度张量和运动方程分别为

$$F(x) = \int_{H_x} [y(x') - y(x)] \otimes [C_N^1(x', x)w(x' - x, \delta)]dV_{x'} \quad (8.38)$$

$$\rho_0 \ddot{u}(x, t) = \int_{H_x} [P(x') + P(x)] \cdot [C_N^1(x', x)w(x' - x, \delta)]dV_{x'} + b(x, t)$$

$$= \int_{H_x} [P(x)C_N^1(x', x)w(x' - x, \delta) - P(x')C_N^1(x, x')w(x - x', \delta)]dV_{x'}$$

$$+ b(x, t) \quad (8.39)$$

8.1.3 非常规态型近场动力学模型

在第 7 章中，通过式 (7.2) 定义了状态缩减算子，并基于能量等效方法推导建立了非常规态型近场动力学的变形梯度张量和积分型运动方程，分别为

$$F(x) = \int_{H_x} [y(x') - y(x)] \otimes [w(x' - x, \delta)K^{-1}\xi]dV_{x'} \quad (8.40)$$

$$\rho_0(x)\ddot{u}(x, t) = \int_{H_x} [wPK^{-1}\xi - w'P'K'^{-1}\xi']dV_{x'} + b(x, t)$$

$$= \int_{H_x} [P(x)wK^{-1}(x)\xi - P(x')w'K^{-1}(x')\xi']dV_{x'} + b(x, t) \quad (8.41)$$

式中，形状张量 $K(x) = \int_{H_x} w(\xi \otimes \xi)dV_{x'}$，影响函数 $w(x' - x, \delta) = w'(x - x', \delta)$，相对位置矢量为 $\xi = x' - x$ 和 $\xi' = x - x'$，P 为第一类 Piola-Kirchhoff 应力张量。

非局部变形梯度张量可以通过 $y' = y(x')$ 关于 $y = y(x)$ 的一阶泰勒级数展开和局部加权积分建立，具体可见式 (7.61) 的推导过程。此外，原非常规态型近场动力学的模型 [14] 可以通过一阶近场动力学微分算子 (peridynamic differential operator, PDDO) 从局部应力平衡方程出发进行构建 [15]。因此，非常规态型近场动力学的变形梯度张量和运动方程可以重构为

$$F(x) = \int_{H_x} [y(x') - y(x)] \otimes g_{N=1}(x, \xi)dV_{x'} \quad (8.42)$$

$$\rho_0(x)\ddot{u}(x, t) = \int_{H_x} [P(x)g_{N=1}(x, \xi) - P(x')g_{N=1}(x', \xi')]dV_{x'} + b(x, t) \quad (8.43)$$

式中，$g_{N=1}(x,\xi) = wK^{-1}\xi$。一般来说，$g$ 向量可由近场动力学函数 $g_N^{p_1\cdots p_M}(\xi)$ 进行构造[16-18]，对于三维问题，$g_{N=1} = \{g_N^{100}(\xi) g_N^{010}(\xi) g_N^{001}(\xi)\}^{\mathrm{T}}$；对于二维问题，$g_{N=1} = \{g_N^{10}(\xi) g_N^{01}(\xi)\}^{\mathrm{T}}$；对于一维问题，$g_{N=1} = g_N^1(\xi)$，其中 $N=1$ 表示泰勒级数展开阶数。值得注意的是，二阶导数对于描述高应变梯度变形是非常重要的，但一阶泰勒级数中并不包含，并且，具有任意形状截断近场范围的边界区域物质点的二阶导数并不为零。

8.1.4 几种无网格法与非常规态型近场动力学方法的对比

基于上述推导分析，以下具体讨论 SPH、CSPH、RKPM、G-RKPM 和 NOSB PD 的异同点，并建立它们的统一积分公式。

将连续形式 SPH 的式 (8.15)、式 (8.16) 与 NOSB PD 的式 (8.40)、式 (8.41) 相比较，当影响函数满足下面的式 (8.44) 时，两者完全等价

$$\frac{\partial w(x'-x,\delta)}{\partial x} = w(x'-x,\delta)K^{-1}\xi \tag{8.44}$$

将连续形式 CSPH 的式 (8.18)、式 (8.20) 与 NOSB PD 的式 (8.40)、式 (8.41) 相比较，当影响函数满足下面的式 (8.45) 或式 (8.46) 时，两者完全等价

$$L^{-1}\frac{\partial w(x'-x,\delta)}{\partial x} = w(x'-x,\delta)K^{-1}\xi \tag{8.45}$$

$$\frac{\partial w(x'-x,\delta)}{\partial x} = w(x'-x,\delta)\xi, \quad L^{-1} = K^{-1} \tag{8.46}$$

将连续形式 RKPM 的式 (8.33)、式 (8.34) 与 NOSB PD 的式 (8.40)、式 (8.41) 相比较，当影响函数满足下面的式 (8.47) 时，两者完全等价

$$\frac{\partial C_N^0 w(x'-x,\delta)}{\partial x} = w(x'-x,\delta)K^{-1}\xi \tag{8.47}$$

将连续形式 G-RKPM 的式 (8.38)、式 (8.39) 与 NOSB PD 的式 (8.40)、式 (8.41) 相比较，当影响函数满足下面的式 (8.48) 时，两者完全等价

$$C_N^1(x',x) = K^{-1}\xi \tag{8.48}$$

RKPM 和 NOSB PD 的零阶导数具有等价性，G-RKPM 和 NOSB PD 的一阶导数 (梯度) 具有等价性。虽然此前未明确提出过，但 NOSB PD 的影响函数需要满足零阶一致性条件，即为

$$\int_{H_\infty} C_N^0 w \mathrm{d}V_{x'} = 1 \quad \text{或} \quad \int_{H_\infty} g_N^{0\cdots 0}(\xi) \mathrm{d}V_{x'} = 1 \tag{8.49}$$

其中，C_N^0 为式 (8.22) 定义的多项式函数，$g_N^{0\cdots0}(\boldsymbol{\xi})$ 为零阶导数的近场动力学函数[16-18]。

于是，SPH、CSPH、RKPM、G-RKPM 和 NOSB PD 的变形梯度张量和运动方程可以写为如下统一积分形式，即

$$\boldsymbol{F}(\boldsymbol{x}) = \int\limits_{H_\infty} [\boldsymbol{y}(\boldsymbol{x}') - \boldsymbol{y}(\boldsymbol{x})] \otimes \boldsymbol{g}_{N=1}(\boldsymbol{x},\boldsymbol{\xi}) \mathrm{d}V_{\boldsymbol{x}'} \quad (8.50)$$

$$\rho_0(\boldsymbol{x})\ddot{\boldsymbol{u}}(\boldsymbol{x},t) = \int\limits_{H_\infty} [\boldsymbol{P}(\boldsymbol{x})\boldsymbol{g}_{N=1}(\boldsymbol{x},\boldsymbol{\xi}) - \boldsymbol{P}(\boldsymbol{x}')\boldsymbol{g}_{N=1}(\boldsymbol{x}',\boldsymbol{\xi}')] \mathrm{d}V_{\boldsymbol{x}'} + \boldsymbol{b}(\boldsymbol{x},t) \quad (8.51)$$

其中，$\boldsymbol{g}$ 向量分别为 $\boldsymbol{g}_{N=1}^{\mathrm{SPH}} = \dfrac{\partial w(\boldsymbol{x}'-\boldsymbol{x},\delta)}{\partial \boldsymbol{x}}$、$\boldsymbol{g}_{N=1}^{\mathrm{CSPH}} = \boldsymbol{L}^{-1}\dfrac{\partial w(\boldsymbol{x}'-\boldsymbol{x},\delta)}{\partial \boldsymbol{x}'}$、$\boldsymbol{g}_{N=1}^{\mathrm{RKPM}} = \dfrac{\partial C_N^0 w(\boldsymbol{x}'-\boldsymbol{x},\delta)}{\partial \boldsymbol{x}}$、$\boldsymbol{g}_{N=1}^{\mathrm{GRKPM}} = \boldsymbol{C}_N^1 w(\boldsymbol{x}'-\boldsymbol{x},\delta)$ 和 $\boldsymbol{g}_{N=1}^{\mathrm{NOSBPD}} = w(\boldsymbol{x}'-\boldsymbol{x},\delta)\boldsymbol{K}^{-1}\boldsymbol{\xi}$。

总体而言，上述方程均可认为是经典局部方程的核近似或积分近似，或是局部空间导数的非局部积分表达，建立的非局部积分型模型被认为是更接近物理实际的数学力学模型。在任意形状支撑域情况下，上述五种方法的连续形式方程具有一阶完备性，但对于边界粒子或非均匀分布粒子，传统的 SPH 方法不能保证一阶完备性甚至零阶完备性，其他四种方法能够至少保证零阶完备性，非均匀或非规则粒子分布会降低五类数值方法计算结果的精度和稳定性。除了传统的 SPH 方法，当选择特定影响函数和修正函数时，其他四种方法是等价的。G-RKPM 与 NOSB PD 的单高斯点积分方法等价，且这两种方法的积分格式能直接计算一阶导数，避免了影响函数的一阶导数计算，可有效降低计算的复杂性和计算量，相比于 SPH、CSPH 和 RKPM 方法，在一定程度上更具有优势。此外，虽然 NOSB PD 和 G-RKPM 的数值计算格式相似，但 G-RKPM 是一阶导数的修正核近似，缺乏明确的物理意义，在求解破坏问题时，与 SPH 和 RKPM 等方法一样，通常采用连续损伤模型描述损伤演化；而 NOSB PD 被认为是一类积分型非局部连续介质力学模型，物理意义清晰，可通过"键"和"断键准则"描述损伤演化和裂纹扩展问题，突出优势是便于处理非连续变形问题。

需要指出的是，带有欧拉核的 NOSB PD 更新拉格朗日格式会面临配点法常见的数值不稳定问题，带有拉格朗日核的 NOSB PD 整体拉格朗日格式能够消除该类应力不稳定现象[19]，但仍然存在零能模式和数值振荡现象，并会导致虚假变形和数值不稳定问题。TL-SPH 方法通常使用小支撑域并配合人工黏性项控制数值振荡问题，但是小支撑域使得数值方法不适用于分析非局部效应显著的动态断裂问题[20]，人工黏性项具有人为性和经验性。

8.2 高阶非常规态型近场动力学模型

如前所述，非常规态型近场动力学模型的计算范式致使其数值计算中出现数值不稳定性现象，而现有的数值不稳定控制措施多存在一定的人为性和经验性。由式 (7.61) 的非局部变形梯度的推导说明和 8.1 节的分析可知，非常规态型近场动力学模型只具有一阶精度，精度相对较低。考虑到近场动力学一阶微分算子可以重构 NOSB PD 的运动方程和变形梯度，但低阶泰勒级数会影响计算精度和稳定性 (见后面的 8.4 节)，本书作者研究团队基于含高阶泰勒级数的近场动力学微分算子 [16-18]，提出新的非局部变形梯度张量和内力密度矢量的定义，建立高阶非常规态型近场动力学 (higher-order NOSB PD, H-NOSB PD) 模型，发展高阶模型的线性隐式和非线性隐式计算方法，从而在不采用人为数值稳定控制措施时，能够提高全局的计算精度，有效抑制数值不稳定现象。

8.2.1 高阶非局部变形梯度和力密度矢量状态

根据近场动力学微分算子 [16-18] (见附录 A)，在三维坐标系中，$\boldsymbol{y}$ 关于 $\boldsymbol{x}$ 的一阶导数的非局部积分表达为

$$\left\{ \frac{\partial y_n}{\partial x_1} \quad \frac{\partial y_n}{\partial x_2} \quad \frac{\partial y_n}{\partial x_3} \right\}^{\mathrm{T}} = \int_{H_x} (y'_n - y_n) \, \boldsymbol{g}_N(\boldsymbol{x}, \boldsymbol{\xi}) \mathrm{d}V_{\boldsymbol{x}'}, \quad n = 1, 2, 3 \tag{8.52}$$

式中，n 为坐标维度，$N = 1, 2, \cdots, \infty$ 是泰勒级数展开的阶数，$\boldsymbol{g}_N$ 向量可由近场动力学微分算子中的 PD 函数 $g_N^{p_1 \cdots p_M}(\boldsymbol{\xi})$ 构造，具体为

$$\begin{cases} \boldsymbol{g}_N = \left\{ g_N^{100}(\boldsymbol{x}, \boldsymbol{\xi}) g_N^{010}(\boldsymbol{x}, \boldsymbol{\xi}) g_N^{001}(\boldsymbol{x}, \boldsymbol{\xi}) \right\}^{\mathrm{T}}, & 三维 \\ \boldsymbol{g}_N = \left\{ g_N^{10}(\boldsymbol{x}, \boldsymbol{\xi}) g_N^{01}(\boldsymbol{x}, \boldsymbol{\xi}) \right\}^{\mathrm{T}}, & 二维 \\ \boldsymbol{g}_N = g_N^{1}(\boldsymbol{x}, \boldsymbol{\xi}), & 一维 \end{cases} \tag{8.53}$$

则非局部变形梯度张量定义如下

$$\boldsymbol{F} = \int_{H_x} [\boldsymbol{y}(\boldsymbol{x}') - \boldsymbol{y}(\boldsymbol{x})] \otimes \boldsymbol{g}_N(\boldsymbol{x}, \boldsymbol{\xi}) \mathrm{d}V_{\boldsymbol{x}'} \tag{8.54}$$

这一修正使得变形梯度张量具有 N 阶误差精度，可称为高阶 (N 阶) 变形梯度张量。于是，变形梯度张量的变分为

$$\delta \boldsymbol{F} = \int_{H_x} [\delta \boldsymbol{y}(\boldsymbol{x}') - \delta \boldsymbol{y}(\boldsymbol{x})] \otimes \boldsymbol{g}_N(\boldsymbol{x}, \boldsymbol{\xi}) \mathrm{d}V_{\boldsymbol{x}'}$$

8.2 高阶非常规态型近场动力学模型

$$= \int_{H_\infty} [\delta \boldsymbol{u}(\boldsymbol{x}') - \delta \boldsymbol{u}(\boldsymbol{x})] \otimes \boldsymbol{g}_N(\boldsymbol{x}, \boldsymbol{\xi}) \mathrm{d}V_{\boldsymbol{x}'} \tag{8.55}$$

以下给出力密度矢量状态和内力密度矢量的推导过程。

类似于文献 [14] 中式 (134)~ 式 (142) 给出的力密度矢量状态推导方法，将经典连续介质力学应变能密度 W_{CCM} 与近场动力学变形能量密度 W_{PD} 进行等效，则 NOSB PD 力密度矢量状态可通过材料变形能密度的弗雷歇导数获得，即

$$\underline{\boldsymbol{T}} = \nabla W_{\mathrm{PD}}(\underline{\boldsymbol{Y}}) = \nabla W_{\mathrm{CCM}}\left(\boldsymbol{F}(\underline{\boldsymbol{Y}})\right) \tag{8.56}$$

通过变形梯度分量 F_{ij} 关于变形矢量状态 $\underline{\boldsymbol{Y}}$ 的增量计算，可以获得变形梯度张量 $\boldsymbol{F}$ 的弗雷歇导数分量 ∇F_{ijk}，即

$$\begin{aligned} F_{ij}(\underline{\boldsymbol{Y}} + \Delta \underline{\boldsymbol{Y}}) &= \int_H (\underline{y}_i \langle \boldsymbol{\xi} \rangle + \Delta \underline{y}_i \langle \boldsymbol{\xi} \rangle) g_j \mathrm{d}V_{\boldsymbol{x}'} \\ &= F_{ij}(\underline{\boldsymbol{Y}}) + \int_H \Delta \underline{y}_i \langle \boldsymbol{\xi} \rangle g_j \mathrm{d}V_{\boldsymbol{x}'} \\ &= F_{ij}(\underline{\boldsymbol{Y}}) + \int_H \Delta \underline{y}_k \langle \boldsymbol{\xi} \rangle \delta_{ik} g_j \mathrm{d}V_{\boldsymbol{x}'} \\ &= F_{ij}(\underline{\boldsymbol{Y}}) + \delta_{ik} g_j \cdot \Delta \underline{y}_k \\ &= F_{ij}(\underline{\boldsymbol{Y}}) + \nabla F_{ijk} \cdot \Delta \underline{y}_k \end{aligned} \tag{8.57}$$

式中，g_j 为向量 $\boldsymbol{g}_N$ 的分量，故有 $\nabla F_{ijk} = \delta_{ik} g_j$。于是，由变形矢量状态 $\underline{\boldsymbol{Y}}$ 增量引起的变形能量密度增量为

$$\begin{aligned} \Delta W_{\mathrm{PD}} &= \frac{\partial W_{\mathrm{CCM}}}{\partial F_{ij}} \Delta F_{ij} = \frac{\partial W_{\mathrm{CCM}}}{\partial F_{ij}} \nabla F_{ijk} \cdot \Delta \underline{y}_k \\ &= P_{ij} \delta_{ik} g_j \cdot \Delta \underline{y}_k \\ &= P_{kj} g_j \cdot \Delta \underline{y}_k \\ &= \nabla_{\underline{y}_k} W_{\mathrm{PD}} \cdot \Delta \underline{y}_k \\ &= \underline{T}_k \cdot \Delta \underline{y}_k \end{aligned} \tag{8.58}$$

根据上式,得到高阶非常规态型近场动力学的力密度矢量状态的分量为 $\underline{T}_k = P_{kj} g_j$，则力密度矢量状态为

$$\underline{\boldsymbol{T}}[\boldsymbol{x}, t]\langle \boldsymbol{\xi} \rangle = \boldsymbol{P} \boldsymbol{g}_N \tag{8.59}$$

将力密度矢量状态代入到近场动力学运动方程中，有

$$\rho_0(\boldsymbol{x})\ddot{\boldsymbol{u}}(\boldsymbol{x},t) = \int_{H_\infty} [\boldsymbol{P}(\boldsymbol{x})\boldsymbol{g}_N(\boldsymbol{x},\boldsymbol{\xi}) - \boldsymbol{P}(\boldsymbol{x}')\boldsymbol{g}_N(\boldsymbol{x}',\boldsymbol{\xi}')]\mathrm{d}V_{\boldsymbol{x}'} + \boldsymbol{b}(\boldsymbol{x},t) \tag{8.60}$$

顺便指出，高阶非常规态型近场动力学中"高阶"的含义是，在建模所需的泰勒级数展开中，包含了 $N(N \geqslant 1)$ 次高阶项，相比于原 NOSB PD 模型的一阶泰勒级数展开，可以包含更为丰富的变形和受力信息，也具有更高的精度。

8.2.2 线性隐式解法

在线弹性小变形假设下，Cauchy 应力、第一类或第二类 Piola-Kirchhoff 应力相等，即 $\boldsymbol{\sigma} = \boldsymbol{T} = \boldsymbol{P}$，则对于静力问题，高阶非常规态型近场动力学的离散平衡方程为

$$\sum_{j=1}^{N^i} \{\boldsymbol{\sigma}_i \boldsymbol{g}(\boldsymbol{\xi}_{ji}) - \boldsymbol{\sigma}_j \boldsymbol{g}(\boldsymbol{\xi}_{ij})\} V_j + \boldsymbol{b}(\boldsymbol{x}_i) = \boldsymbol{0} \tag{8.61}$$

上述方程亦可写为 $\boldsymbol{L}(\boldsymbol{u}_p) + \boldsymbol{b} = \boldsymbol{0}$，其中的内力密度矢量 $\boldsymbol{L}(\boldsymbol{u}_p)$ 可分解为系数矩阵 $\boldsymbol{H}_0$ 与未知位移列向量 $\boldsymbol{u}_p$，即

$$\boldsymbol{H}_0 \boldsymbol{u}_p + \boldsymbol{b}_0 = \boldsymbol{0} \tag{8.62}$$

其中，$\boldsymbol{H}_0$ 为 $n_\mathrm{d} \times n_\mathrm{d}(N^i+1)$ 矩阵，表示单个物质点的劲度系数，$\boldsymbol{u}_p$ 为 $n_\mathrm{d}(N^i+1) \times 1$ 列向量，$\boldsymbol{b}_0$ 为 $n_\mathrm{d} \times 1$ 列向量，n_d 为问题的维度。

进而，整个系统的 PD 平衡关系可组装为代数方程组

$$\boldsymbol{H}\boldsymbol{U} + \boldsymbol{b}^* = \boldsymbol{0} \tag{8.63}$$

式中，$\boldsymbol{U}$ 为 $n_\mathrm{d} N_\mathrm{total} \times 1$ 的列向量，包含系统所有未知位移分量；$\boldsymbol{H}$ 为 $n_\mathrm{d} N_\mathrm{total} \times n_\mathrm{d} N_\mathrm{total}$ 的整体劲度矩阵；$\boldsymbol{b}^*$ 为 $n_\mathrm{d} N_\mathrm{total} \times 1$ 的列向量，包含已知外体力密度。N_total 为系统的物质点总数。

应力边界条件以体力密度的形式施加在靠近边界的一定范围的物质点上，包含在 $\boldsymbol{b}^*$ 向量中；位移边界条件也施加在靠近边界的一定范围的物质点上，其矩阵形式为

$$\boldsymbol{G}\boldsymbol{U} + \boldsymbol{U}^* = \boldsymbol{0} \tag{8.64}$$

式中，已知矩阵 $\boldsymbol{G}$ 为约束方程系数，与未知位移向量 $\boldsymbol{U}$ 相关，$\boldsymbol{U}^*$ 为给定的已知位移值。

约束方程可通过拉格朗日乘子法 (Lagrange multipliers)[16] 或乘大数法等引入到系统控制方程中，形成定解的代数方程组。引入拉格朗日乘子 $\boldsymbol{\lambda}$，系统平衡方程和约束方程可以统一在下述变分中

$$\delta \boldsymbol{U}^\mathrm{T}(\boldsymbol{H}\boldsymbol{U} + \boldsymbol{b}^*) + \delta\left[\boldsymbol{\lambda}^\mathrm{T}(\boldsymbol{G}\boldsymbol{U} + \boldsymbol{U}^*)\right] = 0 \tag{8.65}$$

8.2 高阶非常规态型近场动力学模型

式中，δU 为位移矢量的任意变分。对上式第二项进行一阶变分运算，得到

$$\delta U^{\rm T}(HU+b^*)+\delta\lambda^{\rm T}(GU+U^*)+\delta U^{\rm T}G^{\rm T}\lambda=0 \tag{8.66}$$

上式可重构为

$$\left\{\begin{array}{c}\delta U\\ \delta\lambda\end{array}\right\}^{\rm T}\left\{\left[\begin{array}{cc}H & G^{\rm T}\\ G & 0\end{array}\right]\left(\begin{array}{c}U\\ \lambda\end{array}\right)+\left(\begin{array}{c}b^*\\ U^*\end{array}\right)\right\}=0 \tag{8.67}$$

因此，对任意变分 δU 和 $\delta\lambda$，求解未知量 U 和 λ 的代数方程组为

$$\left[\begin{array}{cc}H & G^{\rm T}\\ G & 0\end{array}\right]\left(\begin{array}{c}U\\ \lambda\end{array}\right)=-\left(\begin{array}{c}b^*\\ U^*\end{array}\right) \tag{8.68}$$

上式左端的系数矩阵具有稀疏特性，但由于近场动力学的非局部特性，其带宽大于传统有限元方法，并依赖于近场范围尺寸选取和粒子编号方法。该方程组可采用 MKL 库提供的大型稀疏方程组求解器 PARDISO 或 GMRES 等进行求解。

要建立完整的线性隐式静力求解方法，关键是构造物质点劲度系数矩阵 H_0，进而组装整体劲度矩阵。以下介绍将高阶 NOSB PD 的本构 $T\langle\xi\rangle=Pg$ 写为矩阵形式，以获得物质点劲度系数矩阵的方法。

小变形条件下的应变张量为 $E=\frac{1}{2}(F+F^{\rm T})-I$，将式 (8.54) 的非局部变形梯度张量代入其中，即 $F(x_i)\approx\sum_j^{N^i}(u_j-u_i)\otimes g(\xi_{ji})V_j+\sum_j^{N^i}\xi_{ji}\otimes g(\xi_{ji})V_j$，可得

$$E_i\approx\frac{1}{2}\sum_{j=1}^{N^i}[(u_j-u_i)\otimes g(\xi_{ji})+g(\xi_{ji})\otimes(u_j-u_i)]V_j$$

$$+\frac{1}{2}\sum_{j=1}^{N^i}[\xi_{ji}\otimes g(\xi_{ji})+g(\xi_{ji})\otimes\xi_{ji}]V_j-I \tag{8.69}$$

上述等式右侧后两项可归并记为 $E_0=\frac{1}{2}\sum_{j=1}^{N^i}[\xi_{ji}\otimes g(\xi_{ji})+g(\xi_{ji})\otimes\xi_{ji}]V_j-I$。

8.2.2.1 三维情况

对于三维问题，将系数矩阵与位移向量解耦，应变张量可重写为如下 Vogit 向量形式

$$\hat{E}=\left[\begin{array}{cccccc}\varepsilon_{11} & \varepsilon_{22} & \varepsilon_{33} & \varepsilon_{23} & \varepsilon_{13} & \varepsilon_{12}\end{array}\right]^{\rm T}=\hat{M}u_p+\hat{E}_0 \tag{8.70}$$

式中, $\hat{\boldsymbol{E}}$ 为 6×1 列向量, $\hat{\boldsymbol{M}}$ 为 $6\times(3N^i+3)$ 矩阵, $\boldsymbol{u}_p$ 为 $(3N^i+3)\times 1$ 列向量, 且有

$$\left[\hat{\boldsymbol{M}}\right]_{6\times(3N^i+3)} = \begin{bmatrix} M^1 & 0 & 0 & \cdots & V_j g_N^{100}(\boldsymbol{\xi}_{ji}) & 0 & 0 & \cdots \\ 0 & M^2 & 0 & \cdots & 0 & V_j g_N^{010}(\boldsymbol{\xi}_{ji}) & 0 & \cdots \\ 0 & 0 & M^3 & \cdots & 0 & 0 & V_j g_N^{001}(\boldsymbol{\xi}_{ji}) & \cdots \\ \frac{1}{2}M^2 & \frac{1}{2}M^1 & 0 & \cdots & \frac{1}{2}V_j g_N^{010}(\boldsymbol{\xi}_{ji}) & \frac{1}{2}V_j g_N^{100}(\boldsymbol{\xi}_{ji}) & 0 & \cdots \\ \frac{1}{2}M^3 & 0 & \frac{1}{2}M^1 & \cdots & \frac{1}{2}V_j g_N^{001}(\boldsymbol{\xi}_{ji}) & 0 & \frac{1}{2}V_j g_N^{100}(\boldsymbol{\xi}_{ji}) & \cdots \\ 0 & \frac{1}{2}M^3 & \frac{1}{2}M^2 & \cdots & 0 & \frac{1}{2}V_j g_N^{001}(\boldsymbol{\xi}_{ji}) & \frac{1}{2}V_j g_N^{010}(\boldsymbol{\xi}_{ji}) & \cdots \end{bmatrix} \tag{8.71}$$

$$[\boldsymbol{u}_p]_{(3N^i+3)\times 1} = \begin{bmatrix} u_{11} & u_{12} & u_{13} & \cdots & u_{N^i 1} & u_{N^i 2} & u_{N^i 3} & u_{(N^i+1)1} & u_{(N^i+1)2} & u_{(N^i+1)3} \end{bmatrix}^{\mathrm{T}} \tag{8.72}$$

$$\left[\hat{\boldsymbol{E}}_0\right]_{6\times 1} = \sum_{j=1}^{N^i} \begin{Bmatrix} \xi_1 g_N^{100}(\boldsymbol{\xi}_{ji}) \\ \xi_2 g_N^{010}(\boldsymbol{\xi}_{ji}) \\ \xi_3 g_N^{001}(\boldsymbol{\xi}_{ji}) \\ \dfrac{\xi_1 g_N^{010}(\boldsymbol{\xi}_{ji}) + \xi_2 g_N^{100}(\boldsymbol{\xi}_{ji})}{2} \\ \dfrac{\xi_1 g_N^{001}(\boldsymbol{\xi}_{ji}) + \xi_3 g_N^{100}(\boldsymbol{\xi}_{ji})}{2} \\ \dfrac{\xi_2 g_N^{001}(\boldsymbol{\xi}_{ji}) + \xi_3 g_N^{010}(\boldsymbol{\xi}_{ji})}{2} \end{Bmatrix} V_j - \begin{Bmatrix} 1 \\ 1 \\ 1 \\ 0 \\ 0 \\ 0 \end{Bmatrix} \tag{8.73}$$

式中, $\hat{\boldsymbol{M}}$ 矩阵前三列的 $M^1 = -\sum_{j=1}^{N^i} V_j g_N^{100}(\boldsymbol{\xi}_{ji})$、$M^2 = -\sum_{j=1}^{N^i} V_j g_N^{010}(\boldsymbol{\xi}_{ji})$ 和 $M^3 = -\sum_{j=1}^{N^i} V_j g_N^{001}(\boldsymbol{\xi}_{ji})$ 对应于源点 $\boldsymbol{x}_i$, 其余各列对应于场点 $\boldsymbol{x}_j$; $g_N^{100}(\boldsymbol{\xi}_{ji})$、$g_N^{010}(\boldsymbol{\xi}_{ji})$ 和 $g_N^{001}(\boldsymbol{\xi}_{ji})$ 是 PD 函数, 具有 N 阶泰勒级数项; u_{11}、u_{12} 和 u_{13} 为源点 $\boldsymbol{x}_i$ 的位移值, 其余各列为场点 $\boldsymbol{x}_j$ 的位移值。

进而，Cauchy 应力和力密度矢量状态分别为

$$\hat{\boldsymbol{\sigma}} = \begin{bmatrix} \sigma_{11} & \sigma_{22} & \sigma_{33} & \sigma_{12} & \sigma_{13} & \sigma_{23} \end{bmatrix}^T = \boldsymbol{C}\hat{\boldsymbol{E}} = \boldsymbol{C}\hat{\boldsymbol{M}}\boldsymbol{u}_p + \boldsymbol{C}\hat{\boldsymbol{E}}_0 \quad (8.74)$$

$$\underline{\boldsymbol{T}}\langle\boldsymbol{\xi}\rangle = \boldsymbol{G}\boldsymbol{C}\hat{\boldsymbol{M}}\boldsymbol{u}_p + \boldsymbol{G}\boldsymbol{C}\hat{\boldsymbol{E}}_0 \quad (8.75)$$

其中，$\boldsymbol{C}$ 为材料的弹性矩阵，如式 (7.33) 所示，$\boldsymbol{G}$ 为 3×6 矩阵，即

$$[\boldsymbol{G}]_{3\times 6} = \begin{bmatrix} g_N^{100}(\boldsymbol{\xi}_{ji}) & 0 & 0 & g_N^{010}(\boldsymbol{\xi}_{ji}) & g_N^{001}(\boldsymbol{\xi}_{ji}) & 0 \\ 0 & g_N^{010}(\boldsymbol{\xi}_{ji}) & 0 & g_N^{100}(\boldsymbol{\xi}_{ji}) & 0 & g_N^{001}(\boldsymbol{\xi}_{ji}) \\ 0 & 0 & g_N^{001}(\boldsymbol{\xi}_{ji}) & 0 & g_N^{100}(\boldsymbol{\xi}_{ji}) & g_N^{010}(\boldsymbol{\xi}_{ji}) \end{bmatrix} \quad (8.76)$$

则物质点的劲度系数矩阵为

$$\boldsymbol{H}_0 = \boldsymbol{G}\boldsymbol{C}\hat{\boldsymbol{M}} \quad (8.77)$$

由此，可组集获得系统整体劲度系数矩阵。式 (8.75) 的第二项 $\boldsymbol{G}\boldsymbol{C}\hat{\boldsymbol{E}}_0$ 需要加到系统代数方程组的右端项中，求解线性代数方程组即可获得解答。

8.2.2.2 二维情况

对于二维问题，将系数矩阵与位移向量解耦，应变张量可重写为如下 Vogit 向量形式

$$\hat{\boldsymbol{E}} = \begin{bmatrix} \varepsilon_{11} & \varepsilon_{22} & \varepsilon_{12} \end{bmatrix}^T = \hat{\boldsymbol{M}}\boldsymbol{u}_p + \hat{\boldsymbol{E}}_0 \quad (8.78)$$

式中，$\hat{\boldsymbol{E}}$ 为 3×1 列向量，$\hat{\boldsymbol{M}}$ 为 3×($2N^i+2$) 矩阵，$\boldsymbol{u}_p$ 为 ($2N^i+2$)×1 列向量，$\hat{\boldsymbol{E}}_0$ 为 3×1 列向量，且有

$$\left[\hat{\boldsymbol{M}}\right]_{3\times(2N^i+2)} = \begin{bmatrix} M^1 & 0 & \cdots & V_j g_N^{10}(\boldsymbol{\xi}_{ji}) & 0 & \cdots \\ 0 & M^2 & \cdots & 0 & V_j g_N^{01}(\boldsymbol{\xi}_{ji}) & \cdots \\ \frac{1}{2}M^2 & \frac{1}{2}M^1 & \cdots & \frac{1}{2}V_j g_N^{01}(\boldsymbol{\xi}_{ji}) & \frac{1}{2}V_j g_N^{10}(\boldsymbol{\xi}_{ji}) & \cdots \end{bmatrix} \quad (8.79)$$

$$[\boldsymbol{u}_p]_{(2N^i+2)\times 1} = \begin{bmatrix} u_{11} & u_{12} & \cdots & u_{N^i 1} & u_{N^i 2} & u_{(N^i+1)1} & u_{(N^i+1)2} \end{bmatrix}^T \quad (8.80)$$

$$\left[\hat{\boldsymbol{E}}_0\right]_{3\times 1} = \sum_{j=1}^{N^i} V_j \begin{bmatrix} \xi_1 g_N^{10}(\boldsymbol{\xi}_{ji}) \\ \xi_2 g_N^{01}(\boldsymbol{\xi}_{ji}) \\ \dfrac{\xi_1 g_N^{01}(\boldsymbol{\xi}_{ji}) + \xi_2 g_N^{10}(\boldsymbol{\xi}_{ji})}{2} \end{bmatrix} - \begin{bmatrix} 1 \\ 1 \\ 0 \end{bmatrix} \quad (8.81)$$

式中，$\hat{M}$ 矩阵前两列的 $M^1 = -\sum_{j=1}^{N^i} V_j g_N^{10}(\boldsymbol{\xi}_{ji})$ 和 $M^2 = -\sum_{j=1}^{N^i} V_j g_N^{01}(\boldsymbol{\xi}_{ji})$ 对应于源点 $\boldsymbol{x}_i$，其余各列对应于场点 $\boldsymbol{x}_j$；$g_N^{10}(\boldsymbol{\xi}_{ji})$ 和 $g_N^{01}(\boldsymbol{\xi}_{ji})$ 是 PD 函数，具有 N 阶泰勒级数项；u_{11} 和 u_{12} 为源点 $\boldsymbol{x}_i$ 的位移值，其余各列为场点 $\boldsymbol{x}_j$ 的位移值。

进而，Cauchy 应力和力密度矢量状态分别为

$$\hat{\boldsymbol{\sigma}} = \begin{bmatrix} \sigma_{11} & \sigma_{22} & \sigma_{12} \end{bmatrix}^{\mathrm{T}} = \boldsymbol{C}\hat{\boldsymbol{E}} = \boldsymbol{C}\hat{\boldsymbol{M}}\boldsymbol{u}_p + \boldsymbol{C}\hat{\boldsymbol{E}}_0 \tag{8.82}$$

$$\underline{\boldsymbol{T}}\langle\boldsymbol{\xi}\rangle = \boldsymbol{G}\boldsymbol{C}\hat{\boldsymbol{M}}\boldsymbol{u}_p + \boldsymbol{G}\boldsymbol{C}\hat{\boldsymbol{E}}_0 \tag{8.83}$$

其中，C 为材料弹性矩阵，如式 (7.42) 所示，G 为 2×3 矩阵，即

$$[\boldsymbol{G}]_{2\times 3} = \begin{bmatrix} g_N^{10}(\boldsymbol{\xi}_{ji}) & 0 & g_N^{01}(\boldsymbol{\xi}_{ji}) \\ 0 & g_N^{01}(\boldsymbol{\xi}_{ji}) & g_N^{10}(\boldsymbol{\xi}_{ji}) \end{bmatrix} \tag{8.84}$$

则物质点的劲度系数矩阵为

$$\boldsymbol{H}_0 = \boldsymbol{G}\boldsymbol{C}\hat{\boldsymbol{M}} \tag{8.85}$$

由此，可组集获得系统整体劲度系数矩阵。式 (8.83) 的第二项 $\boldsymbol{GC}\hat{\boldsymbol{E}}_0$ 需要加到系统代数方程组的右端项中，求解线性代数方程组即可获得解答。

8.2.2.3 一维情况

对于一维问题，可按前述类似方法构造物质点劲度系数矩阵和系统整体劲度系数矩阵，其中涉及的相关物理量表达式如下。

近场动力学函数：$g_N^p(\xi) = w_{ji}(a_1^p \xi_{ji} + a_2^p \xi_{ji}^2 + \cdots + a_N^p \xi_{ji}^N)$，$p = 0, 1, 2$；

变形梯度：$F = \sum_{j=1}^{N^i} (u_j - u_i) g_N^1(\xi_{ji}) V_j + \sum_{j=1}^{N^i} (x_j - x_i) g_N^1(\xi_{ji}) V_j$；

应变：$\varepsilon = \sum_{j=1}^{N^i} (u_j - u_i) g_N^1(\xi_{ji}) V_j + \sum_{j=1}^{N^i} (x_j - x_i) g_N^1(\xi_{ji}) V_j - 1 = \hat{\boldsymbol{N}} \boldsymbol{u}_p + \varepsilon_0$，

其中 $\varepsilon_0 = \sum_{j=1}^{N^i} (x_j - x_i) g_N^1(\xi_{ji}) V_j - 1$；

应力：$\sigma = E\varepsilon = E\hat{\boldsymbol{N}}\boldsymbol{u}_p + E\varepsilon_0$，其中 E 为杨氏弹性模量；

力标量状态：$\underline{T}(x_i) = \sigma(x_i) g_N^1(\xi_{ji}) = E\hat{\boldsymbol{N}}\boldsymbol{u}_p g_N^1(\xi_{ji}) + E\varepsilon_0 g_N^1(\xi_{ji})$，其中的行向量 $\hat{\boldsymbol{N}}_{1\times(N^i+1)} = \begin{bmatrix} N_i^2 & V_{j_1} g_N^1(\xi_{j_1 i}) & \cdots & V_{j_{N^i}} g_N^1(\xi_{j_{N^i} i}) \end{bmatrix}$，$N_i^1 = -\sum_{j=1}^{N^i} V_j g_N^1(\xi_{ji})$；

位移列向量：$[\boldsymbol{u}_p]_{(N^i+1)\times 1} = \begin{bmatrix} u_i & u_{j_1} & \cdots & u_{j_{N^i}} \end{bmatrix}^{\mathrm{T}}$，其中 u_i 为源点 $\boldsymbol{x}_i$ 的位移值，其余为场点 $\boldsymbol{x}_j$ 的位移值。

同样地，根据力标量状态可组集系统整体劲度系数矩阵，且力标量状态的第二项 $E\varepsilon_0 g_2^1(\xi_{ji})$ 需要组装到方程组的右端项中。

8.2.3 非线性隐式解法

与 7.3 节相同，对于静力线性或非线性问题，仍采用 Newton Raphson 迭代法求解增量方程 $\boldsymbol{J}\delta\boldsymbol{u}_p = -(\boldsymbol{L}+\boldsymbol{b})$，获得位移增量解答 $\delta\boldsymbol{u}_p$。

如前所述，对于 NOSB PD 平衡方程，可以得到

$$\mathrm{d}\boldsymbol{L}(\boldsymbol{x}_i) = \boldsymbol{J}\delta\boldsymbol{u}_p = \frac{\partial \boldsymbol{L}}{\partial \boldsymbol{u}_p}\delta\boldsymbol{u}_p = \left[\sum_{j=1}^{N^i}\left(\frac{\partial \underline{\boldsymbol{T}}[\boldsymbol{x}_i]\langle\boldsymbol{\xi}_{ji}\rangle}{\partial \boldsymbol{u}_p} - \frac{\partial \underline{\boldsymbol{T}}[\boldsymbol{x}_j]\langle\boldsymbol{\xi}_{ij}\rangle}{\partial \boldsymbol{u}_p}\right)V_j\right]\delta\boldsymbol{u}_p \tag{8.86}$$

将力密度矢量状态 $\underline{\boldsymbol{T}}[\boldsymbol{x}_i]\langle\boldsymbol{\xi}_{ji}\rangle = \boldsymbol{P}_i \boldsymbol{g}(\boldsymbol{\xi}_{ji})$ 与 $\underline{\boldsymbol{T}}[\boldsymbol{x}_j]\langle\boldsymbol{\xi}_{ij}\rangle = \boldsymbol{P}_j \boldsymbol{g}(\boldsymbol{\xi}_{ij})$ 代入上式，则有

$$\frac{\partial \underline{\boldsymbol{T}}[\boldsymbol{x}_i]\langle\boldsymbol{\xi}_{ji}\rangle}{\partial \boldsymbol{u}_p} = \frac{\partial [\boldsymbol{P}_i \boldsymbol{g}(\boldsymbol{\xi}_{ji})]}{\partial \boldsymbol{u}_p} = \frac{\partial \boldsymbol{P}_i}{\partial \boldsymbol{u}_p}\boldsymbol{g}(\boldsymbol{\xi}_{ji}) = \frac{\partial \boldsymbol{P}_i}{\partial \boldsymbol{F}_i}\frac{\partial \boldsymbol{F}_i}{\partial \boldsymbol{u}_p}\boldsymbol{g}(\boldsymbol{\xi}_{ji}) \tag{8.87}$$

$$\frac{\partial \underline{\boldsymbol{T}}[\boldsymbol{x}_j]\langle\boldsymbol{\xi}_{ij}\rangle}{\partial \boldsymbol{u}_p} = \frac{\partial [\boldsymbol{P}_j \boldsymbol{g}(\boldsymbol{\xi}_{ij})]}{\partial \boldsymbol{u}_p} = \frac{\partial \boldsymbol{P}_j}{\partial \boldsymbol{u}_p}\boldsymbol{g}(\boldsymbol{\xi}_{ij}) = \frac{\partial \boldsymbol{P}_j}{\partial \boldsymbol{F}_j}\frac{\partial \boldsymbol{F}_j}{\partial \boldsymbol{u}_p}\boldsymbol{g}(\boldsymbol{\xi}_{ij}) \tag{8.88}$$

将以上两式代入式 (8.86) 中，有

$$\begin{aligned}\mathrm{d}\boldsymbol{L}(\boldsymbol{x}_i) &= \left[\sum_{j=1}^{N^i}\left(\frac{\partial \boldsymbol{P}_i}{\partial \boldsymbol{F}_i}\frac{\partial \boldsymbol{F}_i}{\partial \boldsymbol{u}_p}\boldsymbol{g}(\boldsymbol{\xi}_{ji}) - \frac{\partial \boldsymbol{P}_j}{\partial \boldsymbol{F}_j}\frac{\partial \boldsymbol{F}_j}{\partial \boldsymbol{u}_p}\boldsymbol{g}(\boldsymbol{\xi}_{ij})\right)V_j\right]\delta\boldsymbol{u}_p \\ &= \sum_{j=1}^{N^i}\left[\frac{\partial \boldsymbol{P}_i}{\partial \boldsymbol{F}_i}\delta\boldsymbol{F}_i\boldsymbol{g}(\boldsymbol{\xi}_{ji})V_j\right] - \sum_{j=1}^{N^i}\left[\frac{\partial \boldsymbol{P}_j}{\partial \boldsymbol{F}_j}\delta\boldsymbol{F}_j\boldsymbol{g}(\boldsymbol{\xi}_{ij})V_j\right] \\ &= \sum_{j=1}^{N^i}\left[\frac{\partial \boldsymbol{P}_i}{\partial \boldsymbol{F}_i}\left(\sum_{k=1}^{N^i}(\delta\boldsymbol{u}_k - \delta\boldsymbol{u}_i)\otimes\boldsymbol{g}(\boldsymbol{\xi}_{ki})V_k\right)\boldsymbol{g}(\boldsymbol{\xi}_{ji})V_j\right] \\ &\quad - \sum_{j=1}^{N^i}\left[\frac{\partial \boldsymbol{P}_j}{\partial \boldsymbol{F}_j}\left(\sum_{k=1}^{N^j}(\delta\boldsymbol{u}_k - \delta\boldsymbol{u}_j)\otimes\boldsymbol{g}(\boldsymbol{\xi}_{kj})V_k\right)\boldsymbol{g}(\boldsymbol{\xi}_{ij})V_j\right]\end{aligned} \tag{8.89}$$

式中，$\boldsymbol{\xi}_{ji} = \boldsymbol{x}_j - \boldsymbol{x}_i$、$\boldsymbol{\xi}_{ij} = \boldsymbol{x}_i - \boldsymbol{x}_j$、$\boldsymbol{\xi}_{ki} = \boldsymbol{x}_k - \boldsymbol{x}_i$ 和 $\boldsymbol{\xi}_{kj} = \boldsymbol{x}_k - \boldsymbol{x}_j$。一旦知晓 $\partial \boldsymbol{P}/\partial \boldsymbol{F}$ 的表达式或具体值，根据上式即可组装物质点和系统的劲度系数矩阵，得到系统的线性代数方程组。

以二维问题为例，为组装物质点和系统整体劲度矩阵，需要将上述公式的 $\frac{\partial P_i}{\partial F_i}\delta F_i g(\xi_{ji})V_j$ 写为如下矩阵形式，以分离系数矩阵 J 和位移增量列向量 δu_p

$$V_j\frac{\partial P_i}{\partial F_i}\delta F_i g(\xi_{ji}) = [G]_{2\times 4}[\delta F]_{4\times 1} = [G]_{2\times 4}[M]_{4\times 2(N^i+1)}[\delta u_p]_{2(N^i+1)\times 1} \quad (8.90)$$

其中，各矩阵如下

$$[\delta F]_{4\times 1} = \begin{bmatrix} \delta F_{11} & \delta F_{22} & \delta F_{12} & \delta F_{21} \end{bmatrix}^\mathrm{T} \quad (8.91)$$

$$[M]_{4\times 2(N^i+1)} = \begin{bmatrix} M^1 & 0 & \cdots & V_j g_N^{10}(\xi_{ji}) & 0 & \cdots \\ 0 & M^2 & \cdots & 0 & V_j g_N^{01}(\xi_{ji}) & \cdots \\ M^2 & 0 & \cdots & V_j g_N^{01}(\xi_{ji}) & 0 & \cdots \\ 0 & M^1 & \cdots & 0 & V_j g_N^{10}(\xi_{ji}) & \cdots \end{bmatrix} \quad (8.92)$$

$$[\delta u_p]_{2(N^i+1)\times 1} = \begin{bmatrix} \delta u_{11} & \delta u_{12} & \cdots & \delta u_{N^i 1} & \delta u_{N^i 2} & \delta u_{(N^i+1)1} & \delta u_{(N^i+1)2} \end{bmatrix}^\mathrm{T} \quad (8.93)$$

$$[G]_{2\times 4} = V_j \begin{bmatrix} g_N^{10}\left(\frac{\partial P}{\partial F}\right)_{11} & g_N^{01}\left(\frac{\partial P}{\partial F}\right)_{12} & g_N^{01}\left(\frac{\partial P}{\partial F}\right)_{11} & g_N^{10}\left(\frac{\partial P}{\partial F}\right)_{12} \\ g_N^{10}\left(\frac{\partial P}{\partial F}\right)_{21} & g_N^{01}\left(\frac{\partial P}{\partial F}\right)_{22} & g_N^{01}\left(\frac{\partial P}{\partial F}\right)_{21} & g_N^{10}\left(\frac{\partial P}{\partial F}\right)_{22} \end{bmatrix} \quad (8.94)$$

式中，矩阵 M 前两列的 $M^1 = -\sum_{j=1}^{N^i} V_j g_N^{10}(\xi_{ji})$ 和 $M^2 = -\sum_{j=1}^{N^i} V_j g_N^{01}(\xi_{ji})$ 对应于源点 x_i，其余各列对应于场点 x_j；δu_{11} 和 δu_{12} 为源点 x_i 的位移增量，其余各列为场点 x_j 的位移增量。至此，获得了组装物质点和系统的劲度系数矩阵的必要矩阵表达。

8.2.4 高阶非局部变形梯度的精度验证

如前所述，物质点的变形梯度直接影响 NOSB PD 模拟结果的精度和稳定性，原非常规态型近场动力学的非局部变形梯度能够精确表示内部物质点的线性变形，也可以在可接受误差内较为准确地描述双线性等二阶变形[12,21]。但对于复杂的变形行为，变形梯度的计算精度较低，尤其是对于边界物质点，且非均匀粒子离散对应的变形梯度的计算精度通常更低。

为验证本章给出的变形梯度高阶非局部积分表达的计算精度，考虑一边长为 2 的方形板，给定的变形构型为 $y_1 = x_1 + 2x_2 + 3x_1 x_2$ 和 $y_2 = x_2$，则变形梯度分

量为 $F_{11}=1+3x_2$、$F_{12}=2+3x_1$、$F_{21}=0$ 和 $F_{22}=1$。分别考虑均匀和非均匀离散两种情形，基于变形梯度的高阶非局部积分表达，求出数值解，并与解析解进行比较。均匀离散点为 400 个，非均匀离散点为 380 个，近场范围 $\delta=0.3$，影响函数 $w(|\boldsymbol{\xi}|)=\delta^3/|\boldsymbol{\xi}|^3$，泰勒级数的阶数为 $N=3$。记变形梯度分量的相对误差为 $(R_{\text{exact}}-R_{\text{PD}})/R_{\text{exact}}$，图 8-2 和图 8-3 给出了均匀离散与非均匀离散构型中所有物质点的 F_{11} 和 F_{12} 的相对误差。结果表明，在均匀或非均匀离散情况下，采用变形梯度的高阶非局部积分表达，均能高精度地描述内部或边界点的变形状态，显著提高原 NOSB PD 模型的计算精度。

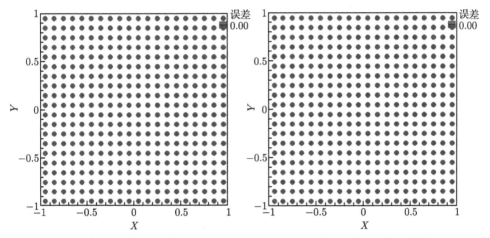

图 8-2　物质点变形梯度分量 F_{11}(左) 和 F_{12}(右) 的相对误差 (均匀离散)

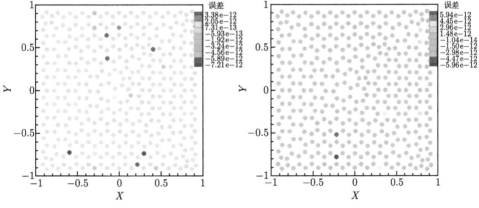

图 8-3　物质点变形梯度分量 F_{11}(左) 和 F_{12}(右) 的相对误差 (非均匀离散)

仍以 7.4.1.3 节中的单点移动为例，进一步说明高阶 NOSB PD 模型中变形梯度的特点。当前构型中，物质点 $\boldsymbol{x}$ 和 $\boldsymbol{x}'$ 的变形梯度张量为 $\boldsymbol{F}_{\mathrm{old}}(\boldsymbol{x})$ 和 $\boldsymbol{F}_{\mathrm{old}}(\boldsymbol{x}')$，此时对物质点 $\boldsymbol{x}$ 施加载荷，使其发生单点移动 $\boldsymbol{u}_{\mathrm{d}}(\boldsymbol{x})$，则物质点 $\boldsymbol{x}$ 和 $\boldsymbol{x}'$ 的变形梯度张量分别为

$$\boldsymbol{F}_{\mathrm{new}}(\boldsymbol{x}) = \int_{H_{\boldsymbol{x}}} [\boldsymbol{y}_{\mathrm{new}}(\boldsymbol{x}') - \boldsymbol{y}_{\mathrm{new}}(\boldsymbol{x})] \otimes \boldsymbol{g}_N(\boldsymbol{x}, \boldsymbol{x}' - \boldsymbol{x}) \mathrm{d}V_{\boldsymbol{x}'}$$

$$= \boldsymbol{F}_{\mathrm{old}}(\boldsymbol{x}) - \boldsymbol{u}_{\mathrm{d}}(\boldsymbol{x}) \otimes \int_{H_{\boldsymbol{x}}} \boldsymbol{g}_N(\boldsymbol{x}, \boldsymbol{x}' - \boldsymbol{x}) \mathrm{d}V_{\boldsymbol{x}'} \quad (8.95)$$

$$\boldsymbol{F}_{\mathrm{new}}(\boldsymbol{x}') = \int_{H_{\boldsymbol{x}'}} [\boldsymbol{y}_{\mathrm{new}}(\boldsymbol{x}'') - \boldsymbol{y}_{\mathrm{new}}(\boldsymbol{x}')] \otimes \boldsymbol{g}_N(\boldsymbol{x}', \boldsymbol{x}'' - \boldsymbol{x}') \mathrm{d}V_{\boldsymbol{x}''}$$

$$= \boldsymbol{F}_{\mathrm{old}}(\boldsymbol{x}') + \boldsymbol{u}_{\mathrm{d}}(\boldsymbol{x}) \otimes \boldsymbol{g}_N(\boldsymbol{x}', \boldsymbol{x} - \boldsymbol{x}') V_{\boldsymbol{x}} \quad (8.96)$$

当采用均匀正交点阵离散构型和球形对称影响函数 w 时，若近场动力学函数的泰勒级数阶数 N 大于等于 2，式 (8.95) 的右端积分项并不等于零，故高阶项的存在一定程度地抑制了零能模式，这也是高阶 NOSB PD 能够降低数值不稳定性的部分原因。

8.2.5 弹性杆拉伸变形的算例分析

如图 8-4 所示，以一维弹性杆单轴拉伸静力变形为例[22]，验证高阶 NOSB PD 模型对计算精度和稳定性的提升作用。分别采用原 NOSB PD 和高阶 NOSB PD 方法求解该问题，杆长 $L = 1\mathrm{m}$、方形横截面面积 $\boldsymbol{A} = h \times h = 1 \times 10^{-6} \mathrm{m}^2$，杨氏模量 $E = 200\mathrm{GPa}$、泊松比 $\nu = 0.2$ 和质量密度 $\rho = 7850\mathrm{kg/m^3}$。杆离散为 N_{total} 个物质点，物质点间距 Δx、体积 $\Delta V = A\Delta x$ 和近场范围 $\delta = m\Delta x (m = 1, 2, \cdots, \infty)$；对杆左端一个近场范围内的物质点施加固定约束，外力 $F = 200\mathrm{N}$ 通过体力密度 $b_x = F/\Delta V$ 施加在最右端的一个物质点上；影响函数 $w(|\boldsymbol{\xi}|) = \delta^3/|\boldsymbol{\xi}|^3$。该问题的解析解为 $u(x) = \dfrac{F}{AE}x = 0.001x$，并根据 $\varepsilon = \dfrac{1}{|u^{(\mathrm{r})}|_{\max}} \sqrt{\dfrac{1}{N_{\mathrm{total}}} \sum\limits_{i=1}^{N_{\mathrm{total}}} \left[u_i^{(\mathrm{r})} - u_i^{(\mathrm{n})} \right]^2}$ 计算全局误差[23]，其中，N_{total} 为物质点总数，上标 r 和 n 分别表示参考解和数值解。

近场动力学的数值结果应满足 m-收敛性要求，其含义是：对于某一近场范围尺寸，当近场范围内物质点 (积分点) 数目不断增多，PD 数值解应收敛于基于 PD 理论的非局部精确解[24,25]。本例中，固定近场范围半径 $\delta = 0.1\mathrm{m}$，改变离散粒子密度，采用不同的近场范围与物质点尺寸的比值 $m = \delta/\Delta x$，进行 m-收敛分析，以获得原 NOSB PD 和 1~4 阶高阶 NOSB PD 的非局部解答。计算条件如

表 8-1 所示。

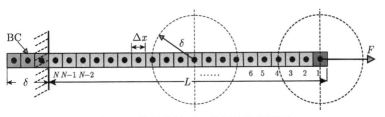

图 8-4 单轴拉伸下一维杆的计算模型

表 8-1 11 组均匀离散构型下 (从粗糙到精细) 单轴拉伸杆的 PD 计算条件

No.	m	Δx/m	$N_{\text{total}} + N_{\text{BC}}$	$\Delta V = A\Delta x$	$b_x/(\text{N/m}^3)$
1	1	1×10^{-1}	10+3	1×10^{-7}	2×10^9
2	2	5×10^{-2}	20+3	5×10^{-8}	4×10^9
3	3	$1/3 \times 10^{-1}$	30+3	$1/3 \times 10^{-7}$	6×10^9
4	4	2.5×10^{-2}	40+4	2.5×10^{-8}	8×10^9
5	5	2×10^{-2}	50+5	2×10^{-8}	1×10^{10}
6	6	$1/6 \times 10^{-1}$	60+6	$1/6 \times 10^{-7}$	1.2×10^{10}
7	8	1.25×10^{-2}	80+8	1.25×10^{-8}	1.6×10^{10}
8	10	1×10^{-2}	100+10	1×10^{-8}	2×10^{10}
9	16	6.25×10^{-3}	160+16	6.25×10^{-9}	3.2×10^{10}
10	20	5×10^{-3}	200+20	5×10^{-9}	4×10^{10}
11	25	4×10^{-3}	250+25	4×10^{-9}	5×10^{10}

图 8-5 给出了基于五种不同的 NOSB PD 模型计算得到的轴向位移 m-收敛性结果与全局误差曲线，表 8-2 给出了全局误差的具体数值。由图表结果可知：① 所有五种 NOSB PD 模型得到的结果均满足 m-收敛要求，表明 NOSB PD 方法能够分析结构的静弹性变形问题，但 PD 解答与传统局部解析解存在一定差异；② 图 8-5(a) 和图 8-5(b) 的轴向位移结果一致，从数值分析的角度证明了基于一阶泰勒级数展开的一阶 NOSB PD 模型与原 NOSB PD 模型等价[15]；③ 二阶 NOSB PD 模型的计算精度低于一阶 NOSB PD 模型，且数值振荡高于一阶模型，其原因在于内部区域物质点通过一阶和二阶泰勒展开获得的近场动力学函数 g_1^1 和 g_2^1 是相等的，但对于边界物质点，近场动力学函数 g_1^1 无额外影响，而 g_2^2 与 g_2^1 是耦合的，这将降低计算精度和放大数值振荡；④ 三阶 NOSB PD 模型显著提高了计算精度和稳定性，而四阶 NOSB PD 模型的计算精度和稳定性低于三阶模型，其原因仍在于近场动力学函数的耦合影响。

当 $m=1$ 并取粒子总数为 $N=10$ 时，观察图 8-5 中二阶至四阶 NOSB PD 模型的对应结果，可以发现计算结果严重偏离解析解，其原因可能是 10 个点的离散模型过于粗糙，不能保证 PD 解答的收敛性；另一方面，$m=1$ 时对应的

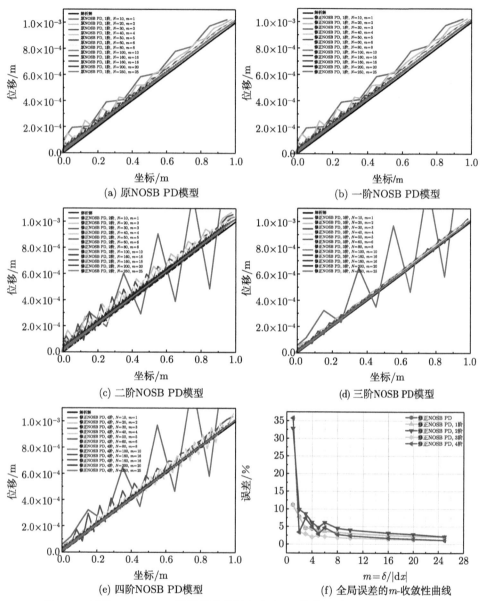

图 8-5 五种 NOSB PD 模型得到的轴向位移 m-收敛性结果与全局误差曲线

近场范围内物质点数目密度过低,不能提供足够的位置信息以准确确定近场动力学函数,导致变形梯度张量和内力密度矢量的计算都存在较大误差。为此,选取粒子总数分别为 $N=10$、$N=20$ 和 $N=40$,给出不同离散情况下 $m=1$ 时三阶 NOSB PD 模型的计算结果,如图 8-6 所示。由图可见,过于粗糙的离散模

8.2 高阶非常规态型近场动力学模型

表 8-2 11 组均匀离散构型下五种 NOSB PD 模型计算结果的全局误差

m	误差/%				
	原 NOSB PD	一阶 NOSB PD	二阶 NOSB PD	三阶 NOSB PD	四阶 NOSB PD
1	11.1888	11.1888	32.8335	35.7976	35.8006
2	7.8123	7.8123	9.8859	3.4177	3.4243
3	4.5460	4.5460	8.6210	2.8978	7.3855
4	4.4030	4.4030	6.1506	2.0310	4.9169
5	4.0323	4.0323	4.7251	2.4340	2.9099
6	3.5902	3.5902	6.1038	2.1356	4.6988
8	3.3684	3.3684	4.4777	1.8578	2.7938
10	3.0572	3.0572	3.9943	1.7025	2.3770
16	2.4930	2.4930	3.1485	1.3904	1.7192
20	2.2693	2.2693	2.8132	1.2567	1.4851
25	2.0709	2.0709	2.1592	1.1545	1.1714

型不能保证 PD 解答的收敛性，合理精细的离散模型是 PD 计算结果合理正确的基本要求；相比原 NOSB PD 模型，高阶 NOSB PD 模型需要较大的近场范围内粒子密度；对于简单变形问题，$m=1$ 即能够获得合理解答，对于复杂变形问题，根据结果收敛性分析，推荐采用基于三阶泰勒级数的 NOSB PD 模型，并配合 $\delta = m\Delta x (m = 2 \sim 8)$ 的近场范围。

近场动力学数值结果还应满足 δ-收敛性要求，即当 m 固定或递增时，随着近场范围尺寸不断递减，PD 数值解应收敛于传统局部理论的精确解[24,25]。选取 6 组不同的 $m(m = 1, 2, 3, 4, 5, 6)$ 值，考虑近场范围内不同的粒子密度进行计算，图 8-7 给出了三阶 NOSB PD 模型全局误差的 δ 收敛性曲线。从图中所示结果可以看出，对于不同的 m 值，总体具有良好的 δ 收敛性，且 $m=2$ 的计算精度最高，其原因是 $m=2$ 既满足了高阶模型需要较大近场范围内的粒子密度要求，也趋于局部化，故与局部解析解的偏差较小。

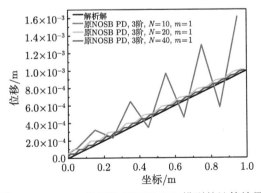

图 8-6 $m=1$ 的三阶 NOSB PD 模型的计算结果

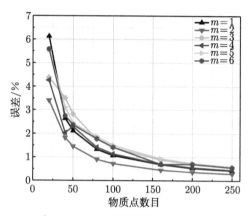

图 8-7　三阶 NOSB PD 模型全局误差的 δ 收敛性曲线

8.2.6　矩形板拉伸变形的算例分析

如图 8-8 所示，受水平均匀单轴拉伸的各向同性矩形板，发生静力弹性变形，采用线弹性 NOSB PD 模型与三阶 NOSB PD 模型的隐式静力方法求解该问题。板的长度、宽度和厚度分别为 $L=1\mathrm{m}$、$W=0.5\mathrm{m}$ 和 $h=\Delta x$；材料的杨氏弹性模量 $E=200\mathrm{GPa}$，泊松比 $\nu=1/3$，质量密度 $\rho=7850\mathrm{kg/m^3}$；上下水平边界自由，水平方向拉伸最大位移值 $d=0.0005\mathrm{m}$，通过线性变化施加左右竖直边界上，边界层数与近场范围半径一致；影响函数 $w(|\boldsymbol{\xi}|)=\mathrm{e}^{-(2|\boldsymbol{\xi}|/\delta)^2}$；物质点均匀离散，$\Delta x=0.005\mathrm{m}$，圆形近场范围半径取为 $\delta=m\Delta x$。选取 5 组不同的 $m(m=1,2,3,4,5)$ 值，通过计算显示高阶 NOSB PD 模型在提高计算精度和降低数值振荡方面的作用，并分析近场范围半径大小对两类 NOSB PD 模型计算精度和稳定性的影响。

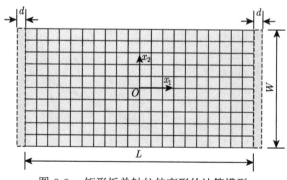

图 8-8　矩形板单轴拉伸变形的计算模型

图 8-9 和图 8-10 给出了不同 m 值时，原 NOSB PD 模型和三阶 NOSB PD

8.2 高阶非常规态型近场动力学模型

模型计算得到的水平位移和竖向位移结果，图 8-11 为根据

$$\varepsilon_{L2} = \sqrt{\sum_{i=1}^{N_{\text{total}}} \left[R_i^{(\text{r})} - R_i^{(\text{n})} \right]^2 } \Big/ \sqrt{\sum_{i=1}^{N_{\text{total}}} \left[R_i^{(\text{r})} \right]^2 }$$

计算[26] 得到的位移分量的全局误差。对于各组 m 值，三阶 NOSB PD 预测结果较原 NOSB PD 更加稳定和精确，证明了对变形梯度张量和键力密度矢量修正的高阶 NOSB PD 模型能显著提高计算精度和降低数值不稳定性。对于本例的典型弹性变形问题，近场范围较小时 ($m=1$ 和 $m=2$)，两类 NOSB PD 计算结果

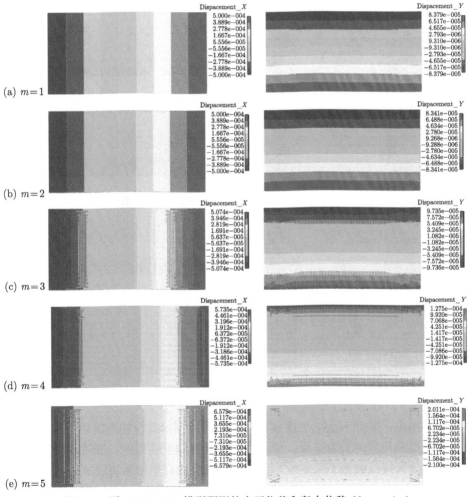

图 8-9　原 NOSB PD 模型预测的水平位移和竖向位移 ($\delta = m\Delta x$)

均接近局部解答，且数值振荡现象轻微，而近场范围较大时 ($m \geqslant 3$)，两类 NOSB PD 均表现出明显的数值振荡现象。但近场范围的减少降低了模型的非局部性，使得计算结果接近局部解答，小近场范围有时不能刻画复杂的变形状态和力状态，选择 $m=3$ 能较好地协调非局部性和计算稳定性两方面的需求。

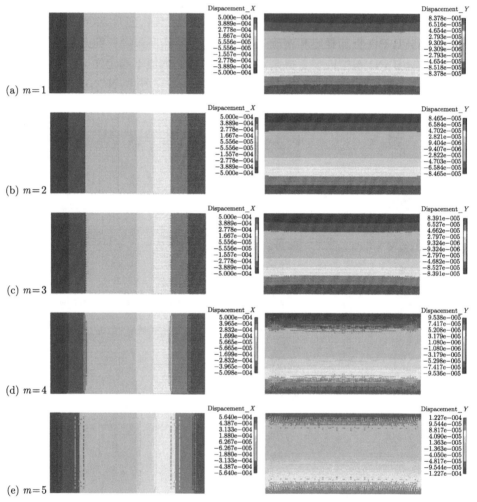

图 8-10　高阶 (三阶)NOSB PD 模型预测的水平位移和竖向位移 ($\delta = m\Delta x$)

另外指出，高阶 NOSB PD 模型要求近场范围内具有较高的粒子数密度，以获得高精度和稳定的数值结果；受相关变量计算精度的影响，高阶模型虽能够提高计算精度和降低数值不稳定性，但仍不能完全消除数值振荡现象。

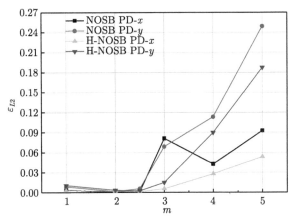

图 8-11 原 NOSB PD 模型和高阶 (三阶)NOSB PD 模型计算结果与解析解的全局误差

8.3 键关联高阶非常规态型近场动力学模型

8.2 节对 7.4.1.1 节提出的非常规态型近场动力学计算精度和数值不稳定问题进行了算例分析。本节基于键关联近场范围的概念,针对 7.4.1.2 节的"键变形协调性"、7.4.1.3 节的点关联非局部变形梯度导致零能模式和伪能量模式问题、7.4.1.4 节的构型变形信息丢失与物质点应力到键力的一对多映射问题,建立键关联非局部变形梯度和力密度矢量状态,构建键关联 (高阶) 非常规态型近场动力学模型 (bond-associated higher-order NOSB PD, BA-H-NOSB PD)[27],给出隐式静力求解方法和验证算例,该模型能够有效解决上述三类问题,但需要较大的计算量。

8.3.1 键关联的非局部变形梯度

如图 8-12 所示,相比于经典连续介质力学的局部变形梯度,点关联的非局部变形梯度引入局部区域加权积分,能包含更多的局部变形信息,表征更为复杂的变形状态,但在一定程度上忽略了局部变形的协调性特征、取向性特征和局部峰值特征,不能严格满足"键变形协调性"。考虑近场动力学建模的基础是两点间的"键",因而提出基于"键关联近场范围"的"键关联高阶非常规态型近场动力学模型",使得建模变量建立在"键层次"的基础上,不出现"点层次"的建模变量,避免了"点到键的一对多映射"问题。

键关联变形梯度张量 $\boldsymbol{F}^{\mathrm{b}}(\boldsymbol{x},\boldsymbol{\xi})$ 定义为 [28,29]

$$\boldsymbol{F}^{\mathrm{b}}(\boldsymbol{x},\boldsymbol{\xi}) = \int_{H_{\boldsymbol{\xi}}} w[\boldsymbol{y}(\boldsymbol{x}') - \boldsymbol{y}(\boldsymbol{x})] \otimes \boldsymbol{\xi} \mathrm{d}V_{\boldsymbol{x}'} (\boldsymbol{K}^{\mathrm{b}}(\boldsymbol{x},\boldsymbol{\xi}))^{-1} \qquad (8.97)$$

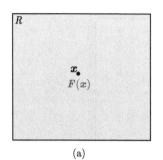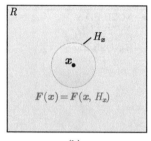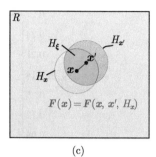

图 8-12　物质点 x 的局部变形梯度 (a)、依赖于近场范围 H_x 的点关联非局部变形梯度 (b) 和依赖于相交近场范围 H_ξ 的键关联非局部变形梯度 (c)

$$F^{\mathrm{b}}(x,\xi) = \int_{H_\xi} [y(x') - y(x)] \otimes g_N(x,\xi) \mathrm{d}V_{x'} \tag{8.98}$$

式中，键关联的非局部形状张量为 $K^{\mathrm{b}}(x,\xi) = \int_{H_\xi} w(\xi \otimes \xi)\mathrm{d}V_{x'}$，键关联的近场范围 (bond-associated horizon) 定义为成键的两物质点 x 和 x' 各自近场范围的交集，即为

$$H_\xi = H_x \cap H_{x'} \tag{8.99}$$

物质点 x 和 x' 通常取相同的近场范围半径，以体现两物质点的同等重要性。对于任意变形状态，在键层次上，$F^{\mathrm{b}}(x,\xi) = F^{\mathrm{b}}(x',-\xi)$ 容易满足，因此 "键变形协调性" 得到保证，即

$$\xi + \eta = F^{\mathrm{b}}(x,\xi)\xi = F^{\mathrm{b}}(x',-\xi)\xi \tag{8.100}$$

进一步定义点关联的变形梯度张量

$$\bar{F}(x) = \int_{H_x} wF^{\mathrm{b}}(x,\xi)(\xi \otimes \xi)\mathrm{d}V_{x'} \left[\int_{H_x} w(\xi \otimes \xi)\mathrm{d}V_{x'}\right]^{-1} \tag{8.101}$$

至此，由键关联非局部变形梯度出发，开展键关联应变、键关联应力和键力密度计算，建模过程都基于"键层次"，不会出现"键变形不协调"、"键变量 (变形状态) 到点变量 (变形梯度) 的多对一映射的信息丢失"和"点变量 (应力) 到键变量 (力密度矢量状态) 的一对多映射"等问题。本书著者认为"键变形不协调""键到点的多对一映射的信息丢失"和"点变量到键变量的一对多映射"是 NOSB PD 模型数值振荡不稳定性的主要来源。同时，由于键关联变形梯度打破了点关联变形梯度的径向对称性，可有效避免零能模式和虚假能量模式问题。

8.3.2 键关联的力密度矢量状态和运动方程

在每根键上，依赖键关联变形梯度的键关联力密度矢量状态可按以下流程建立。

当键关联近场范围 H_{ξ} 发生位移增量时，物质点 $\boldsymbol{x}$ 的键关联的变形能密度增量为 ΔW^{b}，具体为

$$\Delta W^{\mathrm{b}} = \int_{H_{\xi}} \left(\boldsymbol{t}^{\mathrm{b}}\right)^{\mathrm{T}} (\Delta \boldsymbol{y}' - \Delta \boldsymbol{y}) \mathrm{d} V_{\boldsymbol{x}'} \tag{8.102}$$

为获得 ΔW^{b}，需要采用某种假设，此处设键关联近场范围与物质点近场范围的体积比等于键关联的变形能密度增量与物质点应变能密度增量的比值，从而有

$$\Delta W^{\mathrm{b}} = \frac{\int_{H_{\xi}} 1 \mathrm{d} V_{\boldsymbol{x}'}}{\int_{H_{\boldsymbol{x}}} 1 \mathrm{d} V_{\boldsymbol{x}'}} \Delta W \tag{8.103}$$

在经典连续介质力学中，应变能密度增量为

$$\Delta W = \mathrm{tr}((\boldsymbol{P}^{\mathrm{b}})^{\mathrm{T}} \Delta \boldsymbol{F}^{\mathrm{b}}) \tag{8.104}$$

将式 (8.98) 给出的键关联高阶非局部变形梯度张量的变分代入上式，得到

$$\Delta W = \mathrm{tr}\left((\boldsymbol{P}^{\mathrm{b}})^{T} \int_{H_{\xi}} (\Delta \boldsymbol{y}' - \Delta \boldsymbol{y}) \otimes \boldsymbol{g}_{N}(\boldsymbol{x}, \boldsymbol{\xi}) \mathrm{d} V_{\boldsymbol{x}'}\right)$$

$$= \int_{H_{\xi}} (\boldsymbol{P}^{\mathrm{b}} \boldsymbol{g}_{N}(\boldsymbol{x}, \boldsymbol{\xi}))^{\mathrm{T}} (\Delta \boldsymbol{y}' - \Delta \boldsymbol{y}) \mathrm{d} V_{\boldsymbol{x}'} \tag{8.105}$$

将式 (8.103) 和式 (8.105) 代入式 (8.102) 中，有下式成立

$$\frac{\int_{H_{\xi}} 1 \mathrm{d} V_{\boldsymbol{x}'}}{\int_{H_{\boldsymbol{x}}} 1 \mathrm{d} V_{\boldsymbol{x}'}} \int_{H_{\xi}} \left(\boldsymbol{P}^{\mathrm{b}} \boldsymbol{g}_{N}(\boldsymbol{x}, \boldsymbol{\xi})\right)^{\mathrm{T}} (\Delta \boldsymbol{y}' - \Delta \boldsymbol{y}) \mathrm{d} V_{\boldsymbol{x}'} = \int_{H_{\xi}} \left(\boldsymbol{t}^{\mathrm{b}}\right)^{\mathrm{T}} (\Delta \boldsymbol{y}' - \Delta \boldsymbol{y}) \mathrm{d} V_{\boldsymbol{x}'} \tag{8.106}$$

于是，得到键关联的力密度矢量状态

$$\boldsymbol{t}^{\mathrm{b}}(\boldsymbol{x}, t) = \underline{\boldsymbol{T}}^{\mathrm{b}} [\boldsymbol{x}, t] \langle \boldsymbol{\xi} \rangle = \frac{\int_{H_{\xi}} 1 \mathrm{d} V_{\boldsymbol{x}'}}{\int_{H_{\boldsymbol{x}}} 1 \mathrm{d} V_{\boldsymbol{x}'}} \boldsymbol{P}^{\mathrm{b}} \boldsymbol{g}_{N}(\boldsymbol{x}, \boldsymbol{\xi}) \tag{8.107}$$

或

$$t^b(x,t) = \underline{T}^b[x,t]\langle\xi\rangle = \frac{\int_{H_\xi} 1 dV_{x'}}{\int_{H_x} 1 dV_{x'}} w P^b (K^b)^{-1} \xi \qquad (8.108)$$

则近场动力学运动方程为

$$\rho_0(x)\ddot{u}(x,t) = \int_{H_x} [t^b(x,t) - t^b(x',t)] dV_{x'} + b(x,t) \qquad (8.109)$$

当键关联变形梯度满足对称性 $F^b(x,\xi) = F^b(x',-\xi)$ 时，键关联的应力张量和键关联的形状张量分别满足

$$P^b(x,\xi) = P^b(x',-\xi), \quad K^b(x,\xi) = K^b(x',-\xi) \qquad (8.110)$$

此时，键关联的力密度矢量状态满足

$$\underline{T}^b\langle-\xi\rangle = -\underline{T}^b\langle\xi\rangle \qquad (8.111)$$

如图 8-13 所示，键关联的非常规态型近场动力学模型具有大小相等、方向相反、但不平行于键变形方向的力密度矢量状态，介于键型近场动力学和非常规态型近场动力学之间，该模型可视为键型、常规态型和非常规态型之外的第四类近场动力学模型。

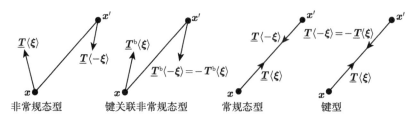

图 8-13　非常规态型、键关联非常规态型、常规态型和键型近场动力学的力密度矢量状态示意图

8.3.3　线性隐式静力解法

键关联的高阶 NOSB PD 模型的线性隐式静力解法与 8.2.2 节的高阶 NOSB PD 的线性隐式解法相似，以下仅给出解法的不同之处。

键关联的高阶非常规态型近场动力学的平衡方程为

$$\int_{H_\infty}\left[\frac{\int_{H_\xi}1\mathrm{d}V_{x'}}{\int_{H_\infty}1\mathrm{d}V_{x'}}P(x,\xi)g_N(x,\xi)-\frac{\int_{H_\xi}1\mathrm{d}V_{x'}}{\int_{H_\infty}1\mathrm{d}V_{x'}}P(x',\xi)g_N(x',\xi')\right]\mathrm{d}V_{x'}+b(x,t)=0 \tag{8.112}$$

物质点 x_i 的离散形式控制方程为

$$\sum_{j=1}^{N^i}\left[\frac{\sum_k^{N^b}V_k}{\sum_k^{N^i}V_k}P(x_i,\xi_{ji})g_N(x_i,\xi_{ji})-\frac{\sum_k^{N^b}V_k}{\sum_k^{N^j}V_k}P(x_j,\xi_{ij})g_N(x_j,\xi_{ij})\right]V_j+b_i=0 \tag{8.113}$$

式中，N^i 和 N^j 分别为物质点 x_i 和 x_j 的近场范围内物质点总数，N^b 表示键关联近场范围内物质点总数，N 为泰勒级数阶数。

不失一般性，将二维问题的系数矩阵与位移向量解耦，应变张量可重写为如下 Vogit 向量形式

$$\hat{E}=\begin{bmatrix}\varepsilon_{11} & \varepsilon_{22} & \varepsilon_{12}\end{bmatrix}^{\mathrm{T}}=\hat{M}u_p+\hat{E}_0 \tag{8.114}$$

式中，$\hat{E}$ 为 3×1 列向量，$\hat{M}$ 为 $3\times(2N^b+2)$ 矩阵，u_p 为 $(2N^b+2)\times 1$ 列向量，$\hat{E}_0$ 为 3×1 列向量，且有

$$\left[\hat{M}\right]_{3\times(2N^b+2)}=\begin{bmatrix}M^1 & 0 & \cdots & V_jg_N^{10}(\xi_{ji}) & 0 & \cdots \\ 0 & M^2 & \cdots & 0 & V_jg_N^{01}(\xi_{ji}) & \cdots \\ \frac{1}{2}M^2 & \frac{1}{2}M^1 & \cdots & \frac{1}{2}V_jg_N^{01}(\xi_{ji}) & \frac{1}{2}V_jg_N^{10}(\xi_{ji}) & \cdots\end{bmatrix} \tag{8.115}$$

$$[u_p]_{(2N^b+2)\times 1}=\begin{bmatrix}u_{11} & u_{12} & \cdots & u_{N^i 1} & u_{N^i 2} & u_{(N^i+1)1} & u_{(N^i+1)2}\end{bmatrix}^{\mathrm{T}} \tag{8.116}$$

$$\left[\hat{E}_0\right]_{3\times 1}=\sum_{j=1}^{N^b}V_j\begin{bmatrix}\xi_1 g_N^{10}(\xi_{ji}) \\ \xi_2 g_N^{01}(\xi_{ji}) \\ \dfrac{\xi_1 g_N^{01}(\xi_{ji})+\xi_2 g_N^{10}(\xi_{ji})}{2}\end{bmatrix}-\begin{bmatrix}1 \\ 1 \\ 0\end{bmatrix} \tag{8.117}$$

其中，$\hat{M}$ 矩阵前两列的 $M^1=-\sum_{j=1}^{N^b}V_jg_N^{10}(\xi_{ji})$ 和 $M^2=-\sum_{j=1}^{N^b}V_jg_N^{01}(\xi_{ji})$ 对应于源点 x_i，其余各列对应于键关联的近场范围内场点 x_j；$g_N^{10}(\xi_{ji})$ 和 $g_N^{01}(\xi_{ji})$ 是

PD 函数，具有 N 阶泰勒级数项；u_{11} 和 u_{12} 为源点 $\boldsymbol{x}_i$ 的位移值，其余各列为键关联近场范围内场点 $\boldsymbol{x}_j$ 的位移值。

进而，Cauchy 应力张量和力密度矢量状态分别为

$$\hat{\boldsymbol{\sigma}} = \begin{bmatrix} \sigma_{11} & \sigma_{22} & \sigma_{12} \end{bmatrix}^{\mathrm{T}} = \boldsymbol{C}\hat{\boldsymbol{E}} = \boldsymbol{C}\hat{\boldsymbol{M}}\boldsymbol{u}_p + \boldsymbol{C}\hat{\boldsymbol{E}}_0 \tag{8.118}$$

$$\underline{\boldsymbol{T}}\langle\boldsymbol{\xi}\rangle = \boldsymbol{G}\boldsymbol{C}\hat{\boldsymbol{M}}\boldsymbol{u}_p + \boldsymbol{G}\boldsymbol{C}\hat{\boldsymbol{E}}_0 \tag{8.119}$$

式中，$\boldsymbol{C}$ 为材料弹性矩阵，$\boldsymbol{G}$ 为 2×3 矩阵，即

$$\boldsymbol{G} = \frac{\sum\limits_{k}^{N^{\mathrm{b}}} V_k}{\sum\limits_{k}^{N^{\mathrm{i}}} V_k} \begin{bmatrix} g_N^{10}(\boldsymbol{\xi}_{ji}) & 0 & g_N^{01}(\boldsymbol{\xi}_{ji}) \\ 0 & g_N^{01}(\boldsymbol{\xi}_{ji}) & g_N^{10}(\boldsymbol{\xi}_{ji}) \end{bmatrix} \tag{8.120}$$

离散形式的内力矢量为

$$\boldsymbol{L}_i = \sum_{j=1}^{N^{\mathrm{i}}} \left[(\boldsymbol{G}_{ji}\boldsymbol{C}\hat{\boldsymbol{M}}_i\boldsymbol{u}_{pi} + \boldsymbol{G}_{ji}\boldsymbol{C}\hat{\boldsymbol{E}}_{0i}) - (\boldsymbol{G}_{ij}\boldsymbol{C}\boldsymbol{M}_j\boldsymbol{u}_{pj} + \boldsymbol{G}_{ij}\boldsymbol{C}\hat{\boldsymbol{E}}_{0j}) \right] V_j \tag{8.121}$$

需要注意，上式的矩阵表达增加了下标 i 和 j，以表示与矩阵有关的物质点。

8.3.4 矩形板拉伸变形的算例分析

采用键关联 NOSB PD 和键关联高阶 NOSB PD 的隐式静力方法求解 8.2.6 节中的矩形板拉伸变形算例。泰勒级数阶数分别为 $N=1$ 和 $N=3$，几何尺寸、材料参数、载荷和离散模型均与前述算例相同，影响函数 $w(|\boldsymbol{\xi}|) = \mathrm{e}^{-(2|\boldsymbol{\xi}|/\delta)^2}$，采用 $\Delta x = 0.005\mathrm{m}$ 物质点均匀离散构型，圆形近场范围半径取为 $\delta = m\Delta x$。如图 8-14 所示，选取 5 组不同的 $m(m=2, 2.5, 3, 4, 5)$ 值，通过计算结果显示键关联 NOSB PD 和键关联高阶 NOSB PD 模型在提高计算精度和降低数值振荡方面的作用，并分析近场范围半径的大小对两类 NOSB PD 模型计算精度和稳定性的影响。

图 8-14 中的红色和绿色圆分别表示红色和绿色物质点的近场范围，相交部分为键关联近场范围。二维键关联 NOSB PD 的最小 m 值为 $\sqrt{2}$[28,29]，以避免相交近场范围内物质点过少而导致矩阵奇异，故本例没有计算 $m=1$ 的情况。当 $\sqrt{2} < m < \sqrt{5}$ 时，计算结果存在数值振荡现象，其原因在于相交近场范围内的物质点过少，不能准确描述变形状态。由于高阶 NOSB PD 需要较大的近场范围[16-18]，故在本例的具体计算中，对于键关联的一阶和三阶 NOSB PD 模型，分别选取 $m \geqslant 2$ 和 $m > 2$。

8.3 键关联高阶非常规态型近场动力学模型

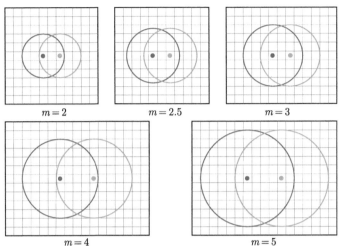

图 8-14 二维离散模型中的不同近场范围示意图 ($\delta = m\Delta x$)

图 8-15 和图 8-16 给出了不同 m 值时，键关联的一阶和三阶 NOSB PD 计算得到的水平位移和竖向位移云图，图 8-17 为根据 $\varepsilon_{L2} = \sqrt{\sum_{i=1}^{N_{\text{total}}} \left[R_i^{(\text{r})} - R_i^{(\text{n})} \right]^2} \Big/ \sqrt{\sum_{i=1}^{N_{\text{total}}} \left[R_i^{(\text{r})} \right]^2}$ 计算[26]得到的位移分量的全局误差。对于各组 m 值，即使选取较大的近场范围，一阶和三阶键关联的 NOSB PD 模型都能获得稳定的计算结果，三阶模型的计算精度略高于一阶模型，没有出现任何数值振荡现象，表明键关联的 NOSB PD 模型能够完全解决数值计算结果不稳定性问题。

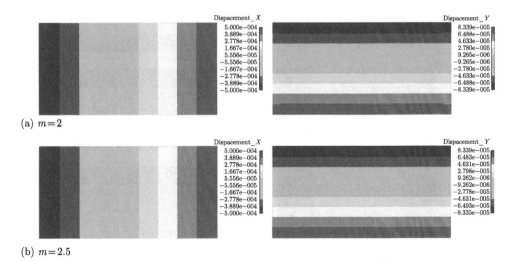

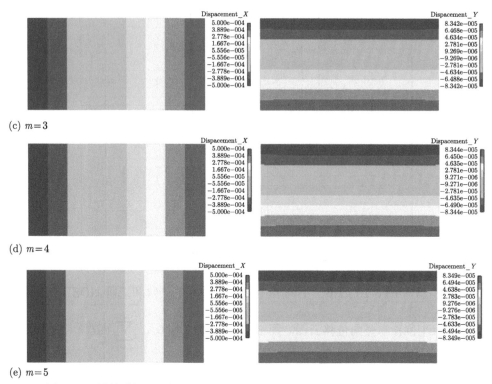

(c) $m=3$

(d) $m=4$

(e) $m=5$

图 8-15 键关联的 (一阶)NOSB PD 模型预测的水平和竖向位移 ($m = \delta/\Delta x$)

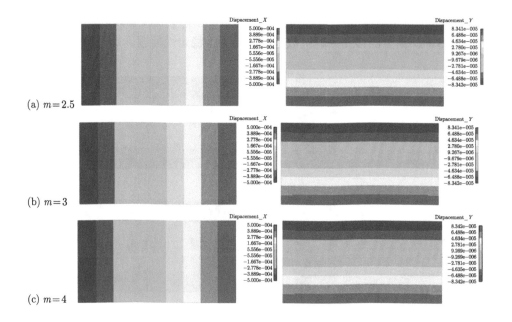

(a) $m=2.5$

(b) $m=3$

(c) $m=4$

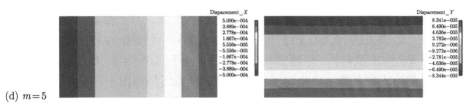

(d) $m=5$

图 8-16 键关联的高阶 (三阶)NOSB PD 模型预测的水平和竖向位移 ($m = \delta/\Delta x$)

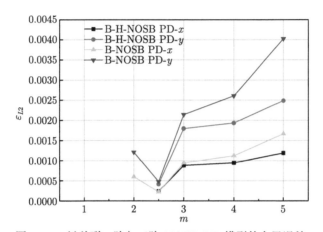

图 8-17 键关联一阶与三阶 NOSB PD 模型的全局误差

8.3.5 三维杆轴压变形的算例分析

如图 8-18 所示，本节通过分析端部受压的三维细长杆静力变形问题，进一步验证键关联的 NOSB PD 在消除数值不稳定性现象的效果。假定均匀各向同性弹性材料的杆长为 1m，方形横截面的边长为 0.1m，杨氏模量为 69.6GPa，泊松比为 0.1454。杆一端固定，另一端受均匀压力 $P = 100$Pa。采用原 NOSB PD 模型和键关联的一阶 NOSB PD 模型的显式动力学拟静力解法进行求解。计算中，均采用立方体物质点均匀离散杆件，物质点边长 Δx，近场范围的半径为 $\delta = m\Delta x$，详细分析近场范围尺寸和离散粒子间距对轴向位移结果的影响。

8.3.5.1 近场范围尺寸对结果精度和稳定性的影响

固定物质点间距为 $\Delta x = 0.01$m，分析不同近场范围尺寸对结果的影响。具体选取 $m = 2.0, 2.5, 3.0, 3.5, 4.0, 4.5, 5.0$。原 NOSB PD 模型计算得到的杆中线轴向位移如图 8-19(a) 所示，呈现出严重的数值振荡现象。采用小近场范围的原 NOSB PD 模型得到的计算结果振幅相对较小，振荡从加载端到固定端逐渐减弱，表明位移的数值振荡现象更容易发生在加载区域；采用较大近场范围 ($4 \leqslant m \leqslant 5$) 的原 NOSB PD 模型进行计算，所得结果在加载端和固定端的振幅都较大，表明除加载区域外,几何边界区域也容易发生数值振荡现象。与此相反,图 8-19(b) 给出了基于一阶键关联 NOSB PD 模型计算得到的杆中线轴向位移分量,从图中所示结果可见,除加载端外,位移曲线平顺光滑,几乎完全消除了数值振荡现象,加载端的位移不均匀现象主要来源于边界区域物质点近场范围不完整。

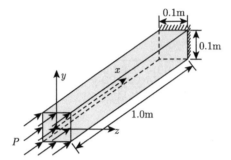

图 8-18　三维细长弹性杆端部受压的计算模型示意图

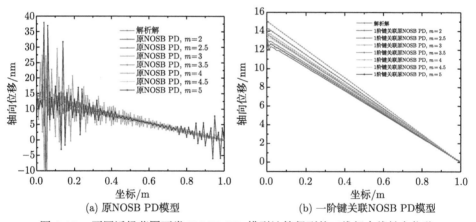

(a) 原NOSB PD模型　　　　(b) 一阶键关联NOSB PD模型

图 8-19　不同近场范围两类 NOSB PD 模型计算得到的三维杆中线轴向位移

图 8-20 给出了 NOSB PD 模型和键关联 NOSB PD 模型计算得到的轴向位移的全局误差 $\left(\varepsilon = \frac{1}{|u^{(r)}|_{\max}} \sqrt{\frac{1}{N} \sum_{i=1}^{N} \left[u_i^{(r)} - u_i^{(n)}\right]^2}\right)$[27]。结果表明,对于 NOSB PD 模型,较大的近场范围导致大振荡和大全局误差,但采用非整数 m 值得到的全局误差小于整数 m 值相应的结果,相似的结论可见文献 [30],这可能是均匀离散导致的。相比而言,键关联的 NOSB PD 模型的计算结果显著降低了全局误差,提高了计算精度;虽然较大的近场范围导致较大的全局误差,但误差来源于非局部解答与局部解答的差异性,而非数值不稳定性;特别地,$m = 2.0$ 和

8.3 键关联高阶非常规态型近场动力学模型

$m = 2.5$ 的全局误差大于 $m = 3.0$ 的全局误差，这是由于过小的近场范围不能准确地描述变形状态和受力状态，故对于键关联的 NOSB PD 模型，推荐采用 $m = 3.0$。

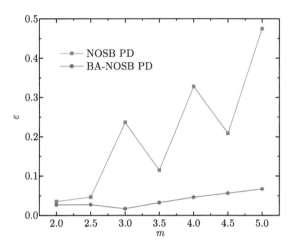

图 8-20 不同近场范围两类 NOSB PD 模型预测的轴向位移全局误差

为直观展示键关联 NOSB PD 模型消除数值不稳定现象的能力，图 8-21 给出了不同近场范围的 NOSB PD 模型和键关联 NOSB PD 模型计算得到的轴向位移云图，从图中所示结果也可以得出类似的认识。

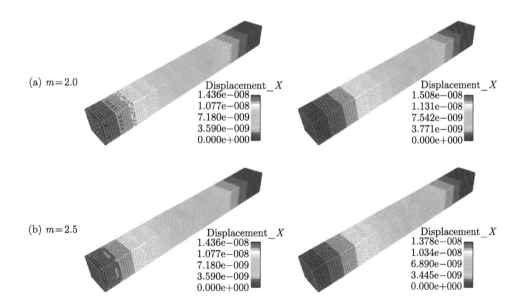

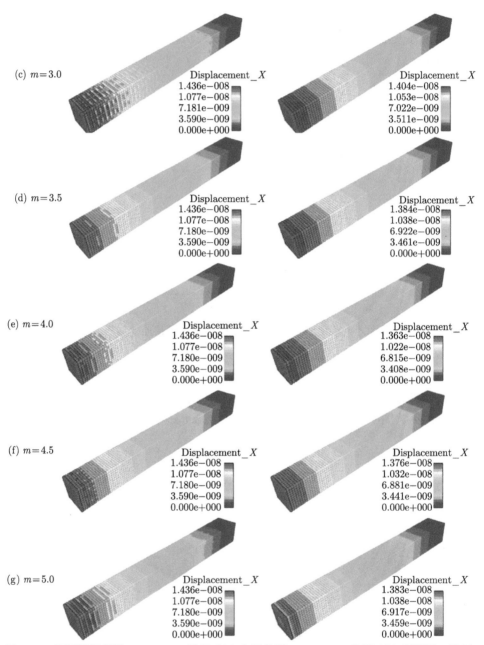

图 8-21 不同近场范围 NOSB PD 模型 (左) 和键关联 NOSB PD 模型 (右) 预测的三维杆轴向位移云图

8.3.5.2 物质点间距对结果精度和稳定性的影响

固定近场范围与物质点间距之比为 $m = 3.0$，分别选取 $\Delta x = 0.005\text{m}, 0.01\text{m}, 0.02\text{m}$，分析不同物质点间距对计算结果的影响。图 8-22 给出了原 NOSB PD 模型和键关联 NOSB PD 模型计算得到的轴向位移分量和全局误差，图 8-23 给出了两类模型计算得到的轴向位移云图。结果显示：精细离散有利于提高计算精度

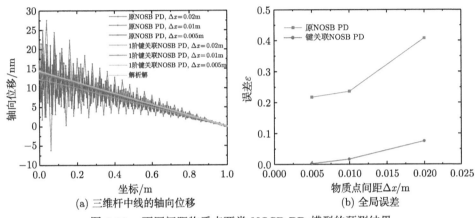

图 8-22 不同间距物质点两类 NOSB PD 模型的预测结果

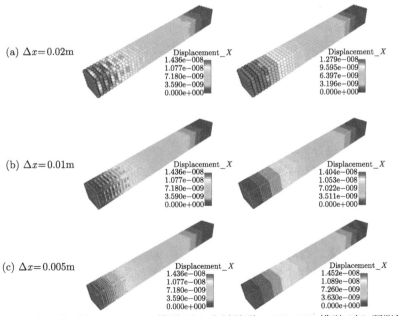

图 8-23 不同间距物质点 NOSB PD 模型 (左) 和键关联 NOSB PD 模型 (右) 预测的三维杆轴向位移云图

和稳定性，基于精细离散模型的原 NOSB PD 模型的计算结果仍存在明显的数值振荡现象，而键关联的 NOSB PD 模型的计算结果没有数值振荡；离散模型越精细，两类模型的计算结果与局部解析解的全局误差均越小，但精细离散将导致物质点数目增多，计算量增大，需要合理选取离散模型，以平衡计算精度、稳定性和计算量，并需要发展非均匀离散和自适应分析技术。

8.3.6 三维杆中准一维弹性波传播的算例分析

本节模拟三维杆中的准一维弹性波传播过程，以验证键关联 NOSB PD 模型分析动力变形的效果。右端固定、左端受压力作用的一维弹性杆中，任意点的位移随时间变化的解析解为 [31]

$$u(x,t) = \frac{P}{E}x - \frac{PL}{E} + \frac{8PL}{E\pi^2}\left[\sum_{n=0}^{\infty}\frac{1}{(2n+1)^2}\cos\frac{(n+1/2)\pi ct}{L}\cos\frac{(n+1/2)\pi x}{L}\right] \quad (8.122)$$

式中，边界压力 $P = P(t)$ 是任意时变函数，x 为杆中任意点坐标位置，t 为时间变量，E 为杨氏模量，L 为杆长，$c = \sqrt{E/\rho}$ 为波速，质量密度为 $\rho = 2530 \text{kg/m}^3$，$n$ 表示幂级数阶次。

杆的几何参数和材料性质均与 8.3.5 节相同，选取物质点尺寸为 $\Delta x = 0.01\text{m}$、近场范围半径为 $\delta = 3\Delta x$，均布压力 $P = 100\text{Pa}$ 在初始时刻突加到杆端并保持不变，即施加矩形脉冲荷载，采用显式动力学方法进行求解。图 8-24 给出了四个不同时刻 ($t = 50\mu\text{s}, 100\mu\text{s}, 150\mu\text{s}, 200\mu\text{s}$) 两类 NOSB PD 模型计算得到的杆中线轴向位移分布曲线。结果显示，原 NOSB PD 模型在不同时刻对应的计算结果都存在数值振荡现象，且从加载端向内部传播，远离加载端位移振荡较小且接近解析解；而键关联的 NOSB PD 模型所得计算结果在不同时刻均无数值振荡现象，且与解析解吻合良好。

图 8-25 展示了杆轴线上 $x = 0\text{m}$ 和 $x = 0.5\text{m}$ 处物质点的轴向位移历史曲线。从图中所示结果可以看出，原 NOSB PD 模型计算得到的加载端点 ($x = 0\text{m}$) 的轴向位移快速偏离解析解，杆中部点 ($x = 0.5\text{m}$) 的轴向位移初始与解析解吻合良好，后逐渐偏离，这是由于加载端点的数值振荡剧烈，并渐渐影响其他物质点的计算结果所致。而对于键关联的 NOSB PD 模型，由于完全消除了数值不稳定性现象，在整个弹性波传播过程中，计算结果与解析解吻合良好。

8.3 键关联高阶非常规态型近场动力学模型

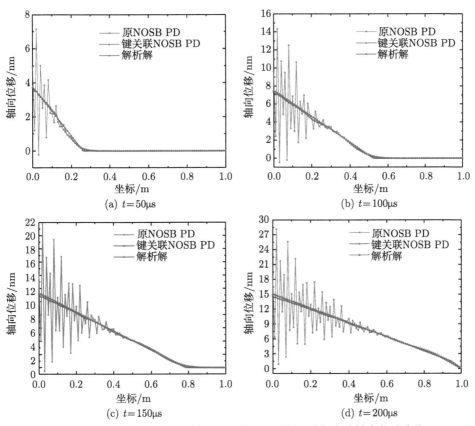

图 8-24 两类 NOSB PD 模型预测的不同时刻三维杆中线轴向位移曲线

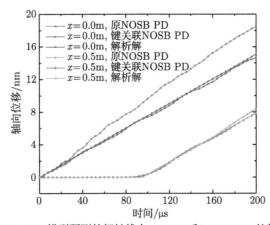

图 8-25 两类 NOSB PD 模型预测的杆轴线上 $x=0\mathrm{m}$ 和 $x=0.5\mathrm{m}$ 处物质点的轴向位移历史曲线

8.4 一个新的非常规态型近场动力学模型

近场动力学微分算子能够将局部微分方程转化为带有长度尺度的非局部积分方程,对于传统的弹性力学 Navier 位移方程,近场动力学微分算子被证明能够构造键型和常规态型近场动力学模型[16,17]。本节基于经典连续介质力学的应力或位移运动方程,采用近场动力学微分算子重构 Silling 博士提出的原 NOSB PD 模型,推导建立一个新的非常规态型近场动力学模型 (refined NOSB PD, RNOSB PD),典型算例分析证明 RNOSB PD 模型能完全避免原 NOSB PD 模型的数值不稳定现象。

8.4.1 基于近场动力学微分算子重构原非常规态型近场动力学模型

基于应力形式的平衡方程,采用近场动力学微分算子重构 Silling 博士提出的非常规态型近场动力学模型。以二维问题为例,应力形式的局部运动方程分量形式为

$$\rho(\boldsymbol{x})\frac{\partial^2 u_i(\boldsymbol{x},t)}{\partial t^2} = \frac{\partial \sigma_{ik}}{\partial x_k} + b_i(\boldsymbol{x},t), \quad i,k=1,2 \tag{8.123}$$

比较局部运动方程与近场动力学运动方程,有下式成立

$$\int_{H_\infty} (t_i(\boldsymbol{x},t) - t_i(\boldsymbol{x}',t))\,\mathrm{d}V_{\boldsymbol{x}'} = \frac{\partial \sigma_{ik}}{\partial x_k}, \quad i,k=1,2 \tag{8.124}$$

对应力分量求导数,应用一阶 PD 微分算子表达式 (见附录 A),则应力导数 $\frac{\partial \sigma_{ik}}{\partial x_k}$ 的 PD 积分表达如下:

$$\left\{\frac{\partial \sigma_{11}(\boldsymbol{x})}{\partial x_1} \frac{\partial \sigma_{11}(\boldsymbol{x})}{\partial x_2}\right\}^{\mathrm{T}} = \frac{1}{\pi h V_1}\left(\int_{H_\infty} w_1(|\boldsymbol{\xi}|)\,(\sigma_{11}(\boldsymbol{x}+\boldsymbol{\xi}) - \sigma_{11}(\boldsymbol{x}))\left\{\begin{array}{c}\xi_1\\ \xi_2\end{array}\right\}\mathrm{d}V_{\boldsymbol{x}'}\right) \tag{8.125}$$

$$\left\{\frac{\partial \sigma_{12}(\boldsymbol{x})}{\partial x_1} \frac{\partial \sigma_{12}(\boldsymbol{x})}{\partial x_2}\right\}^{\mathrm{T}} = \frac{1}{\pi h V_1}\left(\int_{H_\infty} w_1(|\boldsymbol{\xi}|)\,(\sigma_{12}(\boldsymbol{x}+\boldsymbol{\xi}) - \sigma_{12}(\boldsymbol{x}))\left\{\begin{array}{c}\xi_1\\ \xi_2\end{array}\right\}\mathrm{d}V_{\boldsymbol{x}'}\right) \tag{8.126}$$

$$\left\{\frac{\partial \sigma_{21}(\boldsymbol{x})}{\partial x_1} \frac{\partial \sigma_{21}(\boldsymbol{x})}{\partial x_2}\right\}^{\mathrm{T}} = \frac{1}{\pi h V_1}\left(\int_{H_\infty} w_1(|\boldsymbol{\xi}|)\,(\sigma_{21}(\boldsymbol{x}+\boldsymbol{\xi}) - \sigma_{21}(\boldsymbol{x}))\left\{\begin{array}{c}\xi_1\\ \xi_2\end{array}\right\}\mathrm{d}V_{\boldsymbol{x}'}\right) \tag{8.127}$$

8.4 一个新的非常规态型近场动力学模型

$$\left\{\frac{\partial \sigma_{22}(\boldsymbol{x})}{\partial x_1} \quad \frac{\partial \sigma_{22}(\boldsymbol{x})}{\partial x_2}\right\}^{\mathrm{T}} = \frac{1}{\pi h V_1} \left(\int_{H_\infty} w_1(|\boldsymbol{\xi}|) \ (\sigma_{22}(\boldsymbol{x}+\boldsymbol{\xi}) - \sigma_{22}(\boldsymbol{x})) \left\{\begin{array}{c} \xi_1 \\ \xi_2 \end{array}\right\} \mathrm{d}V_{\boldsymbol{x}'} \right) \tag{8.128}$$

式中，$V_n = \int_0^\delta w_n(|\boldsymbol{\xi}|) \ |\boldsymbol{\xi}|^{2n+1} \mathrm{d}|\boldsymbol{\xi}|$，且 $n=1$；影响函数可取 $w_n(|\boldsymbol{\xi}|) = \dfrac{\delta^{n+1}}{|\boldsymbol{\xi}|^{n+1}}$ 或 $w_n(|\boldsymbol{\xi}|) = 1$ 等。

此外，由于影响函数具有球对称性，则以下积分式为零

$$\int_{H_\infty} w_1(|\xi|) \sigma_{11}(\boldsymbol{x}) \xi \mathrm{d}V_{\boldsymbol{x}'} = \sigma_{11}(\boldsymbol{x}) \int_{H_\infty} w_1(|\xi|) \xi \mathrm{d}V_{\boldsymbol{x}'} = 0 \tag{8.129}$$

$$\int_{H_\infty} w_1(|\xi|) \sigma_{12}(\boldsymbol{x}) \xi \mathrm{d}V_{\boldsymbol{x}'} = \sigma_{12}(\boldsymbol{x}) \int_{H_\infty} w_1(|\xi|) \xi \mathrm{d}V_{\boldsymbol{x}'} = 0 \tag{8.130}$$

$$\int_{H_\infty} w_1(|\xi|) \sigma_{21}(\boldsymbol{x}) \xi \mathrm{d}V_{\boldsymbol{x}'} = \sigma_{21}(\boldsymbol{x}) \int_{H_\infty} w_1(|\xi|) \xi \mathrm{d}V_{\boldsymbol{x}'} = 0 \tag{8.131}$$

$$\int_{H_\infty} w_1(|\xi|) \sigma_{22}(\boldsymbol{x}) \xi \mathrm{d}V_{\boldsymbol{x}'} = \sigma_{22}(\boldsymbol{x}) \int_{H_\infty} w_1(|\xi|) \xi \mathrm{d}V_{\boldsymbol{x}'} = 0 \tag{8.132}$$

将式 (8.129)～式 (8.132) 分别加到式 (8.125)～式 (8.128) 的右端项，则有下式成立

$$\begin{bmatrix} \dfrac{\partial \sigma_{11}}{\partial x_1} & \dfrac{\partial \sigma_{12}}{\partial x_2} \\ \dfrac{\partial \sigma_{21}}{\partial x_1} & \dfrac{\partial \sigma_{22}}{\partial x_2} \end{bmatrix}$$

$$= \frac{1}{h\pi V_1}$$

$$\times \begin{bmatrix} \displaystyle\int_{H_\infty} w_1(|\boldsymbol{\xi}|) \left[\sigma_{11}(\boldsymbol{x}+\boldsymbol{\xi})+\sigma_{11}(\boldsymbol{x})\right] \xi_1 \mathrm{d}V_{\boldsymbol{x}'} & \displaystyle\int_{H_\infty} w_1(|\boldsymbol{\xi}|) \left[\sigma_{12}(\boldsymbol{x}+\boldsymbol{\xi})+\sigma_{12}(\boldsymbol{x})\right] \xi_2 \mathrm{d}V_{\boldsymbol{x}'} \\ \displaystyle\int_{H_\infty} w_1(|\boldsymbol{\xi}|) \left[\sigma_{21}(\boldsymbol{x}+\boldsymbol{\xi})+\sigma_{21}(\boldsymbol{x})\right] \xi_1 \mathrm{d}V_{\boldsymbol{x}'} & \displaystyle\int_{H_\infty} w_1(|\boldsymbol{\xi}|) \left[\sigma_{22}(\boldsymbol{x}+\boldsymbol{\xi})+\sigma_{22}(\boldsymbol{x})\right] \xi_2 \mathrm{d}V_{\boldsymbol{x}'} \end{bmatrix} \tag{8.133}$$

类似使用 PD 一阶微分算子推导上式，当定义应力矢量状态为

$$\begin{bmatrix} \sigma_{i1}(\boldsymbol{x}+\boldsymbol{\xi}) + \sigma_{i1}(\boldsymbol{x}) \\ \sigma_{i2}(\boldsymbol{x}+\boldsymbol{\xi}) + \sigma_{i2}(\boldsymbol{x}) \end{bmatrix}$$

利用矢量状态缩减算子[3]，可获得同样的积分形式表达，具体为

$$\begin{bmatrix} \dfrac{\partial \sigma_{i1}(\boldsymbol{x})}{\partial x_1} & \dfrac{\partial \sigma_{i1}(\boldsymbol{x})}{\partial x_2} \\ \dfrac{\partial \sigma_{i2}(\boldsymbol{x})}{\partial x_1} & \dfrac{\partial \sigma_{i2}(\boldsymbol{x})}{\partial x_2} \end{bmatrix}$$

$$= \left(\begin{bmatrix} \sigma_{i1}(\boldsymbol{x}+\boldsymbol{\xi}) + \sigma_{i1}(\boldsymbol{x}) \\ \sigma_{i2}(\boldsymbol{x}+\boldsymbol{\xi}) + \sigma_{i2}(\boldsymbol{x}) \end{bmatrix} * \underline{\boldsymbol{X}} \right) \boldsymbol{K}^{-1}$$

$$= \left[\int_{H_\infty} w_1(|\boldsymbol{\xi}|) \begin{bmatrix} \sigma_{i1}(\boldsymbol{x}+\boldsymbol{\xi}) + \sigma_{i1}(\boldsymbol{x}) \\ \sigma_{i2}(\boldsymbol{x}+\boldsymbol{\xi}) + \sigma_{i2}(\boldsymbol{x}) \end{bmatrix} \begin{bmatrix} \xi_1 & \xi_2 \end{bmatrix} \mathrm{d}V_{\boldsymbol{x}'} \right] \boldsymbol{K}^{-1}$$

$$= \dfrac{1}{h\pi V_1}$$

$$\times \begin{bmatrix} \int_{H_\infty} w_1(|\boldsymbol{\xi}|) [\sigma_{i1}(\boldsymbol{x}+\boldsymbol{\xi}) + \sigma_{i1}(\boldsymbol{x})] \xi_1 \mathrm{d}V_{\boldsymbol{x}'} & \int_{H_\infty} w_1(|\boldsymbol{\xi}|) [\sigma_{i1}(\boldsymbol{x}+\boldsymbol{\xi}) + \sigma_{i1}(\boldsymbol{x})] \xi_2 \mathrm{d}V_{\boldsymbol{x}'} \\ \int_{H_\infty} w_1(|\boldsymbol{\xi}|) [\sigma_{i2}(\boldsymbol{x}+\boldsymbol{\xi}) + \sigma_{i2}(\boldsymbol{x})] \xi_1 \mathrm{d}V_{\boldsymbol{x}'} & \int_{H_\infty} w_1(|\boldsymbol{\xi}|) [\sigma_{i2}(\boldsymbol{x}+\boldsymbol{\xi}) + \sigma_{i2}(\boldsymbol{x})] \xi_2 \mathrm{d}V_{\boldsymbol{x}'} \end{bmatrix}$$

(8.134)

利用式 (8.133) 或式 (8.134) 应力分量的空间导数 $\partial \sigma_{ik}/\partial x_k$ 的 PD 积分表达，得到

$$L_i(\boldsymbol{x}) = \int_{H_\infty} (t_i(\boldsymbol{x},t) - t_i(\boldsymbol{x}',t)) \mathrm{d}V_{\boldsymbol{x}'} = \dfrac{\partial \sigma_{i1}}{\partial x_1} + \dfrac{\partial \sigma_{i2}}{\partial x_2}$$

$$= \dfrac{1}{h\pi V_1} \int_{H_\infty} w_1(|\boldsymbol{\xi}|) [\sigma_{i1}(\boldsymbol{x}+\boldsymbol{\xi}) + \sigma_{i1}(\boldsymbol{x})] \xi_1 \mathrm{d}V_{\boldsymbol{x}'}$$

$$+ \dfrac{1}{h\pi V_1} \int_{H_\infty} w_1(|\boldsymbol{\xi}|) [\sigma_{i2}(\boldsymbol{x}+\boldsymbol{\xi}) + \sigma_{i2}(\boldsymbol{x})] \xi_2 \mathrm{d}V_{\boldsymbol{x}'}$$

$$= \dfrac{1}{h\pi V_1} \int_{H_\infty} w_1(|\boldsymbol{\xi}|) [\sigma_{i1}(\boldsymbol{x}+\boldsymbol{\xi})\xi_1 + \sigma_{i2}(\boldsymbol{x}+\boldsymbol{\xi})\xi_2] \mathrm{d}V_{\boldsymbol{x}'}$$

$$+ \dfrac{1}{h\pi V_1} \int_{H_\infty} w_1(|\boldsymbol{\xi}|) [\sigma_{i1}(\boldsymbol{x})\xi_1 + \sigma_{i2}(\boldsymbol{x})\xi_2] \mathrm{d}V_{\boldsymbol{x}'} \quad (8.135)$$

在二维圆形对称近场范围下，NOSB PD 模型的形状张量为 $\boldsymbol{K} = h\pi V_1 \boldsymbol{I}$，则有 $K_{11}^{-1} = K_{22}^{-1} = 1/h\pi V_1$ 成立，又由于 $\xi_1' = -\xi_1$ 和 $\xi_2' = -\xi_2$，则对于 x_1 与 x_2

方向，式 (8.135) 具体为

$$\left\{\begin{array}{c} L_1 \\ L_2 \end{array}\right\} = \int_{H_x} w_1(|\boldsymbol{\xi}|) \left\{\begin{array}{c} \sigma_{11}(\boldsymbol{x})\xi_1 K_{11}^{-1} + \sigma_{12}(\boldsymbol{x})\xi_2 K_{22}^{-1} \\ \sigma_{21}(\boldsymbol{x})\xi_1 K_{11}^{-1} + \sigma_{22}(\boldsymbol{x})\xi_2 K_{22}^{-1} \end{array}\right\} \mathrm{d}V_{x'}$$

$$- \int_{H_x} w_1(|\boldsymbol{\xi}|) \left\{\begin{array}{c} \sigma_{11}(\boldsymbol{x}+\boldsymbol{\xi})\xi_1' K_{11}^{-1} + \sigma_{12}(\boldsymbol{x}+\boldsymbol{\xi})\xi_2' K_{22}^{-1} \\ \sigma_{21}(\boldsymbol{x}+\boldsymbol{\xi})\xi_1' K_{11}^{-1} + \sigma_{22}(\boldsymbol{x}+\boldsymbol{\xi})\xi_2' K_{22}^{-1} \end{array}\right\} \mathrm{d}V_x \quad (8.136)$$

或

$$\left\{\begin{array}{c} L_1 \\ L_2 \end{array}\right\} = \int_{H_x} w_1(|\boldsymbol{\xi}|) \left[\begin{array}{cc} \sigma_{11}(\boldsymbol{x}) & \sigma_{12}(\boldsymbol{x}) \\ \sigma_{21}(\boldsymbol{x}) & \sigma_{22}(\boldsymbol{x}) \end{array}\right] \left[\begin{array}{cc} K_{11}^{-1} & 0 \\ 0 & K_{22}^{-1} \end{array}\right] \left\{\begin{array}{c} \xi_1 \\ \xi_2 \end{array}\right\} \mathrm{d}V_{x'}$$

$$- \int_{H_x} w_1(|\boldsymbol{\xi}|) \left[\begin{array}{cc} \sigma_{11}(\boldsymbol{x}+\boldsymbol{\xi}) & \sigma_{12}(\boldsymbol{x}+\boldsymbol{\xi}) \\ \sigma_{21}(\boldsymbol{x}+\boldsymbol{\xi}) & \sigma_{22}(\boldsymbol{x}+\boldsymbol{\xi}) \end{array}\right] \left[\begin{array}{cc} K_{11}^{-1} & 0 \\ 0 & K_{22}^{-1} \end{array}\right] \left\{\begin{array}{c} \xi_1' \\ \xi_2' \end{array}\right\} \mathrm{d}V_{x'} \quad (8.137)$$

最终，上述两式可化为下式，即为 NOSB PD 的内力密度矢量项

$$\boldsymbol{L} = \int_{H_x} (\boldsymbol{t}(\boldsymbol{x},t) - \boldsymbol{t}(\boldsymbol{x}',t)) \mathrm{d}V_{x'}$$

$$= \int_{H_x} w_1(|\boldsymbol{\xi}|)\boldsymbol{\sigma}(\boldsymbol{x})\boldsymbol{K}^{-1}(\boldsymbol{x})\boldsymbol{\xi}\mathrm{d}V_{x'} - \int_{H_x} w_1(|\boldsymbol{\xi}|)\boldsymbol{\sigma}(\boldsymbol{x}')\boldsymbol{K}^{-1}(\boldsymbol{x}')\boldsymbol{\xi}'\mathrm{d}V_{x'} \quad (8.138)$$

至此，由传统应力形式的运动方程出发，采用 PD 微分算子重构了非常规态型近场动力学弹性模型的运动方程。比较两种方法定义的形状张量矩阵 $\boldsymbol{K}$ 与形状矩阵 $\boldsymbol{A}$ 的子矩阵 $\boldsymbol{A}_{11}$，以及由 NOSB PD 的矢量状态缩减得到的变形梯度张量与一阶 PD 微分算子得到的变形梯度张量，有

$$\boldsymbol{K} = \int_{H_x} w(|\boldsymbol{\xi}|)(\boldsymbol{\xi}\otimes\boldsymbol{\xi})\mathrm{d}V_{x'} = \int_{H_x} w(|\boldsymbol{\xi}|)\left[\begin{array}{cc} \xi_1^2 & \xi_1\xi_2 \\ \xi_1\xi_2 & \xi_2^2 \end{array}\right]\mathrm{d}V_{x'} = h\pi V_1 \boldsymbol{I} \quad (8.139)$$

$$\boldsymbol{A}_{11} = \int_{H_x} w_1(|\boldsymbol{\xi}|)\left[\begin{array}{cc} \xi_1^2 & \xi_1\xi_2 \\ \xi_1\xi_2 & \xi_2^2 \end{array}\right]\mathrm{d}V_{x'} = h\pi V_1 \boldsymbol{I} \quad (8.140)$$

$$\boldsymbol{F} = \left[\begin{array}{cc} \frac{\partial y_1}{\partial x_1} & \frac{\partial y_1}{\partial x_2} \\ \frac{\partial y_2}{\partial x_1} & \frac{\partial y_2}{\partial x_2} \end{array}\right] = \int_{H_x} w(|\boldsymbol{\xi}|)\left[\begin{array}{cc} (y_1'-y_1)\xi_1 & (y_1'-y_1)\xi_2 \\ (y_2'-y_2)\xi_1 & (y_2'-y_2)\xi_2 \end{array}\right]\mathrm{d}V_{x'}\boldsymbol{K}^{-1} \quad (8.141)$$

$$\boldsymbol{F} = \begin{bmatrix} \dfrac{\partial y_1}{\partial x_1} & \dfrac{\partial y_1}{\partial x_2} \\ \dfrac{\partial y_2}{\partial x_1} & \dfrac{\partial y_2}{\partial x_2} \end{bmatrix} = \dfrac{1}{h\pi V_1} \int_{H_\infty} w_1(|\boldsymbol{\xi}|) \begin{bmatrix} (y_1' - y_1)\xi_1 & (y_1' - y_1)\xi_2 \\ (y_2' - y_2)\xi_1 & (y_2' - y_2)\xi_2 \end{bmatrix} \mathrm{d}V_{\boldsymbol{x}'}$$

(8.142)

可见，两种方法的形状张量和变形梯度张量均一致。综上，从经典连续介质力学的应力平衡方程出发，采用一阶近场动力学微分算子，能够完全重构 Silling 博士提出的非常规态型近场动力学模型，包括运动方程、形状张量和变形梯度张量等。

8.4.2 基于近场动力学微分算子构建新的非常规态型近场动力学模型

仍以二维问题为例，在弹性小变形假设下，Navier 位移运动方程的分量形式为

$$\rho(\boldsymbol{x})\dfrac{\partial^2 u_i(\boldsymbol{x},t)}{\partial t^2} = \dfrac{\partial \sigma_{ik}}{\partial x_k} + b_i(\boldsymbol{x},t) = \mu \dfrac{\partial^2 u_i}{\partial x_k \partial x_k} + (\lambda + \mu)\dfrac{\partial^2 u_k}{\partial x_i \partial x_k} + b_i(\boldsymbol{x},t) \quad (8.143)$$

式中，$i, k = 1, 2$，$\lambda = \dfrac{E\nu}{(1+\nu)(1-2\nu)}$ 和 $\mu = G = \dfrac{E}{2(1+\nu)}$ 为拉梅常数，E 为杨氏模量，ν 为泊松比。注意到 Navier 位移运动方程可以直接退化至平面应变问题，对于平面应力问题，为保持上述方程的形式不变，有 $\lambda = \dfrac{E}{2(1-\nu)} - \mu = \dfrac{E\nu}{1-\nu^2}$。

比较 Navier 位移运动方程与近场动力学运动方程，有下式成立

$$L_i = \int_{H_\infty} (t_i(\boldsymbol{x},t) - t_i(\boldsymbol{x}',t))\,\mathrm{d}V_{\boldsymbol{x}'} = \mu\dfrac{\partial^2 u_i}{\partial x_k \partial x_k} + (\lambda + \mu)\dfrac{\partial^2 u_k}{\partial x_i \partial x_k}, \quad i, k = 1, 2$$

(8.144)

在原 NOSB PD 建模时，将一阶 PD 微分算子应用于应力分量的导数，即 $\partial \sigma_{ik}/\partial x_k$，或者将一阶 PD 微分算子两次应用于位移的导数，即 $\dfrac{\partial}{\partial x_k}\left(\dfrac{\partial u_i}{\partial x_k}\right)$。此外，位移的二阶导数项 $\dfrac{\partial^2 u_i}{\partial x_k \partial x_k}$ 可以直接采用二阶 PD 微分算子进行取代。本节导出一种新的非常规态型近场动力学模型，给出适用于内部区域和边界区域的控制方程，建立新模型的隐式–显式混合求解方法。该模型可以避免传统近场动力学方法"边界效应"问题导致的计算误差，可以施加零或非零应力边界条件，消除传统近场动力学因无法施加零应力边界条件而导致的边界虚假力。

对于结构内部物质点，一般采用对称完整的圆形近场范围，根据附录 A 给出的二阶 PD 微分算子表达式，则各位移分量的二阶微分为

$$\dfrac{\partial^2 u_i}{\partial x_j^2} = \dfrac{1}{\pi h V_2}\int_{H_\infty} w_2(|\boldsymbol{\xi}|)(u_i(\boldsymbol{x}+\boldsymbol{\xi}) - u_i(\boldsymbol{x}))(4\xi_j^2 - \xi_k\xi_k)\,\mathrm{d}V_{\boldsymbol{x}'} \quad (i, j, k = 1, 2)$$

(8.145)

8.4 一个新的非常规态型近场动力学模型

$$\frac{\partial^2 u_i}{\partial x_1 \partial x_2} = \frac{4}{\pi h V_2} \int_{H_\infty} w_2(|\boldsymbol{\xi}|)(u_i(\boldsymbol{x}+\boldsymbol{\xi}) - u_i(\boldsymbol{x}))\xi_1 \xi_2 \mathrm{d}V_{\boldsymbol{x}'} \qquad (8.146)$$

式中，$V_2 = \int_0^\delta w_2(|\boldsymbol{\xi}|)|\boldsymbol{\xi}|^5 \mathrm{d}|\boldsymbol{\xi}|$，影响函数可取 $w_n(|\boldsymbol{\xi}|) = \delta^{n+1}/|\boldsymbol{\xi}|^{n+1}$ 或 $w_n(|\boldsymbol{\xi}|) = 1$ 等形式。

将上述位移分量的二阶 PD 微分算子代入 Navier 位移运动方程中，则有

$$\int_{H_\infty}(t_1(\boldsymbol{x},t) - t_1(\boldsymbol{x}',t))\mathrm{d}V_{\boldsymbol{x}'}$$

$$= (\lambda + 2\mu)\frac{\partial^2 u_1}{\partial x_1 \partial x_1} + \mu\frac{\partial^2 u_1}{\partial x_2 \partial x_2} + (\lambda + \mu)\frac{\partial^2 u_2}{\partial x_1 \partial x_2}$$

$$= \frac{(\mu - \lambda)}{\pi h V_2}\int_{H_\infty} w_2(|\boldsymbol{\xi}|)(u_1(\boldsymbol{x}+\boldsymbol{\xi}) - u_1(\boldsymbol{x}))(\xi_1^2 + \xi_2^2)\mathrm{d}V_{\boldsymbol{x}'}$$

$$+ \frac{4(\lambda + \mu)}{\pi h V_2}\int_{H_\infty} w_2(|\boldsymbol{\xi}|)[(u_1(\boldsymbol{x}+\boldsymbol{\xi}) - u_1(\boldsymbol{x}))\xi_1 + (u_2(\boldsymbol{x}+\boldsymbol{\xi}) - u_2(\boldsymbol{x}))\xi_2]\xi_1 \mathrm{d}V_{\boldsymbol{x}'}$$

$$= \frac{(\mu - \lambda)}{\pi h V_2}\int_{H_\infty} w_2(|\boldsymbol{\xi}|)(u_1(\boldsymbol{x}+\boldsymbol{\xi}) - u_1(\boldsymbol{x}))(\boldsymbol{\xi}\cdot\boldsymbol{\xi})\mathrm{d}V_{\boldsymbol{x}'}$$

$$+ \frac{4(\lambda + \mu)}{\pi h V_2}\int_{H_\infty} w_2(|\boldsymbol{\xi}|)[(\boldsymbol{u}(\boldsymbol{x}+\boldsymbol{\xi}) - \boldsymbol{u}(\boldsymbol{x}))\cdot\boldsymbol{\xi}]\xi_1 \mathrm{d}V_{\boldsymbol{x}'} \qquad (8.147)$$

$$\int_{H_\infty}(t_2(\boldsymbol{x},t) - t_2(\boldsymbol{x}',t))\mathrm{d}V = \mu\frac{\partial^2 u_2}{\partial x_1 \partial x_1} + (\lambda + 2\mu)\frac{\partial^2 u_2}{\partial x_2 \partial x_2} + (\lambda + \mu)\frac{\partial^2 u_1}{\partial x_1 \partial x_2}$$

$$= \frac{(\mu - \lambda)}{\pi h V_2}\int_{H_\infty} w_2(|\boldsymbol{\xi}|)(u_2(\boldsymbol{x}+\boldsymbol{\xi}) - u_2(\boldsymbol{x}))(\boldsymbol{\xi}\cdot\boldsymbol{\xi})\mathrm{d}V_{\boldsymbol{x}'}$$

$$+ \frac{4(\lambda + \mu)}{\pi h V_2}\int_{H_\infty} w_2(|\boldsymbol{\xi}|)[(\boldsymbol{u}(\boldsymbol{x}+\boldsymbol{\xi}) - \boldsymbol{u}(\boldsymbol{x}))\cdot\boldsymbol{\xi}]\xi_2 \mathrm{d}V_{\boldsymbol{x}'}$$

$$(8.148)$$

综合以上两式，可得二维问题内部区域物质点的 RNOSB PD 模型的控制方程

$$\int_{H_\infty}(\boldsymbol{t}(\boldsymbol{x}) - \boldsymbol{t}(\boldsymbol{x}'))\mathrm{d}V_{\boldsymbol{x}'} = \frac{(\mu - \lambda)}{\pi h V_2}\int_{H_\infty} w_2(|\boldsymbol{\xi}|)(\boldsymbol{\xi}\cdot\boldsymbol{\xi})[\boldsymbol{u}(\boldsymbol{x}+\boldsymbol{\xi}) - \boldsymbol{u}(\boldsymbol{x})]\mathrm{d}V_{\boldsymbol{x}'}$$

$$+ \frac{4(\lambda + \mu)}{\pi h V_2} \int_{H_{\boldsymbol{x}}} w_2(|\boldsymbol{\xi}|) \left[(\boldsymbol{u}(\boldsymbol{x}+\boldsymbol{\xi}) - \boldsymbol{u}(\boldsymbol{x})) \cdot \boldsymbol{\xi} \right] \boldsymbol{\xi} \mathrm{d}V_{\boldsymbol{x}'} \tag{8.149}$$

同理，可推导得到三维 RNOSB PD 模型的控制方程

$$\int_{H_{\boldsymbol{x}'}} (\boldsymbol{t}(\boldsymbol{x}) - \boldsymbol{t}(\boldsymbol{x}')) \mathrm{d}V_{\boldsymbol{x}'} = \frac{3(\mu - \lambda)}{4\pi V_2} \int_{H_{\boldsymbol{x}}} w_2(|\boldsymbol{\xi}|)(\boldsymbol{\xi} \cdot \boldsymbol{\xi}) \left[\boldsymbol{u}(\boldsymbol{x}+\boldsymbol{\xi}) - \boldsymbol{u}(\boldsymbol{x}) \right] \mathrm{d}V_{\boldsymbol{x}'}$$
$$+ \frac{15(\lambda + \mu)}{4\pi V_2} \int_{H_{\boldsymbol{x}}} w_2(|\boldsymbol{\xi}|) \left[(\boldsymbol{u}(\boldsymbol{x}+\boldsymbol{\xi}) - \boldsymbol{u}(\boldsymbol{x})) \cdot \boldsymbol{\xi} \right] \boldsymbol{\xi} \mathrm{d}V_{\boldsymbol{x}'} \tag{8.150}$$

值得注意的是，当平面应力问题泊松比为 1/3、平面应变问题和三维问题的泊松比为 1/4 时，上述两式表示的 RNOSB PD 模型等价于键型近场动力学模型。至此，建立了适用于具有完整对称圆形近场范围的内部物质点的 RNOSB PD 模型。比较原 NOSB PD 模型与 RNOSB PD 模型，可以发现：两种 NOSB PD 模型中，物质点的近场范围尺寸都是 δ，但物质点的实际作用范围却不同；具体地，原 NOSB PD 模型的积分计算涉及源点和场点的各自近场范围 δ 内积分，而 RNOSB PD 仅涉及源点的近场范围 δ 内积分，即原 NOSB PD 模型的作用范围为半径 2δ 的区域，而 RNOSB PD 的作用范围半径为 δ。

进而，定义距边界一个近场范围尺寸内的区域为边界区域，其他为内部区域。与其他近场动力学模型相似，上述 RNOSB PD 模型的基础是对称完整的圆形近场范围，但对于边界区域，其近场范围是非对称的，或至少是非完整的圆形，故采用上述 RNOSB PD 方程分析边界区域将会带来误差，而基于非对称非完整的近场范围建立的控制方程能够很好地描述边界区域物质点的运动规律。以下给出边界区域物质点的控制方程，该方程同样适用于全域的一般形式方程。

对于任意形状近场范围 (圆形的、非对称的非完整圆形或任意形状)，根据本书附录 A 中任意形状近场范围下的二阶 PD 微分算子表达式，各位移分量的二阶微分为

$$\left\{ \frac{\partial^2 u_i(\boldsymbol{x})}{\partial x_1^2} \quad \frac{\partial^2 u_i(\boldsymbol{x})}{\partial x_2^2} \quad \frac{\partial^2 u_i(\boldsymbol{x})}{\partial x_1 \partial x_2} \right\}^{\mathrm{T}} = \int_{H_{\boldsymbol{x}}} (u_i(\boldsymbol{x}+\boldsymbol{\xi}) - u_i(\boldsymbol{x})) \left\{ \begin{array}{c} g_2^{20}(\boldsymbol{\xi}) \\ g_2^{02}(\boldsymbol{\xi}) \\ g_2^{11}(\boldsymbol{\xi}) \end{array} \right\} \mathrm{d}V_{\boldsymbol{x}'} \tag{8.151}$$

将上述二阶位移微分代入下述位移平衡方程

$$\int_{H_{\bm{x}}} (t_1(\bm{x},t) - t_1(\bm{x}',t))\,\mathrm{d}V_{\bm{x}'} = (\lambda+2\mu)\frac{\partial^2 u_1}{\partial x_1 \partial x_1} + \mu\frac{\partial^2 u_1}{\partial x_2 \partial x_2} + (\lambda+\mu)\frac{\partial^2 u_2}{\partial x_1 \partial x_2}$$
(8.152)

$$\int_{H_{\bm{x}}} (t_2(\bm{x},t) - t_2(\bm{x}',t))\,\mathrm{d}V_{\bm{x}'} = \mu\frac{\partial^2 u_2}{\partial x_1 \partial x_1} + (\lambda+2\mu)\frac{\partial^2 u_2}{\partial x_2 \partial x_2} + (\lambda+\mu)\frac{\partial^2 u_1}{\partial x_1 \partial x_2}$$
(8.153)

可获得边界区域的 RNOSB PD 运动方程为

$$\int_{H_{\bm{x}}} (\bm{t}(\bm{x}) - \bm{t}(\bm{x}'))\,\mathrm{d}V_{\bm{x}'} = \int_{H_{\bm{x}}} (\mu\mathrm{tr}(\bm{D})\bm{I} + (\lambda+\mu)\bm{D})\,(\bm{u}(\bm{x}') - \bm{u}(\bm{x}))\,\mathrm{d}V_{\bm{x}'} \quad (8.154)$$

其中，tr 为矩阵的迹，矩阵 $\bm{D}$ 为

$$\begin{aligned}
\bm{D} &= \begin{bmatrix} g_2^{200}(|\bm{\xi}|) & g_2^{110}(|\bm{\xi}|) & g_2^{101}(|\bm{\xi}|) \\ g_2^{110}(|\bm{\xi}|) & g_2^{020}(|\bm{\xi}|) & g_2^{011}(|\bm{\xi}|) \\ g_2^{101}(|\bm{\xi}|) & g_2^{011}(|\bm{\xi}|) & g_2^{002}(|\bm{\xi}|) \end{bmatrix}, & \text{三维} \\
\bm{D} &= \begin{bmatrix} g_2^{20}(|\bm{\xi}|) & g_2^{11}(|\bm{\xi}|) \\ g_2^{11}(|\bm{\xi}|) & g_2^{02}(|\bm{\xi}|) \end{bmatrix}, & \text{二维} \\
\bm{D} &= \bm{g}_2^2(\bm{\xi}) = [g_2^2(|\bm{\xi}|)], & \text{一维}
\end{aligned}$$
(8.155)

式中，$g_N^{p_1 p_2 \cdots p_M}(\bm{\xi})$ 为近场动力学微分算子定义的近场动力学函数[16-18]。

8.4.3 新的非常规态型近场动力学模型的隐式-显式混合解法

不考虑惯性项，内部区域和边界区域的二维 RNOSB PD 模型的控制方程为

$$\frac{(\mu-\lambda)}{\pi h V_2} \int_{H_{\bm{x}}} w_2(|\bm{\xi}|)(\bm{\xi}\cdot\bm{\xi})[\bm{u}(\bm{x}+\bm{\xi}) - \bm{u}(\bm{x})]\mathrm{d}V_{\bm{x}'}$$
$$+ \frac{4(\lambda+\mu)}{\pi h V_2} \int_{H_{\bm{x}}} w_2(|\bm{\xi}|)[(\bm{u}(\bm{x}+\bm{\xi}) - \bm{u}(\bm{x}))\cdot\bm{\xi}]\bm{\xi}\mathrm{d}V_{\bm{x}'} + \bm{b}(\bm{x}) = \bm{0} \quad (8.156)$$

$$\int_{H_{\bm{x}}} (\mu\mathrm{tr}(\bm{D})\bm{I} + (\lambda+\mu)\bm{D})\,(\bm{u}(\bm{x}') - \bm{u}(\bm{x}))\,\mathrm{d}V_{\bm{x}'} + \bm{b}(\bm{x}) = \bm{0} \quad (8.157)$$

离散后，可写为矩阵形式

$$\bm{H}_0 \bm{u}_p + \bm{b}_0 = \bm{0} \quad (8.158)$$

式中，对二维问题，H_0 为维度 $2\times(2N^i+2)$ 的物质点劲度系数矩阵，u_p 为维度 $(2N^i+2)\times1$ 的位移列向量，b_0 为维度 2×1 的荷载列向量，N^i 为物质点 x 近场范围内其他物质点的总数。

由此，整个系统的 PD 平衡方程组可组装为代数方程组

$$HU + b^* = 0 \tag{8.159}$$

式中，U 为维度 $2N_{\text{total}}\times1$ 的位移列向量，包含系统所有的未知位移分量；H 为维度 $2N_{\text{total}} \times 2N_{\text{total}}$ 的系统整体劲度系数矩阵；b^* 为维度 $2N_{\text{total}} \times 1$ 的荷载列向量，包含已知的外体力密度，N_{total} 为系统物质点的总数，2 表示二维问题。

RNOSB PD 运动方程的空间求积采用基于无网格粒子离散的高斯积分方法[73]。具体实施时，可采用均匀网格/点阵离散结构，物质点体积为 $V = h\Delta^2$，其中 Δ 为物质点间距，h 为二维问题结构的厚度；圆形近场范围半径为 $\delta = m\Delta$，且 m 通常取 3；边界处的物质点常采用体积修正[4]，以提高积分精度。此外，边界区域可采用均匀或非均匀离散，各物质点体积为实际单元体积；近场范围半径与均匀离散时相同，也可采用变化的近场范围尺寸，只需确保近场范围内有足够的积分点和力守恒。选取无量纲影响函数 $w_n(|\boldsymbol{\xi}|) = \delta^{n+1}/|\boldsymbol{\xi}|^{n+1}$ 或 $w_n(|\boldsymbol{\xi}|) = 1$，$n = 2$。

为使得系统控制方程组成为定解，需要提供边界条件。在传统 PD 方法中，位移边界条件可以直接施加到近场动力学积分方程中；而应力边界条件不能直接施加，需转化为体力密度施加，但该方法不适合零应力边界条件，不能够施加零应力边界条件将导致产生非局部边界鬼力 (如表现为边界应力不为零)。为消除传统近场动力学边界条件的施加方法可能导致的非局部边界鬼力问题，以下将给出位移表示的应力分量，以施加零应力和非零应力边界条件。

如图 8-26 所示，RNOSB PD 求解需补充位移约束条件、零应力边界条件和非零应力边界条件

$$\begin{cases} \boldsymbol{u} = \bar{\boldsymbol{u}}, & \Gamma_{\boldsymbol{u}} \\ \boldsymbol{\sigma n} = \boldsymbol{\tau}, & \Gamma_{\boldsymbol{\tau} \neq 0} \\ \boldsymbol{\sigma n} = \boldsymbol{0}, & \Gamma_{\boldsymbol{\tau} = 0} \end{cases} \tag{8.160}$$

式中，$\boldsymbol{n}$ 为边界物质点的单位外法线向量。采用一阶 PD 微分算子获得位移表示的应力张量，即为

$$\boldsymbol{\sigma}(\boldsymbol{x},t) = \int_{H_{\boldsymbol{x}}} \begin{bmatrix} \lambda \mathrm{tr}\left((\boldsymbol{u}(\boldsymbol{x}') - \boldsymbol{u}(\boldsymbol{x})) \otimes \boldsymbol{g}\right)\boldsymbol{I} \\ +\mu\left((\boldsymbol{u}(\boldsymbol{x}') - \boldsymbol{u}(\boldsymbol{x})) \otimes \boldsymbol{g}\right) + \mu\left(\boldsymbol{g} \otimes (\boldsymbol{u}(\boldsymbol{x}') - \boldsymbol{u}(\boldsymbol{x}))\right) \end{bmatrix} \mathrm{d}V_{\boldsymbol{x}'} \tag{8.161}$$

8.4 一个新的非常规态型近场动力学模型

式中，$\boldsymbol{g} = \left\{ \begin{array}{c} g_2^{10}(|\boldsymbol{\xi}|) \\ g_2^{01}(|\boldsymbol{\xi}|) \end{array} \right\}$。具体地，二维参考坐标系 x_1Ox_2 中的应力分量为

$$\sigma_{11}(x_{1(k)}, x_{2(k)}) = \frac{2\mu}{1-\nu} \sum_{j=1} \left\{ \left(u_1(x_{1(j)}, x_{2(j)}) - u_1(x_{1(k)}, x_{2(k)}) \right) g_2^{10}(\xi_{1(j)(k)}, \xi_{2(j)(k)}) \right.$$
$$\left. + \nu \left(u_2(x_{1(j)}, x_{2(j)}) - u_2(x_{1(k)}, x_{2(k)}) \right) g_2^{01}(\xi_{1(j)(k)}, \xi_{2(j)(k)}) \right\} \Delta A_{(j)}$$

$$\sigma_{22}(x_{1(k)}, x_{2(k)}) = \frac{2\mu}{1-\nu} \sum_{j=1} \left\{ \nu \left(u_1(x_{1(j)}, x_{2(j)}) - u_1(x_{1(k)}, x_{2(k)}) \right) g_2^{10}(\xi_{1(j)(k)}, \xi_{2(j)(k)}) \right.$$
$$\left. + \left(u_2(x_{1(j)}, x_{2(j)}) - u_2(x_{1(k)}, x_{2(k)}) \right) g_2^{01}(\xi_{1(j)(k)}, \xi_{2(j)(k)}) \right\} \Delta A_{(j)}$$

$$\sigma_{12}(x_{1(k)}, x_{2(k)}) = \mu \sum_{j=1} \left\{ \left(u_1(x_{1(j)}, x_{2(j)}) - u_1(x_{1(k)}, x_{2(k)}) \right) g_2^{01}(\xi_{1(j)(k)}, \xi_{2(j)(k)}) \right.$$
$$\left. + \left(u_2(x_{1(j)}, x_{2(j)}) - u_2(x_{1(k)}, x_{2(k)}) \right) g_2^{10}(\xi_{1(j)(k)}, \xi_{2(j)(k)}) \right\} \Delta A_{(j)}$$

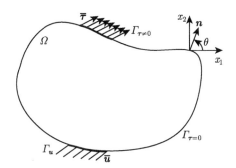

图 8-26 RNOSB PD 模型的位移、零应力与非零应力边界条件

基于 RNOSB PD 模型求解时，施加所有的应力和位移边界条件，且仅施加在最边缘一层物质点上。此时，应力和位移边界条件均是位移的函数，可统一表示为约束方程

$$\boldsymbol{GU} + \boldsymbol{U}^* = \boldsymbol{0} \tag{8.162}$$

其中，矩阵 $\boldsymbol{G}$ 包含约束方程系数，$\boldsymbol{U}^*$ 含有已知位移约束值和应力边界值。通过拉格朗日乘子法施加边界条件，当引入拉格朗日乘子 $\boldsymbol{\lambda}$，平衡方程和约束方程的变分混合形式为

$$\delta \boldsymbol{U}^{\mathrm{T}}(\boldsymbol{HU} + \boldsymbol{b}^*) + \delta \left[\boldsymbol{\lambda}^{\mathrm{T}}(\boldsymbol{GU} + \boldsymbol{U}^*) \right] = 0 \tag{8.163}$$

式中，δU 为位移矢量的任意变分。于是，求解未知量 U 和 λ 的代数方程组为

$$\begin{bmatrix} H & G^{\mathrm{T}} \\ G & 0 \end{bmatrix} \begin{Bmatrix} U \\ \lambda \end{Bmatrix} = - \begin{Bmatrix} b^* \\ U^* \end{Bmatrix} \tag{8.164}$$

对于动力问题，零应力边界条件和位移边界条件需要时刻满足。类似于冲击接触问题的动力有限元的拉格朗日解法[32]，求解带有边界约束方程的 RNOSB PD 积分方程，引入拉格朗日乘子 λ 的运动方程为

$$M\ddot{U} = HU + b^* + G^{\mathrm{T}}\lambda \tag{8.165}$$

式中，M 为集中质量矩阵，矢量 U 为各积分点的未知场变量，H 为整体劲度系数矩阵，b^* 为已知外力矢量。向前增量拉格朗日方法 (forward increment Lagrange multiplier method) 与显式积分方法相匹配，t_{n+1} 时刻的位移约束与 t_n 时刻的拉格朗日乘子关联，则运动方程和约束条件分别为

$$\begin{aligned} M\ddot{U}_n &= HU_n + b_n^* + G_{n+1}^{\mathrm{T}}\lambda_n \\ G_{n+1}U_{n+1} &+ U_{n+1}^* = 0 \end{aligned} \tag{8.166}$$

采用二阶中心差分格式计算速度和加速度，即为

$$\begin{aligned} \dot{U}_n &= \frac{1}{2\Delta t}(U_{n+1} - U_{n-1}) \\ \ddot{U}_n &= \frac{1}{(\Delta t)^2}(U_{n+1} - 2U_n + U_{n-1}) \end{aligned} \tag{8.167}$$

将式 (8.167) 代入式 (8.166) 中，有下式成立

$$U_{n+1} = \underbrace{(\Delta t)^2 M^{-1}(HU_n + b_n^*) + 2U_n - U_{n-1}}_{U_{n+1}^{\mathrm{p}}} + \underbrace{(\Delta t)^2 M^{-1} G_{n+1}^{\mathrm{T}}\lambda_n}_{U_{n+1}^{\mathrm{c}}} \tag{8.168}$$

式中，U_{n+1}^{p} 为未考虑约束方程影响的位移更新。由约束方程可求得 G_{n+1} 矩阵，在本方法中，G 保持不变，则拉格朗日乘子为

$$\lambda_n = - \left[(\Delta t)^2 G_{n+1} M^{-1} G_{n+1}^{\mathrm{T}}\right]^{-1} \left(G_{n+1} U_{n+1}^{\mathrm{p}} + U_{n+1}^*\right) \tag{8.169}$$

由此可计算增量位移 U_{n+1}^{c} 和 t_{n+1} 时刻的最终位移 U_{n+1}。

需要说明的是，在宏观裂纹形成前，采用静力求解方法获得变形构型，再采用显式求解方法分析动力裂纹扩展，是一种优选方案。

8.4.4 矩形板拉伸变形的算例分析

与 8.2.6 节的算例相同，如图 8-27 所示，受水平均匀单轴拉伸的各向同性矩形板，发生静力弹性变形，采用线弹性 NOSB PD 模型和 RNOSB PD 模型的隐式静力分析方法求解该问题。板的长度、宽度和厚度分别为 $L=1\text{m}$、$W=0.5\text{m}$ 和 $h=0.01\text{m}$；材料的杨氏弹性模量 $E=200\text{GPa}$，泊松比 $\nu=1/3$，质量密度 $\rho=7850\text{kg/m}^3$；x_1、x_2 和 x_3 方向分别离散为 100、50 和 1 个物质点，共 5000 个物质点；物质点间距、体积和圆形近场范围半径分别为 $\Delta x=0.01\text{m}$、$\Delta V=(\Delta x)^2 h$ 和 $\delta=3\Delta x$；上下水平边界自由无任何约束，水平向拉伸最大位移值 $d=0.0005\text{m}$，线性变化施加在左右竖直边界上，其中原 NOSB PD 模型施加在三层物质点上，而 RNOSB PD 模型施加在一层物质点上；影响函数取为 $w=1$ 和 $w(|\boldsymbol{\xi}|)=\delta^3/|\boldsymbol{\xi}|^3$ 两种。

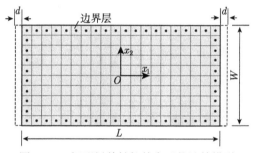

图 8-27 矩形板单轴拉伸变形的计算模型

表 8-3 给出四种计算工况下对应的竖向最大位移值及其与解析解 $\left(d_y^{\max}=\dfrac{W}{L}\nu d_x^{\max}\right)$ 的误差；图 8-28 和图 8-29 分别给出了四种计算工况得到的水平位移和竖向位移云图；图 8-30 给出了不同泊松比情况下基于 RNOSB PD 模型计算得到的板的竖向位移云图。综合计算结果，可获得如下结论：

(1) 对于弹性变形问题，NOSB PD 模型和 RNOSB PD 模型均获得较高精度的解答，竖向最大位移误差较小，表明两种非常规态型近场动力学计算模型的有效性；

(2) 对于弹性变形问题，原 NOSB PD 模型虽然计算精度较高，但计算结果存在明显的数值不稳定性，位移呈现锯齿形振荡，即使递减影响函数 $w(|\boldsymbol{\xi}|)=\delta^3/|\boldsymbol{\xi}|^3$ 能够改善结果的光滑性，但仍然残留位移突变或不连续；

(3) RNOSB PD 模型计算结果光滑平顺，与解析解高度一致，表明 RNOSB PD 方程能够有效消除积分区域不完整带来的计算误差，且 RNOSB PD 模型完全避免了 NOSB PD 模型的数值不稳定问题；

表 8-3 不同 PD 模型得到的竖向最大位移值及其相对误差

工况	模型方法	影响函数	竖向最大位移	误差/%
1	解析解	\	8.3333e−5	0
2	NOSB PD	$w=1$	8.4565e−5	1.478
3	NOSB PD	$w(|\boldsymbol{\xi}|)=\delta^3/|\boldsymbol{\xi}|^3$	8.3838e−5	0.6060
4	RNOSB PD	$w=1$	8.3333e−5	0
5	RNOSB PD	$w(|\boldsymbol{\xi}|)=\delta^3/|\boldsymbol{\xi}|^3$	8.3333e−5	0

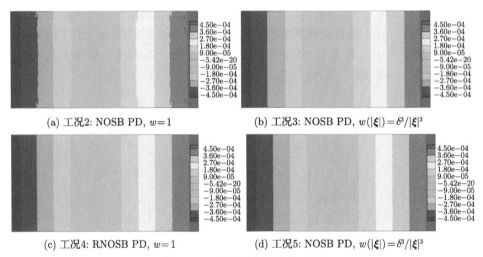

图 8-28 矩形板水平位移云图

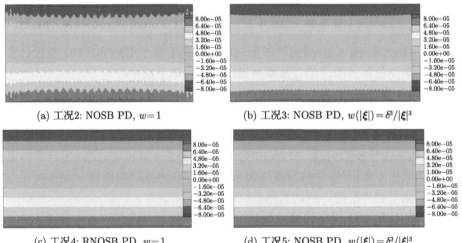

图 8-29 矩形板竖向位移云图

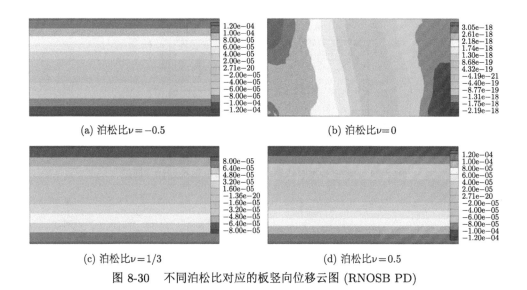

图 8-30 不同泊松比对应的板竖向位移云图 (RNOSB PD)

(4) RNOSB PD 模型能够准确再现弹性变形时材料的泊松比效应，正确描述结构的弹性变形。

8.4.5 含线裂纹矩形板拉伸变形的算例分析

与 7.4.3 节中算例相同，如图 8-31 所示，含有中心预制线裂纹的各向同性矩形板受水平向单轴拉伸载荷作用，发生静力弹性变形，采用线弹性 NOSB PD 和 RNOSB PD 的隐式方法求解该问题。矩形板的长度、宽度和厚度分别为 $L=1\mathrm{m}$、$W=0.5\mathrm{m}$ 和 $h=0.01\mathrm{m}$，中心竖直裂纹长为 $2a=0.1\mathrm{m}$；材料杨氏模量 $E=200\mathrm{GPa}$，泊松比 $\nu=1/3$，质量密度 $\rho=7850\mathrm{kg/m^3}$，对应的 $Q_{11}=225\mathrm{GPa}$、$Q_{12}=75\mathrm{GPa}$ 和 $Q_{44}=75\mathrm{GPa}$。采用 $100\times50\times1$ 点阵离散矩形板，物质点间距、体积和圆形近场范围半径分别为 $\Delta x=0.01\mathrm{m}$、$\Delta V=(\Delta x)^2 h$ 和 $\delta=3\Delta x$，影响函数 $w(|\boldsymbol{\xi}|)=\delta^3\big/|\boldsymbol{\xi}|^3$。矩形板的上下边界自由，水平向单轴拉伸最大位移值为 $d=0.0005\mathrm{m}$，按线性变化施加在左右两端的边界上，对于原 NOSB PD 模型，施加在三层物质点上，对于 RNOSB PD 模型，施加在一层物质点上；应力边界条件施加在一层物质点上。固定矩形板两侧边界中点的竖向位移为 0，以消除刚体位移。

图 8-32 和图 8-33 分别给出 NOSB PD 模型和 RNOSB PD 模型计算得到的水平位移和竖向位移的三维云图。由图中结果可见，NOSB PD 模型的计算结果在裂尖存在显著的数值振荡，而 RNOSB PD 模型的计算结果则非常光滑，完全消除了数值不稳定现象。

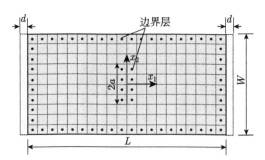

图 8-31 含中心预制线裂纹矩形板的单轴拉伸计算模型

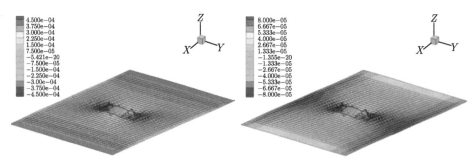

图 8-32 水平位移和竖向位移云图 (NOSB PD)

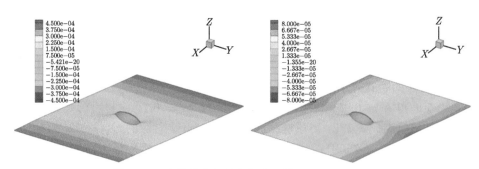

图 8-33 水平位移和竖向位移云图 (RNOSB PD)

8.4.6 含孔板拉伸变形的算例分析

含中心孔洞的各向同性矩形铝合金板受水平向单轴拉伸荷载作用，发生静力弹性变形，如图 8-34 所示。板的长度、宽度和厚度分别为 $L=150\mathrm{mm}$、$W=50\mathrm{mm}$ 和 $h = \Delta\mathrm{mm}$，孔洞直径为 $D = 20\mathrm{mm}$[33]。AA7075-T651 铝合金材料参数为：质量密度 $\rho = 2810\mathrm{kg/m}^3$，拉梅常数 $\lambda = 60.9\mathrm{GPa}$ 和 $\mu = 28.3\mathrm{GPa}$，临界能量释放率 $G_F = 120\mathrm{MPa}$[34]，相应的杨氏模量和泊松比分别为 $E = 75.921\mathrm{GPa}$ 和 $\nu = 0.34137$。左右两端施加位移边界条件 $d = 1.5\times 10^{-5}\mathrm{m}$，使板产生 0.02% 的拉应变。

在平面应力条件下，分别采用对应于传统线弹性模型的 NOSB PD 模型和

8.4 一个新的非常规态型近场动力学模型

RNOSB PD 模型求解该问题。选取圆形近场范围和影响函数 $w(|\boldsymbol{\xi}|) = \delta^3/|\boldsymbol{\xi}|^3$，均匀离散的物质点间距、体积和近场范围分别为 $\Delta x = 0.5\text{mm}$、$\Delta V = (\Delta x)^2 h$ 和 $\delta = 3\Delta x$，物质点总数为 29156。NOSB PD 模型的边界条件施加在三层物质点上，而 RNOSB PD 模型的边界条件施加在一层物质点上。此外，为便于比较分析，采用 ANSYS 有限元软件对该问题进行了求解并获得局部解答 (共离散为 10654 个节点和 10374 个单元)。

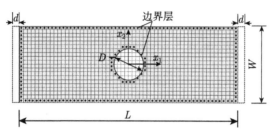

图 8-34 含中心孔洞的各向同性矩形铝板受水平拉伸荷载的计算模型

图 8-35 ~ 图 8-37 分别了给出 FEM、NOSB PD 和 RNOSB PD 对应的解答，包括水平位移和竖向位移以及各应力分量云图；图 8-38 和图 8-39 分别给出了板

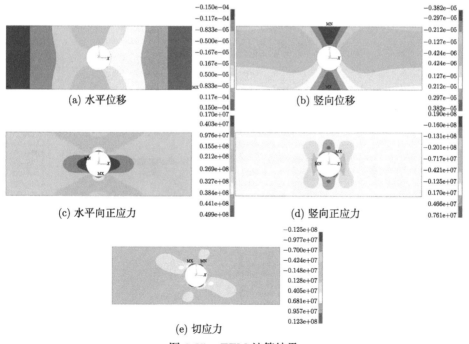

图 8-35 FEM 计算结果

底边和竖直中线的位移结果。比较计算结果可知：两种 PD 模型计算得到的位移分布规律均与 FEM 结果一致；原 NOSB PD 的计算结果表现出位移振荡的数值不稳定现象，在孔边附近区域更甚；RNOSB PD 模型结果没有位移振荡，表明该模型能够完全避免原 NOSB PD 模型的数值不稳定问题。

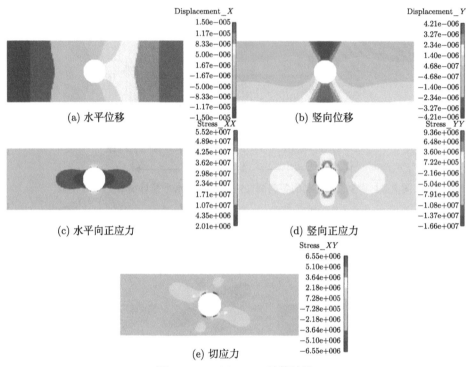

图 8-36　NOSB PD 计算结果

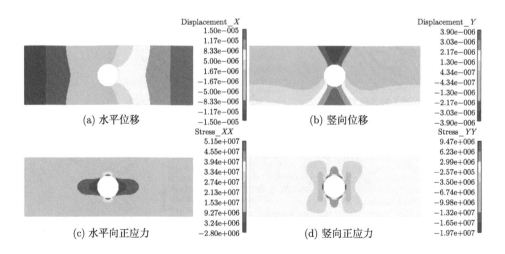

8.4 一个新的非常规态型近场动力学模型 · 289 ·

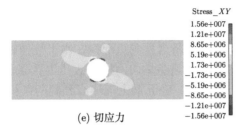

(e) 切应力

图 8-37　RNOSB PD 计算结果

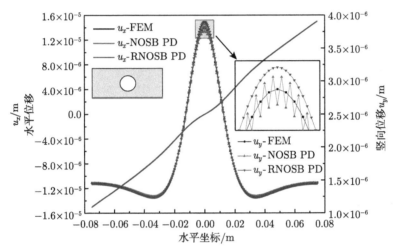

图 8-38　板底边的水平位移 u_x 和竖向位移 u_y 曲线

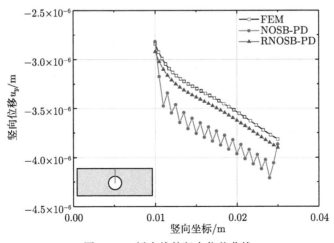

图 8-39　板中线的竖向位移曲线

8.4.7 含孔板裂纹扩展的算例分析

采用 RNOSB PD 方法分析含孔洞矩形板在水平拉伸载荷作用下的裂纹动态扩展问题。如图 8-40 所示,为避免均匀网格离散孔边界造成虚假的数值应力集中以及由此引起的裂纹开裂现象,在孔洞上下边缘预制两条长 2mm 的垂直线裂纹。板的尺寸和材料参数均与 8.4.6 节算例相同,采用 $\Delta x = 2$mm 均匀网格离散得 1896 个物质点。

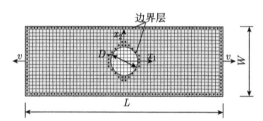

图 8-40 含中心孔和孔边预制裂纹的矩形板受水平拉伸荷载的计算模型

采用隐式-显式混合方法进行求解,在板的两端施加位移增量以实现加载过程。首先,利用隐式方法获得每一个线性增量位移步的变形状态,位移增量为 2.0×10^{-7}m;采用临界伸长率准则判断断键、损伤的产生与累积,当物质点的最大损伤值到达 0.2 时 (该值依据多组数值试验选取,此时静力变形到达较高程度,且尚未形成宏观裂纹),启动显式求解,显式计算的时间步长为 $\Delta t = 2.0 \times 10^{-8}$s,加载速率为 $v = 0.10$m/s。需要注意的是,在形成近场范围内物质点集合时,断开所有穿越孔洞和裂纹的键。

图 8-41 显示了板的损伤累积与裂纹扩展情况;由于应力集中,预制裂纹尖端首先产生损伤,当损伤逐渐累积超过某一临界值时,产生垂直扩展的自相似裂纹,最终扩展到上下水平边界处。图 8-42 给出了板左端第一列物质点所在截面的轴力相对于最大水平位移的变化关系曲线。结果表明,RNOSB PD 模型和相应的隐式-显式混合分析方法能获得合理的裂纹路径与位移载曲线,适合于固体结构的裂纹扩展与断裂分析。

综合算例结果表明,RNOSB PD 模型可以避免传统近场动力学方法"边界效应"问题导致的计算误差,可以施加零或非零应力边界条件,消除传统近场动力

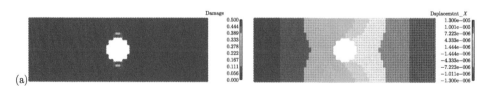

(a)

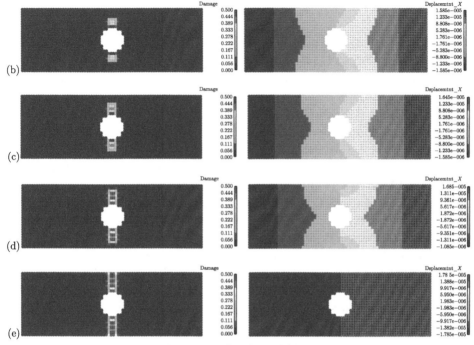

图 8-41 损伤累积与裂纹扩展

学无法施加零应力边界条件导致的边界虚假力，完全消除原 NOSB PD 模型结果的数值不稳定问题，也验证了相应的隐式-显式混合分析方法的有效性。

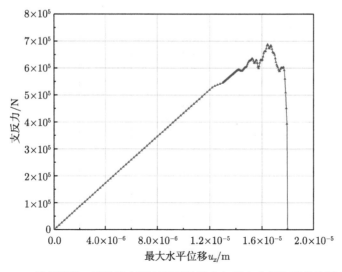

图 8-42 板左端第一列物质点所在截面的轴力与最大水平位移的关系曲线

参 考 文 献

[1] Monaghan J J. Smoothed particle hydrodynamics[J]. Annual Review of Astronomy and Astrophysics, 1992, 30(1):543-574.

[2] Monaghan J J. Smoothed particle hydrodynamics[J]. Reports on Progress in Physics, 2005, 68(8):1703-1759.

[3] Liu M B, Liu G R. Smoothed particle hydrodynamics (SPH): an overview and recent developments[J]. Archives of Computational Methods in Engineering, 2010, 17(1):25-76.

[4] Lee C H, Gil A J, Ghavamian A, et al. A total Lagrangian upwind smooth particle hydrodynamics algorithm for large strain explicit solid dynamics[J]. Computer Methods in Applied Mechanics and Engineering, 2019, 344:209-250.

[5] Bonet J, Kulasegaram S. Correction and stabilization of smooth particle hydrodynamics methods with applications in metal forming simulations[J]. International Journal for Numerical Methods in Engineering, 2000, 47(6):1189-1214.

[6] Belytschko T, Krongauz Y, Dolbow J, et al. On the completeness of meshfree particle methods[J]. International Journal for Numerical Methods in Engineering, 1998, 43(5):785-819.

[7] Liu W K, Jun S, Zhang Y F. Reproducing kernel particle methods[J]. International Journal for Numerical Methods in Fluids, 1995, 20(8-9):1081-1106.

[8] Aluru N R. A point collocation method based on reproducing kernel approximations[J]. International Journal for Numerical Methods in Engineering, 2000, 47(6):1083-1121.

[9] Bonet J, Kulasegaram S. Remarks on tension instability of Eulerian and Lagrangian corrected smooth particle hydrodynamics (CSPH) methods[J]. International Journal for Numerical Methods in Engineering, 2001, 52(11):1203-1220.

[10] Bonet J, Lok T S. Variational and momentum preservation aspects of smooth particle hydrodynamic formulations[J]. Computer Methods in Applied Mechanics and Engineering, 1999, 180(1-2):97-115.

[11] Huerta A, Vidal Y, Bonet J. Updated Lagrangian formulation for corrected smooth particle hydrodynamics[J]. International Journal of Computational Methods, 2006, 3(4):383-399.

[12] Bessa M A, Foster J T, Belytschko T, et al. A meshfree unification: reproducing kernel peridynamics[J]. Computational Mechanics, 2014, 53:1251-1264.

[13] Chi S W, Chen J S, Hu H Y, et al. A gradient reproducing kernel collocation method for boundary value problems[J]. International Journal for Numerical Methods in Engineering, 2013, 93(13):1381-1402.

[14] Silling S A, Epton M, Weckner O, et al. Peridynamic states and constitutive modeling[J]. Journal of Elasticity, 2007, 88(2):151-184.

[15] Gu X, Madenci E, Zhang Q. Revisit of non-ordinary state-based peridynamics[J]. Engineering Fracture Mechanics, 2018, 190:31-52.

[16] Madenci E, Barut A, Futch M. Peridynamic differential operator and its applications[J]. Computer Methods in Applied Mechanics and Engineering, 2016, 304:408-451.

[17] Madenci E, Dorduncu M, Barut A, et al. Numerical solution of linear and nonlinear partial differential equations using the peridynamic differential operator[J]. Numerical Methods for Partial Differential Equations, 2017, 33(5):1726-1753.

[18] Madenci E, Barut A, Dorduncu M. Peridynamic differential operator for numerical analysis[J]. Switzerland: Springer International Publishing, 2019.

[19] Rabczuk T, Belytschko T, Xiao S P. Stable particle methods based on Lagrangian kernels[J]. Computer Methods in Applied Mechanics and Engineering, 2004, 193(12-14):1035-1063.

[20] Gu X, Zhang Q, Madenci E. Non-ordinary state-based peridynamic simulation of elasto-plastic deformation and dynamic cracking of polycrystal[J]. Engineering Fracture Mechanics, 2019, 218:106568.

[21] Ren B, Fan H F, Bergel G L, et al. A peridynamics–SPH coupling approach to simulate soil fragmentation induced by shock waves[J]. Computational Mechanics, 2015, 55(2):287-302.

[22] Madenci E, Oterkus E. Peridynamic Theory and Its Applications[M]. New York: Springer, 2014.

[23] Mukherjee Y X, Mukherjee S. On boundary conditions in the element-free Galerkin method[J]. Computational Mechanics, 1997, 19(4):264-270.

[24] Silling S A, Lehoucq R B. Convergence of peridynamics to classical elasticity theory[J]. Journal of Elasticity, 2008, 93(1):13-37.

[25] Tian X C, Du Q. Asymptotically compatible schemes and applications to robust discretization of nonlocal models[J]. SIAM Journal on Numerical Analysis, 2014, 52(4):1641-1665.

[26] Ren H L, Zhuang X Y, Cai Y C, et al. Dual-horizon peridynamics[J]. International Journal for Numerical Methods in Engineering, 2016, 108(12):1451-1476.

[27] Gu X, Zhang Q, Madenci E, et al. Possible causes of numerical oscillations in non-ordinary state-based peridynamics and a bond-associated higher-order stabilized model[J]. Computer Methods in Applied Mechanics and Engineering, 2019, 357:112592.

[28] Chen H L. Bond-associated deformation gradients for peridynamic correspondence model[J]. Mechanics Research Communications, 2018, 90:34-41.

[29] Chen H L, Spencer B W. Peridynamic bond-associated correspondence model: Stability and convergence properties[J]. International Journal for Numerical Methods in Engineering, 2019, 117(6):713-727.

[30] Ganzenmüller G C, Hiermaier S, May M. Improvements to the prototype micro-brittle model of peridynamics[C].//Meshfree methods for partial differential equations VII Springer, Cham, 2015:163-183.

[31] Gu X, Zhang Q, Yu Y T. An effective way to control numerical instability of a nonordinary state-based peridynamic elastic model[J]. Mathematical Problems in Engineering, 2017, 2017:1-7.

[32] Carpenter N J, Taylor R L, Katona M G. Lagrange constraints for transient finite

element surface contact[J]. International Journal for Numerical Methods in Engineering, 1991, 32(1):103-128.

[33] Silling S A. Stability of peridynamic correspondence material models and their particle discretizations[J]. Computer Methods in Applied Mechanics and Engineering, 2017, 322:42-57.

[34] Littlewood D J. A nonlocal approach to modeling crack nucleation in AA 7075-T651[C]. ASME 2011 International Mechanical Engineering Congress and Exposition, American Society of Mechanical Engineers, 2011: 567-576.

第 9 章 近场动力学方法与有限单元法的混合模型

  基于连续介质力学理论的有限单元法具有计算效率高、通用性好和计算软件商业化程度强等特点，是解决交通运输、机械装备、能源矿业、土木水利、航空航天等几乎所有领域工程问题的主流数值计算方法。但有限单元法在求解固体材料和结构的裂纹扩展等不连续变形破坏问题时存在瓶颈，在模拟尺寸效应或非局部效应显著的力学问题中显现困难。近场动力学理论和方法有望克服经典连续介质力学理论和有限单元法在求解上述问题中的不足。

  显式或隐式数值方法可以用于求解近场动力学方程，但近场动力学的非局部特性决定其计算量要高于有限单元法，计算效率偏低。通常有三种方法提高近场动力学的计算效率，一是近场动力学算法的并行化和高性能计算；二是研究近场动力学的非均匀离散与自适应算法；三是进行近场动力学与以有限单元法为代表的传统数值方法的混合建模[1-3]，以充分利用有限单元法的计算效率和近场动力学求解破坏问题的优势。本章综述了近场动力学与传统有限单元法混合建模的研究进展，介绍各种混合建模方法的基本原理和特点，并重点阐述作者团队在近场动力学与有限单元法混合建模方面的研究工作，提出了两种近场动力学与有限单元法的混合模型，通过典型算例验证了所建模型的有效性。

9.1　近场动力学与有限元混合建模的几种方法

  现有的近场动力学与传统有限单元法的混合建模方法包括位移协调约束、力耦合、混合函数方法以及子模型方法等。可归为两大类：一类是 PD-FEM 并发式分析方法，除子模型方法外，现有 PD-FEM 混合建模方法均可归结为此类，其基本思想是将计算结构划分为近场动力学子域、有限元子域以及两者的交界区域(或连接区域、重叠区域、过渡区域)，即在不连续变形存在或可能产生区域，或非局部效应显著区域主使用近场动力学模型，而在光滑位移场区域选择传统局部模型，交界区域需要满足守恒定律，并注意降低或消除虚假力和虚假应力波问题；另一类是 FEM-PD 顺序分析方法，其中的典型代表是子模型方法，这类方法先采用有限元方法对结构进行连续变形分析，达到临界条件时，将结构全部区域转化为近场动力学物质点集，或将结构非连续变形显现 (或非局部效应显著) 的部分区域转化为物质点集，再进行 PD-FEM 并发式分析。

9.1.1 力耦合方法与镶嵌单元技术

Macek 和 Silling[4] 首先将研究对象划分为近场动力学子域和有限元子域，不同类型子域间设有重叠区域，并将近场动力学的物质点对用有限单元法的杆单元来描述，杆单元的横截面积 A 和弹性模量 E 的计算表达式见式 (4.13) 和式 (4.14)，采用有限元分析软件 ABAQUS 中的镶嵌单元 (embedded element) 技术在显式求解体系下实现了近场动力学与有限单元法的混合建模。

如图 9-1 所示，该混合模型的重叠区域具有一定宽度且采用实体单元离散，杆单元嵌入到实体单元中实现近场动力学物质点与有限元结点的连接作用，单元结点刚度值由杆单元和实体单元叠加而得，为防止镶嵌单元刚度过大，实体镶嵌单元的弹性模量和密度必须取较小的数值，在一定程度上影响混合模型的计算精度和稳定性。该混合模型能够避免有限单元法中的网格扭曲与纠缠等问题，同时兼顾计算效率。Macek 和 Silling[4] 采用所建立的混合模型模拟了拉伸荷载作用下含中心裂纹板的变形、刚性球形弹丸冲击韧性铝板和刚性子弹侵彻金属块等问题，Lall 等 [5,6] 也基于该混合模型研究了电子封装领域无铅合金焊接系统的高速单轴拉伸和冲击破坏问题。

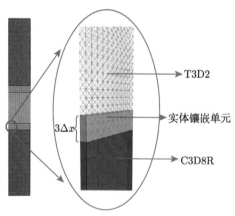

图 9-1　ABAQUS 中镶嵌单元示意图 [5,6]

9.1.2 位移协调约束与位移结合法

Macek 和 Silling[4] 指出可以令重叠区内近场动力学物质点与有限元结点满足位移协调条件，以实现近场动力学和有限元两种数值方法的混合建模。

如图 9-2 所示，在不连续变形和非局部效应显著的区域采用近场动力学模型，其他区域采用有限元模型，两类区域之间设重叠区。在重叠区内，通过有限元结点位移插值的方法确定近场动力学物质点的位移。若三维有限元子区域采用八结

点六面体单元，则有如下关系

$$\boldsymbol{u}_p = \sum_{i=1}^{8} N_i \boldsymbol{u}_i^{(e)} \tag{9.1}$$

其中，$\boldsymbol{u}_p$ 表示重叠区内物质点位移矢量，$\boldsymbol{u}_i$ 表示重叠区内有限元结点位移矢量，$\boldsymbol{u}_i^{(e)}$ 为第 e 个单元中 i 结点的位移，N_i 为形状函数。

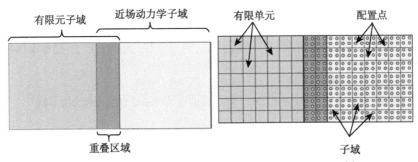

图 9-2 PD 子域、FEM 子域及重叠区域划分示意图 [7]

该混合模型通过形状函数插值的方法实现近场动力学与有限元法结合，重叠区内的物质点满足位移协调条件，无需人为设置重叠区内材料的弹性模量等参数，Kilic 和 Madenci[7] 采用该方法模拟了含中心圆孔矩形板的轴拉变形和破坏问题。但该方法对应的形状函数逆矩阵不存在，无法通过物质点的位移确定有限元结点位移，因此该混合模型只能在有限元子区域施加荷载，以实现有限元区域对近场动力学区域的影响。

9.1.3 力分解法

Liu 和 Hong[8−10] 在近场动力学子区域和有限元子区域间设置界面单元，以力平衡为基本原则，提出了力分解法，以实现近场动力学和有限元法的混合建模。该方法在界面单元内布置足够多的近场动力学物质点，计算界面单元内物质点的受力情况，进而通过形状函数插值，将物质点所受作用力通过两种不同方式分配到界面单元的结点上。当作用力被分解到界面单元的所有结点上时，称为体积 (VL) 分解法，如图 9-3(a) 所示；当作用力被分解到界面单元与近场动力学子域接触的结点上时，称为接触 (CT) 分解法，如图 9-3(b) 所示。

在三维 PD 与 FEM 混合建模中，若有限元子域采用八结点六面体单元，考虑界面单元内物质点受到作用力 $\boldsymbol{f}^{\mathrm{cp}}$，则对于 VL 分解法，采用形状函数插值将作用力分解到界面单元的所有结点上

$$\boldsymbol{f}_i^{\mathrm{cp}} = N_i \boldsymbol{f}^{\mathrm{cp}}, \quad i = 1, \cdots, 8 \tag{9.2}$$

而对于 CT 分解法，作用力被分解到仅与近场动力学子域接触的有限元结点上

$$\bm{f}_i^{\mathrm{cp}} = N_i \bm{f}^{\mathrm{cp}}, \quad i = 3, 4, 7, 8 \tag{9.3}$$

式中，N_i 为界面单元结点 i 的形状函数。

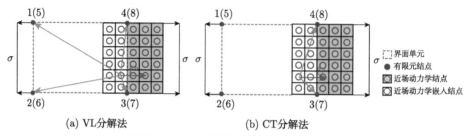

图 9-3　PD 子域与 FEM 子域的界面单元内物质点受力与力的分解

均布荷载作用下双边缺口试件的裂纹扩展模拟[9]显示，CT 分解法模拟结果与试验结果更加吻合。此外，力分解法的作用力计算和分解涉及形状函数，在已知界面单元的结点力求解近场动力学物质点的受力时，还涉及逆等参变换，计算过程较为复杂。

9.1.4　混合函数方法

Han[11,12]、Lubineau[13] 和 Azdoud[14,15] 等发展了一种基于局部和非局部理论混合建模的方法——混合函数方法 (morphing function method)。该方法同样将研究对象划分为局部模型作用区域和非局部模型作用区域，两者之间采用过渡区域连接，局部模型作用区域使用经典连续介质力学本构模型，非局部模型作用区域使用近场动力学模型。在过渡区域，假设应变变化平缓，引入 [0,1] 区间变化的混合函数，将传统局部应变能密度函数与近场动力学应变能函数混合，作为过渡区的应变能密度，进而推导出粗化的局部本构模型。基于 Galerkin 有限元方法实现数值计算，非局部模型作用区域采用非连续 Galerkin 非局部有限元，局部模型作用区域和过渡区采用传统连续 Galerkin 有限元。Seleson 等 [16,17] 利用这种建模方法，采用 [0, 1] 区间变化的混合函数，将经典局部 Navier 弹性方程与键型近场动力学非局部方程相统一，进行了混合方法的误差估计和分片测试。该混合建模方法满足牛顿第三定律，且不出现虚假力情况，但过渡区域选用不同的过渡函数将对计算结果有一定的影响。

9.1.5　子模型方法

PD 与 FEM 混合建模还可通过子模型方法 (submodeling approach) 予以实现。子模型方法可归结为 FEM-PD 顺序分析方法，先采用有限元法进行整体分析，

后采用近场动力学方法对重点区域进行分析。Agwai 等 [18] 和 Oterkus 等 [19-21] 发展了子模型方法，假设子模型区域内的结构细节对结构整体变形结果影响不显著，首先采用有限元方法或解析方法求得一定时间内所研究问题的整体位移解答，固定时间间隔输出一次解答；后通过在预期损伤区域周边设置截断边界，形成子模型，将时间相关的整体位移解答作为加载条件施加在子模型的截断边界上，再基于非局部方法进行损伤分析。

9.2 新的近场动力学与有限元混合模型

9.2.1 重叠模型与接触模型

本书作者团队提出了两类基于力耦合的近场动力学与有限元混合模型及其隐式分析方法，一种称为重叠模型，另一种称为接触模型。具体地，将计算结构划分为近场动力学子域和有限元子域进行离散，在裂纹等不连续现象出现的区域用近场动力学模拟，其他区域使用有限单元法。研究表明，提出的混合模型和求解方法既能有效解决裂纹扩展等不连续问题，又可提高计算效率，为工程结构破坏问题的计算分析提供一种有效方法。

在重叠模型中，有限元子域和近场动力学子域具有重叠区域，重叠区内的物质点不仅与其近场范围内的其他物质点发生作用，还与其所在单元上的有限元结点相互作用，如图 9-4 所示。需要指出的是，类似于镶嵌单元技术的刚度叠加，重叠区内单元结点同时受到单元内物质点与单元结点的刚度贡献，重叠区内的单元存在刚度过大的问题。在接触模型中，有限元子域与近场动力学子域间不再设置重叠区域，在两种子区域的交界面上，采用杆单元连接近场动力学物质点与有限元结点，其中界面上的有限元结点不仅与其所在单元的其他结点发生作用，还通过杆单元与以其为圆心、一定半径的球 (圆) 域的其他物质点相互作用，如图 9-5 所示。

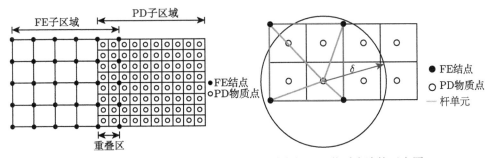

图 9-4　重叠模型子域划分以及 FEM 结点与 PD 物质点连接示意图

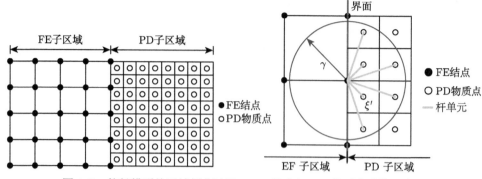

图 9-5　接触模型的子域划分以及 FEM 结点与 PD 物质点连接示意图

无论是重叠模型还是接触模型，均可基于统一的计算框架。采用 3.3.1 节给出的改进 PMB 模型进行分析，微模量函数为

$$c(\boldsymbol{\xi},\delta) = c(0,\delta)g(\boldsymbol{\xi},\delta) \tag{9.4}$$

其中，

$$g(\boldsymbol{\xi},\delta) = \begin{cases} \left(1 - \left(\dfrac{|\boldsymbol{\xi}|}{\delta}\right)^2\right)^2, & |\boldsymbol{\xi}| \leqslant \delta \\ 0, & |\boldsymbol{\xi}| > \delta \end{cases}$$

$$c(0,\delta) = \begin{cases} \dfrac{36E}{\pi\delta^4(1-2\nu)}, & \text{三维} \\ \dfrac{105E}{4\pi h\delta^3(1+\nu)(1-2\nu)}, & \text{平面应变} \\ \dfrac{105E}{4\pi h\delta^3(1-\nu)}, & \text{平面应力} \\ \dfrac{6E}{A\delta^2}, & \text{一维} \end{cases}$$

整体坐标下杆单元劲度矩阵为

$$\underset{\sim}{\boldsymbol{k}} = \dfrac{cV_iV_j}{|\boldsymbol{\xi}_{ji}|}\begin{bmatrix} l^2 & & & & & \\ lm & m^2 & & & \text{对称} & \\ ln & mn & n^2 & & & \\ -l^2 & -lm & -ln & l^2 & & \\ -lm & -m^2 & -mn & lm & m^2 & \\ -ln & -mn & -n^2 & ln & mn & n^2 \end{bmatrix} \tag{9.5}$$

式中，各符号的含义同式 (4.9)。

主要计算步骤如图 9-6 所示，主要包括以下流程。

(1) 读入研究对象的基本信息，包括离散后的物质点数据、有限元结点和单元数据、重叠单元 (或界面单元) 数据、边界条件和材料参数等；

(2) 根据物质点位置信息和给定的近场范围生成每个物质点与其近场范围内其他物质点间的键，计算物质点对和有限元单元劲度矩阵，并计算有连接作用的杆单元劲度矩阵，形成整体劲度矩阵 $\boldsymbol{K}$；

(3) 施加位移、速度和荷载等边界条件，形成整体等效荷载列阵 $\boldsymbol{F}$；

(4) 求解线性方程组 $\boldsymbol{Ku} = \boldsymbol{F}$，得到物质点和有限元结点的位移，计算物质点对间相对伸长率 s，若 $s > s_0$ 即键发生破坏，则更新整体劲度矩阵 $\boldsymbol{K}$，保持整体荷载 $\boldsymbol{F}$ 不变，重新求解线性方程组 $\boldsymbol{Ku} = \boldsymbol{F}$，不断迭代，直至结构达到平衡状态；

(5) 输出计算结果，包括结点和物质点当前位置、位移和损伤情况等，最后采用 ENSIGHT 等后处理软件实现计算结果的可视化。

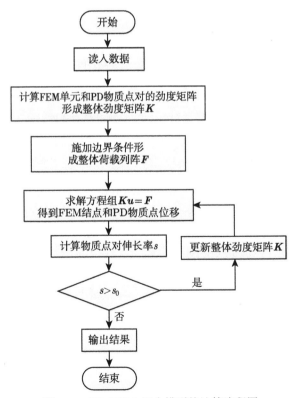

图 9-6　PD-FEM 混合模型的计算流程图

9.2.2 重叠模型的定量分析

如图 9-7 所示,以二维弹性杆轴拉变形问题为例,详细讨论重叠模型中重叠区大小和重叠区近场动力学物质点密度等参数对混合模型计算精度和稳定性的影响。材料弹性模量为 100 GPa,泊松比为 1/3,左端固定,右端施加水平拉伸荷载,$P = 1050$ kN/m。图 9-8 给出了重叠模型子域划分和接触模型子域划分,近场动力学子域均匀离散,物质点间距为 2.5mm×2.5mm,有限元网格采用 10mm×10mm 的四结点等参单元。同时采用 ANSYS 有限元软件求得 FEM 解答 (采用 10mm×10mm 的 SOLID 四边形单元) 作为验证。

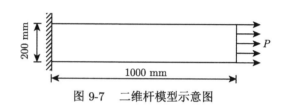

图 9-7 二维杆模型示意图

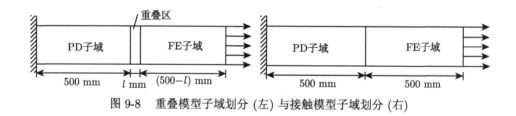

图 9-8 重叠模型子域划分 (左) 与接触模型子域划分 (右)

9.2.2.1 重叠区域的尺寸

取重叠区域尺寸 $l = 10$mm、30mm、60mm,图 9-9 给出了重叠区取不同值时计算得到的杆件水平位移结果 (右图为局部放大的重叠区计算结果)。由计算结果可知,重叠区越大,最大水平位移越小,这是由于重叠区的结点同时受到单元结点与单元内物质点的刚度贡献,当 $l = 10$mm 即重叠区仅为一列单元时,重叠区对整体计算结果的影响可以忽略,重叠区内位移结果偏移也较小,能够保证计算精度要求。需要说明的是,可以通过设置重叠区内单元的弹性模量或物质点微观模量为极小值,使重叠区的弹性模量近似等于材料的弹性模量,以消除重叠区大小对计算精度的影响,但此时必须重新考虑杆单元刚度对计算精度的影响,不利于重叠模型的实际应用,故建议取重叠区为一层单元。

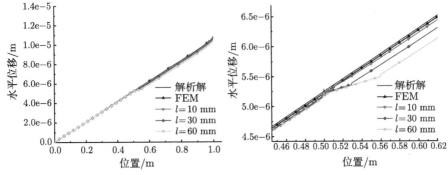

图 9-9 杆件的水平位移曲线 (包括重叠区域的局部放大图)

9.2.2.2 物质点的间距

本小节研究近场动力学中物质点间距对计算精度的影响,有限元网格仍采用 10mm×10mm 的四结点等参单元,物质点间距取 $\Delta x = 5.0$mm、$\Delta x = 2.5$mm 和 $\Delta x = 2.0$mm,则重叠区内一个单元含 4 个、16 个和 25 个物质点。图 9-10 给出了不同物质点间距时计算得到的杆件水平位移曲线。由计算结果可以看出,物质点间距 (近场动力学子域物质点点阵密度) 对重叠模型的计算精度影响较小,当物质点间距取 $\Delta x = 5.0$mm,即重叠区单元含 4 个物质点时,重叠区位移的结果偏移稍大;当物质点间距 $\Delta x \leqslant 2.5$mm 时,计算结果稳定,且精度较高。

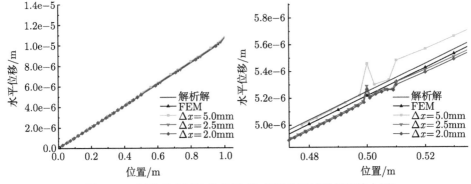

图 9-10 杆件的水平位移曲线 (包括重叠区域的局部放大图)

综上,在使用近场动力学与有限单元法的重叠模型进行计算时,建议取重叠区为一层单元,近场动力学物质点的间距取为有限元单元尺寸的 1/5~1/3。

9.2.3 接触模型的定量分析

9.2.3.1 界面接触点的作用范围

采用接触模型进行有限单元法与近场动力学方法混合建模时，界面上的有限元结点不仅与其所在单元的其他结点发生作用，还通过杆单元与以其为圆心、γ 为半径的球 (圆) 域的其他物质点相互作用，这个 γ 可称为界面接触点的作用范围，是计算时人为选取的参数。

仍以二维弹性杆轴拉变形问题为例，探讨作用范围 γ 对混合模型计算精度和稳定性的影响，取作用范围 $\gamma = 0.25\delta$、$\gamma = 0.50\delta$、$\gamma = 0.75\delta$ 和 $\gamma = 1.0\delta$。图 9-11 给出了基于接触模型计算得到的杆件水平位移曲线。由计算结果可知，当 $\gamma = 0.25\delta$ 时，相对误差最大为 4.38%，且水平位移在界面上有较大幅度的偏移，表明作用范围过小不能取得较好的计算结果；当 $\gamma \geqslant 0.50\delta$ 时，接触模型的计算精度与有限单元法的计算精度相当，且计算结果在界面上偏移较小。

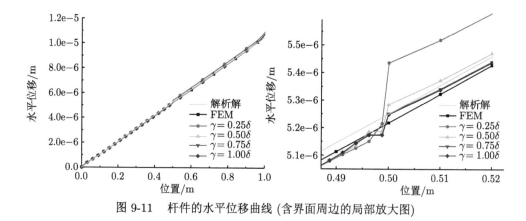

图 9-11　杆件的水平位移曲线 (含界面周边的局部放大图)

9.2.3.2 物质点的间距

本小节考察近场动力学中物质点间距对接触模型计算精度的影响。有限元网格仍采用 10mm×10mm 的四结点等参单元，物质点间距取 $\Delta x = 5.0$mm、$\Delta x = 2.5$mm、$\Delta x = 2.0$mm。图 9-12 为接触模型计算得到的杆件水平位移曲线。计算结果表明，当物质点间距取 $\Delta x = 5.0$mm 时，界面上的水平位移偏移较大；当物质点间距 $\Delta x \leqslant 2.5$mm 时，计算结果稳定，水平位移偏移较小，随着物质点尺寸的减小 (近场动力学子域物质点点阵密度增加)，计算稳定性随之增加。

综上，在使用近场动力学与有限单元法的接触模型进行计算时，建议取界面接触点的作用范围 $\gamma = \delta$，近场动力学中物质点的间距取为有限元单元尺寸的 1/5~1/3。

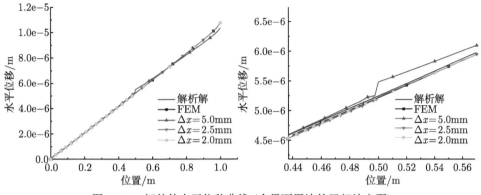

图 9-12 杆件的水平位移曲线 (含界面周边的局部放大图)

9.3 数值算例

基于上述混合模型计算参数的定量分析, 本节应用所建立的两种近场动力学与有限单元法混合模型, 模拟计算二维悬臂梁和简支梁的静力变形、含 I 型和 I-II 复合型裂纹板的裂纹扩展、多裂纹和含切口三点弯曲梁的裂纹扩展等问题, 以验证所建立的混合模型的有效性。

9.3.1 悬臂梁在端部受集中力作用

如图 9-13 所示的悬臂梁, 在端部受集中荷载 $P = 100$ kN 的作用, 材料的弹性模量为 100 GPa、泊松比为 1/3。近场动力学子域均匀离散, 物质点尺寸为 2.5 mm×2.5 mm, 取近场范围 $\delta = 10$ mm; 有限元子域采用 10 mm×10 mm 的四结点等参单元离散。除采用所建立的两种近场动力学与有限单元法混合模型进行计算外, 还基于 ANSYS 软件 (采用 10 mm×10 mm 的 SOLID 四边形单元) 获得有限元解答。在重叠模型中, 重叠区 $l = 10$ mm; 在接触模型中, 界面接触点的作用范围 $\gamma = \delta$。

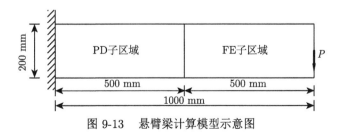

图 9-13 悬臂梁计算模型示意图

表 9-1 给出了不同方法得到的加载点竖向位移计算结果, 图 9-14 为悬臂梁中

性轴的竖向位移曲线。由计算结果可以看出，两种近场动力学与有限单元法混合模型求得的悬臂梁中性轴竖向位移与理论解吻合很好、误差较小。其中，重叠模型计算的加载点竖向位移与解析解的相对误差为 0.1%，接触模型的计算结果与解析解的相对误差为 1.08%，本例中接触模型的精度虽略低于重叠模型，但仍与有限元解的精度相当。

表 9-1 加载点竖向位移计算结果

	解析解	有限元	重叠模型	接触模型
最大位移/m	5.200e-5	5.151e-4	5.195e-4	5.144e-4
相对误差/%	—	0.94	0.10	1.08
计算耗时/s	—	8	40	36

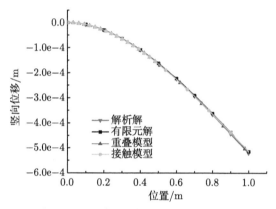

图 9-14 悬臂梁中性轴的竖向位移曲线

9.3.2 简支梁在跨中受集中力作用

如图 9-15 所示的二维简支梁模型，材料的弹性模量为 30 GPa，泊松比为 1/3，该模型被划分为两个有限元子域和一个近场动力学子域，集中力 $P = 10$ kN 施加于 PD 子域，取物质点间距 $\Delta x = 2.5$ mm，近场范围 $\delta = 4\Delta x = 10$ mm，有限元网格采用 10mm×10mm 的四结点等参单元。在重叠模型中，重叠区 $l = 10$ mm，在接触模型中，界面接触点的作用范围 $\gamma = \delta$。

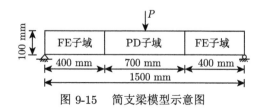

图 9-15 简支梁模型示意图

表 9-2 为不同方法计算得到的中性轴跨中挠度,图 9-16 给出了两种混合模型计算的简支梁中性轴竖向位移与解析解、有限单元法结果的对比。由计算结果可见,重叠模型计算的中性轴跨中挠度为 -2.858×10^{-4} m,与解析解的相对误差为 0.1%;接触模型的计算结果为 -2.852×10^{-4} m,与解析解的相对误差为 0.31%。两种混合模型均具有较高精度。

表 9-2 中性轴跨中挠度的计算结果

	解析解	有限元	重叠模型	接触模型
跨中挠度/m	-2.861×10^{-4}	-2.857×10^{-4}	-2.858×10^{-4}	-2.852×10^{-4}
相对误差/%	—	0.14	0.10	0.31
计算耗时/s	—	7	34	27

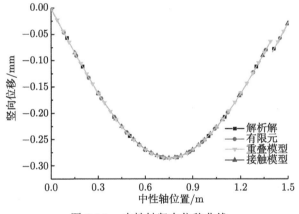

图 9-16 中性轴竖向位移曲线

需要说明的是,在弹性计算中,虽然近场动力学方法的计算效率低于有限单元法,但近场动力学主要用于有限元法难以求解的材料和结构的破坏问题,采用本节提出的近场动力学与有限元混合模型,其精度与有限元相当,计算效率也得到了较大程度的提高,为近场动力学方法应用于工程结构破坏问题的计算分析提供了一种有效途径。

9.3.3 含 I 型裂纹板的裂纹扩展分析

I 型裂纹又称为张开型裂纹,裂纹表面位移的方向垂直于裂纹扩展方向,I 型裂纹扩展是工程结构常见的破坏模式。如图 9-17 所示的中心含单裂纹的正方形平板,裂纹长度 $a = 10$ mm,材料的弹性模量为 30 GPa,泊松比为 1/3,临界伸长率 $s_0 = 8\times 10^{-4}$,为平面应力问题。将该平板划分为两个有限元子域和一个近场动力学子域,取物质点间距 $\Delta x = 0.5$ mm,近场范围 $\delta = 4\Delta x$,有限元网格采用 2 mm×2 mm 的四结点等参单元。加载方式为位移加载，每一荷载步增量为 2×10^{-8} m。重叠模型中重叠区 $l=2$ mm，在接触模型中，界面接触点的作用范围 $\gamma=\delta$。

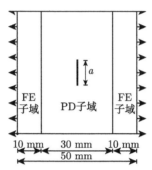

图 9-17　预制裂缝板几何尺寸及子域划分

采用重叠模型计算时，加载到第 256 步 (位移加载值为 5.12×10^{-6} m) 时，平板预制裂纹裂尖出现损伤，此时方板的损伤、水平位移和竖向位移如图 9-18 所示，水平方向位移关于预制裂纹方向对称，重叠区部分的位移光滑平顺，没有波动或突变。图 9-19 给出了重叠模型计算得到的方板裂纹扩展过程，裂尖出现损伤后继续加载，损伤不断累积，当损伤大于 0.40 时，宏观裂纹出现并开始扩展 (第 280 步)，继续加载可以观察到裂纹的扩展和贯穿。

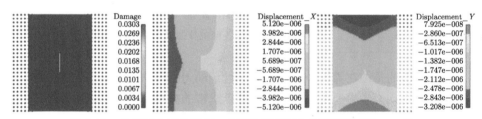

图 9-18　起裂时方板的损伤、水平位移和竖向位移云图 (第 256 步)

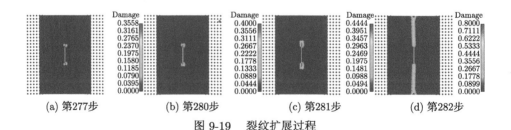

图 9-19　裂纹扩展过程

9.3 数值算例

采用接触模型模拟裂纹扩展，当加载到第 266 步 (位移加载值为 5.32×10^{-6} m) 时，平板预制裂纹裂尖出现损伤，此时方板的损伤、水平位移和竖向位移云图如图 9-20 所示。水平方向位移关于预制裂纹方向对称，有限元子域与近场动力学子域界面部分的位移光滑平顺。图 9-21 给出了方板的裂纹扩展过程，图 9-21(a) 为裂纹萌生阶段，继续加载裂尖损伤不断增加，当加载到第 289 步 (位移加载值为 5.78×10^{-6} m) 时损伤达到 0.40，宏观裂纹出现并开始扩展，加载到第 292 步 (位移加载值为 5.84×10^{-6} m) 时裂纹贯穿平板，结构破坏。值得注意的是，接触模型计算得到的裂纹萌生、扩展和贯穿各阶段的加载步均晚于重叠模型的计算结果，且大致为 10 步，这可能是由于重叠模型重叠区的单元刚度大于材料真实的单元刚度。

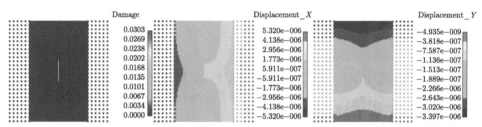

图 9-20 起裂时方板的损伤、水平位移和竖向位移云图 (第 266 步)

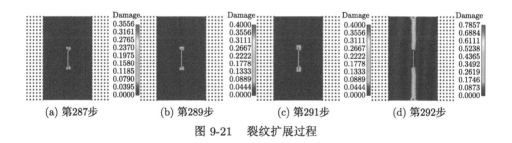

图 9-21 裂纹扩展过程

9.3.4 含 I-II 复合型裂纹板的裂纹扩展分析

I-II 复合型裂纹扩展一直是理论分析、数值模拟和试验研究关注的问题，也是工程结构常见的破坏形式。如图 9-22 所示的中心含单斜裂纹的方形平板，裂纹长度 $a=10$ mm，材料的弹性模量为 30 GPa，泊松比为 1/3，临界伸长率 $s_0 = 8\times10^{-4}$，为平面应力问题。将该平板划分为两个有限元子域和一个近场动力学子域，取物质点间距 $\Delta x = 0.5$ mm，近场范围 $\delta = 4\Delta x$，有限元网格采用 2 mm×2 mm 的四结点等参单元。加载方式为位移加载，每一荷载步增量为 2×10^{-8} m。重叠模型中，重叠区 $l=2$ mm；在接触模型中，界面接触点的作用范围 $\gamma = \delta$。

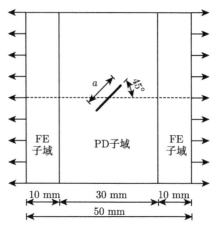

图 9-22 预制裂缝板几何尺寸及子域划分

采用重叠模型计算时，当加载到第 276 步 (位移加载值为 5.52×10^{-6} m) 时预制裂纹裂尖出现损伤，此时方板的损伤、水平位移和竖向位移云图如图 9-23 所示，重叠区部分的位移光滑平顺，没有波动或突变。图 9-24 给出了方板的裂纹扩展过程，裂尖损伤不断增加，当损伤大于 0.40 时，宏观裂纹出现并开始扩展，该裂纹为 I-II 复合型裂纹，重叠模型计算得到的裂纹扩展角度约为 53°，与最大周向应力准则预测的 53.13° [87] 基本一致，加载到第 315 步 (位移加载值为 6.30e−6m) 时裂纹贯穿平板。

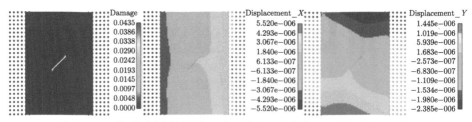

图 9-23 起裂时方板的损伤、水平位移和竖向位移云图 (第 276 步)

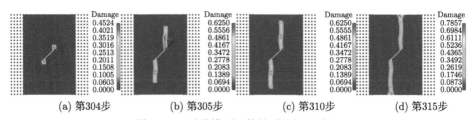

(a) 第304步　　(b) 第305步　　(c) 第310步　　(d) 第315步

图 9-24 重叠模型计算的裂纹扩展过程

采用接触模型模拟裂纹扩展时,加载到第 284 步 (位移加载值为 5.68×10^{-6} m) 时,预制裂纹裂尖出现损伤,此时方板的损伤、水平位移和竖向位移云图如图 9-25 所示,有限元子域与近场动力学子域界面部分的位移光滑平顺。图 9-26 给出了方板的裂纹扩展过程,图 9-26(a) 为裂纹萌生阶段,荷载继续增大,裂尖损伤不断累积,当损伤大于 0.40 时,宏观裂纹出现并开始迅速扩展。接触模型计算得到的裂纹扩展角度也为 53°,裂纹扩展路径也与重叠模型的结果基本一致,但接触模型计算得到的裂纹萌生、扩展和贯穿时刻仍稍晚于重叠模型的结果。

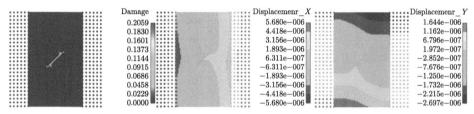

图 9-25　起裂时方板的损伤、水平位移和竖向位移云图 (第 284 步)

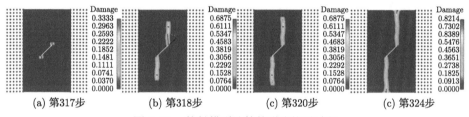

图 9-26　接触模型计算的裂纹扩展过程

9.3.5　多裂纹扩展分析

图 9-27 所示的含双裂纹的正方形平板,裂纹倾角均为 45°,初始裂纹长度 a = 10 mm,间距 b = 10 mm。材料的弹性模量为 30 GPa,泊松比为 1/3,临界伸长率 $s_0 = 8 \times 10^{-4}$,为平面应力问题。将该平板划分为两个有限元子域和一个近场动力学子域,取物质点间距 $\Delta x = 0.5$ mm,近场范围 $\delta = 4\Delta x$,有限元网格采用 2 mm×2 mm 的四结点等参单元。加载方式为位移加载,每一荷载步增量为 2×10^{-8} m。重叠模型中,重叠区 $l = 2$ mm;在接触模型中,界面接触点的作用范围 $\gamma = \delta$。

基于重叠模型计算得到的裂纹扩展过程如图 9-28(a)~(c) 所示,加载到第 279 步 (位移加载值为 5.58×10^{-6} m) 时,预制裂纹外侧裂尖首先出现损伤,内侧裂尖随后出现损伤,裂纹扩展后预制裂纹内侧出现裂纹交汇现象,加载到第 302 步 (位移加载值为 6.04×10^{-6} m) 时,裂纹贯穿平板,试件破坏。

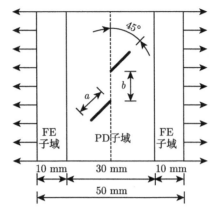

图 9-27　预制裂纹方板几何尺寸及子域划分

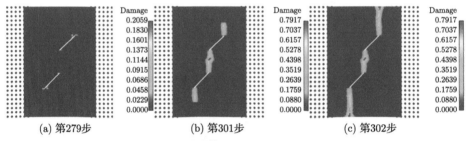

(a) 第279步　　　　　(b) 第301步　　　　　(c) 第302步

图 9-28　重叠模型计算的裂纹扩展过程

基于接触模型计算得到的裂纹扩展过程如图 9-29(a)~(c) 所示,加载到第 288 步 (位移加载值为 5.76×10^{-6} m) 时,预制裂纹外侧裂尖首先出现损伤,内侧裂尖随后出现损伤,裂纹扩展后预制裂纹内侧出现裂纹交汇现象,接触模型计算得到的裂纹交汇形态与重叠模型计算结果一致,加载到第 315 步 (位移加载值为 6.30×10^{-6} m) 时,裂纹贯穿平板,试件破坏。图 9-30 为多维虚内键 (VMIB) 法 [22] 的计算

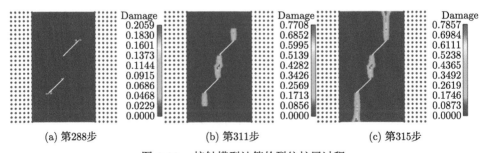

(a) 第288步　　　　　(b) 第311步　　　　　(c) 第315步

图 9-29　接触模型计算的裂纹扩展过程

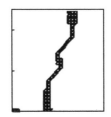

图 9-30 多维虚内键法计算的裂纹路径[22]

结果，比较后不难看出，基于重叠模型和接触模型计算得到的裂纹扩展路径均与多维虚内键法的计算结果相一致。

9.3.6 含切口三点弯曲梁的裂纹扩展分析

在平面应力条件下，研究含切口三点弯曲梁在跨中集中静力载荷作用下的 I 型断裂问题，计算模型如图 9-31 所示。试件长 $L = 228$ mm，高 $D = 76$ mm，跨中切口深度为 $a = 19$ mm。材料杨氏模量为 31.4 GPa，能量释放率为 $G_0 = 31.8$ J/m^2，泊松比为 1/3，通过位移加载方式实现外荷载 F 的施加。模型被划分为两个有限元子域和一个近场动力学子域，近场动力学物质点间距为 $\Delta x = 1$ mm，近场范围 $\delta = 4\Delta x = 4$ mm。重叠模型中，重叠区 $l = 4$ mm；在接触模型中，界面接触点的作用范围 $\gamma = \delta$。有限元单元法计算时采用 4 mm×4 mm 的四结点等参单元。

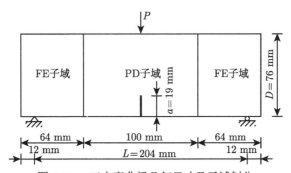

图 9-31 三点弯曲梁几何尺寸及子域划分

重叠模型计算得到的三点弯曲梁破坏过程如图 9-32 所示，当荷载增加到 76.3kN 时，切口位置出现损伤，结构由弹性变形发展到开裂破坏，随着裂纹的不断扩展，结构承载能力降低。接触模型计算得到的三点弯曲梁破坏过程如图 9-33 所示，当荷载达到 74.7kN 时，切口位置出现损伤，随着裂纹的不断扩展，结构承载能力降低。接触模型计算得到的最大荷载略小于重叠模型的计算结果。

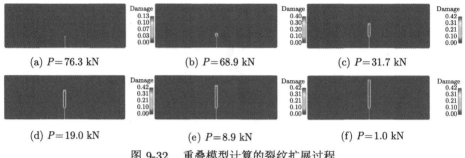

图 9-32 重叠模型计算的裂纹扩展过程

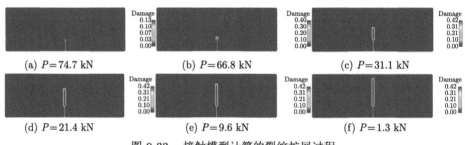

图 9-33 接触模型计算的裂纹扩展过程

含切口三点弯曲梁的线弹性断裂力学解答[23−25]为

$$K_{\mathrm{I}} = \sigma\sqrt{\pi a} \cdot F(\alpha) \tag{9.6}$$

$$\mathrm{CMOD} = \frac{4\sigma a}{E} \cdot V_1(\alpha) \tag{9.7}$$

其中

$$\sigma = \frac{3PL}{2D^2} \tag{9.8}$$

$$F(\alpha) = \frac{1.99 - \alpha(1-\alpha)(2.15 - 3.93\alpha + 2.7\alpha^2)}{\sqrt{\pi}(1+2\alpha)(1-\alpha)^{3/2}} \tag{9.9}$$

$$V_1(\alpha) = 0.76 - 2.28\alpha + 3.87\alpha^2 - 2.04\alpha^3 + \frac{0.66}{(1-\alpha)^2} \tag{9.10}$$

式中，P、L、D、a 的意义如图 9-31 所示，$\alpha = a/D$，CMOD (crack-mouth-opening displacement) 为裂纹口张开位移。由此，重叠模型计算的最大荷载为 76.3 kN，应力强度因子 $K_{\mathrm{I}} = 0.99$ MPa·m$^{1/2}$，接触模型计算的最大荷载为 74.7 kN，应力强度因子 $K_{\mathrm{I}} = 0.97$ MPa·m$^{1/2}$，与试验结果基本一致。

图 9-34 给出两种混合模型计算得到的 P-CMOD 曲线与有限单元法模拟结果和线弹性断裂力学解答的对比，两种混合模型的模拟结果与有限单元法模拟结果基本一致，但在构件未出现裂纹时，接触模型计算的裂纹口张开位移与线弹性断裂力学解答更加接近。

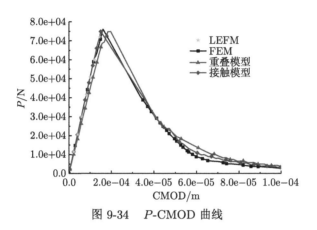

图 9-34 P-CMOD 曲线

上述算例结果表明，所提出的两种近场动力学与有限单元法混合模型在预测裂纹扩展路径、临界应力强度因子和 P-CMOD 曲线等方面均表现良好，能够较为准确地模拟固体材料和结构的渐进破坏全过程 (裂纹的萌生、扩展、汇合等)，且能够有效提高计算效率。需要注意的是，重叠模型由于重叠区内材料参数大于真实材料参数，在模拟裂纹扩展等不连续变形问题时，得到的裂纹扩展路径、角度等结果与接触模型计算结果一致，但重叠模型所预测的裂纹萌生、贯穿等时刻与接触模型不尽一致，采用混合模型进行定量分析时，接触模型计算精度更高。

9.4 近场动力学有限元混合模型在重力坝稳定性分析中的应用

9.4.1 典型重力坝的变形计算和分析

某典型混凝土重力坝尺寸和计算子区域划分如图 9-35 所示，坝体弹性模量为 20 GPa，密度为 2400 kg/m^3，抗拉强度为 1.57 MPa；地基弹性模型为 40 GPa，密度为 2690 kg/m^3，抗拉强度为 3.8 MPa。由于地基的沉陷早已完成，故地基部分只计算刚度而不计算自重。在计算模型中，地基底部采用固定约束，地基两侧采用法向链杆约束，PD 子域的物质点间距 $\Delta x = 1$ m，近场范围 $\delta = 4\Delta x$，有限元网格采用 4 m×4 m 的四结点等参单元。分别采用 PD-FEM 两种混合模型对空库和满库情况进行了计算，并应用 ANSYS 软件获得有限元解答。

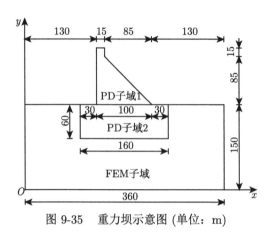

图 9-35 重力坝示意图 (单位：m)

图 9-36 和图 9-37 分别给出了空库时不同方法计算得到的大坝–基础体系的变形情况。从计算结果可以看出，空库时仅考虑坝体的自重，坝体的水平位移倾向上游，竖向位移向下沉降，符合其受力特征和客观认识。有限元计算的坝顶水平位移为 8.591 mm，竖向位移为 10.02 mm；PD-FEM 重叠模型计算的坝顶水平位移为 8.646 mm，竖向位移为 10.05 mm，与有限元结果的相对差异分别为 0.6%和 0.3%；PD-FEM 接触模型计算的坝顶水平位移为 8.663 mm，竖向位移为 10.07 mm，与有限元结果的相对差异分别为 0.8%和 0.5%。

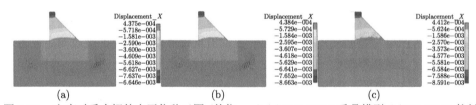

图 9-36 空库时重力坝的水平位移云图 (单位：m) (a) PD-FEM 重叠模型 (b)PD-FEM 接触模型 (c) FEM 计算结果

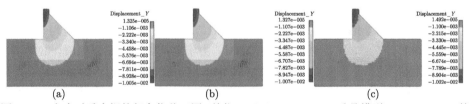

图 9-37 空库时重力坝的竖向位移云图 (单位：m) (a) PD-FEM 重叠模型 (b) PD-FEM 接触模型 (c) FEM 计算结果

在满库情况下，大坝和地基体系除受到坝体自重作用外，还受到水压的作用，图 9-38 和图 9-39 为满库时不同方法计算得到的变形情况。由此可见，由于水压的作用，坝体的水平位移倾向下游，从坝体上部至下部逐渐减小，竖向位移还是向下沉降，也是从坝体上部至下部逐渐减小，符合其受力特征和客观认识。有限元计算的坝顶水平位移为 6.554 mm，竖向位移为 5.606 mm；PD-FEM 重叠模型计算的坝顶水平位移为 6.741 mm，竖向位移为 5.660 mm，与有限元结果的相对差异分别为 2.9% 和 1.0%；PD-FEM 接触模型计算得到的坝顶水平位移为 6.792 mm，竖向位移为 5.611 mm，与有限元结果的相对差异分别为 3.6% 和 0.1%。综合来看，不论是空库还是满库情况，两种近场动力学与有限元混合模型的计算结果与有限元结果具有很好的一致性。

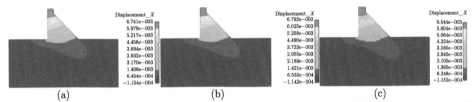

图 9-38　满库时重力坝的水平位移云图 (单位：m) (a) PD-FEM 重叠模型 (b) PD-FEM 接触模型 (c) FEM 计算结果

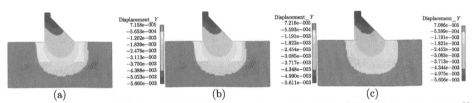

图 9-39　满库时重力坝的竖向位移云图 (单位：m) (a) PD-FEM 重叠模型 (b) PD-FEM 接触模型 (c) FEM 计算结果

9.4.2　典型重力坝的承载力评价

常采用超载法评价大坝的承载能力，超载法认为在某些条件下坝体承受的荷载超过了设计值后坝体或地基发生破坏，破坏时对应的外荷载 P' 与设计荷载 P 的比值为 K，称为超载安全系数

$$K = \frac{P'}{P} \tag{9.11}$$

超载安全系数是评价大坝整体承载能力的重要指标，超载法常采用水容重超载法，该方法保持坝体和地基材料的材料参数不变，逐渐增大水的容重，直至坝

体或地基发生破坏,故超载安全系数也可以表示为

$$K = \frac{P'}{P} = \frac{\gamma'}{\gamma} \tag{9.12}$$

式中,γ',γ 分别为破坏荷载和设计荷载对应的液体容重。

基于所建立的近场动力学与有限元混合模型,采用水容重超载法对上述典型重力坝进行破坏模拟。图 9-40 为 PD-FEM 重叠模型的计算结果,当超载系数 K 达到 1.2 时,坝踵处出现损伤,当超载系数 K 达到 1.8 时,坝踵处出现宏观裂纹并开始扩展,当超载系数 K 达到 2.2 时,坝踵处裂纹向地基继续扩展,同时坝址处也出现了损伤,当超载系数 K 达到 2.8 时,坝踵和坝址处出现大面积损伤破坏,继续加载计算不收敛,表明大坝已失稳,故可认为该典型重力坝的超载安全系数为 2.8。类似地,还采用 PD-FEM 接触模型对该典型重力坝进行超载计算,破坏过程如图 9-41 所示,超载安全系数也为 2.8,与重叠模型的计算结果一致。

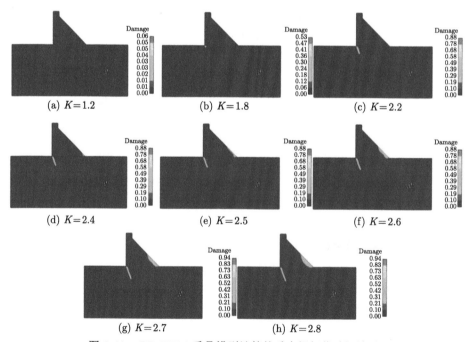

图 9-40 PD-FEM 重叠模型计算的重力坝超载破坏结果

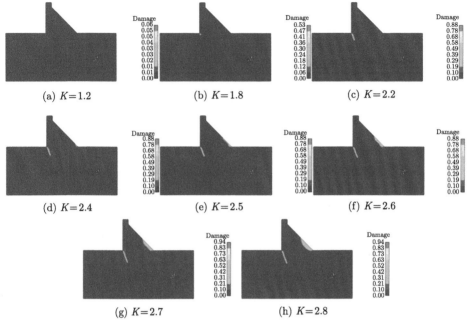

图 9-41 PD-FEM 接触模型计算的重力坝超载破坏结果

参 考 文 献

[1] 章青, 郁杨天, 顾鑫. 近场动力学与有限元的混合建模方法 [J]. 计算力学学报, 2016, 33(4): 441-448.

[2] 郁杨天, 章青, 顾鑫. 近场动力学与有限单元法的混合模型与隐式求解格式 [J]. 浙江大学学报 (工学版), 2017, 51(7): 1324-1330.

[3] Shen F, Yu Y T, Zhang Q, et al. Hybrid model of peridynamics and finite element method for static elastic deformation and brittle fracture analysis[J]. Engineering Analysis with Boundary Elements, 2020, 113: 17-25.

[4] Macek R W, Silling S A. Peridynamics via finite element analysis[J]. Finite Elements in Analysis and Design, 2007, 43(15): 1169-1178.

[5] Lall P, Shantaram S, Panchagade D. Peridynamic-models using finite elements for shock and vibration reliability of leadfree electronics[C]. IEEE Intersociety Conference on Thermal & Thermomechanical Phenomena in Electronic Systems, 2010: 1-12.

[6] Lall P, Shantaram S, Kulkarni M, et al. High strain-rate mechanical properties of SnAgCu leadfree alloys[C]. 2011 IEEE 61st Electronic Components and Technology Conference (ECTC), 2011: 684-700.

[7] Kilic B, Madenci E. Coupling of peridynamic theory and the finite element method[J]. Journal of Mechanics of Materials and Structures, 2010, 5(5): 707-733.

[8] Liu W Y. Discretized bond-based peridynamics for solid mechanics[D]. East Lansing: Michigan State University, 2012.

[9] Liu W Y, Hong J W. A coupling approach of discretized peridynamics with finite element method[J]. Computer Methods in Applied Mechanics and Engineering, 2012, s245-246: 163-175.

[10] Lee J, Liu W Y, Hong J W. Impact fracture analysis enhanced by contact of peridynamic and finite element formulations[J]. International Journal of Impact Engineering, 2016, 87: 108-119.

[11] Han F, Lubineau G. Coupling of nonlocal and local continuum models by the Arlequin approach[J]. International Journal for Numerical Methods in Engineering, 2012, 89(6): 671-685.

[12] Han F, Lubineau G, Azdoud Y, et al. A morphing approach to couple state-based peridynamics with classical continuum mechanics[J]. Computer Methods in Applied Mechanics and Engineering, 2016, 301: 336-358.

[13] Lubineau G, Azdoud Y, Han F, et al. A morphing strategy to couple non-local to local continuum mechanics[J]. Journal of the Mechanics and Physics of Solids, 2012, 60(6): 1088-1102.

[14] Azdoud Y, Han F, Lubineau G. A morphing framework to couple non-local and local anisotropic continua[J]. International Journal of Solids and Structures, 2013, 50(9): 1332-1341.

[15] Azdoud Y, Han F, Lubineau G. The morphing method as a flexible tool for adaptive local/non-local simulation of static fracture[J]. Computational Mechanics, 2014, 54(3): 711-722.

[16] Seleson P, Beneddine S, Prudhomme S. A force-based coupling scheme for peridynamics and classical elasticity[J]. Computational Materials Science, 2013, 66: 34-49.

[17] Seleson P, Ha Y D, Beneddine S. Concurrent coupling of bond-based peridynamics and the navier equation of classical elasticity by blending[J]. International Journal for Multiscale Computational Engineering, 2015, 13(2): 91-113.

[18] Agwai A, Guven I, Madenci E. Drop-shock failure prediction in electronic packages by using peridynamic theory[J]. IEEE Transactions on Components Packaging & Manufacturing Technology, 2012, 2(2): 439-447.

[19] Oterkus E. Peridynamic theory for modeling three-dimensional damage growth in metallic and composite structures[D]. Tucson: The University of Arizona, 2010.

[20] Oterkus E, Barut A, Madenci E. Damage growth prediction from loaded composite fastener holes by using peridynamic theory[C]. 51st AIAA/ASME/ASCE/AHS/ASC Structures, Structural Dynamics, and Materials Conference 18th AIAA/ASME/AHS Adaptive Structures Conference 12th, 2010.

[21] Oterkus E, Madenci E, Weckner O, et al. Combined finite element and peridynamic analyses for predicting failure in a stiffened composite curved panel with a central slot[J]. Composite Structures, 2012, 94(3): 839-850.

[22] 张振南, 陈永泉. 基于 VMIB 多裂纹岩石材料拉伸破坏数值模拟 [J]. 岩土工程学报, 2008, 30(10): 1490-1495.

[23] Jenq Y S, Shah S P. A fracture toughness criterion for concrete[J]. Engineering Fracture Mechanics, 1985, 21(5): 1055-1069.

[24] Jenq Y S, Shah S P. Mixed-mode fracture of concrete[J]. International Journal of Fracture, 1988, 38(2): 123-142.

[25] John R, Shah S P. Mixed-mode fracture of concrete subjected to impact loading[J]. Journal of Structural Engineering, 1990, 116(3): 585-602.

第 10 章　非均匀离散的近场动力学模型与自适应分析

空间离散和积分技术是近场动力学数值计算的基础，目前的研究大多采用均匀正交网格或点阵进行空间离散，并辅之以定常的近场范围 [1,2]，其原因在于这种离散方式和数值计算技术的算法简单，前处理便捷、便于编程实现，且计算误差也容易控制。但近场动力学自身并不要求空间均匀离散，且均匀离散技术在处理具有非规则曲线或曲面的结构体时存在困难。此外，在相同均匀离散条件下，近场动力学的非局部特性 [3] 决定了其计算效率低于传统有限元等方法 [4]，为获得高精度数值解答，均匀离散技术要求全域采用高密度网格或点阵，将导致计算量的急剧增加，计算效率较低。基于上述分析，亟须发展基于非均匀离散和变尺寸近场范围的近场动力学计算技术。本章致力于发展基于 Voronoi 结构图均匀或非均匀离散的近场动力学分析方法以及近场动力学自适应分析方法，并通过二维板弹性变形、二维弹性波传播、二维和三维动力裂纹扩展等典型算例，详细介绍其中的计算细节，验证近场动力学非均匀离散与自适应分析方法的正确性和适用性。

10.1　近场动力学的空间离散方式与自适应分析

自近场动力学问世至今，研究者大多采用均匀正交网格或点阵进行空间离散，并辅之以定常近场范围 [1,2] 开展计算，最近的近场动力学自适应分析文献也多是基于该技术 [5-8]。基于均匀离散，Bobaru 等 [5,6] 首先分析了近场动力学方法的三种收敛性问题：m-收敛、δ-收敛和 δm-收敛，其中 $m = \delta/\Delta x$，δ 为近场范围，Δx 为物质点间距。当 m 值趋于零时，非局部近场动力学解答收敛于传统局部解答 [9]；对于给定近场范围，m 收敛性意味着近场动力学数值解答收敛于近场动力学精确解答。此外，Gerstle 等 [10] 和 Merwe [11] 采用正六边形和正三角形规则分布网格进行结构弹性变形和破坏模拟。除上述规则网格或点阵离散外，Henke 和 Shanbhag [12] 还基于 Voronoi 结构图进行了冲击破坏模拟，并进行了误差分析。Ren 等 [13] 采用非均匀分布的二维三角形和三维四面体网格离散结构体，结合对偶近场范围的概念开展数值计算。Hu 等 [14] 发展了基于非均匀三角形网格离散的近场动力学分析方法。本书作者等 [15] 基于二维和三维 Voronoi 结构图，发展了基于 Voronoi 结构图均匀和非均匀离散的近场动力学分析方法与自适应分析方法。

已有研究指出，近场动力学计算结果具有一定的网格依赖性或网格敏感性 (grid dependent, or mesh dependent, or mesh sensitive)[11,12]，但该观点值得深入探讨。Freimanis 等 [16] 通过改变近场范围和物质点尺寸，研究了近场动力学方法的近场范围依赖性，指出在 $\delta = 3\Delta x$ 和 $\delta = \sqrt{2}\Delta x$ 情况下，近场动力学均能获得接近于局部理论的解答。Dipasquale 等 [17] 指出，保持 $m = 3$ 不变时，裂纹路径依赖于规则分布网格的倾斜角，单一细化均匀网格并不能削弱网格方向的依赖性问题；但当 m 取较大值时 ($m = 6 \sim 8$)，该网格方向依赖性得到削弱，可获得合理可靠、精度较高的解答。由于近场动力学包含描述问题或材料和结构特征长度的尺度参数，所以数值计算中选取的近场范围 δ 需要对应该特征尺度；数值积分的精度要求近场范围内具有足够的物质点，且在各个方向上的分布没有显著差异，这是保证近场动力学数值解答高精度的基本要求。现有文献的计算方案多为：选取 Δx 间距的物质点进行均匀离散，固定近场范围 $\delta = 3\Delta x$。虽然采用上述计算方案求得的弹性问题解答已获得较高精度的验证，但也有值得商榷之处，如：目前选定的近场范围能否真实反映描述问题或材料的特征尺度；在均匀离散下，3 倍物质点间距的近场范围能否保证各直径方向上物质点数量相当或粒子数密度接近，从而避免网格依赖性等等。

Bobaru 等 [5]，Bobaru 和 Ha[6]，Dipasquale 等 [7] 和 Ren 等 [8] 发展了基于初始均匀正交离散网格和局部均匀细化的近场动力学自适应方法，分析了一维弹性波传播、二维弹性变形、二维和三维动力裂纹扩展问题。此外，基于 Dipasquale 等的自适应分析方法，Duzzi 等 [18] 建立了并发多尺度模型，分析了纳米复合材料的裂纹扩展问题。值得注意的是，对于初始均匀或非均匀离散网格或点阵，加密后网格或点阵必然是非均匀的，故要求具体计算时考虑非均匀网格离散的特点，选取与网格密度相匹配的近场范围。

自适应方法需要给定细化因子和细化准则以及相应的局部加密方案。Bobaru 和 Ha[6] 采用物质点应变能密度作为细化因子，Dipasquale 等 [7] 联合采用应变能密度和损伤作为细化因子，当应变能密度达到最大应变能密度的 40%，且损伤到达某一临界值时，细化该物质点。"可视准则" 被用来确定精化区域与原粗糙离散区域的分界，使得细化后物质点不包含在原粗糙物质点的近场范围内 [6,7]。在对初始均匀离散区域进行加密方面，Bobaru 和 Ha[12] 采用四叉树/八叉树方法加密精化区域，Dipasquale 等 [7] 采用简单易行的两点连线中心插点的方法进行区域加密。

变近场范围易导致虚假力或鬼力 (ghost force)、线动量和角动量失衡，将引起虚假波反射问题。数值方法中，鬼力是处于平衡状态的物质点所受的虚假非零力，鬼力可能导致局部轻微的数值振荡或全局计算结果的精度较低，严重情况下将致使计算完全失效，应予以避免或最大限度地削弱 [19]。Bobaru 等 [5,6] 在非均匀离散的研究中注意到了虚假力问题，但由于虚假力对他们所关注的问题影响不大，并没有进行具体处理。Dipasquale 等 [7] 采用增大分界区域粒子近场范围的方法，使得虚假波反射控制在可接受范围内。Silling 等采用偏应力 (partial stress) 方法 [19] 降低变近场范围的虚假力，但该方法损害了近场动力学方法的简明性，此外，他们还采用切片方法 (slice method) 处理分段定常近场范围之间的过渡问题。Ren 等 [8] 发展了对偶作用域的近场动力学方法 (dual-horizon peridynamics)，在变近场范围条件下消除了虚假力问题。Rahman 和 Foster[20] 采用幂律形式的近场动力学核函数，消除了多尺度建模中的虚假波反射问题。王芳 [21] 借鉴 PD 与 FEM 混合建模的思想，通过在带有大近场范围的物质点区域布置虚拟点的方法，降低了虚假力的影响。

10.2 基于 Voronoi 结构图离散的近场动力学自适应方法

基于 Voronoi 结构图离散的近场动力学建模包括三个方面内容：生成结构离散的 Delaunay 三角网和 Voronoi 结构图、确定近场范围尺寸和模型参数的比例关系、消除非定常近场范围导致的虚假力问题。此外，基于 Voronoi 结构图的近场动力学自适应方法还需要明确细化准则、物质点加密方法和相应加密点的变量赋值方法。

10.2.1 均匀/非均匀 Voronoi 胞元离散

均匀正方形或立方体网格易于数值实现，且方便采用四叉树/八叉树或中心点插值方法进行自适应精化分析，但这种均匀离散方法在处理曲线 (曲面) 和非规则边界时存在困难，需要发展非均匀非规则网格的离散方法。非均匀非规则的二维三角形和三维四面体网格是其中的自然选择，比较三角形离散 (主要是 Delaunay 三角形) 和 Voronoi 胞元两种离散方式，可以得到以下认识：① Voronoi 胞元可由其对偶 Delaunay 三角形生成，但生成算法较为复杂；② 在进行近场动力学积分计算时，基于 Voronoi 胞元的积分精度优于三角形网格；③ Voronoi 胞元具有自然邻胞和相对规则的局部影响特性，适合自适应分析时插入新结点。

Delaunay 三角形和 Voronoi 胞元具有对偶关系，即 Delaunay 三角形的外心是 Voronoi 胞元的顶点，Voronoi 胞元的中心是 Delaunay 三角形的顶点，如图 10-1 所示。建模时，首先采用网格划分软件生成结构离散的 Delaunay 三角网或四面体体系，再提取三角网 (或四面体) 顶点生成 Voronoi 胞元，计算各胞元的坐标、面积 (体积) 和自然邻胞等信息，最后绘制生成的 Voronoi 结构图。图 10-2 给出了平面问题方板的离散 Delaunay 三角网与对应 Voronoi 结构图。

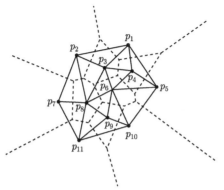

图 10-1 点集的 Voronoi 胞元和 Delaunay 三角剖分

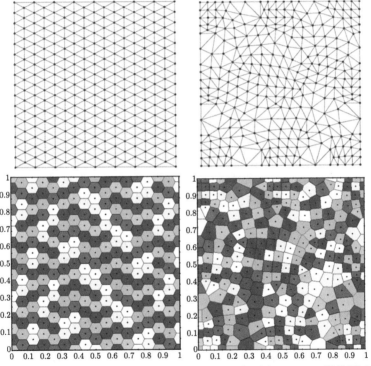

图 10-2 二维方板的离散 Delaunay 三角网 (上) 与对应 Voronoi 结构图 (下)

在近场动力学计算中,每个 Voronoi 胞元被视为物质点,数据文件包含如下信息:总控信息、物质点信息 (坐标、体积、与其自然邻胞的最大距离 Δx、材料编号)、边界条件、加载条件、材料参数,某一物质点的近场范围取为 $\delta = m\Delta x$ ($m = 3\sim 8$)。对于均匀离散网格,在计算近场范围边界物质点的体积时,需要进行体积修正,以提高数值积分精度;对于 Voronoi 非均匀离散网格,在计算近场范围边界物质点体积时，采用胞元的实际体积，此时近场范围的形状不再是规则的圆形或球体，如图 10-3 所示。

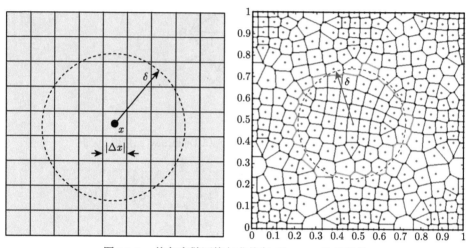

图 10-3　均匀离散网格与非均匀 Voronoi 胞元离散

10.2.2　键型近场动力学模型中参数的比例关系

非均匀离散和自适应分析要求物质点间距和近场范围大小随位置变化，对于不同的近场范围，相应的微观势能函数和微观模量等物理量应该存在比例关系。考察某一物质点，假设具有两个不同的近场范围尺寸 δ 和 ε，则这种比例关系可表示为[6,7]

$$\begin{cases} w_\varepsilon(\boldsymbol{\eta},\boldsymbol{\xi}) = \gamma w_\delta(\gamma\boldsymbol{\eta},\gamma\boldsymbol{\xi}),\ \gamma=\delta/\varepsilon,\quad 一维 \\ w_\varepsilon(\boldsymbol{\eta},\boldsymbol{\xi}) = \gamma^2 w_\delta(\gamma\boldsymbol{\eta},\gamma\boldsymbol{\xi}),\ \gamma=\delta/\varepsilon,\quad 二维 \\ w_\varepsilon(\boldsymbol{\eta},\boldsymbol{\xi}) = \gamma^3 w_\delta(\gamma\boldsymbol{\eta},\gamma\boldsymbol{\xi}),\ \gamma=\delta/\varepsilon,\quad 三维 \end{cases} \quad (10.1)$$

给定变形条件下，由两个近场范围计算的同一点应变能密度必须相同，即

$$\begin{cases} W_\delta(\boldsymbol{x}) = \dfrac{A}{2}\int_{H_\delta} w_\delta(\gamma\boldsymbol{\eta},\gamma\boldsymbol{\xi})\mathrm{d}(\gamma\boldsymbol{\xi}) = \dfrac{A}{2}\int_{H_\varepsilon}\dfrac{1}{\gamma}w_\varepsilon(\boldsymbol{\eta},\boldsymbol{\xi})\gamma\mathrm{d}\boldsymbol{\xi} = W_\varepsilon(\boldsymbol{x}),\quad 一维 \\ W_\delta(\boldsymbol{x}) = \dfrac{1}{2}\int_{H_\delta} w_\delta(\gamma\boldsymbol{\eta},\gamma\boldsymbol{\xi})\mathrm{d}(\gamma^2 A) = \dfrac{1}{2}\int_{H_\varepsilon}\dfrac{1}{\gamma^2}w_\varepsilon(\boldsymbol{\eta},\boldsymbol{\xi})\gamma^2\mathrm{d}A = W_\varepsilon(\boldsymbol{x}),\quad 二维 \\ W_\delta(\boldsymbol{x}) = \dfrac{1}{2}\int_{H_\delta} w_\delta(\gamma\boldsymbol{\eta},\gamma\boldsymbol{\xi})\mathrm{d}(\gamma^3 V) = \dfrac{1}{2}\int_{H_\varepsilon}\dfrac{1}{\gamma^3}w_\varepsilon(\boldsymbol{\eta},\boldsymbol{\xi})\mathrm{d}V = W_\varepsilon(\boldsymbol{x}),\quad 三维 \end{cases}$$

$$(10.2)$$

由 PMB 模型的本构关系式 $f(\pmb{\eta},\pmb{\xi}) = \dfrac{\partial w(\pmb{\eta},\pmb{\xi})}{\partial \pmb{\eta}} = cs\mu\dfrac{\pmb{\xi}+\pmb{\eta}}{|\pmb{\xi}+\pmb{\eta}|}$,可知

$$w = \frac{c(\pmb{x},\delta)|\pmb{\xi}|}{2}s^2 \tag{10.3}$$

将式 (10.3) 代入式 (10.1),则微观模量的比例关系为

$$\begin{cases} c(\pmb{x},\varepsilon) = \gamma^2 c(\pmb{x},\delta),\ \gamma = \delta/\varepsilon,\quad \text{一维} \\ c(\pmb{x},\varepsilon) = \gamma^3 c(\pmb{x},\delta),\ \gamma = \delta/\varepsilon,\quad \text{二维} \\ c(\pmb{x},\varepsilon) = \gamma^4 c(\pmb{x},\delta),\ \gamma = \delta/\varepsilon,\quad \text{三维} \end{cases} \tag{10.4}$$

综合上述分析可知,若近场范围的尺寸改变,本构模型中的相关参数将随之改变。因此,在基于 Voronoi 结构图进行非均匀离散的近场动力学建模时,物质点体积、近场范围尺寸和本构模型的参数都要随位置改变。此外,由于键两端物质点可能具有不同的微观模量,建议使用该点自身的微观模量参数[6],该策略同样适用于对偶双影响域近场动力学计算。

10.2.3 对偶双影响域近场动力学模型

如图 10-4(a) 所示,在传统近场动力学数值计算方案中,采用均匀网格离散和定常的近场范围,不会产生虚假力问题。如果采用分段定常的近场范围进行近场动力学计算,在不同密度网格的界面区,将产生虚假力。如图 10-4(b) 所示,当基于 Voronoi 结构图进行非均匀离散和变化近场范围的传统近场动力学分析时,物质点 $\pmb{x}_i$ 的近场范围包含物质点 $\pmb{x}_j$,而反之不然,将引起全局范围内的虚假力,致使动量失衡,进而产生虚假波反射。动量失衡将可能导致严重的全局计算误差,需要消除或最大限度地限制虚假力。对偶双影响域近场动力学[8] 的建模方法可以完全消除该虚假力问题,适合非均匀离散,如图 10-4(c) 所示。

基于 Voronoi 结构图离散几何模型,并采用对偶双影响域的近场动力学方法进行计算。对于某物质点,近场范围是该源点可见的物质点的集合,而对偶近场范围 (dual-horizon) 是所有可见该源点的场点的集合。对偶双影响域的近场动力学运动方程为

$$\rho\ddot{\pmb{u}}(\pmb{x},t) = \int_{\pmb{x}'\in H'_{\pmb{x}}} \pmb{f}_{\pmb{x}\pmb{x}'}(\pmb{u},\pmb{u}',\pmb{x}',\pmb{x},t)\mathrm{d}V_{\pmb{x}'} + \int_{\pmb{x}'\in H_{\pmb{x}}} -\pmb{f}_{\pmb{x}'\pmb{x}}(\pmb{u},\pmb{u}',\pmb{x}',\pmb{x},t)\mathrm{d}V_{\pmb{x}'} + \pmb{b}(\pmb{x},t) \tag{10.5}$$

式中,$H_{\pmb{x}}$ 为物质点 $\pmb{x}$ 的近场范围内其他物质点 $\pmb{x}'$ 的集合,$H'_{\pmb{x}}$ 为对偶近场范围内的物质点 $\pmb{x}'$ 的集合。点 $\pmb{x}$ 对 $\pmb{x}'$ 施加力密度 $\pmb{f}_{\pmb{x}'\pmb{x}}$,故 $\pmb{x}$ 点受到来自 $\pmb{x}'$ 的反作用力密度 $-\pmb{f}_{\pmb{x}'\pmb{x}}$;同时,对偶近场范围内的物质点 $\pmb{x}'$ 对 $\pmb{x}$ 施加力密度 $\pmb{f}_{\pmb{x}\pmb{x}'}$。与

传统键型近场动力学 PMB 模型对应，键力密度 (bond force density)、对偶键力密度 (dual-bond force density) 和应变能密度的计算公式为

$$\boldsymbol{f}_{\boldsymbol{x}\boldsymbol{x}'} = \frac{1}{2}c(\boldsymbol{x}', \delta_{\boldsymbol{x}'})s\mu\frac{\boldsymbol{\xi}+\boldsymbol{\eta}}{|\boldsymbol{\xi}+\boldsymbol{\eta}|}, \quad \boldsymbol{f}_{\boldsymbol{x}'\boldsymbol{x}} = \frac{1}{2}c(\boldsymbol{x}, \delta_{\boldsymbol{x}})s\mu\frac{\boldsymbol{\xi}+\boldsymbol{\eta}}{|\boldsymbol{\xi}+\boldsymbol{\eta}|} \tag{10.6}$$

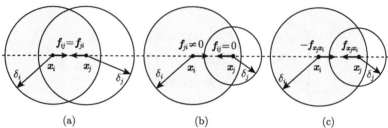

图 10-4 (a) 均匀离散和定常近场范围下，传统 PD 计算无虚假力，(b) 两物质点具有不同近场范围时，传统 PD 计算存在虚假力，(c) 非均匀离散与非定常近场范围下，对偶双影响域 PD 计算无虚假力

和

$$\begin{aligned} W(\boldsymbol{x}) &= \frac{1}{2}\int_{\boldsymbol{x}'\in H'_\infty} w(\boldsymbol{x}', \delta_{\boldsymbol{x}'})\mathrm{d}V_{\boldsymbol{x}'} + \frac{1}{2}\int_{\boldsymbol{x}'\in H_\infty} w(\boldsymbol{x}, \delta_{\boldsymbol{x}})\mathrm{d}V_{\boldsymbol{x}'} \\ &= \frac{1}{2}\int_{\boldsymbol{x}'\in H'_\infty} \frac{c(\boldsymbol{x}', \delta_{\boldsymbol{x}'})}{2}\frac{|\boldsymbol{\xi}|\,s^2}{2}\mathrm{d}V_{\boldsymbol{x}'} + \frac{1}{2}\int_{\boldsymbol{x}'\in H_\infty} \frac{c(\boldsymbol{x}, \delta_{\boldsymbol{x}})}{2}\frac{|\boldsymbol{\xi}|\,s^2}{2}\mathrm{d}V_{\boldsymbol{x}'} \end{aligned} \tag{10.7}$$

式中，所有参数与第 3 章相同。式 (10.7) 意味着一个键储存的能量密度被等分到两个物质点。为消除表面效应，采用实际离散的积分域计算物质点的微观模量 c[22]，因而边界处微观模量增大，表面效应得到削弱，PMB 模型离散形式的微观模量为

$$c_i = \begin{cases} \dfrac{6E}{(1-2\nu)\sum\limits_{j\in H_i}|\boldsymbol{\xi}_{ij}|V_j}, & \text{三维} \\ \dfrac{4E}{(1+\nu)(1-2\nu)\sum\limits_{j\in H_i}|\boldsymbol{\xi}_{ij}|V_j}, & \text{平面应变} \\ \dfrac{4E}{(1-\nu)\sum\limits_{j\in H_i}|\boldsymbol{\xi}_{ij}|V_j}, & \text{平面应力} \end{cases} \tag{10.8}$$

类似地，采用对偶双影响域的概念，修正后的 NOSB PD 运动方程为

$$\rho\ddot{\boldsymbol{u}}(\boldsymbol{x}, t) = \int_{\boldsymbol{x}'\in H'_\infty} \underline{\boldsymbol{T}}\langle\boldsymbol{\xi}\rangle\mathrm{d}V_{\boldsymbol{x}'} + \int_{\boldsymbol{x}'\in H_\infty} -\underline{\boldsymbol{T}}'\langle\boldsymbol{\xi}\rangle\mathrm{d}V_{\boldsymbol{x}'} + \boldsymbol{b}(\boldsymbol{x}, t) \tag{10.9}$$

其中，力密度矢量状态 (force density vector state) 与对偶力密度矢量状态 (dual-force density vector state) 分别为

$$\underline{\boldsymbol{T}}\langle\boldsymbol{\xi}\rangle = w(|\boldsymbol{\xi}|)\boldsymbol{P}\boldsymbol{K}^{-1}\boldsymbol{\xi}, \quad \underline{\boldsymbol{T}}'\langle\boldsymbol{\xi}\rangle = w(|\boldsymbol{\xi}'|)\boldsymbol{P}'\boldsymbol{K}'^{-1}\boldsymbol{\xi}' \quad (10.10)$$

由于同时存在键及其对偶键，发生损伤时两者应分别断开，故定义对偶双影响域近场动力学的损伤为

$$\varphi(\boldsymbol{x},t) = 1 - \frac{\int_{\boldsymbol{x}'\in H'_\infty}\mu(\boldsymbol{x},t,\boldsymbol{\xi}) \mathrm{d}V_{\boldsymbol{x}'} + \int_{\boldsymbol{x}'\in H_\infty}\mu(\boldsymbol{x},t,\boldsymbol{\xi})\mathrm{d}V_{\boldsymbol{x}'}}{\int_{\boldsymbol{x}'\in H'_\infty}\mathrm{d}V_{\boldsymbol{x}'} + \int_{\boldsymbol{x}'\in H_\infty}\mathrm{d}V_{\boldsymbol{x}'}} \quad (10.11)$$

键的断开由源点 $\boldsymbol{x}$ 处的断键准则 (临界伸长率) 确定，对偶键的断开由场点 $\boldsymbol{x}'$ 处的断键准则 (临界伸长率) 确定。至此，可基于 Voronoi 胞元进行结构非均匀离散，采用对偶双影响域近场动力学方法进行计算模拟。

10.2.4 基于 Voronoi 结构图的自适应方案

10.2.4.1 网格细化准则

与文献 [6, 7] 一致，联合采用物质点的应变能密度和损伤值作为自适应分析的细化因子。当物质点的损伤因子 $\varphi(\boldsymbol{x},t)$ 到达某一临界值 $\varphi_0(\boldsymbol{x},t)$ (可取为 0.1)，且应变能密度 W^{nl} 到达最大应变能密度的 40% 时，该物质点需要精化，并称这类物质点为第一类精化物质点，在该类物质点近场范围内的物质点称为第二类精化物质点。细化第一类和第二类精化物质点，其他点不需要细化。此时，第一类精化物质点需满足 "可视准则"[5,6]，即不在原粗糙离散的物质点近场范围内。

10.2.4.2 自适应精化算法

基于 Voronoi 结构图的非均匀离散和对偶双影响域的近场动力学模型，建立相应的自适应分析方法，其数值实现过程如下。

(1) 采用 Voronoi 结构图初始离散构型，建立计算数据文件。

(2) 计算所有物质点的应变能密度和损伤值，确定需要精化的物质点，如果物质点需要精化，进入下一步。

(3) 在两类精化物质点间插入新结点，同时在精化物质点和初始粗物质点间插入新结点。具体做法可借鉴自然单元法 (natural element method) 的精化思想[23]，在物质点 $\boldsymbol{x}_i$ 和其自然邻点 $\boldsymbol{x}_j$ 连线的中点插入新结点，其中的 $\boldsymbol{x}_i$ 和 $\boldsymbol{x}_j$ 至少有一个属于两类精化物质点。该精化方法适用于基于 Voronoi 离散网格的自适应加密，原因在于新结点是 Delaunay 三角形的顶点。在参考构型中，新插入结点的坐标矢量为

$$x_{\text{new}} = \frac{x_i + x_j}{2} \tag{10.12}$$

(4) 基于新物质点的集合，重新生成 Voronoi 结构图，计算 Voronoi 胞元物质点的体积、源点与自然邻接点的最大距离，形成源点的邻接点列表。

(5) 对所有新结点，通过线性插值赋予其各个变量值，包括：参考构型中位置矢量、现时构型中位置矢量、速度、总体力密度等；对于老结点，保持各变量原值不变。

另外指出，在进行自适应精化分析时，需要输出和记录必要的变量值信息和时间步号并重启动 PD 计算。图 10-5 给出了基于 Voronoi 胞元的近场动力学自适应分析流程图。

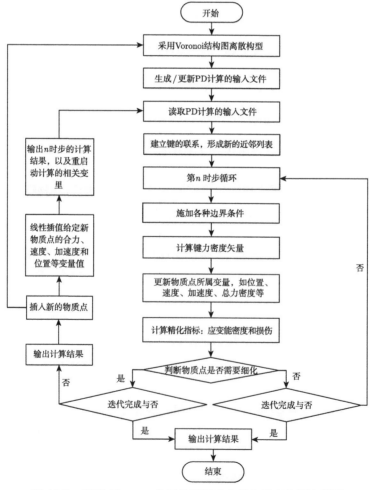

图 10-5　基于 Voronoi 胞元的近场动力学自适应分析流程图

10.3 数值算例

本节通过四个典型算例验证对偶双影响域近场动力学方法能够有效消除虚假力,并检验基于 Voronoi 结构图的非均匀离散近场动力学方法的适用性。

10.3.1 矩形板的拟静力弹性变形

矩形板的几何尺寸为 200mm×100mm,图 10-6 给出矩形板三种离散格式,一种为均匀离散,另两种为基于 Voronoi 胞元的非均匀离散,离散后的物质点总数分别为 20301 个、5894 个和 1989 个。板的左端固定,右端受位移 $d = 1 \times 10^{-4}$ m 的均匀拉伸作用,位移约束均施加在边界处一层物质点上。材料质量密度 $\rho = 2400 \text{ kg/m}^3$,杨氏弹性模量为 $E = 200 \text{ GPa}$,泊松比为 $\nu = 1/3$。采用显式拟静力解法,时间步长 $\Delta t = 1 \times 10^{-7}$ s,人工阻尼系数 $C = 1.5 \times 10^7$,近场范围尺寸 $\delta = 4\Delta x$。分别采用基于均匀网格的有限元方法、基于均匀点阵的传统近场动力学方法和基于 Voronoi 胞元的非均匀离散近场动力学方法进行模拟计算。

图 10-7 分别给出了有限元方法、均匀离散的传统近场动力学方法和基于 Voronoi 胞元的非均匀离散近场动力学方法计算得到的矩形板水平位移、竖直位移和应变能密度云图。结果表明,对于远离边界的区域,四组解答具有较好的一致性,即使采用较少 Voronoi 胞元的模型也能获得足够精度的解答,表明所发展的基于 Voronoi 结构图的近场动力学方法有效性,但该方法求得的解答在近边界区域有一定的误差,误差大小与 Voronoi 胞元的离散密度有关,随着 Voronoi 胞元数目的增加,误差趋于减小。

进一步考察应变能密度结果。图 10-7(c4) 中部分物质点的应变能密度值偏差较大,但对工程分析其精度可以接受,且随着 Voronoi 物质点分布密度的增大,该偏离量快速降低,如图 10-7(c3) 所示。算例表明:均匀荷载作用下的非均匀网格离散结构,应变能密度计算的非均匀性难以避免,但当邻接区域物质点的近场范围尺寸是渐进过渡而非突变,不论 Voronoi 物质点的全局分布密度如何,只需保证近场范围内有足够多的物质点,应变能密度的计算精度是可以接受的。

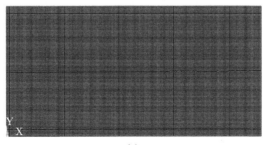

(a)

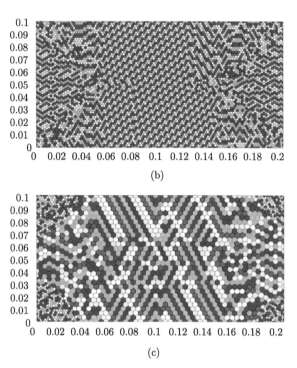

图 10-6 矩形板的三种离散模型：(a) FEM 与 BB PD 分析采用的 1 mm 均匀网格，(b) 带有 5894 个 Voronoi 胞元的非均匀网格，(c) 带有 1989 个 Voronoi 胞元的非均匀网格

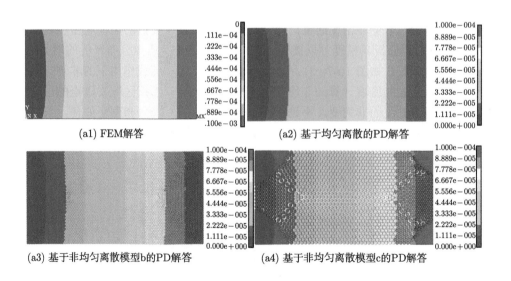

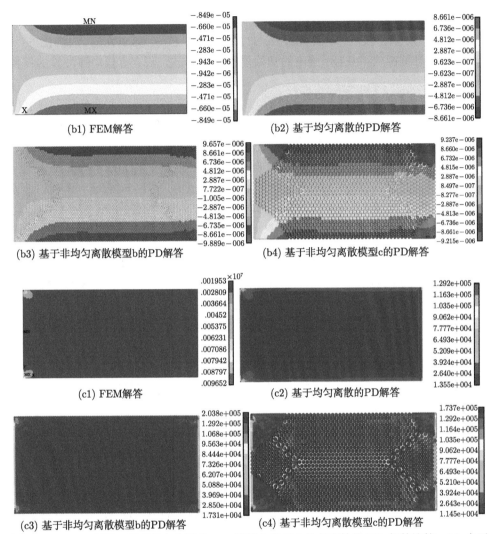

图 10-7 FEM 解答、基于均匀离散的 PD 解答与基于非均匀离散的 PD 解答比较：(a) 水平位移、(b) 竖向位移和 (c) 应变能密度

10.3.2 二维弹性波的传播

边长 30 cm 的方板，具有高斯型函数分布的初始位移，极坐标下的初始位移为[24]

$$u(r, t=0) = \begin{cases} A\exp\left(-\dfrac{r^2}{2r_0^2}\right)\left(1 + b\cos\left(\dfrac{4\pi r}{r_0}\right)\right), & r \leqslant L_c \\ 0, & r > L_c \end{cases} \quad (10.13)$$

式中，$A = 0.0004\text{m}$，$b = 0.002\text{m}$，截断距离为 $L_c = 2r_0$，$r_0 = 0.016\text{m}$。材料参数为：质量密度 $\rho = 2700\text{kg/m}^3$，弹性模量 $E = 68.9\text{GPa}$，泊松比 $\nu = 1/3$。设零时刻由式 (10.13) 给出的初始位移自由释放，研究方板中弹性波的传播规律。

采用近场动力学方法计算上述问题，分别建立了三种不同的离散模型，如图 10-8 所示。第一种为全局 2mm×2mm 的均匀网格离散，共有 22801 个物质点；第二种为中心 10cm×10cm 区域采用 1mm×1mm 均匀网格离散，四周区域采用 2mm×2mm 的均匀网格离散，共有 30000 个物质点；第三种为基于 Voronoi 胞元的全局非均匀离散，共有 14698 个物质点。近场范围尺寸 $\delta = 4\Delta x$，显式动力计算的时间步长 $\Delta t = 1 \times 10^{-7}\text{s}$，计算总时间 $t = 1 \times 10^{-4}\text{s}$。

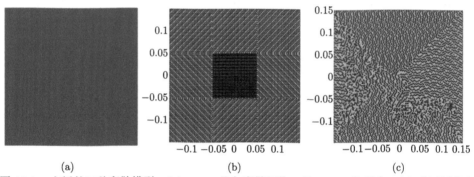

图 10-8　方板的三种离散模型：(a) 2 mm 均匀离散网格，共 22801 物质点；(b) 分区域均匀离散，内部中心 10 cm×10 cm 方形区域采用 1 mm 网格均匀离散，四周采用 2 mm 均匀网格离散；(c) 带有 14698 个 Voronoi 胞元的非均匀离散

图 10-9 给出了基于三种离散模型计算得到的弹性波从方板中心向四周区域的传播历程。图 10-9(a) 显示的径向位移云图为初始波形；图 10-9(b) 是整个弹性波仍在中心区域时的快照，图中两个清晰分离的圆环表示形成两个波峰，外环 (第一个波峰) 的振幅值小于内环 (第二个波峰) 的振幅值；图 10-9(c) 的中间快照显示外环穿过精细与粗糙网格的界面区域，不同密度网格的分界面并未造成位移场的显著不连续，表明非均匀网格下的弹性波能够光滑传播，并无显著的虚假波反射；图 10-9(d) 为内环穿过界面区域的振幅图；图 10-9(e) 和 (f) 显示当整个弹性波穿过两区域界面时，界面区域存在微小残余震荡；图 10-9(g) 给出了弹性波到达边界区域后发生的波反射现象。总体来看，三种离散模型均模拟得到了弹性波沿径向传播扩展时，波振幅随之渐进降低直至消失的过程；此外，由于离散模型 C 是非对称的，在其中心区域会残留微小局部振动，该振动会一定程度地影响反射后的叠加波形。

10.3 数值算例

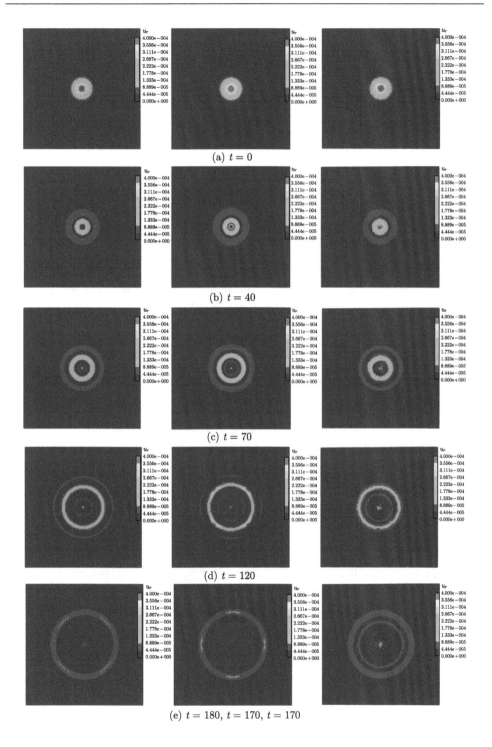

(a) $t = 0$

(b) $t = 40$

(c) $t = 70$

(d) $t = 120$

(e) $t = 180$, $t = 170$, $t = 170$

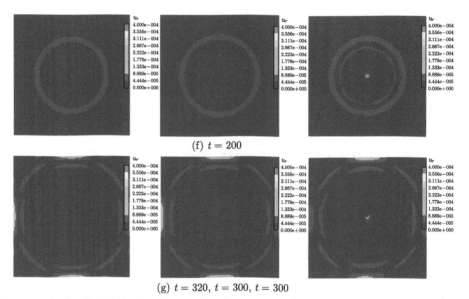

(f) $t = 200$

(g) $t = 320, t = 300, t = 300$

图 10-9　基于三种离散模型的弹性波传播历程：左列为模型 A、中间为模型 B，右列为模型 C (时间单位：Δt)

为定量显示近场动力学模拟的弹性波能有效穿过界面区域，且不伴随显著的波反射和位移不连续现象，考虑几何模型和径向位移分布的对称性，提取模型 A 和模型 B 的水平中线位移 (从中心到边缘 1 条射线)，并提取模型 C 的水平中线和竖直中线位移 (从中心到边缘的 4 条射线)，图 10-10(a)～(e) 分别给出了 5 个不同时刻粒子径向位移的分布曲线。由图可见：① 由模型 B 获得的径向位移在虚线 (两种网格交界面) 左右两侧光滑连续，由模型 C 获得的径向位移在全局具有足够的光滑性，表明粗细不同网格的交界面和相邻 Voronoi 胞元间不发生显著的虚假波反射；② 对于三种离散模型，近场动力学模拟获得的不同时刻波形图 (振幅和相位) 具有良好的一致性；③ 三种离散模型计算得到的低峰弹性波 (第一个圆环) 在相位和幅值上都吻合良好，基于模型 B 求得的高峰弹性波 (第二个圆环) 传播速度大于其余两种模型对应的结果，这是由于近场动力学的内在频散性质所致 [25]，近场动力学的非局部特性决定了 "弹性波在精细网格区域的传播速度高于粗糙网格区域的传播速度"。图 10-10(f) 给出模型 A 在第 180 个时间步与模型 B 在第 170 个时间步时的波形，显示了近场动力学方法的数值频散特性。

图 10-11 和图 10-12 进一步给出了沿与水平线成 30° 角和 45° 角线上 (从中心发出的 30° 射线和 45° 射线各 4 条) 的粒子在不同时刻的径向位移分布曲线。沿不同角度射线的粒子径向位移具有高度相似的传播演化过程，进一步验证了基

于 Voronoi 非均匀离散的近场动力学分析方法的有效性。

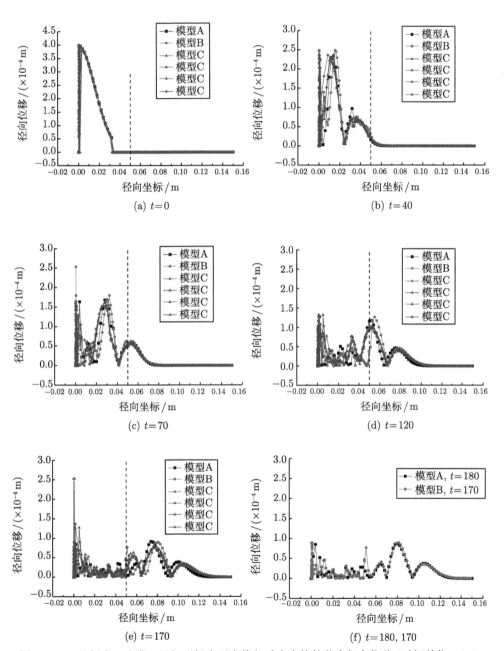

图 10-10　从板中心出发，沿矩形板水平中线与垂直中线的节点径向位移 (时间单位：Δt)

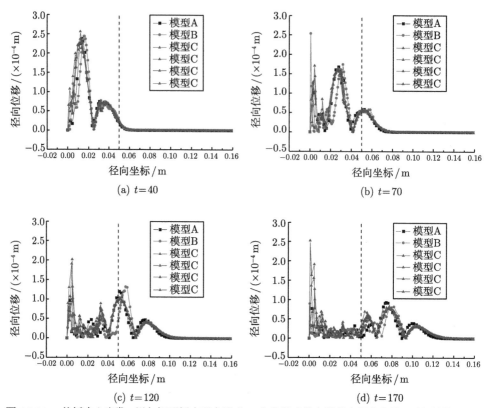

图 10-11 从板中心出发，沿与矩形板水平中线成 30° 角的直线上的节点径向位移 (时间单位：Δt)

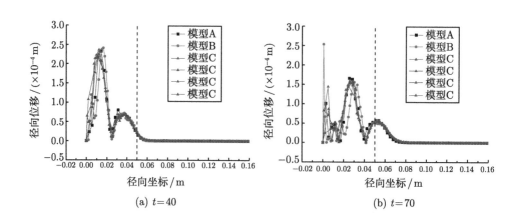

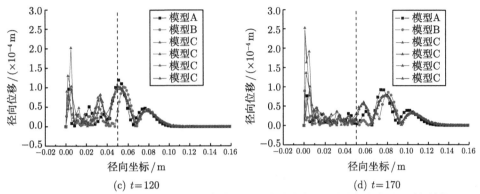

图 10-12 从板中心出发，沿与矩形板水平中线成 45° 角的直线上的节点径向位移 (时间单位：Δt)

10.3.3 含预制裂纹板的动态裂纹扩展

在平面应力假定下，采用基于 Voronoi 结构图的非均匀离散近场动力学方法求解含预制裂纹矩形玻璃板的动态裂纹扩展问题。如图 10-13 所示，板长 100mm、宽 40mm，左端预制裂纹长 50mm。采用断开所有穿过裂纹线段上键的方式，或同时删除裂纹线段上的点和断开穿过裂纹线段上键的方式引入线裂纹。杜兰 50 号玻璃板[26,27]的材料参数为：质量密度 $\rho = 2440 \text{kg/m}^3$、杨氏模量 $E = 72\text{GPa}$、能量释放率 $G_0 = 135\text{MN/m}$，泊松比 $\nu = 0.22$，在键型近场动力学模拟中，泊松比取 $1/3$。板的上下两端在一层物质点上施加均匀拉伸应力载荷 $\sigma = 12\text{MPa}$，持续时间为 $t = 5 \times 10^{-5}\text{s}$，计算时间步长取为 $\Delta t = 5 \times 10^{-8}\text{s}$。

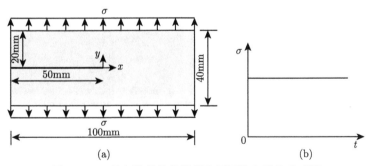

图 10-13 均匀阶越拉伸载荷作用下的含裂纹玻璃板

10.3.3.1 基于均匀离散的传统近场动力学模拟

采用基于均匀正交方形网格离散的传统键型近场动力学进行求解，作为对比分析的基准。根据沿裂纹线段是否布置物质点，形成两种离散模型：一种是沿裂纹线段布置物质点，物质点的层数为奇数；另一种在裂纹线段上下两侧布置物质点，物质点的层数为偶数，如图 10-14(a) 所示。对这两类均匀离散模型，分别采用三组不同的物质点间距进行近场动力学模拟，物质点间距分别为 $\Delta x = 1\text{mm}$、$\Delta x = 0.5\text{mm}$ 和 $\Delta x = 0.25\text{mm}$，近场范围尺寸取 $\delta = 3\Delta x$。

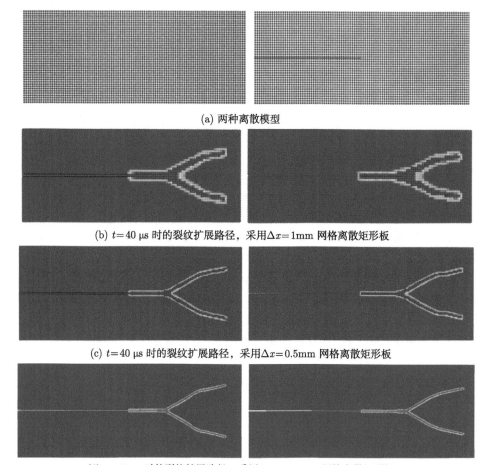

(a) 两种离散模型

(b) $t=40\,\mu\text{s}$ 时的裂纹扩展路径，采用$\Delta x=1\text{mm}$ 网格离散矩形板

(c) $t=40\,\mu\text{s}$ 时的裂纹扩展路径，采用$\Delta x=0.5\text{mm}$ 网格离散矩形板

(d) $t=40\,\mu\text{s}$ 时的裂纹扩展路径，采用$\Delta x=0.25\text{mm}$ 网格离散矩形板

图 10-14　两种离散方式与 $t=40\mu\text{s}$ 时的裂纹扩展路径：沿裂纹布置物质点 (左) 和在裂纹上下两侧布置物质点 (右)

图 10-14(b)~(d) 给出了 6 种条件下，采用近场动力学方法模拟获得的 $t=40\mu\text{s}$ 时刻的裂纹扩展模式，各组裂纹形态 (裂纹分叉和裂纹扩展路径) 非常相似，但物质点间距不同将导致损伤区域的大小有所不同。对于本例而言，基于均匀离散模型，沿裂纹线段是否布置物质点对计算结果没有显著影响；当网格密度 (物

10.3 数值算例

质点间距) 到达一定值后, 单纯均匀加密网格而保持 $m = 3$ 不变, 计算得到的裂纹扩展路径变化很小。

10.3.3.2 基于非均匀离散的近场动力学模拟

采用基于 Voronoi 胞元的非均匀离散近场动力学方法求解该问题。如图 10-15 所示, 考虑三组不同的基于 Voronoi 胞元的非均匀离散模型, 在离散模型构建过程中, 采用 ANSYS 软件生成 Delaunay 三角网, 通过自编的 FORTRAN 90 程序生成 Voronoi 结构图, 并输出 PD 计算所需的数据文件。模型 A 和模型 B 的 Delaunay 三角形的平均边长为 1mm, 只在中心区域具有不同的网格密度, 模型 A 共有 5151 个 Voronoi 胞元, 模型 B 共有 5558 个 Voronoi 胞元; 模型 C 的左半侧 Delaunay 三角形的平均边长为 1.25mm, 右半侧 Delaunay 三角形的平均边长为 0.5mm, 模型 C 共有 10077 个 Voronoi 胞元。

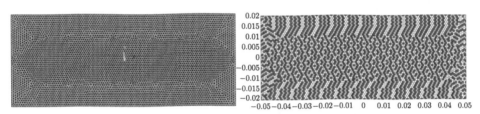

(a) 模型 A, 共 5151 个 Voronoi 胞元

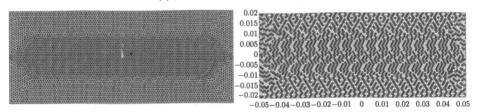

(b) 模型 B, 共 5558 个 Voronoi 胞元

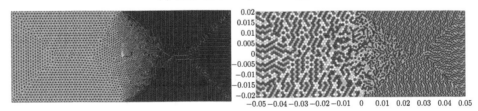

(c) 模型 C, 共 10077 个 Voronoi 胞元

图 10-15 矩形板的 Delaunay 三角形离散与 Voronoi 胞元

选取近场范围为 $\delta = 3\Delta x$, 其中 Δx 为 Voronoi 胞元与其最邻接 Voronoi 胞元之间的最大距离。图 10-16 给出了 $t = 40\mu s$ 时刻板的裂纹形貌, 左列为基

于 Voronoi 胞元非均匀离散的传统近场动力学方法求得的结果，右列为对偶双影响域近场动力学对应的结果。结果表明，在非均匀离散和非定常近场范围条件下，传统近场动力学方法会引起虚假力(鬼力)，虚假力对裂纹动态扩展路径影响显著，模拟得到的裂纹扩展结果完全失真；对偶双影响域近场动力学能够有效消除虚假力，求得相对准确合理的裂纹扩展路径。

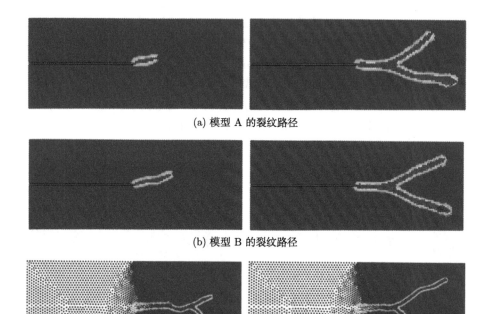

(a) 模型 A 的裂纹路径

(b) 模型 B 的裂纹路径

(c) 模型 C 的裂纹路径

图 10-16　$t = 40\mu s$ 的裂纹路径：传统 PD 模型 (左)，对偶双影响域 PD 模型 (右)

需要注意的是，当 $m = 3$ 时，虽然对偶双影响域的近场动力学方法能够有效消除虚假力，但不同离散模型获得的裂纹路径仍存在较为明显的差异。为避免近场范围内物质点数目过少引起的计算误差，分别取 $m = 4$ 和 $m = 5$ 再对上述问题进行模拟计算，计算结果如图 10-17 所示。结果显示：① 对于三种离散模型，$m = 4$ 和 $m = 5$ 计算获得的裂纹路径非常相似，表明当 m 值到达 4 时，基于 Voronoi 胞元非均匀离散的对偶双影响域近场动力学方法即可求得稳定的裂纹扩展路径；② 离散点阵的粒子分布密度一定时，单一增大 m 值 (增大近场范围)，计算结果的精度不会显著提升；③ 模型 A 和模型 B 获得的裂纹路径非常接近，而模型 C 获得的裂纹路径与模型 A 和模型 B 仍有明显差别，显现出一定的网格敏感性，值得进一步深入研究。

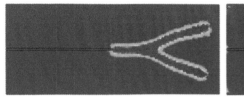 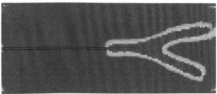

(a) 模型 A 的裂纹路径

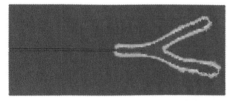 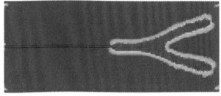

(b) 模型 B 的裂纹路径

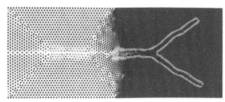 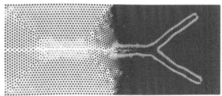

(c) 模型 C 的裂纹路径

图 10-17 对偶双影响域 PD 模型求得的裂纹路径 ($t = 40\mu s$): $m = 4$ (左) 和 $m = 5$ (右)

10.3.3.3 基于非均匀离散的近场动力学自适应分析

采用基于 Voronoi 结构图的近场动力学自适应分析方法，模拟上述含裂纹板的动态裂纹扩展问题。初始离散时，物质点间距为 1mm×1mm，将板均匀离散为 4141 个物质点，初始粗糙网格区域的近场范围为 $\delta = 3\Delta x$，后续细化区域近场范围为 $\delta = 4\Delta x$。为降低精化频率，首先在 $5\mu s$ 时对结构进行精化，以后每隔 $2.5\mu s$ 进行一次精化。

图 10-18 给出了近场动力学实时自适应分析方法计算得到的含裂纹玻璃板动态裂纹扩展过程，基于两组均匀离散模型 ($\Delta x = 1$mm、$\Delta x = 0.5$mm) 的传统近场动力学以及近场动力学自适应分析的裂纹扩展速度如图 10-19 所示。结果显示：实时近场动力学自适应方法能有效分析裂纹渐进扩展过程，获得准确的裂纹扩展模式。需要细化的区域只在损伤和高应变能密度区域附近，自适应加密分析过程中，物质点总数仅从 4141 增加到 7227。此外，传统近场动力学与基于非均匀离散的近场动力学自适应方法获得相近的裂尖扩展速度，可比性良好。

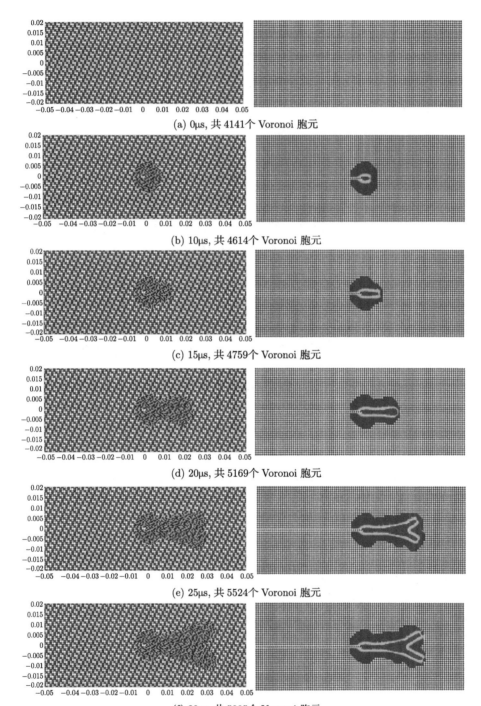

(a) 0μs, 共 4141个 Voronoi 胞元

(b) 10μs, 共 4614个 Voronoi 胞元

(c) 15μs, 共 4759个 Voronoi 胞元

(d) 20μs, 共 5169个 Voronoi 胞元

(e) 25μs, 共 5524个 Voronoi 胞元

(f) 30μs, 共 5905个 Voronoi 胞元

10.3 数值算例

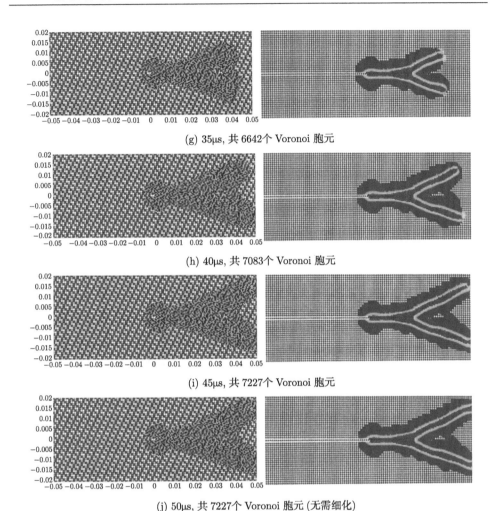

(g) 35µs, 共 6642 个 Voronoi 胞元

(h) 40µs, 共 7083 个 Voronoi 胞元

(i) 45µs, 共 7227 个 Voronoi 胞元

(j) 50µs, 共 7227 个 Voronoi 胞元 (无需细化)

图 10-18　近场动力学实时自适应分析方法计算的含裂纹玻璃板动态裂纹扩展过程

10.3.4 三维马氏体时效钢冲击动力裂纹扩展

选取 Kalthoff-Winkler 的刚性柱体撞击马氏体时效钢靶的试验进行模拟分析，试验概况和计算模型如图 10-20 所示。试验结果表明：刚性柱体撞击马氏体时效钢靶体时产生压缩应力波，并与裂尖相互作用，产生 II 型开裂；在不同的冲击速度下，能观测到不同的裂纹扩展模式，选取恰当的撞击速度 (如 $v_0 = 32\text{m/s}$)，将发生剪切与拉伸混合型开裂，并从裂尖起裂，形成约 68° 的裂纹扩展角。该试验已成为判断数值方法能否模拟脆性材料动力断裂问题的基准考例。

选取与文献 [2] 相同的计算参数：质量密度 $\rho = 8000\text{kg/m}^3$、杨氏弹性模量 $E = 191\text{GPa}$、泊松比 $\nu = 1/4$、材料临界伸长率 $s_0 = 0.01$，刚性柱体弹丸质量

$m=1.57$kg, 初速度 $v_0=-32$m/s。在传统近场动力学模拟中，靶体被均匀离散成 178200 个立方物质点，尺寸 1mm×1mm×1mm；在基于 Voronoi 胞元的近场动力学模拟中，靶体被非均匀离散为 40806 个 Voronoi 胞元物质点。近场范围尺寸为 $\delta=4\Delta x$，时间步长取 $\Delta t=8\times10^{-8}$s，共计算 2000 步。

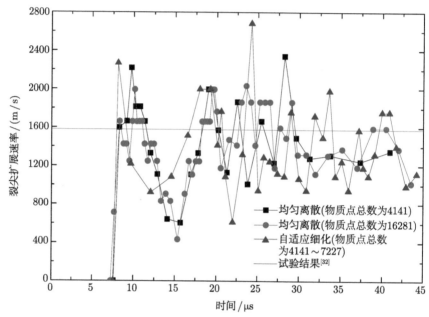

图 10-19 不同模拟方法获得的裂尖扩展速度比较

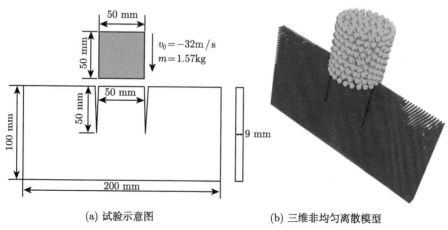

图 10-20 Kalthoff-Winkler 试验概况和计算模型

将损伤值 $\varphi = 0.35$ 的物质点位置作为宏观裂纹，分别采用均匀离散的传统近场动力学方法和基于 Voronoi 胞元的非均匀离散近场动力学方法进行了计算，两种方法的模拟结果如图 10-21 所示。计算结果表明，在冲击荷载作用下，预制裂纹的裂尖产生损伤，损伤累积到一定程度后裂尖起裂，产生裂纹，并在剪切与拉伸的共同作用下，裂纹不断演化发展，形成约 67° 的扩展角；裂纹起裂时间约为 28μs，试验获得的起裂时间略低于 29μs。两种数值模拟方法获得的裂纹起裂时间、扩展角和最终裂纹形态均具有良好的一致性，适用于多裂纹扩展分析。

(a) 基于均匀离散的 PD 模拟　　　(b) 基于非均匀离散的 PD 模拟

图 10-21　两种模拟方法得到的裂纹最终形貌

10.4　主　要　结　论

本章将基于均匀离散的传统近场动力学方法与对偶双影响域近场动力学方法进行了拓展，提出基于 Voronoi 结构图非均匀离散的对偶双影响域近场动力学方法，该方法突破了传统近场动力学方法只能采用均匀离散和定常近场范围的限制，发展了对偶双影响域近场动力学的损伤定义、离散格式和数值积分方式。在此基础上，构建了基于 Voronoi 结构图的近场动力学自适应分析方法，给出了详细的数值计算方案。进行了矩形板的弹性变形、弹性波传播、动态裂纹扩展等四个典型数值算例的计算分析，得到以下结论。

(1) 在数值计算中，当采用非均匀离散和非定常近场范围时，传统近场动力学方法将会引起虚假力、动量失衡和虚假波反射等问题，严重影响动态裂纹扩展分析的正确性，对偶双影响域近场动力学方法能有效避免上述问题。

(2) 采用基于 Voronoi 结构图非均匀离散的近场动力学方法进行了四个典型算例的模拟计算，所得结果与 FEM 方法、传统 PD 方法和试验的结果吻合良好，验证了该方法的有效性。

(3) 基于非均匀离散和自适应加密的近场动力学分析方法在保证计算精度的同时，能有效降低计算负担，提高计算效率。

(4) 一般而言，对于任意的 Voronoi 离散构型，当 $m \geqslant 4$ 时，基于 Voronoi 胞元非均匀离散的对偶双影响域近场动力学方法能获得精度较高的解答，求得稳定的裂纹扩展路径，该方法适用于各种复杂的动态裂纹扩展分析。

(5) 为确保基于 Voronoi 胞元非均匀离散和自适应分析的近场动力学方法的稳健可靠，需要深入开展积分方案精度、误差估计、收敛性分析等方面的研究；需要发展与空间非均匀离散和自适应相匹配的多时间步长积分算法[29]，以进一步提高计算效率；还需要在确保近场范围与材料特征长度相匹配的条件下，在近场范围内布置足够物质点以使各直径方向粒子数密度接近，进而研究近场动力学的网格敏感性问题。

(6) 自适应算法根据中间计算结果，自动选取后续计算所需的网格和离散形式，逐步对误差进行调节，以尽量少的计算量达到所要求的计算精度[30,31]，自适应分析有赖于可靠的误差估计方法[30,31]和功能强大的网格自动生成技术，还处于不断发展和完善中。

参 考 文 献

[1] Silling S A, Askari E. A meshfree method based on the peridynamic model of solid mechanics[J]. Computers & Structures, 2005, 83(17): 1526-1535.

[2] Madenci E, Oterkus E. Peridynamic Theory and Its Applications[M]. New York: Springer, 2014.

[3] Silling S A. Origin and effect of nonlocality in a composite[J]. Journal of Mechanics of Materials and Structures, 2014, 9(2): 245-258.

[4] Macek R W, Silling S A. Peridynamics via finite element analysis[J]. Finite Elements in Analysis and Design, 2007, 43(15): 1169-1178.

[5] Bobaru F, Yang M J, Alves L F, et al. Convergence, adaptive refinement, and scaling in 1D peridynamics[J]. International Journal for Numerical Methods in Engineering, 2009, 77(6): 852-877.

[6] Bobaru F, Ha Y D. Adaptive refinement and multiscale modeling in 2D peridynamics[J]. International Journal for Multiscale Computational Engineering, 2011, 9(6): 635-659.

[7] Dipasquale D, Zaccariotto M, Galvanetto U. Crack propagation with adaptive grid refinement in 2D peridynamics[J]. International Journal of Fracture, 2014, 190(1-2): 1-22.

[8] Ren H L, Zhuang X Y, Cai Y C, et al. Dual-horizon peridynamics[J]. International Journal for Numerical Methods in Engineering, 2016, 108(12): 1451-1476.

[9] Silling S A, Lehoucq R B. Convergence of peridynamics to classical elasticity theory[J]. Journal of Elasticity, 2008, 93(1): 13-37.

[10] Gerstle W, Sau N, Silling S A. Peridynamic modeling of plain and reinforced concrete structures[C]. Proceedings of 18th International Conference on Structural Mechanics in Reactor Technology, 2005.

[11] Merwe C W V D. A peridynamic model for sleeved hydraulic fracture[D]. Stellenbosch: Stellenbosch University, 2014.

[12] Henke S F, Shanbhag S. Mesh sensitivity in peridynamic simulations[J]. Computer Physics Communications, 2014, 185(1): 181-193.

[13] Ren H L, Zhuang X Y, Rabczuk T. Dual-horizon peridynamics: A stable solution to varying horizons[J]. Computer Methods in Applied Mechanics and Engineering, 2017, 318: 762-782.

[14] Hu Y L, Chen H L, Spencer B, et al. Thermomechanical peridynamic analysis with irregular non-uniform domain discretization[J]. Engineering Fracture Mechanics, 2018, 197: 92-113.

[15] Gu X, Zhang Q, Xia X Z. Voronoi-based peridynamics and cracking analysis with adaptive refinement[J]. International Journal for Numerical Methods in Engineering, 2017, 112(13): 2087-2109.

[16] Freimanis A, Paeglitis A. Mesh sensitivity in peridynamic quasi-static simulations[J]. Procedia Engineering, 2017, 172: 284-291.

[17] Dipasquale D, Sarego G, Zaccariotto M, et al. Dependence of crack paths on the orientation of regular 2D peridynamic grids[J]. Engineering Fracture Mechanics, 2016, 160: 248-263.

[18] Duzzi M, Dipasquale D, Sarego G, et al. A concurrent multiscale model to predict crack propagation in nanocomposite materials with the peridynamic theory[C]. Nanotech Italy, 2014.

[19] Silling S A, Littlewood D, Seleson P. Variable horizon in a peridynamic medium[J]. Journal of Mechanics of Materials and Structures, 2015, 10(5): 591-612.

[20] Rahman R, Foster J T. Onto resolving spurious wave reflection problem with changing nonlocality among various length scales[J]. Communications in Nonlinear Science and Numerical Simulation, 2016, 34: 86-122.

[21] 王芳. 双材料界面及其力学性能的近场动力学研究 [D]. 武汉：武汉理工大学, 2015.

[22] Ganzenmüller G C, Hiermaier S, May M. Improvements to the prototype micro-brittle model of peridynamics[C]. International Workshop on Meshfree Methods for Partial Differential Equations, 2015: 163-183.

[23] Yvonnet J, Coffignal G, Ryckelynck D, et al. A simple error indicator for meshfree methods based on natural neighbors[J]. Computers & Structures, 2006, 84(21): 1301-1312.

[24] Tong Q, Li S F. Multiscale coupling of molecular dynamics and peridynamics[J]. Journal of the Mechanics and Physics of Solids, 2016, 95: 169-187.

[25] Gu X, Zhang Q, Huang D, et al. Wave dispersion analysis and simulation method for concrete SHPB test in peridynamics[J]. Engineering Fracture Mechanics, 2016, 160: 124-137.

[26] Ha Y D, Bobaru F. Studies of dynamic crack propagation and crack branching with peridynamics[J]. International Journal of Fracture, 2010, 162(1): 229-244.

[27] Ha Y D, Bobaru F. Characteristics of dynamic brittle fracture captured with peridynamics[J]. Engineering Fracture Mechanics, 2011, 78(6): 1156-1168.

[28] Bobaru F, Hu W K. The meaning, selection, and use of the peridynamic horizon and its relation to crack branching in brittle materials[J]. International Journal of Fracture,

2012, 176(2): 215-222.

[29] Lindsay P, Parks M L, Prakash A. Enabling fast, stable and accurate peridynamic computations using multi-time-step integration[J]. Computer Methods in Applied Mechanics and Engineering, 2016, 306: 382-405.

[30] 王建华, 杨磊. 有限元后验误差估计方法的研究进展 [J]. 力学进展, 2000, 30(2): 175-190.

[31] Du Q, Ju L L, Tian L, et al. A posteriori error analysis of finite element method for linear nonlocal diffusion and peridynamic models[J]. Mathematics of Computation, 2013, 82(284): 1889-1922.

[32] Bowden F P, Brunton J H, Field J E, et al. Controlled fracture of brittle solids and interruption of electrical current[J]. Nature, 1967, 216: 38-42.

第 11 章 冲击侵彻与爆炸问题的近场动力学模拟

冲击和爆炸载荷作用下结构的动力响应与毁伤破坏问题具有重要的学术意义和广泛的应用背景，数值模拟方法已成为该问题研究的主要手段。近场动力学作为一种新兴的积分型非局部连续介质力学理论与数值方法，在进行强动载作用下结构的动力裂纹扩展分析和剧烈破碎毁伤问题模拟时具有独特的优势。本章首先给出刚性体冲击侵彻可变形体的冲击接触算法，介绍一种空中自由爆炸载荷的施加方法；随后详细阐释准脆性材料的 JH-2 动态损伤本构模型；在探讨键型近场动力学方法非局部色散特性的基础上，建立分离式霍普金森压杆 (SHPB) 冲击试验的近场动力学方法，模拟典型混凝土巴西盘的 SHPB 冲击破坏试验；最后采用非常规态型近场动力学方法，进行混凝土冲击层裂与爆炸毁伤破坏问题的计算分析。

11.1 冲击接触算法与爆炸载荷的计算

11.1.1 冲击接触算法

考虑一刚性弹丸冲击靶体 (可变形体)，在采用近场动力学进行该问题计算分析时，需要发展冲击接触算法，以实现刚柔体的接触描述[1]。

设刚性弹丸的初速度为 $\boldsymbol{v}_0$，弹丸与靶体接触后侵入至靶体，靶体的变形遵循近场动力学运动方程。为反映冲击接触的真实物理过程，靶体中与弹丸叠合的物质点需要更新位置，具体实施时，靶体物质点的位置更新滞后于弹丸运动，可将叠合的物质点移动最短距离到弹丸外部，且该物质点与弹丸表面距离保持最近，如图 11-1 所示。该过程反映任意时刻 t 弹丸与靶体间形成接触面，且弹丸与靶体的物质点没有相互叠合，t 时刻弹丸的速度为 $\boldsymbol{v}_t$。

在 t 时刻，某物质点坐标为 $\boldsymbol{x}_k$、位移为 $\boldsymbol{u}_{(k)}^t$、速度为 $\boldsymbol{v}_{(k)}^t$，此时，弹丸与靶体间形成接触面且无叠合现象发生。在 $t+\Delta t$ 时刻，若物质点与弹丸叠合，则启用位置更新算法，计算物质点 $\boldsymbol{x}_k$ 的位移 $\overline{\boldsymbol{u}}_{(k)}^{t+\Delta t}$，再计算物质点速度 $\overline{\boldsymbol{v}}_{(k)}^{t+\Delta t}$，即

$$\overline{\boldsymbol{v}}_{(k)}^{t+\Delta t} = \frac{\overline{\boldsymbol{u}}_{(k)}^{t+\Delta t} - \boldsymbol{u}_{(k)}^t}{\Delta t} \tag{11.1}$$

于是，在 $t+\Delta t$ 时刻，物质点 $\boldsymbol{x}_k$ 对弹丸的作用力为

$$\boldsymbol{F}_{(k)}^{t+\Delta t} = -\rho_{(k)} \frac{\overline{\boldsymbol{v}}_{(k)}^{t+\Delta t} - \boldsymbol{v}_{(k)}^t}{\Delta t} V_{(k)} \tag{11.2}$$

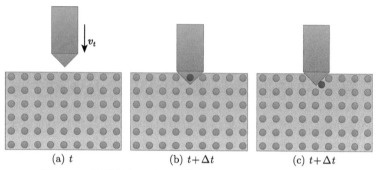

图 11-1　刚性体与可变形体接触中物质点位置更新示意图

此时,弹丸所受的合力则为

$$F^{t+\Delta t} = \sum_{k=1} F_{(k)}^{t+\Delta t} \lambda_{(k)}^{t+\Delta t} \tag{11.3}$$

式中,当靶体的物质点与弹丸发生叠合时,$\lambda_{(k)}^{t+\Delta t}=1$;未发生叠合时,$\lambda_{(k)}^{t+\Delta t}=0$。对于接触区域的物质点,采用上述冲击接触算法,更新物质点的位置和速度等变量,而非接触区域的物质点则依据近场动力学运动方程更新变量。

11.1.2　爆炸载荷施加方法

结构的爆炸毁伤分析涉及炸药爆炸、爆炸冲击波的形成和传播、冲击波或爆炸产物与固体结构的相互作用以及由此引起的结构动态响应四个过程,涉及多相介质在高速、高温、高压等极端条件下的复杂力学行为,也必然会产生复杂的不连续变形,研究难度很大。爆炸是能量在极短时间内迅速释放的过程,爆炸产生的效应主要体现在爆生冲击波上,爆炸点周围瞬间形成的高压火球猛烈向外膨胀、压缩空气形成高压气浪,并且迅速向四周传播,随着距离的增加,传播速度逐渐变慢,压力逐渐变小。

模拟结构爆炸响应的关键是确定爆炸荷载的作用方式和大小。一种简单的处理方法是采用经验公式或爆炸荷载时程曲线,直接将爆炸荷载施加到结构上,以简化爆炸过程的前三个阶段,方便得到问题解答。但这种处理方法不能反映真实爆炸的复杂过程,致使荷载施加不准确,影响结果的真实性。随着对爆轰理论的深入研究和数值模拟水平的提高,研究者愈发地关注结构爆炸响应的全过程,致力于科学地描述炸药爆炸、爆炸冲击波的传播规律、爆炸冲击波或爆炸产物与结构的相互作用,以准确分析结构的爆炸响应[2]。

图 11-2 为空中自由爆炸的冲击波传播示意图,爆炸发生在远离结构的空中,产生的爆炸冲击波传播到离起爆位置较远后才作用到结构上,由于爆源离地面较

远，爆炸冲击波作用于结构之前，尚未传播到地面，爆炸冲击波不会因地面的反射而增强。

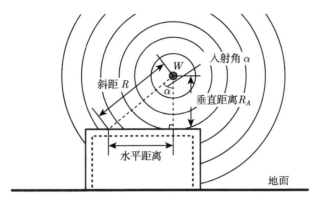

图 11-2 空中自由爆炸的冲击波传播示意图

空中爆炸产生的冲击波典型压力时程曲线如图 11-3 所示，设 P_0 为目标点的大气压，自由空气中的爆源起爆后产生的冲击波经过时间 T_a 到达目标点，目标点处的压力瞬间迅速上升至峰值 P_s。随着冲击波向前传播，目标点处的压力逐渐下降，经时间 T_0 后恢复到大气压。之后，目标点的压强并不是停留在大气压上，而是随着冲击波的向前传播继续下降至负超压，直到降到负的峰值 P_s^-，再逐渐恢复到大气压强，负超压持续时间为 T_0^-。由于该爆炸冲击波的压力时程曲线较为复杂，不利于进行爆炸载荷作用下结构的动力响应和损伤破坏分析，在应用时通常需要对该压力时程曲线进行简化，根据 1979 年 Henrych 出版的 *The Dynamics of Explosion and Its Use* (爆炸动力学及其应用) 专著[4]，简化后的压力时程曲线如图 11-4 所示，本章将利用该压力时程曲线对结构施加空中自由爆炸载荷。

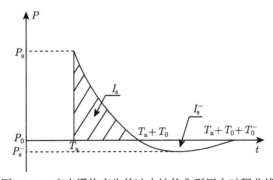

图 11-3 空中爆炸产生的冲击波的典型压力时程曲线

具体地，冲击波波前到达时间 T_a 为

$$T_{\mathrm{a}} = \frac{0.34 R^{1.4} W^{-0.2}}{c_{\mathrm{a}}} \tag{11.4}$$

其中，R 为起爆点与结构之间的距离，W 为 TNT 炸药的质量，c_{a} 为声音在空气中的传播速度，为 340m/s。

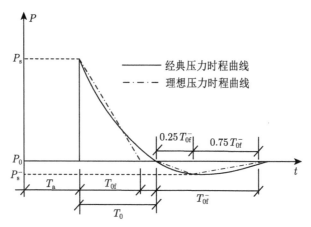

图 11-4 爆炸冲击波的简化压力时程曲线

空中爆炸冲击波的破坏作用主要集中在正超压阶段，正超压阶段的破坏效应主要取决于正超压峰值、正超压作用持续时间和正超压比冲量[3-5]。爆轰超压是指爆轰波前沿压力与大气压之间的压力差 $\Delta P_{\mathrm{s}} = P_{\mathrm{s}} - P_0$，正超压峰值经验公式为

$$\Delta P_{\mathrm{s}} = \begin{cases} \dfrac{1.4072}{z} + \dfrac{0.554}{z^2} - \dfrac{0.0357}{z^3} + \dfrac{0.000625}{z^4}(\mathrm{MPa}), & 0.05 \leqslant z < 0.3 \\ \dfrac{0.6194}{z} - \dfrac{0.033}{z^2} + \dfrac{0.213}{z^3}(\mathrm{MPa}), & 0.3 \leqslant z < 1 \\ \dfrac{0.066}{z} + \dfrac{0.405}{z^2} + \dfrac{0.329}{z^3}(\mathrm{MPa}), & 1 \leqslant z \leqslant 10 \end{cases} \tag{11.5}$$

其中，z 为折合距离，计算式为 $z = \dfrac{R}{W^{1/3}}$。

正超压作用持续时间经验公式为

$$T_0 = \sqrt[3]{W}(0.107 + 0.444z + 0.264z^2 - 0.129z^3 + 0.0335z^4) \times 10^{-3}(\mathrm{s}) \tag{11.6}$$

正超压比冲量经验公式为

$$I_{\mathrm{s}} = \begin{cases} \left(6630 - \dfrac{11150}{z} + \dfrac{6290}{z^2} - \dfrac{1004}{z^3}\right)\sqrt[3]{W}(\mathrm{Pa \cdot s}), & 0.4 \leqslant z < 0.75 \\ \left(-322 + \dfrac{2110}{z} - \dfrac{2160}{z^2} + \dfrac{801}{z^3}\right)\sqrt[3]{W}(\mathrm{Pa \cdot s}), & 0.75 \leqslant z < 3 \end{cases} \tag{11.7}$$

因此，对于简化的压力时程曲线，正超压作用的持续时间为

$$T_{0f} = \frac{2I_s}{\Delta P_s} \tag{11.8}$$

虽然爆炸冲击波的主要破坏力集中在正超压作用阶段，但是在某些特殊情况下，负超压作用阶段也起着一定的破坏作用。需要注意的是对于简化的压力时程曲线，在正超压载荷结束后和负超压载荷开始前有一个时间间隔。简化的负超压载荷部分主要取决于负超压峰值、负超压比冲量，并由此求出负超压作用持续时间。

负超压峰值 $\Delta P_s^- = P_s^- - P_0$ 的经验公式为

$$\Delta P_s^- = \begin{cases} -\dfrac{0.035}{z}(\text{MPa}), & z > 1.6 \\ 0, & z < 1.6 \end{cases} \tag{11.9}$$

负超压比冲量经验公式为

$$I_s^- = I_s\left(1 - \frac{0.5}{z}\right) \quad (\text{Pa}\cdot\text{s}) \tag{11.10}$$

于是，负超压作用的持续时间为

$$T_{0f}^- = \frac{2I_s^-}{-\Delta P_s^-} \tag{11.11}$$

对于简化的负超压载荷曲线，假定由零开始线性下降到负超压峰值的时间为 $0.25T_{0f}^-$，由负超压峰值线性上升到大气压强经历的时间为 $0.75T_{0f}^-$。

综合上述结果，归纳得到简化的爆炸超压公式如下：

$\Delta P(t) =$

$$\begin{cases} 0, & t < T_a \\ \Delta P_s(1 - (t - T_a)/T_{0f}), & T_a \leqslant t < T_a + T_{0f} \\ 0, & T_a + T_{0f} \leqslant t < T_a + T_0 \\ \Delta P_s^-(t - T_a - T_0)/0.25T_{0f}^-, & T_a + T_0 \leqslant t < T_a + T_0 + 0.25T_{0f}^- \\ \Delta P_s^-(1 - (t - T_a - T_0 - 0.25T_{0f}^-)/0.75T_{0f}^-), & T_a + T_0 + 0.25T_{0f}^- \leqslant t < T_a + T_0 + T_{0f}^- \\ 0, & t \geqslant T_a + T_0 + T_{0f}^- \end{cases}$$

$$\tag{11.12}$$

11.2 准脆性材料的 JH-2 本构模型

11.2.1 JH-2 本构关系

针对侵彻和爆炸载荷作用下脆性材料的复杂力学行为，1994 年，Johnson 和 Holmquist[6] 提出了 Johnson-Holmquist-II (JH-2) 动力损伤本构模型，该模型综合考虑了高压、应变率效应和损伤演化对材料本构关系的影响，能较好地描述大应变、高应变率和高压状态下混凝土等脆性材料的大变形、损伤破碎和断裂等问题。JH-2 模型的球应力 (静水压力) 由状态方程控制，用以描述材料的体积变形响应；偏应力由胡克定律控制并通过塑性准则修正，用以描述材料的剪切变形效应。JH-2 模型能很好地反映材料的内能损失与静水压力增量之间的关系，刻画材料从完整状态到断裂状态的损伤过程。与 JH-1 模型相比，JH-2 模型考虑了材料的逐渐软化现象，并通过 Hugoniot 弹性极限 (HEL) 的强度和静水压力进行标定，将一些材料常数无量纲化，便于在试验数据不足的情况下进行估算。

下面简要介绍 JH-2 本构涉及的强度模型、损伤模型和静水压力模型。如图 11-5 所示，归一化的等效应力为

$$\sigma^* = \sigma_i^* - D_{\text{JH-2}}(\sigma_i^* - \sigma_f^*) \tag{11.13}$$

式中，损伤度取值范围为 $0 \leqslant D_{\text{JH-2}} \leqslant 1$，$\sigma_i^*$ 为完好材料的正则化等效应力，σ_f^* 为断裂材料的正则化等效应力。$\sigma^*, \sigma_i^*, \sigma_f^*$ 等正则化等效应力的一般形式为

$$\sigma^* = \sigma/\sigma_{\text{HEL}} \tag{11.14}$$

其中，σ 为实际等效应力，σ_{HEL} 为 Hugoniot 弹性极限下的等效应力。

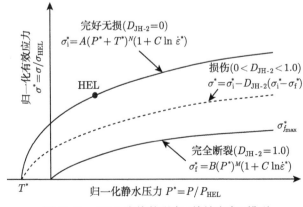

图 11-5 JH-2 本构的强度 (等效应力) 模型

11.2 准脆性材料的 JH-2 本构模型

具体地，完好材料的正则化强度 (等效应力) 为

$$\sigma_i^* = A(P^* + T^*)^N (1 + C \ln \dot{\varepsilon}^*) \tag{11.15}$$

断裂材料的正则化强度 σ_f^* 应小于等于材料最大断裂强度 $\sigma_{f,\max}^*$，可表示为

$$\sigma_f^* = B(P^*)^M (1 + C \ln \dot{\varepsilon}^*) \tag{11.16}$$

式中，$\dot{\varepsilon}^* = \dot{\varepsilon}_{eq}/\dot{\varepsilon}_0$ 为无量纲应变率，$\dot{\varepsilon}_{eq}$ 为实际等效应变率，$\dot{\varepsilon}_0 = 1.0 \text{ s}^{-1}$ 为参考应变率；A、B、C、M、N 为材料强度模型中引入的参数；P^* 为正则化静水压力，T^* 为材料能承受的正则化的最大拉伸球应力

$$\begin{aligned} P^* &= P/P_{\text{HEL}} \\ T^* &= T/P_{\text{HEL}} \end{aligned} \tag{11.17}$$

其中，P 为实际静水压力，T 为材料能承受的最大拉伸球应力 (抗拉强度)，P_{HEL} 为 Hugoniot 弹性极限下的压力。

如图 11-6 所示，考虑材料在断裂过程中损伤的不断累积，有

$$D_{\text{JH-2}} = \sum \Delta \varepsilon_P / \varepsilon_P^f \tag{11.18}$$

式中，$\Delta \varepsilon_P$ 为等效塑性应变增量，ε_P^f 为常压力 P 下断裂时的塑性应变，且有

$$\varepsilon_P^f = D_1 (P^* + T^*)^{D_2} \tag{11.19}$$

其中，D_1 和 D_2 为材料常数，ε_P^f 随 P^* 的增大而增大，当 $P^* = -T^*$ 时，材料不会发生塑性应变。

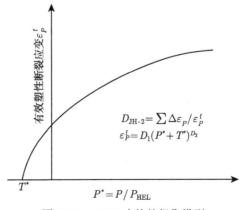

图 11-6　JH-2 本构的损伤模型

如图 11-7 所示，在损伤断裂产生前，$D_{\text{JH-2}} = 0$，静水压力与体应变相关，即

$$P = K_1\mu + K_2\mu^2 + K_3\mu^3 \tag{11.20}$$

式中，K_1 为体积模量，K_2 和 K_3 为材料常数；$\mu = \rho/\rho_0 - 1$，ρ 和 ρ_0 分别为材料当前密度和初始密度。当材料处于拉伸膨胀状态时，即 $\mu < 0$，静水压力根据 $P = K_1\mu$ 计算；当材料开始损伤断裂，即 $D_{\text{JH-2}} > 0$，需在式 (11.20) 的静水压力项中增加额外压力增量，有

$$P = K_1\mu + K_2\mu^2 + K_3\mu^3 + \Delta P \tag{11.21}$$

其中，当损伤 $D_{\text{JH-2}} = 0$ 时，压力增量 $\Delta P = 0$，当损伤 $D_{\text{JH-2}} = 1$ 时，压力增量达到最大值 $\Delta P = \Delta P_{\max}$。

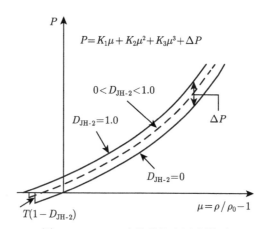

图 11-7　JH-2 本构的静水压力模型

压力增量 ΔP 随损伤度 $D_{\text{JH-2}}$ 的增大而增大。当材料损伤断裂时，材料强度下降导致偏应力减小，偏应力减小继而导致材料的弹性畸变能增量减小，根据能量守恒原理，外力功的作用将更多地转化为内部压力势能，内部压力势能的增加直接体现在压力增量的不断增大。

具体地，畸变能密度或剪切变形能密度可由下式计算

$$U = \sigma^2/(6G) \tag{11.22}$$

其中，σ 为等效塑性应力，G 为剪切模量，则材料的畸变内能损失为

$$\Delta U = U_t - U_{t+\Delta t} \tag{11.23}$$

当能量损失 ΔU 完全转化为压力势能时，内能损失 ΔU 和依赖 ΔP 的压力势能增量满足能量守恒，所以有

$$(\Delta P_{t+\Delta t} - \Delta P_t)\mu_{t+\Delta t} + (\Delta P_{t+\Delta t}^2 - \Delta P_t^2)/2K_1 = \beta \Delta U \tag{11.24}$$

其中，左端第一项为受压 ($\mu > 0$) 时的压力势能，左端第二项为受拉 ($\mu < 0$) 时的压力势能，β 为损失的弹性内能转化为压力势能的比例，随时间更新的 ΔP 由下式计算获得

$$\Delta P_{t+\Delta t} = -K_1\mu_{t+\Delta t} + \sqrt{(K_1\mu_{t+\Delta t} + \Delta P)^2 + 2\beta K_1 \Delta U} \tag{11.25}$$

11.2.2 JH-2 本构的更新算法

速度梯度张量 $\boldsymbol{L}$ 可由非局部变形梯度张量 $\boldsymbol{F}$ 获得，即

$$\boldsymbol{L} = \dot{\boldsymbol{F}}\boldsymbol{F}^{-1} \tag{11.26}$$

式中，$\dot{\boldsymbol{F}}$ 为非局部变形梯度张量的时间导数，$\boldsymbol{F}^{-1}$ 为非局部变形梯度张量的逆。速度梯度张量 $\boldsymbol{L}$ 可以分解为对称张量 $\boldsymbol{D}$ 和反对称张量 $\boldsymbol{W}$ 之和，即

$$\begin{aligned}\boldsymbol{L} &= \boldsymbol{D} + \boldsymbol{W} \\ \boldsymbol{D} &= \frac{1}{2}\left[\boldsymbol{L} + \boldsymbol{L}^{\mathrm{T}}\right] \\ \boldsymbol{W} &= \frac{1}{2}\left[\boldsymbol{L} - \boldsymbol{L}^{\mathrm{T}}\right]\end{aligned} \tag{11.27}$$

其中，对称张量 $\boldsymbol{D}$ 为变形速率张量，非对称张量 $\boldsymbol{W}$ 为旋率张量。

通过变形速率张量 $\boldsymbol{D}$ 可以计算非局部无旋变形率张量 $\boldsymbol{d}$

$$\boldsymbol{d} = \boldsymbol{R}^{\mathrm{T}}\boldsymbol{D}\boldsymbol{R} \tag{11.28}$$

式中，$\boldsymbol{R}$ 为正交转动张量，可以通过如下方法确定。

1) 在第一个时间步 ($n = 0$)，计算 $\boldsymbol{V}_{(0)}$ 和 $\boldsymbol{R}_{(0)}$

根据极分解定理，非局部变形梯度张量 $\boldsymbol{F}$ 存在如下关系

$$\boldsymbol{F} = \boldsymbol{V}\boldsymbol{R} \tag{11.29}$$

其中，$\boldsymbol{V}$ 表示左拉伸张量，利用极分解和转动张量的正交性，有以下公式成立

$$\boldsymbol{V}^2 = \boldsymbol{F}\boldsymbol{F}^{\mathrm{T}} \tag{11.30}$$

计算该式的特征值 $\lambda_1, \lambda_2, \lambda_3$ 以及相应的模态矩阵 $\boldsymbol{N}$，可获得左拉伸张量及其逆

$$\begin{aligned}\boldsymbol{V} &= \boldsymbol{N}^{\mathrm{T}}\mathrm{diag}\begin{bmatrix} \sqrt{\lambda_1} & \sqrt{\lambda_2} & \sqrt{\lambda_3} \end{bmatrix}\boldsymbol{N} \\ \boldsymbol{V}^{-1} &= \boldsymbol{N}^{\mathrm{T}}\mathrm{diag}\begin{bmatrix} \dfrac{1}{\sqrt{\lambda_1}} & \dfrac{1}{\sqrt{\lambda_2}} & \dfrac{1}{\sqrt{\lambda_3}} \end{bmatrix}\boldsymbol{N}\end{aligned} \tag{11.31}$$

因此，初始时刻的左拉伸张量 $\boldsymbol{V}_{(0)}$ 和转动张量 $\boldsymbol{R}_{(0)}$ 分别为

$$\begin{aligned}\boldsymbol{V}_{(0)} &= \boldsymbol{N}_{(0)}^{\mathrm{T}}\mathrm{diag}\begin{bmatrix} \sqrt{\lambda_1} & \sqrt{\lambda_2} & \sqrt{\lambda_3} \end{bmatrix}_{(0)}\boldsymbol{N}_{(0)} \\ \boldsymbol{R}_{(0)} &= \left[\boldsymbol{V}^{-1}\boldsymbol{F}\right]_{(0)}\end{aligned} \tag{11.32}$$

2) 在第 $(n+1)$ 时间步 $(n>0)$，计算当前的 $\boldsymbol{V}_{(n+1)}$ 和 $\boldsymbol{R}_{(n+1)}$

将式 (11.26) 代入到速度梯度张量中，得到

$$\begin{aligned}\boldsymbol{L} &= \dot{\boldsymbol{F}}\boldsymbol{F}^{-1} = (\dot{\boldsymbol{V}}\boldsymbol{R} + \boldsymbol{V}\dot{\boldsymbol{R}})\boldsymbol{R}^{-1}\boldsymbol{V}^{-1} = \dot{\boldsymbol{V}}\boldsymbol{R}\boldsymbol{R}^{-1}\boldsymbol{V}^{-1} + \boldsymbol{V}\dot{\boldsymbol{R}}\boldsymbol{R}^{-1}\boldsymbol{V}^{-1} \\ &= \dot{\boldsymbol{V}}\boldsymbol{V}^{-1} + \boldsymbol{V}\dot{\boldsymbol{R}}\boldsymbol{R}^{-1}\boldsymbol{V}^{-1} = \dot{\boldsymbol{V}}\boldsymbol{V}^{-1} + \boldsymbol{V}\boldsymbol{\Omega}\boldsymbol{V}^{-1}\end{aligned} \tag{11.33}$$

对上式右乘左拉伸张量 $\boldsymbol{V}$，有 $\boldsymbol{L}\boldsymbol{V} = \dot{\boldsymbol{V}}\boldsymbol{V}^{-1}\boldsymbol{V} + \boldsymbol{V}\boldsymbol{\Omega}\boldsymbol{V}^{-1}\boldsymbol{V}$，亦即

$$\dot{\boldsymbol{V}} = \boldsymbol{L}\boldsymbol{V} - \boldsymbol{V}\boldsymbol{\Omega} \tag{11.34}$$

式中，$\boldsymbol{\Omega}$ 为刚体旋转率张量，可由下式计算获得

$$\begin{cases} (1)\ z_i = e_{ikj}D_{jm}V_{mk} \\ (2)\ w_i = -\dfrac{1}{2}e_{ijk}W_{jk} \\ (3)\ \boldsymbol{\omega} = \boldsymbol{w} + [\mathrm{tr}(\boldsymbol{V})\boldsymbol{I} - \boldsymbol{V}]^{-1}\boldsymbol{z} \\ (4)\ \Omega_{ij} = e_{ikj}\omega_k \end{cases} \tag{11.35}$$

其中，$\boldsymbol{\Omega}$ 的矩阵形式为 $\boldsymbol{\Omega} = \begin{bmatrix} 0 & -\omega_3 & \omega_2 \\ \omega_3 & 0 & -\omega_1 \\ -\omega_2 & \omega_1 & 0 \end{bmatrix}$。至此，可获得当前时步的左拉伸张量 $\boldsymbol{V}_{(n+1)}$ 和转动张量 $\boldsymbol{R}_{(n+1)}$

$$\begin{aligned}\boldsymbol{V}_{(n+1)} &= \boldsymbol{V}_{(n)} + \Delta t \dot{\boldsymbol{V}}_{\Delta t} \\ \boldsymbol{R}_{(n+1)} &= \left[\boldsymbol{I} - \dfrac{1}{2}\Delta t\boldsymbol{\Omega}\right]^{-1}\left[\boldsymbol{I} + \dfrac{1}{2}\Delta t\boldsymbol{\Omega}\right]\boldsymbol{R}_{(n)}\end{aligned} \tag{11.36}$$

11.2 准脆性材料的 JH-2 本构模型

如上所述,根据非局部变形梯度张量可求得非局部无旋变形率张量 $\boldsymbol{d}$,进一步可将 JH-2 本构嵌入到非常规态型近场动力学模型框架中,以下给出应变分析和应力分析的过程。

应变张量 $\boldsymbol{\varepsilon}$ 可以分解为偏应变张量 $\boldsymbol{e}$ 和球应变张量 $\boldsymbol{\varepsilon}_\mathrm{m}$

$$\boldsymbol{\varepsilon} = \boldsymbol{e} + \boldsymbol{\varepsilon}_\mathrm{m} = \boldsymbol{e} + \frac{1}{3}\mathrm{tr}(\varepsilon)\boldsymbol{I} \tag{11.37}$$

同样地,无旋变形率张量 $\boldsymbol{d}$ 也可以分解为偏应变率张量 $\dot{\boldsymbol{e}}$ 和球应变率张量 $\dot{\boldsymbol{\varepsilon}}_\mathrm{m}$

$$\boldsymbol{d} = \dot{\boldsymbol{e}} + \dot{\boldsymbol{\varepsilon}}_\mathrm{m} = \dot{\boldsymbol{e}} + \frac{1}{3}\mathrm{tr}(\boldsymbol{d})\boldsymbol{I} \tag{11.38}$$

等效应变率为

$$\dot{\varepsilon}_\mathrm{eq} = \sqrt{\frac{2}{3}}|\dot{\boldsymbol{e}}| = \sqrt{\frac{2}{3}}\sqrt{\dot{e}_{ij}\dot{e}_{ij}} \tag{11.39}$$

根据体应变 $\varepsilon_\mathrm{vol} = 3\varepsilon_\mathrm{m}$ 和球应变率 $\dot{\boldsymbol{\varepsilon}}_\mathrm{m} = \frac{1}{3}\mathrm{tr}(\boldsymbol{d})\boldsymbol{I}$ 的关系式,可以得到增量迭代过程中的体应变表达式

$$[\boldsymbol{\varepsilon}_\mathrm{vol}]_t = [\boldsymbol{\varepsilon}_\mathrm{vol}]_{t-\Delta t} + 3\Delta t\,[\dot{\boldsymbol{\varepsilon}}_\mathrm{m}]_t \tag{11.40}$$

对式 (11.38) 两端同乘以 Δt,将率形式转化为增量形式,则偏应变增量 $\Delta\boldsymbol{e}$ 为

$$\Delta\boldsymbol{e} = \Delta\boldsymbol{\varepsilon} - \frac{1}{3}\mathrm{tr}(\Delta\varepsilon)\boldsymbol{I} = \boldsymbol{d}\Delta t - \frac{1}{3}\mathrm{tr}(\boldsymbol{d}\Delta t)\boldsymbol{I} \tag{11.41}$$

偏应变率张量 $\dot{\boldsymbol{e}}$ 可以分解为弹性偏应变率 $\dot{\boldsymbol{e}}^\mathrm{e}$ 和塑性偏应变率 $\dot{\boldsymbol{e}}^\mathrm{p}$

$$\dot{\boldsymbol{e}} = \dot{\boldsymbol{e}}^\mathrm{e} + \dot{\boldsymbol{e}}^\mathrm{p} \tag{11.42}$$

根据胡克定律,偏应变率张量的弹性部分与偏应力率张量的关系为

$$\dot{\boldsymbol{e}}^\mathrm{e} = \frac{\dot{\boldsymbol{S}}}{2G}, \quad \dot{e}^\mathrm{e}_{ij} = \frac{\dot{S}_{ij}}{2G} \tag{11.43}$$

其中,G 为剪切模量,$\dot{\boldsymbol{S}}$ 为偏应力率张量。根据流动法则,偏应变率张量的塑性部分为

$$\dot{\boldsymbol{e}}^\mathrm{p} = \dot{\lambda}\boldsymbol{Q} = \dot{\lambda}\frac{\boldsymbol{S}}{\sigma_\mathrm{eq}} \tag{11.44}$$

式中，$\dot{\lambda}$ 为待定的标量塑性乘子，$\boldsymbol{Q}$ 为单位偏应力张量，即偏应力张量与等效应力的比值，$\boldsymbol{S}$ 是偏应力张量，$\sigma_{\text{eq}} = \sqrt{\dfrac{3}{2}}|\boldsymbol{S}| = \sqrt{\dfrac{3}{2}S_{ij}S_{ij}}$ 为等效应力。综合式 (11.42)、(11.43) 和 (11.44)，可以获得如下关系

$$\dot{\boldsymbol{e}} = \frac{\dot{\boldsymbol{S}}}{2G} + \dot{\lambda}\boldsymbol{Q} \tag{11.45}$$

上式两边同乘以 Δt，则有如下增量形式成立

$$\Delta \boldsymbol{e} = \frac{\Delta \boldsymbol{S}}{2G} + \Delta\lambda\boldsymbol{Q} \tag{11.46}$$

在时间 $t-\Delta t$ 到时间 t 内，上式表示为

$$\Delta\boldsymbol{e}_{t-\Delta t/2} : \boldsymbol{Q} - \Delta\lambda_{t-\Delta t/2}\boldsymbol{Q} : \boldsymbol{Q} - \frac{1}{2G}[\boldsymbol{S}_t : \boldsymbol{Q} - \boldsymbol{S}_{t-\Delta t} : \boldsymbol{Q}] = 0 \tag{11.47}$$

式中，符号 ":" 表示张量缩并①，下标 $t-\Delta t/2$ 表示 $t-\Delta t$ 到 t 时刻的增量，下标 t 和 $t-\Delta t$ 分别表示 t 时刻和 $t-\Delta t$ 时刻的变量值。若材料在 $t-\Delta t$ 到 t 时段发生屈服，则偏应变张量 $\boldsymbol{S}_t$ 被 $(\sigma_t^* \cdot \sigma_{\text{HEL}})\boldsymbol{Q}$ 取代，继而上式改写为

$$\Delta\boldsymbol{e}_{t-\Delta t/2} : \boldsymbol{Q} - \Delta\lambda_{t-\Delta t/2}\boldsymbol{Q} : \boldsymbol{Q} - \frac{1}{2G}[(\sigma_t^* \cdot \sigma_{\text{HEL}})\boldsymbol{Q} : \boldsymbol{Q} - \boldsymbol{S}_{t-\Delta t} : \boldsymbol{Q}] = 0 \tag{11.48}$$

求解上述两式可获得 $\Delta\lambda_{t-\Delta t/2}$。

等效塑性应变率及其增量形式如下

$$\dot{\varepsilon}_{\text{eq}}^{\text{p}} = \sqrt{\frac{2}{3}}|\dot{\boldsymbol{e}}^{\text{p}}| = \sqrt{\frac{2}{3}}\sqrt{\dot{e}_{ij}^{\text{p}}\dot{e}_{ij}^{\text{p}}} = \dot{\lambda}\sqrt{\frac{2}{3}\boldsymbol{Q} : \boldsymbol{Q}} \tag{11.49}$$

$$\Delta\varepsilon_{\text{eq}}^{\text{p}} = \Delta\lambda\sqrt{\frac{2}{3}\boldsymbol{Q} : \boldsymbol{Q}} \tag{11.50}$$

该等效塑性应变增量可以用于求解 JH-2 损伤模型，一次循环积分过程中等效塑性应变增量为 $\Delta\varepsilon_{\text{p}} = \Delta\varepsilon_{\text{eq}}^{\text{p}} = \Delta\lambda\sqrt{\dfrac{2}{3}\boldsymbol{Q}:\boldsymbol{Q}}$，进而可根据 JH-2 本构关系，计算实际等效应力 σ_{eq} 和实际偏应力张量 $\boldsymbol{S}_t$

$$\begin{aligned}\sigma_{\text{eq}} &= \sigma_t^* \cdot \sigma_{\text{HEL}} \\ \boldsymbol{S}_t &= \sigma_t^* \cdot \sigma_{\text{HEL}}\boldsymbol{Q}\end{aligned} \tag{11.51}$$

---

① 二阶张量 $\boldsymbol{A}$ 和 $\boldsymbol{B}$ 的缩并运算为 $\boldsymbol{A} : \boldsymbol{B} = \sum\limits_{i=1}^{n}\sum\limits_{i=1}^{n}A_{ij}B_{ij}$.

其中，σ_t^* 由式 (11.13) 获得。

假设在时间 $t-\Delta t$ 到时间 t 的过程中为弹性变形，则在 t 时刻，试偏应力的大小为

$$\boldsymbol{S}_t^{\text{tr}} = \boldsymbol{S}_{t-\Delta t} + 2G\Delta \boldsymbol{e}_{t-\Delta t/2} \tag{11.52}$$

弹性等效应力 σ'_{eq} 和弹性偏应力张量 $\boldsymbol{S}'_t$ 分别为

$$\begin{aligned}\sigma'_{\text{eq}} &= \sqrt{\frac{3}{2}}|\boldsymbol{S}'_t| = \sqrt{\frac{3}{2}S'_{ij}S'_{ij}} \\ \boldsymbol{S}'_t &= \boldsymbol{S}_{t-\Delta t} + 2G(\Delta \boldsymbol{e} - \Delta \boldsymbol{e}^{\text{p}})_{t-\Delta t/2} = \boldsymbol{S}_t^{\text{tr}} - 2G\Delta \boldsymbol{e}^{\text{p}}_{t-\Delta t/2}\end{aligned} \tag{11.53}$$

式中，偏应变张量增量的塑性部分 $\Delta \boldsymbol{e}^{\text{p}}$ 为

$$\Delta \boldsymbol{e}^{\text{p}} = \dot{\boldsymbol{e}}^{\text{p}}\Delta t = \dot{\lambda}\boldsymbol{Q}\Delta t = \Delta\lambda\boldsymbol{Q} \tag{11.54}$$

在通常情况下，σ'_{eq} 和 σ_{eq} 并不相等，这是由于材料弹性内能损失引起的，则增量内能损失为

$$\Delta U = (\sigma'_{\text{eq}})^2/6G - (\sigma_{\text{eq}})^2/6G \tag{11.55}$$

根据能量守恒原理，增量内能损失 ΔU 会转化为静水压力增量 ΔP，即式 (11.25)，静水压力 P_t 可根据式 (11.21) 计算。

于是，可求出 t 时刻无旋 Cauchy 应力和有旋 Cauchy 应力

$$\begin{aligned}\boldsymbol{\tau}_t &= \boldsymbol{S}_t + P_t \boldsymbol{I} \\ \boldsymbol{\sigma}_t &= \boldsymbol{R}_t \boldsymbol{\tau}_t \boldsymbol{R}_t^{\text{T}}\end{aligned} \tag{11.56}$$

至此，可以计算得到非常规态型近场动力学的力密度矢量状态

$$\underline{\boldsymbol{T}_t}\langle\boldsymbol{\xi}\rangle = w(|\boldsymbol{\xi}|)\det(\boldsymbol{F}_t)\boldsymbol{\sigma}_t\boldsymbol{F}_t^{-\text{T}}\boldsymbol{K}_t^{-1}\boldsymbol{\xi} \tag{11.57}$$

11.2.3 JH-2 本构的基准验证

为验证基于 JH-2 本构的非常规态型近场动力学模型的正确性，现重现 Johnson 和 Holmquist 早年提出的基准算例。

基准算例的计算对象为边长为 1.0m 的立方体，底面固定约束，四个侧面受法向方向位移约束，顶面自由，并承受外力载荷。在加载过程中，垂直向下载荷 $\boldsymbol{F}$ 逐渐增加，一旦顶面向下位移达到 0.05m，外力逐渐释放直至构型到达零应力状态。由于变形膨胀，构型最终体积大于其初始体积。

图 11-8 为几何模型示意图，表 11-1 为相应的本构模型参数，共考虑了三种情况。在采用非常规态型近场动力学模型进行模拟计算时，将立方体离散为 8 个物

质点,如图 11-9 所示。底部 4 点的所有位移分量为零,顶部 4 点除了竖向位移分量外,其余位移分量皆为零。外载荷 F 通过位移加载实现,在每一加载步中,对顶部 4 点施加垂直向下的位移增量 10^{-5}m;当向下总位移达到 0.05m 后,在每一加载步中,给顶部 4 点施加垂直向上的位移增量 10^{-5}m,直至达到零应力状态。每一加载计算步中,通过粒子应力状态可计算等效外力载荷 F,也即 $F = -\sigma_z$。

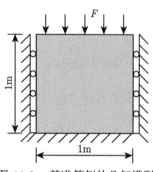

图 11-8 基准算例的几何模型

表 11-1 JH-2 模型验证算例的不同材料本构参数

	工况 A	工况 B	工况 C	单位
密度 ρ	3700	3700	3700	kg·m^{-3}
杨氏模量 E	220	220	220	GPa
泊松比 ν	0.22	0.22	0.22	/
A	0.93	0.93	0.93	/
B	0	0	0.31	/
C	0	0	0	/
M	0	0	0.6	/
N	0.6	0.6	0.6	/
σ_{HEL}	2.0	2.0	2.0	GPa
P_{HEL}	1.46	1.46	1.46	GPa
T	0.2	0.2	0.2	GPa
$\dot{\varepsilon}_0$	1.0	1.0	1.0	s^{-1}
D_1	0	0.005	0.005	/
D_2	0	1.0	1.0	/
K_1	130.95	130.95	130.95	GPa
K_2	0	0	0	GPa
K_3	0	0	0	GPa
β	1.0	1.0	1.0	/

下面针对三种不同的 JH-2 本构参数进行计算结果分析。

1) 计算工况 A

对于计算工况 A,损伤模型参数 D_1 和 D_2 均为零,表明材料不产生塑性变形,即 $\varepsilon_{\text{f}}^{\text{p}} = 0$,当应力状态达到屈服强度,材料会瞬时发生断裂。材料强度参数

B 和 M 也都为零，表明断裂材料没有强度，即 $\sigma_f = 0$，一旦完全断裂破坏，材料不再能够承受任何外载，应力状态始终为零。图 11-10 给出了计算得到的工况 A 等效应力–压强历史曲线和力–竖向位移曲线。值得注意的是，图中点②到点③是瞬间断裂变化过程，此时材料弹性内能损失完全转化为压力势能（取 $\beta = 1.0$），该过程导致膨胀压力增量为 $\Delta P = 0.560 \text{GPa}$。

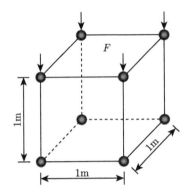

图 11-9 近场动力学模拟计算的示意图

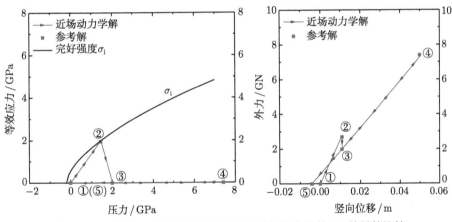

图 11-10 工况 A 的近场动力学计算结果与参考文献 [6] 结果的比较

2) 计算工况 B

计算工况 B 与工况 A 不同的是损伤模型参数不再为零，取为 $D_1 = 0.005$ 和 $D_2 = 1.0$，也即在材料开裂过程中，有塑性应变产生和累积。材料强度参数 B 和 M 也都为零，即 $\sigma_f = 0$，一旦完全断裂破坏，材料不再能够承受任何外载，应力状态始终为零。图 11-11 给出了算例 B 的应力–压力历史曲线和力–竖向位移曲线。值得注意的是，图中从点②的损伤产生到点③的完全破坏是一个渐变过程，材料弹性内能损失完全转化为压力势能 (取 $\beta = 1.0$),该过程导致膨胀压力增量逐渐增加到最大值,即 $\Delta P = 0.715\text{GPa}$。

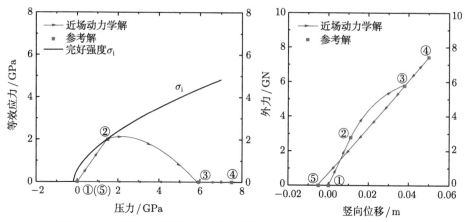

图 11-11　工况 B 的近场动力学计算结果与参考文献 [6] 结果的比较

3) 计算工况 C

计算工况 C 与工况 B 相似,但允许材料在完全开裂后仍有一定强度。在工况 C 中,其他参数与工况 B 相同,只是材料强度参数取 $B = 0.31$ 和 $M = 0.6$,即完全断裂的材料仍具有一定的断裂强度 $\sigma_\text{f}^* = 0.31(P^*)^{0.6}$。图 11-12 给出了算例 C 的应力-压力历史曲线和力-竖向位移曲线。在整个损伤断裂过程中,材料弹性内能损失完全转化为压力势能 (取 $\beta = 1.0$),该过程导致膨胀压力增量逐渐增加到最大值,即 $\Delta P = 0.649\text{GPa}$。

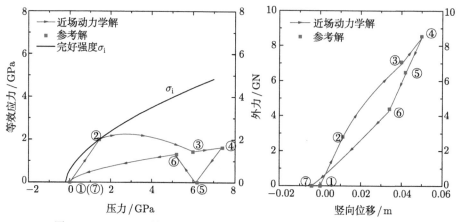

图 11-12　工况 C 的近场动力学计算结果与参考文献 [6] 结果的比较

11.3 近场动力学的非局部色散特性与霍普金森压杆冲击试验的模拟

11.3.1 键型近场动力学的非局部色散特性

一般来说，均匀材料的弹性波传播显现低频均匀传播、高频色散传播的特性，而非均匀材料的弹性波传播具有色散现象。在高频成分较多的冲击载荷作用下，弹性波通常具有色散特性。由于近场动力学的非局部特性，其具有分析弹性波传播和结构冲击破坏问题的潜力。但近场动力学的计算方案会带来一定的数值频散，导致计算得到的波速和裂纹扩展速度有一定的失真，需要研究近场动力学的内禀色散与数值色散特性，完善近场动力学的计算模型。

分别采用键型近场动力学的 PMB 模型和改进的 PMB 模型，对一维杆中波传播的近场动力学数值频散特性进行分析。不计体力和损伤，将本构模型代入近场动力学运动方程中，有

$$\begin{aligned}
\rho \ddot{u}(\boldsymbol{x}, t) &= \int_{H_\infty} c(0,\delta) g(\boldsymbol{\xi},\delta) \frac{|\boldsymbol{\xi}+\boldsymbol{\eta}|-|\boldsymbol{\xi}|}{|\boldsymbol{\xi}|} \frac{\boldsymbol{\xi}+\boldsymbol{\eta}}{|\boldsymbol{\xi}+\boldsymbol{\eta}|} \mathrm{d}\boldsymbol{x}' \\
&= \int_{H_\infty} \frac{c(0,\delta) g(\boldsymbol{\xi},\delta)}{|\boldsymbol{\xi}|} \left[\boldsymbol{\xi}+\boldsymbol{\eta} - \frac{|\boldsymbol{\xi}|}{|\boldsymbol{\xi}+\boldsymbol{\eta}|}(\boldsymbol{\xi}+\boldsymbol{\eta}) \right] \mathrm{d}\boldsymbol{x}' \\
&= \int_{H_\infty} \frac{c(0,\delta) g(\boldsymbol{\xi},\delta)}{|\boldsymbol{\xi}|} \boldsymbol{\eta} \mathrm{d}\boldsymbol{x}' \\
&= \int_{H_\infty} \frac{c(0,\delta) g(\boldsymbol{\xi},\delta)}{|\boldsymbol{x}'-\boldsymbol{x}|} (\boldsymbol{u}'-\boldsymbol{u}) \mathrm{d}\boldsymbol{x}' \\
&= \int_{H_\infty} C(\boldsymbol{x}'-\boldsymbol{x})(\boldsymbol{u}'-\boldsymbol{u}) \mathrm{d}\boldsymbol{x}' \quad (11.58)
\end{aligned}$$

一维运动方程的通解为

$$u = u_0 \mathrm{e}^{\mathrm{i}(kx-\omega t)} = u_0[\cos(kx-\omega t) + \mathrm{i}\sin(kx-\omega t)] \quad (11.59)$$

式中，u_0 为振幅，$\mathrm{i}=\sqrt{-1}$ 为虚数单位，k 为波数，ω 为角频率，经典一维波动方程的弹性波速为 $v = \dfrac{\omega}{k} = \sqrt{\dfrac{E}{\rho}}$。

将式 (11.59) 代入式 (11.58) 中，得到如下频散关系[7-9]

$$\omega^2(\boldsymbol{x}, k) = \int_{H_\infty} [1-\cos k(\boldsymbol{x}'-\boldsymbol{x})] \frac{C(\boldsymbol{x}'-\boldsymbol{x})}{\rho} \mathrm{d}\boldsymbol{x}' \quad (11.60)$$

设一维杆横截面积为单位 1，采用一维改进的 PMB 模型和原 PMB 模型，式 (11.60) 成为

$$\omega^2(\boldsymbol{x},k) = \int_{H_\infty} [1-\cos k(\boldsymbol{x}'-\boldsymbol{x})] \frac{6E\left(1-\left(\frac{\boldsymbol{x}'-\boldsymbol{x}}{\delta}\right)^2\right)^2}{\rho\delta^2|\boldsymbol{x}'-\boldsymbol{x}|} \mathrm{d}\boldsymbol{x}', \quad \text{改进 PMB 模型}$$
$$\omega^2(\boldsymbol{x},k) = \int_{H_\infty} [1-\cos k(\boldsymbol{x}'-\boldsymbol{x})] \frac{2E}{\rho\delta^2|\boldsymbol{x}'-\boldsymbol{x}|} \mathrm{d}\boldsymbol{x}', \quad \text{原 PMB 模型}$$

(11.61)

令近场范围 $\delta = m\Delta x$ ($m = 1, 2, 3, 4, 5$)，则式 (11.61) 的离散求和形式为

$$\omega^2(\boldsymbol{x},k) = \frac{2v^2}{m^2(\Delta x)^2} \sum_{j\in[-m,m], j\neq 0} \frac{3\left(1-\left(\frac{j}{m}\right)^2\right)^2 [1-\cos kj\Delta x]}{|j|}, \quad \text{改进 PMB 模型}$$
$$\omega^2(\boldsymbol{x},k) = \frac{2v^2}{m^2(\Delta x)^2} \sum_{j\in[-m,m], j\neq 0} \frac{[1-\cos kj\Delta x]}{|j|}, \quad \text{原 PMB 模型}$$

(11.62)

近场动力学离散运动方程显示，物质点间距 Δx、近场范围尺寸 δ 和时间步长将影响近场动力学的数值频散特性。式 (11.62) 不含有时间变量，故认为满足稳定性条件的时间步长对近场动力学数值频散特性的影响可以忽略，而物质点间距和近场范围尺寸是两个重要影响因素。为此，考虑不同的近场范围 ($\delta = m\Delta x$, $m = 1, 2, 3, 4$) 和物质点间距 (Δx 为 0.00005m、0.0001m、0.0002m、0.0005m、0.001m、0.002m 和 0.005m)，根据式 (11.62)，研究归一化的角频率 $\omega/(E/\rho)^{1/2}$ 与波数 k 的关系，得到一系列频散曲线，如图 11-13 所示。

图 11-13(a) 给出了物质点间距 $\Delta x = 0.001$m 时不同近场范围 (m 值) 对应的频散曲线。结果表明，① 对某一确定波数和物质点间距，随着 m 的增大，两种模型的 PD 频散曲线与理论解的偏差加剧，说明近场范围对 PD 的频散特性有重要影响；② 当波数较小，即波长较长、频率较低时，PD 方法的频散程度较小，当频率增大时，频散程度加剧；③ 改进的 PMB 模型在 $m = 1$ 时，频率始终为零，不能分析波频散现象，但 PD 的非局部性决定 m 始终大于 1，故不影响改进的 PMB 模型在波频散和开裂破坏分析中的应用。

图 11-13(b) 比较了改进 PMB 模型和原 PMB 模型的波频散程度。从图中结果可以看出，① 原 PMB 模型在 $m = 1$ 时趋向于理论解，表明当近场范围趋近于零时，PD 解趋近于传统解；② 改进的 PMB 模型在 $m = 2$ 或 $m = 3$ 时，误差较小，且精度均高于原 PMB 模型对应的结果，说明改进的 PMB 模型显著降

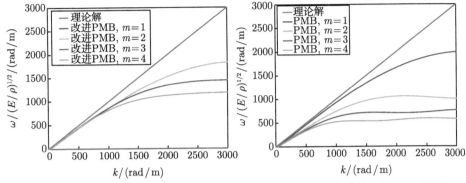

(a) 近场范围变化的频散关系(Δx=0.001m, 左: 改进PMB模型, 右: 原PMB模型)

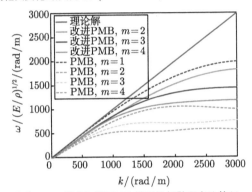

(b) 改进PMB模型与原PMB模型的频散程度比较(Δx=0.001m)

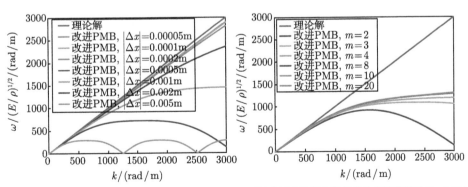

(c) 物质点间距变化的频散关系 ($m=3$)　　(d) 近场范围内物质点数目变化的频散关系(δ=0.004m)

图 11-13　归一化的角频率-波数的频散关系曲线

低了原 PMB 模型的频散特性。

图 11-13(c) 给出了 $m=3$ 时不同物质点间距对应的频散关系。结果表明, ① 当波数较低时 ($k<150$), 物质点间距增大到毫米量级造成的波频散程度较低;

当波数较高时,物质点间距增大会显著增强波频散,PD 数值解与局部理论解偏差较大,反映了非局部近场动力学中角频率对于波数的依赖关系,即式 (11.61) 或 (11.62);② 对于某种材料,可根据频率或波数,结合频散曲线确定物质点间距,以确保精度和减小计算量;③ 物质点间距过大时,出现某波数对应的频率为零情况,说明数值离散模型可能存在遮蔽结构动力特性的缺陷。

图 11-13(d) 给出了近场范围尺寸 $\delta = 0.004$m 时不同 m 值对应的波频散关系。从图中结果可以看出,① 当 m 增大,即近场范围内物质点增多时,波频散程度减弱;近场范围内物质点越密集,PD 数值解越趋近于 PD 精确解;② 当 $m \geqslant 3$ 时,PD 频散误差较精确解已较小,可以较好地体现 PD 方法的波频散特征;③ 选取合适的近场范围后,取 $m = 3 \sim 4$ 计算,则波传播分析的误差较小,进行固体破坏分析是可靠的。采用近场动力学方法分析高频荷载作用时,可以根据频散曲线选定物质点间距和近场范围尺寸。

11.3.2 分离式霍普金森压杆冲击巴西圆盘的近场动力学模拟

分离式霍普金森压杆 (split Hopkinson pressure bar,SHPB) 试验技术是测量高应变率条件下材料动态力学特性的主要方法。本节建立 SHPB 试验的近场动力学模拟方法,模拟混凝土巴西圆盘的冲击破坏。

如图 11-14 所示,将 SHPB 装置杆件视为一维弹性杆,试件可以是二维或三维,将装置杆与试件进行空间离散,离散的物质点遵循近场动力学运动方程。在建模和模拟计算中,对各部件之间和内部的作用效应采用如下方法进行处理:

(1) 采用近场动力学的本构力函数描述杆件或试件内部的作用效应;

(2) 对于撞击杆撞击入射杆的冲击接触作用效应,采用排斥力模型[10] 进行描述;

(3) 将杆端与混凝土试件接触的物质点视为刚性体,通过刚性体与可变形体的接触算法 (见 11.1.1 节) 描述入射杆与试件、试件与透射杆之间的接触作用效应;

(4) 试件出现破坏后,采用 Parks[17] 的排斥力模型描述破碎的物质点间以及破碎的物质点与试件完整部分的作用效应。

在计算模拟中,记录入射杆及透射杆上两个应变测量点的应变-时间曲线,通过专门程序分析应变-时间曲线,可求得试件内部应力-时间、应力-应变关系变化过程。

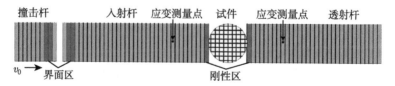

图 11-14 SHPB 冲击试验的近场动力学模拟器示意图

11.3 近场动力学的非局部色散特性与霍普金森压杆冲击试验的模拟

利用近场动力学方法模拟 SHPB 试验时，实现撞击杆撞击入射杆的冲击加载过程十分重要，拟采用两种方法模拟该冲击加载过程。方法一类似于近场动力学中的力边界条件，计算体系中只包含入射杆而不含撞击杆，入射杆边界所受的撞击压力 σ 与作用时间 t_0 可根据经典理论确定：具有初始速度 v_0 的杆件完美匹配撞击目标杆，有 $\sigma = v_0 \rho c/2$ 和 $t_0 = 2L_s/c$ 两式成立，其中弹性波速 $c = \sqrt{E/\rho}$，L_s 为撞击杆长度，t_0 时刻后保持外力为零。方法二类似于近场动力学中速度边界条件，计算体系中同时包含撞击杆与入射杆，两杆均采用近场动力学方法进行离散，给定撞击杆初速度 v_0，入射杆初始静止，通过可变形体之间的冲击接触模型描述撞击杆撞击入射杆的过程。此处，完美匹配撞击是指撞击杆和入射杆直径相同、接触杆端完全匹配，接触后两杆可视为一根完整杆件，则在采用 Parks 的排斥力模型描述杆件接触作用之外 (考虑图 11-14 中绿色物质点的接触排斥作用)，方法二还可采用本构力函数描述杆件间的接触作用 (考虑图 11-14 中蓝色绿色物质点的非局部相互作用)，但只在两杆挤压时存在排斥力。

在本例中，混凝土巴西圆盘的直径为 74 mm，混凝土材料质量密度 $\rho = 2400 \text{kg/m}^3$，弹性模量 $E = 30$ GPa，泊松比 $\nu = 0.25$，临界拉伸伸长率为 $s_0 = 0.0018$。撞击杆、入射杆及透射杆均为 347 不锈钢材料，其长度分别为 600 mm、3200 mm 和 1800 mm，杆直径均为 74 mm；杆的弹性模量 $E = 193$ GPa，质量密度 $\rho = 8027 \text{kg/m}^3$。入射杆与透射杆的应变测量点与试件距离均为 1m。采用改进的 PMB 模型进行模拟计算，杆件按一维问题进行离散，试件按二维问题进行离散，离散试件的物质点间距为 $\Delta x = 0.0002$m，离散杆件的物质点间距为 $\Delta x = 0.0005$m，近场范围尺寸均为 $\delta = 3\Delta x$，时间步长 $\Delta t = 5 \times 10^{-8}$s，杆件离散 11202 个物质点，圆盘离散 107501 个物质点。

图 11-15 给出撞击杆速度 $v_0 = 10$m/s 时基于近场动力学方法计算得到的入

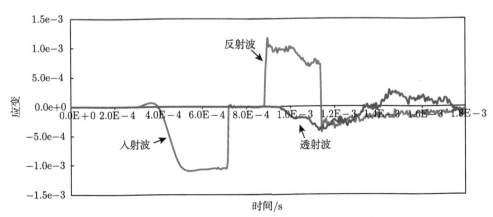

图 11-15 近场动力学模拟得到的 SHPB 冲击混凝土巴西盘的应变--时间曲线 ($v_0 = 10$m/s)

射杆和透射杆应变测量点处的应变–时间曲线，经分析后可得到入射波、反射波和透射波的应力–时间曲线、应力–应变曲线等。图 11-16 给出冲击速度 10m/s 时混凝土巴西圆盘的渐进破坏过程。从图示结果可以看出，在冲击加载初期，混凝土巴西圆盘在与入射杆和透射杆接触区域发生轻微损伤破碎；随后产生一条主裂纹，并从试件两端沿径向加载方向向中部扩展；在主裂纹的发展演化过程中，试件出现较大范围的损伤，并有多处出现微裂纹形核与扩展，以消耗更多的能量；最后主裂纹出现贯通破坏，试件最终劈裂为两半。图 11-17 对比了文献 [11] 中试验与近场动力学模拟的最终裂纹模式，两者具有良好的一致性，近场动力学模拟再现了混凝土试件渐进破坏过程和最终裂纹模式。

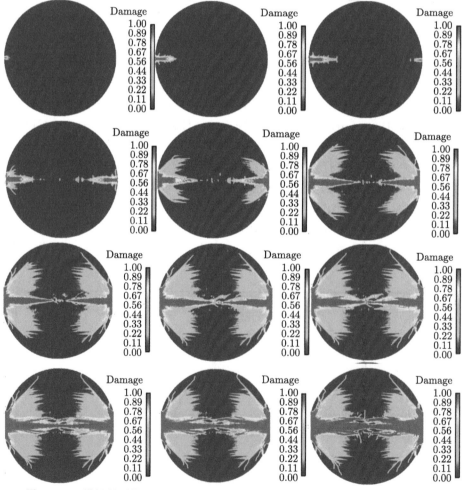

图 11-16　模拟得到的 SHPB 冲击混凝土巴西圆盘的渐进破坏过程 ($v = 10\text{m/s}$)

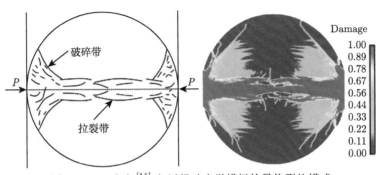

图 11-17　试验 [11] 与近场动力学模拟的最终裂纹模式

11.4　混凝土层裂与多重层裂的近场动力学模拟

在爆炸冲击和其他复杂的强动力荷载作用下，混凝土结构内部会产生压缩波传播。受到某一压缩脉冲作用时，混凝土结构可能不会出现损伤，但压缩脉冲会在混凝土中持续传播，并在自由表面发生反射，产生一个拉伸脉冲，该拉伸波易导致混凝土产生拉伸损伤，甚至出现断裂破坏。这种压缩脉冲在自由表面反射引起的动态断裂通常称为层裂。一旦混凝土结构发生层裂剥落，将形成一个新的自由表面，连续入射的压缩脉冲会再次在新的自由表面上发生反射，导致二次层裂甚至多次层裂剥落。因此，研究层裂问题具有重要的现实意义。

本节采用键关联的非常规态型近场动力学 (BA-NOSB PD) 方法实现 JH-2 本构仿真建模，并对混凝土杆的层裂破坏问题进行模拟分析 [12]。

11.4.1　矩形冲击波作用下混凝土杆的单层层裂模拟

如图 11-18 所示，三维混凝土杆的横截面为 $0.1\mathrm{m}\times 0.1\mathrm{m}$、长度为 $1.0\mathrm{m}$，杆一端 ($x=0$) 受矩形冲击波 $\boldsymbol{F}$ 作用，作用时长为 T_0，杆另一端自由。混凝土断裂能密度为 $G_\mathrm{F}=175\mathrm{N/m}$，相应的临界伸长率为 $s_0=\sqrt{5G_\mathrm{F}/9K\delta}=4.67\times 10^{-4}$，混凝土所采用的 JH-2 本构模型相关参数 [13,14] 见表 11-2。

在近场动力学模拟中，采用均匀正交物质点进行离散，离散粒子间距为 $\Delta x=0.01\mathrm{m}$，共有 12221 个粒子，近场范围尺寸为 $\delta=3\Delta x$。值得注意的是，本节计算模拟的对象为方形截面杆，而非试验研究中常用的圆形截面杆，主要是基于以下几点原因：① 采用均匀正交的物质点离散时，点阵无论粗细，离散后的截面都不能与理想的圆截面完全匹配；② 粗颗粒离散化既不能很好地逼近圆形边界，也不能准确地模拟层裂现象，虽然极细粒子离散化能较好地逼近圆形截面形状，提高模拟精度，但将带来严重的计算负担；③ 细长杆的截面形状对一维应力波沿轴向的传播、波在自由表面的反射以及入射波与反射波的相互作用影响不大。

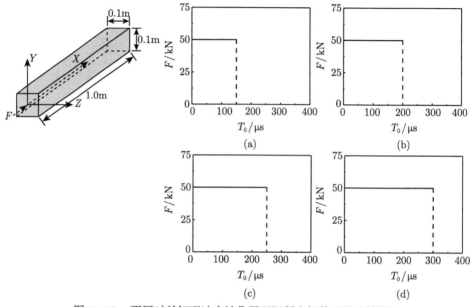

图 11-18　不同时长矩形冲击波作用下混凝土杆的层裂计算模型

表 11-2　JH-2 模型的材料参数[13,14]

参数	数值	参数	数值
密度 $\rho/(\mathrm{kg\cdot m^{-3}})$	2400	$\sigma_{\mathrm{HEL}}/\mathrm{GPa}$	0.03
剪切模量 G/GPa	14.86	$P_{\mathrm{HEL}}/\mathrm{GPa}$	0.016
泊松比 ν	0.1454	T/GPa	0.004
A	0.90	K_1/GPa	16.0
B	1.30	K_2/GPa	0
C	0.007	K_3/GPa	0
M	0.62	$\dot{\varepsilon}_0$	1.0
N	0.62	β	1.0
D_1	0.04	D_2	1.00

假定层裂是瞬时产生的,王礼立[15]详细讨论了一维杆件在波长 λ 的矩形冲击波作用下的层裂特性,指出当矩形入射波从自由面反射后,一旦导致层裂,不论矩形冲击力幅值有多大,都不会发生多重层裂,层裂厚度为 $h = \lambda/2$。为验证嵌入 JH-2 本构的 BA-NOSB PD 方法能够正确模拟混凝土层裂的结果,选取四个不同持续时间的矩形冲击波进行模拟计算,即 $T_0 = 150\mu s, 200\mu s, 250\mu s, 300\mu s$,如图 11-18 所示,冲击波幅值为 $F = 5 \times 10^4 \mathrm{N}$,时间步长 $1\mu s$,总计算时长为 $600\mu s$。

图 11-19 给出了不同持续时长矩形冲击波作用下混凝土杆层裂的损伤特征和轴向位移。通过测量层裂处至自由端的长度,可以得到层裂段的厚度。表 11-3 给出了四组模拟的定量结果,包括层裂位置、层裂段厚度、冲击波长、层裂厚度与冲击波长之比。结果表明,模拟的层裂厚度均接近于经典理论解析解,即 $h = $

11.4 混凝土层裂与多重层裂的近场动力学模拟

$\lambda/2$。以矩形冲击波作用时长 $T_0 = 200\mu s$ 为例,本例混凝土杆的理论波速或矩形冲击波传播的速度为 $c = \sqrt{E/\rho} = \sqrt{2(1+\nu)G/\rho} = 3766 \mathrm{m \cdot s^{-1}}$,矩形冲击波波长为

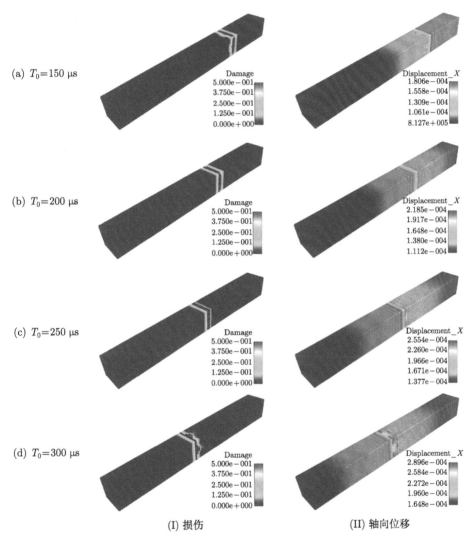

(I) 损伤　　　　　　　　(II) 轴向位移

图 11-19　不同持续时长矩形冲击波作用下混凝土杆层裂的损伤特征和轴向位移

表 11-3　不同时长矩形冲击波作用下混凝土层裂模拟的定量结果

	$T_0 = 150\mu s$	$T_0 = 200\mu s$	$T_0 = 250\mu s$	$T_0 = 300\mu s$
层裂位置 X/m	0.715	0.625	0.525	0.435
层裂段厚度 h'/m	0.285	0.375	0.475	0.565
冲击波长 λ/m	0.565	0.753	0.942	1.130
层裂厚度/冲击波长	0.504	0.498	0.504	0.500

$\lambda = cT_0 = 0.7532\text{m}$，理论层裂段厚度为 $h = \lambda/2 = 0.3766\text{m}$，近场动力学数值模拟得到的层裂厚度为 0.375m，与理论层裂厚度的相对误差仅为 0.42%。模拟结果验证了与 JH-2 本构对应的 BA-NOSB PD 方法在混凝土层裂问题模拟中的有效性。

11.4.2 三角冲击波作用下混凝土杆的多重层裂模拟

11.4.2.1 三角冲击波作用下混凝土杆的层裂特征

经典层裂理论与矩形冲击波下混凝土的层裂模拟已经证明矩形冲击波不会产生多重层裂，本节研究三角卸载冲击波作用下混凝土杆的层裂特征。如图 11-20 所示，三维混凝土杆的横截面仍为 0.1m×0.1m、长度为 1.0m，杆一端 ($x = 0$) 受三角冲击波 F 作用，杆的另一端自由，载荷峰值为 $F = 90\text{ kN}$，冲击波作用时间为 $T_0 = 360\text{ μs}$，混凝土采用的 JH-2 本构模型参数与上例相同，混凝土断裂能密度为 $G_\text{F} = 175\text{ MN/m}$。近场动力学模拟采用均匀正交物质点离散，离散粒子间距为 $\Delta x = 0.01\text{m}$，共有 12221 个粒子，时间步长 1 μs，总计算时长为 600 μs。

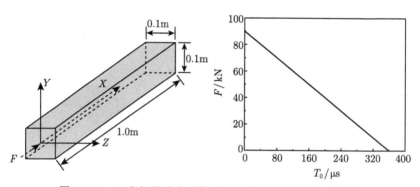

图 11-20　三角卸载冲击波作用下混凝土杆的层裂计算模型

图 11-21 显示了混凝土杆在给定的三角冲击波作用下几个典型时刻的损伤特征和轴向位移。在冲击作用 $t = 400\text{μs}$ 之前，混凝土杆完好无损，但 $X = 0.7\text{m}$ 截面两侧物质点的轴向位移差已经较大，表明该截面处于高拉应力状态。当 $t = 414\text{μs}$，在 $X = 0.705\text{m}$ 处发生第一次层裂，位于层裂段厚度为 $h' = 0.295\text{m}$。当 $t = 453\text{μs}$，在 $X = 0.495\text{m}$ 处发生第二次层裂，第二次层裂位置到新生成自由面间的层裂段厚度为 $h' = 0.210\text{m}$；此外，在第二次层裂产生的过程中，在第一次层裂新生成自由面附近出现了结构表面损伤，校核该处损伤与位移结果，发现此处未形成贯穿裂纹面。当 $t = 479\text{μs}$，在 $X = 0.375\text{m}$ 处发生第三次层裂，第三次层裂位置到新生成自由面间的层裂段厚度为 $h' = 0.120\text{m}$。当 $t = 500\text{μs}$，混凝土杆损伤状态与 $t = 479\text{μs}$ 时的损伤状态基本一致，仅发生裂缝处表面损伤的轻微加

11.4 混凝土层裂与多重层裂的近场动力学模拟

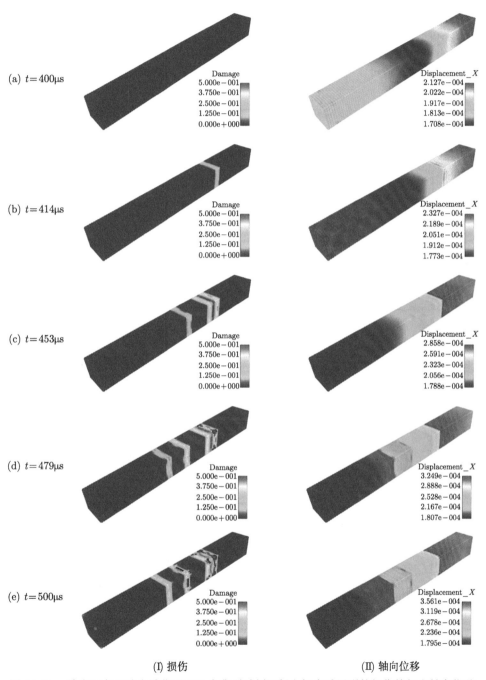

(I) 损伤 (II) 轴向位移

图 11-21 给定三角形冲击波作用下几个典型时刻混凝土杆多重层裂的损伤特征和轴向位移

剧,并不再有新的层裂段产生。模拟结果表明,在三角形卸载冲击波作用下,混凝土杆将发生多重层裂现象。

11.4.2.2 相同作用时间、不同峰值的三角形冲击波对混凝土杆层裂特性的影响

本节将采用近场动力学方法,研究相同作用时间、不同峰值的三角形卸载冲击波对混凝土杆层裂特性的影响。如图 11-22 所示,三角形卸载冲击波的持续时间固定为 $T_0 = 200\mu s$,冲击力初始峰值分别为 $F = 50\,kN$、$60kN$、$70kN$、$80kN$ 和 $90kN$。PD 空间离散与时间离散格式同前所述。

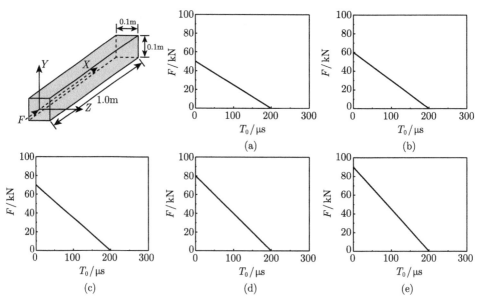

图 11-22　相同作用时间、不同峰值的三角形冲击波作用下混凝土杆层裂的计算模型

图 11-23 给出了五种三角卸载冲击载荷作用下混凝土杆层裂后的损伤特征和轴向位移分布。从图中所示结果可以看出:

(1) 当混凝土杆受到图 11-22(a) 所示的三角卸载冲击波作用时,由于该载荷峰值的冲击波不能激发超过抗拉强度的应力状态,混凝土杆不产生损伤和层裂现象。

(2) 当混凝土杆受到图 11-22(b) 所示的三角卸载冲击波作用时,只产生一次层裂,层裂位置为 $X = 0.695m$,自由面到第一次层裂位置间的层裂段厚度为 $h' = 0.305m$。

(3) 当混凝土杆受到图 11-22(c) 和 (d) 所示的三角卸载冲击波作用时,混凝土杆层裂两次。对于峰值载荷 $F = 70kN$ 的情况,第一次层裂和第二次层裂分别

11.4 混凝土层裂与多重层裂的近场动力学模拟 · 379 ·

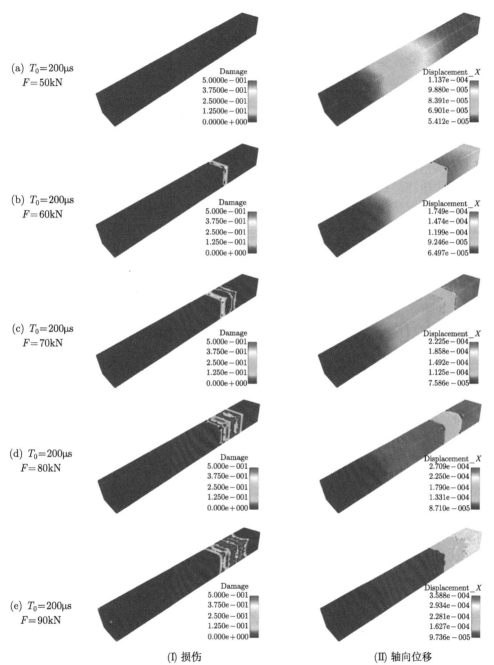

(I) 损伤   (II) 轴向位移

图 11-23 相同作用时间、不同峰值的三角形卸载冲击波作用下混凝土杆多重层裂的损伤和轴向位移

发生在 $X = 0.765$m 和 $X = 0.675$m 位置，则自由面到第一次层裂位置间的层

裂段厚度为 $h' = 0.235$m，第一次层裂新生成自由面位置到第二次层裂位置间的层裂段厚度为 $h' = 0.090$m；对于峰值载荷 $F = 80$kN 的情况，第一次层裂和第二次层裂分别发生在 $X = 0.815$m 和 $X = 0.665$m 位置，相应的层裂段厚度为 $h' = 0.185$m 和 $h' = 0.150$m。

(4) 当混凝土杆受到图 11-22(e) 所示的三角卸载冲击波作用时，强烈冲击引起的损伤较为剧烈，尤其是层裂裂缝附近的杆表面产生额外损伤，损伤连成一片首次层裂段厚度较小，受限于较粗的离散物质点尺寸，较难以确定多次层裂的位置和各层裂段厚度。

综上所述，当受到三角卸载冲击波作用时，随着冲击力初始峰值的增大或三角卸荷脉冲卸荷速率的增加，混凝土中的层裂次数增加，第一次层裂段厚度逐渐减小。

另外指出，除第一次层裂的主断裂裂纹外，层裂段中可能会出现一些额外损伤，特别是在第二次层裂段或中间层裂段，如图 11-23(d) 所示的第二次剥落体的损伤，并且，中间层裂段中的严重损伤使得难以清晰区分所有层裂位置，如图 11-23(e) 所示。这些额外的损伤无法基于经典层裂理论进行解释，因为经典层裂理论假设层裂是瞬时完成的，但对于实际问题，从轻微损伤到完全断裂形成新的自由面，这个过程需要一段时间，理论上应随层裂体剥落消失的应力波并没有立即消失，从而在新的自由表面左侧产生过大的应力状态，会引起一些额外损伤，这个特征也被一些试验所证实。

11.4.2.3 相同卸载斜率、不同峰值的三角形冲击波对混凝土杆层裂特性的影响

本节将采用近场动力学方法，进一步研究三角形卸荷冲击波对混凝土层裂的影响。如图 11-24 所示，考虑冲击波具有相同的卸载斜率、不同的冲击峰值和作用时间，三角形卸载冲击波的斜率为 2.5×10^8N/s，冲击力初始峰值分别为 $F = 60$kN、70kN、80kN 和 90kN，相应的作用时间分别是 $T_0 = 240$μs、280μs、320μs 和 360μs。PD 空间离散与时间离散格式同前所述。

图 11-25 给出了四种三角卸载冲击载荷作用下混凝土层裂后的损伤特征和轴向位移分布。从图中结果可以看出：

(1) 当冲击力峰值为 $F = 60$kN 时，发生一次层裂，层裂位置为 $X = 0.665$m，层裂段厚度为 $h' = 0.335$m；

(2) 当冲击力峰值为 $F = 70$kN 时，在 $X = 0.665$m 和 $X = 0.565$m 位置相继发生层裂，两次的层裂段厚度分别为 $h' = 0.335$m 和 $h' = 0.100$m。

(3) 当冲击力峰值为 $F = 80$kN 时，也发生两次层裂，层裂位置分别为 $X = 0.705$m 和 $X = 0.495$m，层裂段厚度分别为 $h' = 0.295$m 和 $h' = 0.210$m。

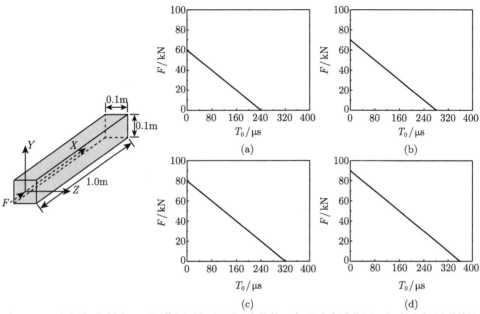

图 11-24　相同卸载斜率、不同作用时间和不同峰值的三角形冲击波作用下混凝土杆层裂的计算模型

(4) 当冲击力峰值为 $F = 90\text{kN}$ 时，发生三次层裂，层裂位置分别为 $X = 0.705\text{m}$、$X = 0.495\text{m}$ 和 $X = 0.375\text{m}$，相应的层裂段厚度分别为 $h' = 0.295\text{m}$、$h' = 0.210\text{m}$ 和 $h' = 0.120\text{m}$。需指出的是，对于这种情况，图 11-25(d) 的第二层

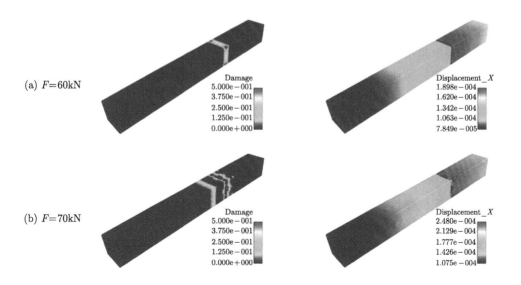

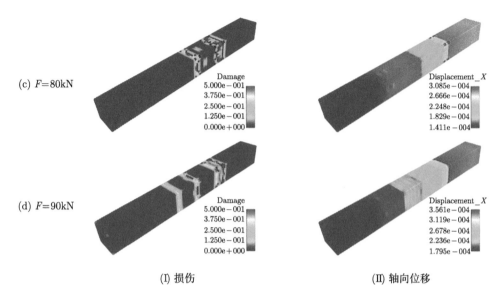

图 11-25 相同卸载斜率、不同作用时间和不同峰值的三角形卸载冲击波作用下混凝土杆多重层裂的损伤与轴向位移

裂段中也存在类似层裂的额外损伤，但对应的轴向位移并不说明这些额外损伤处出现层裂现象。

从模拟结果还可以看出，$F=60\text{kN}$ 和 $F=70\text{kN}$ 冲击波作用下，混凝土杆的第一段层裂厚度相等，$F=80\text{kN}$ 和 $F=90\text{kN}$ 冲击波作用下，混凝土杆的两次层裂厚度也都相同；四组模拟中的第一层裂段厚度基本一致，表明在相同卸载速率下，不同冲击波强度对首次层裂的影响很小。

综上所述，在三角形卸荷冲击波作用下，混凝土杆的层裂次数主要取决于冲击波初始峰值，并随着冲击力的增大而逐渐增多。在连续多次层裂过程中，层裂段厚度越来越小。影响层裂厚度的因素很多，其中最重要的是冲击波卸载速率或斜率。如果卸荷速率过大，则层裂段厚度太小，以至于无法区分层裂位置和层裂数目。

11.5 钢筋混凝土板空中爆炸毁伤的近场动力学模拟

11.5.1 问题描述

本节研究爆炸冲击载荷对混凝土结构的破坏过程，选用文献 [16] 中爆炸冲击作用下钢筋混凝土板破坏的现场试验作为参考，如图 11-26 所示。钢筋混凝土板的长和宽均为 1000 mm，厚度为 40 mm，钢筋分布在混凝土板的抗拉区域，布置的钢筋在长度和宽度方向均为 75 mm，钢筋的直径为 6 mm，见图 11-27。混凝土的弹性模量为 28.3 GPa，抗压强度为 39.5 MPa，抗拉强度为 4.2 MPa，断裂能量释放率为 $G_F = 155$ N/m；钢筋的弹性模量为 200 GPa，屈服强度为 600 MPa。TNT 炸药量为 0.31 kg，炸药位于钢筋混凝土板中心的正上方，起爆距离为 400 mm。

图 11-26　爆炸载荷下钢筋混凝土板破坏的现场试验[16]

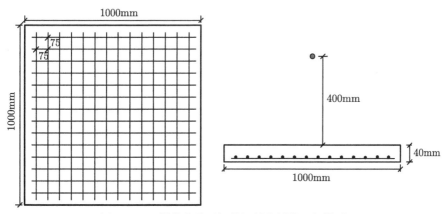

图 11-27　爆炸载荷下钢筋混凝土板的几何模型

采用基于 JH-2 本构的键关联非常规态型近场动力学方法仿真模拟爆炸载荷对结构的响应，再现爆炸冲击作用下钢筋混凝土板试验的破坏过程。采用 JH-2 本构描述混凝土力学响应，材料参数见表 11-4，采用弹性本构模型刻画钢筋行为。计算中，对钢筋混凝土板进行均匀离散，离散间距为 $\Delta x = 5$mm，物质点总数为

363609，近场范围取为 $\delta = 2\Delta x$，影响函数采用 $\omega(|\boldsymbol{\xi}|) = \mathrm{e}^{-|\boldsymbol{\xi}|^2/\delta^2}$，时间步长为 $\Delta t = 1\mu s$，总计算时长为爆炸冲击波作用到混凝土上表面后的 $600\mu s$。TNT 爆炸荷载通过 11.1.2 节中的爆压时程曲线经验公式施加在钢筋混凝土板上表面，施加过程中需要计算爆源与上表面各物质点间的爆距，爆压的方向沿各物质点与爆源的连线方向。在现场试验中，钢筋混凝土板左右两边被角钢紧紧夹住，为了防止夹住的部位出现损伤破坏，在角钢和钢筋混凝土板之间，辅以与角钢相同宽度和长度的木条，以均匀地固定支撑试件 (相当于板的夹支边或固定边)。在对应的数值模拟中，在板的左右两边各固定三排物质点，作为固定边的边界条件，并设置这六排物质点为无损区域。这样处理不仅与边界条件的实际情况相符合，也可以避免在模拟过程中钢筋混凝土板的边界过早破坏，致使边界条件不准确而影响模拟计算的结果。

表 11-4 混凝土 JH-2 模型的材料参数

参数/单位	数值	参数	数值
密度 $\rho/(\mathrm{kg} \cdot \mathrm{m}^{-3})$	2400	A	0.90
剪切模量 G/GPa	12.4	B	1.30
泊松比 ν	0.1454	C	0.007
$\sigma_{\mathrm{HEL}}/\mathrm{GPa}$	0.025	M	0.62
$P_{\mathrm{HEL}}/\mathrm{GPa}$	0.0133	N	0.62
T/GPa	0.004	$\dot{\varepsilon}_0$	1.0
K_1/GPa	13.3	D_1	0.04
K_2/GPa	0	D_2	1.00
K_3/GPa	0	β	1.0

11.5.2 计算结果与分析

根据计算结果，TNT 炸药起爆后产生冲击波，经过 $T = 350\mu s$ 后，首先到达钢筋混凝土板的迎爆面中心处，所引起的钢筋混凝土板压应力为 1.99MPa，未超过混凝土的强度，钢筋混凝土板未出现损伤。

图 11-28 给出了在 0.31kg 的 TNT 炸药起爆后引起的冲击载荷作用下，钢筋混凝土板几个典型时刻的损伤破坏情况。计算结果表明，当 $T = 480\mu s$ 时，板的背爆面首先开始产生损伤，而此时迎爆面还未发生损伤，其原因可归结为：当冲击波刚传播到迎爆面时，冲击波作用力的垂直分量为压力，迎爆面混凝土主要承受压应力，由于混凝土抗压强度相对较高，迎爆面没有出现损伤；当冲击波传播到背爆面时，会反射形成拉应力，冲击波短时间内在迎爆面和背爆面之间反复传播叠加，在背爆面中心处产生较大的拉应力，首先出现损伤现象。

11.5 钢筋混凝土板空中爆炸毁伤的近场动力学模拟

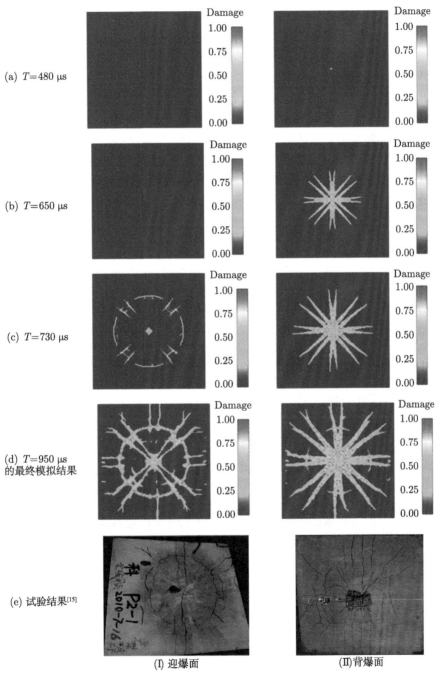

(a) $T=480\ \mu s$
(b) $T=650\ \mu s$
(c) $T=730\ \mu s$
(d) $T=950\ \mu s$ 的最终模拟结果
(e) 试验结果[15]

(I) 迎爆面   (II) 背爆面

图 11-28 爆炸载荷作用下钢筋混凝土板的破坏过程

当 $T=650\mu s$ 时，迎爆面才开始发生损伤，而此时背爆面已经出现了震塌开裂现象，发生较为严重的破坏，并形成多条裂纹，向四周延伸扩展。当 $T = 730\mu s$ 时，迎爆面开始出现环向裂纹，经分析，这种环向裂纹主要是由于爆炸冲击波的水平分量传播至边界发生反射所致，也与不同方向应力波之间的叠加效应有关；此时，背爆面的裂纹继续延伸扩展。

$T = 730\mu s$ 以后，钢筋混凝土板已经出现较为严重的损伤破坏，之后的应力波非常复杂，主要还是驱使已经产生的裂纹不断演化发展。图 11-28(d) 给出了 $T = 950\mu s$ 钢筋混凝土板的最终破坏形态，比较图 11-28(e) 所示的现场试验结果可以看出，数值模拟预测的裂纹分布特征与实际试验结果具有较好的一致性，也表明基于 JH-2 本构的键关联非常规态型近场动力学方法求解爆炸冲击破坏问题的有效性。

参 考 文 献

[1] Madenci E, Oterkus E. Peridynamic Theory and Its Applications[M]. New York: Springer, 2014.

[2] 顾鑫, 章青. 爆炸荷载作用下大坝破坏分析的数值模拟研究进展 [J]. 河海大学学报 (自然科学版), 2017, 45(001): 45-55.

[3] Technical Manual (TM5-1300). To resist the effect of accidental explosions[R]. Washington, DC: Department of the Army, Navy and the Air force, 1990.

[4] Henrych J. The Dynamics of Explosion and Its Use[M]. Amsterdam: Elseviser Scientific Publishing Company, 1979.

[5] Wu C, Hao H. Modeling of simultaneous ground shock and airblast pressure on nearby structures from surface explosions[J]. International Journal of Impact Engineering, 2005, 31(6): 699-717.

[6] Johnson G R, Holmquist T J. An improved computational constitutive model for brittle materials[C]. AIP Conference Proceedings, American Institute of Physics, 1994, 309(1): 981-984.

[7] Silling S A. Reformulation of elasticity theory for discontinuities and long-range forces[J]. Journal of the Mechanics and Physics of Solids, 2000, 48(1): 175-209.

[8] Weckner O, Silling S A. Determination of nonlocal constitutive equations from phonon dispersion relations[J]. International Journal for Multiscale Computational Engineering, 2011, 9(6): 623-634.

[9] Wildman R A, Gazonas G A. A finite difference-augmented peridynamics method for reducing wave dispersion[J]. International Journal of Fracture, 2014, 190(1-2): 39-52.

[10] Gu X, Zhang Q, Huang D, et al. Wave dispersion analysis and simulation method for concrete SHPB test in peridynamics[J]. Engineering Fracture Mechanics, 2016, 160: 124-137.

[11] Zhou Z, Li X, Zou Y, et al. Dynamic Brazilian tests of granite under coupled static and dynamic loads[J]. Rock Mechanics and Rock Engineering, 2014, 47(2): 495-505.

[12] Yang S Y, Gu X, Zhang Q, et al. Bond-associated non-ordinary state-based peridynamic model for multiple spalling simulation of concrete[J]. Acta Mechanica Sinica, 2021, 37(7): 1104-1135.

[13] 王高辉. 极端荷载作用下混凝土重力坝的动态响应行为和损伤机理 [D]. 天津：天津大学, 2014.

[14] 柴传国. 异形头部弹体对混凝土靶的侵彻效应研究 [D]. 北京：北京理工大学, 2014.

[15] 王礼立. 应力波基础 [M]. 北京: 国防工业出版社, 2005.

[16] Wang W, Zhang D, Lu F, et al. Experimental study on scaling the explosion resistance of a one-way square reinforced concrete slab under a close-in blast loading[J]. International Journal of Impact Engineering, 2012, 49: 158-164.

[17] Parks M L, Lehoucq R B, Plimpton S J, et al. Implementing peridynamics within a molecular dynamics code [J]. Computer Physics Communications, 2008, 179(11): 777-783.

第 12 章 热传导与热-力耦合问题的近场动力学模拟

能量和物质传输是通过各种粒子承载发生的，本质上是一种非局部过程。在传输过程中，粒子携带自身及其他点的能量 (或质量) 到达空间中的某一点，如果粒子的平均自由程 (即粒子携带的多余能量消失前所运动的平均距离) 远小于驱动力变化的距离尺度 (如温度差的变化距离)，则经典局部理论 (傅里叶定律、菲克定律等) 足以描述该过程，如果粒子的平均自由程大于驱动力变化的距离尺度，则需要借助非局部传输理论。非局部效应比较显著的典型传输扩散过程有：多孔介质中的扩散过程、微纳米尺度器件的热扩散过程以及超快速扩散过程等。诸多学者已在非傅里叶热传导、非局部传输理论、非局部热力学和非局部热弹性理论等领域进行了大量研究 [1-6]，为求解与非局部效应有关的各类扩散问题和多物理场耦合问题做出了重要贡献。鉴于近场动力学的积分非局部特性，本章将简要阐述热传导问题的近场动力学模型，基于近场动力学微分算子建立具有一般性的近场动力学热传导模型和热-力耦合模型，并给出相应的数值求解算法，进行混合热边界条件下完好或含裂纹结构的一维、二维和三维热传导分析，发展混凝土材料和结构导热分析以及热-力耦合分析的近场动力学方法，研究裂纹对于混凝土导热性能的影响，进行混凝土厚壁圆筒的热冲击破裂模拟。

12.1 热传导问题的近场动力学模型

在热传导的傅里叶定律中，热流总是从温度高的区域流向温度低的区域，温度差是热流流动的驱动力，单位时间内通过单位截面的热量与温度梯度成正比，即 [7]

$$\boldsymbol{q}(\boldsymbol{x},t) = -k_{\mathrm{T}}\nabla T(\boldsymbol{x},t) \tag{12.1}$$

式中，$\boldsymbol{x}$ 为物质点的位置矢量；t 为时间，单位 s；$\boldsymbol{q}$ 为热通量矢量，单位是 $J/(s \cdot m^2)$；k_{T} 为导热系数，单位是 $J/(s \cdot m \cdot K)$；T 为温度，单位 K；∇ 为梯度算子。

经典的局部热传导方程可以通过傅里叶定律得到 [7]

$$\rho c_v \frac{\partial T(\boldsymbol{x},t)}{\partial t} = -\nabla \cdot \boldsymbol{q}(\boldsymbol{x},t) + S_{\mathrm{T}}(\boldsymbol{x},t) = k_{\mathrm{T}}\nabla^2 T(\boldsymbol{x},t) + S_{\mathrm{T}}(\boldsymbol{x},t) \tag{12.2}$$

式中，ρ 为密度，单位是 kg/m³；c_v 为比热容，单位是 J/(kg·K)；S_T 为体积热源项，单位是 J/(m³·s)，表示单位时间单位体积吸收或放出的热量；∇^2 为拉普拉斯算子。

12.1.1 键型近场动力学热传导模型

将局部热传导方程中温度的二阶空间导数用热响应函数 f_T 的积分形式替换，可得到键型近场动力学的热传导方程[8,9]

$$\rho c_v \frac{\partial T(\boldsymbol{x},t)}{\partial t} = \int_{H_{\boldsymbol{x}}} f_T(\boldsymbol{x},\boldsymbol{x}',T,T') \mathrm{d}V_{\boldsymbol{x}'} + S_T(\boldsymbol{x},t) \tag{12.3}$$

其中，热响应函数 f_T 是描述物质点 $\boldsymbol{x}$ 与 $\boldsymbol{x}'$ 相互作用的热流密度函数，单位为 J/(s·m⁶)；其他参数含义同公式 (12.2)。在键型近场动力学热传导方程中，两点间的热响应函数与其他物质点对无关。

表 12-1 给出了几种热响应函数形式，其中 κ_T 为微导热系数，微导热系数形式取决于热响应函数的形式，它可以通过在施加了简单线性温度场的物质点处，将近场动力学热势与经典热势等效[10,11]或者将近场动力学热流通量和经典热流通量等效[8,9]得到。

表 12-1　三种热响应函数及其不同维度的微导热系数

热响应函数	微导热系数		
	一维	二维	三维
$f_T = \kappa_T(T' - T)$	$\kappa_T = \dfrac{3k_T}{A\delta^3}$	$\kappa_T = \dfrac{8k_T}{\pi h \delta^4}$	$\kappa_T = \dfrac{8k_T}{\pi \delta^5}$
$f_T = \kappa_T \dfrac{T' - T}{|\boldsymbol{\xi}|}$	$\kappa_T = \dfrac{2k_T}{A\delta^2}$	$\kappa_T = \dfrac{6k_T}{\pi h \delta^3}$	$\kappa_T = \dfrac{6k_T}{\pi \delta^4}$
$f_T = \kappa_T \dfrac{T' - T}{|\boldsymbol{\xi}|^2}$	$\kappa_T = \dfrac{k_T}{A\delta}$	$\kappa_T = \dfrac{4k_T}{\pi h \delta^2}$	$\kappa_T = \dfrac{4k_T}{\pi \delta^3}$

注：A 是一维杆的截面面积，h 是二维板的厚度。

以下以一维问题为例，简要说明基于热势等效确定近场动力学微导热系数的方法。

对于一维问题，假设简单线性温度场为 $T(x) = ax + b$，其中 a、b 为常量，则两物质点 $\boldsymbol{x}$ 与 $\boldsymbol{x}'$ 的温度差为

$$\tau = T(x') - T(x) = a(x' - x) = a\xi \tag{12.4}$$

近场动力学物质点间的热流密度与微热势满足下述关系

$$f_T = \frac{\partial z}{\partial \tau} \tag{12.5}$$

当假设热响应函数为 $f_\mathrm{T} = \kappa_\mathrm{T}(T' - T)$ 时，热响应函数对温度差积分，则可以得到微热势的表达式

$$z = \frac{1}{2}\kappa_\mathrm{T}\tau^2 = \frac{1}{2}\kappa_\mathrm{T}(T' - T)^2 \tag{12.6}$$

对于线性温度场 $T(x) = ax + b$，有 $z = \frac{1}{2}\kappa_\mathrm{T}(a\xi)^2$。将 "热键" 的微热势函数代入近场动力学物质点的热势中，有

$$Z_\mathrm{PD} = \frac{1}{2}\int_{H_\infty} z(x,x')\mathrm{d}V_{x'} = \frac{1}{2}\int_{-\delta}^{\delta}\frac{1}{2}\kappa_\mathrm{T}(a\xi)^2 A\mathrm{d}\xi = \frac{1}{6}\kappa_\mathrm{T}a^2 A\delta^3 \tag{12.7}$$

对于线性温度场 $T(x) = ax + b$，有 $\nabla T = a$，则经典局部热传导理论的热势为

$$Z_\mathrm{CT} = \frac{1}{2}\nabla T \cdot k_\mathrm{T}\nabla T = \frac{1}{2}k_\mathrm{T}a^2 \tag{12.8}$$

令 $Z_\mathrm{PD} = Z_\mathrm{CT}$，即式 (12.7) 与式 (12.8) 相等，可以建立微导热系数与材料宏观导热系数的关系

$$\kappa_\mathrm{T} = \frac{3k_\mathrm{T}}{A\delta^3} \tag{12.9}$$

此即为表 12-1 给出的第一种热响应函数的一维微导热系数值。值得注意的是，表 12-1 中的热响应函数是针对具有完整对称圆形或球形近场域的物质点导出的，对于边界区域物质点，由于其近场域不完整且非对称，需进行微导热系数的表面修正以减小数值误差。

12.1.2 态型近场动力学热传导模型

态型近场动力学的热传导控制方程为 [11,12]

$$\rho c_v \frac{\partial T(\boldsymbol{x},t)}{\partial t} = \int_{H_\infty} (\underline{h}[\boldsymbol{x},t]\langle \boldsymbol{x}' - \boldsymbol{x}\rangle - \underline{h}[\boldsymbol{x}',t]\langle \boldsymbol{x} - \boldsymbol{x}'\rangle)\mathrm{d}V_{\boldsymbol{x}'} + S_\mathrm{T}(\boldsymbol{x},t) \tag{12.10}$$

式中，$\underline{h}$ 为标量热流状态。近场动力学状态是一个包含着与该物质点相关的所有相互作用的全部信息的无限维数组，方括号 $[\boldsymbol{x},t]$ 代表空间和时间位置，角括号内包含的是相互作用的物质点对。

运用标量状态减缩算子的定义 [11,13]，假设近场动力学热势变化量与经典热势变化量相等，可以得到近场动力学热流状态与经典热流通量之间的关系

$$\underline{h}[\boldsymbol{x},t]\langle \boldsymbol{x}' - \boldsymbol{x}\rangle = w\boldsymbol{q}^T m^{-1}\boldsymbol{\xi} \tag{12.11}$$

$$\underline{h}[\boldsymbol{x},t]\langle \boldsymbol{x}'-\boldsymbol{x}\rangle = w\boldsymbol{q}^T\boldsymbol{K}^{-1}\boldsymbol{\xi} \tag{12.12}$$

式中，w 为影响函数，$\boldsymbol{\xi} = \boldsymbol{x}'-\boldsymbol{x}$ 是两个物质点的相对位置矢量，m 和 $\boldsymbol{K}$ 分别是标量加权量和形状张量，具体形式如下

$$m = \int_{H_\infty} w|\boldsymbol{\xi}|\cdot|\boldsymbol{\xi}|\mathrm{d}V_{\boldsymbol{x}'} \tag{12.13}$$

$$\boldsymbol{K} = \int_{H_\infty} w\boldsymbol{\xi}\otimes\boldsymbol{\xi}\mathrm{d}V_{\boldsymbol{x}'} \tag{12.14}$$

对于具有完整对称的圆形或球形近场范围的物质点，当选定合适的影响函数时，式 (12.13) 和式 (12.14) 是等价的。

此外，热流通量也可以通过标量状态减缩运算符得到，即根据式 (7.2) ($\Re\{\underline{\boldsymbol{A}}\} = \int_{H_\infty} w(|\boldsymbol{\xi}|)(\underline{\boldsymbol{A}}\otimes\underline{\boldsymbol{X}})\mathrm{d}V_{\boldsymbol{x}'}\boldsymbol{K}^{-1}$)，有

$$\boldsymbol{q}(\boldsymbol{x},t) = -k_\mathrm{T}\int_{H_\infty} w\underline{\tau}\langle\boldsymbol{\xi}\rangle\boldsymbol{I}\otimes\boldsymbol{\xi}\mathrm{d}V_{\boldsymbol{x}'}(m^{-1}\boldsymbol{I}) \tag{12.15}$$

$$\boldsymbol{q}(\boldsymbol{x},t) = -k_\mathrm{T}\int_{H_\infty} w\underline{\tau}\langle\boldsymbol{\xi}\rangle\boldsymbol{I}\otimes\boldsymbol{\xi}\mathrm{d}V_{\boldsymbol{x}'}\boldsymbol{K}^{-1} \tag{12.16}$$

其中

$$\underline{\tau}[\boldsymbol{x},t]\langle\boldsymbol{x}'-\boldsymbol{x}\rangle = T'-T \tag{12.17}$$

$\underline{\tau}\langle\boldsymbol{\xi}\rangle$ 表示在 t 时刻 $\boldsymbol{x}$ 点处作用于相对位置矢量 $\boldsymbol{x}'-\boldsymbol{x}$ 上的温度标量状态 $\underline{\tau}$。

将式 (12.15) 和式 (12.11) 代入式 (12.10)，或将式 (12.16) 和式 (12.12) 代入式 (12.10)，得到具有双重积分的近场动力学热传导方程，可称之为两阶段建模方法，所得到的方程在形式上与在近场范围内进行一次积分的键型近场动力学运动方程有所不同。

12.2 基于近场动力学微分算子的热传导模型

将式 (12.2) 给出的局部热传导方程改写为笛卡儿直角坐标下的分量形式，对于三维、二维和一维热传导问题，分别有

$$\rho c_v \frac{\partial T(\boldsymbol{x},t)}{\partial t} = k_\mathrm{T}\left(\frac{\partial^2 T(\boldsymbol{x},t)}{\partial x_1^2}+\frac{\partial^2 T(\boldsymbol{x},t)}{\partial x_2^2}+\frac{\partial^2 T(\boldsymbol{x},t)}{\partial x_3^2}\right)+S_\mathrm{T}(\boldsymbol{x},t) \tag{12.18}$$

$$\rho c_v \frac{\partial T(\boldsymbol{x},t)}{\partial t} = k_{\mathrm{T}} \left(\frac{\partial^2 T(\boldsymbol{x},t)}{\partial x_1^2} + \frac{\partial^2 T(\boldsymbol{x},t)}{\partial x_2^2} \right) + S_{\mathrm{T}}(\boldsymbol{x},t) \qquad (12.19)$$

$$\rho c_v \frac{\partial T(x,t)}{\partial t} = k_{\mathrm{T}} \frac{\partial^2 T(x,t)}{\partial x^2} + S_{\mathrm{T}}(x,t) \qquad (12.20)$$

本节采用近场动力学微分算子取代局部空间导数, 构建具有一般性的键型近场动力学非局部热传导模型[14]。近场动力学微分算子[15,16] 能够将任意阶局部偏微分转化为对应的非局部积分表达式, 从数学上避免了不连续处导数不存在的奇异性问题 (见附录 A)。根据附录 A 中式 (A.54)、(A.93) 和 (A.112) 的二阶局部微分的非局部积分表达式, 得到三维、二维和一维问题中具有完整对称近场域的物质点温度二阶导数的积分表达式, 即

$$\left\{ \begin{array}{c} \dfrac{\partial^2 T(\boldsymbol{x})}{\partial x_1^2} \\ \dfrac{\partial^2 T(\boldsymbol{x})}{\partial x_2^2} \\ \dfrac{\partial^2 T(\boldsymbol{x})}{\partial x_3^2} \end{array} \right\} = \frac{3}{4\pi V_2^{\mathrm{3D}}} \int\limits_{H_\infty} w_2(|\boldsymbol{\xi}|)(T(\boldsymbol{x}+\boldsymbol{\xi}) - T(\boldsymbol{x})) \left\{ \begin{array}{c} 4\xi_1^2 - \xi_2^2 - \xi_3^2 \\ -\xi_1^2 + 4\xi_2^2 - \xi_3^2 \\ -\xi_1^2 - \xi_2^2 + 4\xi_3^2 \end{array} \right\} \mathrm{d}V_{\boldsymbol{x}'}$$

$$(12.21)$$

$$\left\{ \begin{array}{c} \dfrac{\partial^2 T(\boldsymbol{x})}{\partial x_1^2} \\ \dfrac{\partial^2 T(\boldsymbol{x})}{\partial x_2^2} \end{array} \right\} = \frac{1}{h\pi V_2^{\mathrm{2D}}} \int\limits_{H_\infty} w_2(|\boldsymbol{\xi}|)(T(\boldsymbol{x}+\boldsymbol{\xi}) - T(\boldsymbol{x})) \left\{ \begin{array}{c} 3\xi_1^2 - \xi_2^2 \\ -\xi_2^2 + 3\xi_1^2 \end{array} \right\} \mathrm{d}V_{\boldsymbol{x}'}$$

$$(12.22)$$

$$\frac{\partial^2 T(x)}{\partial x^2} = \frac{1}{A V_2^{\mathrm{1D}}} \int\limits_{H_\infty} w_2(|\xi|)(T(x+\xi) - T(x))\xi^2 \mathrm{d}V_{\boldsymbol{x}}' \qquad (12.23)$$

式中, 物质点相对位置矢量为 $\boldsymbol{\xi} = (\xi_1, \xi_2, \xi_3)$, 且有 $\xi_1 = x_1' - x_1$、$\xi_2 = x_2' - x_2$ 和 $\xi_3 = x_3' - x_3$; 对于三维问题, $V_2^{\mathrm{3D}} = \int_0^\delta w_2(|\boldsymbol{\xi}|)|\boldsymbol{\xi}|^6 \mathrm{d}|\boldsymbol{\xi}|$; 对于二维问题, $V_2^{\mathrm{2D}} = \int_0^\delta w_2(|\boldsymbol{\xi}|)|\boldsymbol{\xi}|^5 \mathrm{d}|\boldsymbol{\xi}|$, h 为二维板厚度; 对于一维问题, $V_2^{\mathrm{1D}} = \int_0^\delta w_2(|\xi|)|\xi|^4 \mathrm{d}|\xi|$, A 为一维杆件横截面积。

将上述温度导数的积分式代入局部热传导方程中, 得到如下三维、二维和一

维问题的近场动力学热传导方程

$$\rho c_v \frac{\partial T(\boldsymbol{x},t)}{\partial t} = \frac{3k_\mathrm{T}}{2\pi V_2^{3\mathrm{D}}} \int_{H_\infty} w_2(|\boldsymbol{\xi}|)(T(\boldsymbol{x}+\boldsymbol{\xi}) - T(\boldsymbol{x})) |\boldsymbol{\xi}|^2 \, \mathrm{d}V_{\boldsymbol{x}'} + S_\mathrm{T}(\boldsymbol{x},t) \quad (12.24)$$

$$\rho c_v \frac{\partial T(\boldsymbol{x},t)}{\partial t} = \frac{2k_\mathrm{T}}{h\pi V_2^{2\mathrm{D}}} \int_{H_\infty} w_2(|\boldsymbol{\xi}|)(T(\boldsymbol{x}+\boldsymbol{\xi}) - T(\boldsymbol{x})) |\boldsymbol{\xi}|^2 \, \mathrm{d}V_{\boldsymbol{x}'} + S_\mathrm{T}(\boldsymbol{x},t) \quad (12.25)$$

$$\rho c_v \frac{\partial T(x,t)}{\partial t} = \frac{k_\mathrm{T}}{A\pi V_2^{1\mathrm{D}}} \int_{H_\infty} w_2(|\xi|)(T(x+\xi) - T(x)) |\xi|^2 \, \mathrm{d}V_{\boldsymbol{x}}' + S_\mathrm{T}(x,t)$$

$$= \frac{k_\mathrm{T}}{V_2^{1\mathrm{D}}} \int_{-\delta}^{\delta} w_2(|\xi|)(T(x+\xi) - T(x)) |\xi|^2 \, \mathrm{d}|\xi| + S_\mathrm{T}(x,t) \quad (12.26)$$

与现有的键型近场动力学热传导模型不同的是，上述基于近场动力学微分算子的热传导模型不含有微导热系数，自然也不需要率定，而是直接使用物质宏观导热系数，并且也不需要假定热响应函数形式。当选定合适影响函数时，构建的新模型能够退化为现有近场动力学热传导模型，例如，当一维热传导公式 (12.26) 取影响函数为 $w_2(\xi) = 1/|\xi|^4$ 时，该方程与现有键型近场动力学中热响应函数取 $f_\mathrm{T} = \kappa_\mathrm{T}(T'-T)/|\xi|^2$ 及微导热系数 $\kappa_\mathrm{T} = \dfrac{k_\mathrm{T}}{A\delta}$ 相一致。

当物质点位于边界区域时，物质点的近场域是不完整非对称的，则可以根据附录 A 中近场动力学微分算子公式 (A.44)、(A.83) 和 (A.108)，得到三维、二维和一维问题温度的二阶导数积分表达式为

$$\left\{\begin{array}{c} \dfrac{\partial^2 T(\boldsymbol{x})}{\partial x_1^2} \\ \dfrac{\partial^2 T(\boldsymbol{x})}{\partial x_2^2} \\ \dfrac{\partial^2 T(\boldsymbol{x})}{\partial x_3^2} \end{array}\right\} = \int_{H_\infty} (T(\boldsymbol{x}+\boldsymbol{\xi}) - T(\boldsymbol{x})) \left\{\begin{array}{c} g_2^{200}(\boldsymbol{\xi}) \\ g_2^{020}(\boldsymbol{\xi}) \\ g_2^{002}(\boldsymbol{\xi}) \end{array}\right\} \mathrm{d}V_{\boldsymbol{x}'} \quad (12.27)$$

$$\left\{\begin{array}{c} \dfrac{\partial^2 T(\boldsymbol{x})}{\partial x_1^2} \\ \dfrac{\partial^2 T(\boldsymbol{x})}{\partial x_2^2} \end{array}\right\} = \int_{H_\infty} (T(\boldsymbol{x}+\boldsymbol{\xi}) - T(\boldsymbol{x})) \left\{\begin{array}{c} g_2^{20}(\boldsymbol{\xi}) \\ g_2^{02}(\boldsymbol{\xi}) \end{array}\right\} \mathrm{d}V_{\boldsymbol{x}'} \quad (12.28)$$

$$\frac{\partial^2 T(x)}{\partial x^2} = \int_{H_\infty} (T(x+\xi) - T(x)) g_2^2(\xi) \mathrm{d}V_{\boldsymbol{x}}' \quad (12.29)$$

式中，$g_N^{p_1p_2\cdots p_M}(\boldsymbol{\xi})$ 为近场动力学函数，具体构建过程可见附录 A。

类似地，将上述温度导数的积分式代入局部热传导方程中，得到如下近场动力学热传导方程

$$\rho c_v \frac{\partial T(\boldsymbol{x},t)}{\partial t} = k_{\mathrm{T}} \int_{H_{\boldsymbol{x}}} w_2(|\boldsymbol{\xi}|)(T(\boldsymbol{x}+\boldsymbol{\xi})-T(\boldsymbol{x}))\mathrm{tr}(\boldsymbol{g}_2^2(\boldsymbol{\xi}))\mathrm{d}V_{\boldsymbol{x}'} + S_{\mathrm{T}}(\boldsymbol{x},t) \quad (12.30)$$

式中，tr 表示矩阵的迹，且有

$$\boldsymbol{g}_2^2(\boldsymbol{\xi}) = \begin{cases} \mathrm{diag}(g_2^{200}, g_2^{020}, g_2^{002}), & \text{三维} \\ \mathrm{diag}(g_2^{20}, g_2^{02}), & \text{二维} \\ \mathrm{diag}(g_2^{2}), & \text{一维} \end{cases} \quad (12.31)$$

式 (12.30) 也适用于构型内部具有完整对称近场范围的物质点。该式还可以统一改写为

$$\rho c_v \frac{\partial T(\boldsymbol{x},t)}{\partial t} = \sum_{i=1}^{n_{\mathrm{d}}} S_{\mathrm{T}_{x_i}} + S_{\mathrm{T}}(\boldsymbol{x},t) \quad (12.32)$$

式中，n_d 为空间维数，分量 $S_{\mathrm{T}_{x_i}}$ 为热量密度，表示单位时间内构型自身 x_i 方向的热传导对单位体积物质点的热量贡献，具体表达式如下：

$$S_{\mathrm{T}_{x_1}} = k_{\mathrm{T}} \int_{H_{\boldsymbol{x}}} (T(\boldsymbol{x}+\boldsymbol{\xi})-T(\boldsymbol{x}))g_2^{200}(\boldsymbol{\xi})\mathrm{d}V_{\boldsymbol{x}'} \quad (12.33)$$

$$S_{\mathrm{T}_{x_2}} = k_{\mathrm{T}} \int_{H_{\boldsymbol{x}}} (T(\boldsymbol{x}+\boldsymbol{\xi})-T(\boldsymbol{x}))g_2^{020}(\boldsymbol{\xi})\mathrm{d}V_{\boldsymbol{x}'} \quad (12.34)$$

$$S_{\mathrm{T}_{x_3}} = k_{\mathrm{T}} \int_{H_{\boldsymbol{x}}} (T(\boldsymbol{x}+\boldsymbol{\xi})-T(\boldsymbol{x}))g_2^{002}(\boldsymbol{\xi})\mathrm{d}V_{\boldsymbol{x}'} \quad (12.35)$$

最后，将傅里叶定律中的一阶导数项用近场动力学微分算子替换，得到近场动力学热流通量表达式

$$\boldsymbol{q}(\boldsymbol{x},t) = -k_{\mathrm{T}} \int_{H_{\boldsymbol{x}}} (T(\boldsymbol{x}+\boldsymbol{\xi})-T(\boldsymbol{x}))\boldsymbol{g}_2^1(\boldsymbol{\xi})\mathrm{d}V_{\boldsymbol{x}'} \quad (12.36)$$

且有

$$\boldsymbol{g}_2^1(\boldsymbol{\xi}) = \begin{cases} (g_2^{100}(\boldsymbol{\xi}), g_2^{010}(\boldsymbol{\xi}), g_2^{001}(\boldsymbol{\xi})), & \text{三维} \\ (g_2^{10}(\boldsymbol{\xi}), g_2^{01}(\boldsymbol{\xi})), & \text{二维} \\ (g_2^1(\boldsymbol{\xi})), & \text{一维} \end{cases} \quad (12.37)$$

12.3 基于近场动力学微分算子的热-力耦合模型

热-力耦合问题通常是指热传导方程中包含考察体变形引起的加热或冷却效应，而考察体运动方程中涉及的本构定律需考虑热效应的作用。本节将结合经典热-力耦合方程和近场动力学微分算子，构建基于近场动力学微分算子的热-力耦合模型。

由应力应变表示的经典热-力耦合方程为[17]

$$\rho c_v \dot{T} = \nabla \cdot (k_\mathrm{T} \nabla T) - \alpha(3\lambda + 2\mu) T_0 \dot{\varepsilon}_{kk} + S_\mathrm{T} \quad (12.38)$$

$$\rho \ddot{\boldsymbol{u}} = \nabla \cdot \boldsymbol{\sigma} + \boldsymbol{b} \quad (12.39)$$

式中，$\dot{\varepsilon}_{kk}$ 为体应变率，$\boldsymbol{u}$ 为位移矢量，$\alpha(3\lambda+2\mu)$ 为热模量，α 为热膨胀系数，T 和 T_0 分别为当前构型的温度和参考温度，$\boldsymbol{b}$ 为体力密度矢量，∇ 为梯度算子。在线弹性热力学中，Cauchy 应力张量 $\boldsymbol{\sigma}$ 的分量形式为

$$\sigma_{ij} = 2\mu\varepsilon_{ij} + (\lambda\varepsilon_{kk} - \alpha(3\lambda + 2\mu)(T - T_0))\delta_{ij} \quad (12.40)$$

式中，$\lambda = \dfrac{E\nu}{(1+\nu)(1-2\nu)}$ 和 $\mu = \dfrac{E}{2(1+\nu)}$ 为拉梅常数，ε_{ij} 为应变张量的分量，ε_{kk} 为体积应变，δ_{ij} 为 Kronecker 符号。

根据线弹性应变与位移的关系，将式 (12.40) 中的应变分量用位移分量表示，有

$$\sigma_{ij} = \mu(u_{i,j} + u_{j,i}) + (\lambda u_{k,k} - \alpha(3\lambda + 2\mu)(T - T_0))\delta_{ij} \quad (12.41)$$

将上述关系代入到式 (12.39) 中，并注意到式 (12.38) 中体应变率可表示为 $\dot{\varepsilon}_{kk} = \dfrac{\partial}{\partial t}\left(\dfrac{\partial u_k}{\partial x_k}\right)$，得到以位移分量表示的经典热-力耦合方程

$$\rho c_v \frac{\partial T}{\partial t} = k_\mathrm{T} \frac{\partial^2 T}{\partial x_j \partial x_j} - \alpha(3\lambda + 2\mu) T_0 \frac{\partial}{\partial t}\left(\frac{\partial u_k}{\partial x_k}\right) + S_\mathrm{T} \quad (12.42)$$

$$\rho \frac{\partial^2 u_i}{\partial t^2} = \mu \frac{\partial^2 u_i}{\partial x_k \partial x_k} + (\mu + \lambda) \frac{\partial^2 u_k}{\partial x_k \partial x_i} - \alpha(3\lambda + 2\mu) \frac{\partial(T - T_0)}{\partial x_i} + b_i \quad (12.43)$$

对于任意的物质点，不论其处于考察体的体内或边界，也不论该物质点的近场范围是完整对称的或是任意形状的，根据附录 A 中公式 (A.43)、(A.44)、(A.83)、(A.84)、(A.107) 和 (A.108) 给出的一阶、二阶局部微分的积分表达式，可以得到任意物质点的近场动力学热-力耦合方程

$$\rho c_v \frac{\partial T(\boldsymbol{x},t)}{\partial t} = k_\mathrm{T} \int_{H_\infty} (T(\boldsymbol{x}+\boldsymbol{\xi}) - T(\boldsymbol{x}))\mathrm{tr}(\boldsymbol{g}_2^2(|\boldsymbol{\xi}|))\mathrm{d}V_{\boldsymbol{x}'}$$

$$- \alpha(3\lambda+2\mu)T_0 \frac{\partial}{\partial t}\left(\int_{H_\infty}(\boldsymbol{u}(\boldsymbol{x}+\boldsymbol{\xi})-\boldsymbol{u}(\boldsymbol{x}))\boldsymbol{g}_2^1(\boldsymbol{\xi})\mathrm{d}V_{\boldsymbol{x}'}\right)$$

$$+ S_\mathrm{T}(\boldsymbol{x},t) \qquad (12.44)$$

$$\rho\frac{\partial^2 \boldsymbol{u}(\boldsymbol{x},t)}{\partial t^2} = \int_{H_\infty}(\mu\mathrm{tr}(\boldsymbol{D})\boldsymbol{I}+(\lambda+\mu)\boldsymbol{D})(\boldsymbol{u}(\boldsymbol{x}+\boldsymbol{\xi})-\boldsymbol{u}(\boldsymbol{x}))\mathrm{d}V_{\boldsymbol{x}'}$$

$$- \alpha(3\lambda+2\mu)\int_{H_\infty}(T(\boldsymbol{x}+\boldsymbol{\xi})-T(\boldsymbol{x}))\boldsymbol{g}_2^1(\boldsymbol{\xi})\mathrm{d}V_{\boldsymbol{x}'} + \boldsymbol{b}(\boldsymbol{x},t) \qquad (12.45)$$

式中，$\boldsymbol{g}_2^1(\boldsymbol{\xi})$ 和 $\boldsymbol{g}_2^2(\boldsymbol{\xi})$ 分别如式 (12.37) 和式 (12.31) 所示，对于不同维度的问题，矩阵 $\boldsymbol{D}$ 分别为

$$\boldsymbol{D} = \begin{bmatrix} g_2^{200}(\boldsymbol{\xi}) & g_2^{110}(\boldsymbol{\xi}) & g_2^{101}(\boldsymbol{\xi}) \\ g_2^{110}(\boldsymbol{\xi}) & g_2^{020}(\boldsymbol{\xi}) & g_2^{011}(\boldsymbol{\xi}) \\ g_2^{101}(\boldsymbol{\xi}) & g_2^{011}(\boldsymbol{\xi}) & g_2^{002}(\boldsymbol{\xi}) \end{bmatrix}, \quad 三维 \qquad (12.46)$$

$$\boldsymbol{D} = \begin{bmatrix} g_2^{20}(\boldsymbol{\xi}) & g_2^{11}(\boldsymbol{\xi}) \\ g_2^{11}(\boldsymbol{\xi}) & g_2^{02}(\boldsymbol{\xi}) \end{bmatrix}, \quad 二维 \qquad (12.47)$$

$$\boldsymbol{D} = \boldsymbol{g}_2^2(\xi) = [g_2^2(\xi)], \quad 一维 \qquad (12.48)$$

对于考察体内部非边缘区域的物质点，当其具有完整球形或圆形对称近场范围时，近场动力学微分算子见式 (A.53)、(A.54)、(A.92)、(A.93)、(A.111) 和 (A.112)。采用完整对称近场范围下的近场动力学非局部微分算子取代经典热-力耦合方程 (12.42) 和 (12.43) 中的局部微分，可以获得积分型非局部热-力耦合方程。

12.3 基于近场动力学微分算子的热-力耦合模型

具体地，对于三维问题，积分型非局部热-力耦合方程为

$$\rho c_v \frac{\partial T(\boldsymbol{x},t)}{\partial t} = \frac{3k_\mathrm{T}}{2\pi V_2^{\mathrm{3D}}} \int_{H_\infty} w_2(|\boldsymbol{\xi}|)(T(\boldsymbol{x}+\boldsymbol{\xi})-T(\boldsymbol{x}))|\boldsymbol{\xi}|^2 \mathrm{d}V_{\boldsymbol{x}'}$$

$$- \alpha(3\lambda+2\mu)T_0 \frac{\partial}{\partial t}\left(\frac{3}{4\pi V_1^{\mathrm{3D}}} \int_{H_\infty} w_1(|\boldsymbol{\xi}|)\boldsymbol{\xi}\cdot(\boldsymbol{u}(\boldsymbol{x}+\boldsymbol{\xi})-\boldsymbol{u}(\boldsymbol{x}))\mathrm{d}V_{\boldsymbol{x}'}\right)$$

$$+ S_\mathrm{T}(\boldsymbol{x},t) \tag{12.49}$$

$$\rho \frac{\partial^2 \boldsymbol{u}(\boldsymbol{x},t)}{\partial t^2} = \frac{3(\mu-\lambda)}{4\pi V_2^{\mathrm{3D}}} \int_{H_\infty} w_2(|\boldsymbol{\xi}|)(\boldsymbol{\xi}\cdot\boldsymbol{\xi})(\boldsymbol{u}(\boldsymbol{x}+\boldsymbol{\xi})-\boldsymbol{u}(\boldsymbol{x}))\mathrm{d}V_{\boldsymbol{x}'}$$

$$+ (\lambda+\mu)\frac{15}{4\pi V_2^{\mathrm{3D}}} \int_{H_\infty} w_2(|\boldsymbol{\xi}|)((\boldsymbol{u}(\boldsymbol{x}+\boldsymbol{\xi})-\boldsymbol{u}(\boldsymbol{x}))\cdot\boldsymbol{\xi})\boldsymbol{\xi}\mathrm{d}V_{\boldsymbol{x}'}$$

$$- \alpha(3\lambda+2\mu)\frac{3}{4\pi V_1^{\mathrm{3D}}} \int_{H_\infty} w_1(|\boldsymbol{\xi}|)(T(\boldsymbol{x}+\boldsymbol{\xi})-T(\boldsymbol{x}))\boldsymbol{\xi}\mathrm{d}V_{\boldsymbol{x}'}$$

$$+ \boldsymbol{b}(\boldsymbol{x},t) \tag{12.50}$$

类似地，对于二维问题，相应的热-力耦合方程为

$$\rho c_v \frac{\partial T(\boldsymbol{x},t)}{\partial t} = \frac{2k_\mathrm{T}}{h\pi V_2^{\mathrm{2D}}} \int_{H_\infty} w_2(|\boldsymbol{\xi}|)(T(\boldsymbol{x}+\boldsymbol{\xi})-T(\boldsymbol{x}))|\boldsymbol{\xi}|^2 \mathrm{d}V_{\boldsymbol{x}'}$$

$$- \alpha(3\lambda+2\mu)T_0 \frac{\partial}{\partial t}\left(\frac{1}{\pi h V_1^{\mathrm{2D}}} \int_{H_\infty} w_1(|\boldsymbol{\xi}|)\boldsymbol{\xi}\cdot(\boldsymbol{u}(\boldsymbol{x}+\boldsymbol{\xi})-\boldsymbol{u}(\boldsymbol{x}))\mathrm{d}V_{\boldsymbol{x}'}\right)$$

$$+ S_\mathrm{T}(\boldsymbol{x},t) \tag{12.51}$$

$$\rho \frac{\partial^2 \boldsymbol{u}(\boldsymbol{x},t)}{\partial t^2} = \frac{(\mu-\lambda)}{\pi h V_2^{\mathrm{2D}}} \int_{H_\infty} w_2(|\boldsymbol{\xi}|)(\boldsymbol{\xi}\cdot\boldsymbol{\xi})(\boldsymbol{u}(\boldsymbol{x}+\boldsymbol{\xi})-\boldsymbol{u}(\boldsymbol{x}))\mathrm{d}V_{\boldsymbol{x}'}$$

$$+ (\lambda+\mu)\frac{4}{\pi h V_2^{\mathrm{2D}}} \int_{H_\infty} w_2(|\boldsymbol{\xi}|)\left[(\boldsymbol{u}(\boldsymbol{x}+\boldsymbol{\xi})-\boldsymbol{u}(\boldsymbol{x}))\cdot\boldsymbol{\xi}\right]\boldsymbol{\xi}\mathrm{d}V_{\boldsymbol{x}'}$$

$$- \alpha(3\lambda+2\mu)\frac{1}{\pi h V_1^{\mathrm{2D}}} \int_{H_\infty} w_1(|\boldsymbol{\xi}|)(T(\boldsymbol{x}+\boldsymbol{\xi})-T(\boldsymbol{x}))\boldsymbol{\xi}\mathrm{d}V_{\boldsymbol{x}'}$$

$$+ \boldsymbol{b}(\boldsymbol{x},t) \tag{12.52}$$

一维问题对应的积分型非局部热-力耦合方程为

$$\rho c_v \frac{\partial T(x,t)}{\partial t} = \frac{k_T}{AV_2} \int_{H_x} w_2(|\xi|)(T(x+\xi) - T(x))|\xi|^2 \mathrm{d}V_{x'}$$

$$- \alpha \mu T_0 \frac{\partial}{\partial t} \left(\frac{1}{2AV_1} \int_{H_x} w_1(|\xi|)(u(x+\xi) - u(x))\xi \mathrm{d}V_{x'} \right)$$

$$+ S_T(x,t) \qquad (12.53)$$

$$\rho \frac{\partial^2 u(x,t)}{\partial t^2} = \mu \frac{1}{2AV_2} \int_{H_x} w_2(|\xi|)(u(x+\xi) - u(x))\xi^2 \mathrm{d}V_{x'}$$

$$- \alpha \mu \frac{1}{AV_1} \int_{H_x} w_1(|\xi|)(T(x+\xi) - T(x))\xi \mathrm{d}V_{x'}$$

$$+ b(x,t) \qquad (12.54)$$

在一维问题的理想模型中，泊松比 ν 为零，拉梅常数 $\lambda = 0$ 和 $\mu = E/2$。

相比于 Agwai[10] 于 2011 年建立的近场动力学热-力耦合方程，本节给出的热-力耦合模型无需假设热响应函数形式，可以直接使用材料的导热系数，并不需要引入键的微导热系数，避免了基于完整对称近场范围的微导热系数的率定过程。此外，在变形引起的温度项中，不再使用伸长量的时间导数，而是直接采用位移的微分表示。

12.4 热传导和热-力耦合问题近场动力学模型的数值计算

12.4.1 初始条件与边界条件

热传导问题的计算需要知晓初始条件和热边界条件。三类热边界条件为：狄利克雷 (Dirichlet) 边界条件或本质边界条件、诺依曼 (Neumann) 边界条件或自然边界条件、罗宾 (Robin) 边界条件或对流边界条件。

对于初始条件，构型中所有物质点都被赋予一个初始温度值

$$T(\boldsymbol{x}, t=0) = T^*(\boldsymbol{x}) \qquad (12.55)$$

对于狄利克雷边界条件，构型边界温度是已知的，只需将温度直接施加在一层边界点上

$$T = T_{\mathrm{bc}}(\boldsymbol{x}, t) \qquad (12.56)$$

式中，$T_{bc}(\boldsymbol{x},t)$ 为在时间 t 时预设的边界温度。

对于热流边界条件或诺依曼边界条件，构型热流密度已知或与外部介质存在热交换，当使用均匀离散时，每个物质点所占空间体积相同，且截面积也是一个常数[19]，计算单位时间内通过热传导面的热流密度，通过施加体积热源的方式施加在物质点上，可以表示为

$$S_{\mathrm{T}} = -\frac{\boldsymbol{q}_{bc}(\boldsymbol{x},t) \cdot \boldsymbol{n}}{\Delta_{\mathrm{th}}} \tag{12.57}$$

或者用分量表达形式

$$S_{\mathrm{T}_{x_i}} = -\frac{q_{bc}(\boldsymbol{x},t)_{x_i}}{\Delta_{\mathrm{th}}} \tag{12.58}$$

式中，S_{T} 为体积热源项，热量密度分量 $S_{\mathrm{T}_{x_i}}$ 表示单位时间内构型自身 x_i 方向的热传导对单位体积物质点的热量贡献，$\boldsymbol{q}_{bc}$ 为指定的热流量向量，$\boldsymbol{n}$ 为热传导面的单位法向量，$q_{bc}(\boldsymbol{x},t)_{x_i}$ 为沿 x_i 方向的热流量分量，Δ_{th} 是热流边界层厚度，即 $\Delta_{\mathrm{th}} = \Delta x$。

对于给定的对流边界条件，也可以通过体积热源项施加在一层边界物质点上，即

$$\boldsymbol{q}_{bc} \cdot \boldsymbol{n} = h_c(T_{bc} - T_\infty) \tag{12.59}$$

$$S_{\mathrm{T}} = -\frac{\boldsymbol{q}_{bc}(\boldsymbol{x},t) \cdot \boldsymbol{n}}{\Delta_{\mathrm{th}}} = \frac{h_c(T_\infty - T_{bc})}{\Delta_{\mathrm{th}}} \tag{12.60}$$

其中，$T_{bc}(\boldsymbol{x},t)$ 为 t 时刻的边界物质点温度，T_∞ 为环境温度，h_c 为固体介质表面与环境介质之间的热对流系数，单位为 $\mathrm{J/(m^2 \cdot s \cdot K)}$。

12.4.2 方程离散与求解

采用显式无网格粒子法求解近场动力学热传导和热–力耦合方程。无量纲的影响函数 $w_n(|\boldsymbol{\xi}|)$ 应能反映物质点间相互作用强度随距离增加而递减的效应，可以选取

$$w_n(|\boldsymbol{\xi}|) = \frac{\delta^{n+1}}{|\boldsymbol{\xi}|^{n+1}} \tag{12.61}$$

不失一般性，以完整圆形对称近场范围下的二维热–力耦合方程求解为例。考虑上述影响函数，二维热–力耦合问题的热传导方程可以被离散为

$$\rho(\boldsymbol{x}_i) c_v(\boldsymbol{x}_i) \frac{\partial T^n(\boldsymbol{x}_i)}{\partial t} = \frac{6k_{\mathrm{T}}}{h\pi\delta^3} \sum_{j=1}^{N_j} \frac{T(\boldsymbol{x}_j) - T(\boldsymbol{x}_i)}{|\boldsymbol{\xi}_{ji}|} V_j$$

$$-\alpha(3\lambda+2\mu)T_0\frac{\partial}{\partial t}\left(\frac{2}{h\pi\delta^2}\sum_{j=1}^{N_i}\frac{1}{|\boldsymbol{\xi}_{ji}|^3}\boldsymbol{\xi}_{ji}\cdot(\boldsymbol{u}(\boldsymbol{x}_j)-\boldsymbol{u}(\boldsymbol{x}_i))V_j\right)$$
$$+S_{\mathrm{T}}(\boldsymbol{x}_i,t) \tag{12.62}$$

式中，n 为时间步号，$\boldsymbol{x}_j$ 为考察点 $\boldsymbol{x}_i$ 近场范围内的物质点，N_i 为 $\boldsymbol{x}_i$ 近场范围内的物质点总数，V_j 为物质点 $\boldsymbol{x}_j$ 的积分体积，采用正交均匀离散时，三维问题的物质点实际体积为 $\bar{V}_j=(\Delta x)^3$，二维问题的物质点实际体积为 $\bar{V}_j=h(\Delta x)^2$，一维问题的物质点体积为 $\bar{V}_j=A\Delta x$，Δx 为物质点长度，h 为二维问题的板厚，A 为一维问题的杆横截面积。圆形近场范围半径为 $\delta=m\Delta x$，在一般情况下，近场范围的边界点可以采用体积修正[1]；当采用非完整对称近场范围下的近场动力学热传导和热力耦合模型时，近场范围的边界点不必做体积修正。

采用逐步积分法完成热传导过程计算 (对于单纯的热传导问题，在式 (12.62) 右端略去变形场的影响即可)，逐步积分法可通过隐式向后差分格式实现，或者显式向前差分格式实现。当采用显式向前差分格式时，从第 n 步到第 $n+1$ 步的温度为

$$T(\boldsymbol{x}_i,t^{n+1})=T(\boldsymbol{x}_i,t^n)+\Delta t\cdot\dot{T}(\boldsymbol{x}_i,t^n) \tag{12.63}$$

式中，Δt 为热传导计算的时间步长。显式向前差分法是条件稳定的，为保证数值计算的收敛性，必须对时间步长进行限制，给出稳定性条件[19]。

含耦合效应的热传导方程稳定性分析较为复杂，以下仅以单纯的一维热传导为例，采用冯·诺依曼稳定性分析方法阐释热传导问题的稳定性条件。对于无热源的一维热传导方程进行空间离散，再采用向前差分格式对时间离散，得到

$$\rho c_v \frac{T_i^{n+1}-T_i^n}{\Delta t}=\frac{k_{\mathrm{T}}}{AV_2}\sum_{j=-\infty}^{\infty}w_2(|\xi_{ji}|)(T_j^n-T_i^n)|\xi_{ji}|^2 V_j \tag{12.64}$$

式中，j 表示 i 点两侧近场范围内物质点的编号。

根据冯·诺依曼稳定性分析方法，假设温度场的形式为

$$T_i^n=\zeta^n \mathrm{e}^{(\kappa i\sqrt{-1})} \tag{12.65}$$

式中，κ 为正实数，ζ 为复数。解的稳定性要求对于任意正实数 κ，确定稳定时间步长 Δt，满足 $|\zeta|\leqslant 1$，也即保证解答不会无限增长。与近场动力学一维热传导方程 (式 (12.64)) 相对应的经典一维热传导方程 (式 (12.20)) 的齐次方程解析解为 $T=A\mathrm{e}^{-wt}$，其中 A 为任意常数，w 总为正值，该解析解单调趋于零，也即满足 $|\zeta|\leqslant 1$。

12.4 热传导和热-力耦合问题近场动力学模型的数值计算

将式 (12.65) 代入式 (12.64)，有

$$\frac{\rho c_v}{\Delta t}(\zeta^{n+1}\mathrm{e}^{(\kappa i\sqrt{-1})} - \zeta^n\mathrm{e}^{(\kappa i\sqrt{-1})})$$

$$= \frac{k_\mathrm{T}}{AV_2}\sum_{j=-\infty}^{\infty} w_2(|\xi_{ji}|)(\zeta^n\mathrm{e}^{(\kappa j\sqrt{-1})} - \zeta^n\mathrm{e}^{(\kappa i\sqrt{-1})})\,|\xi_{ji}|^2\,V_j$$

$$\Rightarrow \frac{\rho c_v}{\Delta t}(\zeta - 1) = \frac{k_\mathrm{T}}{AV_2}\sum_{j=-\infty}^{\infty} w_2(|\xi_{ji}|)(\mathrm{e}^{(\kappa(j-i)\sqrt{-1})} - 1)\,|\xi_{ji}|^2\,V_j$$

$$\Rightarrow \frac{\rho c_v}{\Delta t}(\zeta - 1) = \frac{k_\mathrm{T}}{AV_2}\sum_{j=-\infty}^{\infty} w_2(|\xi_{ji}|)[\cos(\kappa(j-i))$$

$$+ \sqrt{-1}\sin(\kappa(j-i)) - 1]\,|\xi_{ji}|^2\,V_j$$

$$\Rightarrow \frac{\rho c_v}{\Delta t}(\zeta - 1) = \frac{2k_\mathrm{T}}{AV_2}\sum_{j=1}^{\infty} w_2(|\xi_{ji}|)[\cos(\kappa(j-i)) - 1]\,|\xi_{ji}|^2\,V_j$$

$$\Rightarrow \zeta = 1 + \frac{2\Delta t k_\mathrm{T}}{\rho c_v AV_2}\sum_{j=1}^{\infty} w_2(|\xi_{ji}|)[\cos(\kappa(j-i)) - 1]\,|\xi_{ji}|^2\,V_j \tag{12.66}$$

在上述推导中，利用了欧拉公式 $\mathrm{e}^{x\sqrt{-1}} = \cos x + \sqrt{-1}\sin x$，以及 $\sin x$ 的奇函数性质。欲使条件 $|\zeta| \leqslant 1$ 恒成立，则有

$$-1 < 1 + \frac{2\Delta t k_\mathrm{T}}{\rho c_v AV_2}\sum_{j=1}^{\infty} w_2(|\xi_{ji}|)[\cos(\kappa(j-i)) - 1]\,|\xi_{ji}|^2\,V_j < 1 \tag{12.67}$$

考虑到 $\sum_{j=1}^{\infty} w_2(|\xi_{ji}|)[\cos(\kappa(j-i)) - 1]\,|\xi_{ji}|^2\,V_j \geqslant -\sum_{j=1}^{\infty} 2w_2(|\xi_{ji}|)\,|\xi_{ji}|^2\,V_j$，进而有

$$0 < \Delta t < \frac{\rho c_v AV_2}{2k_\mathrm{T}\sum_{j=1}^{\infty} w_2(|\xi_{ji}|)\,|\xi_{ji}|^2\,V_j} = \Delta t_{\mathrm{cr}} \tag{12.68}$$

进而，采用拟静力法求解准静态热-力耦合运动方程，引入阻尼项，则空间离散的近场动力学运动方程为

$$\rho \ddot{\boldsymbol{u}}_i + c_n \dot{\boldsymbol{u}}_i = \boldsymbol{G}_i + \boldsymbol{b}_i \tag{12.69}$$

其中

$$\boldsymbol{G}_i = \frac{3(\mu - \lambda)}{\pi h \delta^3}\sum_{j=1}^{N_i} \frac{(\boldsymbol{\xi}_{ji} \cdot \boldsymbol{\xi}_{ji})}{|\boldsymbol{\xi}_{ji}|^3}(\boldsymbol{u}(\boldsymbol{x}_j) - \boldsymbol{u}(\boldsymbol{x}_i))V_j$$

$$+ (\lambda + \mu) \frac{12}{\pi h \delta^3} \sum_{j=1}^{N_i} \frac{(\boldsymbol{u}(\boldsymbol{x}_j) - \boldsymbol{u}(\boldsymbol{x}_i)) \cdot \boldsymbol{\xi}_{ji}}{|\boldsymbol{\xi}_{ji}|^3} \boldsymbol{\xi}_{ji} V_j$$

$$- \alpha(3\lambda + 2\mu) \frac{2}{\pi h \delta^2} \sum_{j=1}^{N_i} \frac{(T(\boldsymbol{x}_j) - T(\boldsymbol{x}_i))}{|\boldsymbol{\xi}_{ji}|^3} \boldsymbol{\xi}_{ji} V_j \quad (12.70)$$

式中，c_n 为人工阻尼系数项。

采用显式中心差分[20]对运动方程进行时间积分。忽略热传导对变形的影响，根据稳定性分析 (推导过程见 5.2.2 节)，可得到近场动力学运动方程的稳定时间步长条件，热传导方程的稳定时间步长远大于运动方程的稳定时间步长。

对于热–力耦合问题，可采用交错迭代方法进行数值计算。如图 12-1 所示，首先进行热传导分析，在经过一个时间步的热传导计算后，将得到的当前构型温度状态设置为一个 "假定稳态"，然后整个系统进入一个变形场分析的循环，这个循环会在变形场达到稳态时停止，然后进入下一个温度分析步。在变形场分析循环中，引入收敛准则判断当前构型变形是否达到稳态，当一个物质点将要达到稳态时，它在一个时间步内的位移增量接近于零；在每步变形场分析后，计算构型物质点位移的均方根 $\mathrm{Re} = \sqrt{\sum_{m=1}^{N_{\mathrm{total}}} (u_m^{i,j} - u_m^{i,j-1})^2 \Big/ N_{\mathrm{total}}}$，其中 N_{total} 为构型物质点总数；然后将其与小量 Ω 作比较，当 $\mathrm{Re} > \Omega$ 时，认为系统是不稳定的并进入 $j+1$ 步；否则，认定系统达到稳定状态并进入 $i+1$ 步。

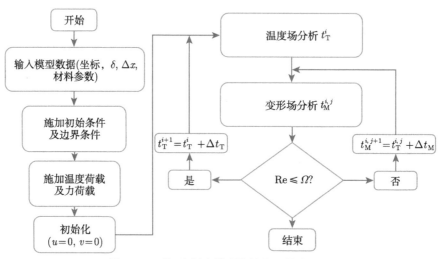

图 12-1　热–力耦合模型的数值计算流程图

12.4.3 损伤区域导热系数的修正

当"机械键"断裂时，键力被释放即等于零，但热传导并未完全停止。尽管两个物质点间的距离逐渐远离，但只要两点仍在彼此的近场范围内，热传导过程就依然存在，即"热键"并没有断裂。由于两个物质点相距越来越远，对各自的影响逐渐减弱，当距离超过近场范围尺寸 δ 时，认为空气完全进入"机械键"断裂后形成的裂隙，成为一种新的传导介质，空气的热学参数被视为已断键的热学参数。为此，对于断开的"机械键"，可将导热系数修正为

$$k_{\mathrm{T}}^{\mathrm{B}} = \begin{cases} k_{\mathrm{T}}\left(1 - \dfrac{|\boldsymbol{\xi}|}{\delta}\right), & |\boldsymbol{\xi}| < \delta \\ k_{\mathrm{A}}, & |\boldsymbol{\xi}| \geqslant \delta \end{cases} \tag{12.71}$$

式中，$k_{\mathrm{T}}^{\mathrm{B}}$ 是断开的"机械键"的导热系数，k_{A} 是空气的导热系数。

另外说明，空气的导热系数通常远小于其他材料的导热系数，这也是许多研究者将裂纹边界视为绝热的原因。

12.5 热传导问题近场动力学模型的算例分析

本节采用热传导问题的近场动力学模型研究混合热边界条件下，完好或含裂纹结构的一维、二维和三维热传导问题[8,9,11]，验证所发展的模型和方法的有效性。在此基础上，建立混凝土材料和结构导热分析的细观近场动力学方法，研究裂纹对于混凝土导热性能的影响。

12.5.1 一维杆件热传导问题的算例分析

长度 $L = 0.1\mathrm{m}$、横截面积 $A = 2.5 \times 10^{-5}\mathrm{m}^2$ 的一维杆件，初始温度 $100\,^\circ\mathrm{C}$，杆两端温度突降并保持在 $0\,^\circ\mathrm{C}$，杆的材料质量密度 $\rho = 1000\mathrm{kg/m}^3$，比热容 $c_v = 1\mathrm{J/(kg \cdot K)}$，导热系数 $k_{\mathrm{T}} = 11.4\mathrm{J/(kg \cdot K)}$，相应的导温系数为 $\alpha = k_{\mathrm{T}}/(\rho c_v) = 0.0114\mathrm{m}^2/\mathrm{s}$[8]。

经典局部热传导理论对该问题的描述为

$$\begin{cases} \dfrac{\partial T(x,t)}{\partial t} = \alpha \dfrac{\partial^2 T(x,t)}{\partial x^2}, & 0 \leqslant x \leqslant L, 0 \leqslant t < \infty \\ T(0,t) = T(L,t) = 0, & 0 \leqslant t < \infty \\ T(x,0) = 100\,^\circ\mathrm{C}, & 0 \leqslant x \leqslant L \end{cases} \tag{12.72}$$

基于分离变量方法，可以获得温度的级数形式解析解

$$T(x,t) = T(0,t) + (T(L,t) - T(0,t))\frac{x}{L} + \sum_{n=1,3,5,\cdots}^{\infty} \frac{4T(x,0)}{n\pi} \sin\frac{n\pi x}{L} e^{-\alpha t(n^2\pi^2/L^2)}$$
(12.73)

此处取级数 $n = 55$ 计算解析解。

在近场动力学模拟中,时间步长取为 $\Delta t = 10^{-6}$s 以保证计算稳定性,Dirichlet 边界条件分别施加在杆两端的一个物质点上,分别采用 50、100、200、400 和 800 个物质点离散杆件,相应的物质点间距 $\Delta x = 2 \times 10^{-3}$m、$1 \times 10^{-3}$m、$5 \times 10^{-4}$m、$2.5 \times 10^{-4}$m 和 1.25×10^{-4}m。

如前所述,δ 为近场范围半径,m 为近场范围半径与物质点间距的比值,即 $\delta = m\Delta x$;对于给定的近场范围半径 δ,当比值 $m \to \infty$,近场动力学数值解将会收敛于近场动力学非局部精确解;对于固定的 m,当 $\delta \to 0$,近场动力学数值解将会收敛于经典连续介质力学的局部解。在 m-收敛性和 δ-收敛性分析中,通过全局误差评价计算结果的收敛性,全局误差采用 $\varepsilon = \dfrac{1}{|T^{(r)}|_{\max}}\sqrt{\dfrac{1}{N_{\text{total}}}\sum_{i=1}^{N_{\text{total}}}\left[T_i^{(r)} - T_i^{(n)}\right]^2}$,其中 N_{total} 为物质点总数,上标 r 和 n 分别表示参考解与数值解。

给定近场范围 $\delta = 2 \times 10^{-3}$m,考虑 5 个不同的比值 $m = 1, 2, 4, 8, 16$,进行 m-收敛性分析。图 12-2(a) 显示了 $t = 0.01$s 时 PD 预测温度与经典解析解的全局误差,5 组近场动力学模拟对应的全局误差分别为 2.7899%、1.3838%、0.6901%、0.3457% 和 0.1743%,表明近场动力学计算的 m-收敛性良好,选取 $m = 4$ 可以较好地兼顾计算精度和计算效率。固定 $m = 4$,选取不同离散网格进行 5 组近场动力学模拟,$t = 0.01$s 时的全局误差如图 12-2(b) 所示,其收敛速度是线性的,表明

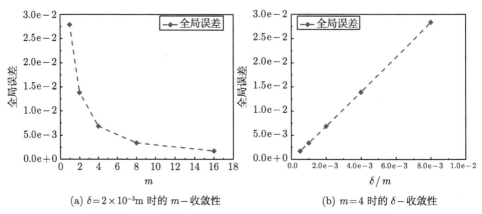

(a) $\delta = 2\times 10^{-3}$m 时的 $m-$收敛性   (b) $m = 4$ 时的 $\delta-$收敛性

图 12-2 $t = 0.01$s 时 PD 预测值与经典解析解的全局误差

近场动力学计算的 δ-收敛性良好。图 12-3 给出了 $t=0.001\mathrm{s}$、$t=0.01\mathrm{s}$、$t=0.04\mathrm{s}$ 和 $t=0.08\mathrm{s}$ 四个时刻局部理论的解析解与近场动力学求得的杆中温度分布曲线，在近场动力学模拟中，物质点间距和近场范围半径分别为 $\Delta x = 2.5 \times 10^{-4}\mathrm{m}$ 和 $\delta = 4\Delta x$。图中所示的计算结果表明，近场动力学计算结果与解析解高度一致。

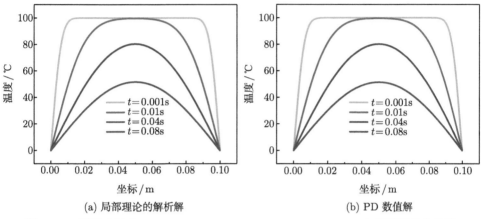

图 12-3 不同时刻 ($t=0.001\mathrm{s}$、$t=0.01\mathrm{s}$、$t=0.04\mathrm{s}$、$t=0.08\mathrm{s}$) 的温度分布曲线

12.5.2 二维方板热传导问题的算例分析

图 12-4 所示为边长 $L = W = 0.01\mathrm{m}$ 的二维方板，初始温度 $T = 0°\mathrm{C}$，板左边 $T = 1°\mathrm{C}$，其他三边为绝热边界，即热流通量为零。杆的材料质量密度 $\rho = 1000\mathrm{kg/m^3}$，比热容 $c_v = 1\mathrm{J/(kg \cdot K)}$，导热系数 $k_\mathrm{T} = 1000\mathrm{J/(s \cdot m \cdot K)}$，相应的导温系数 $\alpha = k_\mathrm{T}/(\rho c_v) = 1\mathrm{m^2/s}$[9]。

经典局部热传导理论对该问题的描述为

$$\begin{cases} \dfrac{\partial T(\boldsymbol{x},t)}{\partial t} = \dfrac{\partial^2 T(\boldsymbol{x},t)}{\partial x_1^2} + \dfrac{\partial^2 T(\boldsymbol{x},t)}{\partial x_2^2}, & 0 \leqslant x_1 \leqslant 1, 0 \leqslant x_2 \leqslant 1, 0 \leqslant t < \infty \\ T(x_1,x_2,0) = 0, & 0 \leqslant x_1 \leqslant 1, 0 \leqslant x_2 \leqslant 1 \\ T(0,x_2,t) = 1, \quad T_{,x_1}(1,x_2,t) = 0, & 0 \leqslant t < \infty \\ T_{,x_2}(x_1,0,t) = T_{,x_2}(x_1,1,t) = 0, & 0 \leqslant t < \infty \end{cases}$$

(12.74)

沿方板水平中线 ($x_2 = 0.005\mathrm{m}$) 的温度分布解析解如下：

$$T(x_1, x_2 = 0.005, t) = 1 + \sum_{n=1,2,\cdots}^{\infty} \tilde{b}_{0n} \mathrm{e}^{(-\alpha t \beta_n^2/L^2)} \sin(\beta_n x_1/L) \quad (12.75)$$

式中，$\tilde{b}_{0n} = \dfrac{\cos(\beta_n) - 1}{\left(1/2 - \dfrac{\sin(2\beta_n)}{4\beta_n}\right)\beta_n}$ 和 $\beta_n = \dfrac{(2n-1)\pi}{2}$。在计算级数形式的解析解时，取 $n = 20$。

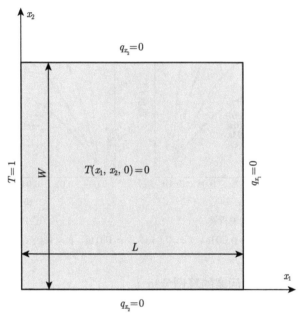

图 12-4　Dirichlet 和 Neumann 边界条件下二维方板热传导的计算模型示意图

在近场动力学模拟中，物质点尺寸 $\Delta x = 1 \times 10^{-4}$m，共有 $100 \times 100 = 10000$ 个物质点，近场范围半径取 $\delta = 3\Delta x$ 和 $\delta = 4\Delta x$，时间步长取 $\Delta t = 1 \times 10^{-8}$s，Dirichlet 边界条件施加在左侧边界一层物质点上，诺依曼边界条件通过体积热源项施加在边界一层物质点上，即 $S_{\mathrm{T}_{x_1}}(x_1 = L, x_2, t) = S_{\mathrm{T}_{x_2}}(x_1, x_2 = 0, t) = S_{\mathrm{T}_{x_2}}(x_1, x_2 = L, t) = 0, 0 \leqslant t < \infty$。

图 12-5 给出 6×10^{-6}s、1×10^{-5}s、2×10^{-5}s、4×10^{-5}s 和 1×10^{-4}s 五个时刻的局部热传导理论和近场动力学方法求得的方板水平中线 ($x_2 = 0.005$m) 温度分布曲线，从图中给出的计算结果可以看出，$\delta = 3\Delta x$ 和 $\delta = 4\Delta x$ 对应的近场动力学计算结果均与解析解吻合良好；由于近场动力学的非局部特性和色散行为，近场动力学计算的热传导过程滞后于局部热传导过程，且近场范围越大，这种滞后性愈加突出。图 12-6 给出了 $\delta = 3\Delta x$ 近场动力学模拟得到的不同时刻方板的温度分布云图，由此可以了解温度的时空演化情况，符合客观认知。

12.5 热传导问题近场动力学模型的算例分析

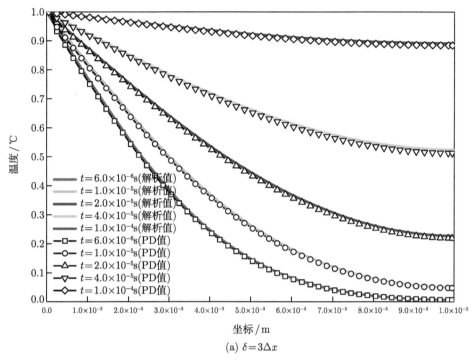

(a) $\delta = 3\Delta x$

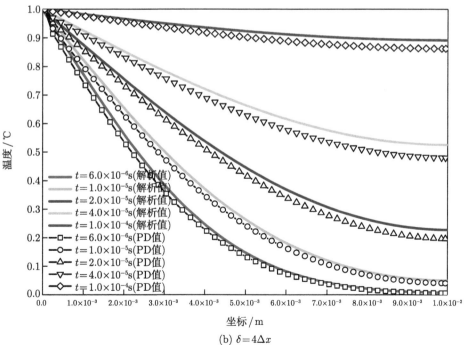

(b) $\delta = 4\Delta x$

图 12-5 理论解和近场动力学方法求得的不同时刻方板水平中线温度分布曲线

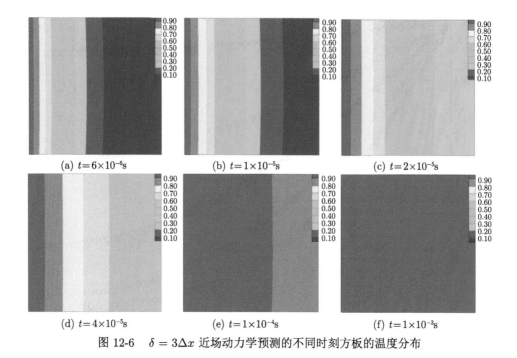

图 12-6 $\delta = 3\Delta x$ 近场动力学预测的不同时刻方板的温度分布

12.5.3 三维厚板热传导问题的算例分析

本节采用近场动力学方法研究混合热边界条件下含两条对称斜裂纹的三维厚板热传导问题。如图 12-7 所示，上下表面为常温边界，左右表面为热对流边界，前后表面为绝热边界[11]。板的长度、宽度和厚度分别为 $L = 2\text{cm}$、$W = 2\text{cm}$ 和 $H = 0.2\text{cm}$，裂纹边界视为绝热，裂纹长度 $2a = 0.6\text{cm}$，与水平面夹角 $\theta = 60°$，两裂纹中点距离 $2e = 0.66\text{cm}$。板的材料质量密度、比热容和导热系数分别为 $\rho = 1\text{kg/cm}^3$、$c_v = 1\text{J/(kg·K)}$ 和 $k_T = 1.14\text{J/(s·cm·K)}$。板的初始温度 $T = 0°C$，厚板的热边界条件如下

$$\begin{cases} T(x_1, x_2 = 0, x_3, t) = -100°C, \quad T(x_1, x_2 = W, x_3, t) = 100°C, \quad 0 \leqslant t < \infty \\ -k_T T_{,x_1}(x_1 = 0, x_2, x_3, t) = h_c(T_\infty - T_{bc}), \quad 0 \leqslant t < \infty \\ k_T T_{,x_1}(x_1 = L, x_2, x_3, t) = h_c(T_\infty - T_{bc}), \quad 0 \leqslant t < \infty \\ T_{,x_3}(x_1, x_2, x_3 = 0, t) = T_{,x_3}(x_1, x_2, x_3 = H, t) = 0, \quad 0 \leqslant t < \infty \end{cases}$$

(12.76)

其中，热对流系数 $h_c = 10\text{J/(s·cm}^2\text{·K)}$，环境温度 $T_\infty = 0°C$，$T_{bc}(\boldsymbol{x}, t)$ 为 t 时刻的边界物质点温度。

12.5 热传导问题近场动力学模型的算例分析

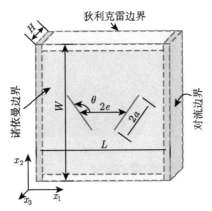

图 12-7 混合热边界条件下含两条斜裂纹的三维厚板热传导问题

在近场动力学模拟中，物质点间距和近场范围分别为 $\Delta x = 0.02\text{cm}$ 和 $\delta = 3\Delta x$，物质点总数为 $100\times100\times10 = 10^5$，时间步长为 $\Delta t = 1\times10^{-4}\text{s}$，所有热边界条件均施加在一层物质点上，左右两边的热对流边界条件和前后表面的绝热边界条件通过体力热源项施加，即

$$\begin{cases} S_\text{T}(x_1=0,x_2,x_3,t) = S_\text{T}(x_1=L,x_2,x_3,t) = \dfrac{h_\text{c}(T_\infty - T_{bc})}{\Delta x} \\ S_{\text{T}_{x_3}}(x_1,x_2,x_3=0,t) = S_{\text{T}_{x_3}}(x_1,x_2,x_3=H,t) = 0 \end{cases}$$

裂纹表面的绝热条件是通过移除穿过裂纹面的"热键"实现的。

基于近场动力学方法计算得到 $t=0.05\text{s}$、0.10s、0.15s、0.20s、0.25s、0.30s、0.35s、0.40s 和 0.45s 等不同时刻的温度分布，如图 12-8 和图 12-9 所示，本例预

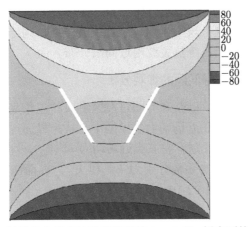

图 12-8 近场动力学方法计算得到的 $t=0.45\text{s}$ 板中面的温度分布

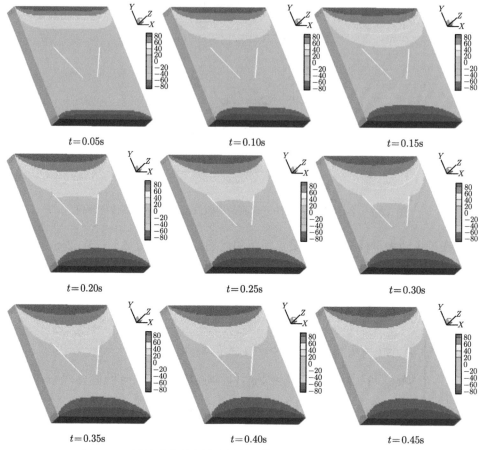

图 12-9 近场动力学方法计算得到的不同时刻厚板的温度分布

测的 $t=0.45$s 时温度分布与文献 [11] 一致。绝热斜裂纹阻碍热流传播并改变热流传播方向，热流在裂缝上表面累积、绕过裂缝传播，则绝热裂纹迎流面的温度高于背流面温度，温度场在绝热裂纹附近间断分布。

12.5.4 混凝土试件热传导问题的细观分析

将混凝土视为由骨料、砂浆和界面过渡区组成的三相复合材料，由于界面过渡区很薄，为方便起见，在近场动力学模型中，认为界面过渡区厚度为零，物质点只位于砂浆或骨料内部，则存在三类近场动力学"键"，即砂浆键、骨料键和界面键，设砂浆和骨料的导热系数分别为 k_T^A 和 k_T^m，界面键的导热系数简单设为 $k_T^{\text{ITZ}} = (k_T^A + k_T^m)/2$，如图 12-10 所示。

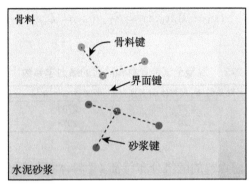

图 12-10　混凝土中物质点间的近场动力学"热键"示意图

12.5.4.1　二维方形混凝土试件的近似一维热传导问题

研究二维方形混凝土标准试件的近似一维热传导问题。如图 12-11 所示，混凝土试件边长 $L = 150\text{mm}$，二级配混凝土骨料体积分数为 46.998%，骨料形状假设为圆形，包括直径 30mm 的中石 6 块和直径 12mm 小石 56 块。试件初始温度为 20℃；左右两边为绝热边界，热流量为零，即 $q_{\text{left}} = q_{\text{right}} = 0$；上下两边为温度已知的边界，具体为 $T_{\text{top}} = 35$℃ 和 $T_{\text{bottom}} = 25$℃。混凝土各个组分的热力学参数见表 12-2[21,22]。在近场动力学模拟中，采用间距 $\Delta x = 1\times 10^{-3}\text{m}$ 的点阵均匀离散方形试件，共有 $150\times 150 = 22500$ 个物质点，近场范围 $\delta = 3\Delta x$，时间步长 $\Delta t = 1\times 10^{-1}\text{s}$。温度已知的边界条件施加在边界单层物质点上，左右两边的绝热边界

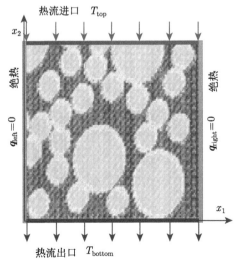

图 12-11　二维混凝土方形试件热传导问题的细观分析模型

条件通过体积热源 $S_{T_{x_1}}(x_1=0,x_2,t) = S_{T_{x_2}}(x_1=L,x_2=0,t) = 0, 0 \leqslant t < \infty$ 进行施加。

表 12-2 混凝土试件骨料和砂浆的热力学参数 [21,22]

材料组分	导热系数 $k_T/(J/(s·m·K))$	密度 $\rho/(kg/m^3)$	比热容 $c_v/(J/(kg·K))$
骨料	2.77	2733	710
砂浆	1.20	2078	1175

具体计算时，考虑骨料的随机分布特征 (骨料体积分数均为 46.998%)，形成三组随机骨料模型。采用近场动力学方法对上述三组随机骨料模型进行热传导模拟，图 12-12~ 图 12-14 分别给出了这三组模型对应的混凝土试件温度和竖向热流通量的分布云图。从图中所示结果可以看出，由于骨料导致的非均质性，混凝土试件的等温线呈现出一定的锯齿状和非光滑性 (均质材料为平行等温线)；竖向热流密度均为负值，与热量从上向下传导的客观事实一致，由于骨料的导热系数大于砂浆，其热流密度也大于砂浆中的热流密度，在骨料间形成 "热桥"[23]。

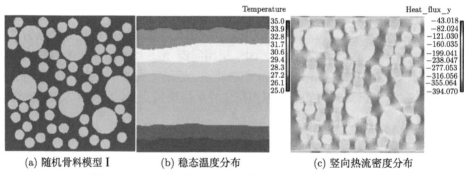

(a) 随机骨料模型 I   (b) 稳态温度分布   (c) 竖向热流密度分布

图 12-12 二维混凝土方形试件热传导问题的细观模拟

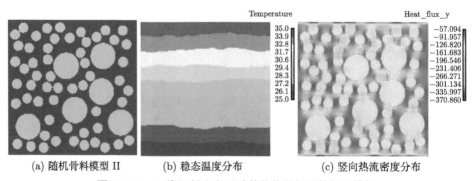

(a) 随机骨料模型 II   (b) 稳态温度分布   (c) 竖向热流密度分布

图 12-13 二维混凝土方形试件热传导问题的细观模拟

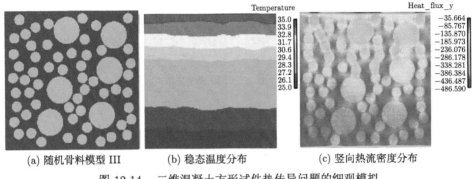

(a) 随机骨料模型 III　　(b) 稳态温度分布　　(c) 竖向热流密度分布

图 12-14　二维混凝土方形试件热传导问题的细观模拟

当热传导过程趋于稳态时，可以通过下式得到混凝土的有效导热系数

$$k_{\text{T}} = -\frac{q_{x_2}^{\text{ave}}}{T_{\text{top}} - T_{\text{bottom}}} \cdot L \tag{12.77}$$

式中，$q_{x_2}^{\text{ave}}$ 是混凝土试件竖向热流密度的平均值，T_{top} 和 T_{bottom} 分别为上下表面的温度，L 是试样的侧边长度。

在混凝土的热学试验中，通常采用防护热板法，基于式 (12.77) 测定混凝土试件的宏观有效导热系数。同理，根据 PD 计算结果按式 (12.77) 求得混凝土试件的有效导热系数分别为 1.7690J/(s·m·K)、1.7592J/(s·m·K) 和 1.7620J/(s·m·K)，平均值为 $k_{\text{T}}^{\text{eff}} = 1.7634$J/(s·m·K)，试验测定的结果为 $k_{\text{T}}^{\text{eff}} = 1.7$J/(s·m·K)[21]，两者非常接近，也验证了所建立的非均质材料的近场动力学热扩散模型的有效性。

12.5.4.2　绝热裂纹对混凝土试件导热过程的影响

本小节采用所建立的混凝土试件热传导问题的近场动力学细观模型，研究裂纹对混凝土试件导热过程的影响。在研究中，假设裂纹是绝热的，即没有穿越裂纹的热辐射和热对流，分别考虑不同长度的水平裂纹 (垂直于热流方向) 和不同方位的斜裂纹 (长度不变) 对于导热过程的影响效应。

如图 12-15 所示，设水平裂纹和斜裂纹的长度均为 $2a$，斜裂纹的倾斜角为 θ，通过断开跨越裂纹的 "热键" 引入预制裂纹。在近场动力学模拟中，水平裂纹长度 $2a$ 分别取 0mm、10mm、20mm、30mm、40mm、50mm 和 60mm，长度 40mm 的裂纹倾斜角 θ 分别为 0°、15°、30°、45°、60°、75° 和 90°，研究水平绝热裂纹长度和裂纹倾斜角对混凝土有效导热系数的影响。

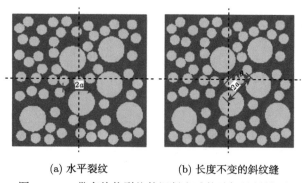

(a) 水平裂纹　　　　　　(b) 长度不变的斜纹缝

图 12-15　带有绝热裂纹的混凝土试件随机骨料模型

计算条件和计算参数与 12.5.4.1 节中的算例完全相同，图 12-16 和图 12-17 分别给出两种情形下混凝土试件的稳态温度分布云图，图 12-18 给出了预制裂纹长度和角度对混凝土试件宏观有效导热系数的影响。结果表明，混凝土试件中的裂纹长度和倾角对温度分布有很大影响，具体表现为：绝热裂纹阻碍热流传播并改变热流传播方向，热流在裂缝表面累积、绕过裂缝传播，导致水平裂纹上下表面的温度梯度最大；斜裂纹使热流传播方向发生局部改变；垂直热流方向的裂纹投影长度越大，热流受阻现象越严重，混凝土试件的宏观导热系数也越小。

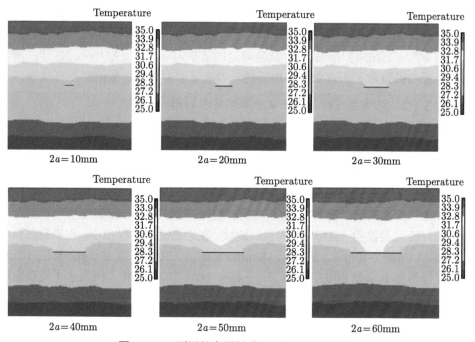

图 12-16　不同长度预制裂纹对应的温度分布

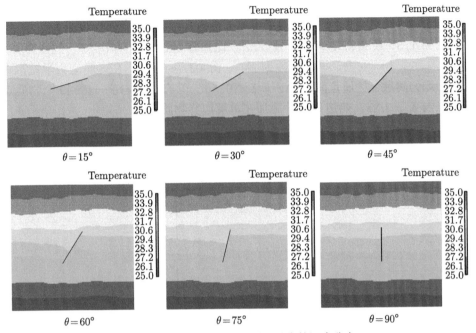

图 12-17 不同倾角预制裂纹对应的温度分布

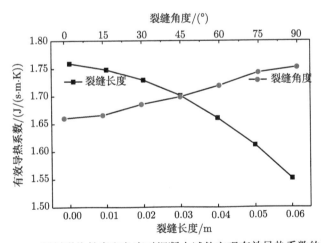

图 12-18 预制裂纹长度和角度对混凝土试件宏观有效导热系数的影响

12.6 热-力耦合问题近场动力学模型的算例分析

12.6.1 二维方板四边受给定温度荷载作用

如图 12-19 所示,边长 1m 的二维方板的四角固定,初始温度为 $T_0 = 0°C$,四

边受温度荷载 $T' = 0.15℃$ 的作用 [24]，采用热力单向弱耦合模型分析该问题，即热传导不考虑变形效应影响，而变形采用热弹性本构描述。板的弹性模量为 200GPa，泊松比为 1/3，质量密度为 $\rho = 1\text{kg/m}^3$，比热容为 $c_v = 1\text{J/(kg·K)}$，线热膨胀系数为 $\alpha = 5 \times 10^{-5}\text{K}^{-1}$，导热系数为 $k_\text{T} = 1000\text{J/(s·m·K)}$。在近场动力学模拟中，方板被均匀离散，物质点间距 $\Delta x = 1 \times 10^{-2}\text{m}$，共有 $100 \times 100 = 10000$ 个物质点，近场范围 $\delta = 3\Delta x$，热边界条件施加在单层边界物质点上。

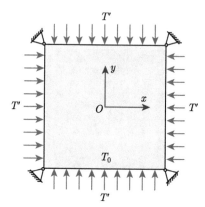

图 12-19　四角固定的平面板受到四面的热荷载

对该问题还采用商业有限元软件 ABAQUS 进行了计算，单元类型为八结点热-力耦合单元 C3D8T。图 12-20 为方板某一时刻的近场动力学和有限元计算的温度、水平位移和竖向位移云图。从图中所示结果可见，两种方法得到的计算结果几乎一致，且等温线的形状由外到内逐渐由方变圆；由于平板的四角固定，外部温度高于内部温度，最大位移发生在板边上。

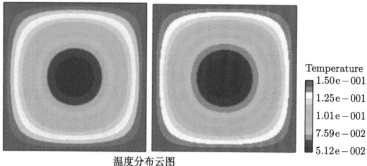

温度分布云图

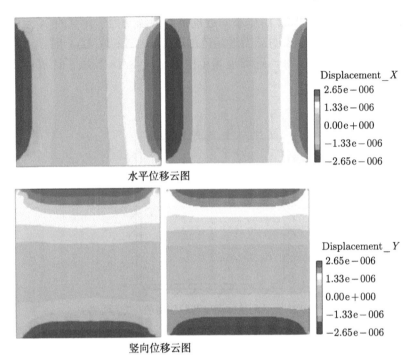

水平位移云图

竖向位移云图

图 12-20　近场动力学 (右) 和有限元 (左) 计算得到的温度场和位移场

12.6.2　二维方板三边绝热和一边受给定温度荷载作用

如图 12-21 所示，边长 1m 的方板三边受法向位移约束而上边界自由，初始温度为 $T_0 = 0$℃，上边界受温度荷载 $T' = 1.0$℃，其余三边均绝热。板材料的弹性模量为 1Pa，泊松比为 1/3，质量密度为 $\rho = 1\text{kg/m}^3$，比热容为 $c_v = 1\text{J/(kg·K)}$，导热系数为 $k_T = 1\text{J/(s·m·K)}$，线热膨胀系数为 $\alpha = 2 \times 10^{-2}\text{K}^{-1}$。当仅考虑热力单向弱耦合时，即温度场对变形场有贡献，而结构变形不影响热传导过程，则该问题的温度和竖向位移的局部解析解分别为[25,26]

$$T(y,t) = 1 - \frac{4}{\pi} \sum_{n=0}^{\infty} \frac{(-1)^n}{2n+1} \exp\left(-\frac{(2n+1)\pi^2 kt}{4L^2}\right) \cos\left(\frac{(2n+1)\pi y}{2L}\right) \quad (12.78)$$

$$u_y(y,t) = \frac{(1+\nu)}{(1-\nu)} \alpha \int_0^y T(y,t) \mathrm{d}y \quad (12.79)$$

在热力单向弱耦合的近场动力学模拟中，正交均匀的物质点间距 $\Delta x = 1 \times 10^{-2}\text{m}$，共离散 $100 \times 100 = 10000$ 个物质点，分别取近场范围尺寸 $\delta = 2\Delta x, 3\Delta x, 4\Delta x$。选取方板竖直对称轴上的三点 A、B 和 C (分别位于上边界、板中心和下边界)作为考察点，图 12-22 给出不同近场范围近场动力学计算的 A 点竖向位移与时间的关系曲线。结果表明，当近场范围取 $\delta=3\Delta x$ 或 $4\Delta x$ 时，近场动力学解答与局部解析解吻合良好，选取近场范围为 $\delta=3\Delta x$ 兼顾了计算精度和计算效率。

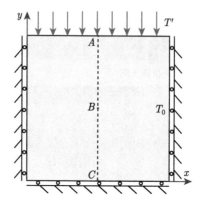

图 12-21 受到单面温度荷载作用的正方形平板

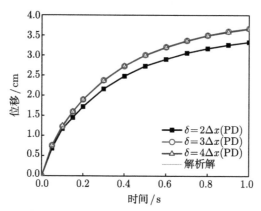

图 12-22 不同近场范围下 A 点竖向位移的 PD 解与解析解比较

为了分析温度场和变形场的耦合作用效应，分别采用热-力耦合和非耦合的近场动力学模型求解上述问题，计算得到了 A、B 点的竖向位移以及 B、C 点的温度随时间的变化曲线，如图 12-23 所示。结果表明，在温度荷载作用的早期，热-力耦合作用效应对温度和变形的影响很小；随着时间的不断增加，热-力耦合作用效应对温度场的影响较为明显，考虑热-力耦合作用效应使 B、C 点的温度有所下降，但对 A、B 点的竖向位移影响很小，其原因可能是在温度荷载作用下，不论是否考虑热-力耦合效应，变形场的运动方程都是相同的，均含有温度的作用项。

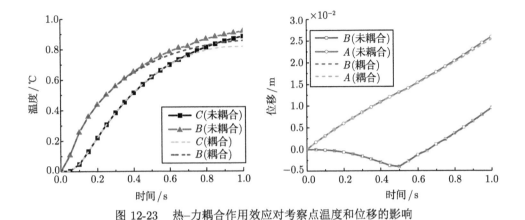

图 12-23　热–力耦合作用效应对考察点温度和位移的影响

12.6.3　混凝土厚壁圆筒内壁温升后裂纹扩展的二维模拟

图 12-24 为含中心芯棒的混凝土厚壁圆筒，内外半径分别为 25mm 和 76mm，圆筒受到中心芯棒加热产生的热载荷作用，考虑热–力耦合作用效应，采用近场动力学方法对该问题进行模拟计算，研究混凝土厚壁圆筒的热致变形和裂纹扩展特征。

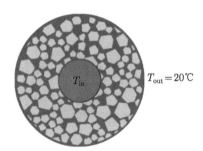

图 12-24　含中心芯棒的混凝土厚壁圆筒

混凝土圆筒的初始温度 $T_0 = 20°C$，圆筒外边界始终保持为 20°C，即 $T_{out} = 20°C$。芯棒的温度与时间有关，设为 $T_{in} = (2t_T + 20)°C$。假设混凝土随机骨料模型中的骨料形状为多边形，体积分数为 80.0%，二级配混凝土骨料的最大和最小粒径分别为 10mm 和 4mm。混凝土骨料、砂浆和界面过渡区的热力学参数如表 12-3 所示，芯棒的弹性模量、泊松比和热膨胀率分别为 $E_{cen} = 80\text{GPa}$、$\nu_{cen} = 0.2$ 及 $\alpha_{cen} = 2.7 \times 10^{-5} \text{K}^{-1}$，空气的导热系数、密度及比热容分别为 $k_A = 0.024 \text{J}/(\text{s}\cdot\text{m}\cdot\text{K})$、$\rho_A = 1.29 \text{kg/m}^3$ 及 $c_{v_A} = 1100 \text{J}/(\text{kg}\cdot\text{K})$[27]。

作为二维算例，混凝土圆筒被均匀离散为 113365 个物质点，物质点间距 $\Delta x = 2 \times 10^{-4}\text{m}$，近场范围尺寸 $\delta = 3\Delta x$。温度场及变形场的计算时间步长分别为 $\Delta t_T = 1 \times 10^{-2}\text{s}$ 和 $\Delta t_M = 1 \times 10^{-8}\text{s}$。图 12-25 给出了混凝土圆筒在 26s、32s、

表 12-3 混凝土各组分的热力学参数[27,28]

参数	骨料	砂浆	界面过渡区
密度 $\rho/(\text{kg/m}^3)$	2400	1800	1800
杨氏模量 E/GPa	55.4	25.7	23.6
泊松比 ν	0.2	0.2	0.2
比热容 $c_v/(\text{J}/(\text{kg}\cdot\text{K}))$	920	1688	1688
导热系数 $k_T/(\text{J}/(\text{s}\cdot\text{m}\cdot\text{K}))$	3.1	1.0	0.5
热胀系数 α/K^{-1}	8e−6	1.8e−5	8e−6

38s 和 44s 的温度分布云图；图 12-26 和图 12-27 分别为混凝土圆筒在对应时刻的径向位移和环向位移云图，图中，径向位移向外为正，环向位移逆时针方向为正；图 12-28 为相应时刻的损伤及裂纹扩展情况。从图中所示的温度分布云图结果可以看出，芯棒加热后，靠近内边缘的混凝土首先吸收热量，开始升温，热流沿径向向试件外边缘传播；由于混凝土的导热系数比较小，远离内边缘区域的温度较低且温度梯度也较小，混凝土圆筒内外壁的温度保持一个较大的温差；当芯棒加热时间达到 32s 时，在圆筒内侧已出现裂纹，由于裂纹对热流传导路径的影响，温度分布产生非均匀特征，并且这种非均匀特征随时间的进行愈加明显。此外，由于空气进入裂纹，裂纹处的导热系数会变小，热量会在裂纹处聚集，使得裂纹处的温度略高于其他位置的温度，该结果也与试验结果[29]一致。

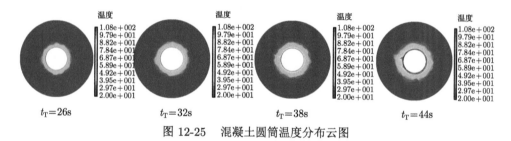

图 12-25 混凝土圆筒温度分布云图

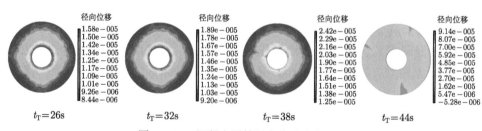

图 12-26 混凝土圆筒径向位移分布云图

对于径向位移，由于圆筒内边缘处的温度较高，热膨胀效应导致内部区域的径向位移大于外部区域；当 $t_T \geq 38\text{s}$ 时，由于裂纹萌生和扩展，裂纹附近物点间的键断开并释放键力,使得裂纹附近的径向位移小于相同半径其他位置的位移值。对于环向位移,其主要源自混凝土的非均质性,骨料的弹性模量大于砂浆的弹性模量,致使骨料两侧的环向位移有一定的差别;当 $t_T \geqslant 32s$ 时,由于微裂纹萌生和演化,裂纹两侧的环向位移方向相反,并随着裂纹两侧物质点间距的持续增大,环向位移的变化更为显著。

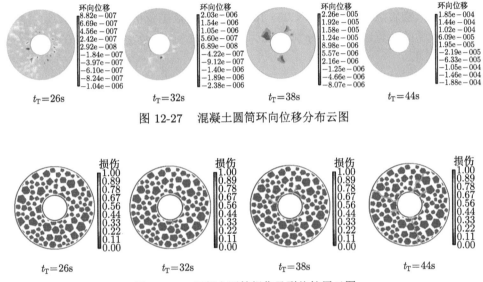

图 12-27 混凝土圆筒环向位移分布云图

图 12-28 混凝土圆筒损伤及裂纹扩展云图

图 12-28 表明,当芯棒升温、热流向外传播时,由于内缘温度高于外缘温度,将在内缘附近产生拉应力,当拉应力超过混凝土的抗拉极限后,便会产生微裂纹。随着时间的推移,温度继续升高,微裂纹不断产生和汇集,形成宏观裂纹,并贯穿圆筒,导致圆筒破裂。与图 12-29 所示的试验结果相比较,计算模拟得到的圆筒裂纹扩展路径以及最终破坏形式与试验结果[30]吻合良好。

图 12-29 Abdalla 等的试验结果[30]

12.6.4 混凝土厚壁圆筒内壁温升后裂纹扩展的三维模拟

本节对上例按三维情形进行模拟计算，圆筒长度取为 100mm，除材料的泊松比改为 0.25 外，圆筒的横截面形状、初始温度、外边界温度、中心芯棒的升温速度以及芯棒和混凝土各组分的材料参数等均与二维算例一致。但在三维混凝土随机骨料模型中，骨料形状设为球形，体积分数为 46.0%，骨料的最大和最小粒径分别为 15mm 和 5mm，如图 12-30 所示。计算中，混凝土圆筒被均匀离散为 232400 个物质点，物质点间距 $\Delta x = 1.25\times 10^{-3}$m，近场范围依旧取为 $\delta = 3\Delta x$，温度场和变形场的计算时间步长分别为 $\Delta t_\mathrm{T} = 1 \times 10^{-2}$s 和 $\Delta t_\mathrm{M} = 1 \times 10^{-8}$s，与二维算例相同。

图 12-30　含中心芯棒的三维厚壁混凝土圆筒

图 12-31 为三维混凝土圆筒在 15s、25s、32s 和 40s 时的温度分布云图。得到的温度分布特征与二维情形下得到的结果基本一致，裂纹处的温度分布符合此前描述的热流聚集升温的现象。

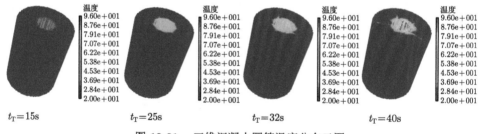

图 12-31　三维混凝土圆筒温度分布云图

图 12-32～图 12-34 分别为三维混凝土圆筒在 15s、25s、32s 和 40s 时的径向位移、环向位移及轴向位移云图。其中，柱坐标系的原点位于圆柱中心点，高度坐标轴沿圆柱轴线，向上为正，径向位移与环向位移的符号规定与之前的二维算例相同。从径向位移的云图可见，圆筒的上下表面自由，受热后产生的径向位移普遍大于圆筒中部区域，且温度高的区域对应的位移也较大，裂纹处的位移要

明显小于相邻区域。从环向位移的云图可见，裂纹面两侧的位移方向相反，且随着裂纹的不断演化和扩展愈加明显。从轴向位移的云图可见，圆筒的轴向位移关于中间截面呈现出较好的对称性，在圆筒的两个端部，轴向位移数值最大，但方向相反。总体来看，圆筒的热致变形符合规律。

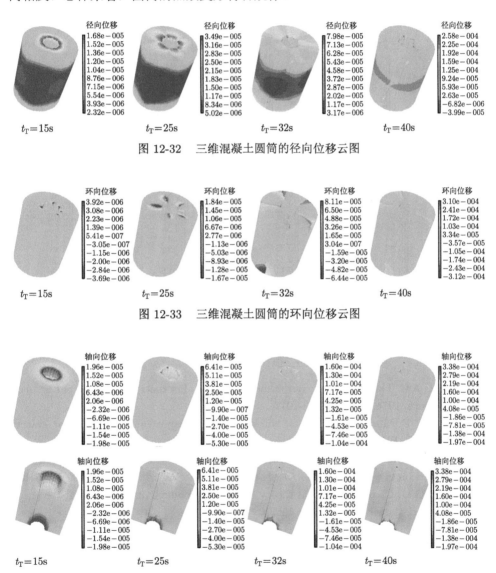

图 12-32　三维混凝土圆筒的径向位移云图

图 12-33　三维混凝土圆筒的环向位移云图

图 12-34　三维混凝土圆筒轴向位移云图及其剖面图

图 12-35 给出了混凝土圆筒在 15s、25s、32s 和 40s 时热致损伤的外观及内部情况。从图示结果可以看到，在圆筒的上下表面近内壁区域首先产生损伤，然

后,随着时间的推演,温度不断升高,损伤区域也在不断发展,并形成裂纹;裂纹从圆筒的端部沿径向和轴向逐渐向筒身发展,最后贯穿筒体。对比 Abdalla 的试验结果,三维计算模拟得到的圆筒裂纹扩展路径和破坏形式与试验结果具有较好的相似性,进一步验证了近场动力学热-力耦合模型可以很好地模拟混凝土材料和结构在温度荷载作用下的断裂破坏行为。

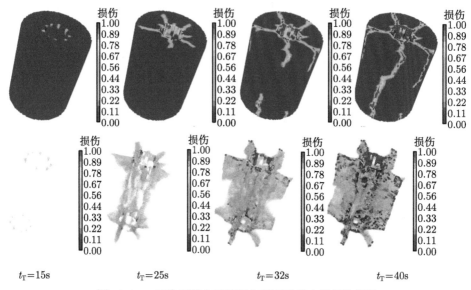

图 12-35　三维混凝土圆筒损伤云图及其内部损伤情况

参 考 文 献

[1] Eringen A C. Theory of nonlocal thermoelasticity[J]. International Journal of Engineering Science, 1974, 12(12): 1063-1077.

[2] Luciani J F, Mora P, Virmont J. Nonlocal heat transport due to steep temperature gradients[J]. Physical Review Letters, 1983, 51(18): 1664-1667.

[3] Koch D L, Brady J F. A non-local description of advection-diffusion with application to dispersion in porous media[J]. Journal of Fluid Mechanics, 1987, 180(180): 387-403.

[4] Mahan G D, Claro F. Nonlocal theory of thermal conductivity[J]. Physical Review B Condensed Matter, 1988, 38(3): 1963.

[5] Grmela M, Lebon G. Finite-speed propagation of heat: a nonlocal and nonlinear approach[J]. Physica A Statistical Mechanics & Its Applications, 1998, 248(3-4): 428-441.

[6] Sobolev S L. Nonlocal diffusion models: Application to rapid solidification of binary mixtures[J]. International Journal of Heat and Mass Transfer, 2014, 71(1): 295-302.

[7] Hahn D W, Özisik M N. Heat Conduction[M]. Chichester: John Wiley & Sons, 2012.

[8] Bobaru F, Duangpanya M. The peridynamic formulation for transient heat conduction[J]. International Journal of Heat and Mass Transfer, 2010, 53(19-20): 4047-4059.

[9] Bobaru F, Duangpanya M. A peridynamic formulation for transient heat conduction in bodies with evolving discontinuities[J]. Journal of Computational Physics, 2012, 231(7): 2764-2785.

[10] Agwai A. A peridynamic approach for coupled fields[D]. Tucson: University of Arizona, 2011.

[11] Oterkus S, Madenci E, Agwai A. Peridynamic thermal diffusion[J]. Journal of Computational Physics, 2014, 265: 71-96.

[12] Liao Y, Liu L S, Liu Q W, et al. Peridynamic simulation of transient heat conduction problems in functionally gradient materials with cracks[J]. Journal of Thermal Stresses, 2017, 40(12): 1484-1501.

[13] Silling S A, Epton M, Weckner O, et al. Peridynamic states and constitutive modeling[J]. Journal of Elasticity, 2007, 88(2): 151-184.

[14] Gu X, Zhang Q, Madenci E. Refined bond-based peridynamics for thermal diffusion[J]. Engineering Computations, 2019, 36(8): 2557-2587.

[15] Madenci E, Barut A, Futch M. Peridynamic differential operator and its applications[J]. Computer Methods in Applied Mechanics and Engineering, 2016, 304, 408-451.

[16] Madenci E, Barut A, Dorduncu M. Peridynamic Differential Operator for Numerical Analysis[M]. Switzerland: Springer International Publishing, 2019.

[17] Maugin G A. Non-classical Continuum Mechanics[M]. Singapore: Springer Verlag, 2017.

[18] Oterkus S, Madenci E, Agwai A. Fully coupled peridynamic thermomechanics[J]. Journal of the Mechanics and Physics of Solids, 2014, 64: 1-23.

[19] Silling S A, Askari E. A meshfree method based on the peridynamic model of solid mechanics[J]. Computers & Structures, 2005, 83(17-18): 1526-1535.

[20] Kilic B, Madenci E. Peridynamic Theory for Thermomechanical Analysis[J]. IEEE Transactions on Advanced Packaging, 2010, 33(1): 97-105.

[21] Zhang W, Min H, Gu X, et al. Mesoscale model for thermal conductivity of concrete[J]. Construction and Building Materials, 2015, 98: 8-16.

[22] Zhao J, Zheng J J, Peng G F, et al. A meso-level investigation into the explosive spalling mechanism of high-performance concrete under fire exposure[J]. Cement and Concrete Research, 2014, 65: 64-75.

[23] Shen L, Ren Q W, Xia N, et al. Mesoscopic numerical simulation of effective thermal conductivity of tensile cracked concrete[J]. Construction and Building Materials, 2015, 95: 467-475.

[24] Wang H L, Oterkus E, Celik S, et al. Thermomechanical analysis of porous solid oxide fuel cell by using peridynamics[J]. AIMS Energy, 2017, 5(4): 585-600.

[25] Timoshenko S P, Goodier J N. Theory of Elasticity[M]. New York: McGraw-Hill, 1951.

[26] Carslaw H S, Jaeger J C. Conduction of Heat in Solids[M]. Oxford: Clarendon press, 1992.

[27] 唐世斌, 唐春安, 梁正召, 等. 混凝土热传导与热应力的细观特性及热开裂过程研究 [J]. 土木工程学报, 2012, 45(2): 11-19.

[28] Zhang N L, Guo X M, Zhu B B, et al. A mesoscale model based on Monte-Carlo method for concrete fracture behavior study[J]. Science China (Technological Sciences), 2012, 55(12): 3278-3284.

[29] Wu B, Xiong W, Wen B. Thermal fields of cracked concrete members in fire[J]. Fire Safety Journal, 2014, 66: 15-24.

[30] Abdalla H. Concrete cover requirements for FRP reinforced members in hot climates[J]. Composite Structures, 2006, 73(1): 61-69.

第 13 章 混凝土材料与结构破坏的近场动力学建模分析

混凝土作为重要的建筑材料已有百余年的历史，由于其经济性和便于施工等一系列优点，被广泛应用于各种工程领域，混凝土材料与结构的损伤过程和破坏机制的研究，是学术界与工程界长期关注的热点和难点问题。基于连续介质力学理论框架下的传统数值方法在模拟混凝土材料与结构的损伤累积和渐进破坏问题时面临网格重构等诸多困难，且难以从根本上予以消除。基于非局部积分思想的近场动力学理论和方法，彻底避开传统方法中采用局部微分方程求解不连续问题时的奇异性，通过空间积分型支配方程的离散和求解，能够模拟混凝土材料与结构中微缺陷的发展、损伤累积、宏观裂纹萌生、裂纹扩展、局部断裂直至结构整体失稳的渐进破坏全过程，有望揭示混凝土材料与结构的破坏机制，为混凝土结构的安全性评估和寿命预测提供科学依据。

本章基于近场动力学方法，发展混凝土材料与结构破坏分析的微观、细观和宏观等不同尺度的计算模型，进行相应的数值模拟研究。具体包括：考虑水泥水化微结构复杂的多相非均质性与孔隙结构特征，提出水泥水化产物拉伸破坏的近场动力学建模方法；构建混凝土细观破坏的近场动力学模型，进行典型混凝土试件的细观破坏分析；建立热–水–力多物理场耦合作用下混凝土冻融循环分析的近场动力学计算模型；开展混凝土重力坝的冲击侵彻与爆炸毁伤分析；采用近场动力学–有限元混合方法，进行向家坝水电站典型坝段的抗滑稳定性研究。

13.1 水泥水化产物的近场动力学建模与拉伸破坏分析

混凝土是由粗骨料、细骨料、水泥水化产物、未水化水泥颗粒和孔隙等组成的非均质复合材料，水泥水化产物在很大程度上决定了水泥砂浆的物理力学性能，进而影响混凝土的力学特性。水泥水化产物具有多相多孔的复杂微观结构，不同的水胶比和添加材料等都会影响水泥水化产物的微结构特征，从而导致水泥水化产物的物理、化学、热学和力学性能的不同。

确定水泥水化产物的宏观力学性能通常需要进行大量的试验，耗费巨大的人力和物力，而基于水化微观结构信息的微观力学建模与数值模拟方法 (microstructure informed micromechanical modeling method)[1] 有助于预测水泥水化产物的宏观力学性能，系统深入地研究水泥水化产物的损伤断裂行为。本节基于近场动力学理论和方法，考虑水泥水化微结构复杂的多相非均质性和孔隙结构特征，进行水泥水化产物的建模分析，为水泥净浆的裂纹扩展和宏观力学性能预测提供一个有效的计算方法，也可为水泥砂浆乃至混凝土力学性能的研究提供科学依据。

13.1.1 水泥水化产物微结构的生成

水泥的水化过程是指水泥中的各种熟料矿物与水发生一系列化学反应，形成具有黏结性和可塑性的水泥浆体，并最终凝结和硬化变成水泥石的过程。水泥的主要矿物成分包括硅酸三钙 (C_3S)、硅酸二钙 (C_2S)、铝酸三钙 (C_3A) 和铁铝酸四钙 (C_4AF)；水泥中一般都需加入适量石膏，石膏分为二水石膏 (GYPSUM)、半水石膏 (HEMIHYD) 和无水石膏 (ANHYDRITE)，在水泥水化过程中充当缓凝剂，以调节水化反应与凝结速度；水泥中主要掺合料和外加剂有硫铝酸盐矿物外加剂 (ASG)、非活性掺料 (INERT)、活性火山灰掺料 (POZZOLAN) 和矿渣掺料 (SLAG)。水泥水化的主要产物包括水化硅酸钙凝胶 (CSH)、氢氧化钙 (CH)、水化硫铝酸钙 (AFM、Aft、ETTR)、水化铝酸钙 (C3AH6)、水化铁酸钙 (CFH)、氢氧化铁 (FH3) 等。由于水泥组分及掺合料的差异性，矿渣、粉煤灰、硅灰、填料和其他次要相的存在，使得复合水泥中的水化产物组分相数急剧增加，水泥水化产物的组分还包括氯化钙 (CACL2)、碳酸钙 (CACO3)、硅铝酸钙 (CAS2)、富铁钙钒石 (ETTRC4AF)、弗里德尔盐 (FRIEDEL)、火山灰水化硅酸钙 (POZZCSH)、矿渣水化硅酸钙 (SLAGCSH)、单碳型水化碳铝酸钙 (AFMC)、水铝黄长石 (STRAT)、石膏 (GYPSUMS、ABSGYP)、细骨料 (AGGREGATE)、空孔隙 (EMPTYP) 等。

目前，CEMHYD3D、μic、HYMOSTRUC 是三种代表性的描述水泥水化过程和浆体微结构变化的数值模拟软件。其中，CEMHYD3D 软件[2-5] 是美国国家标准与技术研究院 (National Institute of Standards and Technology, NIST) 于 20 世纪 90 年代初开始研发，2000 年于 NIST 网站发布 CEMHYD3D2.0 版开源代码，2005 年发布 CEMHYD3D3.0 版开源代码。CEMHYD3D 软件将水泥水化过程视为一个多输入和多输出的非线性动态系统，利用物理化学定律和数学模型建立的一个计算机虚拟系统。利用 CEMHYD3D 软件进行数值试验，可以预测材料的化学反应过程及各种性能，模拟得到水化热、水化程度、水化过程中主要反应物和产物的变化情况，并与试验结果具有较好的一致性。CEMHYD3D 软件根据水泥粒径分布 (particle size distribution, PSD) 和背散射电子 (back scattered electron, BSE) 图像建模，模拟系统由代表水泥矿物的大量像素构成，采用元胞自动机技术将水化规则算法化，操纵像素进行溶解、扩散和水化反应等。生成的水泥水化产物微结构一般是 100μm 立方体，1μm 为最小单元或像素点。

CEMHYD3D3.0 版开源软件采用 C 语言编写，共 70 个子程序 (函数)，包含 3 个主要模块：初始微结构构建模块、物相分配模块和水化模拟模块，按序依次执行，生成水泥水化产物微结构。以下简述基于 CEMHYD3D 软件生成水泥水化产物微结构的过程。

13.1.1.1 建立初始未分相的水泥三维微结构

选取 NIST 数据库中"水泥和混凝土参考实验室 (cement and concrete reference laboratory, CCRL)"的 115 号 I 型普通硅酸盐水泥，由勃氏 (Blaine) 法或透气法测定的水泥细度为 363 m^2/kg，该水泥主要成分硅酸三钙 (C_3S)、硅酸二钙 (C_2S)、铝酸三钙 (C_3A) 和铁铝酸四钙 (C_4AF) 的体积分数见表 13-1。

表 13-1　CCRL 115 水泥的主要组分比例

矿物成分	C_3S	C_2S	C_3A	C_4AF
体积分数	0.6338	0.2316	0.0334	0.1011

按以下步骤建立 CCRL 115 号水泥未分相的三维微结构：

(1) 基于试验获得水泥的背散射电子图像与 X 射线能谱图，进行数字图像处理，得到 BSE/X-ray 图像和相应的数据文件。本例图像的大小为 512μm×400μm，将系统中的像素划分成 C_3S、C_2S、C_3A、C_4AF 和石膏，其中红色为 C_3S、蓝色为 C_2S、绿色为 C_3A、黄色为 C_4AF、浅绿色为石膏、黑色为孔隙，石膏含量为 6.0%，如图 13-1 所示。

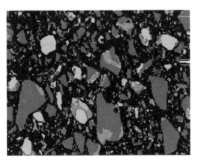

图 13-1　CCRL 115 号水泥的 BSE/X-ray 二维图像

(2) 对 (1) 中获得的含数字化标记的水泥图像进行统计分析，得到水泥的粒径分布，如图 13-2 所示。根据体视学原理，颗粒的体积分数和表面积分数分别等于平面图中颗粒的面积分数和周长分数，基于 CEMHYD3D 软件的 statsimp.c 子程序，求出 CCRL 115 号水泥的体积分数与表面积分数，见表 13-2；再通过 CEMHYD3D 软件的子程序 corrcalc.c 和 corrxy2r.c，计算得到各矿物的自相关函数信息。上述工作为基于水泥的二维 BSE/X-ray 图像构建三维初始微结构提供基础。

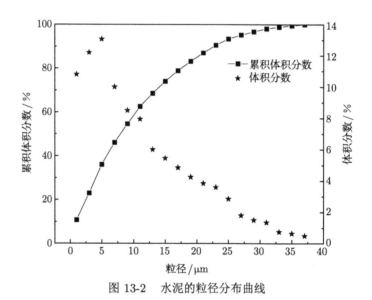

图 13-2 水泥的粒径分布曲线

表 13-2 CCRL 115 水泥的体积分数与表面积分数

组分	体积分数 (面积分数)	表面积分数 (周长分数)
C_3S	0.6338	0.5831
C_2S	0.2316	0.2763
C_3A	0.0334	0.0522
C_4AF	0.1011	0.0885

(3) 分步执行 CEMHYD3D 相关软件，生成未分相、未水化的水泥微结构。建立体积为 $100\mu m^3$ 的未分相、未水化的不同粒径水泥颗粒的空间分布模型，此时所生成的三维微结构仅包含两种物质，即 C_3S 颗粒和石膏颗粒。

设定水灰比为 $W/C = 0.39$，该值决定了 $100\mu m^3$ 空间中的 C_3S 像素点总数。假定水泥颗粒为球形，每个小球采用 $1\mu m^3$ 立方体像素作为最小堆积单元堆积而成，相应直径的水泥颗粒所含像素点数目见表 13-3。根据水泥颗粒的质量分布数据 PSD 和水灰比，可计算出给定体积的相应粒径的水泥颗粒数目等信息；给定石膏的总质量分数 6%，其中二水石膏、半水石膏和无水石膏占石膏总量的比例分别为 44.4%、51.5%和 4.1%，即三类石膏的质量分数分别为 2.664%、3.09%和 0.246%。

基于上述参数信息，运行 CEMHYD3D 软件子程序 genpartnew.c，实现三维微观结构的建模并输出数据。采用改编的 oneimage.c 程序提取 100 张切面图，并导入图像处理软件 IMAGEJ 中，形成一个 stack 块，再通过 Plugins 中的 3D Volume Viewer 命令，生成三维微结构图像。图 13-3 为体积 $100\mu m^3$ 的初始未分相、未水化的不同粒径水泥颗粒的空间分布模型。

表 13-3　颗粒直径及颗粒内所含像素数目的对应关系

颗粒直径 (1μm)	像素数目	颗粒直径 (1μm)	像素数目
1	1	29	12893
3	19	31	15515
5	81	33	18853
7	179	35	22575
9	389	37	26745
11	739	39	31103
13	1189	41	36137
15	1791	43	41851
17	2553	45	47833
19	3695	47	54435
21	4945	49	61565
23	6403	51	69599
25	8217	61	119009
27	10395	73	203965

图 13-3　体积 $100\mu m^3$ 的初始未分相、未水化的不同粒径水泥颗粒的空间分布模型

13.1.1.2 进行初始未分相三维微结构的物相分配

对上述形成的初始未分相(仅含 C_3S 和石膏两相)的三维微结构进行分相。基于 CEMHYD3D 软件中的 distrib3d.c 模块,不断调整水泥熟料中各矿物的体积分数和表面积分数,进行物相划分,建立分相未水化水泥微结构的三维模型。将空隙像素点的材料属性设为水,得到水分的空间分布(此时尚未发生水化反应),生成加水后的水泥初始三维微结构的数据文件。再采用改编的 CEMHYD3D 子程序 oneimage.c,提取 100 张切面图,使用图像处理软件 IMAGEJ 生成三维微结构图像,也可以得到微结构的截面图像,如图 13-4 所示。图中的彩色圆球表示水泥熟料,灰色代表石膏,黑色区域是没有被水泥占据的孔隙位置,水化后为充满水的空间。

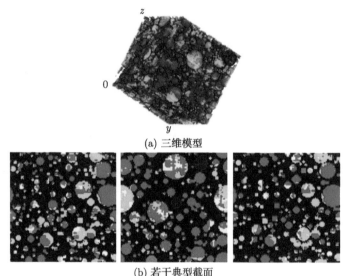

(a) 三维模型

(b) 若干典型截面
图 13-4　分相未水化水泥初始微结构
红色为 C_3S、蓝色为 C_2S、绿色为 C_3A、黄色为 C_4AF、浅灰色为石膏、黑色为孔隙

需要指出的是,为提高水泥组分分相的计算效率,在这个阶段,物相分配并没有考虑各组分 1 个像素的颗粒;在水化模拟阶段,将根据子程序 distrib3d.c 生成的各相的像素数量,计算出各组分 1 个像素的颗粒数目,在 disrealnew.c 子程序中实现投放。

13.1.1.3 基于分相的三维微结构的水泥水化模拟

CEMHYD3D 软件采用元胞自动机方法将水泥水化规则算法化,操纵分相未水化的水泥微结构中所有像素点,进行溶解、扩散和水化反应,实现对像素点的水化演变。基于生成的分相(C_3S、C_2S、C_3A、C_4AF 和 GYPSUM 等 5 相)未

13.1 水泥水化产物的近场动力学建模与拉伸破坏分析

水化水泥初始微结构，运行 CEMHYD3D 软件中的 disrealnew.c 子程序，调用 burn3d.c、burnset.c、complex.c、hydrealnew.c、nrutil.c、parthyd.c、pHpred.c 和 ran1.c 等子程序，模拟水泥水化过程。程序读取的主要参数包括：各组分 1 个像素的颗粒数目、循环次数、每一次水化循环的最大扩散次数、各水化产物的结晶系数、初始温度养护温度、各物相的活化能等。

此外，CEMHYD3D 软件的输出结果是不同循环次数下水泥水化微结构的组分信息，因此，需要基于模拟与试验中水泥烧蚀量相等的原则，确定水化循环次数与实际水化时间的对应关系。CCRL 115 号水泥的水化循环次数与实际水化时间之间的关系为 $t = BN^2$，其中 t 为实际水化时间（小时），$B = 0.00035$ 小时/循环次数^2 为经验系数，N 为水泥水化的循环次数，则水泥水化 7 天需要循环计算 693 次，水化 28 天需要循环计算 1386 次。系统初始温度设为 20°C，养护条件设置为 20°C 的恒温及饱水条件。

水泥水化后形成的水化产物即为水泥净浆，水泥净浆由水泥水化微结构组成，也是水泥砂浆的基质。为生成 28 天龄期的 100μm 水泥净浆微结构，采用 CEMHYD3D 软件进行 1386 次循环计算，获得水泥水化 28 天后的各个水化产物的体积分数 (表 13-4)，再通过 CEMHYD3D 软件中的 oneimage.c 子程序提取 100 张切面图，采用 IMAGEJ 软件导入 100 张图片，生成水泥净浆微结构三维图

表 13-4 水泥水化 28 天后各水化产物的体积分数

物相成分	ID	像素个数	体积分数/%	物相成分	ID	像素个数	体积分数/%
Porosity	0	119779	11.98	ETTR	16	34901	3.49
C_3S	1	61280	6.13	ETTRC$_4$AF	17	33106	3.31
C_2S	2	88568	8.86	AFM	18	32074	3.21
C_3A	3	2819	0.28	FH$_3$	19	6053	0.61
C_4AF	4	28608	2.86	POZZCSH	20	0	0
GYPSUM	5	2691	0.27	SLAGCSH	21	0	0
HEMIHYD	6	315	0.032	CACL$_2$	22	0	0
ANHYDRITE	7	38	0.0038	FREIDEL	23	0	0
POZZOLAN	8	0	0	STRAT	24	0	0
INERT	9	0	0	GYPSUMS	25	2662	0.27
SLAG	10	0	0	CACO$_3$	26	0	0
ASG	11	0	0	AFMC	27	0	0
CAS2	12	0	0	AGG	28	0	0
CH	13	145754	14.58	ABSGYP	29	4013	0.40
CSH	14	401229	40.12	EMPTYP	45	17161	1.72
C_3AH_6	15	18949	1.89				

像和二维图像，如图 13-5 所示。最后输出各像素点的坐标、材料组分等信息，即可进行水泥水化微结构的力学分析。

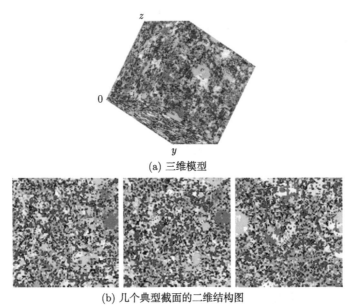

(a) 三维模型

(b) 几个典型截面的二维结构图

图 13-5　生成的 28 天水化龄期的水泥净浆微结构

13.1.2　水泥水化微结构的近场动力学计算模型

13.1.1 节已经生成水灰比为 $W/C = 0.39$、28 天水化龄期的 $100\mu m^3$ 水泥净浆的几何微观结构，共有一百万像素点。目前，$50\mu m^3$ 立方体已被证明是水泥净浆的合理代表体积单元[6,7]，为提高计算效率，从 $100\mu m^3$ 的水泥水化微观结构中随机抽取 4 个 $50\mu m^3$ 立方体，形成模型 Ⅰ、Ⅱ、Ⅲ、Ⅳ，如图 13-6 所示。

图 13-6　$50\mu m^3$ 水泥水化微结构立方体的 4 个模型 (颜色代表不同的材料组分)

在这四个模型中，每个像素被视为一个近场动力学物质点，四个模型的物质点数目分别为 108127、107272、108947 和 108143。基于近场动力学理论和方法，进行水泥水化微结构立方体的静态和动态单向拉伸模拟计算，相应的边界条件为：

13.1 水泥水化产物的近场动力学建模与拉伸破坏分析

立方体底面受竖向固定约束，水平各方向自由；顶面受垂直位移荷载作用，水平各方向自由；四个侧面均为自由面。采用本书第 3 章介绍的基于改进 PMB 模型的隐式-显式混合解法进行求解，对于静态问题，采用多级加载方式，位移增量为 $\Delta u = 3/400 \mu m$；对于动态问题，每级施加不同的位移增量，以实现不同的加载速率。时间步长取为 $\Delta t = 5.0 \times 10^{-11} s$，近场范围半径取为 $\delta = 3\Delta x$。

根据参考文献 [6-8]，水化产物的力学参数 (杨氏模量、泊松比、质量密度和断裂能释放率) 见表 13-5。在构建水泥净浆微结构的近场动力学模型时，若键两端物质点材料的属性不同，材料属性设置为两种材料的平均值。

表 13-5 28 天水化龄期的水泥净浆各组分的力学性能参数 [6-8]

物相组分	ID	杨氏模量/GPa	泊松比	质量密度/(kg/m^3)	断裂能量释放率/(J/m^2)
C$_3$S	1	137.4	0.25	3210	65.8
C$_2$S	2	135.5	0.25	3280	56.3
C$_3$A	3	145.2	0.25	3030	94.3
C$_4$AF	4	150.8	0.25	2320	72.5
GYPSUM	5	44.5	0.25	3000	1.0
HEMIHYD	6	132.0	0.25	3000	1.4
ANHYDRITE	7	88.8	0.25	3000	4.6
POZZ	8	23.8	0.25	3000	6.1
INERT	9	71.0	0.25	3000	3.3
SLAG	10	137.4	0.25	3000	65.8
ASG	11	71.7	0.25	3000	73.4
CAS2	12	71.7	0.25	3000	73.4
CH	13	43.5	0.25	3000	6.1
CSH	14	23.8	0.25	3000	5.9
C$_3$AH$_6$	15	93.8	0.25	3000	55.6
ETTR	16	24.1	0.25	3000	3.2
ETTRC$_4$AF	17	24.1	0.25	3000	3.2
AFM	18	43.2	0.25	3000	77.3
FH$_3$	19	22.4	0.25	3000	178.5
POZZCSH	20	23.8	0.25	3000	5.9
SLAGCSH	21	23.8	0.25	3000	5.9
CACL$_2$	22	30	0.25	3000	20
FREIDEL	23	22.4	0.25	3000	12.2
STRAT	24	22.4	0.25	3000	225.4
GYPSUMS	25	44.5	0.25	3000	1.0
CACO$_3$	26	97.0	0.25	3000	2.4
AFMC	27	79.0	0.25	3000	40.9
AGG	28	30	0.25	3000	20
ABSGYP	29	88.8	0.25	3000	4.6
EMPTYP	45	0	0	0	0

采用三种方式 (I、II、III) 施加本质边界条件，以研究边界条件对模拟结果的影响。在方式 I 和方式 II 中，在图 13-6 所示的立方体底部和顶部分别添加三层物

质点用于施加位移约束，四个模型的物质点总数分别为 123127、122272、123947 和 123143，新增边界区域的材料参数取整个微观结构的体积加权平均值，且设为无损区域。在方式 I 中，每个加载步的位移增量按线性分布施加于三层物质点上(由内向外逐渐增大)，在方式 II 中，位移增量均匀施加于三层物质点上，在方式 III 中，位移约束直接施加于微观结构上下表面的一层物质点上。

13.1.3 水泥水化微结构单向拉伸的计算结果与分析

13.1.3.1 荷载处理方式的影响

基于改进的近场动力学 PMB 模型及其隐式-显式混合求解方法，进行水泥水化微结构立方体的单向拉伸模拟计算。在隐式静力求解阶段，采用上述方式 I 施加位移荷载，显式求解阶段的加载速率为 $L_{\text{rate}} = \dfrac{\Delta u}{\Delta t} = \dfrac{3 \times 10^{-12} \text{m}}{5 \times 10^{-11} \text{s}} = 0.06 \text{m/s}$。计算得到水泥水化微结构四种模型的最终损伤特征、裂纹分布情况和试样的应力-应变曲线，如图 13-7 所示。其中，为获取试件的应力-应变曲线，采用近场动力学

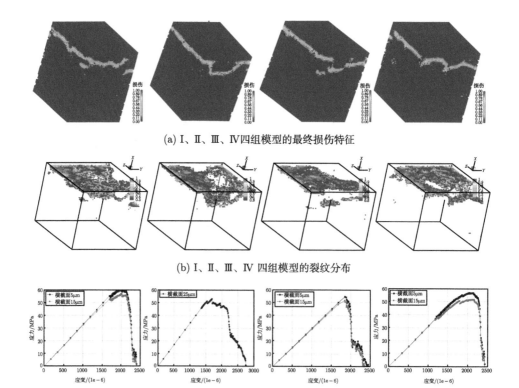

(a) I、II、III、IV 四组模型的最终损伤特征

(b) I、II、III、IV 四组模型的裂纹分布

(c) I、II、III、IV 四组模型的应力-应变曲线

图 13-7 方式 I 加载下水泥水化微结构的近场动力学预测结果

微分算子对材料的弹性本构关系进行处理，选取距试件顶面 5μm、15μm、25μm、35μm 和 45μm 处的横截面，根据计算得到的位移场，求出这些截面的应变和应力，得到相应的力-加载位移曲线和应力-应变曲线。

从图 13-7 中所示结果可以看出，四个模型得到的应力-应变曲线基本一致，表明水泥水化后不同的微结构特征对其应力-应变关系影响不大，损伤主要发生在边界加载区域附近，这是由于多孔水泥水化微结构和增加的连续边界区域之间的刚度不匹配所致。

图 13-8 给出了基于改进的近场动力学 PMB 模型，采用方式 II 施加位移荷载得到的计算结果。在这种情况下，微裂纹主要从试件侧面孔隙边缘开始萌生，并沿着孔隙率较高的薄弱区域扩展，最终在试件中部形成较大范围的贯穿性破坏区，从而导致水泥水化微结构断裂，且增加的连续边界区域与微观结构之间仅在部分位置发生了轻微损伤；此外，四组模型得到的应力-应变曲线具有较好的一致性。模拟结果表明，将位移约束均匀施加在增加的边界区域上，可以获得可靠的损伤破坏特征和水泥水化产物的应力-应变曲线等力学性能。

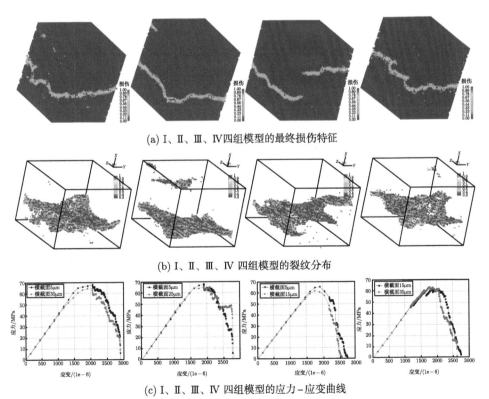

(a) I、II、III、IV四组模型的最终损伤特征

(b) I、II、III、IV 四组模型的裂纹分布

(c) I、II、III、IV 四组模型的应力-应变曲线

图 13-8　方式 II 加载下水泥水化微结构的近场动力学预测结果

图 13-9 显示了基于改进的近场动力学 PMB 模型，采用方式 Ⅲ 施加位移荷载得到的计算结果。计算结果表明，四种模型在加载方式 Ⅱ 和方式 Ⅲ 下的损伤破坏特征非常相似，得到的应力-应变曲线也具有较好的一致性，同时加载方式 Ⅲ 不存在刚度失配问题及其诱发的虚假损伤问题。

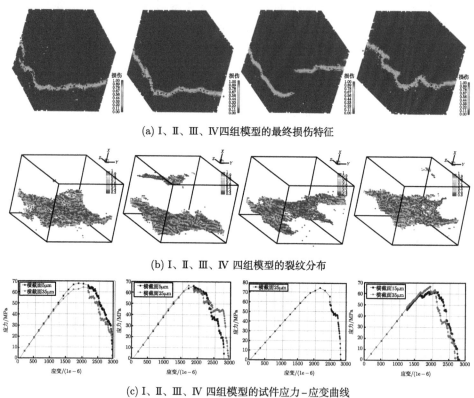

(a) Ⅰ、Ⅱ、Ⅲ、Ⅳ 四组模型的最终损伤特征

(b) Ⅰ、Ⅱ、Ⅲ、Ⅳ 四组模型的裂纹分布

(c) Ⅰ、Ⅱ、Ⅲ、Ⅳ 四组模型的试件应力-应变曲线

图 13-9　方式 Ⅲ 加载下水泥水化微结构的近场动力学预测结果

根据图 13-9(c) 的四组应力-应变曲线，可以得到试件材料的相关力学参数，如峰值应力、最大拉伸应变、杨氏模量和断裂能量释放率，如图 13-10 所示。

13.1.3.2　加载速率的影响

选取模型 Ⅱ，采用加载方式 Ⅲ，动态加载速率分别取 0.006m/s、0.06m/s、0.6m/s 和 6m/s，研究显式计算阶段的加载速率对计算结果的影响。图 13-11 给出了四种加载速率下水泥净浆试样的最终损伤特征、裂纹分布情况和应力-应变曲线。由应力-应变曲线的直线上升段可见，静态加载阶段的应力-应变响应几乎完全一致，但在动态加载阶段，由于动态加载速率的不同，损伤演化模式、贯穿裂纹表面和应力-应变曲线有很大差异。具体体现在：① 在 0.006m/s 和 0.06m/s

13.1 水泥水化产物的近场动力学建模与拉伸破坏分析 · 439 ·

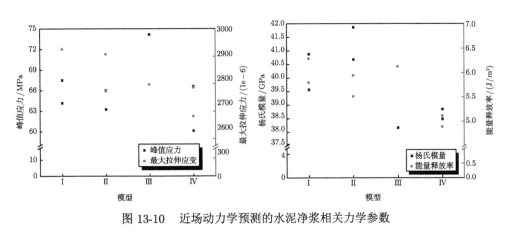

图 13-10 近场动力学预测的水泥净浆相关力学参数

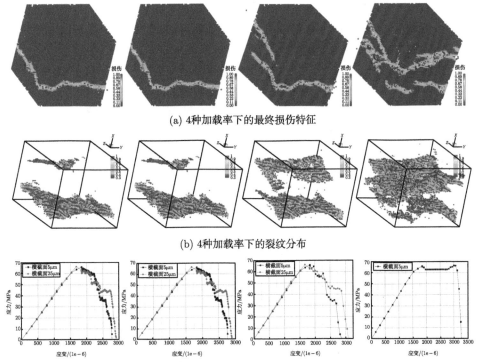

(a) 4种加载率下的最终损伤特征

(b) 4种加载率下的裂纹分布

(c) 4种加载率下水泥水化微结构的应力-应变曲线

图 13-11 不同加载速率 0.006m/s、0.06m/s、0.6m/s、6m/s 下近场动力学的预测结果

加载速率下，接近准静态加载，两种情况下的最终损伤特征和应力-应变曲线高度一致，表明较小的加载速率并未引发水泥净浆过大的动力效应；② 较大加载速率下，不再是一个准静态断裂问题，水泥水化微结构的整体损伤更为严重，不同加

载速率下的应力–应变曲线呈现不同的下降段趋势；③ 在 0.6m/s 和 6m/s 加载速率下，与准静态加载相比，动态加载导致峰后材料强度和断裂韧性提高；④ 动态载荷引发由边界向内传播的应力波，多相多孔特性使波的传播行为更加复杂，从而导致更为严重的损伤和裂纹分叉。模拟结果表明，加载速率显著影响水泥水化微结构的损伤模式和应力–应变关系。

13.1.3.3 近场动力学计算模型的完善

在上述计算中，对于模型中穿过孔洞的物质点之间的"键"，并未进行初始处理，即没有消除这些"键"的作用，这在一定程度上影响计算结果的精度。本节完善前述近场动力学计算模型，不计穿过孔隙的"键"的作用效应，具体采用断"键"的方式进行处理，简称为"完善模型"。采用加载方式 III 进行四组模型的准静态 (加载速率 0.06m/s) 单轴拉伸模拟。考虑到水泥水化微结构具有复杂的孔隙结构，将可能导致近场范围内物质点数目过少，使得整体劲度矩阵出现病态或奇异现象，而扩大近场范围并不能从本质上解决该问题。因此，一旦出现上述情况，将直接由隐式求解转换为显式求解。

图 13-12 给出了基于近场动力学方法计算得到的水泥净浆试样的最终损伤特

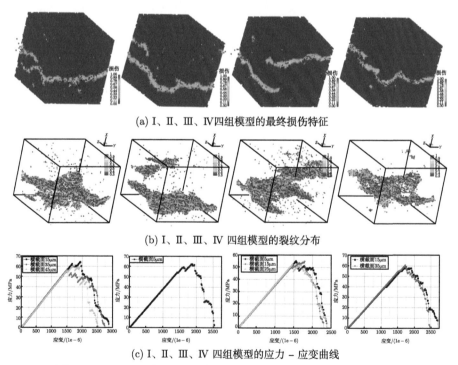

(a) I、II、III、IV 四组模型的最终损伤特征

(b) I、II、III、IV 四组模型的裂纹分布

(c) I、II、III、IV 四组模型的应力 – 应变曲线

图 13-12　近场动力学"完善模型"(加载速率 0.06m/s) 的预测结果

征、裂纹分布情况和应力–应变曲线,并由此获得峰值应力、最大拉伸应变、杨氏弹性模量和断裂能释放率等力学参数,如图 13-13 所示。计算结果表明,在考虑和不考虑穿过孔隙的"键"作用效应下,获得的损伤模式和裂纹分布情况总体相似,但不计穿过孔隙的"键"作用效应时,材料断裂的阻力有所降低,裂纹萌生时间更早,损伤更容易累积。表 13-6 列出了"完善模型"与原模型预测得到的力学参数平均值,以及文献中相关参数的取值,结果表明,完善模型所预测的峰值应力、最大拉伸应变、杨氏弹性模量和断裂能量释放率均低于原模型的对应的结果,且预测值数据范围合理可靠[9]。

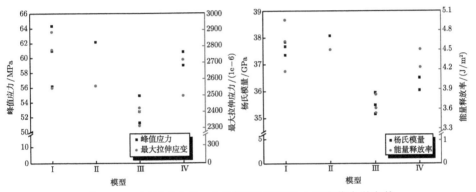

图 13-13 近场动力学"完善模型"预测的水泥净浆力学参数

表 13-6 近场动力学模型预测的水泥净浆相关力学参数

	峰值应力/MPa	最大拉伸应变	杨氏模量/GPa	能量释放率/(J/m²)
原模型结果	65.97	2.829e-3	39.78	5.669
完善模型结果	58.04	2.570e-3	36.68	4.207
文献取值范围	几 ~ 几百 [6,7]	0.15e-3~3.0e-3 [6,11]	15~40 [11]	1.72~20 [7,9]

13.2 混凝土细观破坏的近场动力学建模分析

混凝土通常可视为由骨料、砂浆及两者之间的界面过渡区组成的三相复合材料。为了充分考虑混凝土不同组分材料的本构特征,进行混凝土材料和结构的细观力学分析,首先需根据混凝土的级配要求,建立混凝土的随机骨料模型,再构建混凝土材料和结构的近场动力学细观数值计算模型。

13.2.1 混凝土骨料的富勒级配理论

骨料是混凝土的重要组成部分，在混凝土工程中，根据粒径大小将骨料分为粗骨料和细骨料。根据我国建设部行业标准《普通混凝土用砂、石质量及检验方法标准 (附条文说明)》(JGJ52—2006)，骨料粒径在 0.15~5mm 之间称为细骨料，粒径大于 5mm 的骨料统称为粗骨料。水工混凝土中，将 85% 以上的质量通过 5mm 筛孔的骨料称为细骨料，85% 以上的质量留在 5mm 筛孔上的骨料称为粗骨料。普通混凝土中所用细骨料，一般是由天然岩石长期风化等自然条件形成的天然砂，根据产源不同，可分为河砂、海砂和山砂三类；粗骨料按种类分为卵石、碎石、卵石和碎石的混合物等。粗骨料按粒径分类，可以分为小石 (5~20mm)、中石 (20~40mm)、大石 (40~80mm)、特大石 (80~150mm)，它们依次称为一、二、三、四配。骨料的级配，是指骨料中不同粒径颗粒的分布情况。当混凝土配比中包含这四种级配时，称为全级配混凝土。通常将只包含一、二级配的混凝土称为小骨料混凝土，将只包含三、四级配的称为大骨料混凝土。常用的四级配混凝土中的小石:中石:大石:特大石比例为 2:2:3:3，三级配混凝土中的小石:中石:大石的比例为 3:3:4，二级配混凝土中的小石:中石为 5.5:4.5。良好的级配应当能使骨料的空隙率和总表面积均较小，不仅能减少所需的水泥浆量，而且还可以提高混凝土的密实度、强度及其他性能。

20 世纪初，富勒 (Fuller) 等美国学者经过大量的试验工作，依靠筛分试验结果，提出最大密度的理想级配曲线，如图 13-14 所示。Fuller 理想级配曲线是将混凝土材料的骨料按颗粒度的大小，粗细合理搭配，有规则地组合排列，成为密度最大、

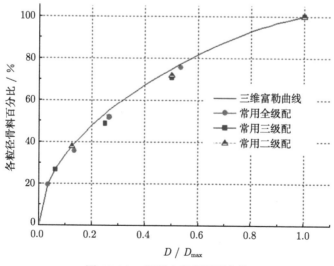

图 13-14 混凝土理想级配曲线

空隙率最小的混合物。该曲线将细骨料以下的颗粒级配用抛物线表示，其方程为

$$P = 100\sqrt{\frac{D}{D_{\max}}} \tag{13.1}$$

式中，P 表示骨料通过筛孔 D 的骨料质量百分比，D 为筛孔直径，$D_{\max}$ 为骨料的最大粒径。

从图 13-14 中可以看出，三种常用的混凝土级配曲线均与 Fuller 理想级配曲线吻合较好。按照 Fuller 理想级配曲线设计的混凝土 (混凝土骨料大小和配合比) 具有强度高、结构密实性好、抗渗性好等特点，同时可以减少水泥的用量，节约成本[13]。

13.2.2 基于蒙特卡罗法和随机游走法的骨料生成与投放

蒙特卡罗方法亦被称为随机模拟方法，或随机抽样技术或统计试验方法，该方法出现在 20 世纪 40 年代中期，是数学、概率统计和计算机技术结合形成的数值计算方法。蒙特卡罗方法利用随机数进行统计试验，用于求解具有随机性的非确定性问题，以求得的统计特征值作为待求解问题的数值解。该方法的缺点是计算量大，收敛速度较慢，但随着计算机技术的高速发展，计算机的运算能力大幅提高，蒙特卡罗方法在各个领域研究中得到广泛的应用。

蒙特卡罗方法使用一组随机数来代替某些随机过程中的变量。在数值试验中，最基本的随机数是在 [0,1] 区间上均匀分布的随机数，其概率密度函数为

$$f(x) = \begin{cases} 1, & x \in [0,1] \\ 0, & x \notin [0,1] \end{cases} \tag{13.2}$$

其中，x 为区间 [0,1] 上均匀分布的随机变量。在具体计算时，可产生随机抽样序列 $\{x_n\}$，x_n 称为区间 [0,1] 上均匀分布的随机变量 x 的随机数，这些随机数必须相互独立，且无相关性。在区间 $[a,b]$ 上均匀分布的随机数 x'，可由下式变换得到

$$x' = a + (b-a)x \tag{13.3}$$

在建立混凝土随机骨料模型时，根据混凝土结构的几何尺寸和骨料级配，基于 Fuller 最大密度理想级配曲线求得骨料粒径和数目；认为骨料的位置满足均匀随机分布，采用 Monte-Carlo 方法生成与骨料位置对应的随机数，直接在混凝土结构的几何区域内确定各骨料的位置，随后便可进行骨料投放，完成随机骨料模型的构建[14]。

13.2.2.1 二维圆形骨料的生成与投放方法

为生成混凝土随机骨料模型，除了采用 Monte-Carlo 方法在混凝土结构区域内直接投放骨料，还可以采用随机游走方法实现骨料投放。

假定某一级配混凝土的骨料粒径可以表示为 $(d_{11},d_{12}),\cdots,(d_{i1},d_{i2}),\cdots,(d_{n1},d_{n2})$，其中，$d_{i1}$ 和 d_{i2} 分别为第 i 个粒径段中的最小粒径和最大粒径。根据下式

$$\pi\frac{D_i^2}{4}(d_{i2}-d_{i1})=\frac{1}{4}\int_{d_{i1}}^{d_{i2}}\pi r^2 \mathrm{d}r \tag{13.4}$$

可获得第 i 个级配段中骨料面积特征粒径 D_i 为

$$D_i=\sqrt{\frac{d_{i1}^2+d_{i1}d_{i2}+d_{i1}^2}{3}} \tag{13.5}$$

记 Q_i 为第 i 个级配段的骨料质量百分比，S_i 为第 i 个级配段的骨料数量百分比，则根据各级配骨料密度一致原则，也即各级骨料的质量百分比的比例关系 $(Q_1:Q_2:\cdots:Q_i:\cdots:Q_n)$ 与面积百分比的比例关系 $(S_1D_1^2:S_2D_2^2:\cdots:S_iD_i^2:\cdots:S_nD_n^2)$ 相等原则，有下式成立

$$\frac{S_1D_1^2}{Q_1}=\cdots\frac{S_iD_i^2}{Q_i}=\cdots=\frac{S_nD_n^2}{Q_n},\quad i=1,\cdots,n \tag{13.6}$$

进而，可以确定骨料的累积质量百分比

$$S_{iM}=\sum_{i=1}^{M}S_i=\sum_{i=1}^{M}\frac{\dfrac{Q_i}{D_i^2}}{\sum_{j=1}^{n}\dfrac{Q_j}{D_j^2}} \tag{13.7}$$

其中，S_{iN} 表示骨料的累积质量百分比，$M(1,2,\cdots,n)$ 表示前 M 级的混凝土骨料。

在骨料的生成阶段，采用随机数生成算法生成 0~1 的均匀分布的随机数 random，比较随机数与 $S_{i(M-1)}$ 和 S_{iM} 大小，即可确定该圆形骨料的半径区间，例如 $0<\text{random}<S_{i1}$，则生成第一级粒径的骨料，如 $S_{i(M-1)}<\text{random}<S_{iM}$，则生成第 M 级粒径的骨料。在特定粒径段的骨料生成阶段，仍然采用蒙特卡罗方法，随机生成粒径为 r 的骨料，随后进行投放游走，直至满足预定的骨料投放率。

13.2 混凝土细观破坏的近场动力学建模分析

以下以二维问题的矩形计算区域为例, 说明如何在有限空间区域内, 通过分批投放和随机移动生成混凝土随机骨料模型[15,16]。如图 13-15 所示, 矩形四个顶点坐标分别为 $(x_{\min}, y_{\min})$、$(x_{\min}, y_{\max})$、$(x_{\max}, y_{\min})$ 和 $(x_{\max}, y_{\max})$。在计算区域外, 设置骨料初始投放区域, 每次投放骨料个数设为 N, 在初始投放区域内生成混凝土圆形骨料, 圆形骨料的圆心坐标根据初始投放区域位置确定, 即第 i 个混凝土骨料圆心坐标为

$$\begin{cases} x_i = x_{\min} + \left(2 \bmod \left(\dfrac{i}{N}\right) - 1\right) \\ y_i = y_{\max} + \dfrac{D_{\max}}{2} \end{cases} \tag{13.8}$$

其中, $N = \mathrm{int}\left(\dfrac{x_{\max} - x_{\min}}{D_{\max}}\right)$ 为每次投放骨料个数, $D_{\max}$ 为骨料最大直径, mod 为取余数函数。

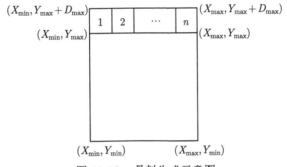

图 13-15 骨料生成示意图

为使骨料充分占据计算区域, 使骨料在有限区域内随机移动, 每次随机移动后更新圆形骨料的圆心坐标, 则第 i 个骨料的圆心坐标为

$$\begin{cases} x'_i = x_i + \mathrm{random1} \times \dfrac{D_{\max}}{2} \\ y'_i = y_i - \mathrm{random2} \times \dfrac{D_{\max}}{2} \end{cases} \tag{13.9}$$

其中, random1 和 random2 分别为 $[-1, 1]$ 和 $[0, 1]$ 中的随机数。

骨料随机移动后, 需要判断该次骨料移动的有效性, 即移动后骨料之间不能相交, 亦即任意两个圆形骨料的圆心距离须大于两个圆的半径之和, 可表示为

$$S(i', j) > r_{i'} + r_j \tag{13.10}$$

其中，$S(i',j) = \sqrt{(x_{i'} - x_j)^2 + (y_{i'} + y_j)^2}$ 为第 i' 个骨料和第 $j(j=1,\cdots,n)$ 个骨料的圆心距离，$i' \neq j$，n 表示当前投放骨料的总数，$r_{i'}$、r_j 为圆形骨料的半径。

随机移动的骨料还要满足几何方面的限制条件，即骨料不能移动到计算对象几何区域的外部，对于矩形，需满足

$$\begin{cases} x_{\max} - r_{i'} \geqslant x_i' \geqslant x_{\min} + r_{i'} \\ y_i' \geqslant y_{\min} + r_i \end{cases} \tag{13.11}$$

满足骨料不相交条件和几何限制条件后，骨料的随机移动即是有效运动，此时有 $x_i = x_i', y_i = y_i'$。如果不满足上述条件，则该骨料不移动，维持原有位置不变，进行第 $i+1$ 个骨料的随机移动，直到完成所有已投放骨料的移动。可以设置随机移动的次数，使骨料更加密实；完成规定次数的随机移动后，再进行下次骨料投放。

重复进行骨料的投放和随机移动，直到满足骨料的投放率，则停止投放。此时，有些骨料可能会留在初始投放区域和计算区域的边界处，需要进行剔除，即骨料坐标需满足 $y_i' \leqslant y_{\max} - r_i, i = 1, \cdots, n$。具体实施时，还可调节骨料之间的距离，以使骨料在相同的投放率下保持均匀。

图 13-16 展示了二级配骨料在边长 150mm 方形区域内的生成与投放过程，其中，两种骨料的半径范围分别为 (5~20mm) 和 (20~30mm)，小石：中石的质量比为 0.50:0.50，设定的目标投放率为 70%。投放最大次数为 500 次，随机游走次数设为 40 次，则投放 144 次后的骨料投放率为 74.31%。

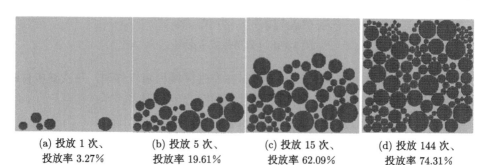

(a) 投放 1 次、投放率 3.27%　(b) 投放 5 次、投放率 19.61%　(c) 投放 15 次、投放率 62.09%　(d) 投放 144 次、投放率 74.31%

图 13-16　圆形骨料投放结果

13.2.2.2　二维任意多边形骨料的生成与投放方法

确定多边形骨料初始中心位置的算法与圆形骨料的相应算法完全相同，与圆形骨料相比，多边形骨料顶点坐标的生成、随机移动、骨料不相交条件和几何限制条件的算法有所不同，下面具体介绍二维多边形骨料的生成和投放过程。

13.2 混凝土细观破坏的近场动力学建模分析

首先，确定第 i 个多边形骨料的中心位置，为

$$\begin{cases} c_{x_i} = x_{\min} + \left(2\bmod\left(\dfrac{i}{N}\right) - 1\right) \\ c_{y_i} = y_{\max} + \dfrac{D_{\max}}{2} \end{cases} \quad (13.12)$$

式中，$N = \mathrm{int}\left(\dfrac{x_{\max} - x_{\min}}{D_{\max}}\right)$ 为每次投放骨料个数，$D_{\max}$ 为骨料最大直径，mod 为取余数函数。

其次，生成多边形骨料各顶点 p_i 的坐标，为

$$\begin{cases} x_i = c_{x_i} + r_i \cos\alpha_i \\ y_i = c_{y_i} + r_i \sin\alpha_i \end{cases} \quad (13.13)$$

其中，(c_{x_i}, c_{y_i}) 为骨料中心点坐标，$i = 1, 2, \cdots, n$，n 为多边形顶点数，为使起点和终点重合，使 $p_{n+1} = p_1$，r_i 为顶点 i 距骨料中心点的长度，α_i 为 r_i 与 x 轴的夹角。

为确定多边形骨料的顶点坐标，令

$$\alpha_1 = 0, \; \alpha_i = \alpha_{i-1} + 45 + 45 \times \mathrm{random3}, \; i = 2, 3, \cdots \quad (13.14)$$

当 $\alpha_k > 360°$ 时，则停止顶点坐标的生成，此时取 $n = k - 1$。

骨料粒径段为 $(d_{\min}, d_{\max})$ 的骨料顶点 i 距其中心点的长度可表示为

$$r_i = r_{\min} + (r_{\max} - r_{\min}) \times \mathrm{random4}, \; i = 1, 2, 3, \cdots, n \quad (13.15)$$

其中，$r_{\min} = d_{\min}/2$，$r_{\max} = d_{\max}/2$，random3 和 random4 表示 (0, 1) 间的随机数。

多边形骨料的随机移动包括平移和转动两种运动状态，当骨料发生随机平移时，骨料中心点的坐标为

$$\begin{cases} c'_{x_i} = c_{x_i} + r_{\max}\mathrm{random5} \\ c'_{y_i} = c_{y_i} + r_{\max}\mathrm{random6} \end{cases} \quad (13.16)$$

当骨料发生随机转动时，骨料顶点 P_{ij}(i 表示第 i 个骨料，j 表示第 i 个骨料的第 j 个顶点) 的转动角度为

$$\alpha'_{ij} = \alpha_{ij} + \theta \times \mathrm{random7} \quad (13.17)$$

其中，random5 表示 $(-1,1)$ 中的随机数，random6 和 random7 表示 $(0,1)$ 中的随机数，θ 为最大转动角度，可取为 $45°$。

对于随机移动的任意多边形骨料，必须满足几何方面的限制条件和骨料之间互不相交条件，如下所述。

(1) 几何限制条件：对于随机移动后的多边形骨料 i，其顶点 P'_{ij} 的坐标 (x'_{ij}, y'_{ij}) 需要满足 $x_{\min} \leqslant x'_{ij} \leqslant x_{\max}$，$y_{\min} \leqslant y'_{ij}$。

(2) 骨料互不相交条件：对于随机移动后的多边形骨料 i，不与其他未移动骨料相交。具体处理时，可以首先剔除与骨料 i 相距较远的骨料点，以减少计算工作量，如果满足

$$\sqrt{(x_{i,c}-x_{j,c})^2+(y_{i,c}-y_{j,c})^2} \geqslant 2r_{\max} \tag{13.18}$$

或者

$$\sqrt{(x_{i,c}-x_{j,c})^2+(y_{i,c}-y_{j,c})^2} \geqslant r_{i\max}+r_{j\max} \tag{13.19}$$

则表明两骨料必不相交，否则需判断两骨料的相交性，具体有下面两种情况。

① 判断骨料 i 的顶点 $P_{i,k}$ 是否在骨料 j 内。如图 13-17(a) 所示，判断面积 $S_{C_j P_{j,l} P_{j,l+1}}$ 与 S 是否相等 (其中，$S = S_{C_j P_{i,k} P_{j,l+1}} + S_{C_j P_{j,l} P_{i,k}} + S_{P_{i,k} P_{j,l} P_{j,l+1}}$，$(i,j=1,2,\cdots,n))$，如果相等，则 $P_{i,k}$ 在骨料 j 内；如果不等，则 $P_{i,k}$ 不在骨料 j 内，需要进行下一步判断。

② 判断骨料 i 的边界是否与骨料 j 边界相交。如图 13-17(b) 所示，如果两线段 P_1P_2、Q_1Q_2 相交，则向量 $\overrightarrow{P_1Q_1}$ 和向量 $\overrightarrow{P_2Q_2}$ 必然在向量 $\overrightarrow{Q_1Q_2}$ 两侧，有

$$(\overrightarrow{P_1Q_1} \times \overrightarrow{Q_1Q_2}) \cdot (\overrightarrow{P_2Q_2} \times \overrightarrow{Q_1Q_2}) < 0 \tag{13.20}$$

上述两个条件需要同时满足。

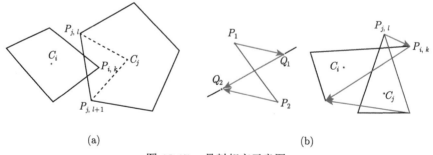

图 13-17 骨料相交示意图

重复骨料投放和随机移动过程，当满足投放率时，则停止投放。根据 $y_{i,k} \leqslant y_{\max}$ 的条件，对位于初始区域和几何构型边界处的骨料进行剔除，同样也可以通过控制骨料之间的距离保持骨料投放的均匀性。

图 13-18 给出了二级配骨料在边长 150mm 方形区域内的生成与投放过程,其中任意多边形骨料半径范围分别为 (5~20mm) 和 (20~30mm),小石:中石的质量比为 0.50:0.50,设定的目标投放率为 70%。投放最大次数为 500 次,随机游走次数设为 40 次,则投放 135 次后的骨料投放率为 70.05%。

(a) 投放 1 次、投放率 4.61%　　(b) 投放 5 次、投放率 17.90%　　(c) 投放 15 次、投放率 55.07%　　(d) 投放 135 次、投放率 70.05%

图 13-18　任意多边形骨料投放结果

13.2.3　数值算例与分析

13.2.3.1　基于随机骨料模型的混凝土试件单轴压缩模拟

边长 150mm 的混凝土立方体试件,在竖直方向承受压载荷。按照平面应变问题分析,采用二级配圆形骨料,生成的随机骨料模型包括 6 颗粒径 30mm 的中骨料和 56 颗粒径 12mm 的小骨料,骨料体积分数为 46.998%[17,18],如图 13-19(a) 所示。

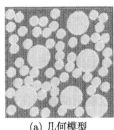 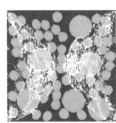 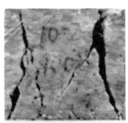

(a) 几何模型　　(b) PD模拟结果　　(c) 试验结果

图 13-19　混凝土的单轴受压破坏结果

在近场动力学计算模型中,含有骨料和砂浆两类物质点,三种"键"的作用,分别为骨料键、砂浆键及界面键,它们具有不同的力学性能,需要赋予相应的材料参数,以反映混凝土的非均匀特性。骨料的力学参数为:弹性模量 50GPa、泊松比 0.2、质量密度 2700kg/m³、抗拉强度为 10.0MPa;水泥砂浆的力学参数为:弹性模量 25GPa、泊松比 0.2、质量密度 2100kg/m³、抗拉强度 4.0MPa;界面过渡区的材料参数为:弹性模量 25GPa、泊松比 0.2、抗拉强度 4.0MPa。

基于前述近场动力学模型计算得到的试件破坏形态如图 13-19(b) 所示。由图可见，破坏形态呈 X 型分布，剪胀现象明显，与混凝土单轴压缩的试验结果 (图 13-19(c))[19] 具有较好的一致性。此外，采用近场动力学模型计算得到的试件极限荷载为 34.98MPa，文献 [18] 基于有限元方法得到的极限荷载均值为 28.5MPa，试验给出的抗压强度为 37.1MPa[20]，本例的计算结果与试验值更为接近。

13.2.3.2 基于真实骨料分布的混凝土试件单轴压缩模拟

除前述的随机骨料模型外，还可以通过对实际的混凝土试件进行切片、CT 扫描和数字图像处理，得到真实的骨料形状和分布特征，再据此构建混凝土细观数值模型。这类模型基于混凝土真实的骨料分布情况，能够更好地反映混凝土的细观结构组成 [21]。图 13-20(a) 为边长 150mm 的某四级配混凝土立方试件一个截面的 CT 扫描图片，在进行该 CT 扫描图片的图像处理过程中，忽略其中的微空隙和微裂隙，进行模型重构，得到的计算模型如图 13-20(b) 所示。

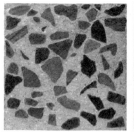
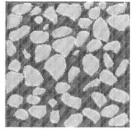

(a) 扫描图像[21]　　　　(b) 图像重构

图 13-20　混凝土试件的某截面的扫描图像和重构

基于近场动力学方法对上述混凝土细观模型进行竖向单轴压缩模拟，采用与 13.2.3.1 节中相同的材料参数，计算得到的混凝土试件破坏过程以及应力-应变曲线分别如图 13-21 和图 13-22 所示。模拟结果表明，混凝土单轴受压破坏的模拟

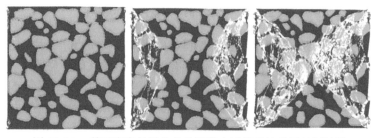

图 13-21　基于真实骨料分布的混凝土试件单轴受压破坏过程

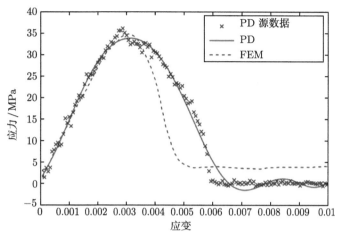

图 13-22 混凝土试件单轴压缩的应力–应变曲线

结果与试验结果相符，也与随机骨料模型得到的结果相吻合，基于近场动力学方法计算得到的应力–应变曲线与有限元结果[21]具有较好的一致性。

13.2.3.3 含初始裂纹混凝土三点弯曲梁的断裂模拟

考虑如图 13-23 所示的含初始裂纹混凝土三点弯曲梁，在梁中间截面的底部预置初始裂纹，裂纹长度 80mm、宽度为 8mm，梁中部承受集中载荷 P。混凝土梁的砂浆力学参数为：弹性模量 E=25GPa、泊松比 ν=0.19、质量密度 ρ=2000kg/m³、能量释放率 G=80J/m²；骨料力学参数为：弹性模量 E=80Gpa、泊松比 ν=0.24、质量密度 ρ=3500kg/m³、能量释放率 G=120J/m²。对梁进行均匀离散，物质点总数为 50481。采用临界伸长率作为断键损伤判据，骨料与砂浆的界面的临界伸长率采用骨料和砂浆临界伸长率最小值的 0.8 倍[13,15]。

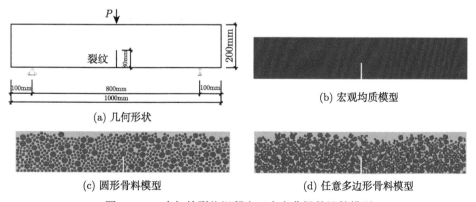

图 13-23 含初始裂纹混凝土三点弯曲梁的计算模型

基于近场动力学方法，分别建立了宏观均质计算模型、骨料形状为圆形和任意多边形的非均质细观数值模型，二级配非均质模型的骨料投放率均为60%，骨料粒径为 (10~15mm) 和 (25~30mm)，两种骨料质量百分比为 1:1。

图 13-24 给出了基于近场动力学方法计算得到的三种模型的裂纹演化过程与扩展路径。由计算结果可知，裂纹均是从预置裂纹尖端产生，对于宏观均质模型，裂纹按预置裂纹的方向，沿直线扩展至梁顶部，符合问题的对称性；对非均质细观数值模型，裂纹没有严格沿裂纹的方向竖直向上扩展，而是在裂纹尖端附近弯曲延伸，总体向加载点方向扩展，裂纹扩展路径受骨料分布的影响较大，与试验结果[22]相吻合。

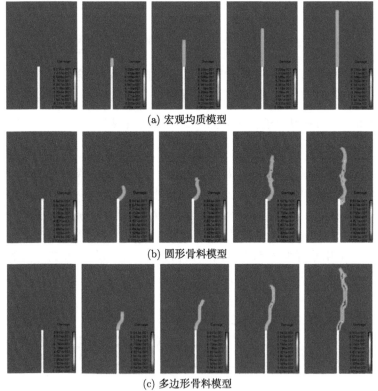

(a) 宏观均质模型

(b) 圆形骨料模型

(c) 多边形骨料模型

图 13-24　三种混凝土梁模型的裂纹扩展过程

13.3　混凝土冻融循环热–水–力耦合问题的近场动力学模型

冻融破坏是寒区水工混凝土建筑物最主要的病害，给水工混凝土结构的安全运行带来严重威胁，是工程界和学术界长期关注的重要课题。混凝土冻融破坏是指已硬化的混凝土在浸水饱和或潮湿状态下，由于环境温度正负交替变化，使混凝土内部孔隙水形成冻结膨胀压力、渗透压力和结晶压力等，造成混凝土内部损伤开裂或表面成层剥蚀，并导致混凝土力学性能降低的一种破坏现象。

混凝土的冻融循环效应强烈依赖于砂浆基体中的孔隙结构，在冻融循环过程中，孔隙水发生相变，导致水分迁移，引起温度重分布和结构变形。本节阐述水泥砂浆的孔隙结构特征和孔隙水的冻结规律，考虑孔隙水相变潜热、孔隙压力、结冰速率和水冰压力差等的影响，运用近场动力学微分算子重构连续介质力学局部理论的温度场、渗流场和应力场方程，建立积分形式的近场动力学热–水–力耦合模型和相应的数值算法，对典型水泥砂浆试件进行冻融循环模拟计算。

13.3.1 多孔介质中孔隙水冻结规律与孔隙压力

对于不同水灰比的水泥砂浆，其孔隙含量和分布大不相同，水泥砂浆中的孔隙主要分为凝胶孔、毛细孔和非毛细孔三类，孔径大小差别可达 6 个数量级。多孔介质的孔径分布可通过压汞法或氮气吸附法测得。

多孔介质的总孔隙率 n(孔隙体积占总体积的分数) 定义为

$$n = \lim_{r \to 0} \int_r^\infty \frac{\mathrm{d}\varphi}{\mathrm{d}r} \mathrm{d}r \tag{13.21}$$

式中，φ 为半径大于 r 的孔隙总含量，当 r 趋近于零时，则 $n = \varphi$。在低温条件下，饱和砂浆孔隙中始终只存在冰和水两相，即满足 $S_l + S_s = 1$，其中的 S_l 和 S_s 分别为水和冰占孔隙率的百分比。

温度降低将导致水泥砂浆的孔隙水冻结，是引发混凝土低温冻害的重要原因。与开放状态的水不同，孔隙水的冰点与孔径有关，随孔径减小而不断降低，且与水的相变潜热、水和冰的表面张力以及水的密度相关。基于热力学理论，可建立能结冰的最小孔隙半径与温度和上述因素的关系式[23,24]

$$R_c = -\frac{2\gamma T_0}{\rho_w L T} \tag{13.22}$$

式中，R_c 为温度降至 T 时能结冰的最小孔隙半径，单位为 nm；T_0 为水的冰点，T_0=273.15K；γ 为水与冰之间的表面张力，取γ=0.039N/m；水的密度ρ_w = 1000kg/m^3；L 为水的相变潜热，即单位质量水冻结 (或冰融化) 放出 (或吸收) 的热量，$L = 333.5$kJ/kg。

研究发现，在极低温度下，水泥砂浆孔隙壁上厚度约为 $\delta = 1.97 |T|^{-1/3}$ nm 的吸附水膜也不会冻结，故需考虑水膜厚度对孔隙半径进行校正，如图 13-25 所

示。因此，温度为 T 时，将式 (13.22) 中相关参数代入，得到校正后结冰的最小孔隙半径 [23,24]

$$R_{\mathrm{cr}} = R_{\mathrm{c}} + \delta = \frac{64}{|T|} + 1.97\,|T|^{-1/3} \tag{13.23}$$

考虑水泥砂浆完全饱和，温度 T 时水变为冰的体积为 $V_{\mathrm{s}} = \varphi(R_{\mathrm{cr}}(T)) - V_\delta(T)$，其中 $V_\delta(T)$ 为温度 T 时的吸附层体积，φ 为半径大于 R_{cr} 的孔隙总含量。

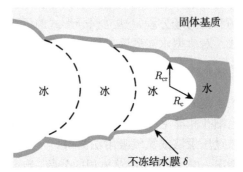

图 13-25　孔隙结冰时的最小孔隙半径与吸附层厚度示意图

水泥砂浆的冻融循环分析可以从微观和宏观两个尺度开展。若已知水泥砂浆的孔隙微结构，则可以将孔隙压力施加于孔壁上，开展微观计算分析；若采用宏观代表性体积单元开展数值计算，则需要计算平均孔隙压力。

作用在孔壁上的平均孔隙压力可以定义为 [25]

$$p^* = \sum_{\mathrm{REV}} p_r \frac{V_r}{V_{\mathrm{REV}}} \tag{13.24}$$

式中，V_r 为半径为 r 的孔隙的体积，p_r 为作用在体积为 V_r 的孔隙孔壁上的压力，V_{REV} 为代表性体积单元 REV 体积。根据总孔隙率的定义，可知 $\dfrac{V_r}{V_{\mathrm{REV}}} = \dfrac{\mathrm{d}\varphi}{\mathrm{d}r}\mathrm{d}r$，则平均孔隙压力 p^* 为

$$p^* = \frac{1}{n}\int_0^\infty p_r \frac{\mathrm{d}\varphi}{\mathrm{d}r}\mathrm{d}r \tag{13.25}$$

对于半径 r 的孔隙，若孔隙水已经冻结，则 $p_r = \pi_r$，π_r 为孔隙吸附层的水压力；若未冻结，则 $p_r = p_1$，p_1 为未冻结孔的水压力。所以，对于某一时刻 t 和温度 T，平均孔隙压力为

$$p^*(t) = \frac{1}{n}\int_0^{R_{\mathrm{cr}}} p_1(t) \frac{\mathrm{d}\varphi}{\mathrm{d}r}\mathrm{d}r + \frac{1}{n}\int_{R_{\mathrm{cr}}}^\infty \pi_r(t) \frac{\mathrm{d}\varphi}{\mathrm{d}r}\mathrm{d}r \tag{13.26}$$

式中，等号右端第一项为未冻结孔的压力，第二项为冻结孔的压力。

如图 13-26 所示，在冻结孔与未冻结孔联通处的水冰交界面，由于毛细孔表面张力的作用，存在平衡关系

$$p_s(t) - p_l(t) = \gamma(t)\kappa_c(t) \tag{13.27}$$

式中，p_s 为固相冰压力，κ_c 为冻结孔与未冻结孔联通处水冰接触面的曲率，γ 为冰与水之间的表面张力。

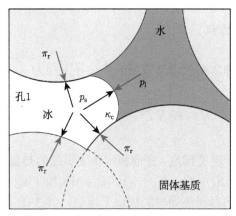

图 13-26　孔隙结冰时的孔隙压力示意图

由于吸附层的存在，对于半径为 r 的结冰孔，冰的半径变为 $r-\delta$，κ 为结冰孔内冰与吸附层水接触面的曲率。基于平衡法则，水冰表面及作用在孔壁上的压力存在如下关系：

$$p_s(t) = \pi_r(t) + \gamma(t)\kappa(r-\delta, t) \tag{13.28}$$

将式 (13.38) 代入式 (13.37)，有

$$\pi_r = p_l(t) + \gamma(t)\left(\kappa_c(t) - \kappa(r-\delta, t)\right) \tag{13.29}$$

在目前的研究中，通常假设冻结孔与未冻结孔联通处的水冰接触面为球形，结冰孔内冰与吸附层水的接触面为圆柱形，则有 $\kappa_c = 2/R_c(t)$、$\kappa(r-\delta, t) = 1/(r-\delta(t))$，故上式可写为

$$\pi_r = p_l(t) + \gamma(t)\left(\frac{2}{R_c(t)} - \frac{1}{r-\delta(t)}\right) \tag{13.30}$$

则最终得到平均孔隙压力

$$p^*(t) = p_l(t) + \frac{1}{n}\int_{R_{cr}}^{\infty} \gamma(t)\left(\frac{2}{R_c(t)} - \frac{1}{r-\delta(t)}\right)\frac{d\varphi}{dr}dr \tag{13.31}$$

式中，右端第二项与孔隙冻结状态相关，记为 X，表明水泥砂浆的平均孔隙压力是液体压力与孔隙冻结状态相关项的总和。

13.3.2 多孔介质热-水-力耦合问题的近场动力学模型

采用近场动力学微分算子重构连续介质力学的局部控制方程，建立积分型非局部近场动力学热-水-力耦合方程，包括考虑孔隙水相变潜热的热传导方程、多孔介质体系中的渗流方程和考虑孔隙中平均孔隙压力的运动方程。

13.3.2.1 热传导方程

考虑变形热效应和水相变潜热效应的热传导方程为

$$\rho C \frac{\partial T(\boldsymbol{x},t)}{\partial t} = \nabla \cdot (K_k \nabla T) - \alpha(3\lambda + 2\mu)T_0 \dot{\varepsilon}_{kk} + L\dot{w}_s \tag{13.32}$$

等式右边分别表示由于温度梯度、变形和水相变引起的热量。式中，$C = (nS_lC_l + nS_sC_s + C_m)/(nS_l + nS_s + 1)$，$K_k = (nS_lk_l + nS_sk_s + k_m)/(nS_l + nS_s + 1)$，分别表示多孔体系的比热容和热导率的加权平均值，其中的 C_l, C_s, C_m 分别为液相水、固相冰和混凝土基质的比热容，k_l, k_s, k_m 分别为液相水、固相冰及混凝土基质的热导率；ρ、α、λ 和 μ 分别为多孔材料的密度、线热胀系数和拉密常数；L 为水的相变潜热；$\dot{w}_s$ 为结冰速率，可由结冰体积与温度的关系，并结合具体问题中的温度随时间变化率得到。

采用近场动力学微分算子将方程 (13.32) 改写为非局部积分形式，对于三维问题，有

$$\begin{aligned}\rho C \frac{\partial T}{\partial t} =& \frac{3k_T}{2\pi V_2^{3D}} \int_{H_x} w_2 \left(T(\boldsymbol{x}') - T(\boldsymbol{x})\right) |\boldsymbol{\xi}|^2 \, \mathrm{d}V_{\boldsymbol{x}'} \\ & - \alpha(3\lambda + 2\mu)T_0 \frac{\partial}{\partial t} \left(\frac{3}{4\pi V_1^{3D}} \int_{H_x} w_1 \boldsymbol{\xi} \cdot (\boldsymbol{u}(\boldsymbol{x}') - \boldsymbol{u}(\boldsymbol{x})) \, \mathrm{d}V_{\boldsymbol{x}'} \right) + L\dot{w}_s \end{aligned} \tag{13.33}$$

式中，$V_2^{3D} = \int_0^\delta w_2(|\boldsymbol{\xi}|)|\boldsymbol{\xi}|^6 \mathrm{d}|\boldsymbol{\xi}|$，$V_1 = \int_0^\delta w_1(|\boldsymbol{\xi}|)|\boldsymbol{\xi}|^4 \mathrm{d}|\boldsymbol{\xi}|$，$\boldsymbol{\xi} = \boldsymbol{x}' - \boldsymbol{x}$ 是两物质点的相对位置矢量，$w_1(|\boldsymbol{\xi}|)$ 和 $w_2(|\boldsymbol{\xi}|)$ 为近场动力学影响函数，$\boldsymbol{u}$ 为物质点位移。

13.3.2.2 渗流方程

液相质量平衡方程是一个扩散方程，液相质量守恒方程为

$$\beta \frac{\partial p_l(\boldsymbol{x},t)}{\partial t} = \nabla \cdot \left(\frac{D}{\eta}\nabla p_l\right) + S_p - b\dot{\varepsilon}_V \tag{13.34}$$

式中，$\beta = \frac{nS_l}{K_l} + \frac{nS_s}{K_s} + \frac{b-n}{K_m}$，$S_p = \left(\frac{1}{\rho_s} - \frac{1}{\rho_l}\right)\dot{w}_s + \bar{\alpha}\dot{T} - \frac{b-n}{K_m}\dot{X} - \frac{nS_s}{K_s}\gamma\dot{\kappa}_c$，$\bar{\alpha} = nS_l\alpha_l + nS_s\alpha_s + (b-n)\alpha_0$，$D$ 为渗透系数，η 为水的动力黏滞系数，ε_V 为体应变，K_l、K_s 为水和冰的压缩体积模量，$\alpha_l, \alpha_s, \alpha_0$ 为水、冰和基质的体膨胀系数，ρ_s 为冰的密度，$b = 1 - K_0/K_m$ 为 Biot 系数，K_0 和 K_m 分别为多孔介质和基体骨架的弹性体积模量。右端项分别为孔隙水压梯度、源项和体积改变导致的液相转移。源项 S_p 由四部分组成：结冰速率相关项、温度膨胀效应相关项、平均孔隙压力与水压力差的相关项，第四项由水冰表面张力系数和曲率半径组成，与水冰间压力差相关。研究表明，这四项的大小与温度有关，当温度低于 $-10°C$ 后，后两项的影响变得非常小，可忽略不计。

采用近场动力学微分算子将方程 (13.44) 改写为非局部积分形式，在三维情况下，相应的渗流方程为

$$\beta\frac{\partial p_l}{\partial t} = \frac{3k_p}{2\pi V_2^{3D}}\int_{H_x} w_2\left(p_l(\boldsymbol{x}') - p_l(\boldsymbol{x})\right)|\boldsymbol{\xi}|^2\,\mathrm{d}V_{\boldsymbol{x}'}$$

$$- b\frac{\partial}{\partial t}\left(\frac{3}{4\pi h V_1^{3D}}\int_{H_x} w_1\boldsymbol{\xi}\cdot(\boldsymbol{u}(\boldsymbol{x}') - \boldsymbol{u}(\boldsymbol{x}))\,\mathrm{d}V_{\boldsymbol{x}'}\right) + S_p(\boldsymbol{x},t) \tag{13.35}$$

式中，$k_p = D/\eta$，其他参数同前所述。

13.3.2.3 运动方程

为反映热–水–力耦合效应的影响，引入平均孔隙压力，则 Cauchy 应力分量为 $\sigma_{ij} = \sigma'_{ij} - bp^*\delta_{ij}$，其中的 $\sigma'_{ij} = C_{ijkl}(\varepsilon_{kl} - \alpha(T-T_0)\delta_{kl})$ 为有效应力分量，C_{ijkl} 为弹性张量分量，ε_{ij} 为应变分量，δ_{ij} 为克罗内克符号，T_0 为参考温度。对于理想弹性体，Cauchy 应力分量为

$$\sigma_{ij} = 2\mu\varepsilon_{ij} + (\lambda\varepsilon_{kk} - \alpha(3\lambda+2\mu)(T-T_0) - bp^*)\delta_{ij} \tag{13.36}$$

将上式中的应变分量通过几何方程用位移分量表示，并代入运动方程 $\rho\ddot{\boldsymbol{u}} = \nabla\cdot\boldsymbol{\sigma} + \boldsymbol{f}$，得到位移微分形式的运动方程，再使用近场动力学微分算子将其改写

为非局部积分形式，得到引入平均孔隙压力的近场动力学运动方程。三维问题的控制方程为

$$\rho\frac{\partial^2 \boldsymbol{u}(\boldsymbol{x},t)}{\partial t^2} = \frac{3(\mu-\lambda)}{4\pi V_2^{3D}}\int_{H_\infty} w_2(|\boldsymbol{\xi}|)(\boldsymbol{\xi}\cdot\boldsymbol{\xi})(\boldsymbol{u}(\boldsymbol{x}+\boldsymbol{\xi})-\boldsymbol{u}(\boldsymbol{x}))\mathrm{d}V_{\boldsymbol{x}'}$$

$$+(\lambda+\mu)\frac{15}{4\pi V_2^{3D}}\int_{H_\infty} w_2(|\boldsymbol{\xi}|)((\boldsymbol{u}(\boldsymbol{x}+\boldsymbol{\xi})-\boldsymbol{u}(\boldsymbol{x}))\cdot\boldsymbol{\xi})\boldsymbol{\xi}\mathrm{d}V_{\boldsymbol{x}'}$$

$$-\alpha(3\lambda+2\mu)\frac{1}{4\pi V_1^{3D}}\int_{H_\infty} w_1(|\boldsymbol{\xi}|)(T(\boldsymbol{x}+\boldsymbol{\xi})-T(\boldsymbol{x}))\boldsymbol{\xi}\mathrm{d}V_{\boldsymbol{x}'}$$

$$-b\frac{1}{4\pi V_1^{3D}}\int_{H_\infty} w_1(p^*(\boldsymbol{x}')-p^*(\boldsymbol{x}))\boldsymbol{\xi}\mathrm{d}V_{\boldsymbol{x}'}+\boldsymbol{f}(\boldsymbol{x},t) \quad (13.37)$$

至此，构建了多孔介质热-水-力耦合问题的近场动力学模型，待求未知函数包括温度、液体压力和位移，三者相互影响。

13.3.3 水泥砂浆试件的冻融循环分析

考虑一尺寸 40mm×40mm×160mm 的长方体均质水泥砂浆试件，受温降作用，根据问题的对称性，可取 1/4 区域进行分析，以 O 为坐标原点选取坐标系，考察点 A 的坐标为 (20,20,80)，如图 13-27 所示。边界条件为：底面和两个对称面的法向位移为零，表面、底面和两个外侧边界降温，降温范围为 3~20℃，试件的初始温度为 3℃。表 13-7 列出了水泥砂浆的材料参数。

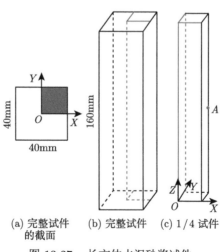

(a) 完整试件的截面　(b) 完整试件　(c) 1/4 试件

图 13-27　长方体水泥砂浆试件

13.3 混凝土冻融循环热-水-力耦合问题的近场动力学模型

图 13-28 为文献 [26] 测定的试件孔隙率及孔隙分布，并给出相应的拟合曲线及对应公式。

表 13-7 水泥砂浆的材料参数

参数	砂浆	水	冰
密度 $\rho/(\mathrm{kg/m^3})$	2140	1000	961
弹性模量 E/GPa	10.1	—	—
体积模量 E/GPa	5.61	—	—
压缩模量 E/GPa	—	2	8
泊松比 ν	1/3	—	—
比热容 $c_v/(\mathrm{J/(kg\cdot K)})$	840	4220	2110
热导系数 $k_T/(\mathrm{J/(s\cdot m\cdot K)})$	0.93	0.55	2.2
热膨胀系数 α/K^{-1}	9.87×10^{-5}	—	5.167×10^{-5}
表面张力系数 $\lambda/(\mathrm{N/m})$		3.9×10^{-2}	
相变潜热 $L/(\mathrm{J/kg})$		3.335×10^{5}	
渗透系数 D/m^2		8.343×10^{-21}	
水的动力黏滞系数 $\eta/(\mathrm{Pa\cdot s})$		$1.38\times 10^{-7}\times e^{\frac{2590}{T+273.15}}$	

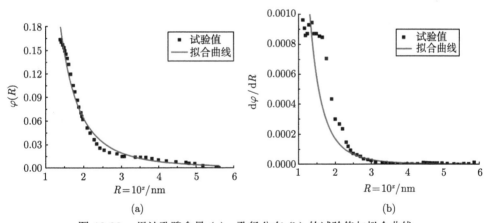

图 13-28 累计孔隙含量 (a)、孔径分布 (b) 的试验值与拟合曲线

根据图 13-28 中的试验结果，可以得到累计孔隙含量与孔径分布的拟合公式，分别为 $\varphi = \dfrac{0.1861}{1+\left(\dfrac{\lg R}{2.0783}\right)^{3.8059}}$ 和 $\dfrac{\mathrm{d}\varphi}{\mathrm{d}R} = A_2 + \dfrac{A_1}{1+\left(\dfrac{R}{R_0}\right)^B}$，其中，参数 $A_1 = 9.366\times 10^{-4}$，$A_2 = 4.9386\times 10^{-6}$，$B = 2.27832$，$R_0 = 95.80985$。

在近场动力学计算中，将试件均匀离散为 135876 个物质点，物质点间距为

$\Delta x=0.8\text{mm}$,近场范围为 $\delta = 3\Delta x$,温度场计算时间步长为 $\Delta t_T = 0.1\text{s}$,渗流场计算时间步长为 $\Delta t_p = 10^{-4}\text{s}$,位移场计算时间步长为 $\Delta t_u = 10^{-7}\text{s}$。以 12℃/h 的降温速率从表面、底面和两个外侧边界面开始降温。计算得到 A 点在 z 方向的正应变,并与试验数据进行对比,见表 13-8,计算结果与试验结果吻合良好。

表 13-8 试验测定的不同温度下试件的应变差

温度/℃	z 向正应变	
	试验值	PD 模拟值
−5	1.98×10^{-4}	1.88×10^{-4}
−10	3.09×10^{-4}	2.53×10^{-4}
−15	3.48×10^{-4}	3.10×10^{-4}

依旧以 12℃/h 的降温及升温速率进行数值计算,正负温度峰值分别为 3℃ 和 −20℃,在降温或升温到峰值时,即开始升温或者降温。图 13-29 为计算得到的 6 个降升温循环过程中 A 点的 z 方向位移随时间变化图。结果表明,A 点 z 方向位移主要向下,这是限制了底面 z 向位移和外界温度变化始终低于初始温度所致,随着冻融循环的不断进行,A 点的 z 向位移有向上的趋势,符合物理实际。

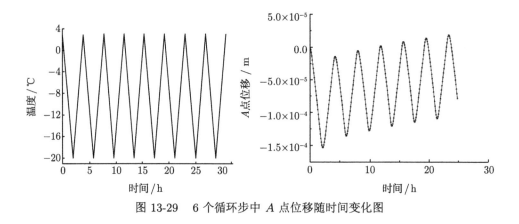

图 13-29 6 个循环步中 A 点位移随时间变化图

图 13-30 为试件在 2500s、4000s 和 5500s 时的内部温度分布图,从图中所示结果可以看出,在此三个时刻,模拟得到试件中心温度分别为 −4.943℃、−9.985℃ 和 −15.011℃,试件表面温度分别为 −8.333℃、−13.333℃ 和 −18.333℃,试件中心温度与环境温度大致相差 3.35℃,试件表面温度稍低于试件内部温度。这是由于试件尺寸较小,且降温速率较慢,试件内部温度有足够时间下降到与外表温度相近值,也表明所建立的热−水−力耦合问题的近场动力学模型和数值方法的有

效性。

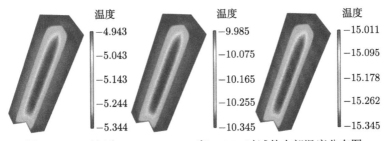

图 13-30 时间为 2500s、4000s 和 5500s 时试件内部温度分布图

13.4 混凝土重力坝的侵彻与爆炸毁伤分析

混凝土坝具有体型简单和适应各种地质条件等特点，是挡水建筑物的主要坝型，大坝的安全防护已成为工程界长期关注的重点问题。混凝土坝在服役期内除受到自重、水压等常规荷载作用外，还可能遭受爆炸冲击等极端外荷载作用。大坝一旦发生事故，不仅会对工程自身造成影响，更可能会引发连锁反应导致严重的次生灾害，对国家安全、社会经济稳定和人民日常生活带来巨大影响，因此，开展爆炸冲击荷载下的混凝土坝毁伤破坏研究具有重要的意义。

13.4.1 混凝土重力坝的侵彻毁伤分析

研究如图 13-31 所示的典型混凝土重力坝受到钻地弹作用的侵彻毁伤问题。坝高 85m，坝头宽 15m，上游坝面垂直，下游坝面坡度比为 1∶0.60，水位高度为 83m。考虑 GBU-28 制导炸弹垂直撞击大坝顶面的中心位置，球形弹头的直径 0.37m，长度 5.84m，弹壳厚度 0.01m，质量密度 7896kg/m³、质量 2130kg；高能炸药装药直径 0.35m、高度 1.71m、质量密度 1860kg/m³、装药质量 306kg[27]；刚性弹丸垂直撞击的速度分别为 400m/s、200m/s、100m/s。混凝土坝体和岩体坝基均假设为均质弹性材料，坝体混凝土的弹性模量、泊松比和质量密度分别为 $E=26\text{GPa}$、$\nu=0.167$ 和 $\rho=2450\text{kg/m}^3$；基岩岩体的弹性模量、泊松比和质量密度分别为 $E=35\text{GPa}$、$\nu=0.230$ 和 $\rho=2700\text{kg/m}^{3[28-30]}$。混凝土和岩石的能量释放率分别取为 $G_F=0.15\text{MN/m}$ 和 $G_F=0.12\text{MN/m}$。

基于近场动力学方法进行计算时，上下游坝基水平方向位移均固定为零，而坝基底部受竖向约束。采用空间均匀正交点阵离散坝体和坝基，物质点尺寸和近场范围半径为 Δx 和 $\delta=3\Delta x$，物质点体积为 $\Delta V=(\Delta x)^2$。坝体受到的静水压力通过边界点的体力密度施加，不计坝体重力作用。

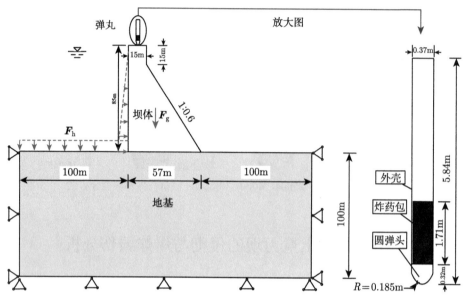

图 13-31　GBU-28 制导导弹撞击混凝土重力坝坝顶面的计算模型

13.4.1.1　拟静力变形分析

分别采用共轭键型近场动力学方法和商用软件 ANSYS 进行静水压力作用下混凝土重力坝的拟静力变形分析[31]。有限元分析采用 28450 四节点平面单元，单元结点总数为 28878。在近场动力学计算中，采用 $\Delta x=0.25\mathrm{m}$ 的均匀正交网格离散计算模型，物质点总数为 455092，拟静力解法的时间步长为 $\Delta t=5\times 10^{-5}\mathrm{s}$、人工阻尼系数为 $C=1\times 10^{5}$，界面键参数取为两物质点材料参数的算数平均值，考虑表面效应修正算法修正径向键微观模量。图 13-32 和图 13-33 分别给出 PD

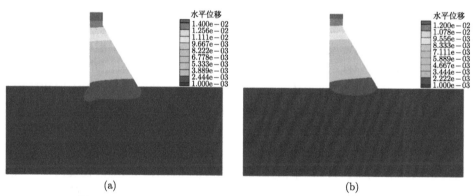

图 13-32　PD(a) 和 FEM(b) 预测的水平位移云图 (单位：m)

13.4 混凝土重力坝的侵彻与爆炸毁伤分析

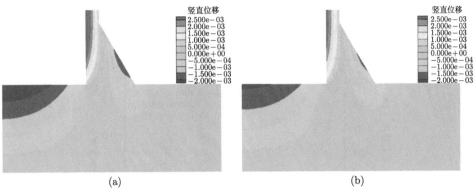

图 13-33 PD(a) 和 FEM(b) 预测的竖直位移云图 (单位：m)

与 FEM 计算的水平位移和竖直位移结果，其中对 PD 解进行了 δ-收敛性分析 (Δx=2m、Δx=1m、Δx=0.5m 和 Δx=0.25m)，均接近有限元解答。两种模型的计算结果整体吻合良好，表明改进的共轭键型近场动力学方法的有效性。

13.4.1.2 冲击侵彻毁伤分析

将静态变形解作为动力分析的初始条件，动力计算的时间步长为 $\Delta t = 1\times 10^{-6}$s。在动力分析开始时，即 $t = 0$s，钻地弹以 400m/s 的速度垂直撞击混凝土重力坝顶面中心位置。图 13-34 显示了弹丸侵彻过程中混凝土重力坝的损伤演化过程，虽然可以在一定程度描述弹丸的侵彻过程，但 Δx=0.25m 的物质点间距相比于弹丸直径 0.37m 过于粗糙，影响刚性体--变形体接触碰撞算法的精度，使得损伤演化的模拟不够精准。

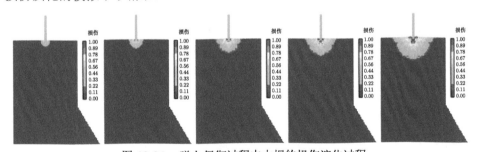

图 13-34 弹丸侵彻过程中大坝的损伤演化过程

为了更准确地研究弹丸侵彻大坝的过程，并兼顾计算效率，下面将仅选取坝体作为计算对象，采用物质点间距 Δx=0.0625m 进行精细离散，相应的物质点总数为 702608。图 13-35 给出了不同初始撞击速度下，基于 PD 方法模拟计算得到的弹丸垂直速度和侵彻深度随时间的变化过程。结果可见，垂直速度的历史曲线基本呈双线性分布特征，在图中竖直方向虚线前为第一减速阶段，虚线后为第二减速阶段,对于各撞击速度,第一阶段减速更为显著,说明这个阶段大坝混凝土具有较强的抗侵彻能力。此外,高速弹丸的侵彻深度明显大于低速弹丸的侵彻深度。

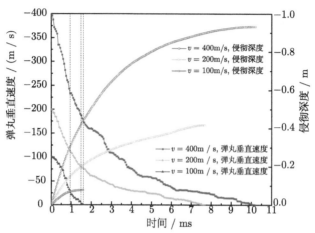

图 13-35　弹丸垂直速度和侵彻深度随时间变化过程

图 13-36～图 13-38 给出了不同撞击速度下大坝在侵彻区附近的损伤破坏演

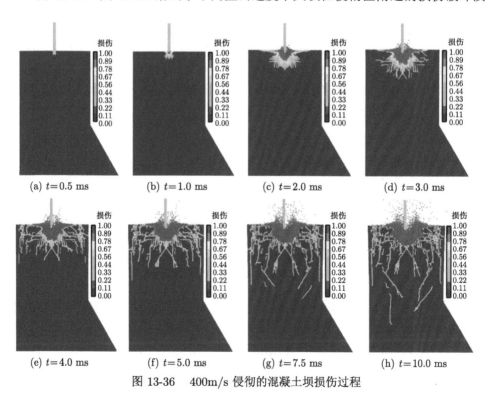

图 13-36　400m/s 侵彻的混凝土坝损伤过程

13.4 混凝土重力坝的侵彻与爆炸毁伤分析

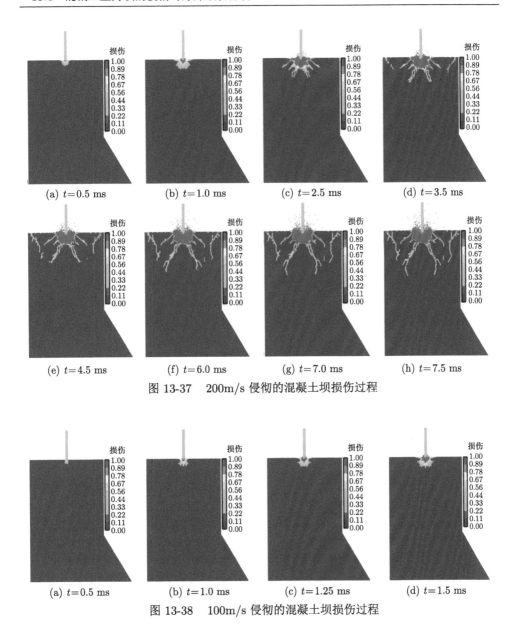

图 13-37 200m/s 侵彻的混凝土坝损伤过程

图 13-38 100m/s 侵彻的混凝土坝损伤过程

化过程。从图中所示结果可以看出，弹丸侵彻后，弹头周围的混凝土材料受到严重挤压，使混凝土破碎，在大坝撞击区域形成弹坑；随着弹丸侵彻的不断加深，损伤不断演化，弹坑逐渐扩大，并有一些混凝土碎块从弹坑中飞溅出去。

13.4.2 空中爆炸致使混凝土重力坝的毁伤分析

研究空中爆炸致使混凝土重力坝的毁伤破坏过程。图 13-39 为中国西南地区某混凝土重力坝的几何结构与空中爆炸示意图，坝高 142m，满库时水深 138m。建立直角坐标系，顺河流方向为 X 轴，自上游指向下游，竖直方向为 Y 轴，竖直向上为正。大坝混凝土的弹性模量 $E=34$GPa，泊松比 $\nu=0.1454$，密度 $\rho=2400$kg/m^3，能量释放率 $G_F=150$N/m，混凝土 JH-2 本构模型的相关参数见表 11-2。TNT 炸药的爆炸位置如图 13-39 所示，距离坝底的高度固定为 123.33m，距离大坝下游侧的水平距离为 L，取 $L=10$m。

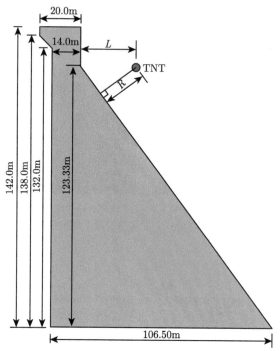

图 13-39　混凝土重力坝的几何结构与空中爆炸示意图

TNT 炸药在坝体的下游侧空中起爆后，爆生冲击波在空气介质中传播，然后作用于混凝土坝的下游面，坝体下游面受到的爆炸超压载荷根据本书的式 (11.12) 计算。基于近场动力学模拟时，采用均匀正交离散，物质点间距为 $\Delta x=0.2$m，物质点总数为 193996 个。

13.4.2.1　毁伤过程分析

质量 $W=2000$kg 的 TNT 炸药爆炸产生的冲击波经过 $t=4$ms 后到达混凝土坝下游侧，图 13-40 为几个典型时刻混凝土坝等效应力分布云图。当爆生冲击波作用到混凝土坝下游坝面时，应力波会迅速从大坝下游侧向坝体内部传播，到达坝体边界面时发生反射，并从压应力波反射成拉应力波，再向坝体内传播，与后续从下游侧传播的压应力波相互叠加，形成新的应力波，如图 13-40(b) 所示。当拉应力波继续传播时，在坝体边界面继续发生反射，形成压应力波，再反射出的压应力波会和之前的应力波继续发生叠加，如图 13-40(c) 所示，如此反复循环。

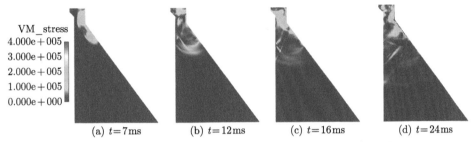

图 13-40　混凝土坝未破坏前几个典型时刻的等效应力云图

经过应力波传播过程中的多次反射叠加，产生图 13-40(d) 所示的应力状态，在混凝土坝下游侧变坡处出现了等效应力值较大的区域，达到了混凝土材料的破坏条件，开始发生损伤破坏，具体破坏过程如图 13-41 所示。坝体的损伤首先出现在下游侧的斜坡和竖直面的交接处，然后开始出现小面积的破坏现象，紧接着破坏区域向四周扩散，形成较大范围的损伤破坏区。

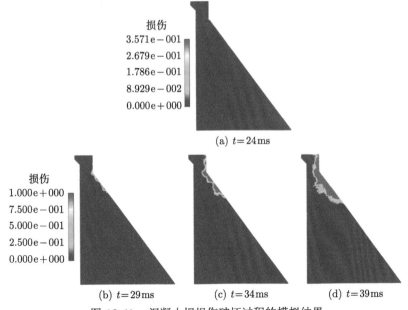

图 13-41　混凝土坝损伤破坏过程的模拟结果

13.4.2.2 炸药起爆位置的影响

下面分析 TNT 炸药起爆位置对混凝土坝毁伤破坏的影响规律。TNT 炸药的质量 $W=2000\mathrm{kg}$，炸药爆炸的位置距离混凝土坝坝底的高度固定为 $123.33\mathrm{m}$，距离大坝下游侧的水平距离为 L，分别取 $L=5.0\mathrm{m}$、$10.0\mathrm{m}$、$15.0\mathrm{m}$ 和 $20.0\mathrm{m}$。不同爆炸距离对应的混凝土坝最终毁伤的模拟结果如图 13-42 所示。结果显示，相同质量 TNT 炸药的爆距越小，混凝土坝的损伤破坏越严重，主要破坏位置位于大坝的下游侧靠近坝顶的区域，而在混凝土坝的下部和底部区域，各工况均未发生损伤破坏现象。

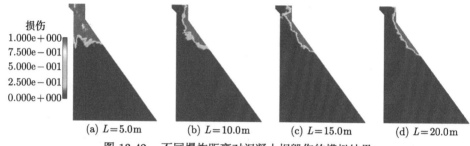

图 13-42 不同爆炸距离对混凝土坝毁伤的模拟结果

值得注意的是，当 TNT 炸药的起爆位置距离混凝土坝下游侧的水平距离为 $L=5.0\mathrm{m}$ 时，坝顶附近上、下游区域均已经完全破坏，如图 13-42(a) 所示。为更清晰分析这种情况下的破坏特征，图 13-43 给出 $L=5.0\mathrm{m}$ 时，三个典型时刻混凝土坝的损伤破坏云图。从中可以看出，在 $L=5.0\mathrm{m}$ 工况下，由于爆距较小，坝体内的应力波传播、反射和叠加更为频繁，坝体当 $t=29\mathrm{ms}$ 时，除坝体下游侧区域发生损伤破坏外，上游侧区域也出现了局部损伤破坏，随着爆炸时间的推移，上游侧损伤破坏区域逐渐增大，并与下游侧的损伤破坏区域相连，致使坝顶部分基本全部损伤破坏。

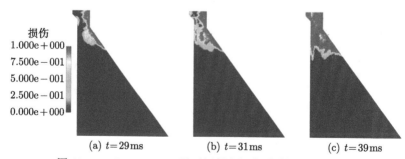

图 13-43 $L=5.0\mathrm{m}$ 工况下混凝土坝典型时刻的损伤云图

13.4.2.3 炸药质量的影响

以下分析 TNT 炸药质量对混凝土坝毁伤破坏的影响规律。炸药的位置距离混凝土坝坝底的高度仍然固定为 123.33m，距离大坝下游侧的水平距离为 $L = 10.0$m，TNT 炸药的质量分别取为 1000kg、2000kg、3000kg 和 4000kg。不同 TNT 炸药质量对混凝土坝最终毁伤的模拟结果如图 13-44 所示，结果表明，当 TNT 炸药的爆距固定时，TNT 炸药的质量越大，混凝土坝的损伤破坏越严重，且破坏位置仍然是靠近爆炸中心的下游侧和坝顶区域，而大坝下部和底部区域在四种炸药质量工况下均未发生损伤破坏。

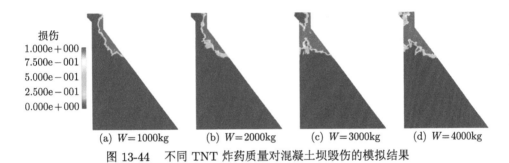

图 13-44　不同 TNT 炸药质量对混凝土坝毁伤的模拟结果

13.5　向家坝水电站泄洪坝段的抗滑稳定分析

13.5.1　工程概况

向家坝水电站是金沙江梯级开发战略中最后一级水电站，是西电东送计划的重要电源点。向家坝水电站是一等大 (1) 型工程，由通航建筑物、挡水建筑物、位于右岸的发电系统和位于河床中部的泄洪排沙建筑物等组成。拦河坝是混凝土重力坝，为 I 级建筑物，最大坝高为 162m，坝顶长度为 909.26m，坝体混凝土量约为 896 万立方米。

向家坝水电站基岩以泥岩、砂岩和含煤地层为主，岩性和岩相的变化均较大，层理交错发育，岩体质量主要为 II~III 类岩体，局部存在 III$_2$ 和 IV 类岩体。坝基地质条件对大坝的抗滑稳定性产生很大的不利影响，主要包括以下三个方面：一是受膝状挠曲构造的影响，基岩层产状明显倾向于下游，仅有少数坝段的坝基岩层产状较平缓；二是坝基岩体内顺岩层方向存在多条软弱夹层；三是坝基内部和坝后存在较大规模的陡倾破碎带。

13.5.2　泄④坝段近场动力学-有限元混合模型

选取向家坝水电站泄④坝段剖面图作为研究对象，采用本书介绍的近场动力学-有限元混合模型进行该坝段的抗滑稳定分析。如图 13-45 所示，泄④坝段坝基岩体质量主要为 II、III$_1$ 和 III$_2$ 类，并含有大量软弱夹层。具体建模时，考虑了不同类别的岩体和倾向下游的一级软弱结构面 T_3^{2-3} 和 T_3^{2-5} 等主要软弱夹层的影响，将坝体和近坝区域的地基采用 PD 离散，其他区域采用有限元离散，如图 13-46 所示。采用的材料参数为：坝体弹性模量为 20GPa，密度为 2400kg/m^3，抗拉强度为 1.57MPa；II 类岩石的弹性模量为 20GPa，密度为 2690kg/m^3，抗拉强度为 3.3MPa；III$_1$ 类岩石的弹性模量为 13GPa，密度为 2690kg/m^3，抗拉强度为 2.7MPa；III$_2$ 类岩石的弹性模量为 10GPa，密度为 2690kg/m^3，抗拉强度为 2.0MPa；软弱夹层的弹性模量为 0.72GPa，密度为 2300kg/m^3。大坝基岩底部采用固定约束，基岩两侧采用法向链杆约束。

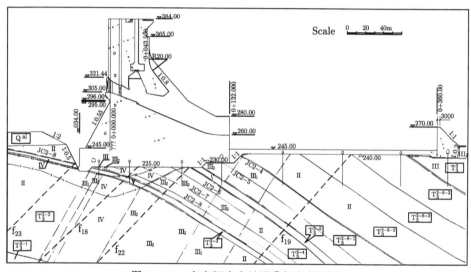

图 13-45 向家坝水电站泄④坝段剖面图

13.5.3 计算结果与分析

由于在正常蓄水位情况下，大坝处于安全状态，为了评价大坝的承载能力，学术界和工程界通常采用超载法进行分析，即定义超载系数 K 为计算库水容重和实际库水容重的比值，将放大的计算库水容重作为外荷载进行计算分析。图 13-47 为基于 PD-FEM 混合模型计算得到的向家坝水电站泄④坝段的超载破坏过程。

由计算结果可见，当超载系数 K 达到 1.9 时，坝踵出现损伤，但不影响坝体正常工作。当超载系数 K 达到 2.3 时，坝踵裂纹开始扩展；当超载系数 K 达到 2.5 时，坝踵裂纹扩展至软弱结构面 T_3^{2-3}；当超载系数 K 达到 3.3 时，坝踵裂纹沿软弱结构面 T_3^{2-3} 扩展，坝趾处出现损伤；当超载系数 K 不断增加，坝趾处损伤区继续扩大,坝踵裂纹不断扩展出现损伤;当超载系数 K 达到 4.4 时,计算结果不收敛,表明坝体-地基体系出现较大的失稳破坏区,超载安全系数可以认为是 4.4。

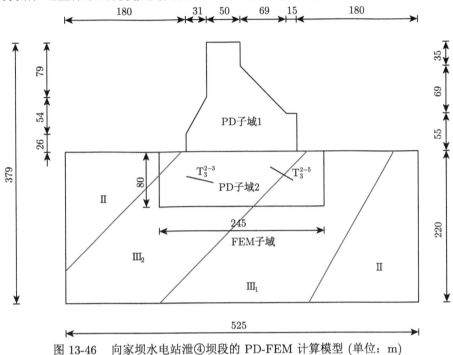

图 13-46　向家坝水电站泄④坝段的 PD-FEM 计算模型 (单位: m)

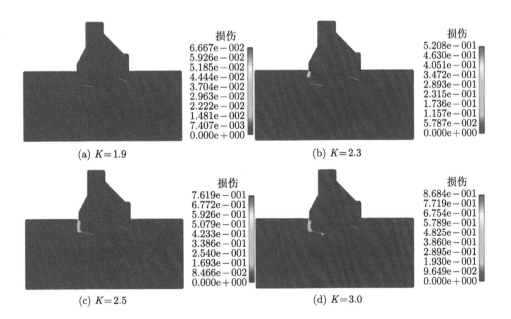

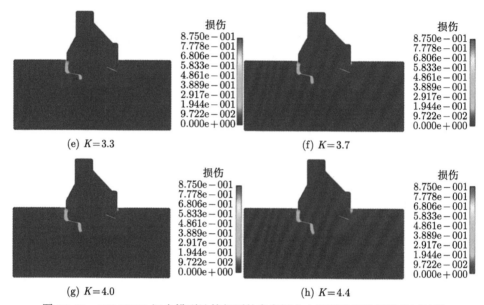

图 13-47 PD-FEM 混合模型计算得到的向家坝水电站泄④坝段超载破坏过程

参 考 文 献

[1] Zhang H, Xu Y, Gan Y, et al. Microstructure informed micromechanical modelling of hydrated cement paste: Techniques and challenges[J]. Construction and Building Materials, 2020, 251:118983.

[2] Bentz D P. Three-dimensional computer simulation of Portland cement hydration and microstructure development[J]. Journal of the American Ceramic Society, 1997, 80(1):3-21.

[3] Bentz D P. Modelling cement microstructure: pixels, particles, and property prediction[J]. Materials and structures, 1999, 32(217):187-195.

[4] Bentz D P. CEMHYD3D: A three-dimensional cement hydration and microstructure development modelling package. Version 2.0[R]. US Department of Commerce, National Institute of Standards and Technology, 2000.

[5] 黄宝华. 水泥石微观结构力学性能模拟 [M]. 武汉：武汉理工大学，2013.

[6] Bernard F, Kamali-Bernard S, Prince W. 3D multi-scale modelling of mechanical behaviour of sound and leached mortar[J]. Cement and Concrete Research, 2008, 38(4):449-458.

[7] Rhardane A, Grondin F, Alam S Y. Development of a micro-mechanical model for the determination of damage properties of cement pastes[J]. Construction and Building Materials, 2020, 261:120514.

[8] Mazaheripour H, Faria R, Ye G, et al. Microstructure-based prediction of the elastic behaviour of hydrating cement pastes[J]. Applied Sciences, 2018, 8(3):442.

[9] Gu X, Li X, Xia X Z, et al. A robust peridynamic computational framework for predicting mechanical properties of porous quasi-brittle material[J]. Composite Structures, 2023, 303:116245.

[10] Němeček J, Králík V, Šmilauer V, et al. Tensile strength of hydrated cement paste phases assessed by micro-bending tests and nanoindentation[J]. Cement and Concrete Composites, 2016, 73:164-173.

[11] Hou DS, Zhang W, Wang P, et al. Microscale peridynamic simulation of damage process of hydrated cement paste subjected to tension[J]. Construction and Building Materials, 2019, 228:117053.

[12] Li X, Gu X, Xia X Z, et al. Effect of water-cement ratio and size on tensile damage in hardened cement paste: Insight from peridynamic simulation[J]. Construction and Building Materials, 2022, 356:129256.

[13] 沈峰. 混凝土材料和结构损伤破坏的近场动力学模拟 [D]. 南京：河海大学，2014.

[14] 夏晓舟. 混凝土细观数值仿真及宏细观力学研究 [D]. 南京：河海大学，2007.

[15] 郭士强. 自适应近场动力学方法及其在混凝土宏细观破坏分析中的应用 [D]. 南京：河海大学，2019.

[16] 宋晓刚，杨智春. 混凝土多边形骨料投放的随机游走算法 [J]. 应用力学学报，2009, 26(4): 808-811.

[17] 李天一，章青，夏晓舟，等. 考虑混凝土材料非均质特性的近场动力学模型 [J]. 应用数学和力学，2018, 39(8):913-924.

[18] 马怀发，陈厚群，黎保琨. 混凝土试件细观结构的数值模拟 [J]. 水利学报，2004, 35(10): 27-35.

[19] 闫东明. 混凝土动态力学性能试验与理论研究 [D]. 大连：大连理工大学，2006.

[20] 吴锋. 混凝土单轴抗拉应力—应变全曲线的试验研究 [D]. 长沙：湖南大学，2006.

[21] 吴纹达. 基于数字图像处理的混凝土非线性多尺度有限元分析 [D]. 南京：河海大学，2015.

[22] 张廼龙. 基于细观结构特征的混凝土破坏过程试验研究和数值模拟 [D]. 南京：东南大学，2013.

[23] Bažant Z P, Chern J C, Rosenberg A M, et al. Mathematical model for freeze-thaw durability of concrete[J]. Journal of the American Ceramic Society, 1988, 71(9):776-783.

[24] Penttala V. Freezing-induced strains and pressures in wet porous materials and especially in concrete mortar[J]. Advanced Cement Based Material, 1998, 7: 8-19.

[25] Zuber B, Marchand J. Predicting the volume instability of hydrated cement systems upon freezing using poro-mechanics and local phase equilibria[J]. Materials and structures, 2004, 37(4):257-270.

[26] 曾强，李克非. 冻融情况下降温速率对水泥基材料变形和损伤的影响 [J]. 清华大学学报：自然科学版，2008, 48(9):1390-1394.

[27] Zhang S R, Wang G H. Study on the antiknock performance and measures of concrete gravity dam[J]. Shuili Xuebao(Journal of Hydraulic Engineering), 2012, 43(10):1202-1213.

[28] Yu T T. Dynamical response simulation of concrete dam subjected to underwater contact explosion load[C]. 2009 WRI world congress on computer science and information engineering, IEEE, 2009, 1:769-774.

[29] Chen J, Liu X, Xu Q. Numerical simulation analysis of damage mode of concrete gravity dam under close-in explosion[J]. KSCE Journal of Civil Engineering, 2017, 21(1):397-407.

[30] Yang D, Dong W, Liu X F, et al. Investigation on mode-I crack propagation in concrete using bond-based peridynamics with a new damage model[J]. Engineering Fracture Mechanics, 2018, 199:567-581.

[31] Gu X, Zhang Q. A modified conjugated bond-based peridynamic analysis for impact failure of concrete gravity dam[J]. Meccanica, 2020, 55(3):547-566.

附录 A 近场动力学微分算子

2016 年，美国亚利桑那大学 Madenci 教授提出了近场动力学微分算子 (peridynamic differential operator, PDDO) 的概念，并由此开展了系列研究，包括：散点数据的数值微分计算、各类型常 (偏) 微分方程求解、积分方程求解、图像压缩与复原等 [1,2]。2019 年，出版了专著 *Peridynamic differential operator for numerical analysis* [3]，该书的中译本也于 2021 年由科学出版社出版 [4]。采用 PD 微分算子，可由近场动力学运动方程推导重现经典连续介质力学运动方程，也可由连续介质力学运动方程出发推导得出非常规态型、常规态型和键型近场动力学运动方程 [1,5,6]。

近场动力学微分算子的本质功能是：在物质点的局部空间域或时间域上，定义该物质点任意阶局部导数的非局部积分形式表达，采用一种统一的方式将任意常 (偏) 微分方程转化为对应积分方程，从而在数学概念上避免了不连续处导数不存在或不唯一的奇异性问题。值得指出的是，在近场动力学微分算子中，各个物质点的近场半径不再要求相同，近场范围形状也不再限于圆形或球体，可为任意形状，适用于非均匀离散和自适应分析；各个物质点的影响函数可以不同，以表征不同的非局部作用程度；此外，近场动力学微分算子能够考虑时间进程的非对称性和空间离散的非均匀性，自然描述其中的非局部特性。本附录将系统给出任意形状近场范围和完整对称近场范围下，近场动力学微分算子的表达式，为将任意阶局部微分转化为非局部积分以及近场动力学的理论建模奠定基础。

A.1 多维空间任意形状近场范围的任意阶近场动力学微分算子

M 维空间中，对含有多个变量的标量函数 $f(\boldsymbol{x}') = f(\boldsymbol{x}+\boldsymbol{\xi})$ 进行泰勒级数展开，引入并构造近场动力学函数 (peridynamic function) $g_N^{p_1 p_2 \cdots p_M}$，利用近场动力学函数的正交性质可以获得任意阶偏导数的非局部积分表达，构造得到近场动力学微分算子。本节介绍 M 维空间中，物质点 $\boldsymbol{x}$ 在其近场范围内任意位置时，任意阶近场动力学微分算子的一般数学表述。

近场动力学微分算子有两种表达形式，一种是将函数 $f(\boldsymbol{x})$ 的任意阶局部导数用 $f(\boldsymbol{x}+\boldsymbol{\xi})$ 的非局部积分形式表示，称为 PD 微分算子的绝对表达式；另一种是将函数 $f(\boldsymbol{x})$ 的零阶以上任意阶局部导数用 $f(\boldsymbol{x}+\boldsymbol{\xi}) - f(\boldsymbol{x})$ 的非局部积分形式表示，称为 PD 微分算子的相对表达式。

A.1.1 近场动力学微分算子的绝对表达式

M 维空间中，函数 $f(\boldsymbol{x}') = f(\boldsymbol{x}+\boldsymbol{\xi})$ 的 N 阶泰勒级数展开式为

$$f(\boldsymbol{x}+\boldsymbol{\xi}) = \sum_{n_1=0}^{N}\sum_{n_2=0}^{N-n_1}\cdots\sum_{n_M=0}^{N-n_1\cdots-n_{M-1}} \frac{1}{n_1!n_2!\cdots n_M!}\xi_1^{n_1}\xi_2^{n_2}\cdots\xi_M^{n_M}$$
$$\cdot \frac{\partial^{n_1+n_2+\cdots+n_M}f(\boldsymbol{x})}{\partial x_1^{n_1}\partial x_2^{n_2}\cdots\partial x_M^{n_M}} + R(N,\boldsymbol{x}) \tag{A.1}$$

其中，$\boldsymbol{\xi}=\boldsymbol{x}'-\boldsymbol{x}$ 为物质点间的相对位置；n_i 为 $f(\boldsymbol{x})$ 对变量 x_i 的微分阶数，其中 $i=1,\cdots,M$，且有 $n_i=0,\cdots,N$，此处 n_i 可同时取为零；$R(N,\boldsymbol{x})$ 为残余项，假设残余量足够小，则可忽略其对函数 $f(\boldsymbol{x}+\boldsymbol{\xi})$ 的贡献。

引入并构造如下形式的近场动力学函数 $g_N^{p_1p_2\cdots p_M}(\boldsymbol{\xi})$

$$g_N^{p_1p_2\cdots p_M}(\boldsymbol{\xi}) = \sum_{q_1=0}^{N}\sum_{q_2=0}^{N-q_1}\cdots\sum_{q_M=0}^{N-q_1\cdots-q_{M-1}} a_{q_1q_2\cdots q_M}^{p_1p_2\cdots p_M} w_{q_1q_2\cdots q_M}(|\boldsymbol{\xi}|) \xi_1^{q_1}\xi_2^{q_2}\cdots\xi_M^{q_M} \tag{A.2}$$

其中，p_i 表示对变量 x_i ($i=1,\cdots,M$) 的微分阶数；$p_i,q_i=0,\cdots,N$；$w_{q_1q_2\cdots q_M}(|\boldsymbol{\xi}|)$ 是影响函数，其与 PD 函数的多项式中的每一项 $\xi_1^{q_1}\xi_2^{q_2}\cdots\xi_M^{q_M}$ 相关联，每一项可有不同形式的影响函数，表示泰勒级数展开中的每一项可以有相同或不同的影响程度，但为简化分析，常采用相同形式的影响函数。

近场动力学函数具有正交性质，即为

$$\int_{H_\infty} \xi_1^{n_1}\xi_2^{n_2}\cdots\xi_M^{n_M}\, g_N^{p_1p_2\cdots p_M}(\boldsymbol{\xi})\mathrm{d}V_{\boldsymbol{x}'} = n_1!n_2!\cdots n_M!\, \delta_{n_1p_1}\delta_{n_2p_2}\cdots\delta_{n_{M-1}p_{M-1}}\delta_{n_Mp_M} \tag{A.3}$$

其中，$\delta_{n_ip_i}$ 为克罗内克符号。将函数 $f(\boldsymbol{x}+\boldsymbol{\xi})$ 的泰勒展开式乘以 PD 函数，并施加该正交性质，则函数 $f(\boldsymbol{x})$ 任意阶偏导数的 PD 微分算子绝对表达式为

$$\frac{\partial^{p_1+p_2+\cdots+p_M}f(\boldsymbol{x})}{\partial x_1^{p_1}\partial x_2^{p_2}\cdots\partial x_M^{p_M}} = \int_{H_\infty} f(\boldsymbol{x}+\boldsymbol{\xi})\, g_N^{p_1p_2\cdots p_M}(\boldsymbol{\xi})\mathrm{d}V_{\boldsymbol{x}'} \tag{A.4}$$

至此，只需要知晓每点的 PD 函数 $g_N^{p_1p_2\cdots p_M}(\boldsymbol{\xi})$，也即只需要知晓未知系数 $a_{q_1q_2\cdots q_M}^{p_1p_2\cdots p_M}$，则可由上述非局部积分形式确定函数 $f(\boldsymbol{x})$ 的任意阶偏导数。由 PD 函数的正交性质，可知系数 $a_{q_1q_2\cdots q_M}^{p_1p_2\cdots p_M}$ 是下式的解答

$$\sum_{q_1=0}^{N}\sum_{q_2=0}^{N-q_1}\cdots\sum_{q_M=0}^{N-q_1\cdots-q_{M-1}} A_{(n_1n_2\cdots n_M)(q_1q_2\cdots q_M)} a_{q_1q_2\cdots q_M}^{p_1p_2\cdots p_M} = b_{n_1n_2\cdots n_M}^{p_1p_2\cdots p_M} \Leftrightarrow \boldsymbol{Aa}=\boldsymbol{b} \tag{A.5}$$

其中，$\boldsymbol{A}$ 为形状矩阵，$\boldsymbol{a}$ 为 PD 函数未知系数矩阵，$\boldsymbol{b}$ 为已知矩阵，且有

$$A_{(n_1n_2\cdots n_M)(q_1q_2\cdots q_M)} = \int_{H_{\boldsymbol{x}}} w_{q_1q_2\cdots q_M}(|\boldsymbol{\xi}|)\xi_1^{n_1+q_1}\xi_2^{n_2+q_2}\cdots\xi_M^{n_M+q_M} \mathrm{d}V_{\boldsymbol{x}'} \tag{A.6}$$

$$b_{n_1n_2\cdots n_M}^{p_1p_2\cdots p_M} = n_1!n_2!\cdots n_M! \, \delta_{n_1p_1}\delta_{n_2p_2}\cdots\delta_{n_Mp_M} \tag{A.7}$$

需要指出的是，当近场范围 $H_{\boldsymbol{x}}$ 尺寸减小或 $g_N^{p_1p_2\cdots p_M}(\boldsymbol{\xi})$ 函数的项数增加时，非局部程度随着近场范围尺寸减小或泰勒级数项数的增多而减小，近场动力学微分算子能够重现局部微分值。此外，为求得 N 阶局部偏导数的 PD 微分算子绝对表达式，需要保证泰勒级数展开的阶数大于等于 N。

A.1.2　近场动力学微分算子的相对表达式

若函数 $f(\boldsymbol{x})$ 已知，则不必采用近场动力学微分算子表示零阶偏导数，即函数自身，仅需关注 $f(\boldsymbol{x})$ 的一阶及一阶以上偏导数的计算。由于物质点之间的相互作用是近场动力学建模的核心，以函数相对差值 $f(\boldsymbol{x}+\boldsymbol{\xi}) - f(\boldsymbol{x})$ 的非局部积分形式表示的近场动力学微分算子更符合建模预期。因此，可以构造与 $f(\boldsymbol{x}+\boldsymbol{\xi}) - f(\boldsymbol{x})$ 相匹配的 PD 函数 $g_N^{p_1p_2\cdots p_M}(\boldsymbol{\xi})$，此时零阶导数与二阶导数的耦合消失，则 n_i, p_i 和 q_i 系列的取值不同时为 0，也即近场动力学函数中不含有常数项，具体推导过程如下。

将式 (A.1) 中隐含的 $f(\boldsymbol{x})$ 移至等式左边，函数 $f(\boldsymbol{x}') = f(\boldsymbol{x}+\boldsymbol{\xi})$ 的 N 阶泰勒级数展开式为

$$f(\boldsymbol{x}+\boldsymbol{\xi}) - f(\boldsymbol{x}) = \sum_{n_1=0}^{N}\sum_{n_2=0}^{N-n_1}\cdots\sum_{n_M=0}^{N-n_1\cdots-n_{M-1}} \frac{1}{n_1!n_2!\cdots n_M!}$$
$$\cdot \xi_1^{n_1}\xi_2^{n_2}\cdots\xi_M^{n_M} \frac{\partial^{n_1+n_2+\cdots+n_M} f(\boldsymbol{x})}{\partial x_1^{n_1}\partial x_2^{n_2}\cdots\partial x_M^{n_M}} + R(N,\boldsymbol{x}) \tag{A.8}$$

注意，上式与式 (A.1) 不同的是需满足条件 $n_1+n_2+\cdots+n_M \neq 0$。

引入并构造如下形式的近场动力学函数 $g_N^{p_1p_2\cdots p_M}(\boldsymbol{\xi})$

$$g_N^{p_1p_2\cdots p_M}(\boldsymbol{\xi}) = \sum_{q_1=0}^{N}\sum_{q_2=0}^{N-q_1}\cdots\sum_{q_M=0}^{N-q_1\cdots-q_{M-1}} a_{q_1q_2\cdots q_M}^{p_1p_2\cdots p_M} w_{q_1q_2\cdots q_M}(|\boldsymbol{\xi}|)\xi_1^{q_1}\xi_2^{q_2}\cdots\xi_M^{q_M}$$
$$\tag{A.9}$$

其中，$w_{q_1q_2\cdots q_M}(|\boldsymbol{\xi}|)$ 是影响函数，p_i 表示对变量 x_i $(i=1,\cdots,M)$ 的微分阶数；$p_i, q_i = 0,\cdots,N$，且 $p_1+p_2+\cdots+p_M \neq 0$，$q_1+q_2+\cdots+q_M \neq 0$。

近场动力学函数具有正交性质，即为

$$\int_{H_\infty} \xi_1^{n_1}\xi_2^{n_2}\cdots\xi_M^{n_M}\, g_N^{p_1p_2\cdots p_M}(\boldsymbol{\xi})\mathrm{d}V_{\boldsymbol{x}'} = n_1!n_2!\cdots n_M!\, \delta_{n_1p_1}\delta_{n_2p_2}\cdots\delta_{n_{M-1}p_{M-1}}\delta_{n_Mp_M} \tag{A.10}$$

其中，$\delta_{n_ip_i}$ 为克罗内克符号。将函数 $f(\boldsymbol{x}+\boldsymbol{\xi})$ 的泰勒展开式乘以 PD 函数，并施加该正交性质，则函数 $f(\boldsymbol{x})$ 任意高于零阶偏导数的 PD 微分算子相对表达式为

$$\frac{\partial^{p_1+p_2+\cdots+p_M} f(\boldsymbol{x})}{\partial x_1^{p_1}\partial x_2^{p_2}\cdots\partial x_M^{p_M}} = \int_{H_\infty} (f(\boldsymbol{x}+\boldsymbol{\xi})-f(\boldsymbol{x}))\, g_N^{p_1p_2\cdots p_M}(\boldsymbol{\xi})\mathrm{d}V_{\boldsymbol{x}'} \tag{A.11}$$

至此，只需要知晓每点的 PD 函数 $g_N^{p_1p_2\cdots p_M}(\boldsymbol{\xi})$，也即知晓未知系数 $a_{q_1q_2\cdots q_M}^{p_1p_2\cdots p_M}$，就可以确定函数 $f(\boldsymbol{x})$ 的任意阶偏导数。根据 PD 函数的正交性质，可知系数 $a_{q_1q_2\cdots q_M}^{p_1p_2\cdots p_M}$ 是下式的解答

$$\boldsymbol{Aa} = \boldsymbol{b} \Rightarrow \sum_{q_1=0}^{N}\sum_{q_2=0}^{N-q_1}\cdots\sum_{q_M=0}^{N-q_1\cdots-q_{M-1}} A_{(n_1n_2\cdots n_M)(q_1q_2\cdots q_M)}\, a_{q_1q_2\cdots q_M}^{p_1p_2\cdots p_M} = b_{n_1n_2\cdots n_M}^{p_1p_2\cdots p_M} \tag{A.12}$$

其中

$$A_{(n_1n_2\cdots n_M)(q_1q_2\cdots q_M)} = \int_{H_\infty} w_{q_1q_2\cdots q_M}(|\boldsymbol{\xi}|)\xi_1^{n_1+q_1}\xi_2^{n_2+q_2}\cdots\xi_M^{n_M+q_M}\mathrm{d}V_{\boldsymbol{x}'} \tag{A.13}$$

$$b_{n_1n_2\cdots n_M}^{p_1p_2\cdots p_M} = n_1!n_2!\cdots n_M!\, \delta_{n_1p_1}\delta_{n_2p_2}\cdots\delta_{n_Mp_M} \tag{A.14}$$

A.2　三维球形对称近场范围的二阶近场动力学微分算子

A.2.1　近场动力学微分算子的绝对表达式

三维情况下，函数 $f(\boldsymbol{x}+\boldsymbol{\xi})$ 关于变量 $x_i(i=1,2,3)$ 的二阶泰勒级数展开为

$$\begin{aligned}f(\boldsymbol{x}+\boldsymbol{\xi}) =& f(\boldsymbol{x}) + \xi_1\frac{\partial f(\boldsymbol{x})}{\partial x_1} + \xi_2\frac{\partial f(\boldsymbol{x})}{\partial x_2} + \xi_3\frac{\partial f(\boldsymbol{x})}{\partial x_3} + \frac{1}{2!}\xi_1^2\frac{\partial^2 f(\boldsymbol{x})}{\partial x_1^2} + \frac{1}{2!}\xi_2^2\frac{\partial^2 f(\boldsymbol{x})}{\partial x_2^2} \\ &+ \frac{1}{2!}\xi_3^2\frac{\partial^2 f(\boldsymbol{x})}{\partial x_3^2} + \xi_1\xi_2\frac{\partial^2 f(\boldsymbol{x})}{\partial x_1\partial x_2} + \xi_1\xi_3\frac{\partial^2 f(\boldsymbol{x})}{\partial x_1\partial x_3} + \xi_2\xi_3\frac{\partial^2 f(\boldsymbol{x})}{\partial x_2\partial x_3}\end{aligned} \tag{A.15}$$

此时，构造如下形式的 PD 函数

$$g_2^{p_1p_2p_3}(\boldsymbol{\xi}) = a_{000}^{p_1p_2p_3}\, w_{000}(|\boldsymbol{\xi}|) + a_{100}^{p_1p_2p_3}\, w_{100}(|\boldsymbol{\xi}|)\xi_1$$

A.2 三维球形对称近场范围的二阶近场动力学微分算子

$$+ a_{010}^{p_1p_2p_3} w_{010}(|\boldsymbol{\xi}|)\xi_2 + a_{001}^{p_1p_2p_3} w_{001}(|\boldsymbol{\xi}|)\xi_3$$

$$+ a_{200}^{p_1p_2p_3} w_{200}(|\boldsymbol{\xi}|)\xi_1^2 + a_{020}^{p_1p_2p_3} w_{020}(|\boldsymbol{\xi}|)\xi_2^2 + a_{002}^{p_1p_2p_3} w_{002}(|\boldsymbol{\xi}|)\xi_3^2$$

$$+ a_{110}^{p_1p_2p_3} w_{110}(|\boldsymbol{\xi}|)\xi_1\xi_2 + a_{101}^{p_1p_2p_3} w_{101}(|\boldsymbol{\xi}|)\xi_1\xi_3 + a_{011}^{p_1p_2p_3} w_{011}(|\boldsymbol{\xi}|)\xi_2\xi_3 \tag{A.16}$$

则函数 $f(\boldsymbol{x})$ 各阶局部偏导数的 PD 微分算子绝对表达式为

$$f(\boldsymbol{x}) = \int_{H_{\infty}} f(\boldsymbol{x} + \boldsymbol{\xi}) g_2^{000}(\boldsymbol{\xi}) \mathrm{d}V_{\boldsymbol{x}'} \tag{A.17}$$

$$\left\{ \frac{\partial f(\boldsymbol{x})}{\partial x_1} \quad \frac{\partial f(\boldsymbol{x})}{\partial x_2} \quad \frac{\partial f(\boldsymbol{x})}{\partial x_3} \right\}^{\mathrm{T}} = \int_{H_{\infty}} f(\boldsymbol{x} + \boldsymbol{\xi}) \left\{ \begin{array}{c} g_2^{100}(\boldsymbol{\xi}) \\ g_2^{010}(\boldsymbol{\xi}) \\ g_2^{001}(\boldsymbol{\xi}) \end{array} \right\} \mathrm{d}V_{\boldsymbol{x}'} \tag{A.18}$$

$$\left\{ \frac{\partial^2 f(\boldsymbol{x})}{\partial x_1^2} \quad \frac{\partial^2 f(\boldsymbol{x})}{\partial x_2^2} \quad \frac{\partial^2 f(\boldsymbol{x})}{\partial x_3^2} \quad \frac{\partial^2 f(\boldsymbol{x})}{\partial x_1 \partial x_2} \quad \frac{\partial^2 f(\boldsymbol{x})}{\partial x_1 \partial x_3} \quad \frac{\partial^2 f(\boldsymbol{x})}{\partial x_2 \partial x_3} \right\}^{\mathrm{T}}$$

$$= \int_{H_{\infty}} f(\boldsymbol{x} + \boldsymbol{\xi}) \left\{ \begin{array}{c} g_2^{200}(\boldsymbol{\xi}) \\ g_2^{020}(\boldsymbol{\xi}) \\ g_2^{002}(\boldsymbol{\xi}) \\ g_2^{110}(\boldsymbol{\xi}) \\ g_2^{101}(\boldsymbol{\xi}) \\ g_2^{011}(\boldsymbol{\xi}) \end{array} \right\} \mathrm{d}V_{\boldsymbol{x}'} \tag{A.19}$$

其中，未知系数 $\boldsymbol{a}$ 由下式获得，进而可获知所有的 PD 函数。

$$\boldsymbol{A}\boldsymbol{a} = \boldsymbol{b} \Rightarrow \begin{bmatrix} \boldsymbol{A}_{00} & \boldsymbol{A}_{01} & \boldsymbol{A}_{02} \\ \boldsymbol{A}_{10} & \boldsymbol{A}_{11} & \boldsymbol{A}_{12} \\ \boldsymbol{A}_{20} & \boldsymbol{A}_{21} & \boldsymbol{A}_{22} \end{bmatrix} \begin{bmatrix} \hat{\boldsymbol{a}}_{00} & \hat{\boldsymbol{a}}_{01} & \hat{\boldsymbol{a}}_{02} \\ \hat{\boldsymbol{a}}_{10} & \hat{\boldsymbol{a}}_{11} & \hat{\boldsymbol{a}}_{12} \\ \hat{\boldsymbol{a}}_{20} & \hat{\boldsymbol{a}}_{21} & \hat{\boldsymbol{a}}_{22} \end{bmatrix} = \begin{bmatrix} \hat{\boldsymbol{b}}_{00} & \boldsymbol{0} & \boldsymbol{0} \\ \boldsymbol{0} & \hat{\boldsymbol{b}}_{11} & \boldsymbol{0} \\ \boldsymbol{0} & \boldsymbol{0} & \hat{\boldsymbol{b}}_{22} \end{bmatrix}$$

$$\tag{A.20}$$

当影响函数取同一形式时，即 $w_{q_1q_2q_3}(|\boldsymbol{\xi}|) = w(|\boldsymbol{\xi}|)$，有

$$\boldsymbol{A} = \int_{H_{\infty}} w(|\boldsymbol{\xi}|)$$

$$\times \begin{bmatrix} 1 & \xi_1 & \xi_2 & \xi_3 & \xi_1^2 & \xi_2^2 & \xi_3^2 & \xi_1\xi_2 & \xi_1\xi_3 & \xi_2\xi_3 \\ \xi_1 & \xi_1^2 & \xi_1\xi_2 & \xi_1\xi_3 & \xi_1^3 & \xi_1\xi_2^2 & \xi_1\xi_3^2 & \xi_1^2\xi_2 & \xi_1^2\xi_3 & \xi_1\xi_2\xi_3 \\ \xi_2 & \xi_1\xi_2 & \xi_2^2 & \xi_2\xi_3 & \xi_1^2\xi_2 & \xi_2^3 & \xi_2\xi_3^2 & \xi_1\xi_2^2 & \xi_1\xi_2\xi_3 & \xi_2^2\xi_3 \\ \xi_3 & \xi_1\xi_3 & \xi_2\xi_3 & \xi_3^2 & \xi_1^2\xi_3 & \xi_3\xi_2^2 & \xi_3^3 & \xi_1\xi_2\xi_3 & \xi_1\xi_3^2 & \xi_2\xi_3^2 \\ \xi_1^2 & \xi_1^3 & \xi_1^2\xi_2 & \xi_1^2\xi_3 & \xi_1^4 & \xi_1^2\xi_2^2 & \xi_1^2\xi_3^2 & \xi_1^3\xi_2 & \xi_1^3\xi_3 & \xi_1^2\xi_2\xi_3 \\ \xi_2^2 & \xi_1\xi_2^2 & \xi_2^3 & \xi_3\xi_2^2 & \xi_1^2\xi_2^2 & \xi_2^4 & \xi_2^2\xi_3^2 & \xi_1\xi_2^3 & \xi_1\xi_2^2\xi_3 & \xi_2^3\xi_3 \\ \xi_3^2 & \xi_1\xi_3^2 & \xi_2\xi_3^2 & \xi_3^3 & \xi_1^2\xi_3^2 & \xi_2^2\xi_3^2 & \xi_3^4 & \xi_1\xi_2\xi_3^2 & \xi_1\xi_3^3 & \xi_2\xi_3^3 \\ \xi_1\xi_2 & \xi_1^2\xi_2 & \xi_1\xi_2^2 & \xi_1\xi_2\xi_3 & \xi_1^3\xi_2 & \xi_1\xi_2^3 & \xi_1\xi_2\xi_3^2 & \xi_1^2\xi_2^2 & \xi_1^2\xi_2\xi_3 & \xi_1\xi_2^2\xi_3 \\ \xi_1\xi_3 & \xi_1^2\xi_3 & \xi_1\xi_2\xi_3 & \xi_1\xi_3^2 & \xi_1^3\xi_3 & \xi_1\xi_2^2\xi_3 & \xi_1\xi_3^3 & \xi_1^2\xi_2\xi_3 & \xi_1^2\xi_3^2 & \xi_1\xi_2\xi_3^2 \\ \xi_2\xi_3 & \xi_1\xi_2\xi_3 & \xi_2^2\xi_3 & \xi_2\xi_3^2 & \xi_1^2\xi_2\xi_3 & \xi_2^3\xi_3 & \xi_2\xi_3^3 & \xi_1\xi_2^2\xi_3 & \xi_1\xi_2\xi_3^2 & \xi_2^2\xi_3^2 \end{bmatrix} dV_{x'}$$

(A.21)

$$\boldsymbol{a} = \begin{bmatrix} a_{000}^{000} & a_{000}^{100} & a_{000}^{010} & a_{000}^{001} & a_{000}^{200} & a_{000}^{020} & a_{000}^{002} & a_{000}^{110} & a_{000}^{101} & a_{000}^{011} \\ a_{100}^{000} & a_{100}^{100} & a_{100}^{010} & a_{100}^{001} & a_{100}^{200} & a_{100}^{020} & a_{100}^{002} & a_{100}^{110} & a_{100}^{101} & a_{100}^{011} \\ a_{010}^{000} & a_{010}^{100} & a_{010}^{010} & a_{010}^{001} & a_{010}^{200} & a_{010}^{020} & a_{010}^{002} & a_{010}^{110} & a_{010}^{101} & a_{010}^{011} \\ a_{001}^{000} & a_{001}^{100} & a_{001}^{010} & a_{001}^{001} & a_{001}^{200} & a_{001}^{020} & a_{001}^{002} & a_{001}^{110} & a_{001}^{101} & a_{001}^{011} \\ a_{200}^{000} & a_{200}^{100} & a_{200}^{010} & a_{200}^{001} & a_{200}^{200} & a_{200}^{020} & a_{200}^{002} & a_{200}^{110} & a_{200}^{101} & a_{200}^{011} \\ a_{020}^{000} & a_{020}^{100} & a_{020}^{010} & a_{020}^{001} & a_{020}^{200} & a_{020}^{020} & a_{020}^{002} & a_{020}^{110} & a_{020}^{101} & a_{020}^{011} \\ a_{002}^{000} & a_{002}^{100} & a_{002}^{010} & a_{002}^{001} & a_{002}^{200} & a_{002}^{020} & a_{002}^{002} & a_{002}^{110} & a_{002}^{101} & a_{002}^{011} \\ a_{110}^{000} & a_{110}^{100} & a_{110}^{010} & a_{110}^{001} & a_{110}^{200} & a_{110}^{020} & a_{110}^{002} & a_{110}^{110} & a_{110}^{101} & a_{110}^{011} \\ a_{101}^{000} & a_{101}^{100} & a_{101}^{010} & a_{101}^{001} & a_{101}^{200} & a_{101}^{020} & a_{101}^{002} & a_{101}^{110} & a_{101}^{101} & a_{101}^{011} \\ a_{011}^{000} & a_{011}^{100} & a_{011}^{010} & a_{011}^{001} & a_{011}^{200} & a_{011}^{020} & a_{011}^{002} & a_{011}^{110} & a_{011}^{101} & a_{011}^{011} \end{bmatrix}$$

(A.22)

$$\boldsymbol{b} = \begin{bmatrix} \hat{\boldsymbol{b}}_{00} & \boldsymbol{0} & \boldsymbol{0} \\ \boldsymbol{0} & \hat{\boldsymbol{b}}_{11} & \boldsymbol{0} \\ \boldsymbol{0} & \boldsymbol{0} & \hat{\boldsymbol{b}}_{22} \end{bmatrix} = \begin{bmatrix} 1 & 0 & 0 & 0 & 0 & 0 & 0 & 0 & 0 & 0 \\ 0 & 1 & 0 & 0 & 0 & 0 & 0 & 0 & 0 & 0 \\ 0 & 0 & 1 & 0 & 0 & 0 & 0 & 0 & 0 & 0 \\ 0 & 0 & 0 & 1 & 0 & 0 & 0 & 0 & 0 & 0 \\ 0 & 0 & 0 & 0 & 2 & 0 & 0 & 0 & 0 & 0 \\ 0 & 0 & 0 & 0 & 0 & 2 & 0 & 0 & 0 & 0 \\ 0 & 0 & 0 & 0 & 0 & 0 & 2 & 0 & 0 & 0 \\ 0 & 0 & 0 & 0 & 0 & 0 & 0 & 1 & 0 & 0 \\ 0 & 0 & 0 & 0 & 0 & 0 & 0 & 0 & 1 & 0 \\ 0 & 0 & 0 & 0 & 0 & 0 & 0 & 0 & 0 & 1 \end{bmatrix}$$

(A.23)

由于三维球形近场范围的对称性，式 (A.20) 中矩阵 $\boldsymbol{A}$ 的一些非对角子矩阵

A.2 三维球形对称近场范围的二阶近场动力学微分算子

项消失，即奇数阶导数项和偶数阶导数项解耦，故式 (A.20) 简化为

$$\begin{bmatrix} \boldsymbol{A}_{00} & \boldsymbol{0} & \boldsymbol{A}_{02} \\ \boldsymbol{0} & \boldsymbol{A}_{11} & \boldsymbol{0} \\ \boldsymbol{A}_{20} & \boldsymbol{0} & \boldsymbol{A}_{22} \end{bmatrix} \begin{bmatrix} \hat{\boldsymbol{a}}_{00} & \boldsymbol{0} & \hat{\boldsymbol{a}}_{02} \\ \boldsymbol{0} & \hat{\boldsymbol{a}}_{11} & \boldsymbol{0} \\ \hat{\boldsymbol{a}}_{20} & \boldsymbol{0} & \hat{\boldsymbol{a}}_{22} \end{bmatrix} = \begin{bmatrix} \hat{\boldsymbol{b}}_{00} & \boldsymbol{0} & \boldsymbol{0} \\ \boldsymbol{0} & \hat{\boldsymbol{b}}_{11} & \boldsymbol{0} \\ \boldsymbol{0} & \boldsymbol{0} & \hat{\boldsymbol{b}}_{22} \end{bmatrix} \tag{A.24}$$

此时，$\boldsymbol{A}$、$\boldsymbol{a}$ 和 $\boldsymbol{b}$ 的各子矩阵如下

$$\boldsymbol{A}_{00} = \int_{H_{\boldsymbol{x}}} w_0(|\boldsymbol{\xi}|) \mathrm{d}V_{\boldsymbol{x}'} = 4\pi V_0 \tag{A.25}$$

$$\boldsymbol{A}_{02} = \boldsymbol{A}_{20}^{\mathrm{T}} = \int_{H_{\boldsymbol{x}}} w_1(|\boldsymbol{\xi}|) \begin{bmatrix} \xi_1^2 & \xi_2^2 & \xi_3^2 & \xi_1\xi_2 & \xi_1\xi_3 & \xi_2\xi_3 \end{bmatrix} \mathrm{d}V_{\boldsymbol{x}'}$$

$$= \frac{4\pi V_1}{3}[1,1,1,0,0,0] \tag{A.26}$$

$$\boldsymbol{A}_{11} = \int_{H_{\boldsymbol{x}}} w_1(|\boldsymbol{\xi}|) \begin{bmatrix} \xi_1^2 & \xi_1\xi_2 & \xi_1\xi_3 \\ \xi_1\xi_2 & \xi_2^2 & \xi_2\xi_3 \\ \xi_1\xi_3 & \xi_2\xi_3 & \xi_3^2 \end{bmatrix} \mathrm{d}V_{\boldsymbol{x}'} = \frac{4\pi V_1}{3} \begin{bmatrix} 1 & 0 & 0 \\ 0 & 1 & 0 \\ 0 & 0 & 1 \end{bmatrix} \tag{A.27}$$

$$\boldsymbol{A}_{22} = \int_{H_{\boldsymbol{x}}} w_2(|\boldsymbol{\xi}|) \begin{bmatrix} \xi_1^4 & \xi_1^2\xi_2^2 & \xi_1^2\xi_3^2 & \xi_1^3\xi_2 & \xi_1^3\xi_3 & \xi_1^2\xi_2\xi_3 \\ \xi_1^2\xi_2^2 & \xi_2^4 & \xi_2^2\xi_3^2 & \xi_1\xi_2^3 & \xi_1\xi_2^2\xi_3 & \xi_2^3\xi_3 \\ \xi_1^2\xi_3^2 & \xi_2^2\xi_3^2 & \xi_3^4 & \xi_1\xi_2\xi_3^2 & \xi_1\xi_3^3 & \xi_2\xi_3^3 \\ \xi_1^3\xi_2 & \xi_1\xi_2^3 & \xi_1\xi_2\xi_3^2 & \xi_1^2\xi_2^2 & \xi_1^2\xi_2\xi_3 & \xi_1\xi_2^2\xi_3 \\ \xi_1^3\xi_3 & \xi_1\xi_2^2\xi_3 & \xi_1\xi_3^3 & \xi_1^2\xi_2\xi_3 & \xi_1^2\xi_3^2 & \xi_1\xi_2\xi_3^2 \\ \xi_1^2\xi_2\xi_3 & \xi_2^3\xi_3 & \xi_2\xi_3^3 & \xi_1\xi_2^2\xi_3 & \xi_1\xi_2\xi_3^2 & \xi_2^2\xi_3^2 \end{bmatrix} \mathrm{d}V_{\boldsymbol{x}'}$$

$$= \frac{4\pi V_2}{15} \begin{bmatrix} 3 & 1 & 1 & 0 & 0 & 0 \\ 1 & 3 & 1 & 0 & 0 & 0 \\ 1 & 1 & 3 & 0 & 0 & 0 \\ 0 & 0 & 0 & 1 & 0 & 0 \\ 0 & 0 & 0 & 0 & 1 & 0 \\ 0 & 0 & 0 & 0 & 0 & 1 \end{bmatrix}$$

$$\tag{A.28}$$

$$\boldsymbol{A}_{11}^{-1} = \frac{3}{4\pi V_1}\begin{bmatrix} 1 & 0 & 0 \\ 0 & 1 & 0 \\ 0 & 0 & 1 \end{bmatrix}, \quad \boldsymbol{A}_{22}^{-1} = \frac{3}{8\pi V_2}\begin{bmatrix} 4 & -1 & -1 & 0 & 0 & 0 \\ -1 & 4 & -1 & 0 & 0 & 0 \\ -1 & -1 & 4 & 0 & 0 & 0 \\ 0 & 0 & 0 & 10 & 0 & 0 \\ 0 & 0 & 0 & 0 & 10 & 0 \\ 0 & 0 & 0 & 0 & 0 & 10 \end{bmatrix} \tag{A.29}$$

$$\hat{\boldsymbol{b}}_{00} = 1, \quad \hat{\boldsymbol{b}}_{11} = \begin{bmatrix} 1 & 0 & 0 \\ 0 & 1 & 0 \\ 0 & 0 & 1 \end{bmatrix}, \quad \hat{\boldsymbol{b}}_{22} = \begin{bmatrix} 2 & 0 & 0 & 0 & 0 & 0 \\ 0 & 2 & 0 & 0 & 0 & 0 \\ 0 & 0 & 2 & 0 & 0 & 0 \\ 0 & 0 & 0 & 1 & 0 & 0 \\ 0 & 0 & 0 & 0 & 1 & 0 \\ 0 & 0 & 0 & 0 & 0 & 1 \end{bmatrix} \tag{A.30}$$

$$\hat{\boldsymbol{a}}_{00} = \left[\boldsymbol{A}_{00} - \boldsymbol{A}_{02}\boldsymbol{A}_{22}^{-1}\boldsymbol{A}_{20}\right]^{-1}\hat{\boldsymbol{b}}_{00} \tag{A.31}$$

$$\hat{\boldsymbol{a}}_{11} = \boldsymbol{A}_{11}^{-1}\hat{\boldsymbol{b}}_{11} \tag{A.32}$$

$$\hat{\boldsymbol{a}}_{20} = -\boldsymbol{A}_{22}^{-1}\boldsymbol{A}_{20}\hat{\boldsymbol{a}}_{00} \tag{A.33}$$

$$\hat{\boldsymbol{a}}_{22} = \left[\boldsymbol{A}_{22} - \boldsymbol{A}_{20}\boldsymbol{A}_{00}^{-1}\boldsymbol{A}_{02}\right]^{-1}\hat{\boldsymbol{b}}_{22} \tag{A.34}$$

$$\hat{\boldsymbol{a}}_{02} = -\boldsymbol{A}_{00}^{-1}\boldsymbol{A}_{02}\hat{\boldsymbol{a}}_{22} \tag{A.35}$$

式中

$$V_n = \int_0^\delta w_n(|\boldsymbol{\xi}|)\,|\boldsymbol{\xi}|^{2n+2}\,\mathrm{d}\,|\boldsymbol{\xi}|, \quad n=1,2 \tag{A.36}$$

当选定以下影响函数时，其值分别为 $V_0 = \delta^3/2$、$V_1 = \delta^5/3$ 和 $V_2 = \delta^7/4$

$$w_n(|\boldsymbol{\xi}|) = \frac{\delta^{n+1}}{|\boldsymbol{\xi}|^{n+1}} \tag{A.37}$$

其中，$n=(a+b)/2$，且 a 和 b 为子矩阵 $\boldsymbol{A}_{ab}$ 的下标。需要指出的是，影响函数形式多样，如 $w_n(|\boldsymbol{\xi}|) = 1$ 等，但以能够反映长程相互作用的空间分布特征为佳。

A.2 三维球形对称近场范围的二阶近场动力学微分算子

从以上各式出发，式 (A.17)～式 (A.19) 表示的各阶偏导数可化为

$$f(\boldsymbol{x}) = \int_{H_{\boldsymbol{x}}} \hat{\boldsymbol{a}}_{00}^{\mathrm{T}} \{1\} w_0(|\boldsymbol{\xi}|) f(\boldsymbol{x}+\boldsymbol{\xi}) \mathrm{d}V_{\boldsymbol{x}'} + \int_{H_{\boldsymbol{x}}} \hat{\boldsymbol{a}}_{20}^{\mathrm{T}} \begin{Bmatrix} \xi_1^2 \\ \xi_2^2 \\ \xi_3^2 \\ \xi_1\xi_2 \\ \xi_1\xi_3 \\ \xi_2\xi_3 \end{Bmatrix} w_0(|\boldsymbol{\xi}|) f(\boldsymbol{x}+\boldsymbol{\xi}) \mathrm{d}V_{\boldsymbol{x}'} \tag{A.38}$$

$$\left\{ \begin{array}{ccc} \dfrac{\partial f(\boldsymbol{x})}{\partial x_1} & \dfrac{\partial f(\boldsymbol{x})}{\partial x_2} & \dfrac{\partial f(\boldsymbol{x})}{\partial x_3} \end{array} \right\}^{\mathrm{T}} = \int_{H_{\boldsymbol{x}}} \hat{\boldsymbol{a}}_{11}^{\mathrm{T}} w_1(|\boldsymbol{\xi}|) f(\boldsymbol{x}+\boldsymbol{\xi}) \left\{ \begin{array}{ccc} \xi_1 & \xi_2 & \xi_3 \end{array} \right\}^{\mathrm{T}} \mathrm{d}V_{\boldsymbol{x}'} \tag{A.39}$$

$$\begin{Bmatrix} \dfrac{\partial^2 f(\boldsymbol{x})}{\partial x_1^2} \\ \dfrac{\partial^2 f(\boldsymbol{x})}{\partial x_2^2} \\ \dfrac{\partial^2 f(\boldsymbol{x})}{\partial x_3^2} \\ \dfrac{\partial^2 f(\boldsymbol{x})}{\partial x_1 \partial x_2} \\ \dfrac{\partial^2 f(\boldsymbol{x})}{\partial x_1 \partial x_3} \\ \dfrac{\partial^2 f(\boldsymbol{x})}{\partial x_2 \partial x_3} \end{Bmatrix} = \int_{H_{\boldsymbol{x}}} \hat{\boldsymbol{a}}_{02}^{\mathrm{T}} \{1\} w_2(|\boldsymbol{\xi}|) f(\boldsymbol{x}+\boldsymbol{\xi}) \mathrm{d}V_{\boldsymbol{x}'}$$

$$+ \int_{H_{\boldsymbol{x}}} \hat{\boldsymbol{a}}_{22}^{\mathrm{T}} w_2(|\boldsymbol{\xi}|) f(\boldsymbol{x}+\boldsymbol{\xi}) \begin{Bmatrix} \xi_1^2 \\ \xi_2^2 \\ \xi_3^2 \\ \xi_1\xi_2 \\ \xi_1\xi_3 \\ \xi_2\xi_3 \end{Bmatrix} \mathrm{d}V_{\boldsymbol{x}'} \tag{A.40}$$

A.2.2 近场动力学微分算子的相对表达式

当函数 $f(\boldsymbol{x})$ 已知时，可构造与 $f(\boldsymbol{x}+\boldsymbol{\xi})-f(\boldsymbol{x})$ 相匹配的 PD 函数 $g_N^{p_1 p_2 \cdots p_M}(\boldsymbol{\xi})$ 以及近场动力学微分算子的相对表达式，具体推导过程如下。

三维情况下，函数 $f(\boldsymbol{x}+\boldsymbol{\xi})$ 的二阶泰勒级数展开式可写为

$$f(\boldsymbol{x}+\boldsymbol{\xi})-f(\boldsymbol{x})=\xi_1\frac{\partial f(\boldsymbol{x})}{\partial x_1}+\xi_2\frac{\partial f(\boldsymbol{x})}{\partial x_2}+\xi_3\frac{\partial f(\boldsymbol{x})}{\partial x_3}+\frac{1}{2!}\xi_1^2\frac{\partial^2 f(\boldsymbol{x})}{\partial x_1^2}+\frac{1}{2!}\xi_2^2\frac{\partial^2 f(\boldsymbol{x})}{\partial x_2^2}$$

$$+\frac{1}{2!}\xi_3^2\frac{\partial^2 f(\boldsymbol{x})}{\partial x_3^2}+\xi_1\xi_2\frac{\partial^2 f(\boldsymbol{x})}{\partial x_1 \partial x_2}+\xi_1\xi_3\frac{\partial^2 f(\boldsymbol{x})}{\partial x_1 \partial x_3}+\xi_2\xi_3\frac{\partial^2 f(\boldsymbol{x})}{\partial x_2 \partial x_3}$$

(A.41)

此时，构造如下形式的 PD 函数

$$g_2^{p_1p_2p_3}(\boldsymbol{\xi})=a_{100}^{p_1p_2p_3}w_{100}(|\boldsymbol{\xi}|)\xi_1+a_{010}^{p_1p_2p_3}w_{010}(|\boldsymbol{\xi}|)\xi_2+a_{001}^{p_1p_2p_3}w_{001}(|\boldsymbol{\xi}|)\xi_3$$

$$+a_{200}^{p_1p_2p_3}w_{200}(|\boldsymbol{\xi}|)\xi_1^2+a_{020}^{p_1p_2p_3}w_{020}(|\boldsymbol{\xi}|)\xi_2^2+a_{002}^{p_1p_2p_3}w_{002}(|\boldsymbol{\xi}|)\xi_3^2$$

$$+a_{110}^{p_1p_2p_3}w_{110}(|\boldsymbol{\xi}|)\xi_1\xi_2+a_{101}^{p_1p_2p_3}w_{101}(|\boldsymbol{\xi}|)\xi_1\xi_3+a_{011}^{p_1p_2p_3}w_{011}(|\boldsymbol{\xi}|)\xi_2\xi_3$$

(A.42)

则函数 $f(\boldsymbol{x})$ 一阶和二阶局部偏导数的 PD 微分算子相对表达式为

$$\left\{\begin{array}{ccc}\frac{\partial f(\boldsymbol{x})}{\partial x_1} & \frac{\partial f(\boldsymbol{x})}{\partial x_2} & \frac{\partial f(\boldsymbol{x})}{\partial x_3}\end{array}\right\}^{\mathrm{T}}$$

$$=\int_{H_{\boldsymbol{x}}}(f(\boldsymbol{x}+\boldsymbol{\xi})-f(\boldsymbol{x}))\left\{\begin{array}{ccc}g_2^{100}(\boldsymbol{\xi}) & g_2^{010}(\boldsymbol{\xi}) & g_2^{001}(\boldsymbol{\xi})\end{array}\right\}^{\mathrm{T}}\mathrm{d}V_{\boldsymbol{x}'}$$

$$=\int_{H_{\boldsymbol{x}}}w_1(|\boldsymbol{\xi}|)(f(\boldsymbol{x}+\boldsymbol{\xi})-f(\boldsymbol{x}))\hat{\boldsymbol{a}}_{11}^{\mathrm{T}}\left\{\begin{array}{c}\xi_1\\ \xi_2\\ \xi_3\end{array}\right\}\mathrm{d}V_{\boldsymbol{x}'}$$

$$+\int_{H_{\boldsymbol{x}}}w_2(|\boldsymbol{\xi}|)(f(\boldsymbol{x}+\boldsymbol{\xi})-f(\boldsymbol{x}))\hat{\boldsymbol{a}}_{21}^{\mathrm{T}}\left\{\begin{array}{c}\xi_1^2\\ \xi_2^2\\ \xi_3^2\\ \xi_1\xi_2\\ \xi_1\xi_3\\ \xi_2\xi_3\end{array}\right\}\mathrm{d}V_{\boldsymbol{x}'} \qquad (\text{A.43})$$

$$\left\{\begin{array}{cccccc}\frac{\partial^2 f(\boldsymbol{x})}{\partial x_1^2} & \frac{\partial^2 f(\boldsymbol{x})}{\partial x_2^2} & \frac{\partial^2 f(\boldsymbol{x})}{\partial x_3^2} & \frac{\partial^2 f(\boldsymbol{x})}{\partial x_1 \partial x_2} & \frac{\partial^2 f(\boldsymbol{x})}{\partial x_1 \partial x_3} & \frac{\partial^2 f(\boldsymbol{x})}{\partial x_2 \partial x_3}\end{array}\right\}^{\mathrm{T}}$$

A.2 三维球形对称近场范围的二阶近场动力学微分算子

$$= \int_{H_x} (f(\boldsymbol{x}+\boldsymbol{\xi}) - f(\boldsymbol{x})) \begin{Bmatrix} g_2^{200}(\boldsymbol{\xi}) \\ g_2^{020}(\boldsymbol{\xi}) \\ g_2^{002}(\boldsymbol{\xi}) \\ g_2^{110}(\boldsymbol{\xi}) \\ g_2^{101}(\boldsymbol{\xi}) \\ g_2^{011}(\boldsymbol{\xi}) \end{Bmatrix} \mathrm{d}V_{x'}$$

$$= \int_{H_x} \hat{\boldsymbol{a}}_{12}^{\mathrm{T}} \begin{Bmatrix} \xi_1 \\ \xi_2 \\ \xi_3 \end{Bmatrix} w_1(|\boldsymbol{\xi}|)(f(\boldsymbol{x}+\boldsymbol{\xi}) - f(\boldsymbol{x}))\mathrm{d}V_{x'} + \int_{H_x} \hat{\boldsymbol{a}}_{22}^{\mathrm{T}} w_2(|\boldsymbol{\xi}|)(f(\boldsymbol{x}+\boldsymbol{\xi})$$

$$- f(\boldsymbol{x})) \begin{Bmatrix} \xi_1^2 \\ \xi_2^2 \\ \xi_3^2 \\ \xi_1\xi_2 \\ \xi_1\xi_3 \\ \xi_2\xi_3 \end{Bmatrix} \mathrm{d}V_{x'} \tag{A.44}$$

其中，系数 $\boldsymbol{a}$ 由下式获得

$$\boldsymbol{A}\boldsymbol{a} = \boldsymbol{b} \Rightarrow \begin{bmatrix} \boldsymbol{A}_{11} & \boldsymbol{A}_{12} \\ \boldsymbol{A}_{21} & \boldsymbol{A}_{22} \end{bmatrix} \begin{bmatrix} \hat{\boldsymbol{a}}_{11} & \hat{\boldsymbol{a}}_{12} \\ \hat{\boldsymbol{a}}_{21} & \hat{\boldsymbol{a}}_{22} \end{bmatrix} = \begin{bmatrix} \hat{\boldsymbol{b}}_{11} & \boldsymbol{0} \\ \boldsymbol{0} & \hat{\boldsymbol{b}}_{22} \end{bmatrix} \tag{A.45}$$

其中，$\hat{\boldsymbol{a}}_{11} = (\boldsymbol{A}_{11} - \boldsymbol{A}_{12}\boldsymbol{A}_{22}^{-1}\boldsymbol{A}_{21})^{-1}\hat{\boldsymbol{b}}_{11}$、$\hat{\boldsymbol{a}}_{22} = (\boldsymbol{A}_{22} - \boldsymbol{A}_{21}\boldsymbol{A}_{11}^{-1}\boldsymbol{A}_{12})^{-1}\hat{\boldsymbol{b}}_{22}$，$\hat{\boldsymbol{a}}_{12} = -\boldsymbol{A}_{11}^{-1}\boldsymbol{A}_{12}\hat{\boldsymbol{a}}_{22}$ 和 $\hat{\boldsymbol{a}}_{21} = -\boldsymbol{A}_{22}^{-1}\boldsymbol{A}_{21}\hat{\boldsymbol{a}}_{11}$。且当影响函数取同一形式 $w_{q_1q_2q_3}(|\boldsymbol{\xi}|) = w(|\boldsymbol{\xi}|)$ 时，有

$$\boldsymbol{A}\boldsymbol{a} = \boldsymbol{b} \Rightarrow \begin{bmatrix} \boldsymbol{A}_{11} & \boldsymbol{A}_{12} \\ \boldsymbol{A}_{21} & \boldsymbol{A}_{22} \end{bmatrix} \begin{bmatrix} \hat{\boldsymbol{a}}_{11} & \hat{\boldsymbol{a}}_{12} \\ \hat{\boldsymbol{a}}_{21} & \hat{\boldsymbol{a}}_{22} \end{bmatrix} = \begin{bmatrix} \hat{\boldsymbol{b}}_{11} & \boldsymbol{0} \\ \boldsymbol{0} & \hat{\boldsymbol{b}}_{22} \end{bmatrix} \tag{A.46}$$

其中

$$\boldsymbol{A} = \int_{H_x} w(|\boldsymbol{\xi}|)$$

$$\times \begin{bmatrix} \xi_1^2 & \xi_1\xi_2 & \xi_1\xi_3 & \xi_1^3 & \xi_1\xi_2^2 & \xi_1\xi_3^2 & \xi_1^2\xi_2 & \xi_1^2\xi_3 & \xi_1\xi_2\xi_3 \\ \xi_1\xi_2 & \xi_2^2 & \xi_2\xi_3 & \xi_1^2\xi_2 & \xi_2^3 & \xi_2\xi_3^2 & \xi_1\xi_2^2 & \xi_1\xi_2\xi_3 & \xi_2^2\xi_3 \\ \xi_1\xi_3 & \xi_2\xi_3 & \xi_3^2 & \xi_1^2\xi_3 & \xi_3\xi_2^2 & \xi_3^3 & \xi_1\xi_2\xi_3 & \xi_1\xi_3^2 & \xi_2\xi_3^2 \\ \xi_1^3 & \xi_1^2\xi_2 & \xi_1^2\xi_3 & \xi_1^4 & \xi_1^2\xi_2^2 & \xi_1^2\xi_3^2 & \xi_1^3\xi_2 & \xi_1^3\xi_3 & \xi_1^2\xi_2\xi_3 \\ \xi_1\xi_2^2 & \xi_2^3 & \xi_3\xi_2^2 & \xi_1^2\xi_2^2 & \xi_2^4 & \xi_2^2\xi_3^2 & \xi_1\xi_2^3 & \xi_1\xi_2^2\xi_3 & \xi_2^3\xi_3 \\ \xi_1\xi_3^2 & \xi_2\xi_3^2 & \xi_3^3 & \xi_1^2\xi_3^2 & \xi_2^2\xi_3^2 & \xi_3^4 & \xi_1\xi_2\xi_3^2 & \xi_1\xi_3^3 & \xi_2\xi_3^3 \\ \xi_1^2\xi_2 & \xi_1\xi_2^2 & \xi_1\xi_2\xi_3 & \xi_1^3\xi_2 & \xi_1\xi_2^3 & \xi_1\xi_2\xi_3^2 & \xi_1^2\xi_2^2 & \xi_1^2\xi_2\xi_3 & \xi_1\xi_2^2\xi_3 \\ \xi_1^2\xi_3 & \xi_1\xi_2\xi_3 & \xi_1\xi_3^2 & \xi_1^3\xi_3 & \xi_1\xi_2^2\xi_3 & \xi_1\xi_3^3 & \xi_1^2\xi_2\xi_3 & \xi_1^2\xi_3^2 & \xi_1\xi_2\xi_3^2 \\ \xi_1\xi_2\xi_3 & \xi_2^2\xi_3 & \xi_2\xi_3^2 & \xi_1^2\xi_2\xi_3 & \xi_2^3\xi_3 & \xi_2\xi_3^3 & \xi_1\xi_2^2\xi_3 & \xi_1\xi_2\xi_3^2 & \xi_2^2\xi_3^2 \end{bmatrix} \mathrm{d}V_{x'} \tag{A.47}$$

$$\boldsymbol{a} = \begin{bmatrix} a_{100}^{100} & a_{100}^{010} & a_{100}^{001} & a_{100}^{200} & a_{100}^{020} & a_{100}^{002} & a_{100}^{110} & a_{100}^{101} & a_{100}^{011} \\ a_{010}^{100} & a_{010}^{010} & a_{010}^{001} & a_{010}^{200} & a_{010}^{020} & a_{010}^{002} & a_{010}^{110} & a_{010}^{101} & a_{010}^{011} \\ a_{001}^{100} & a_{001}^{010} & a_{001}^{001} & a_{001}^{200} & a_{001}^{020} & a_{001}^{002} & a_{001}^{110} & a_{001}^{101} & a_{001}^{011} \\ a_{200}^{100} & a_{200}^{010} & a_{200}^{001} & a_{200}^{200} & a_{200}^{020} & a_{200}^{002} & a_{200}^{110} & a_{200}^{101} & a_{200}^{011} \\ a_{020}^{100} & a_{020}^{010} & a_{020}^{001} & a_{020}^{200} & a_{020}^{020} & a_{020}^{002} & a_{020}^{110} & a_{020}^{101} & a_{020}^{011} \\ a_{002}^{100} & a_{002}^{010} & a_{002}^{001} & a_{002}^{200} & a_{002}^{020} & a_{002}^{002} & a_{002}^{110} & a_{002}^{101} & a_{002}^{011} \\ a_{110}^{100} & a_{110}^{010} & a_{110}^{001} & a_{110}^{200} & a_{110}^{020} & a_{110}^{002} & a_{110}^{110} & a_{110}^{101} & a_{110}^{011} \\ a_{101}^{100} & a_{101}^{010} & a_{101}^{001} & a_{101}^{200} & a_{101}^{020} & a_{101}^{002} & a_{101}^{110} & a_{101}^{101} & a_{101}^{011} \\ a_{011}^{100} & a_{011}^{010} & a_{011}^{001} & a_{011}^{200} & a_{011}^{020} & a_{011}^{002} & a_{011}^{110} & a_{011}^{101} & a_{011}^{011} \end{bmatrix} \tag{A.48}$$

$$\boldsymbol{b} = \begin{bmatrix} \hat{\boldsymbol{b}}_{11} & \boldsymbol{0} \\ \boldsymbol{0} & \hat{\boldsymbol{b}}_{22} \end{bmatrix} = \begin{bmatrix} 1 & 0 & 0 & 0 & 0 & 0 & 0 & 0 & 0 \\ 0 & 1 & 0 & 0 & 0 & 0 & 0 & 0 & 0 \\ 0 & 0 & 1 & 0 & 0 & 0 & 0 & 0 & 0 \\ 0 & 0 & 0 & 2 & 0 & 0 & 0 & 0 & 0 \\ 0 & 0 & 0 & 0 & 2 & 0 & 0 & 0 & 0 \\ 0 & 0 & 0 & 0 & 0 & 2 & 0 & 0 & 0 \\ 0 & 0 & 0 & 0 & 0 & 0 & 1 & 0 & 0 \\ 0 & 0 & 0 & 0 & 0 & 0 & 0 & 1 & 0 \\ 0 & 0 & 0 & 0 & 0 & 0 & 0 & 0 & 1 \end{bmatrix} \tag{A.49}$$

当近场范围为三维球形对称时，式 (A.46) 退化为

A.2 三维球形对称近场范围的二阶近场动力学微分算子

$$\begin{bmatrix} \boldsymbol{A}_{11} & \boldsymbol{0} \\ \boldsymbol{0} & \boldsymbol{A}_{22} \end{bmatrix} \begin{bmatrix} \hat{\boldsymbol{a}}_{11} & \boldsymbol{0} \\ \boldsymbol{0} & \hat{\boldsymbol{a}}_{22} \end{bmatrix} = \begin{bmatrix} \hat{\boldsymbol{b}}_{11} & \boldsymbol{0} \\ \boldsymbol{0} & \hat{\boldsymbol{b}}_{22} \end{bmatrix} \tag{A.50}$$

其中

$$\hat{\boldsymbol{a}}_{11} = \boldsymbol{A}_{11}^{-1}\hat{\boldsymbol{b}}_{11} = \frac{3}{4\pi V_1} \begin{bmatrix} 1 & 0 & 0 \\ 0 & 1 & 0 \\ 0 & 0 & 1 \end{bmatrix} \tag{A.51}$$

$$\hat{\boldsymbol{a}}_{22} = \boldsymbol{A}_{22}^{-1}\hat{\boldsymbol{b}}_{22} = \frac{3}{4\pi V_2} \begin{bmatrix} 4 & -1 & -1 & 0 & 0 & 0 \\ -1 & 4 & -1 & 0 & 0 & 0 \\ -1 & -1 & 4 & 0 & 0 & 0 \\ 0 & 0 & 0 & 5 & 0 & 0 \\ 0 & 0 & 0 & 0 & 5 & 0 \\ 0 & 0 & 0 & 0 & 0 & 5 \end{bmatrix} \tag{A.52}$$

将以上两式代入式 (A.43) 和式 (A.44) 中，则函数 $f(\boldsymbol{x})$ 的一阶和二阶局部偏导数的 PD 微分算子相对表达式为

$$\left\{ \begin{array}{ccc} \dfrac{\partial f(\boldsymbol{x})}{\partial x_1} & \dfrac{\partial f(\boldsymbol{x})}{\partial x_2} & \dfrac{\partial f(\boldsymbol{x})}{\partial x_3} \end{array} \right\}^{\mathrm{T}} = \frac{3}{4\pi V_1} \int_{H_{\boldsymbol{x}}} w_1(|\boldsymbol{\xi}|)(f(\boldsymbol{x}+\boldsymbol{\xi})-f(\boldsymbol{x})) \left\{ \begin{array}{c} \xi_1 \\ \xi_2 \\ \xi_3 \end{array} \right\} \mathrm{d}V_{\boldsymbol{x}'} \tag{A.53}$$

$$\left\{ \begin{array}{c} \dfrac{\partial^2 f(\boldsymbol{x})}{\partial x_1^2} \\ \dfrac{\partial^2 f(\boldsymbol{x})}{\partial x_2^2} \\ \dfrac{\partial^2 f(\boldsymbol{x})}{\partial x_3^2} \\ \dfrac{\partial^2 f(\boldsymbol{x})}{\partial x_1 \partial x_2} \\ \dfrac{\partial^2 f(\boldsymbol{x})}{\partial x_1 \partial x_3} \\ \dfrac{\partial^2 f(\boldsymbol{x})}{\partial x_2 \partial x_3} \end{array} \right\} = \frac{3}{4\pi V_2} \int_{H_{\boldsymbol{x}}} w_2(|\boldsymbol{\xi}|)(f(\boldsymbol{x}+\boldsymbol{\xi})-f(\boldsymbol{x})) \left\{ \begin{array}{c} 4\xi_1^2 - \xi_2^2 - \xi_3^2 \\ -\xi_1^2 + 4\xi_2^2 - \xi_3^2 \\ -\xi_1^2 - \xi_2^2 + 4\xi_3^2 \\ 5\xi_1\xi_2 \\ 5\xi_1\xi_3 \\ 5\xi_2\xi_3 \end{array} \right\} \mathrm{d}V_{\boldsymbol{x}'} \tag{A.54}$$

A.3 二维圆形对称近场范围的二阶近场动力学微分算子

A.3.1 近场动力学微分算子的绝对表达式

二维情况下，函数 $f(\boldsymbol{x}+\boldsymbol{\xi})$ 的二阶泰勒级数展开式为

$$f(\boldsymbol{x}+\boldsymbol{\xi})=f(\boldsymbol{x})+\xi_1\frac{\partial f(\boldsymbol{x})}{\partial x_1}+\xi_2\frac{\partial f(\boldsymbol{x})}{\partial x_2}+\frac{1}{2!}\xi_1^2\frac{\partial^2 f(\boldsymbol{x})}{\partial x_1^2}+\frac{1}{2!}\xi_2^2\frac{\partial^2 f(\boldsymbol{x})}{\partial x_2^2}+\xi_1\xi_2\frac{\partial^2 f(\boldsymbol{x})}{\partial x_1\partial x_2} \tag{A.55}$$

此时，构造如下形式的 PD 函数

$$\begin{aligned}g_2^{p_1p_2}(\boldsymbol{\xi})=&a_{00}^{p_1p_2}w_{00}(|\boldsymbol{\xi}|)+a_{10}^{p_1p_2}w_{10}(|\boldsymbol{\xi}|)\xi_1+a_{01}^{p_1p_2}w_{01}(|\boldsymbol{\xi}|)\xi_2\\&+a_{20}^{p_1p_2}w_{20}(|\boldsymbol{\xi}|)\xi_1^2+a_{02}^{p_1p_2}w_{02}(|\boldsymbol{\xi}|)\xi_2^2+a_{11}^{p_1p_2}w_{11}(|\boldsymbol{\xi}|)\xi_1\xi_2\end{aligned} \tag{A.56}$$

类似地，函数 $f(\boldsymbol{x})$ 各阶局部偏导数的 PD 微分算子绝对表达式为

$$f(\boldsymbol{x})=\int_{H_{\boldsymbol{x}}}f(\boldsymbol{x}+\boldsymbol{\xi})g_2^{00}(\boldsymbol{\xi})\mathrm{d}V_{\boldsymbol{x}'} \tag{A.57}$$

$$\left\{\begin{array}{cc}\dfrac{\partial f}{\partial x_1} & \dfrac{\partial f}{\partial x_2}\end{array}\right\}^{\mathrm{T}}=\int_{H_{\boldsymbol{x}}}f(\boldsymbol{x}+\boldsymbol{\xi})\left\{\begin{array}{cc}g_2^{10}(\boldsymbol{\xi}) & g_2^{01}(\boldsymbol{\xi})\end{array}\right\}^{\mathrm{T}}\mathrm{d}V_{\boldsymbol{x}'} \tag{A.58}$$

$$\left\{\begin{array}{ccc}\dfrac{\partial^2 f}{\partial x_1^2} & \dfrac{\partial^2 f}{\partial x_2^2} & \dfrac{\partial^2 f}{\partial x_1\partial x_2}\end{array}\right\}^{\mathrm{T}}=\int_{H_{\boldsymbol{x}}}f(\boldsymbol{x}+\boldsymbol{\xi})\left\{\begin{array}{ccc}g_2^{20}(\boldsymbol{\xi}) & g_2^{02}(\boldsymbol{\xi}) & g_2^{11}(\boldsymbol{\xi})\end{array}\right\}^{\mathrm{T}}\mathrm{d}V_{\boldsymbol{x}'} \tag{A.59}$$

其中，系数 $\boldsymbol{a}$ 由下式获得

$$\boldsymbol{A}\boldsymbol{a}=\boldsymbol{b}\Rightarrow\begin{bmatrix}A_{00} & A_{01} & A_{02}\\ A_{10} & A_{11} & A_{12}\\ A_{20} & A_{21} & A_{22}\end{bmatrix}\begin{bmatrix}\hat{a}_{00} & \hat{a}_{01} & \hat{a}_{02}\\ \hat{a}_{10} & \hat{a}_{11} & \hat{a}_{12}\\ \hat{a}_{20} & \hat{a}_{21} & \hat{a}_{22}\end{bmatrix}=\begin{bmatrix}\hat{b}_{00} & 0 & 0\\ 0 & \hat{b}_{11} & 0\\ 0 & 0 & \hat{b}_{22}\end{bmatrix} \tag{A.60}$$

当影响函数取同一形式时，即 $w_{q_1q_2}(|\boldsymbol{\xi}|)=w(|\boldsymbol{\xi}|)$，有

$$\boldsymbol{A}=\int_{H_{\boldsymbol{x}}}w(|\boldsymbol{\xi}|)\begin{bmatrix}1 & \xi_1 & \xi_2 & \xi_1^2 & \xi_2^2 & \xi_1\xi_2\\ \xi_1 & \xi_1^2 & \xi_1\xi_2 & \xi_1^3 & \xi_1\xi_2^2 & \xi_1^2\xi_2\\ \xi_2 & \xi_1\xi_2 & \xi_2^2 & \xi_1^2\xi_2 & \xi_2^3 & \xi_1\xi_2^2\\ \xi_1^2 & \xi_1^3 & \xi_1^2\xi_2 & \xi_1^4 & \xi_1^2\xi_2^2 & \xi_1^3\xi_2\\ \xi_2^2 & \xi_1\xi_2^2 & \xi_2^3 & \xi_1^2\xi_2^2 & \xi_2^4 & \xi_1\xi_2^3\\ \xi_1\xi_2 & \xi_1^2\xi_2 & \xi_1\xi_2^2 & \xi_1^3\xi_2 & \xi_1\xi_2^3 & \xi_1^2\xi_2^2\end{bmatrix}\mathrm{d}V_{\boldsymbol{x}'} \tag{A.61}$$

A.3 二维圆形对称近场范围的二阶近场动力学微分算子

$$\boldsymbol{a} = \begin{bmatrix} a_{00}^{00} & a_{00}^{10} & a_{00}^{01} & a_{00}^{20} & a_{00}^{02} & a_{00}^{11} \\ a_{10}^{00} & a_{10}^{10} & a_{10}^{01} & a_{10}^{20} & a_{10}^{02} & a_{10}^{11} \\ a_{01}^{00} & a_{01}^{10} & a_{01}^{01} & a_{01}^{20} & a_{01}^{02} & a_{01}^{11} \\ a_{20}^{00} & a_{20}^{10} & a_{20}^{01} & a_{20}^{20} & a_{20}^{02} & a_{20}^{11} \\ a_{02}^{00} & a_{02}^{10} & a_{02}^{01} & a_{02}^{20} & a_{02}^{02} & a_{02}^{11} \\ a_{11}^{00} & a_{11}^{10} & a_{11}^{01} & a_{11}^{20} & a_{11}^{02} & a_{11}^{11} \end{bmatrix} \tag{A.62}$$

$$\boldsymbol{b} = \begin{bmatrix} 1 & 0 & 0 & 0 & 0 & 0 \\ 0 & 1 & 0 & 0 & 0 & 0 \\ 0 & 0 & 1 & 0 & 0 & 0 \\ 0 & 0 & 0 & 2 & 0 & 0 \\ 0 & 0 & 0 & 0 & 2 & 0 \\ 0 & 0 & 0 & 0 & 0 & 1 \end{bmatrix} \tag{A.63}$$

由于二维圆形近场范围的对称性，式 (A.60) 中矩阵 $\boldsymbol{A}$ 的一些非对角子矩阵项消失，即奇数阶导数项和偶数阶导数项解耦，故式 (A.60) 简化为

$$\begin{bmatrix} \boldsymbol{A}_{00} & \boldsymbol{0} & \boldsymbol{A}_{02} \\ \boldsymbol{0} & \boldsymbol{A}_{11} & \boldsymbol{0} \\ \boldsymbol{A}_{20} & \boldsymbol{0} & \boldsymbol{A}_{22} \end{bmatrix} \begin{bmatrix} \hat{\boldsymbol{a}}_{00} & \boldsymbol{0} & \hat{\boldsymbol{a}}_{02} \\ \boldsymbol{0} & \hat{\boldsymbol{a}}_{11} & \boldsymbol{0} \\ \hat{\boldsymbol{a}}_{20} & \boldsymbol{0} & \hat{\boldsymbol{a}}_{22} \end{bmatrix} = \begin{bmatrix} \hat{\boldsymbol{b}}_{00} & \boldsymbol{0} & \boldsymbol{0} \\ \boldsymbol{0} & \hat{\boldsymbol{b}}_{11} & \boldsymbol{0} \\ \boldsymbol{0} & \boldsymbol{0} & \hat{\boldsymbol{b}}_{22} \end{bmatrix} \tag{A.64}$$

此时，$\boldsymbol{A}$、$\boldsymbol{a}$ 和 $\boldsymbol{b}$ 的各子矩阵如下

$$\boldsymbol{A}_{00} = \int_{H_x} w_0(|\boldsymbol{\xi}|) \mathrm{d}V_{x'} = 2h\pi V_0 \tag{A.65}$$

$$\boldsymbol{A}_{02} = \boldsymbol{A}_{20}^{\mathrm{T}} = \int_{H_x} w_1(|\boldsymbol{\xi}|) \begin{bmatrix} \xi_1^2 & \xi_2^2 & \xi_1\xi_2 \end{bmatrix} \mathrm{d}V_{x'} = h\pi V_1 \begin{bmatrix} 1 & 1 & 0 \end{bmatrix} \tag{A.66}$$

$$\boldsymbol{A}_{11} = \int_{H_x} w_1(|\boldsymbol{\xi}|) \begin{bmatrix} \xi_1^2 & \xi_1\xi_2 \\ \xi_1\xi_2 & \xi_2^2 \end{bmatrix} \mathrm{d}V_{x'} = h\pi V_1 \begin{bmatrix} 1 & 0 \\ 0 & 1 \end{bmatrix} \tag{A.67}$$

$$\boldsymbol{A}_{22} = \int_{H_x} w_2(|\boldsymbol{\xi}|) \begin{bmatrix} \xi_1^4 & \xi_1^2\xi_2^2 & \xi_1^3\xi_2 \\ \xi_1^2\xi_2^2 & \xi_2^4 & \xi_1\xi_2^3 \\ \xi_1^3\xi_2 & \xi_1\xi_2^3 & \xi_1^2\xi_2^2 \end{bmatrix} \mathrm{d}V_{x'} = \frac{h\pi V_2}{4} \begin{bmatrix} 3 & 1 & 0 \\ 1 & 3 & 0 \\ 0 & 0 & 1 \end{bmatrix} \tag{A.68}$$

$$\boldsymbol{A}_{11}^{-1} = \frac{1}{h\pi V_1} \begin{bmatrix} 1 & 0 \\ 0 & 1 \end{bmatrix}, \quad \boldsymbol{A}_{22}^{-1} = \frac{1}{2h\pi V_2} \begin{bmatrix} 3 & -1 & 0 \\ -1 & 3 & 0 \\ 0 & 0 & 8 \end{bmatrix} \tag{A.69}$$

$$\hat{\boldsymbol{b}}_{00} = 1, \quad \hat{\boldsymbol{b}}_{11} = \begin{bmatrix} 1 & 0 \\ 0 & 1 \end{bmatrix}, \quad \hat{\boldsymbol{b}}_{22} = \begin{bmatrix} 2 & 0 & 0 \\ 0 & 2 & 0 \\ 0 & 0 & 1 \end{bmatrix} \tag{A.70}$$

$$\hat{\boldsymbol{a}}_{00} = \left[\boldsymbol{A}_{00} - \boldsymbol{A}_{02} \boldsymbol{A}_{22}^{-1} \boldsymbol{A}_{20} \right]^{-1} \hat{\boldsymbol{b}}_{00} \tag{A.71}$$

$$\hat{\boldsymbol{a}}_{11} = \boldsymbol{A}_{00}^{-1} \hat{\boldsymbol{b}}_{11} \tag{A.72}$$

$$\hat{\boldsymbol{a}}_{20} = -\boldsymbol{A}_{22}^{-1} \boldsymbol{A}_{20} \hat{\boldsymbol{a}}_{00} \tag{A.73}$$

$$\hat{\boldsymbol{a}}_{22} = \left[\boldsymbol{A}_{22} - \boldsymbol{A}_{20} \boldsymbol{A}_{00}^{-1} \boldsymbol{A}_{02} \right]^{-1} \hat{\boldsymbol{b}}_{22} \tag{A.74}$$

$$\hat{\boldsymbol{a}}_{02} = -\boldsymbol{A}_{00}^{-1} \boldsymbol{A}_{02} \hat{\boldsymbol{a}}_{22} \tag{A.75}$$

式中

$$V_n = \int_0^\delta w_n(|\boldsymbol{\xi}|) |\boldsymbol{\xi}|^{2n+1} \, \mathrm{d}|\boldsymbol{\xi}|, \quad n=1,2 \tag{A.76}$$

当选定以下影响函数时，其值分别为 $V_0 = \delta^2$、$V_1 = \delta^4/2$ 和 $V_2 = \delta^6/3$

$$w_n(|\boldsymbol{\xi}|) = \frac{\delta^{n+1}}{|\boldsymbol{\xi}|^{n+1}} \tag{A.77}$$

其中，$n = (a+b)/2$，且 a 和 b 为子矩阵 $\boldsymbol{A}_{ab}$ 的下标；h 为板厚度。

从以上各式出发，式 (A.57)~式 (A.59) 表示的各阶偏导数为

$$f(\boldsymbol{x}) = \int_{H_\infty} \hat{\boldsymbol{a}}_{00}^{\mathrm{T}} \{1\} w_0(|\boldsymbol{\xi}|) f(\boldsymbol{x}+\boldsymbol{\xi}) \mathrm{d}V_{\boldsymbol{x}'} + \int_{H_\infty} \hat{\boldsymbol{a}}_{20}^{\mathrm{T}} \begin{Bmatrix} \xi_1^2 \\ \xi_2^2 \\ \xi_1 \xi_2 \end{Bmatrix} w_0(|\boldsymbol{\xi}|) f(\boldsymbol{x}+\boldsymbol{\xi}) \mathrm{d}V_{\boldsymbol{x}'} \tag{A.78}$$

$$\begin{Bmatrix} \dfrac{\partial f}{\partial x_1} & \dfrac{\partial f}{\partial x_2} \end{Bmatrix}^{\mathrm{T}} = \int_{H_\infty} \hat{\boldsymbol{a}}_{11}^{\mathrm{T}} w_1(|\boldsymbol{\xi}|) f(\boldsymbol{x}+\boldsymbol{\xi}) \begin{Bmatrix} \xi_1 & \xi_2 \end{Bmatrix}^{\mathrm{T}} \mathrm{d}V_{\boldsymbol{x}'} \tag{A.79}$$

$$\begin{Bmatrix} \dfrac{\partial^2 f}{\partial x_1^2} \\ \dfrac{\partial^2 f}{\partial x_2^2} \\ \dfrac{\partial^2 f}{\partial x_1 \partial x_2} \end{Bmatrix} = \int_{H_\infty} \hat{\boldsymbol{a}}_{02}^{\mathrm{T}} \{1\} w_2(|\boldsymbol{\xi}|) f(\boldsymbol{x}+\boldsymbol{\xi}) \mathrm{d}V_{\boldsymbol{x}'}$$

$$+ \int_{H_{\infty}} \hat{\boldsymbol{a}}_{22}^{\mathrm{T}} w_2(|\boldsymbol{\xi}|) f(\boldsymbol{x}+\boldsymbol{\xi}) \left\{ \begin{array}{c} \xi_1^2 \\ \xi_2^2 \\ \xi_1 \xi_2 \end{array} \right\} \mathrm{d}V_{\boldsymbol{x}'} \qquad (\text{A.80})$$

A.3.2 近场动力学微分算子的相对表达式

当函数 $f(\boldsymbol{x})$ 已知时,可构造与 $f(\boldsymbol{x}+\boldsymbol{\xi})-f(\boldsymbol{x})$ 相匹配的 PD 函数 $g_N^{p_1 p_2 \cdots p_M}(\boldsymbol{\xi})$ 以及近场动力学微分算子的相对表达式,具体推导过程如下。

二维情况下,函数 $f(\boldsymbol{x}+\boldsymbol{\xi})$ 的二阶泰勒级数展开式可改写为

$$f(\boldsymbol{x}+\boldsymbol{\xi})-f(\boldsymbol{x}) = \xi_1 \frac{\partial f(\boldsymbol{x})}{\partial x_1} + \xi_2 \frac{\partial f(\boldsymbol{x})}{\partial x_2} + \frac{1}{2!}\xi_1^2 \frac{\partial^2 f(\boldsymbol{x})}{\partial x_1^2} + \frac{1}{2!}\xi_2^2 \frac{\partial^2 f(\boldsymbol{x})}{\partial x_2^2} + \xi_1\xi_2 \frac{\partial^2 f(\boldsymbol{x})}{\partial x_1 \partial x_2}$$
(A.81)

此时,构造如下形式的 PD 函数

$$\begin{aligned} g_2^{p_1 p_2}(\boldsymbol{\xi}) =\, & a_{10}^{p_1 p_2} w_{10}(|\boldsymbol{\xi}|)\xi_1 + a_{01}^{p_1 p_2} w_{01}(|\boldsymbol{\xi}|)\xi_2 \\ & + a_{20}^{p_1 p_2} w_{20}(|\boldsymbol{\xi}|)\xi_1^2 + a_{02}^{p_1 p_2} w_{02}(|\boldsymbol{\xi}|)\xi_2^2 + a_{11}^{p_1 p_2} w_{11}(|\boldsymbol{\xi}|)\xi_1\xi_2 \end{aligned} \qquad (\text{A.82})$$

类似地,函数 $f(\boldsymbol{x})$ 的一阶和二阶局部偏导数的 PD 微分算子相对表达式为

$$\begin{aligned} \left\{ \begin{array}{cc} \dfrac{\partial f}{\partial x_1} & \dfrac{\partial f}{\partial x_2} \end{array} \right\}^{\mathrm{T}} &= \int_{H_{\infty}} (f(\boldsymbol{x}+\boldsymbol{\xi})-f(\boldsymbol{x})) \left\{ \begin{array}{c} g_2^{10}(\boldsymbol{\xi}) \\ g_2^{01}(\boldsymbol{\xi}) \end{array} \right\} \mathrm{d}V_{\boldsymbol{x}'} \\ &= \int_{H_{\infty}} w_1(|\boldsymbol{\xi}|)(f(\boldsymbol{x}+\boldsymbol{\xi})-f(\boldsymbol{x}))\hat{\boldsymbol{a}}_{11}^{\mathrm{T}} \left\{ \begin{array}{c} \xi_1 \\ \xi_2 \end{array} \right\} \mathrm{d}V_{\boldsymbol{x}'} \\ &\quad + \int_{H_{\infty}} w_2(|\boldsymbol{\xi}|)(f(\boldsymbol{x}+\boldsymbol{\xi})-f(\boldsymbol{x}))\hat{\boldsymbol{a}}_{21}^{\mathrm{T}} \left\{ \begin{array}{c} \xi_1^2 \\ \xi_2^2 \\ \xi_1\xi_2 \end{array} \right\} \mathrm{d}V_{\boldsymbol{x}'} \end{aligned} \qquad (\text{A.83})$$

$$\begin{aligned} \left\{ \begin{array}{ccc} \dfrac{\partial^2 f}{\partial x_1^2} & \dfrac{\partial^2 f}{\partial x_2^2} & \dfrac{\partial^2 f}{\partial x_1 \partial x_2} \end{array} \right\}^{\mathrm{T}} &= \int_{H_{\infty}} (f(\boldsymbol{x}+\boldsymbol{\xi})-f(\boldsymbol{x})) \left\{ \begin{array}{c} g_2^{20}(\boldsymbol{\xi}) \\ g_2^{02}(\boldsymbol{\xi}) \\ g_2^{11}(\boldsymbol{\xi}) \end{array} \right\} \mathrm{d}V_{\boldsymbol{x}'} \\ &= \int_{H_{\infty}} \hat{\boldsymbol{a}}_{12}^{\mathrm{T}} \left\{ \begin{array}{c} \xi_1 \\ \xi_2 \end{array} \right\} w_1(|\boldsymbol{\xi}|)(f(\boldsymbol{x}+\boldsymbol{\xi})-f(\boldsymbol{x}))\mathrm{d}V_{\boldsymbol{x}'} \end{aligned}$$

$$+ \int_{H_\infty} \hat{a}_{22}^{\mathrm{T}} w_2(|\boldsymbol{\xi}|)(f(\boldsymbol{x}+\boldsymbol{\xi}) - f(\boldsymbol{x})) \left\{ \begin{array}{c} \xi_1^2 \\ \xi_2^2 \\ \xi_1\xi_2 \end{array} \right\} \mathrm{d}V_{\boldsymbol{x}'} \tag{A.84}$$

其中，系数 $\boldsymbol{a}$ 由下式获得

$$\boldsymbol{Aa} = \boldsymbol{b} \Rightarrow \left[\begin{array}{cc} \boldsymbol{A}_{11} & \boldsymbol{A}_{12} \\ \boldsymbol{A}_{21} & \boldsymbol{A}_{22} \end{array} \right] \left[\begin{array}{cc} \hat{\boldsymbol{a}}_{11} & \hat{\boldsymbol{a}}_{12} \\ \hat{\boldsymbol{a}}_{21} & \hat{\boldsymbol{a}}_{22} \end{array} \right] = \left[\begin{array}{cc} \hat{\boldsymbol{b}}_{11} & \boldsymbol{0} \\ \boldsymbol{0} & \hat{\boldsymbol{b}}_{22} \end{array} \right] \tag{A.85}$$

其中，$\hat{a}_{11} = \left(\boldsymbol{A}_{11} - \boldsymbol{A}_{12}\boldsymbol{A}_{22}^{-1}\boldsymbol{A}_{21}\right)^{-1}\hat{\boldsymbol{b}}_{11}$、$\hat{a}_{22} = \left(\boldsymbol{A}_{22} - \boldsymbol{A}_{21}\boldsymbol{A}_{11}^{-1}\boldsymbol{A}_{12}\right)^{-1}\hat{\boldsymbol{b}}_{22}$、$\hat{a}_{12} = -\boldsymbol{A}_{11}^{-1}\boldsymbol{A}_{12}\hat{a}_{22}$ 和 $\hat{a}_{21} = -\boldsymbol{A}_{22}^{-1}\boldsymbol{A}_{21}\hat{a}_{11}$。且当影响函数取同一形式 $w_{q_1q_2}(|\boldsymbol{\xi}|) = w(|\boldsymbol{\xi}|)$ 时，有

$$\boldsymbol{A} = \int_{H_\infty} w(|\boldsymbol{\xi}|) \left[\begin{array}{ccccc} \xi_1^2 & \xi_1\xi_2 & \xi_1^3 & \xi_1\xi_2^2 & \xi_1^2\xi_2 \\ \xi_1\xi_2 & \xi_2^2 & \xi_1^2\xi_2 & \xi_2^3 & \xi_1\xi_2^2 \\ \xi_1^3 & \xi_1^2\xi_2 & \xi_1^4 & \xi_1^2\xi_2^2 & \xi_1^3\xi_2 \\ \xi_1\xi_2^2 & \xi_2^3 & \xi_1^2\xi_2^2 & \xi_2^4 & \xi_1\xi_2^3 \\ \xi_1^2\xi_2 & \xi_1\xi_2^2 & \xi_1^3\xi_2 & \xi_1\xi_2^3 & \xi_1^2\xi_2^2 \end{array} \right] \mathrm{d}V_{\boldsymbol{x}'} \tag{A.86}$$

$$\boldsymbol{a} = \left[\begin{array}{ccccc} a_{10}^{10} & a_{10}^{01} & a_{10}^{20} & a_{10}^{02} & a_{10}^{11} \\ a_{01}^{10} & a_{01}^{01} & a_{01}^{20} & a_{01}^{02} & a_{01}^{11} \\ a_{20}^{10} & a_{20}^{01} & a_{20}^{20} & a_{20}^{02} & a_{20}^{11} \\ a_{02}^{10} & a_{02}^{01} & a_{02}^{20} & a_{02}^{02} & a_{02}^{11} \\ a_{11}^{10} & a_{11}^{01} & a_{11}^{20} & a_{11}^{02} & a_{11}^{11} \end{array} \right] \tag{A.87}$$

$$\boldsymbol{b} = \left[\begin{array}{ccccc} 1 & 0 & 0 & 0 & 0 \\ 0 & 1 & 0 & 0 & 0 \\ 0 & 0 & 2 & 0 & 0 \\ 0 & 0 & 0 & 2 & 0 \\ 0 & 0 & 0 & 0 & 1 \end{array} \right] \tag{A.88}$$

当近场范围为二维对称圆形时，式 (A.85) 退化为

$$\left[\begin{array}{cc} \boldsymbol{A}_{11} & \boldsymbol{0} \\ \boldsymbol{0} & \boldsymbol{A}_{22} \end{array} \right] \left[\begin{array}{cc} \hat{\boldsymbol{a}}_{11} & \boldsymbol{0} \\ \boldsymbol{0} & \hat{\boldsymbol{a}}_{22} \end{array} \right] = \left[\begin{array}{cc} \hat{\boldsymbol{b}}_{11} & \boldsymbol{0} \\ \boldsymbol{0} & \hat{\boldsymbol{b}}_{22} \end{array} \right] \tag{A.89}$$

其中

$$\hat{\boldsymbol{a}}_{11} = \boldsymbol{A}_{11}^{-1}\hat{\boldsymbol{b}}_{11} = \frac{1}{h\pi V_1}\begin{bmatrix} 1 & 0 \\ 0 & 1 \end{bmatrix} \tag{A.90}$$

$$\hat{\boldsymbol{a}}_{22} = \boldsymbol{A}_{22}^{-1}\hat{\boldsymbol{b}}_{22} = \frac{1}{2h\pi V_2}\begin{bmatrix} 3 & -1 & 0 \\ -1 & 3 & 0 \\ 0 & 0 & 8 \end{bmatrix}\begin{bmatrix} 2 & 0 & 0 \\ 0 & 2 & 0 \\ 0 & 0 & 1 \end{bmatrix} = \frac{1}{h\pi V_2}\begin{bmatrix} 3 & -1 & 0 \\ -1 & 3 & 0 \\ 0 & 0 & 4 \end{bmatrix} \tag{A.91}$$

将以上两式代入式 (A.83) 和式 (A.84) 中，则可得到函数 $f(\boldsymbol{x})$ 的一阶和二阶局部偏导数的 PD 微分算子相对表达式

$$\left\{ \frac{\partial f}{\partial x_1} \quad \frac{\partial f}{\partial x_2} \right\}^{\mathrm{T}} = \frac{1}{\pi h V_1}\int_{H_{\boldsymbol{x}}} w_1(|\boldsymbol{\xi}|)(f(\boldsymbol{x}+\boldsymbol{\xi})-f(\boldsymbol{x}))\left\{ \begin{matrix} \xi_1 \\ \xi_2 \end{matrix} \right\}\mathrm{d}V_{\boldsymbol{x}'} \tag{A.92}$$

$$\left\{ \frac{\partial^2 f}{\partial x_1^2} \quad \frac{\partial^2 f}{\partial x_2^2} \quad \frac{\partial^2 f}{\partial x_1 \partial x_2} \right\}^{\mathrm{T}}$$
$$= \frac{1}{h\pi V_2}\int_{H_{\boldsymbol{x}}} w_2(|\boldsymbol{\xi}|)(f(\boldsymbol{x}+\boldsymbol{\xi})-f(\boldsymbol{x}))\begin{bmatrix} 3\xi_1^2 - \xi_2^2 \\ -\xi_1^2 + 3\xi_2^2 \\ 4\xi_1\xi_2 \end{bmatrix}\mathrm{d}V_{\boldsymbol{x}'} \tag{A.93}$$

A.4　一维对称近场范围的二阶近场动力学微分算子

在一维情况下，函数 $f(x+\xi)$ 关于 x 的二阶泰勒展开式和 PD 函数分别为

$$f(x+\xi) = f(x) + \xi \frac{\partial f(x)}{\partial x} + \frac{1}{2!}\xi^2 \frac{\partial^2 f(x)}{\partial x^2} \tag{A.94}$$

$$g_2^p(\xi) = a_0^p w_0(|\xi|) + a_1^p w_1(|\xi|)\xi + a_2^p w_2(|\xi|)\xi^2, \quad p = 0,1,2 \tag{A.95}$$

相应的形状矩阵 $\boldsymbol{A}$、PD 函数的未知系数矩阵 $\boldsymbol{a}$ 和已知系数矩阵 $\boldsymbol{b}$ 分别为

$$\boldsymbol{A} = \begin{bmatrix} A_{00} & A_{01} & A_{02} \\ A_{10} & A_{11} & A_{12} \\ A_{20} & A_{21} & A_{22} \end{bmatrix} = \int_{H_x} w(|\xi|)\begin{bmatrix} 1 & \xi & \xi^2 \\ \xi & \xi^2 & \xi^3 \\ \xi^2 & \xi^3 & \xi^4 \end{bmatrix}\mathrm{d}V_{x'} \tag{A.96}$$

$$\boldsymbol{a} = \begin{bmatrix} a_0^0 & a_0^1 & a_0^2 \\ a_1^0 & a_1^1 & a_1^2 \\ a_2^0 & a_2^1 & a_2^2 \end{bmatrix} \tag{A.97}$$

$$\boldsymbol{b} = \begin{bmatrix} b_0 & 0 & 0 \\ 0 & b_1 & 0 \\ 0 & 0 & b_2 \end{bmatrix} = \begin{bmatrix} 1 & 0 & 0 \\ 0 & 1 & 0 \\ 0 & 0 & 2 \end{bmatrix} \quad (A.98)$$

最终，函数 $f(x)$ 及其一阶和二阶局部偏导数的 PD 微分算子绝对表达式分别为

$$f(x) = \int_{H_x} f(x+\xi) g_2^0(\xi) \mathrm{d}V_{x'} \quad (A.99)$$

$$\frac{\partial f(x)}{\partial x} = \int_{H_x} f(x+\xi) g_2^1(\xi) \mathrm{d}V_{x'} \quad (A.100)$$

$$\frac{\partial^2 f(x)}{\partial x^2} = \int_{H_x} f(x+\xi) g_2^2(\xi) \mathrm{d}V_{x'} \quad (A.101)$$

类似地，在一维情况下，可以求得函数 $f(x)$ 的一阶和二阶局部偏导数的 PD 微分算子相对表达式，具体为

$$f(x+\xi) - f(x) = \xi \frac{\partial f(x)}{\partial x} + \frac{1}{2!} \xi^2 \frac{\partial^2 f(x)}{\partial x^2} \quad (A.102)$$

$$g_2^p(\xi) = a_1^p w_1(|\xi|)\xi + a_2^p w_2(|\xi|)\xi^2, \quad p=1,2 \quad (A.103)$$

$$\boldsymbol{A} = \begin{bmatrix} A_{11} & A_{12} \\ A_{21} & A_{22} \end{bmatrix} = \int_{H_x} w(|\xi|) \begin{bmatrix} \xi^2 & \xi^3 \\ \xi^3 & \xi^4 \end{bmatrix} \mathrm{d}V_x' \quad (A.104)$$

$$\boldsymbol{a} = \begin{bmatrix} a_1^1 & a_1^2 \\ a_2^1 & a_2^2 \end{bmatrix} \quad (A.105)$$

$$\boldsymbol{b} = \begin{bmatrix} b_1 & 0 \\ 0 & b_2 \end{bmatrix} = \begin{bmatrix} 1 & 0 \\ 0 & 2 \end{bmatrix} \quad (A.106)$$

$$\frac{\partial f(x)}{\partial x} = \int_{H_x} (f(x+\xi) - f(x))\, g_2^1(\xi) \mathrm{d}V_{x'} \quad (A.107)$$

$$\frac{\partial^2 f(x)}{\partial x^2} = \int_{H_x} (f(x+\xi) - f(x)) g_2^2(\xi) \mathrm{d}V_{x'} \quad (A.108)$$

当近场范围为对称线段时，则有

$$\begin{bmatrix} A_{11} & 0 \\ 0 & A_{22} \end{bmatrix} \begin{bmatrix} a_{11} & 0 \\ 0 & a_{22} \end{bmatrix} = \begin{bmatrix} b_{11} & 0 \\ 0 & b_{22} \end{bmatrix} \tag{A.109}$$

式中，$A_{11} = 2AV_1$，$A_{22} = 2AV_2$，$a_{11} = b_{11}/A_{11} = \dfrac{1}{2AV_1}$，$a_{22} = b_{22}/A_{22} = \dfrac{1}{AV_2}$，且有

$$V_n = \int_0^\delta w_n(|\xi|)\,|\xi|^{2n}\,\mathrm{d}|\xi|, \quad n = 1, 2 \tag{A.110}$$

其中，$n = (a+b)/2$，且 a 和 b 为 A_{ab} 的下标；A 为一维杆件的截面积。则函数 $f(x)$ 的一阶和二阶局部偏导数的 PD 微分算子相对表达式为

$$\frac{\partial f(x)}{\partial x} = \frac{1}{2AV_1} \int_{H_x} w_1(|\xi|)(f(x+\xi) - f(x))\,\xi\,\mathrm{d}V_{x'} \tag{A.111}$$

$$\frac{\partial^2 f(x)}{\partial x^2} = \frac{1}{AV_2} \int_{H_x} w_2(|\xi|)(f(x+\xi) - f(x))\xi^2\,\mathrm{d}V_{x'} \tag{A.112}$$

参 考 文 献

[1] Madenci E, Barut A, Futch M. Peridynamic differential operator and its applications [J]. Computer Methods in Applied Mechanics and Engineering, 2016, 304:408-451.

[2] Madenci E, Dorduncu M, Barut A, et al. Numerical solution of linear and nonlinear partial differential equations using the peridynamic differential operator [J]. Numerical Methods for Partial Differential Equations, 2017, 33(5):1726-1753.

[3] Madenci E, Barut A, Dorduncu M. Peridynamic Differential Operator for Numerical Analysis [M]. Switzerland: Springer International Publishing, 2019.

[4] Madenci E, Barut A, Dorduncu M. 近场动力学微分算子——在数值分析中的应用 [M]. 韩非, 张玲, 译. 北京: 科学出版社, 2021.

[5] 顾鑫. 非常规态型近场动力学建模及其微分算子重构 [D]. 南京: 河海大学, 2018.

[6] Gu X, Zhang Q, Madenci E. Refined bond-based peridynamics for thermal diffusion [J]. Engineering Computations, 2019, 36(8):2557-2587.

附录 B　近场动力学的显式动力学算法程序

本书正文部分阐述了近场动力学的理论模型与数值方法，第 5 章具体介绍了近场动力学的显式动力无网格粒子算法。为了使读者更好地理解近场动力学理论和方法，也便于读者开展近场动力学研究，本附录给出与第 5 章算法配套的 FORTRAN 源代码，并包含必要的注释；此外，还给出输入数据文件格式说明，供读者参考使用。

B.1　键型近场动力学的显式动力学算法 FORTRAN 源程序

本程序包括一个公用模块、一个主程序和 9 个子程序。公用模块定义全局变量，主程序 master 依次调用子程序 data_input、element_form、add_load_displacement、add_load_bc_force、add_load_impact_rigid、equation_solve、model_PMB、model_short_range_force、data_output 等，实现键型近场动力学 PMB 模型的显式动力学时程积分计算。

B.1.1　公用模块：Module

本段程序包含 globalParameters、scalars、serialArrays、impact_parameters 等程序段，实现全局常数、标量和矢量数组的定义，具体程序段及参数含义如下。

```
module scalars
    integer::load_type, gravity_type, nchek
    integer::npart, ipart, jpart, nstep, istep, i_out, n_mat, i_mat,
        n_bt
    integer::nBC_dis, iBC_dis, nBC_force, iBC_force, nload_velocity,
        iload_velocity,& nload_dis, iload_dis, nload_force,
        iload_force
!******************************************************************
!   load_type：加载方式，0为速度加载，1为位移加载，2为力加载(2为线性
    加载，20为瞬时加载)，*
!                     3为一般刚性体冲击非刚性体(30为刚性球冲击)，*
!                     4为非刚性体冲击非刚性体，*
!                     5为爆炸荷载/应力波加载,6为混合加载方式，  *
!   gravity_type：计算时，是否计入重力作用：0为不计重力，1为计重力  *
!******************************************************************
```

```fortran
!     nchek: 收敛判断，0为已收敛，1为未收敛
!     npart: 物质点总数
!     ipart: 第i个物质点
!     jpart: 第j个物质点
!     nstep: 总迭代时间步数
!     istep: 第i个时间步
!     i_out: 每隔一定步数，输出结果
!     n_mat: 材料类型总数
!     i_mat: 第i种材料
!     n_bt: 键类型数
!     nBC_dis: 位移边界物质点总数
!     iBC_dis: 第i个位移边界物质点
!     nBC_force: 应力边界物质点总数
!     iBC_force: 第i个应力边界物质点
!     nload_velocity: 速度加载物质点总数
!     iload_velocity: 第i个速度加载物质点
!     nload_dis: 位移加载物质点总数
!     iload_dis: 第i个位移加载物质点
!     nload_force: 力加载物质点总数
!     iload_force: 第i个力加载物质点
      real(8)::lat_c, half_lat, mod_c, str_c, vol_1, tstep, pehor, &
          pehsq, vfrac_scale, Cv, velocity_initial
!     lat_c: 物质点间距
!     half_lat: 物质点尺寸的一半
!     mod_c: 物质点微观模量c
!     str_c: 物质点临界伸长率s0
!     vol_1: 物质点体积(或面积)
!     tstep: 迭代时间步长
!     pehor: 近场尺寸
!     pehsq: 近场尺寸的平方
!     vfrac_scale: 近场附近物质点体积加权系数
!     Cv: 局部阻尼系数
!     velocity_initial: 构型的初速度
      real(8)::toler
!     toler: 为收敛容差
      integer::ii,iii  !用以临时计数
      integer,dimension(:),allocatable::num_temp   !计算裂尖速度
      real(8),allocatable::damage0(:)   !只用于计算裂纹传播速度
      real(8)::t1,t2  !起裂前后两次的时间
end module scalars
```

```fortran
!*******************************************************************
module serialArrays
   integer,allocatable::bc_dis_id(:),bc_dis_code(:), bc_force_id(:),
       load_velocity_id(:), &
   load_dis_id(:), load_force_id(:)
!  bc_dis_id(:)：为位移边界物质点编号数组
!  bc_dis_code(:)：为位移边界条件类型数组：取值为001，
   010,100,011,101,110,111；分别对应x,y,z三个方向，其中0表示自由，1
   表示固定约束；
!  bc_force_id(:)：为力边界物质点编号数组
!  load_velocity_id(:)：为速度加载物质点编号数组
!  load_dis_id(:)：为位移加载物质点编号数组
!  load_force_id(:)：为力加载物质点编号数组
   integer,allocatable::temp_n(:), ele_b0(:,:), ele_b(:,:), ele_n0
       (:), ele_n(:), free_p(:)
!  temp_n(:)：
!  ele_b0(:,:),ele_b(:,:)：初始和计算时间步后，近场范围内粒子编号
!  ele_n0(:),ele_n(:)：初始和计算时间步后，近场范围内粒子总数
!  free_p(:)：物质点标记数组，特别标出位移边界区域物质点
   real(8),allocatable::coord(:,:),bc_xdis_value(:),bc_ydis_value
       (:),bc_zdis_value(:),&
                        bc_xforce_value(:),bc_yforce_value(:),
                            bc_zforce_value(:),&
                        load_xvelocity_value(:),
                            load_yvelocity_value(:),
                            load_zvelocity_value(:),&
                        load_xdis_value(:),load_ydis_value(:),
                            load_zdis_value(:),&
                        load_xforce_value(:),load_yforce_value(:)
                            ,load_zforce_value(:),&
                        x_old(:),x_now(:),y_old(:),y_now(:),z_old
                            (:),z_now(:),&
                        b_fx(:),b_fy(:),b_fz(:),b_fx_old(:),
                            b_fy_old(:),b_fz_old(:),&
                        x_dis(:),y_dis(:),z_dis(:),v_old(:,:),
                            v_middle(:,:),v_now(:,:),&
                        b_len(:,:),mat(:,:),damage(:),c_mod(:),
                            c_str(:),s0(:),dens(:)
!  coord(:,:)：物质点坐标数组
```

!   bc_xdis_value(:),bc_ydis_value(:),bc_zdis_value(:): 分别为位移边界物质点的x,y,z方向位移值
!   bc_xforce_value(:),bc_yforce_value(:),bc_zforce_value(:): 分别为应力边界条件物质点x,y方向荷载值
!   load_xdis_value(:),load_ydis_value(:): 分别为位移加载区域物质点x,y,z方向位移增量(加载总位移值)
!   load_xforce_value(:),load_yforce_value(:),load_zforce_value(:): 分别为力加载区域物质点x,y,z方向力值(每个物质点上受到的力的总值,2线性加载,20瞬时加载)
!   b_fx(:),b_fy(:),b_fz(:): 物质点所受合力
!   b_fx_old(:),b_fy_old(:),b_fz_old(:): 物质点所受合力,Velocity-Verlet算法中使用
!   x_old(:),x_now(:),y_old(:),y_now(:),z_old(:),z_now(:): 分别对应差分算法中三个时刻物质点位置
!   v_old(:),v_middle(:),v_now(:): 分别对应Velocity_Verlet差分格式中三个时刻物质点速度
!   b_len(:,:): 物质点近场范围内点对距离
!   mat(:,:): 材料信息
!   damage(:): 物质点损伤值
!   x_dis(:),y_dis(:),z_dis(:): 物质点位移值
!   c_mod(:): 材料微观模量数组
!   c_str(:): 材料临界拉伸数组,初始定义
!   s0(:): 每一时步的临界伸长率数组,记录所有物质点的临界伸长率
!   dens(:): 材料密度数组
end module serialArrays

!**
module globalParameters
    real(8),parameter::n_delta=3.015d0
    integer,parameter::ndime=3,max_nb=343
    real(8),parameter::pi=3.14159265358979323846264338327 95d0,g=9.80d0
!   n_delta: 近场范围与物质点尺寸的比值
!   ndime: 问题维度
!   max_nb: 近场范围内最大物质点数,三维7^3=343
end module globalParameters
!**

module impact_parameters
!一般刚性体、刚性球、非刚性体共享参数:

```fortran
    integer::npart_projectile        !弹丸离散粒子总数
    real(8)::Volume_projectile,dens_projectile      !弹丸体积，密度
    real(8),dimension(3)::Velocity_projectile       !弹丸初速度
    real(8)::Xcoor_centroid,Ycoor_centroid,Zcoor_centroid      !弹丸质
        心初始坐标，始终保持不变
    real(8)::Xcoor_centroid0,Ycoor_centroid0,Zcoor_centroid0     !弹丸
        质心初始坐标，不断更新
    real(8)::x_dis_centroid,y_dis_centroid,z_dis_centroid      !弹丸质
        心的位移值
    real(8),dimension(:,:),allocatable::coord_projectile,          &
        coord_projectile_old     !弹丸物质点坐标数组：更新、固定不变
    real(8)::X_acceleration,Y_acceleration,Z_acceleration,          &
        X_acceleration_old,Y_acceleration_old,Z_acceleration_old     !
        弹丸加速度
!一般刚性体参数：
    character(len=15)::rigid_shape      !刚性体形状：3D:球体、柱体、
        立方体；2D:圆盘、矩形、圆锥曲线形
    real(8)::radius_projectile,heigh_projectile,length_projectile,   &
        width_projectile    !弹丸几何参数，用于判断侵彻是否发生：球、
        柱、立方体
!刚性球冲击薄板参数：
    real(8),parameter::Ks=1.e17        !力模型参数： Ks
!非刚性体冲击非刚性体参数
end module impact_parameters
!*****************************************************************
```

B.1.2 主程序：Master

本程序以键型近场动力学 PMB 模型的三维情形为基础编写，定义整体坐标系满足右手螺旋定则，水平向右为 x 轴正向，竖直向上为 y 轴正向，垂直纸面向外是 z 轴正向。

```fortran
program master
    use scalars
    use serialArrays
    use globalParameters
    use impact_parameters
    implicit none
    open(5,file='PD_input_Kalthoff_Winkler.dat',status='old')
    open(6,file='target_crack_damage_observe.dat',status='old')
    open(7,file='target_dis_observe.dat',status='old')
```

```fortran
open(8,file='target_velocity_observe.dat',status='old')
open(9,file='target_crack_velocity_observe.dat',status='old')
open(1,file='projectile_dis_observe.dat',status='old')
open(2,file='projectile_Velocity_observe.dat',status='old')
open(3,file='projectile_acceleration_observe.dat',status='old')
 !调用数据读入子程序
call data_input
write(*,*)'end data_input'
 !调用形成物质点的近场域内粒子列表的子程序
call element_form()
write(*,*)'end element_form'
i_out=5      !i_out:输出时间步间隔
!****************************************
ii=0   !用以临时计数
allocate(num_temp(npart),damage0(npart))
num_temp=0
damage0=0
iii=0
!****************************************
 !显式时间积分法的迭代求解过程
do istep=0,nstep
    !输出数据
    if(mod(istep,i_out)==0) then
        call data_output()
    endif
    !观察计算过程
    write(*,*)istep
    !******************************************
    b_fx=0.0d0        !物质点间作用力赋初值为0
    b_fy=0.0d0
    b_fz=0.0d0
    dens=mat(1,4)
    Cv=0.0d0          !阻尼系数,通过给定阻尼值,实现静力和拟静力模
                      拟
    !******************************************
    !调用施加位移边界条件子程序
    call add_load_displacement()
    !调用刚性体冲击非刚性体子程序
    call add_load_impact_rigid()
    !调用施加外力子程序
```

```
        !call add_load_bc_force()
        !调用Velocity_Verlet差分格式求解子程序
        call equation_solve()
    enddo
     stop
end program master
```

B.1.3 子程序1: data_input

子程序 data_input 实现 PD 计算数据文件的读取和相关参数的计算。

```
subroutine data_input()
    use scalars
    use serialArrays
    use globalParameters
    use impact_parameters
    implicit none
    integer::i,idime
    !读入计算总控信息数据
    read(5,*) npart,nBC_dis,nBC_force,nload_velocity,nload_dis,
        nload_force,lat_c
    read(5,*) load_type,gravity_type,tstep,nstep
    read(5,*) velocity_initial
    allocate(coord(npart,ndime),x_old(npart),x_now(npart),y_old(
        npart),y_now(npart),&
            z_old(npart),z_now(npart),v_old(npart,ndime),v_middle(
                npart,ndime),v_now(npart,ndime),&
            b_fx(npart),b_fy(npart),b_fz(npart),b_fx_old(npart),
                b_fy_old(npart),b_fz_old(npart),&
            ele_b0(npart,max_nb),ele_b(npart,max_nb),ele_n0(npart),
                ele_n(npart),b_len(npart,max_nb),&
            damage(npart),x_dis(npart),y_dis(npart),z_dis(npart),
                free_p(npart),s0(npart),dens(npart))
    if(nBC_dis/=0) allocate(bc_dis_id(nBC_dis),bc_dis_code(nBC_dis),
        bc_xdis_value(nBC_dis), &
        bc_ydis_value(nBC_dis), bc_zdis_value(nBC_dis))
    if(nBC_force/=0) allocate(bc_force_id(nBC_force),
        bc_xforce_value(nBC_force), &
        bc_yforce_value(nBC_force), bc_zforce_value(nBC_force))
    if(nload_velocity/=0) allocate(load_velocity_id(nload_velocity),
        load_xvelocity_value(nload_velocity), &
```

```fortran
            load_yvelocity_value(nload_velocity), load_zvelocity_value(
                nload_velocity))
    if(nload_dis/=0)   allocate(load_dis_id(nload_dis),
        load_xdis_value(nload_dis),&
            load_ydis_value(nload_dis), load_zdis_value(nload_dis))
    if(nload_force/=0) allocate(load_force_id(nload_force),
        load_xforce_value(nload_force), &
            load_yforce_value(nload_force), load_zforce_value(
                nload_force))
    !读入几何模型信息(冲击情况下为：靶体几何模型信息)
    do ipart=1,npart
        read(5,*) i,(coord(ipart,idime),idime=1,ndime)
    enddo
    !读入约束与荷载信息(冲击情况下为：靶体约束与荷载信息)
    if(nBC_dis/=0) then
        do iBC_dis=1,nBC_dis
            read(5,*) bc_dis_id(iBC_dis), bc_dis_code(iBC_dis),
                bc_xdis_value(iBC_dis),&
                    bc_ydis_value(iBC_dis),bc_zdis_value(iBC_dis)
        enddo
    end if
    if(nBC_force/=0) then
        do iBC_force=1,nBC_force
            read(5,*) bc_force_id(iBC_force), bc_xforce_value(
                iBC_force),&
                    bc_yforce_value(iBC_force), bc_zforce_value(
                        iBC_force)
        end do
    end if
    if(nload_velocity/=0) then
        do iload_velocity=1,nload_velocity
            read(5,*) load_velocity_id(iload_velocity),
                load_xvelocity_value(iload_velocity),&
                    load_yvelocity_value(iload_velocity),
                        load_zvelocity_value(iload_velocity)
        enddo
    endif
    if(nload_dis/=0) then
        do iload_dis=1,nload_dis
            read(5,*) load_dis_id(iload_dis), load_xdis_value(
```

```fortran
            iload_dis),&
            load_ydis_value(iload_dis), load_zdis_value(iload_dis)
        end do
    end if
    if(nload_force/=0) then
        do iload_force=1,nload_force
            read(5,*) load_force_id(iload_force), load_xforce_value(
                iload_force),&
                load_yforce_value(iload_force), load_zforce_value(
                iload_force)
        end do
    end if
    !处理坐标信息，避免微小误差
    call initial_coord()
    !读入材料信息
    read(5,*)n_mat
    n_bt=(n_mat+1)*n_mat/2
    allocate(mat(n_mat,4),c_mod(n_bt),c_str(n_bt))
    do i=1,n_mat
        read(5,*) mat(i,1:4) ! mat()每行为弹性模量、泊松比、能量释放
            率、材料密度
    enddo
    !读入冲击弹丸信息
    if(load_type==3) then       !一般刚性体可在此处修改
        read(5,*)npart_projectile,dens_projectile   !弹丸物质点综述、
            质量密度
        read(5,*)rigid_shape        !弹丸形状
        if(trim(rigid_shape)=='cylinder')then   !柱形弹丸
            read(5,*)radius_projectile,heigh_projectile,
                Xcoor_centroid,Ycoor_centroid,Zcoor_centroid
        elseif(trim(rigid_shape)=='sphere')then   !球形弹丸
            read(5,*)radius_projectile,Xcoor_centroid,Ycoor_centroid
                ,Zcoor_centroid
        elseif(trim(rigid_shape)=='cubic')then   !长方体弹丸
            read(5,*)length_projectile, width_projectile,
                heigh_projectile, &
                    Xcoor_centroid, Ycoor_centroid, Zcoor_centroid
        endif
        read(5,*)(Velocity_projectile(i),i=1,ndime)   !弹丸初速度
        allocate(coord_projectile(npart_projectile,ndime),
```

```fortran
                coord_projectile_old(npart_projectile,ndime))
        do ipart=1,npart_projectile
            read(5,*)i,(coord_projectile(ipart,idime),idime=1,ndime)
                    !弹丸物质点坐标
        enddo
    end if
    if(load_type==31)then      !刚性球撞击薄板
        read(5,*)npart_projectile,dens_projectile
        read(5,*)radius_projectile,Xcoor_centroid,Ycoor_centroid,
            Zcoor_centroid
        read(5,*)(Velocity_projectile(i),i=1,ndime)
        allocate(coord_projectile(npart_projectile,ndime),
            coord_projectile_old(npart_projectile,ndime))
        do ipart=1,npart_projectile
            read(5,*)i,(coord_projectile(ipart,idime),idime=1,ndime)
        enddo
    endif
    if(load_type==4)then       !非刚性体冲击非刚性体,可在此处修改
        continue
    endif
    if(npart_projectile/=0) call initial_projectile()    !处理坐标信
        息,避免微小误差
    return
end subroutine data_input

!   本子程序用于修正网格划分引起的误差,四舍五入到小数点后一定位数
subroutine initial_coord()
    use scalars
    use serialArrays
    use globalParameters
    implicit none
    real(8)::x_coord,y_coord,z_coord
    do ipart=1,npart
        x_coord=coord(ipart,1)*1.0d16
        y_coord=coord(ipart,2)*1.0d16
        z_coord=coord(ipart,3)*1.0d16
        x_old(ipart)=(dble(anint(x_coord)))/1.0d16
        y_old(ipart)=(dble(anint(y_coord)))/1.0d16
        z_old(ipart)=(dble(anint(z_coord)))/1.0d16
        x_now(ipart)=x_old(ipart)
```

```
            y_now(ipart)=y_old(ipart)
            z_now(ipart)=z_old(ipart)
            coord(ipart,1)=x_old(ipart)
            coord(ipart,2)=y_old(ipart)
            coord(ipart,3)=z_old(ipart)
        enddo
        return
end subroutine initial_coord
```

! 本子程序用于修正刚性弹丸网格划分引起的误差，四舍五入到小数点后一定位数

```
subroutine initial_projectile()
    use scalars
    use serialArrays
    use globalParameters
    use impact_parameters
    implicit none
    real(8)::x_coord,y_coord,z_coord
    do ipart=1,npart_projectile
        x_coord=coord_projectile(ipart,1)*1.0d16
        y_coord=coord_projectile(ipart,2)*1.0d16
        z_coord=coord_projectile(ipart,3)*1.0d16
        coord_projectile(ipart,1)=(dble(anint(x_coord)))/1.0d16
        coord_projectile(ipart,2)=(dble(anint(y_coord)))/1.0d16
        coord_projectile(ipart,3)=(dble(anint(z_coord)))/1.0d16
    enddo
    coord_projectile_old=coord_projectile
    !数据文件中直接给出质心坐标
    Xcoor_centroid0=Xcoor_centroid      !弹丸质心坐标，不断改变
    Ycoor_centroid0=Ycoor_centroid
    Zcoor_centroid0=Zcoor_centroid
    return
end subroutine initial_projectile
```

B.1.4 子程序 2：element_form

子程序 element_form 的功能是生成物质点近场域内的粒子链表，为后续非局部相互作用及积分求和计算奠定基础。

```
subroutine element_form()
    use scalars
    use serialArrays
```

B.1 键型近场动力学的显式动力学算法 FORTRAN 源程序

```fortran
    use globalParameters
    use impact_parameters
    implicit none
    integer::j_num,flag
    real(8)::cutsq,delx0,dely0,delz0,rsq0
    pehor=n_delta*lat_c
    pehsq=pehor*pehor
    cutsq=pehsq+1.0d-15    ! cutsq:在截断半径的平方上,增加一个小量'1
        d-15'减小系统误差
    vol_1=lat_c*lat_c*lat_c
    b_len=0.0d0;ele_b=0;ele_b0=0

    do ipart=1,npart
        j_num=0
        do jpart=1,npart
            if(ipart.eq.jpart) go to 200
            delx0= x_old(jpart) - x_old(ipart)
            dely0= y_old(jpart) - y_old(ipart)
             delz0= z_old(jpart) - z_old(ipart)
              rsq0=delx0*delx0+dely0*dely0+delz0*delz0
            if(rsq0.le.cutsq) then
                j_num=j_num+1
                ele_b(ipart,j_num)=jpart
                ele_b0(ipart,j_num)=jpart
                b_len(ipart,j_num)=sqrt(rsq0)
            endif
200         flag=1
        enddo
        ele_n0(ipart)=j_num
        ele_n(ipart)=j_num
    enddo

    b_fx_old=0.0d0    !给出Velocity-Verlet差分格式的初始值
    b_fy_old=0.0d0
    b_fz_old=0.0d0
    v_old=velocity_initial
    v_middle=velocity_initial
    v_now=velocity_initial
    X_acceleration_old=0.0d0
    Y_acceleration_old=0.0d0
```

```
    Z_acceleration_old=0.0d0
    return
end subroutine element_form
```

B.1.5 子程序 3：add_load_displacement

子程序 add_load_displacement 的功能是实现位移边界加载，施加位移增量或总量。

```
subroutine add_load_displacement()
    use scalars
    use serialArrays
    use globalParameters
    implicit none
    integer::idot
    real(8)::delt_t,v_scale

    !位移加载，每步施加一给定位移增量
    if (load_type==1) then
        delt_t=1.0d0    ! delt_t取[0~1]范围内的值，现取为0.8；在
            delt_t时间内位移加载完成，其后不再加载
    if(dble(istep)<delt_t*nstep) then
        v_scale=1.0d0/(delt_t*dble(nstep))
    else
        v_scale=0.0d0
    end if
    do iload_dis=1,nload_dis
        idot=load_dis_id(iload_dis)
        x_old(idot)=x_old(idot)+load_xdis_value(iload_dis)*v_scale
        y_old(idot)=y_old(idot)+load_ydis_value(iload_dis)*v_scale
        z_old(idot)=z_old(idot)+load_zdis_value(iload_dis)*v_scale
        end do
    end if
    return
end subroutine add_load_displacement
```

B.1.6 子程序 4：add_load_bc_force

子程序 add_load_bc_force 的功能是实现外荷载的施加，包括恒定不变的外力，线性或突加的面力荷载和体力荷载。

```
subroutine add_load_bc_force()
```

```fortran
    use scalars
    use serialArrays
    use globalParameters
    implicit none
    real::v_scale,delt_t
!**********************************************************
! 力边界区域物质点体力:力边界条件已经施加,并一直存在*
!**********************************************************
if(nBC_force/=0) then
    do ibc_force=1,nBC_force
        b_fx(bc_force_id(iBC_force))=b_fx(bc_force_id(iBC_force))
            +bc_xforce_value(iBC_force)/vol_1
        b_fy(bc_force_id(iBC_force))=b_fy(bc_force_id(iBC_force))
            +bc_yforce_value(iBC_force)/vol_1
        b_fz(bc_force_id(iBC_force))=b_fz(bc_force_id(iBC_force))
            +bc_zforce_value(iBC_force)/vol_1
    enddo
end if


!****************
! 力加载区域体力*
!****************
if(load_type==2) then   !比例加载:delt_t,[0~1]时间内,线性加载完
    毕;剩余1-detl_t时间步保持不变加载
    delt_t=0.1d0    ! 可根据加载方式,进行修改
    if(dble(istep)<delt_t*dble(nstep)) then
        v_scale=1.0d0/delt_t*dble(istep)/dble(nstep)
    else
        v_scale=1.0d0
    end if
    do iload_force=1,nload_force
b_fx(load_force_id(iload_force))=b_fx(load_force_id(iload_force)
    ) +& v_scale*load_xforce_value(iload_force)/vol_1
b_fy(load_force_id(iload_force)) =b_fy(load_force_id(iload_force
    )) +&
                        v_scale*load_yforce_value(
                            iload_force)/vol_1
b_fz(load_force_id(iload_force))=b_fz(load_force_id(iload_force)
    ) +&
                        v_scale*load_zforce_value(
```

```
                                        iload_force)/vol_1
        enddo
    elseif(load_type==20) then     !瞬时加载
        v_scale=1.0d0
        do iload_force=1,nload_force
    b_fx(load_force_id(iload_force))=b_fx(load_force_id(iload_force)&
            ) + &
                            v_scale*load_xforce_value(&
                                iload_force)/vol_1
    b_fy(load_force_id(iload_force))=b_fy(load_force_id(iload_force)&
            ) +&
                            v_scale*load_yforce_value(&
                                iload_force)/vol_1
    b_fz(load_force_id(iload_force))=b_fz(load_force_id(iload_force)&
            ) +&
                            v_scale*load_zforce_value(&
                                iload_force)/vol_1
        enddo
    endif
    !*************
    !是否考虑自重*
    !*************
    if(gravity_type==1) then
        b_fx=b_fx
        b_fy=b_fy
        b_fz=b_fz-g*dens
    elseif(gravity_type==0) then
        b_fx=b_fx;b_fy=b_fy;b_fz=b_fz
end if

    return
end subroutine add_load_bc_force
```

B.1.7 子程序 5: add_load_impact_rigid

子程序 add_load_impact_rigid 的功能是实现刚性体冲击可变形体的接触过程模拟。

```
subroutine add_load_impact_rigid()
    use scalars
    use serialArrays
    use globalParameters
```

```fortran
use impact_parameters
implicit none
real(8)::sum_fx,sum_fy,sum_fz,mass
real(8)::deltx,delty,deltz,rijsq,rij
real(8)::x_temp,y_temp,z_temp,v_xtemp,v_ytemp,v_ztemp,f_xtemp, &
    f_ytemp,f_ztemp
integer::i

if(load_type==3)then
!采用Velocity-Verlet差分格式更新弹丸质心速度和位置
Xcoor_centroid0=Xcoor_centroid0+(velocity_projectile(1)+0.5d0* &
    X_acceleration_old*tstep)*tstep
Ycoor_centroid0=Ycoor_centroid0+(velocity_projectile(2)+0.5d0* &
    Y_acceleration_old*tstep)*tstep
Zcoor_centroid0=Zcoor_centroid0+(velocity_projectile(3)+0.5d0* &
    Z_acceleration_old*tstep)*tstep
!更新弹丸整体坐标信息
x_dis_centroid=Xcoor_centroid0-Xcoor_centroid
y_dis_centroid=Ycoor_centroid0-Ycoor_centroid
z_dis_centroid=Zcoor_centroid0-Zcoor_centroid
coord_projectile(:,1)=coord_projectile_old(:,1)+x_dis_centroid
coord_projectile(:,2)=coord_projectile_old(:,2)+y_dis_centroid
coord_projectile(:,3)=coord_projectile_old(:,3)+z_dis_centroid
!更新弹丸合力和加速度
! Madenci 2014年专著中的冲击接触模型，表示冲击过程中弹丸与颗粒的
!    接触作用
sum_fx=0.0d0     !刚性体所受合力
sum_fy=0.0d0
sum_fz=0.0d0
!刚性体为柱体
if(trim(rigid_shape)=="cylinder")then
mass=pi*radius_projectile**2*heigh_projectile*dens_projectile
mass=3.0d0
do ipart=1,npart
    deltx=Xcoor_centroid0-x_old(ipart)     !注意数据文件坐标轴与计
        算程序一致
    delty=Ycoor_centroid0-y_old(ipart)
    deltz=Zcoor_centroid0-z_old(ipart)
    rijsq=deltx*deltx+delty*delty
    rij=sqrt(rijsq)
```

```fortran
        if(rij<radius_projectile+0.5d0*lat_c .and. abs(deltz)<0.5d0*
           heigh_projectile+0.5d0*lat_c) then !判断靶体与弹丸是否叠
           合
            !step1.将物质点移到最近的边界面:x_temp、y_temp、z_temp
            x_temp=x_old(ipart)
            y_temp=y_old(ipart)
            z_temp=z_old(ipart)-(0.5d0*heigh_projectile+0.5d0*lat_c-
               deltz)
            !step2:更新物质点速度:v_xtemp、v_ytemp、v_ztemp
            v_xtemp=(x_temp-x_old(ipart))/tstep
            v_ytemp=(y_temp-y_old(ipart))/tstep
            v_ztemp=(z_temp-z_old(ipart))/tstep
            !step3:计算物质点由于弹丸冲击所受合力:f_xtemp、f_ytemp、
               f_ztemp
            f_xtemp=dens(ipart)*vol_1*(v_xtemp-v_old(ipart,1))/tstep
            f_ytemp=dens(ipart)*vol_1*(v_ytemp-v_old(ipart,2))/tstep
            f_ztemp=dens(ipart)*vol_1*(v_ztemp-v_old(ipart,3))/tstep
            !step4:累加,计算弹丸所受合力
            sum_fx=sum_fx+f_xtemp
            sum_fy=sum_fy+f_ytemp
            sum_fz=sum_fz+f_ztemp
            !step5:更新物质点位置、速度,用于后续迭代求解
            x_old(ipart)=x_temp
            y_old(ipart)=y_temp
            z_old(ipart)=z_temp
            v_old(ipart,1)=v_xtemp
            v_old(ipart,2)=v_ytemp
            v_old(ipart,3)=v_ztemp
            b_fx(ipart)=b_fx(ipart)+f_xtemp
            b_fy(ipart)=b_fy(ipart)+f_ytemp
            b_fz(ipart)=b_fz(ipart)+f_ztemp
        else
            continue
        endif
    enddo
    endif

    sum_fx=-sum_fx
    sum_fy=-sum_fy
    sum_fz=-sum_fz
```

B.1 键型近场动力学的显式动力学算法 FORTRAN 源程序

```fortran
    !计算弹丸加速度
    X_acceleration=sum_fx/mass
    Y_acceleration=sum_fy/mass
    Z_acceleration=sum_fz/mass        !考虑弹丸自重：-g
    !更新弹丸质心速度
    Velocity_projectile(1)=Velocity_projectile(1)+0.5d0*tstep*(
        X_acceleration_old+X_acceleration)
    Velocity_projectile(2)=Velocity_projectile(2)+0.5d0*tstep*(
        Y_acceleration_old+Y_acceleration)
    Velocity_projectile(3)=Velocity_projectile(3)+0.5d0*tstep*(
        Z_acceleration_old+Z_acceleration)
    X_acceleration_old=X_acceleration
    Y_acceleration_old=Y_acceleration
    Z_acceleration_old=Z_acceleration
    !输出弹丸位置、速度、加速度变化信息
    if(mod(istep,i_out)==0) then
        write(1,'(1x,i10,3(2x,e12.5))') istep,x_dis_centroid,
            y_dis_centroid,z_dis_centroid
        write(2,'(1x,i10,3(2x,e12.5))') istep,(Velocity_projectile(i
            ),i=1,3)
        write(3,'(1x,i10,3(2x,e12.5))') istep,X_acceleration_old,
            Y_acceleration_old,Z_acceleration_old
    endif
    endif
    return
end subroutine add_load_impact_rigid
```

B.1.8 子程序 6：equation_solve

子程序 equation_solve 的功能是根据 Verlet-velocity 差分格式，进行近场动力学运动方程的时程积分计算，实现物质点受力、加速度、速度和位置的更新。

```fortran
subroutine equation_solve()
    use scalars
    use serialArrays
    use globalParameters
    implicit none
    real(8)::tstsq,coe1
    tstsq=tstep*tstep
    coe1=0.5d0*Cv*tstep
    !更新物质点位置
    do ipart=1,npart
```

```
    v_middle(ipart,1)=v_old(ipart,1)*(dens(ipart)-coe1)/dens(ipart)
       +0.5d0*tstep*b_fx_old(ipart)/dens(ipart)
    v_middle(ipart,2)=v_old(ipart,2)*(dens(ipart)-coe1)/dens(ipart)
       +0.5d0*tstep*b_fy_old(ipart)/dens(ipart)
      v_middle(ipart,3)=v_old(ipart,3)*(dens(ipart)-coe1)/dens(
         ipart)+0.5d0*tstep*b_fz_old(ipart)/dens(ipart)
      x_now(ipart)=x_old(ipart)+v_middle(ipart,1)*tstep
      y_now(ipart)=y_old(ipart)+v_middle(ipart,2)*tstep
      z_now(ipart)=z_old(ipart)+v_middle(ipart,3)*tstep
      !计算位移
      x_dis(ipart)=x_now(ipart)-coord(ipart,1)
      y_dis(ipart)=y_now(ipart)-coord(ipart,2)
      z_dis(ipart)=z_now(ipart)-coord(ipart,3)
   enddo
   !调用力场计算子程序
   call model_short_range_force()
   call model_PMB()
   !更新物质点速度
   do ipart=1,npart
   v_now(ipart,1)=v_middle(ipart,1)*dens(ipart)/(dens(ipart)+coe1)+
      &
              0.5d0*tstep*b_fx(ipart)/(dens(ipart)+coe1)
      v_now(ipart,2)=v_middle(ipart,2)*dens(ipart)/(dens(ipart)+
         coe1)+ &
              0.5d0*tstep*b_fy(ipart)/(dens(ipart)+coe1)
      v_now(ipart,3)=v_middle(ipart,3)*dens(ipart)/(dens(ipart)+
         coe1)+ &
              0.5d0*tstep*b_fz(ipart)/(dens(ipart)+coe1)
   enddo
   !更新下一次迭代数据
   b_fx_old=b_fx
   b_fy_old=b_fy
   b_fz_old=b_fz
   v_old=v_now
   x_old=x_now
   y_old=y_now
   z_old=z_now
   return
end subroutine equation_solve
```

B.1.9 子程序 7: model_PMB

子程序 model_PMB 的功能是根据键型近场动力学 PMB 模型, 实现物质点间非局部相互作用力 f 及物质点所受合力 L 的计算。

```fortran
subroutine model_PMB()
    use scalars
    use serialArrays
    use globalParameters
    implicit none
    real(8)::delx,dely,delz,rsq,r,dr,stret,fbond
    real(8)::cos_x, cos_y, cos_z
    integer::i_jn,jnum,flag
    integer::ibond,num_b   !记录损伤
    real(8)::max_damage,sum_damage
    real(8)::max_damage0,sqrt_dis1,sqrt_dis2,crack_vel1,crack_vel2 !
        裂纹传播速度
    integer::i,count_min,sqrt_tempnum       !以下三者记录可能的裂尖点编
        号、距离
    integer,dimension(:),allocatable::damage_min
    real(8),dimension(:),allocatable::sqrt_dis
    real(8)::xx1,xx2,yy1,yy2   !起裂前后两次的物质点位置
    real(8)::xxx1,xxx2,yyy1,yyy2
    integer,dimension(1)::crack_point_1,crack_point_2  !起裂点编号
    !三维PMB模型
    !c_mod(1)=6.0d0*mat(1,1)/(pi*pehsq*pehsq*(1.0d0-2.0d0*mat(1,2)))
    !c_str(1)=sqrt(5.0d0*mat(1,3)*(1.0d0-2.0d0*mat(1,2))/(3.0d0*mat
        (1,1)*pehor))
    !三维改进PMB模型
    c_mod(1)=36.0d0*mat(1,1)/(pi*pehsq*pehsq*(1.0d0-2.0d0*mat(1,2)))
    !c_str(1)=sqrt(35.0d0*mat(1,3)*(1.0d0-2.0d0*mat(1,2))/(34.0d0*
        mat(1,1)*pehor))
    c_str(1)=sqrt(5.0d0*mat(1,3)*(1.0d0-2.0d0*mat(1,2))/(3.0d0*mat
        (1,1)*pehor))
    mod_c=c_mod(1)
    !str_c=c_str(1)
    str_c=0.015d0
    !s0=c_str(1)
    s0=0.015d0
    half_lat=0.5d0*lat_c
    !计算所有物质点所受键力
```

```
do ipart=1,npart
    i_jn=ele_n0(ipart)
    do jnum=1,i_jn
        jpart=ele_b(ipart,jnum)
        if(jpart.eq.0) goto 400
        delx= x_now(jpart) - x_now(ipart)
        dely= y_now(jpart) - y_now(ipart)
        delz= z_now(jpart) - z_now(ipart)
        rsq=delx*delx+dely*dely+delz*delz
        r=sqrt(rsq)
        dr=r-b_len(ipart,jnum)
        cos_x= delx/r    !当前构型中，键的方向向量
        cos_y= dely/r
        cos_z= delz/r
        !体积修正
        if(abs(r-pehor)<half_lat) then
            vfrac_scale=(pehor-r)/lat_c+0.5d0
        elseif(pehor-r>half_lat) then
            vfrac_scale=1.0d0
        else
            vfrac_scale=0.0d0
        endif

        if(abs(dr).lt.2.0e-20) dr=0.0d0
        if(b_len(ipart,jnum) .ne. 0.d00)   stret=dr/b_len(ipart,
            jnum)
        str_c=min(s0(ipart),s0(jpart))   !对称化操作
        fbond=0.0d0
        if(stret .gt. str_c) then
            ele_b(ipart,jnum)=0
            fbond=0.0d0
        elseif(stret .lt. str_c) then
            mod_c= c_mod(1)    ! PMB模型
            mod_c=c_mod(1)*(1.0d0-rsq/pehsq)**2    !改进PMB模型
            fbond=mod_c*stret*vol_1*vfrac_scale
        endif
        b_fx(ipart)=b_fx(ipart)+ fbond*cos_x
        b_fy(ipart)=b_fy(ipart)+ fbond*cos_y
        b_fz(ipart)=b_fz(ipart)+ fbond*cos_z
        flag=1
```

B.1 键型近场动力学的显式动力学算法 FORTRAN 源程序

```fortran
        enddo
    enddo
    !*****************************
    !     计算最大损伤值、总损伤值*
    !*****************************
    do ipart=1,npart
        ibond=0
        i_jn=ele_n0(ipart)
        do jpart=1,i_jn
            if(ele_b(ipart,jpart).ne.0) ibond=ibond+1
        enddo
        ele_n(ipart)=ibond
        damage(ipart)=1.0d0-dble(ele_n(ipart))/dble(ele_n0(ipart))
    enddo
    max_damage=maxval(damage)
    sum_damage=sum(damage)/dble(npart)
    write(6,101) istep,max_damage,sum_damage    !某一特定输出步的最大
        损伤值、总损伤值
101 format(1x,i10,2e12.5)

    return
end subroutine model_PMB
```

B.1.10 子程序 8：model_short_range_force

子程序 model_short_range_force 的功能是实现 Park 等[1]提出的短程排斥力模型，防止完全损伤的物质点与结构发生非物理重叠变形。

```fortran
subroutine model_short_range_force()
    use scalars
    use serialArrays
    use globalParameters
    implicit none
    integer::j_num,i_num,flag
    real(8):: deltx,delty,deltz,rijsq,rij,d_ij,drij,fpair
    real(8):: cos_x,cos_y, cos_z
    c_mod(1)=6.0d0*mat(1,1)/(pi*pehsq*pehsq*(1.0d0-2.0d0*mat(1,2)))
            !三维PMB模型的微模量
```

---

[1] Parks M L, Lehoucq R B, Plimpton S J, et al. Implementing peridynamics within a molecular dynamics code [J]. Computer Physics Communications, 2008, 179(11):777-783.

```fortran
        c_str(1)=sqrt(5.0d0*mat(1,3)*(1.0d0-2.0d0*mat(1,2))/(3.0d0*mat
           (1,1)*pehor))
        mod_c=c_mod(1)
        str_c=c_str(1)
        s0=c_str(1)
        half_lat=0.5d0*lat_c

        do ipart=1,npart
            j_num=ele_n0(ipart)
            do i_num=1,j_num
                jpart=ele_b0(ipart,i_num)
                deltx= x_now(jpart) -x_now(ipart)
                delty= y_now(jpart)-y_now(ipart)
                deltz= z_now(jpart)-z_now(ipart)
                rijsq=deltx*deltx+delty*delty+deltz*deltz
                rij=sqrt(rijsq)
                cos_x= deltx/rij
                cos_y= delty/rij
                cos_z= deltz/rij
                d_ij=min((0.9d0*b_len(ipart,i_num)),(1.35d0*lat_c))    !
                    短程力作用距离
                if(rij.lt.d_ij) then
                    drij=rij-d_ij
                     if(rij.lt.2.0d-20) rij=2.0d-20   !处理两质点非常近
                        的情况，即rij=0的情况
                    fpair=0.0d0
                    if(rij.gt.0) fpair=15.0d0*mod_c*drij/pehor*vol_1
                    b_fx(ipart)=b_fx(ipart)+cos_x*fpair
                    b_fy(ipart)=b_fy(ipart)+ cos_y*fpair
                     b_fz(ipart)=b_fz(ipart)+cos_z*fpair
                endif
            enddo
        enddo
        return
end subroutine model_short_range_force
```

B.1.11 子程序 9: data_output

子程序 data_output 的功能是输出计算结果，包括位移、速度、损伤等信息，按照可视化软件 ENSIGHT 规定的格式输出。

```fortran
subroutine data_output()
```

B.1 键型近场动力学的显式动力学算法 FORTRAN 源程序

```fortran
    use scalars
    use serialArrays
    use globalParameters
    use impact_parameters
    implicit none
!   按ensight软件格式输出几何信息、位移、损伤
!*******************************************
    if(load_type==3.or.load_type==31.or.load_type==4) then
        call impact_output()
    else
        call uniformity_output()
    end if
    return
end subroutine data_output

subroutine impact_output()
    use scalars
    use serialArrays
    use globalParameters
    use impact_parameters
    implicit none
    character  (len=20):: cstr
    character  (len=20):: cFmt
    integer::i
    real(8)::a
    i=int(istep/i_out)
    !------------------------!
    !写靶体物质点坐标
    write( cStr , '(i5)' ) i
    Open( 10 , File = 'point.geo' // AdjustL(Trim( cStr ) ) )
    write(10,'(a)')"enSight Model Geometry File"
    write(10,'(a)')"enSight 7.4.1"
    write(10,'(a)')"node id assign"
    write(10,'(a)')"element id assign"
    write(10,'(a)')"extents"
    write(10,'(a)')"-1.00000e+03 1.00000e+03"
    write(10,'(a)')"-1.00000e+03 1.00000e+03"
    write(10,'(a)')"-1.00000e+03 1.00000e+03"
    write(10,'(a)')"part"
    write(10,'(a)')"            1"
```

```
write(10,'(a)')"point"
write(10,'(a)')"coordinates"
write(10,'(I10)') npart
do ipart=1,npart
   a=x_now(ipart)
   if(a.ge.0) then
      cFmt="(1x,e11.5)"
    else
      cFmt="(e12.5)"
    end if
    write( 10 , cFmt ) a
end do
do ipart=1,npart
   a=y_now(ipart)
   if(a.ge.0) then
      cFmt="(1x,e11.5)"
    else
      cFmt="(e12.5)"
    end if
    write( 10 , cFmt ) a
enddo
do ipart=1,npart
   a=z_now(ipart)
   if(a.ge.0) then
      cFmt="(1x,e11.5)"
    else
      cFmt="(e12.5)"
    end if
    write( 10 , cFmt ) a
enddo
!---------------------!
!写靶体损伤
write( cStr , '(i5)' ) i
open( 11 , File = 'point.Damage' // AdjustL(Trim( cStr ) ) )
write(11,'(a)')"Damage"
write(11,'(a)')"part"
write(11,'(a)')"           1"
write(11,'(a)')"coordinates"
do ipart=1,npart
   a=damage(ipart)
```

```fortran
      if(a.ge.0) then
         cFmt="(1x,e11.5)"
      else
         cFmt="(e12.5)"
      endif
      write(11,cFmt ) a
   enddo
!------------------!
!写靶体位移x
write( cStr , '(i5)' ) i
open(12,File= 'point.Displacement-X' // AdjustL(Trim(cStr) ) )
write(12,'(a)')"Displacement-X"
write(12,'(a)')"part"
write(12,'(a)')"              1"
write(12,'(a)')"coordinates"
   do ipart=1,npart
      a=x_dis(ipart)
      if (a.ge.0) then
      cFmt="(1x,e11.5)"
      else
      cFmt="(e12.5)"
      end if
      write(12,cFmt ) a
   enddo
!写靶体位移y
write( cStr , '(i5)' ) i
open(13,File= 'point.Displacement-Y' // AdjustL(Trim( cStr ) ) )
write(13,'(a)')"Displacement-Y"
write(13,'(a)')"part"
write(13,'(a)')"              1"
write(13,'(a)')"coordinates"
   do ipart=1,npart
      a=y_dis(ipart)
      if(a.ge.0) then
      cFmt="(1x,e11.5)"
      else
      cFmt="(e12.5)"
      end if
      write(13,cFmt) a
   enddo
```

```
!写靶体位移z
write( cStr , '(i5)' ) i
open(14,File= 'point.Displacement-Z' // AdjustL(Trim( cStr ) ) )
write(14,'(a)')"Displacement-Z"
write(14,'(a)')"part"
write(14,'(a)')"          1"
write(14,'(a)')"coordinates"
    do ipart=1,npart
       a=z_dis(ipart)
       if(a.ge.0) then
       cFmt="(1x,e11.5)"
       else
       cFmt="(e12.5)"
       end if
       write(14,cFmt) a
   enddo
!------------!
!写靶体速度x
write( cStr , '(i5)' ) i
open( 15 , File = 'point.Velocity-X' // AdjustL(Trim( cStr ) ) )
write(15,'(a)')"Velocity-X"
write(15,'(a)')"part"
write(15,'(a)')"          1"
write(15,'(a)')"coordinates"
    do ipart=1,npart
       a=v_now(ipart,1)
       if (a.ge.0) then
       cFmt="(1x,e11.5)"
       else
       cFmt="(e12.5)"
       end if
       write(15,cFmt ) a
    enddo
!写靶体速度y
write( cStr , '(i5)' ) i
open( 16 , File = 'point.Velocity-Y' // AdjustL(Trim( cStr ) ) )
write(16,'(a)')"Velocity-Y"
write(16,'(a)')"part"
write(16,'(a)')"          1"
write(16,'(a)')"coordinates"
```

```fortran
   do ipart=1,npart
      a=v_now(ipart,2)
      if(a.ge.0) then
      cFmt="(1x,e11.5)"
      else
      cFmt="(e12.5)"
      end if
      write(16,cFmt) a
   enddo
!写靶体速度z
write( cStr , '(i5)' ) i
open( 17 , File = 'point.Velocity-Z' // AdjustL(Trim( cStr ) ) )
write(17,'(a)')"Velocity-Z"
write(17,'(a)')"part"
write(17,'(a)')"            1"
write(17,'(a)')"coordinates"
   do ipart=1,npart
      a=v_now(ipart,3)
      if(a.ge.0) then
      cFmt="(1x,e11.5)"
      else
      cFmt="(e12.5)"
      end if
      write(17,cFmt) a
   enddo

!------------------------!
!写弹丸坐标
write(10,'(a)')"part"
write(10,'(a)')"            2"
write(10,'(a)')"point"
write(10,'(a)')"coordinates"
write(10,'(I10)') npart_projectile
do ipart=1,npart_projectile
   a=coord_projectile(ipart,1)
   if(a.ge.0) then
      cFmt="(1x,e11.5)"
    else
      cFmt="(e12.5)"
    end if
```

```fortran
         write( 10 , cFmt ) a
      end do
      do ipart=1,npart_projectile
         a=coord_projectile(ipart,2)
         if(a.ge.0) then
            cFmt="(1x,e11.5)"
         else
            cFmt="(e12.5)"
         end if
         write( 10 , cFmt ) a
      enddo
      do ipart=1,npart_projectile
         a=coord_projectile(ipart,3)
         if(a.ge.0) then
            cFmt="(1x,e11.5)"
         else
            cFmt="(e12.5)"
         end if
         write( 10 , cFmt ) a
      enddo
      !----------------------!
      !写弹丸损伤
      write(11,'(a)')"part"
      write(11,'(a)')"           2"
      write(11,'(a)')"coordinates"
      do ipart=1,npart_projectile
         a=0.
         if(a.ge.0) then
            cFmt="(1x,e11.5)"
         else
            cFmt="(e12.5)"
         endif
         write(11,cFmt ) a
      enddo
      !------------------------!
      !写弹丸位移x
      write(12,'(a)')"part"
      write(12,'(a)')"           2"
      write(12,'(a)')"coordinates"
         do ipart=1,npart_projectile
```

```fortran
            a=x_dis_centroid
            if(a.ge.0) then
               cFmt="(1x,e11.5)"
            else
               cFmt="(e12.5)"
            end if
            write(12,cFmt ) a
         enddo
!写弹丸位移y
      write(13,'(a)')"part"
      write(13,'(a)')"         2"
      write(13,'(a)')"coordinates"
         do ipart=1,npart_projectile
            a=y_dis_centroid
            if(a.ge.0) then
               cFmt="(1x,e11.5)"
            else
               cFmt="(e12.5)"
            endif
            write(13,cFmt) a
         enddo
!写弹丸位移z
      write(14,'(a)')"part"
      write(14,'(a)')"         2"
      write(14,'(a)')"coordinates"
         do ipart=1,npart_projectile
            a=z_dis_centroid
            if(a.ge.0) then
               cFmt="(1x,e11.5)"
            else
               cFmt="(e12.5)"
            endif
            write(14,cFmt) a
         enddo
!------------------------!
!写弹丸速度x
      write(15,'(a)')"part"
      write(15,'(a)')"         2"
      write(15,'(a)')"coordinates"
         do ipart=1,npart_projectile
```

```fortran
            a=Velocity_projectile(1)
            if(a.ge.0) then
                cFmt="(1x,e11.5)"
            else
                cFmt="(e12.5)"
            end if
            write(15,cFmt) a
        enddo
    !写弹丸速度y
    write(16,'(a)')"part"
    write(16,'(a)')"           2"
    write(16,'(a)')"coordinates"
        do ipart=1,npart_projectile
            a=Velocity_projectile(2)
            if(a.ge.0) then
               cFmt="(1x,e11.5)"
            else
               cFmt="(e12.5)"
            endif
            write(16,cFmt) a
        enddo
    !写弹丸速度z
    write(17,'(a)')"part"
    write(17,'(a)')"           2"
    write(17,'(a)')"coordinates"
        do ipart=1,npart_projectile
            a=Velocity_projectile(3)
            if(a.ge.0) then
               cFmt="(1x,e11.5)"
            else
               cFmt="(e12.5)"
            endif
            write(17,cFmt) a
        enddo
    return
end subroutine impact_output
```

B.2 输入数据文件

B.2.1 输入数据文件的一般格式

输入数据文件包含的信息：计算总控信息，物体的坐标信息、材料信息、约束信息、加载信息，对于冲击问题，还包括冲击弹丸的坐标信息、材料信息、加载信息，冲击接触算法信息等。具体如下：

```
npart, nBC_dis, nBC_force, nload_velocity, nload_dis,
    nload_force, lat_c    !粒子总数、位移约束粒子总数、力边界粒子
        总数、速度加载粒子总数、位移加载粒子总数、力加载粒子总数、粒
        子间距
load_type, gravity_type, tstep, nstep         !加载类型、是否考虑
        重力、时间步长、总迭代步数
velocity_initial                              !构型初始速度
1       x(1)        y(1)        z(1)          !靶体物质点编号ID, 坐标位
    置x, y, z, 共npart组
2       x(2)        y(3)        z(2)
3       x(3)        y(3)        z(3)
4       x(4)        y(4)        z(4)
5       x(5)        y(5)        z(5)
6       x(6)        y(6)        z(6)
⋮       ⋮           ⋮           ⋮
⋮       ⋮           ⋮           ⋮
npart-1 x(npart-1)  y(npart-1)  z(npart-1)
npart   x(npart)    y(npart)    z(npart)
bc_dis_id(iBC_dis), bc_dis_code(iBC_dis), bc_xdis_value(iBC_dis),
    bc_ydis_value(iBC_dis), bc_zdis_value(iBC_dis)    !位移约束
        粒子编号，约束标记：001,010,100,011,101,110,111, 其中1表示该
        方向有约束、0表示无约束，x,y,z方向的位移约束值，共nBC_dis行
        数据
bc_force_id(iBC_force), bc_xforce_value(iBC_force),
    bc_yforce_value(iBC_force), bc_zforce_value(iBC_force) !力约
        束粒子编号，x,y,z方向的力约束值，共nBC_ force行数据
load_velocity_id(iload_velocity), load_xvelocity_value(
    iload_velocity), load_yvelocity_value(iload_velocity),
    load_zvelocity_value(iload_velocity) !速度加载粒子编号，x,y,
    z方向的加载速度值，共nload_velocity行数据
load_dis_id(iload_dis), load_xdis_value(iload_dis),
```

```
        load_ydis_value(iload_dis), load_zdis_value(iload_dis) !位移
            加载粒子编号，x,y,z方向的加载位移值，共nload_dis行数据
    load_force_id(iload_force),load_xforce_value(iload_force),
        load_yforce_value(iload_force), load_zforce_value
        (iload_force) !力加载粒子编号，x,y,z方向的加载力值，共nload_
            force行数据
    n_mat                  !材料总数
    mat(i_mat,1:4)         !材料参数：每行分别为弹性模量、泊松比、能量释
            放率、材料密度，共n_mat行
    npart_projectile, dens_projectile  !弹丸信息：粒子总数、密度
    rigid_shape                        !弹丸形状:cylinder、sphere、cubic等
    radius_projectile, heigh_projectile, Xcoor_centroid,
        Ycoor_centroid, Zcoor_centroid    !柱形弹丸的半径、高度、中
            心坐标x,y,z
    Velocity_projectile(1:ndime)       !弹丸初速度的x,y,z速度分量，
        ndime表示问题维度
    ID(1)       x(1)         y(1)          z(1)     !弹丸物质点编号ID，位置坐
        标x,y,z，共npart_projectile组
    ID(2)       x(2)         y(3)          z(2)
    ID(3)       x(3)         y(3)          z(3)
    ID(4)       x(4)         y(4)          z(4)
    ID(5)       x(5)         y(5)          z(5)
     ⋮           ⋮            ⋮             ⋮
     ⋮           ⋮            ⋮             ⋮
    ID(npart_projectile)     x(npart_projectile)
     y(npart_projectile)     z(npart_projectile)
```

B.2.2 马氏体时效钢冲击破坏试验模拟的输入数据文件

以 5.5.3 节中 Kalthoff-Winkler 冲击试验模拟为例。马氏体时效钢靶体长 200mm、高 100mm、厚 9mm，采用各方向间距均为 1mm 的物质点离散构型，共有 178200 个物质点，靶体材料弹性模量 $E=191$GPa，泊松比 $\nu=1/4$，质量密度 $\rho=8000$kg/m^3，临界伸长率 $s_0=0.015$，初速度为零。刚性圆柱体弹丸半径 50mm、高度 50mm，质量 $m=1.57$kg，质量密度 8000kg/m^3，竖向初速度 $v_0=-32$m/s，为便于显示，离散为 579 个物质点。显式动力学计算的时间步长 $\Delta t = 8 \times 10^{-8}$s，共计算 4000 步。则相应计算所需的数据文件如下所示：

```
178200   0    0    0    0    0    1.00E-03  !粒子总数、位移约束粒子
    总数、力边界粒子总数、速度加载粒子总数、位移加载粒子总数、力加载
    粒子总数、粒子尺寸
```

B.2 输入数据文件

```
3       0      8.0e-8      4e3          !加载类型、是否考虑重力、时间步长、总迭代步数
0     !构型初始速度
 1       0.00000000000         0.00000000000         0.00000000000       ! 粒子编号ID，坐标x，y，z
 2       0.199000000000        0.00000000000         0.00000000000
 3       0.100000000000E-02    0.00000000000         0.00000000000
 4       0.200000000000E-02    0.00000000000         0.00000000000
 5       0.300000000000E-02    0.00000000000         0.00000000000
 6       0.400000000000E-02    0.00000000000         0.00000000000
 7       0.500000000000E-02    0.00000000000         0.00000000000
 8       0.600000000000E-02    0.00000000000         0.00000000000
 9       0.700000000000E-02    0.00000000000         0.00000000000
10       0.800000000000E-02    0.00000000000         0.00000000000
11       0.900000000000E-02    0.00000000000         0.00000000000
12       0.100000000000E-01    0.00000000000         0.00000000000
13       0.110000000000E-01    0.00000000000         0.00000000000
14       0.120000000000E-01    0.00000000000         0.00000000000
15       0.130000000000E-01    0.00000000000         0.00000000000
16       0.140000000000E-01    0.00000000000         0.00000000000
17       0.150000000000E-01    0.00000000000         0.00000000000
18       0.160000000000E-01    0.00000000000         0.00000000000
19       0.170000000000E-01    0.00000000000         0.00000000000
  ⋮             ⋮                    ⋮                     ⋮
178196   0.710000000000E-01    0.980000000000E-01    0.300000000000E-02
178197   0.710000000000E-01    0.980000000000E-01    0.400000000000E-02
178198   0.710000000000E-01    0.980000000000E-01    0.500000000000E-02
178199   0.710000000000E-01    0.980000000000E-01    0.600000000000E-02
178200   0.710000000000E-01    0.980000000000E-01    0.700000000000E-02
1    !材料总数
1.91E+11    0.25    4.25e4    8000       !弹性模量、泊松比、断裂能密度、密度
579    8000    !弹丸信息：粒子总数、密度
cylinder             !弹丸形状
0.025    0.05    0.0995    0.12450    0.004    !刚性体几何信息
0    -32.0    0     ! x,y,z速度分量
178201    0.0995    0.0996    0.029    !弹丸物质点坐标信息：编号、x,y,z位置坐标
```

178202	0.1245	0.0996	0.004
178203	0.105970476	0.0996	0.028148146
⋮	⋮	⋮	⋮
⋮	⋮	⋮	⋮
178777	0.118222445	0.129224464	0.004014068
178778	0.102050051	0.119672962	0.003666348
178779	0.088298731	0.129482925	0.018914234

近场动力学主要术语的中英文对照表

（按英文字母次序排列，括号内为部分术语的缩写词）

Bond	键
Bond breakage criterion	断键准则
Bond-associated horizon	键关联的近场范围
Bond-associated higher-order non-ordinary state-based peridynamics (BA-H-NOSB PD)	键关联高阶非常规态型近场动力学
Bond-based peridynamics (BB PD)	键型近场动力学
Bond force density vector	键力密度矢量
Conjugated bond-based peridynamics (conjugated BB PD)	共轭键型近场动力学
Constitutive force density vector	本构力密度矢量
Critical stretch	临界伸长率
Deformation vector state	变形矢量状态
Dilatation	膨胀量或体应变
Double state	双状态
Dual-horizon	对偶近场范围
Dual-horizon peridynamics (dual-PD)	对偶作用域的近场动力学
Elastoplastic peridynamic solid (EPS) model	（常规态型）近场动力学弹塑性模型
Extension scalar state	拉伸标量状态
Family of material point	物质点近场域内的点簇
Fictitious material layer (FML)	虚拟材料层
Force density vector state	力密度矢量状态
Fréchet derivative	弗雷歇导数
Higher-order non-ordinary state-based peridynamics (H-NOSB PD)	高阶非常规态型近场动力学
Horizon	近场范围
Linearized peridynamic solid (LPS) model	（常规态型）近场动力学线弹性模型
Material point	物质点
Micro-conductivity	微导热系数
Micro-modulus tensor	微模量张量
Microplastic (MP) model	微观塑性模型
Micropolar peridynamics	微极近场动力学
Morphing function method	混合函数方法

Nonlocal shape tensor	非局部形状张量
Non-ordinary state-based peridynamics (NOSB PD)	非常规态型近场动力学
Ordinary state-based peridynamics (OSB PD)	常规态型近场动力学
Pairwise potential function	点对势能函数
Peridynamics (PD)	近场动力学
Peridynamic correspondence material model (PD-CMM)	近场动力学对应材料模型
Peridynamic differential operator (PDDO)	近场动力学微分算子
Peridynamic function	近场动力学函数
Peridynamic laminate model	近场动力学层压板模型
Peridynamic least-squares minimization (PD-LSM)	近场动力学最小二乘法
Peridynamics-based finite element method (PeriFEM)	近场有限元法
Peridynamics in Large-scale Atomic/Molecular Massively Parallel Simulator (PDLAMMPS)	大规模原子分子并行模拟软件中的近场动力学模块
Peristatics	近场静力学
Point-associated horizon	点关联的近场范围
Point-associated nonlocal deformation gradient	点关联的非局部变形梯度
Prototype microelastic brittle (PMB) model	经典微弹脆性模型
Refined non-ordinary state-based peridynamics (RNOSB PD)	新的非常规态型近场动力学
Scalar state	标量状态
State-based peridynamics (SB PD)	态型近场动力学
State-based peridynamic lattice model (SPLM)	常规态型近场动力学点阵模型
State operator	状态算子
State reduction operator	状态缩减算子
Vector state	矢量状态
Visco-elastoplastic peridyamic solid (VES) model	(常规态型)近场动力学黏弹塑性模型

索　引

(按汉语拼音字母次序排列)

A

爱因斯坦求和约定, 216

B

本构关系, 3
变形率张量, 61
变形梯度张量, 16, 59, 185
表面效应, 12, 96, 99
并矢积运算符号, 59

C

参考构型, 58, 66, 69
冲击接触算法, 351

D

代表性体积单元, 3
单向弱耦合模型, 416
Drucker-Prager 弹塑性模型, 162
Delaunay 三角形, 324
等效塑性应变, 214, 357
狄利克雷边界条件, 398
第一类 Piola-Kirchhoff 应力, 66, 73, 186
动态松弛法, 140
多时间步长积分算法, 348

F

非局部边界条件, 32, 73, 201
非局部理论, 1
非连续 Galerkin 有限元法, 32
菲克定律, 388

分离式霍普金森压杆, 34, 359, 370
分切应力, 211
分子动力学, 1
冯·诺依曼稳定性分析方法, 143
傅里叶定律, 388, 394

G

高阶龙格-库塔差分法, 142
格子玻尔兹曼方法, 18
更新拉格朗日格式, 18
光滑粒子流体动力学, 16, 25, 228
广义连续介质力学, 2, 32
广义最小残量法, 196
鬼力, 280, 323

H

核函数近似, 229
横观各向同性弹性材料, 130
回映算法, 10

J

积分型非局部理论, 32
基尔霍夫-勒夫板, 14
Kirchhoff 应力, 65
畸变能密度, 93
级联跨尺度建模, 83
极分解定理, 65
剪切变形能密度, 358
键变形协调性, 199, 201, 258
角动量守恒的相容性条件, 76
紧支撑性, 75

紧支域, 229
浸入边界法, 18
晶体塑性有限元, 217
径向基函数, 10

K

抗旋转微模量, 106
Cauchy-Born 准则, 107
Cauchy 应力张量, 65
克罗内克符号, 93
扩展有限单元法, 28

L

拉格朗日乘子, 127, 189, 242
拉格朗日-格林应变张量, 189
拉普拉斯算子, 28
离散单元法, 18
裂尖扩展速度, 343
裂纹口张开位移, 314
零能模式, 15, 199, 239
罗宾边界条件, 398

M

蒙特卡罗法, 443

N

黏聚力模型, 28
能量释放率, 79, 90, 129
牛顿-拉夫逊迭代法, 194
诺依曼边界条件, 398

P

频散关系, 27, 149, 367

Q

屈服准则, 13

R

热力学定律, 4
人工阻尼, 16, 118, 145, 331
瑞利波波速, 151

S

沙漏模式, 202
施密特张量, 211
实时自适应, 343
斯托克斯定理, 85
四叉树/八叉树方法, 323
松弛构型, 170
速度梯度张量, 61
塑性变形体积不可压缩, 170
塑性乘子, 170
塑性剪切变形, 211
塑性流动法则, 170
塑性增量理论, 172
随机骨料模型, 412
缩减弹性矩阵, 193

T

泰勒级数展开, 27, 59, 86, 198
弹塑性乘法分解, 210
弹塑性加法分解, 170
梯度再生核质点法, 17

W

完全拉格朗日格式, 18
伪能量模式, 199
位移梯度张量, 60
沃罗诺伊图, 220

X

线动量守恒的相容性条件, 75
向前增量拉格朗日方法, 282
形状函数, 122
旋率张量, 61
旋转弹簧, 104

Y

雅可比矩阵, 119
一致性条件, 13, 170, 213
移动粒子半隐式方法, 18
应变能密度, 5, 66, 91, 102

应变偏张量, 93, 165
应力强度因子, 13
预测-校正法, 143

Z

再生核质点法, 17
正交性质, 476
中心差分法, 142
转换矩阵, 116
自然单元法, 329
最大周向应力准则, 310
左拉伸张量, 359

其他

δ-收敛性, 253
J 积分, 78
J_2 弹塑性模型, 162
JH-2 动力损伤本构模型, 356
Kalthoff-Winkler 试验, 28
Lax-Wendroff 差分法, 142
Navier 位移运动方程, 276
Velocity-Verlet 差分法, 142
Vogit 向量, 190